PROCEEDINGS OF THE NINTH INTERNATIONAL SYMPOSIUM ON CYCLODEXTRINS

Proceedings of the Ninth International Symposium on Cyclodextrins

Santiago de Compostela, Spain, May 31–June 3, 1998

Edited by

J.J. Torres Labandeira

and

J.L. Vila-Jato

Department of Pharmacy and Pharmaceutical Technology,
University of Santiago de Compostela,
Spain

SPRINGER-SCIENCE+BUSINESS MEDIA, B.V.

A C.I.P. Catalogue record for this book is available from the Library of Congress.

ISBN 978-94-010-5971-8 ISBN 978-94-011-4681-4 (eBook)
DOI 10.1007/978-94-011-4681-4

Printed on acid-free paper

The cover illustration displays the solid-state conformation of α-cycloaltrin (α-CA) in different forms: the molecule adopts a conformation of 3-fold rotational symmetry with a unique alternating sequence of $^4C_1$ and $^1C_4$ altropyranose chair geometries (upper left model, space group $P6_3$), resulting in banana-shaped disaccharide units, of which one is represented as a space-filling CPK-type model (lower left entry). In the crystal lattice the compact molecules are stacked in transposed layers (upper right, water molecules left off for clarity). Each α-CA molecule is embedded into a matrix of 21 water molecules (lower right, view along the hexagonal *c*-axis). The space filling models reveal α-CA to be devoid of a central 'through-going' cavity. The graphics were generated using Brickmann's MOLCAD molecular modeling program; for further information see the webpages at 'http://caramel.oc.chemie.tudarmstadt.de/immel/molcad/Gallery.html'or 'S. Immel, G.E. Schmitt, and F.W. Lichtenthaler: α-Cycloaltrin: Conformation and Properties in the Solid-State and Aqueous Solution.', pp. 41–48 of this volume.

Dedicated to

Dr. Tsuneji Nagai and Dr. Josef Szejtli

on their 65th Birthday

Contents

4. CYCLODEXTRINS IN CHEMISTRY

5. CYCLODEXTRINS IN ENVIRONMENT SCIENCES

6. CYCLODEXTRINS AS CHIRAL SELECTORS

7. CYCLODEXTRIN COMPLEXES AND ITS CHARACTERIZATION

8. CYCLODEXTRINS IN FOOD AND COSMETICS

Dear Colleagues and Friends,

WELLCOME TO SANTIAGO DE COMPOSTELA

Welcome to this small and wonderful town in the Westcorner of Spain. After the discovery of the tomb of the Apostle St. James, the city has become the goal of thousands of pilgrims who came from all over Europe to venerate the Apostle and looking for a renew of their soul and life. This has not changed nowadays, but Santiago has become a place for cultural and scientific pilgrimage. It's University celebrated its 500th anniversary in 1995, and Santiago became a scientific meetingpoint for the thousands of researchers that participated in the Conferences and Meetings.

This is the ***9th International Symposium on Cyclodextrins***. Previous Cyclodextrin Meetings were very successful and we wish you all the best for this one. Cyclodextrin technology has grown up a lot since the discovery of these substances at the beginning of the Century. Many people from all over the world have contributed to this development, looking for its characterization and for new applications. This is not the end of the way and our participation in this Symposium proves it. We have here the opportunity to continue the trail, renew our knowledge and share with our friends the scientific uneasiness.

Have a nice stage in Santiago de Compostela and we look forward to being able to welcome you again some other time in future!

ACKNOWLEDGEMENTS

The generous financial support of the following sponsors made it possible to organize this Symposium:

UNIVERSITY OF SANTIAGO DE COMPOSTELA. (Spain)

Secretaría de Estado de Universidades e Investigación (Spain)

XUNTA DE GALICIA. Dirección Xeral de Investigación (Spain)

The Nagai Foundation (Japan)

CYCLOLAB Ltd. (Hungary)

Cyclodextrin Technology Development Inc. CTD Inc. (USA)

CyDex, L.C. (USA)

Wacker-Chemie GmbH (Germany)

Asahi Foods Corporation (Japan)

Nihon Shokuhin Kako Co., Ltd (Japan)

Mercian Corporation (Japan)

Ensuiko (Bio Research Corp. of Yokohama. Japan)

Cyclodextrin Society of Japan

GlaxoWellcome, S.A.(Spain)

Chiesi Farmaceutici S.p.A. (Italy)

Kluwer Academic Publishers (The Netherlands)

****

Special thanks to Mr. Carlos M. Varela Ulla, your help made it possible.

Chapter 1

SEPARATION AND CHARACTERIZATION OF CYCLODEXTRINS

DEVELOPMENT OF A FLUORIMETRIC-FIA METHOD TO MONITOR CYCLODEXTRIN PRODUCTION FROM STARCH AND THE FORMATION OF INCLUSION COMPLEXES.

JOSÉ V. CARBONELL Y JOSÉ Mª. SENDRA.
Instituto de Agroquímica y Tecnología de Alimentos, Consejo Superior de Investigaciones Científicas, P.O. Box 73, 46100-Burjassot, Valencia, Spain

Introduction

Native amylose is known to be composed of mostly linear homopolymeric chains of (1→4)-α-D-linked glucoses, with different degrees of polymerisation. In aqueous solution, amylose chain takes on the conformation of a flexible random coil, with non-helical segments connecting extended, irregular and labile sequences of helical structures. Amylopectin is a branched homopolymer of glucose residues. Its structure is composed of an amylose-like backbone carrying clusters of ramified (1→4)-α-D-glucan branches which are connected among them and to the backbone through (1→6)-α-D-linkages. In aqueous solution, the (1→4)-α–D-linked segments of amylopectin can also take on a helical conformation. Both amylose and amylopectin can form a complex with some guest molecules which modify the initial conformation of the corresponding chain by inducing a coil-helix transition. The driving forces involved in the formation of the complex are not yet completely understood but it is assumed that hydrophobic interactions play a very important role.

2-*p*-Toluidinylnaphthalene-6-sulfonate (2,6-TNS) is barely fluorescent in pure aqueous solution but its fluorescence is strongly enhanced when the environment becomes hydrophobic, thus it has been extensively used as a fluorescent hydrophobic probe. The addition of cyclodextrins to 2,6-TNS aqueous solutions greatly enhances its fluorescence (Kondo et al., 1976), as does the addition of amylose, amylopectin and glycogen, albeit to a lesser extent (Nakatani et al., 1984). It has been found that the increase in the fluorescence intensity of 2,6-TNS when it forms a complex with (1→4)-α-D-glucans depends on the average molecular weight of the latter, hence the fluorescence of 2,6-TNS decreases as the enzymic depolymerization of the (1→4)-α-D-glucan proceeds (Nakatani et al., 1977). A model to explain this interaction proposes the inclusion of 2,6-TNS within the hydrophobic microenvironment thus occupying the central core of the helical (1→4)-α-D-glucan segments.

In this work, a new FIA methodology, using 2,6-TNS as the fluorescent probe, has been developed for the determination of: (1) the enzymic activities of enzymes degrading linear and ramified (1→4)-α-D-glucans; (2) the yield of cyclodextrins from amylose and amylopectin using cyclodextrin glycosyltransferase (CGTase); and (3) the yield of inclusion complexes between β- and γ-cyclodextrins and some apolar guest molecules.

Experimental

Figure 1 shows a schematic representation of the FIA system. It includes a high-pressure pump, an injection valve, a mixing coil reactor and a fluorescence detector connected to an integrator/plotter. The general operating conditions were: an excitation wavelength of 358 nm; an emission wavelength of 455 nm; a slit width of 25 nm. Other operating conditions were chosen according to the characteristics and concentrations of the substrates and the end products.

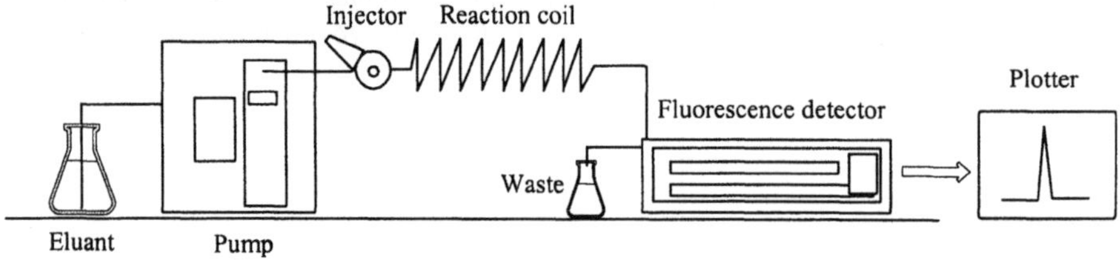

Fig. 1. Scheme of the fluorimetric-FIA system

Results

DEPOLYMERIZATION AND CYCLATION PROCESSES

Figure 2 shows the response peaks corresponding to the injection (in duplicate) of different concentrations of amylose. As can be seen, an excellent correlation between the peak heights and the corresponding injected amylose concentrations was obtained. The delay time between the injection of the sample and the corresponding peak apex time was approximately 9 seconds thus making this a potentially useful tool for real time monitoring of the processes.

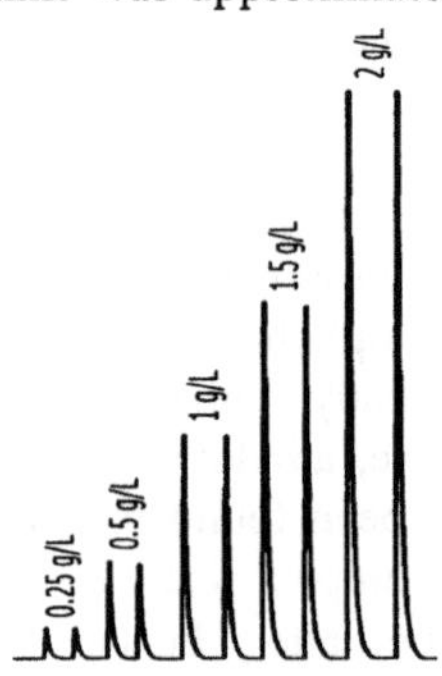

Fig. 2. Plot corresponding to the injection of different amylose concentrations.

Figure 3 shows the responses of amylopectin, α–, β– and γ–cyclodextrins, and different amyloses and dextrins in the FIA system as a function of their concentrations (mg/L). As can be seen, in all cases the calibration response curves were straight lines for the range of concentrations assayed. Figure 4 shows the dependence of fluorescence intensity on the degree of polymerisation (DP) of the amylose chain, at constant concentration (w/v). As can be seen, at low DP's an increase in DP leads to an increase in the response level. However, for DP's higher than 300 the response level reaches practically its maximum and becomes independent of DP.

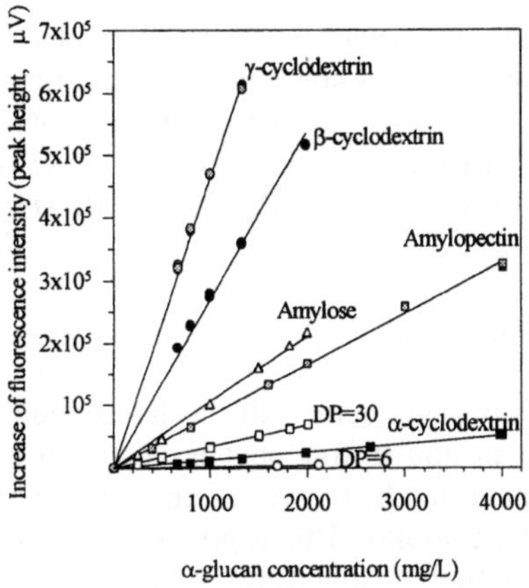

Fig 3. Peak heights of different α-glucans.

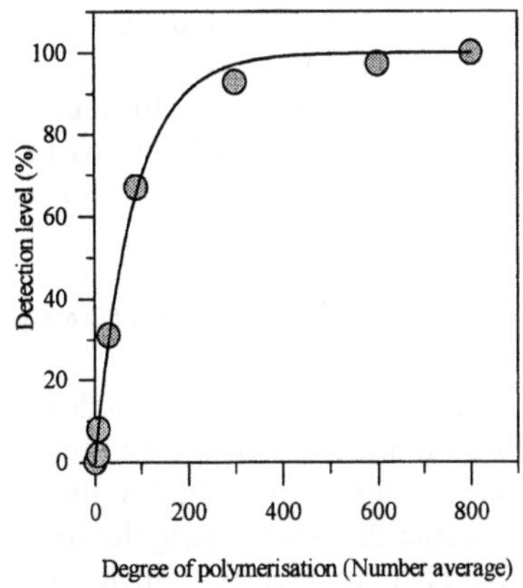

Fig. 4. Response curve of the detection system to the degree of polymerisation of amylose samples

The fluorimetric-FIA technique may thus be very useful for monitoring the kinetics of polymerisation as well as the processes of depolymerisation of (1→4)-α-D-glucans induced by specific enzymes. Figure 5 shows the time evolution of substrate concentration during the depolymerisation of amylose and amylopectin by *B. subtilis* α-amylase.

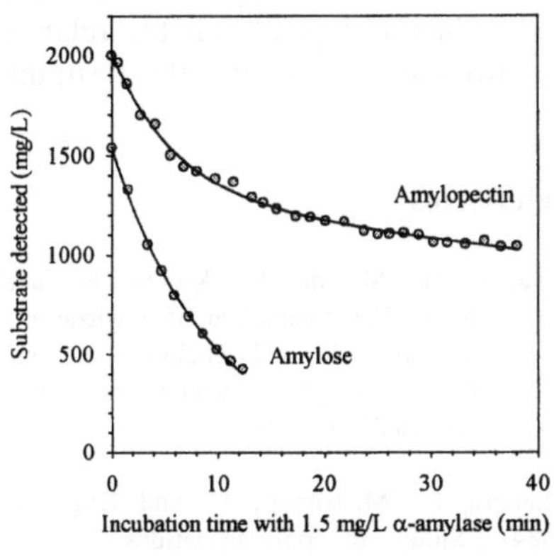

Fig. 5. Amylose and amylopectin detected in depolymerisation processes with α-amylase.

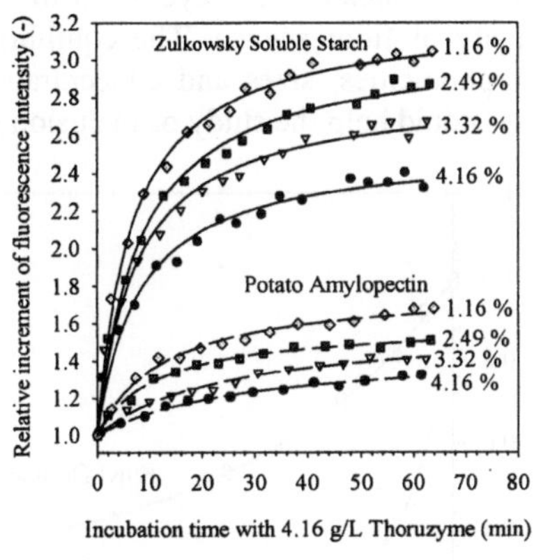

Fig. 6. Increase of fluorescence intensity due to the formation of β– and γ-cyclodextrins

Figure 6 shows the time evolution of fluorescence intensity during the incubation of potato amylopectin and Zulkowsky's soluble starch with CGTase. The increase in fluorescence intensity with incubation time can be attributed to the production of β- and γ-cyclodextrins which over-compensate the decrease of the signal due to the depolymerisation of the substrate and the production of α-cyclodextrin. From Figure 6, it is apparent that Zulkowsky's starch is a much better substrate than potato amylopectin for the production of β- and γ-cyclodextrins. In fact, the increase in fluorescence due to the cyclation of a given concentration of Zulkovsky's starch is approximately four times that of the increase due to the cyclation of the same concentration of potato amylopectin. However half of this increment results from the lower detection level of Zulkowsky's starch (about 50 % of that of potato amylopectin) in the fluorimetric-FIA system.

INCLUSION PROCESSES IN ß- AND γ-CYCLODEXTRINS

The internal cavities of cyclodextrins are empty cylinders with a hydrophobic microenvironment. When a molecule containing an apolar fragment is added to an aqueous cyclodextrin solution the apolar fragment tends to take refuge in these hydrophobic cavities forming the so-called inclusion complex. This is precisely the case for 2,6-TNS molecules which contain a hydrophobic toluidinyl moiety. When molecules of 2,6-TNS are excited by light of 358 nm, only those molecules included in the hydrophobic cavity emit a strong signal at 455 nm, which can then be measured with a suitable fluorescence detector. In excess of 2,6-TNS molecules, all the cyclodextrin cavities are occupied and hence the intensity of the emitted signal is directly proportional to the concentration of cyclodextrin molecules. Other hydrophobic molecules added to this solution can compete with 2,6-TNS for inclusion in the cyclodextrin internal cavity thus leading to a decrease in the fluorescent signal.
Figure 7 shows that the addition of increasing amounts of sodium dodecylsulfate (SDS) or ethyl decanoate to a β-cyclodextrin solution leads to a corresponding decrease of the intensity of fluorescence. The equilibrium of this exchange depends on the relative hydrophobicities, sizes and concentrations of the different solutes, and the resulting curves could help the study of inclusion processes.

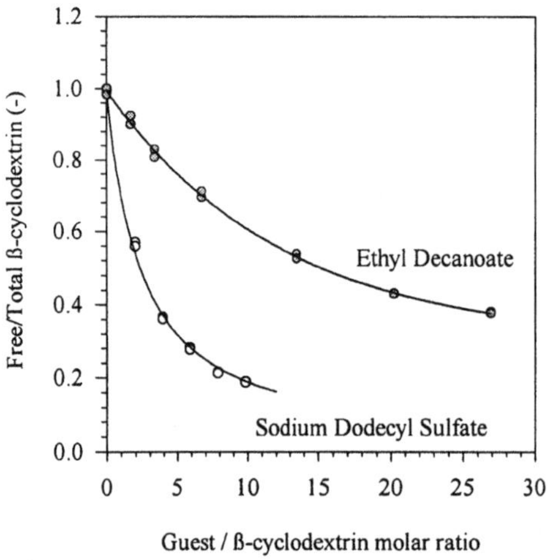

Fig. 7. Decrease of the concentration of free ß-cyclodextrin due to the formation of complexes with sodium dodecyl sulfate and ethyl decanoate.

References

Nakatani, H., Shibata, K., Kondo, H., and Hiromi, K. (1977). Interaction of amylose and other α-glucans with hydrophobic fluorescent probe (2-p-toluidinylnaphthalene-6-sulfonate). *Biopolymers* **16** 2363-2370.

Kitamura, S., Matsumori, S., and Kuge, T. (1984). Study of polysaccharides by the fluorescence method. V. Interaction of 2-p-toluidinylnaphthalene-6-sulfonate with amylose and its related compounds in aqueous solution. *Journal of Inclusion Phenomena* 2, 725-735.

Kondo, H., Nakatani, H., and Hiromi, K.(1976). Analysis of mixtures of α- and β-cyclodextrins using fluorescent dyes. *Carbohydrate Research* 52:1-10.

DEVELOPMENT OF ENZYME IMMUNOASSAYS SPECIFIC FOR NATURAL β- AND γ-CYCLODEXTRINS AND SYNTHETIC DERIVATIVES. PHARMACOKINETIC ANALYSIS OF CYCLODEXTRINS IN RAT.

C. CRÉMINON [a], F. DJEDAINI-PILARD [b], C. PÉAN [b], R. VIENET [a], B. PERLY [b], J. GRASSI [a], P. PRADELLES [a]

a: CEA, DRM/SPI, CE-Saclay, F-91191 Gif sur Yvette, (France)
b :CEA, DRECAM/SCM, CE-Saclay, F-91191 Gif sur Yvette, (France)

1. INTRODUCTION

Pharmacological applications of CDs are mainly based on an increase of the solubility of the active molecule entrapped. The development of pharmaceutical uses requires quantitative measurements of CD in biological material. We have developed different enzyme immunoassays (EIAs) specific for β-cyclodextrin (β-CD), di-2,6-O-methyl-β--cyclodextrin (di-Me-β-CD), tri-2,3,6-O-methyl-β-cyclodextrin (per-Me-β-CD) and γ--cyclodextrin (γ-CD). Specific β- and γ-CD EIAs allowed us to perform qualitative and quantitative measurements of intact mono-6-amino-6-deoxy-β-CD (βCD-NH_2) (after i.v. and oral administrations) and γ-cd (after i.v. administration) in rat biological fluids (plasma and urine). To our knowledge, this is the first pharmacokinetic study performed for these molecules using such specific immunological methods.

2. MATERIALS AND METHODS

2.1 Materials

β-CD and γ-CD were obtained from Roquette Fréres and Wacker respectively. Mono-6-amino-6-deoxy-β or γ-CD were prepared as previously describes (1). Mono-6-amino-6-deoxy-heptakis-2-β-methyl-hexakis-6-*O*-methyl-β-CD and Mono-6-amino-6-deoxy-heptakis-2,3-O-methyl-hexakis-6-O-methyl-β-CD were synthesized using modified methods already describes (2) and purified by HPLC. Pure acetylcholinesterase (ACHE,EC 3.1.1.7) was prepared from the electric organs of the electric eel *Electrophorus electricus* (3).

2.2 Competitive EIA procedure

For each molecule, competitive EIA was performed as previously describes using specific polyclonal antibodies raised against mono-6-amino-CD synthetic derivatives and enzymatic tracers corresponding to these molecules covalently coupled to ACHE (4).

2.3 Animal experirnents and pharmacokinetic evaluation

The experiments were performed on 10 weeks old male rats weighting 325 ± 60 g. The animals received a single intravenous or oral 25 mg/kg dose solubilized in saline solution of either $\beta CD\text{-}NH_2$ or γ-CD. Blood was sampled at various time between 5 min to 24 h at the arterial catheter. Blood was centrifuged, plasma kept frozen and pellets containing red blood cells reinjected. The rats were housed individually in glass metabolic cages allowing separate collection of urine and faeces. Urine sample collections were made 4 h, 8 h and 24 h after dosing.
Pharmacokinetic parameters were evaluated using non compartimental approach (5) and processed with Siphar computer program.

3. RESULTS AND DISCUSSION

3.1 Enzyme immunoassays

The different immunoassays are sensitive (Fig. 1 and Table 1) and the precision is also very satisfactory as a coefficient of variation less than 10% is observed in the 0.2-50 ng/ml range. Moreover these assays are quite specific since for natural CD assays neither linear analog nor others (α- or γ-CD with β-CD assay for instance) were recognized while di-Me-β-CD assay do not allow detection of β-CD or per-Me-β-CD and per-Me-β-CD assay fails to recognize di-Me-β-CD or β-CD.

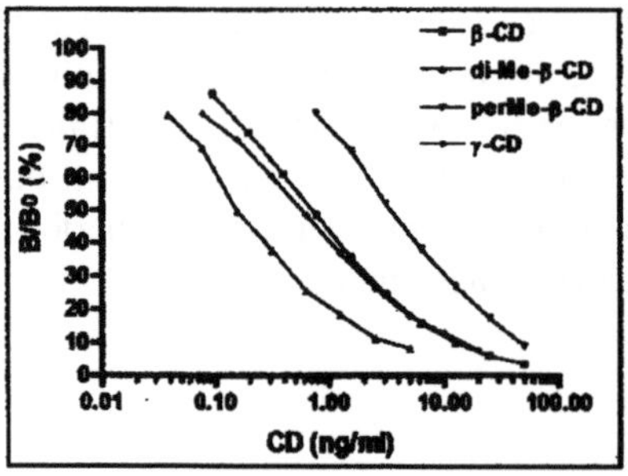

Compound	Minimum detectable concentration (pg/ml)
β-CD	120
di-Me-β-CD	40
perMe-β-CD	800
γ-CD	80

Fig. 1: Standard curves for the different CD assays; Table 1: Sensitivity of the different CD assays.

To characterize the specificity of the di-Me-β-CD antibodies, standard curves were performed with various randomized dimethyled derivatives obtained from Fluka, Acros, Eras or Waker (RAMEB). These derivatives showed a full cross-reactivity thus demonstrating that the presence of many isomers does not preclude to the recognition by this assay. However it should be noted that RAMEB exhibited a lower cross-reactivity since this derivative contains less than 1% of the pure di-Me-β-CD. In this case the sensitivity of the assay is lower but stays satisfactory.

3.2 Pharmacokinetic analysis

After single intravenous injection at 25 mg/kg, immunoreactive CD is detected above 10 ng/ml until 4 h in plasma (Fig. 2). In urine, values are highly significant for all collection time including 8-24 h where concentrations are up to 10 μg/ml (inset, Fig. 2).

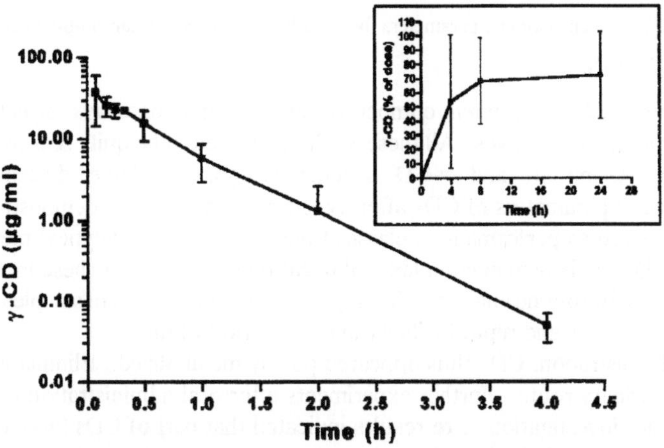

Fig. 2: Pharmacokinetic of γ-CD in plasma. Inset: Pharinacokinetic of γ-CD in urine.

Main pharmacokinetics parameters are fisted in Table 2. 1.v. administered CDs are subjected to rapid disappearance from plasma as demonstrated by short half-life and MRT. The volume of distribution at steady state is up to those of blood and extracellular fluids indicating that CDs are certainly distributed in tissues. Most of the administered CD is eliminated in urine with a apparent urinary clearance close to the systemic clearance. This clearance value is near the renal glomerular filtration. CDs are thus rapidly cleared from plasma and appeared poorly metabolized. For oral administration, preliminary experiments showed that about 1% of the dose is observes in plasma and eliminated in urine.

	β-CD-NH_2 (mean ± SD)	γ-CD (mean ± SD)
T ½ (min)	21±2	28 ±4
MRT (min)	28±5	31±13
Cls ($ml.min^{-1}.kg^{-1}$)	6.9±2.1	16.9 ±2.6
AUC 0-inf ($\mu g.ml^{-1}.h$)	69.3±17.2	24.99 ± 3.63
Vdss ($ml.kg^{-1}$)	202±56	687 ± 211
Qel 0-24h (% dose)	100±11	72 ± 31
Clu 0-24h ($ml.min^{-1}.kg^{-1}$)	7.1±3.1	14.4 ± 5.9

Table 2: Pharmacokinetic parameters for i.v. administration of 6-amino-β-CD and γ-CD.

4. CONCLUSION

We have developed four different competitive enzyme immunoassays for natural CDs as well as synthetic derivatives. All these methods proved to be quite sensitive, precise and specific. Using β- and γ-CD specific assays, we have determined the pharmacokinetic parameters of CDs after i.v. administration, thus demonstrating that immunoassays are as performant as classical physicochemical analytical methods for analysis of CD metabolism in complex biological fluids. Moreover these methods are choice methods for routine analysis, allowing assay of large number of samples with high sensitivity, specificity and reproducibility in short period of time.
After i.v. administration, CDs thus appeared poorly metabolized. Elimination mainly occurred by urinary route. Further experiments after oral administration of CDs are presently under investigation. Pre results indicated that part of CDs (about 1%) was observed in plasma and eliminated in urine.

This work was supported by the European Commission (DGXII) under the FAIR Programme CT95-0300

5. REFERENCES

(1) F. Djedaini-Pilard, N. Azaroual-Bellanger, M. Gosnat, D. Vemet and B. Perly, *J. Chem. Soc. Perkin Trans. II*, 723-730 (1995)
(2) C.M. Spencer, J.F. Stoddart, R. Zarzycki *J Chem. Soc. Perkin Trans II*, 1323-1330 (1987). J. Boger, J.Corcoran, J.M. Lehn *Helv. Chim. Acta*, **61** (6), 2190-2218 (1978)
(3) J. Massoulié; S. Bon; *Eur. J Biochem.* **68**, 531-539 (1976)
(4) C. Créminon, F. Djedaïni-Pilard, J. Grassi, B. Perly and P. Pradelles; *Carbohyd. Res..* **63**, 207-213 (1994).
(5) M. Gibaldi and D. Perrier; in *Pharmacokinetics in drugs and the pharmaceutical sciences*, Marcel Dekker Publishers, New York (1982).

THE ROLE OF HOST - GUEST INTERACTION IN THICKENING AND PHASE SEPARATION OF B-CYCLODEXTRIN POLYMER AND POLYMERS BEARING CYCLIC HYDROPHOBIC SIDE GROUPS.

C. AMIEL*, B. SEBILLE*, E. RENARD*, S. PICAVET*, R. GREF [£]
** Laboratoire de Recherche sur les Polymères, UMR C7581 CNRS, 2-8 rue H. Dunant, 94320 Thiais France, [£] Laboratoire de Chimie-Physique Macromoléculaire, URA 494 CNRS, BP 451, 54001 Nancy France.*

1. Introduction

A large variety of architectures can be obtained from macromolecular assemblies involving inclusion complexes between polymers in aqueous solutions. Mixtures of hydrophically end-capped PEO with β-cyclodextrin polymers have already been described in previous papers [1 - 3]. It was shown that the hydrophobic moieties were interacting by host-guest association with the β-cyclodextrin cavities and this was at the origin at the formation of intermolecular aggregates. The strength of the interaction between the two polymers appeared to be mostly determined by the number of hydrophobe per PEO chain. In the present paper, the hydrophobic molar ratio has been varied by using "star" PEO of constant arm size (5000) but different number of arms (3, 4, 8). The PEO stars were terminated with adamantyl end groups which are known to make very stable complexes with β-cyclodextrin [4]. The β-cyclodextrin polymers (β-CD/EP) are formed by polycondensation with epichlorohydrin under strongly alkaline conditions [5]. High molecular weights but water soluble polymers were obtained by conducting the reaction close to the gelation threshold.

2. Phase diagrams

PEO and β-CD/EP are incompatible polymers. The phase diagrams (figure 1) are qualitatively comparable to those obtained with PEO / Dextran systems [6]. At total concentration higher than 20% w/w, mixtures PEO 20000 / β-CD/EP exhibit liquid-liquid phase separation with the top phase enriched in PEO and the bottom phase enriched in β-CD/EP. This segregative behaviour, which is very common in binary polymer mixtures, is indicative that there is no specific affinity between the two polymers but repulsive effective interactions.

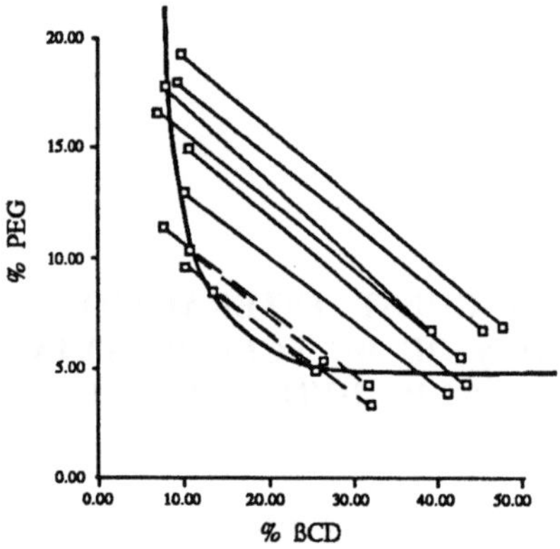

Figure 1: Segregative phase separation in aqueous mixtures of b-CD/E¨P polymers and PEO (linear, Mw= 20000).

The two incompatible polymer backbones can be compatibilized by introduction of the hydrophobic end-groups: due to inclusion complexes between hydrophobic moieties and β-cyclodextrin cavities, additional attractive interactions are experienced by the two polymers. When the number of hydrophobic moieties per chain is two (linear PEO derivatives) , mixtures give homogeneous one phase systems independantly of the PEO chain length, even in concentration domain where two phases are obtained in mixtures with unmodified PEO.

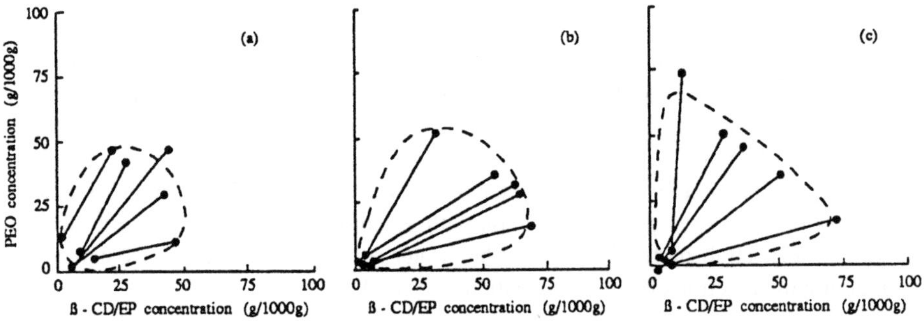

Figure 2: Associating phase diagrams of β-cyclodextrin-epichlorohydrin copolymers and adamantyl terminated multi-arm branched PEO with 3 arms (a), 4 arms (b) or 8 arms (c).

As soon as the number of hydrophobes per chain reaches 3, a completely different pattern is observed. As shown in figure 2, associative phase separations are observed for mixtures involving PEO with 3, 4 or 8 arms. Two phases systems are obtained in a relatively low concentration range, 0 to 10 % w/w, where a polymer rich phase is topped by a polymer poor phase.

The class of associating systems presented in this work gives the opportunity to observe for the first time associative phase separations (also denoted as complex coacervation) [7] mediated by host-guest complexation interactions between two neutral polymers in aqueous media [8].

3. Rheology

Mixtures of modified star PEO with β-CD/EP polymers do show thickening properties. Viscosity enhancements, which can be as high as 4000 (table 2), are much higher than those obtained with linear PEO. This tendency has also been observed in associative PEO thickeners where better thickening properties were obtained as the number of alkyl groups per chain was increased from 2 to 3 [9] . This has been interpreted as an increase of the percentage of intermolecular over intramolecular links. However, in our case, no further enhancements are observed when the number of adamantyl groups is increased from 3 to 8. Viscosities seem to be rather insensitive to the number of end groups. This could be attributed to an interplay between autoassociation and inclusion complexes of the hydrophobic moieties.

TABLE 2: Low shear viscosities of monophasic, mixtures with PEO compounds, concentration 70 g/l, and β-CD/EP polymers.

Sample	C(β-CD/EP) (g/l)	η (Pa.s)
PEO unmodified	30	$5\ 10^{-3}$
PEO-Ad3	30	20
PEO-Ad4	10	$9.5\ 10^{-2}$
PEO-Ad4	20	1.4
PEO-Ad4	30	25
PEO-Ad8	30	16

Oscillatory shear measurements have shown a viscoelastic liquid behavior of the monophasic mixtures: G' is found much smaller than G" in the whole frequency range and at any concentrations. In the current models of reversible polymer networks [10, 11], the frequency at which G' crosses G" is generally identified as the inverse of an average relaxation time characteristic of the transient network. Moreover, this relaxation time should be related to the lifetime of the bonds (inclusion complexes). Two possible mechanisms could explain this behavior: - kinetics of dissociation of the complexes fast compared to the rate of dissociation of alkyl groups in hydrophobic aggregates in the case of associative PEO thickeners; - An alternative explanation of the viscous fluid behavior of our concentrated mixtures could be that aggregates of limiting size are formed instead of percolating structures. This could be attributed to the repulsive interactions experienced by the two polymer backbones (PEO and β-CD/EP, see part 2).

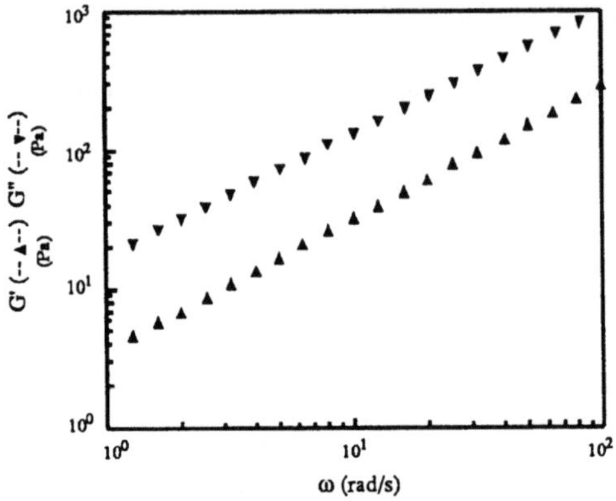

Figure 3: Plots of G' (Δ) and G" (∇) vs frequency for homogeneous mixture of PEO-Ad3 (3 arms) and β-CD/EP at a total concentration 100g/l.

4. Conclusion

An original associating system has been presented in this paper. It involves the mixing of two polymers : hydrophobically end-capped PEO and β-cyclodextrin copolymer. Inclusion complexes between the hydrophobic moieties and β-cyclodextrin cavities are at the origin of the attraction between the two polymers. The presence of these complexes which act as reversible bonds, leads to two physical phenomena : - intermolecular clusters growing eventually to transient percolating structures as the concentration is increased and - associative phase separation. These two phenomena have been observed in our system. The interplay between them has been modified by using branched PEO of variable number of arms: the number of hydrophobic end-groups per PEO chain was varied from two to eight.

5. References

1. C. Amiel, A. Sandier, B. Sebille, P. Valat, V. Wintgens, *Int. J. Polymer Analysis & Characterization*, 1 (1995) 289.
2. C. Amiel, B. Sebille, Journal of Inclusion Phenomena, 25 (1996) 61.
3. C. Amiel, B. Sebille, submitted to *Advances in Colloid and Interface Science*.
4. M. R. Eftink, M. L. Andy, K. Bystrom, H. D. Perlmutter, D. S. Kristol, *J. Am. Chem. Soc.*, 111 (1989) 6765.
5. E. Renard, A. Deratani, G. Volet, B. Sebille, *Eur. Polym. J.*, 33 (1997) 49..
6. P. A. Albertsson, Partition of Cell Particles and Macromolecules, 3rd ed., Wiley & Sons, New York 1986.
7. H. G. Bungenberg de Jong, in H. R. Kruyt (Ed.) Colloid Science, Elsevier, Amsterdam, 1949, Ch 10, 235.
8. N. M. Gosselet, C. Borie, C. Amiel, B. Sebille, *Journal of Dispersion Science and Technology*, 19 (1998), in press.
9. B. Xu, L. Li, K. Zhang, P. M. Macdonald, M. A.Winnik, *Langmuir*, 13 (1997) 6896.
10. F. Tanaka, S. F. Edwards, *J. Non-Newtonian Fluid Mech.*, 43 (1992) 247.
11. T. Annable, R. Buscall, R. Ettelaie, D. Whittlestone, *J. Rheol.*, 37 (1993) 695.

SPECIFIC BINDING AND PRECIPITATION OF γ-CYCLODEXTRIN WITH MONOSACCHARIDE IN AN AQUEOUS SOLUTION

HIDEO IMATA AND KEIKO TAKAHASHI § *, *Sagami Keisoku Co., Ltd., 1892-1 Shimomizo, Sagamihara, Kanagawa 229-0015, JAPAN,*
§Department of Industrial Chemistry, Faculty of Engineering , Tokyo Institute of Polytechnics, 1583 Iiyama Atsugi, Kanagawa 243-0239 JAPAN

1.Introduction

Molecular recognition of carbohydrates is characterized by the confluence of many polar and nonpolar intermolecular forces. However, the understanding of the contribution of these forces in carbohydrate recognition is probably more intriguing due to their molecular complexity. Cyclodextrin(CD) has several hydroxy groups at both rims of the cavity. The early stages in CD chemistry presented a wide range of chemists with a very impressive concept, that of a host-guest complex which has been described as a ball in a bottomless pail. Actually, the ball or guest is not always deeply embedded in the pail cavity. Other interactions play sufficiently significant roles that they recognize the ball. The so-called "rims" of the pail must also be considered. The γ-CD is characterized by a large sized cavity and high solubility for water. We now report the direct evidence that γ-CD recognizes D-fucose using the "rims "of the cavity.

2.Experimental

α-, β- and γ-CDs were obtained from Japan Maize Co.. Saccharides were commercially available guaranteed reagent grade and were used without further purification. NMR spectra were taken using a JEOL Lambda 500 spectrometer.

3.Results and Discussion

3.1. COMPLEX FORMATION BETWEEN CDS AND MONOSACCHARIDES

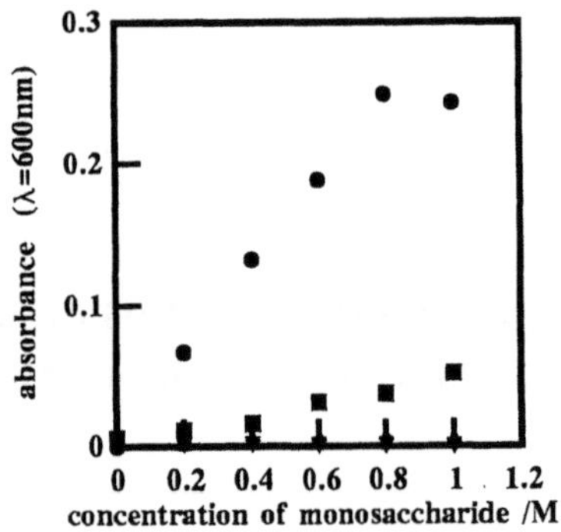

Figure 1. Dependence of the turbidity at 600 nm of the γ-CD solution(pH 6.84) on the concentration of the monosaccharides D-fucose(●), D-arabinose(■), D-sucrose(▼), D-garactose(▲) and D-glucose(◆) at 3°C.[γ-CD]=$1.2x10^{-2}$ mol/l, [monosaccharide]=0.0-1.0 mol/l.

Fig. 1 shows the dependence of A_{600} of a pH 6.84 γ-CD solution on the concentration of the monosaccharides, such as D-fucose, D-arabinose, D-sucrose, D-garactose and D-glucose. Only the addition of D-fucose to the solution increased A_{600} to saturate in high concentrations. When D-fucose was added to the α- or β-CD solution instead of the γ-CD solution, no change was observed. These results suggested that an insoluble compound was formed in the solution. γ-CD($2.4x10^{-2}$mmol) and D-fucose ($1x10^{-2}$mol) were dissolved in 10ml of pH 6.84 phosphate buffer, and the clear solution for 24 hr at 3°C was maintained and a 11.8mg of a white precipitate was produced . The ^{1}H NMR spectra of the precipitate in DMSO-d_6 gave direct evidence that the precipitate is a mixture composed of only γ-CD and D-fucose with molar ratio of 1:1-8.

3.2 PRECIPITATION OF FUCOSE AND γ-CD IN THE PRESENCE OF 1-ANILINO-8-NAPHTHYLSULFONIC ACID

1-Anilino-8-naphthylsulfonic acid(ANS) is known to be included in γ-CD and forms a 1:1 host-guest complex (K=246 M^{-1})[1]. The ROESY method has also indicated that both aromatic groups of ANS, such as the phenyl and naphtyl groups, were included in the hydrophobic cavity. The fluorescence intensity of ANS is increased by the hydrophobic environment. Interestingly, the precipitate was also observed in the γ-CD and D-fucose system even in the presence of ANS. Based on the NMR spectrum of the precipitate, it was composed of fucose, γ-CD and ANS. Fig. 2 shows the aromatic ring proton region of the 500MHz NMR spectra of ANS in the precipitate, in buffer solution and in the γ-CD cavity. The chemical shift of ANS in the CD cavity is different from those in bulk

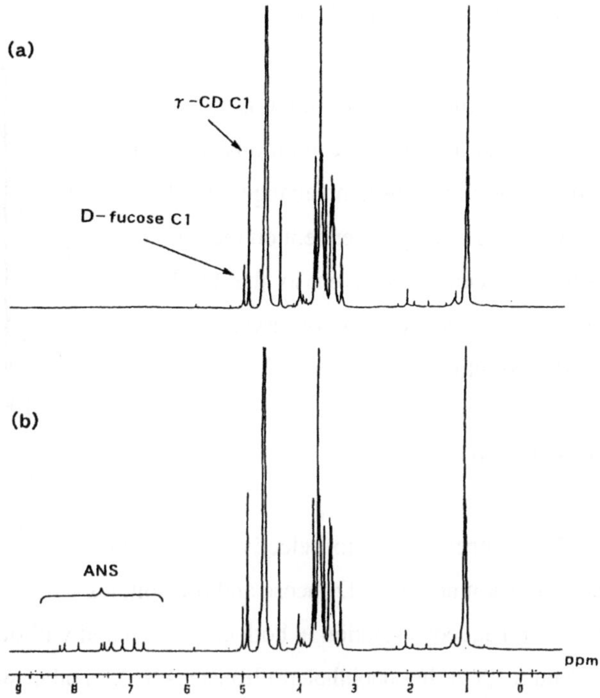

Figure 2. 400 MHz ^{1}H NMR spectra of the precipitate obtaining from the γ-CD and D-fucose solution in the absence(a) and presence of the ANS(b) in D_2O.

solution. The chemical shift during the precipitation is almost the same as that in the CD cavity. These observations suggest that the ANS molecule is embedded in the cavity. Induced circular dichroism spectra of the ANS also indicated that the ANS molecule was in the cavity along with fucose. The fluorescence intensity of ANS was increased by the addition of fucose (Fig. 3). We have already reported that the fluorescence intensity of the CD-ANS complex depended on the kinds of saccharide and remarkable increases were observed in the case of the γ-CD-ANS-sucrose system[2]. Saccharides in the system seem to stay near the rims of the CD cavity due to hydrogen bonding between the hydroxyl groups of the saccharides and those on the CD rims. The γ-CD-ANS-fucose system seem to be same as the above system. These results suggest that the fucose molecules are not included in the cavity of the CD, but stay near the rims of the cavity. The orientation of the hydroxy groups on the CD ring may be fit the orientation of the hydroxy groups on fucose.

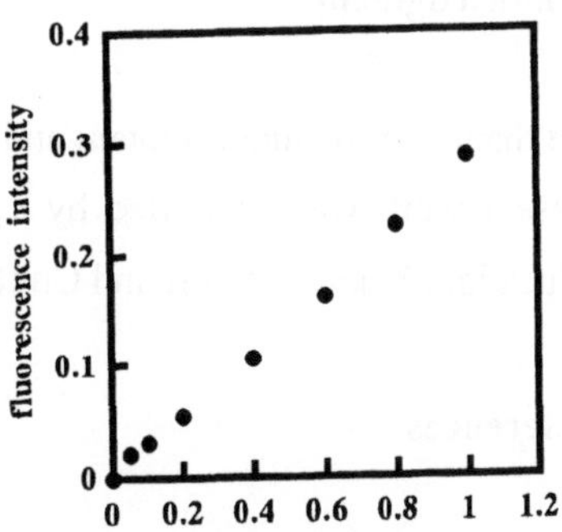

Figure 3. Dependence of the fluorescence intensity on the concentration of D-fucose in pH6.84 phosphate buffer at 25.0±1.0°C. λex=350 nm, λem=500nm, [γ-CD]=1.0×10^{-2} mol/l, [monosaccharide]=0.0-1.0 mol/l, [ANS]=1.0×10^{-4} mol/l.

3.3 ENENTIOSELECTIVITY

By adding D-fucose to the solution, A_{600} increases and saturates at high concentrations, whereas L-fucose addition produces no change. As shown in Fig. 4, A_{600} is clearly dependent on the ratio of the D-form concentration even in the D- L-mixed solution. No changes were observed using the α- or β-CD solution.

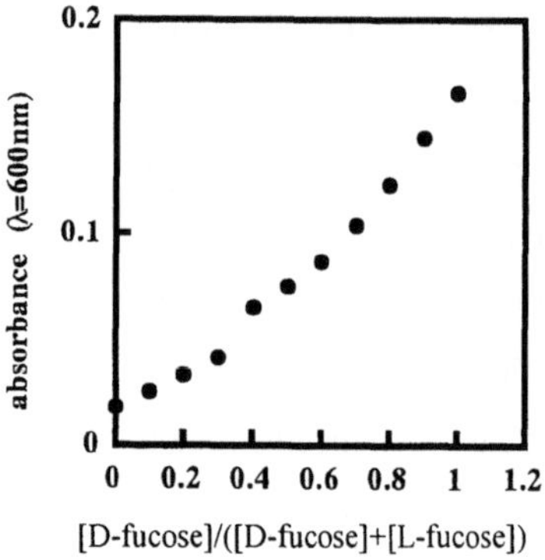

Figure 4. Dependence of the turbidity at 600 nm of the γ-CD solution(pH 6.84) on the ratio of the concentration of D-fucose in total fucose concentration at 3°C.[γ-CD]=$1.2x10^{-2}$ mol/l.

4.Conclusion

γ-CD has been found to selectively bind with excess molar amounts of D-fucose and precipitates in a neutral aqueous solution in both cases with and without other guest molecule, such as sodium 1,8-anilinonaphthalensulfonate(ANS). In other words, γ-CD discriminates the optical isomers of D-fucose. Further designs of CD-saccharide complexes and the detailed mechanism elucidation are now currently under investigation.

Ackowledgment

We thank Mr. Kouichi Imotani and Mr. Mikio Izawa for their experimental help. The present work was supported by a Grant-in-Aid No.09680579 from the Ministry of Education, Science, Sport and Culture.

References

1.Takahashi, K., Koizumi,U. and Hattori,K., (1990) Inclusion Behavior of Methylated Cyclodextrins with Fluorescent Probes, *The Academic Reports of the Faculty of Engineering, Tokyo Institute of Polytechnics,* 13, 18-22.

2.Imata,H., Takahashi,K. and Hattori,K., (1995) Recognition of Mono- and Disaccharides with 1,8-Anilinonaphthalenesulsonic Acid as Fluorescence Probe in Cyclodextrin Cavities, *KOUBUNSHI RONBUNSHU,* 52, 639-643.

ANALYSIS OF HYDROXYPROPYL CYCLODEXTRINS BY LIQUID CHROMATOGRAPHY AND EVAPORATIVE LIGHT SCATTERING DETECTION

N. RABEARIMONJY, S. OUNNAR, M. RIGHEZZA AND M.DREUX
Université d'Orléans. ICOA. BP 6759. 45067 Orléans Cedex 2. France

Introduction

Hydroxylated cyclodextrins have been used successfully in various fields such as pharmaceutical applications as drug delivery to increase the aqueous solubility and stability of drugs, or enantiomeric separations as chiral additive. The hydrophilic interactions given by the hydroxypropyl substituent confer to the molecule a high solubility in aqueous surroundings. Hydroxypropyl cyclodextrins (HP-CD) studied are derived of α-, β- or γ-cyclodextrins. The HP-CD are mixtures of homologous in which the number of HP substituents ranges from 1 to 9. The distribution of the degree of substitution depends on the preparation[1]. Previous works[2-4] have been published on the analysis of complexes mixtures of HP-CD. Determination of the distribution of the sample is often limited by the detection system[4] and a mass sensitive detector seems to be preferable to improve the characterization of complex mixtures.
The aim of the work is to study the composition of derivatives HP-CD by liquid chromatography, evaporative light scattering detection and mass spectroscopy and to determine the influence of the substituent on the retention. LC method using a mass sensitive detector shows a separation of homologous according to the degree of substitution and provides a pattern analysis of the homologous distribution.

Experimental

The equipment includes an L6200 pump (Merck-Hitachi, Darmstadt, Germany), an evaporative light scattering detector (SEDERE, France), the temperature of the drift tube is set to 50°C, and an D2500M integrator (Merck-Hitachi, Darmstadt, Germany). The compounds are studied with two columns used specially for complex carbohydrates, an AX (ColoChrom, Gagny, France) (250x4.6 mm ID) 5µm and a Polyglycoplex (PolyLC Inc., Columbia, USA) (200x4.6 mm ID) 5µm, and an amino bonded stationary phase, Hypersil (SFCC, Neuilly-Plaisance, France) (150 x 4.6 mm ID) 5µm column. They are used with an oven, model 655A-52 (Merck-Hitachi, Darmstadt, Germany), thermostated at 30°C or 80°C. The flow rate was 1 ml/min and the volume of injection 20µl.
The eluents consist of mixtures of acetonitrile Ultra Gradient Grade (Mallinckrodt

Baker, Deventer, Holland) and water for HPLC (Elga). The exact eluent composition of the chromatographic systems chosen was acetonitrile/water 75/25 (V/V) for the carbohydrates columns and 85/15 (V/V) for the amino column.
Structural analysis was carried out with triple quadripole LC/MS/MS spectrometer API 300 (Perkin-Elmer Sciex Instruments).
The HP-CD were obtained from Wacker Chimie (Lyon, France). The relative degree of substitution (DS) equals 0.6 for α- and γ-cyclodextrins and 0.9 for β-cyclodextrins.

Results and Discussion

Volatile eluents based on acetonitrile-water mixtures are used with an evaporative light scattering detector (ELSD). The three stationary phases, devoted to complex carbohydrates separation, show a good separation of HP-α-CD derivatives at room temperature. The separation and resolution between homologous are maximised on decreasing water concentration till an acetonitrile-water mixture of 75/25 % (V/V) with the AX or Polyglycoplex columns and a mixture of 85/15 % (V/V) with the amino bonded phase (figure 1).

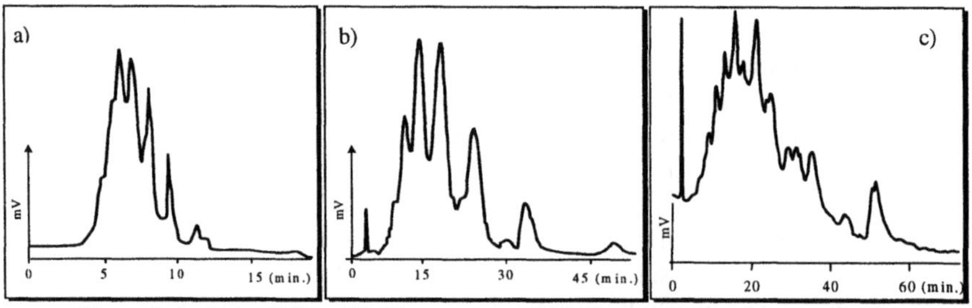

Figure 1. Analysis of HP-α-CD. Temperature : 30°C. Detector ELSD. Columns a) AX, b) Polyglycoplex, c) amino bonded phase

The isocratic elution of the HP-α-CD derivatives is better with the amino bonded phase than with the two other columns. To improve the separation, the elution is studied with both AX and Polyglycoplex columns (figure 2.a) at 30°C. With the two best chromatographic systems , amino bonded phase and coupled AX and Polyglycoplex columns, the temperature of columns is set at 80°C. The selectivity is decreased with amino bonded phase (figure 2.b) and increased with the two coupled columns (figure 2.c). The coupled system is used to study the separation of the HP-β-CD and HP-γ-CD derivatives (figure 3). With the ELSD, the relative intensity of the peaks can be directly link to the relative concentration of the homologous. With HP-β-CD, distribution of the homologous HP-CD is centred on the mean of the retention times. With HP-α-CD and HP-γ-CD, the distribution is centred on the low retention times. To identify the homologous HP-CD and to explain the influence of the substituent on retention, fractions of eluted sample are collected and analysed by MS/MS.

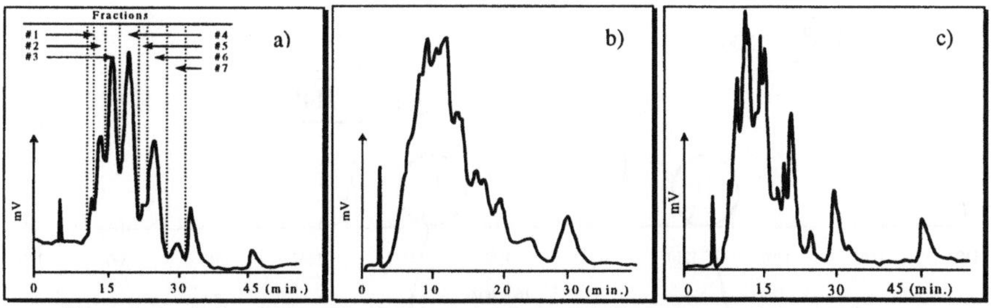

Figure 2. Analysis of HP-α-CD. Columns a) AX and Polyglycoplex coupled at 30°C and b) amino bonded phase at 80°C., c) AX and Polyglycoplex coupled at 80°C.

Seven fractions of the eluted HP-α-CD (figure 2.a) are collected at the outlet of the coupled columns. MS is performed for each fraction. From the MS spectra, a MS/MS analysis is done for each significant signal. An example of MS and MS/MS results of fraction #3 is given figure 4. The fraction #3 is a mixture of two homologous for which the number of HP substituents equals 4 and 5 characterised by the masses 1206 and 1264 (m/z amu) respectively. The MS/MS of the two signals indicate the presence of

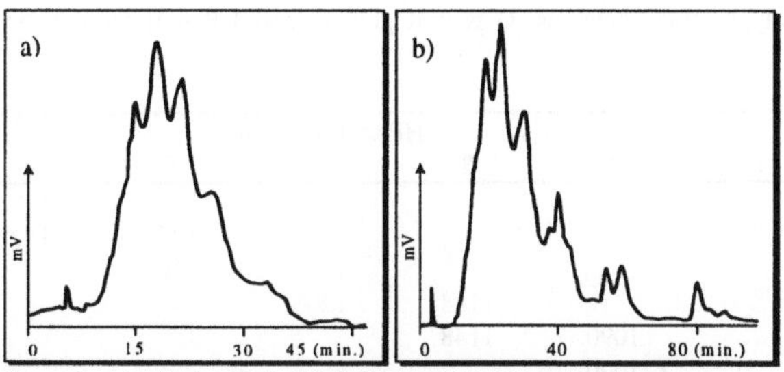

Figure 3 : Analysis of a) HP-β-CD and b) HP-γ-CD. Temperature : 30°C. Detector : ELSD. Columns AX and Polyglycoplex coupled.

mono, di or three substituted glucose or condensed HP (i.e. 1HP-1G, 2HP-1G and 3HP-1G). All MS analysis are summarised table I. From the MS analysis, it appears that the retention of HP-α-CD decreases by increasing the number of HP substituents. Considering a homologous compound at a given mass (m/z), isomers can be found in two successive fractions. This is confirmed by the MS/MS analysis which indicates different spectrum for the two samples. The relative intensities of the MS/MS signals, suggest that the retention increases when the number of HP substituent per glucose unit increases.

In conclusion, the LC and MS/MS analyses of HP-α-CD preparation which have a low relative degree of substitution (DS equals 0.6), show a wide variety of homologous or isomers. The number of HP substituent per glucose unit can be range from zero up to three. A condensation of HP group has been found. The diversity of the molecular substitution suggests the use of LC/MS technique with ELSD for structural and pattern determinations.

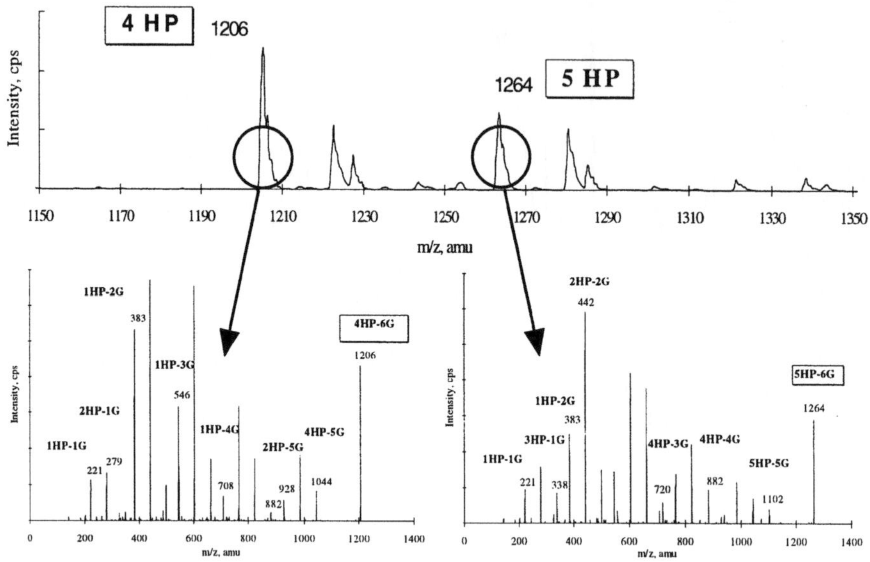

Figure 4. MS and MS/MS of the fraction #3 of a mixture of HP-α-CD eluted on coupled AX and Polyglycoplex columns. G is unit glucose and HP unit for the hydroxypropyl substituent

fraction	HP unit (m/z amu)						
#	1	2	3	4	5	6	7
1						1322	1380
2					1264	1322	
3				1206	1264		
4			1148	1206			
5		1090	1148				
6		1090					
7	1032	1090					

Table I. MS of the isolated fraction of HP-α-CD eluted on coupled AX and Polyglycoplex columns.

References

1. Pitha, J., Milecki, J., Fales, H., Pannell, L., Uekama, K. (1986) hydroxypropyl β-cyclodextrin : Preparation and characterization ; effects on solubility of drugs, International Journal of Pharmaceutics, 29, 73-82.
2. Liu, G., Goodall, D.M., Loran, J.S. (1993) Quantification of trace levels of β-cyclodextrin and substitution patterns in hydroxypropyl β-cyclodextrin using high performance liquid chromatography with polarimetric detection, Chirality. 5, 220-223.
3. Jindrich, J., Pitha, J., Lindberg, B. (1995) Separation of cyclodextrins and their derivatives by thin-layer and preparative column chromatography, Carbohydrate research, 275, 1-7.
4. Bielejewska, A., Kozbial, M., Nowakowski, R., Duszczyk, K., Sybilska, D. (1995) Studies of the behaviour of α-, β- and γ-cyclodextrins and some derivatives under reversed-phase liquid chromatographic conditions, Analytica Chimica Acta, 300, 201-206

LC-ELSD OR (AND) LC-MS : A WELL SUITED TECHNIQUE FOR CHARACTERIZATION OF THE DIFFERENCES IN THE COMMERCIALLY AVAILABLE DM-β-CDS

I. CARON, P. CHAIMBAULT, C. ELFAKIR, M. DREUX
Institut de Chimie Organique et Analytique, Université d'Orléans, CNRS UPRES-A 6005, BP 6759, 45067 Orléans cedex 2, France

1. Introduction

Heptakis (2,6-di-O-methyl)-β-cyclodextrin (DM-β-CD) is one of the most potent solubilizers of apolar guest molecules. However, all the methods used to obtain DM-β-CD give mixtures of partially methylated β-CDs with under- and overmethylated homologues.

Analyses which provide accurate characterization of commercially available DM-β-CDs are therefore necessary to control the quality of these compounds widely used in various fields such as in cosmetics, food technology, the pharmaceutical industry and in the production of agrochemicals.

The analytical methods currently used are thin layer chromatography [1-3], gas chromatography [1, 4, 5], and liquid chromatography [1, 2, 4-6].

In this work, our aim was to develop fast and effective methods for comparing the composition of some commercial DM-β-CDs and to offer LC systems which can be directly coupled with mass spectrometry and performed on the preparative scale to obtain methylated derivatives in high purity.

Furthermore, evaporative light scattering detection, which has proved useful for cyclodextrin determination [7-10], has been used before a complementary LC-MS analysis to detect the differences in the commercially available DM-β-CDs.

2. Material

DM-β-CDs come from different commercial sources. Pure DM-β-CD E is provided by Cyclolab (Budapest, Hungary).

3. Results and discussion

3.1 LC-ELSD

Greater knowledge of the interactions involved in the retention mechanism of methylated β-CDs [9, 10] allowed us to optimize two LC systems in reversed (RP-LC) and normal (NP-LC) phase modes.

Figures 1 and 2 show the chromatographic profiles of five commercially available DM-β-CDs in these two LC modes. It can be seen from the chromatograms that from one manufacturer to another the mixture composition is different.

In these chromatographic conditions, heptakis (2,6-di-O-methyl)-β-CD (2,6 DiMe) and hexakis (2,6-di-O-methyl)-mono(2,3,6-tri-O-methyl)-β-CD (15 OCH_3) gave respectively a chromatographic peak at tr = 17.2 min and tr = 19.2 min in RP-LC and a peak at tr = 15.5 min and tr = 14 min in NP-LC. This revealed that, in the two LC systems the elution order of methylated β-CDs is reversed. In NP-LC, the more methylated the CD was, the shorter the retention time. We also show that 2,6 DiMe and 15 OCH_3 are the major products in most of the commercial samples (DM-β-CDs B, C, D and E).

DM-β-CD A was the most complex mixture as it can be separated into 8-10 components depending on the chromatographic system and contains higher amounts of overmethylated homologues than the other samples (this result confirmed its higher DS = 2.21).

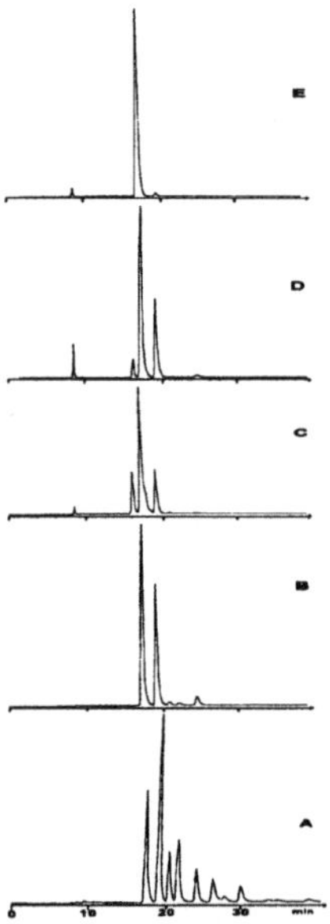

Figure 1 : Chromatograms of five commercial DM-β-CDs in RP-LC on Nucleosil 50-5-C8 (250 x 4 mm I.D.) column, 50 °C, eluent : ACN/water (34 : 66), flow rate : 1 $ml.min^{-1}$, detection ELSD.

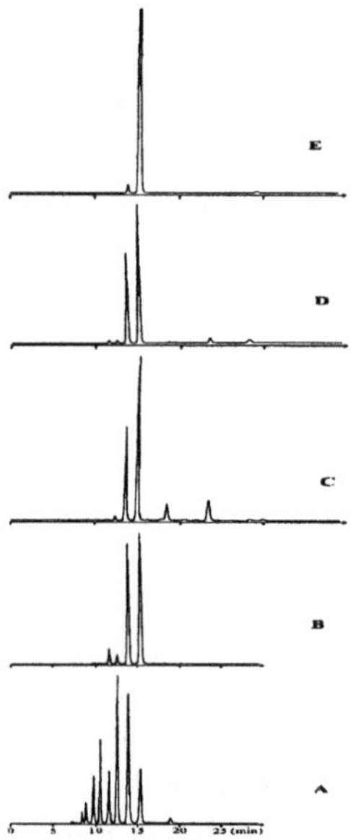

Figure 2 : Chromatograms of five commercial DM-β-CDs in NP-LC on Nucleosil 100-5-CN (250 x 4 mm I.D.) column, 20 °C, eluent : heptane/$CHCl_3$/MeOH (85 : 10 : 5), flow rate : 1 $ml.min^{-1}$, detection ELSD.

DM-β-CD E seemed to be the purest sample : only one major peak corresponding to 2,6 DiMe and two minor peaks are obtained in the two chromatographic systems. It is in good agreement with the manufacturer's claim that the amount of 2,6 DiMe in this sample was greater than 80 %. We also confirmed that DM-β-CD D from the same manufacturer was less purified than batch E.

It also underscored the fact that the two LC-ELSD systems differed from each other in selectivity.

In fact, DM-β-CDs C and D presented similar chromatographic profiles in RP-LC, whereas in NP-LC higher amounts of minor impurities were detected in the sample D. We also observed that the three positional isomers of pentakis (2,6-di-O-methyl)-tri (2,3,6-tri-O-methyl)-β-CD (16 OCH_3), which are present in sample B, gave two chromatographic peaks in NP-LC against three peaks in RP-LC.
The two LC-ELSD systems therefore provide complementary and characteristic fingerprints of each sample of DM-β-CD and can be used favorably as complementary techniques in the isolation and purification of a given isomer in a commercial mixture.

3.2 LC-MS

In order to obtain additional information on the composition of the two samples C and D for which the component distribution was found to be similar on Nucleosil 50-5-C8 column, the chromatographic systems were directly coupled with electrospray mass spectrometry (figures 3 and 4). Post-column addition of ammonium acetate ions was used to produce positive $[M+NH_4]^+$ ions.

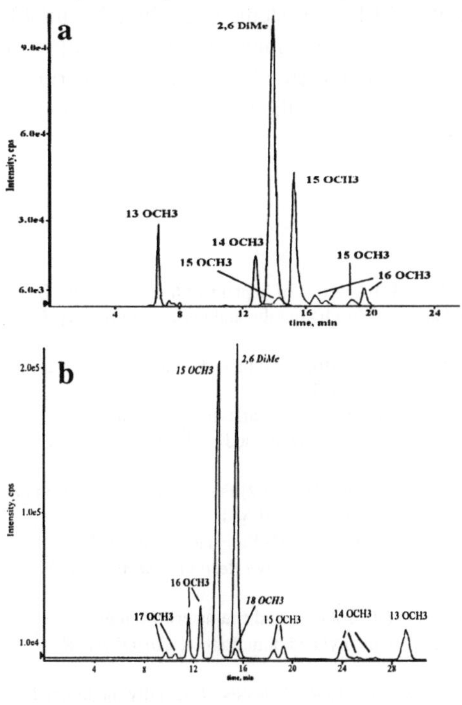

Figure 3 : mass spectrum of DM-β-CD C on a) Nucleosil 50-5-C8 column (eluent : ACN/water 34 : 66) and b) Nucleosil 100-5-CN (eluent : heptane/chloroform/methanol 85 : 10: 5)

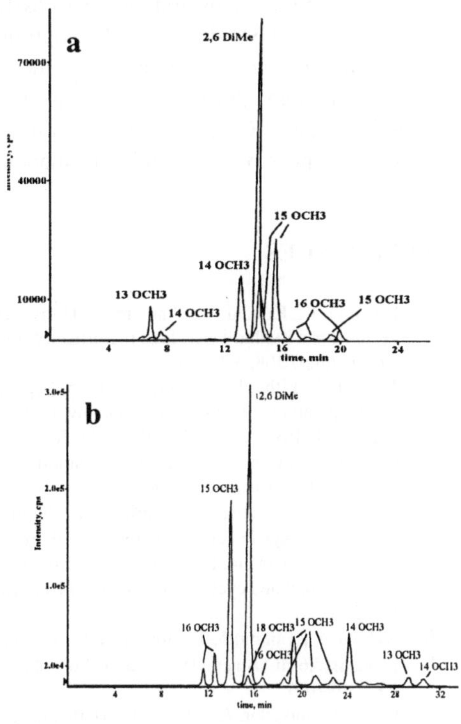

Figure 4 : mass spectrum of DM-β-CD D on a) Nucleosil 50-5-C8 column (eluent : ACN/water 34 : 66) and b) Nucleosil 100-5-CN (eluent : heptane/chloroform/methanol 85 : 10: 5)

LC-MS exacerbated the different selectivities obtained in RP-LC and NP-LC for each DM-β-CD sample analysis and the differences in composition between the two samples

C and D. The component eluted in front of the 2,6 DiMe peak in RP-LC for the two samples appeared to be a substitution isomer of 2,6 DiMe (it contains methyl groups in the C(3) position).

Other substitution isomers of 15 OCH_3 can also be observed in the two samples in various proportions. LC-MS showed that the selectivity between the different substitution isomers was higher in NP-LC than in RP-LC. In RP-LC, the coelution of a substitution isomer of 15 OCH_3 with 2,6 DiMe explained the non symetrical peak obtained for 2,6 DiMe in the sample D in figure1.

LC-MS analyses allowed to obtain complementary informations on the composition of the studied CDs, particularly, the presence of different substitution isomers in the samples. This system allowed too, to verify the quality and efficiency of the different optimized chromatographic systems.

4. Conclusion

Each mixture composition was successfully and rapidly verified by using two LC systems with ELSD detection. The optimized LC-ELSD systems were directly suitable to LC-MS. Using a combination of these two chromatographic systems will allow to isolate in high purity a given isomer in a commercial mixture. Furthermore, complementary LC-MS analyses provided informations on the molecular weight and the presence of possible substitution isomers in the mixture.

REFERENCES

1. Koizumi, K., Kubota, Y., Utamura, T., Horiyama, S. (1986) Analysis of heptakis(2,6-di-O-methyl)-β-cyclodextrin by thin-layer, high performance liquid and gas chromatograhy and mass spectrometry *J. Chromatogr.* 368, 329-337.
2. Jindrich, J., Pitha, J., Lindberg, B. (1995) Separation of cyclodextrins and their derivatives by thin-layer and preparative column chromatography, *Carbohydr. Res.* 275, 1-7.
3. Szeman, J., Roos, N., Csabai, K. (1997) Ruggedness of enantiomaeric separation by capillary electrophoresis and high-performance liquid chromatography with methylated cyclodextrins as chiral selectors, *J. Chromatogr. A* 763, 139-147.
4. Schomburg, G., Deege, A., Hinrichs, H., Hübinger, E., Husmann, H. (1992) Preparation, purification and analysis of alkylated cyclodextrins, *J. High Resolut. Chromatogr.* 15, 579-584.
5. Deege, A., Husmann, H., Hübinger, E., Kobor, F., Schomburg, G. (1993) Purification and analysis of partially alkylated cyclodextrins by liquid and gas chromatography, *J. High Resolut. Chromatogr.* 16, 587-593.
6. Kubota, Y., Tanimoto, T., Horiyama, S., Koizumi, K. (1989) Isolation and characterization of heptakis(2,6-di-O-methyl)-β-cyclomaltoheptaose and over-methylated homologues, *Carbohydr. Res.* 192, 159-166.
7. Caron, I., Salvador, A., Elfakir, C., Herbreteau, B., Dreux, M. (1996) Analysis of partially methylated cyclodextrins by subcritical fluid and liquid chromatography, *J. Chromatogr. A* 746, 103-108.
8. Caron, I., Elfakir, C., Dreux, M. (1997) Advantages of evaporative light scattering detection for the purity control of commercial cyclodextrins, *J. Liq. Chrom. & Rel. Technol.* 20, 1015-1035.
9. Caron, I., Elfakir, C., Dreux, M. (1998) Correlation between the chromatographic properties of silica phenyl packing materials and the retention behaviour of methylated β-cyclodextrins, *Chromatographia* 47, 383-390.
10. Caron, I., Elfakir, C., Dreux, M. Partially methylated β-cyclodextrin analysis : a systematic approach for a proper RP column selection, *J. High Resolut. Chromatogr.*, submitted.

THERMAL DEGRADATION OF CYCLODEXTRINS

F.TROTTA[1],V.MORONE[2], M. ZANETTI[1] and G.CAMINO[1]

1. *Dipartimento di Chimica IFM dell'Università di Torino, Via P. Giuria 7, 10125 Torino, ITALY*
2. *ISRIM, Località Pentima Bassa 21, 05100 Terni, ITALY*

1. Introduction

Owing to environmental issues, there is now a very important request for halogen-free fire retardant polymer materials.
A large research and development activity is concerned with the so called "intumescent" systems which char and blow on exposure to heat. The resulting multicellular char protects the underlying material from the action of flames.
Polysaccharides are most common components of such intumescent systems in which they act as the char source[1]. Cyclodextrins have been used in an intumescent system developed for polyethylene in which however their fire protection showed to be too poor[2].
In this work substituted cyclodextrins are employed. They show a strong relationship between nature of sobstituents and charring behaviour. An example is reported here concerning the mono-6-deoxy-6-amino-β-cyclodextrin (β-CD-NH_2). These results may be relevant for the development of effective intumescent systems based on cyclodextrins.

2. Materials and methods

β-cyclodextrin was kindly supplied from Roquette-Italia (Cassano Spinola-Italy). Other substituted cyclodextrins were synthetized according to the reported methods. All solvents and reagents employed (ACS grade) were bought from Aldrich (U.S.A.) and used without further purification.

2.1. THERMOGRAVIMETRIC ANALYSIS

Thermogravimetric analysis (TGA) have been performed on a Du Pont Instruments TGA 2950 thermobalance, with a heating ramp of 10°C/min to 800°C in nitrogen or air (60 cm^3/min).

2.2. ANALYSIS OF THE SOLID CHARS

The samples have been degraded in a thermobalance, interrupting the heating at various preset temperatures and the obtained solid chars have been analysed by infrared analysis (Perkin Elmer 2000 FTIR) in KBr pellet.

3. Results and discussion

Degradations carried out in nitrogen allow the comparison with previous work with cellulose[3],where the thermal degradations have been performed under vacuum.
The process of weigth loss of the analized cyclodextrins consists of three stages:

- the first at a temperature lower then 100°C, probably due to loss of absorbed water ($\leq$10%);
- the second stage develops in the temperature range between 250°C and 350°C and it is associated to a weight loss of 70÷80%, with the formation of residue ("char");
- the last stage is concerned with a relatively slow process of thermal degradation of the char.

Temperature of degradation, yield and thermal stability of the char depend on type of substituent in cyclodextrin.
As an example, figure 1 compares the thermogravimetric curves of β-cyclodextrin (β-CD) and of β-CD-NH_2.

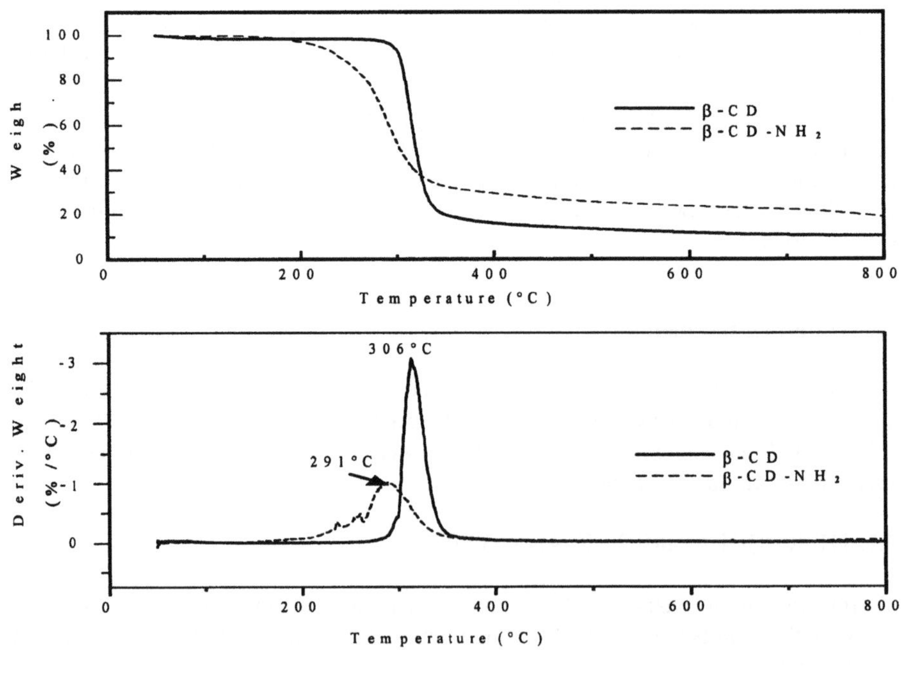

Figure 1

Infrared spectra of residues of heating of figure 2 confirm that in the first stage of weight loss the chemical structure of cyclodextrin is preserved. In the second stage, corresponding to the major weight loss (250÷355°C, weigth loss 80%) the glucosidic rings are decomposed as indicated by decrease of their tipical absorptions (1200÷900 cm^{-1}) while carbonyl (1704 cm^{-1}) and carbon-carbon double bonds (1631 cm^{-1}) are simultaneously created in the resulting thermally stable residue ("char", 355 °C). Water elimination is involved in this degradation step as shown by the strong decrease of OH and CH absorption at 3430 and 2920 cm^{-1} respectively.
Decomposition of the glucosidic structure as well as OH and CH concentration decrease in the residue seem more pronounced in the case of cyclodextrin than in cellulose[3].

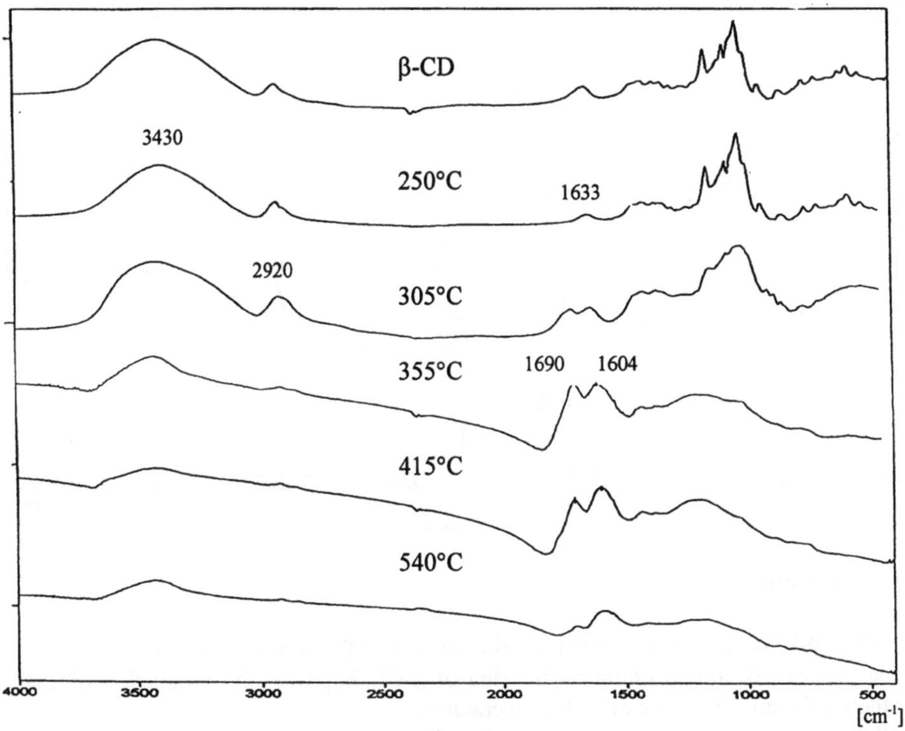

Figure 2

The amino-cyclodextrin begins to decompose to volatile products at lower temperature than β-cyclodextrin (175°C and 250°C respectively) giving however as much as twice the amount of residual char stable up to 800°C given by the latter one (20% instead 10%). Figure 3 shows that in the major weight loss stage (175°C, 260°C, 310°C, 420°C and 600°C) the glucosidic structure is decomposed as in the case of β-CD (figure 2).

However, figure 3 shows that while this process occurs, the C—NH_2 groups evolve to a nitrile structure (2230 cm^{-1}). At high temperatures (≥ 420 °C) there is evidence of new chemical bonds created on heating (1461, 962, 624 and 514 cm^{-1}) which are stable at least to 600°C and are responsible for increasing char yield in this amino modified cyclodextrin.

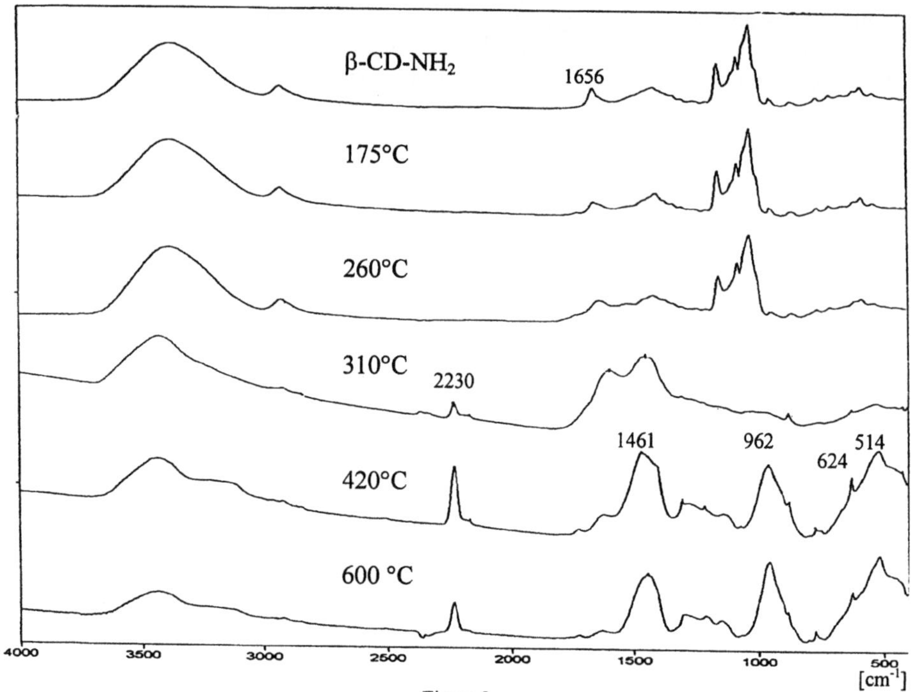

Figure 3

4. Conclusions

In this work we have studied the thermal behaviour of β-CD and β-CD-NH_2.
The presence of amino group on the ring of β-CD improves the production of char, which is thermically stable at high temperatures.
All that grants to cyclodextrins a primal interest for:

- Models for the study of thermal degradation mechanisms.
- Development of effective fire retardant polymer materials based on intumescent systems.

---

[1] Vandersall, H. L. (1971) Intumescent coating sistems, their development and chemistry, *J. Fire and Flammability* 2, 97.

[2] Le Bras, M., Bourbigot, S., Le Tallec, Y.& Laureyns, J. (1997) Synergy in intumescence-application to β-cyclodextrin carbonisation agent in intumescent additives for fire retardant polyethilene formulations, *Polymer Degradation and Stability* 56, 11-21.

[3] Soares, S., Camino, G., Levchik, S. (1995) Comparative study of the thermal decomposition of pure cellulose and pulp paper, *Polymer Degradation and Stability* 49, 275-283.

Chapter 2

CYCLODEXTRIN DERIVATIVES

SELECTIVE MODIFICATIONS OF CYCLODEXTRINS

ABDUL RAUF KHAN, PETER FORGO, KEITH J. STINE AND VALERIAN T. D'SOUZA*, *Department of Chemistry, University of Missouri-St. Louis, St. Louis, MO 63121, USA*

It is indeed a pleasure and an honor to present this talk at the beautiful and ancient city of Santiago de Compostela and I would like to thank the organizers for giving me an opportunity to do so. We all love cyclodextrins and we like to study them, use them, play with them and even make some money out of them. The question is, if they are so incredible molecules, why do we need to change them? The answer is that in spite of their extraordinary properties, limitations of available functional groups and rigidity of their structure restrict our ability to utilize them. A number of very prominent chemists have attempted to rectify this situation and have provided us with methods to modify them.[1] In this talk, I will present our contribution to this effort. I will first introduce the chemistry involved in selective modification of cyclodextrin, then I will discuss mono-modification of cyclodextrins in which we have attached the same group (4-methylamino-3-nitro-benzyl group) selectively at the 2-, 3- and the 6-positions. I will then present our contribution in per-modification of cyclodextrins.

Chemistry Involved in the Modification of Cyclodextrins: A Cyclodextrin molecule contains hydroxyl groups at the 2-, 3- and 6-positions and a hydrophobic cavity. Thus, the nucleophilicity of the hydroxyl groups and the ability of cyclodextrins to form complexes with the reagents used are the important factors to be considered in selective modification of cyclodextrins. Of the three types of hydroxyl groups present in cyclodextrins, those at the 6-position are the most basic (and often most nucleophilic), those at the 2-position are the most acidic and those at the 3-position the most inaccessible.[2,3] Thus, under normal circumstances, an electrophilic reagent attacks the 6-position. This situation can be reversed by the use of a strong base. Since the hydroxyl groups at the 2-position are the most acidic, they will be the first to get deprotonated.[4] The oxy-anion thus formed is more nucleophilic than the non-deprotonated hydroxyl groups at the 6-position. An interesting factor affecting the chemistry of the hydroxyl groups is provided by the ability of cyclodextrins to form complexes. If the electrophilic reagent forms a complex with cyclodextrin, then the orientation of the reagent within the complex introduces an additional factor in determining the nature of the product.

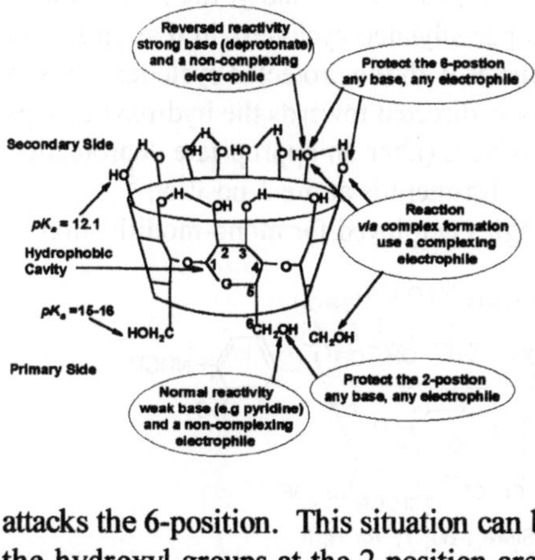

Mono-modification of Cyclodextrins: In support of the discussion presented above, we present syntheses of three derivatives in which the same group [*N*-methyl-4-chloromethyl-2-nitroaniline (**2**)][5] has been attached to three different (2-, 3- and 6-) positions of cyclodextrin.

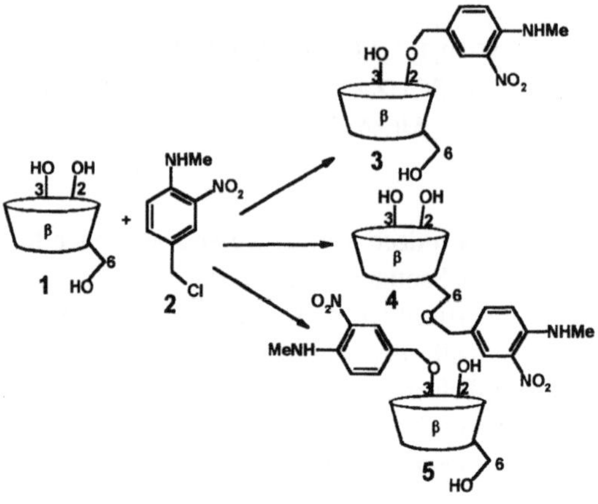

The substitution at the 2-position (**3**) was achieved by reacting cyclodextrin with a strong base (NaH) and reacting the resultant oxy-anion with the electrophile **2** to give desired 2-substituted cyclodextrin.[4] However, this electrophile does not exclusively afford the 6-substituted (**4**) product under conditions where the normal nucleophilicity would be retained[6] but always gives a mixture of both the 2- and 6-substituted products. We can explain this only by suggesting that it forms a complex with cyclodextrin such that the chloromethyl group of the reagent is oriented towards the secondary side of the host. The desired exclusively 6-substituted product (**4**) can only be formed after protection of hydroxyl groups at the 2-position with TBDMS groups. In this protection strategy it is estimated that on an average[7] six out of the seven hydroxyl groups are silylated which is sufficient prevent the incoming electrophile from attacking the secondary side and direct it to the 6-position.[8] Finally, the 3-substituted derivative (**5**) is obtained by reacting **2** with 6-per-silylated cyclodextrin **6** as shown in below. In this strategy, the electrophile **1** complexes with the protected cyclodextrin **6** in orientation where the reactive benzyl functionality is directed towards the hydroxyl groups at the 3-position. The subsequent reaction produces (after an appropriate deprotection step) a derivative of cyclodextrin in which the substituent is at the 3-position.[9]

Per-modification of Cyclodextrins: The strategies employed for mono-modification of

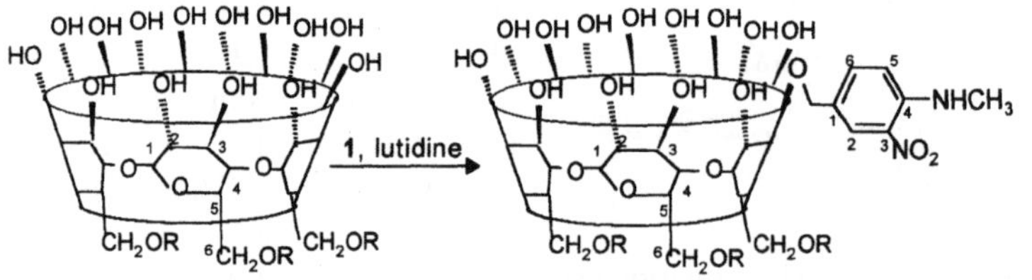

R= SiMe$_2$*t*-Bu: **6** R= SiMe$_2$*t*-Bu: **7**; R= H: **8**

cyclodextrins are not applicable in per-modification of cyclodextrins. We have developed and present here strategies for per-modification of cyclodextrins at the secondary as well as the primary side.

Per-modification of the Secondary Side of Cyclodextrins: The strategy used for per-modification of the secondary side is to protect the primary side and then react the protected cyclodextrin with an excess of the desired electrophile.

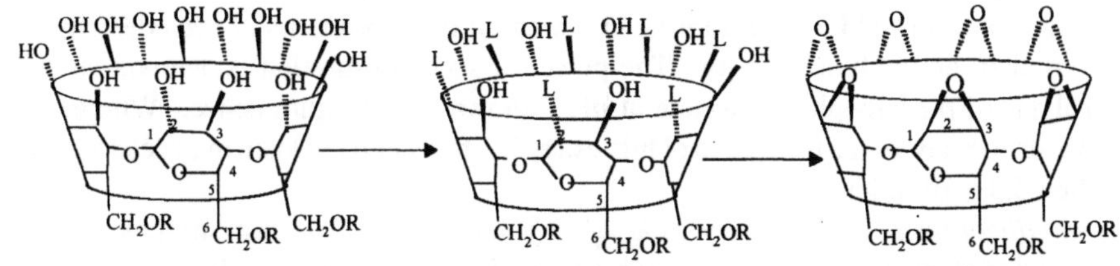

1 R= H;
6 R= $SiMe_2t$-Bu;

9 R= $SiMe_2t$-Bu;
10 R= H;
L = leaving group, $-OSO_2Ar$

11 R= $SiMe_2t$-Bu;
12 R= H;

Thus, per-6-silylated cyclodextrin **6** is reacted with excess of benzene sulfonyl chloride to yield per-2-sulfonated-6-silylated cyclodextrin **9**. Upon deprotection of the silyl groups, one obtains per-2-sulfonated cyclodextrin **10**. This derivative can be further reacted with a base to per-2,3-epoxy cyclodextrin **12** which is a valuable intermediate for the synthesis of a variety of per-3-substituted cyclodextrins. These per-2,3-epoxycyclodextrins can be obtained in a single pot without isolation of the intermediate by reacting **6** with benzene

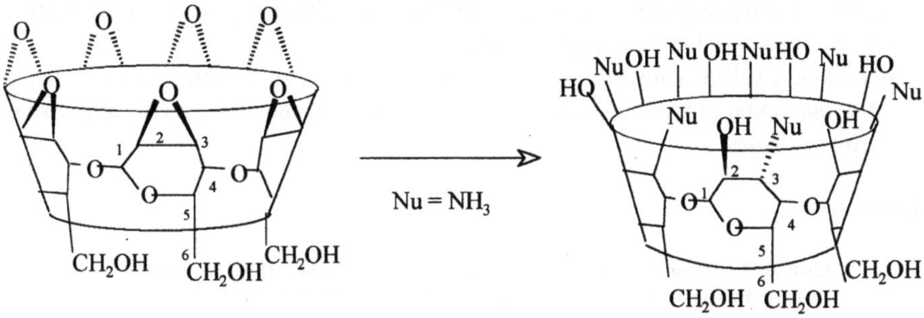

sulfonyl chloride in the presence of excess of a strong base (NaH) to give directly **11** which on deprotection gives **12**.[10]

As mentioned before **12** is an important intermediate for synthesis of a variety of 3-substituted cyclodextrins by a ring opening reaction as shown in the figure above. Thus, any nucleophile can react at the 3-position of **12** to give a cyclodextrin derivative in which all the 3-positions are substituted with the desired functionality.[11] It is important to note that this is not a

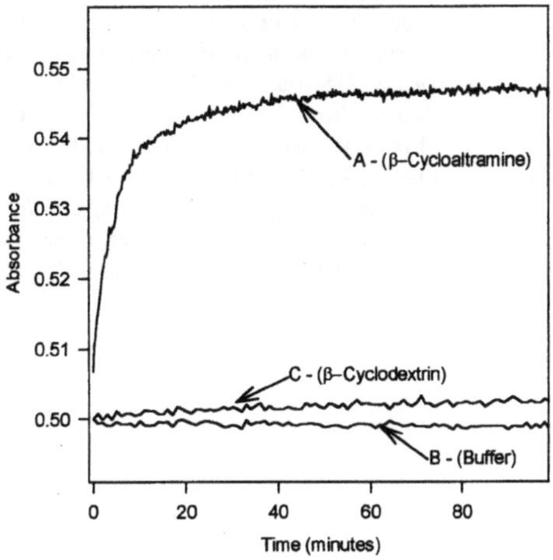

true cyclodextrin derivative but because of stereochemical inversions at both the 2- and the 3-positions, it is a derivative of altrose and is called cycloaltrin. We have reacted **12** with ammonia to yield cycloaltramine in which all the 3-positions of cyclodextrins are converted an amine functionality. The power of such substitution is demonstrated by hydrolysis of *p-t*-butylphenyl acetate at *p*H 7 shown in the figure above. Whereas, β-cyclodextrin catalyzes the hydrolysis of this ester, 7.2 folds and cycloaltrin accelerates this reaction 5.3 x 10^4 folds.[12]

Per-modification of the Primary Side of Cyclodextrins: The best strategy for per-modification of cyclodextrins on the primary side is through their halides. Simplest and the most convenient method for synthesis per-6-halogenated cyclodextrin is by its reaction with sulfonyl chloride in DMF and in the presence of imidazole.[13] Under these conditions, per-6-chlorocyclodextrin is formed in very high yield (>90%) and in high purity. This reaction is believed to go through Vilsmeier-Haack type complex. These chloro compounds are important intermediates in the synthesis of cyclodextrins derivatized at the 6-position. A variety of cyclodextrin derivatives can be synthesized by reacting these with appropriate nucleophiles. We have reacted these with thiols to give 6-thio derivatives which have been used in monolayer studies.[14]

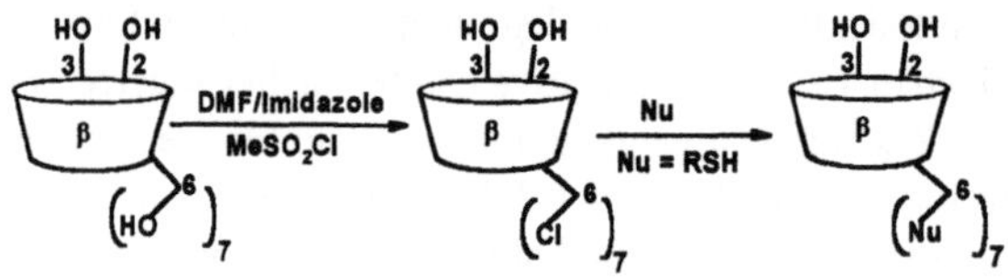

Conclusion: Cyclodextrins are limited in their uses because of lack of availability of desirable functional groups. However, these limitations can be overcome by appropriate synthetic strategies

References

1. Khan, A. R..; Forgo, P.; Stine, K. J. and D'Souza, V. T. *Chem Rev.* **1998**, *98*, in press.
2. Hybl, A.; Rundle, R. E.; William, D. E. *J. Am. Chem. Soc.* **1965**, *87*, 2779.
3. Saenger, W.; Noltemeyer, M.; Manor, P. C.; Himgerty, B.; Klar, B. *Bioorg. Chem.* **1976**, *5*, 187.
4. Rong, D.; D'Souza, V. T. *Tetrahedron Lett.* **1990**, *31*, 4275.
5. Rong, D.; Ye, H.; Boehlow, T. R.; D'Souza, V. T. *J. Org. Chem.* **1992**, *57*, 163.
6. Reaction conditions for these reactions include use a) pyridine or lutidine as solvent/base and b) silver oxide, silver carbonate or proton sponge as a catalyst in dry DMF
7. We suggest that the substitution pattern is a range from 4-7 with an average of approximately six.
8. Tian, S.; D'Souza, V. T. *Tetrahedron Lett.* **1994**, *35*, 9339.
9. Tian, S.; Forgo, P.; D'Souza, V. T. *Tetrahedron Lett.* **1996**, *37*, 8309.
10. Khan, A. R.; Barton, L., D'Souza, V. T. *J. Org. Chem.*, **1996**, *61*, 8301.
11. Khan, A. R.; Tong, W.; D'Souza, V.T. *Supramole. Chem.* **1995**, *4*, 243.
12. Groaning, M. D..; D'Souza, V.T. *Supramole. Chem.* **1996**, *8*, 5.
13. Khan, A. R.; D'Souza, V. T. *J. Org. Chem.* **1994**, *59*, 7492.
14. Parazak, D. P.; Khan, A. R.; D'Souza, V. T.; Stine, K. J. *Langmuir* **1996**, *12*, 4046.

CYCLODEXTRIN DERIVATIVES WITH ENHANCED SOLUBILIZING POWER AND LOWER TOXICITY

JOSEF PITHA, *417 Anglesea Street, Baltimore, MD 21224. USA*

1. Introduction

Mixtures of partially substituted cyclodextrins (CDs) contain potentially millions of compounds, each with its own individuality. When the large number of compounds is considered, along with our inability to fully analyze complex mixtures, it is likely that various preparations with the same collective name (e.g., partial methyl ethers of cyclodextrin) contain groups of compounds with little overlap. Thus, in the search for new compositions with better properties, study of derivatives with previously used substituents may be as fruitful as that of derivatives containing sustituents not yet evaluated. Furthermore, CD derivatives with better properties are needed since many applications with theoretical potential have only marginal performance when tested.

This presentation addresses three groups of partial CD derivatives: (1) methyl ethers, (2) glycerol cyclic ethers (i.e., derivatives which have a 1,4-dioxane ring fused to glucose residues) and (3) tertiary butyl ethers. In this work, we actively used rules (shown in Table 1) which govern the substitution patterns in CDs.

	Acid Medium	Neutral	Basic Medium
Mechanism	CD Reacts with Cation of Alkylating Agent $CD\text{-}OH + (CH_3)_3C^+ \rightarrow CD\text{-}O\text{-}C(CH_3)_3 + H^+$	**No Reaction**	CD Forms Anions Which React with Alkylating Agent $CD\text{-}O^- + CH_3I \rightarrow CD\text{-}OCH_3 + I^-$
Substitution	The Most Accessible Hydroxyls (Primary) Are Substituted		Low Basicity: Only a Few Hydroxyls (Secondary) Form Anions; These Hydroxyls Are Substituted High Basicity: Many Hydroxyls Form Anions (Primary & Secondary); Those Most Accessible are Substituted

Table 1. Rules Governing Substitution of Cyclodextrins

To prepare the methyl and glycerol cyclic ethers, we used conditions of low basicity to maximize the selectivity of substitution. It is important to note that the basicity should not be decreased below a certain level because then reactions proceed too sluggishly. To denote these principles, we call these conditions "minimal effective basicity" and realize them, for example, by using an excess of suspended calcium hydroxide as a base. Aqueous solutions over suspended calcium hydroxide have a pH close to 12.4; if the pH in such suspensions decreases due to the progress of alkylation, more calcium hydroxide is dissolved and the pH remains the same. Similar buffering systems were developed on the basis of alkali salts of water insoluble acids, for example, sodium aluminate. In this case, water insoluble aluminum hydroxide may precipitate as the alkylation progresses. When using these systems, the final results are given by the combination of low basicity and coordination of CD anions. Consequently, these systems yield somewhat different compositions of products.

2. Methyl Ethers of CDs

Although methyl ethers of CDs have been intensively investigated, our data show that improvements in their properties are still possible using conditions of minimal effective basicity. Figure 1 shows data on solubilization power. Solubilizations of hydrocortisone by hydroxypropyl beta-CD (HPBCD), commercial randomly methylated beta-CD (RAMEB) and heptakis-2,6-dimethyl beta-cyclodextrin (DIMEB) are compared to that of several methyl beta-CDs prepared in conditions of minimal effective basicity (denoted by numbers in Figure 1). The new derivatives consistently outperformed HPBCD, RAMEB and DIMEB.

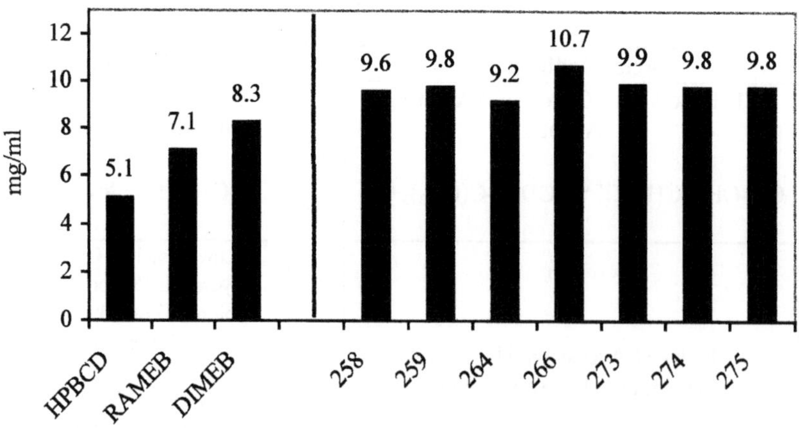

Figure 1. Solubilization of Hydrocortisone in 5% (w/w) Solutions of Methyl Beta-CDs Prepared in Conditions of Minimal Effective Basicity Versus Other Ethers of Beta-CD

Not only were the new preparations better solubilizers, they also were less toxic that DIMEB. Figure 2 shows that none of the mice receiving ip. injections of minimal basicity methyls died over a five-day period, whereas all of those receiving DIMEB did.

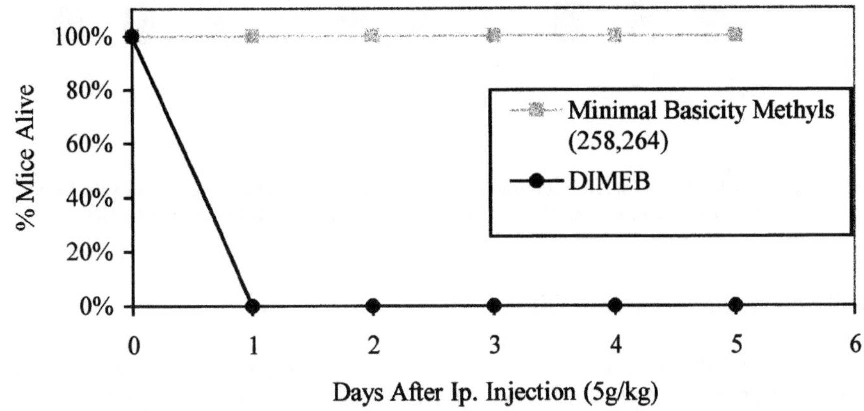

Figure 2. Toxicity of Methyl Beta-CDs Prepared in Conditions of Minimal Effective Basicity Versus DIMEB

3. CD-Glycerol Cyclic Ethers

Many studies have focused on reactions of CDs with alkylating agents having two alkylating moieties, but only recently we found that cyclic ethers of CDs can be formed when conditions are favorable. For example, in the reaction of CD with epichlorohydrin, cyclic ethers of glycerol and CD were formed in which the secondary hydroxyls of a glucose residue of CD become bridged by a $-CH_2.CH(CH_2OH)-$ group. The problem we faced here was that epichlorohydrin, in addition to forming these elegant bridges, also crosslinks different CD molecules to oligomers and polymers; additionally, the bridge formed from epichlorohydrin may not close and remain open. CD then is substituted with a dihydroxypropyl group. The early preparations of this type, in spite of having these undesired components present in detectable amounts, had promise. They were studied by researchers from Cyclolab, by Drs. Irie and Donova (who also studied some of the new methyl derivatives) and by Drs. Masson and Loftsson. Recently, we found the conditions to suppress these byproducts. Figure 3 shows that the CD-glycerol cyclic ethers (denoted by numbers in Figure 3) had superior solubilization power as compared to HPBCD, DIMEB and dihydroxypropyl beta-CD (DHPBCD). Solubility comparisons with DHPBCD show the advantages of closing the bridges.

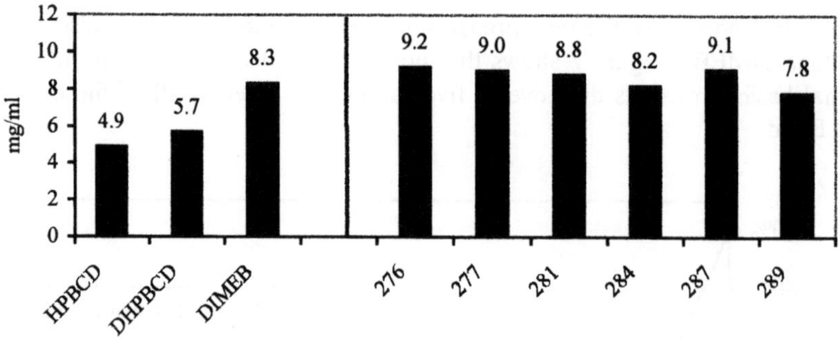

Figure 3. Solubilization of Hydrocortisone in 5% (w/w) Solutions of Beta-CD-Glycerol Cyclic Ethers Prepared in Conditions of Minimal Effective Basicity Versus Other Ethers of Beta-CD

4. Tertiary-Butyl Ethers of CD

In collaboration with Dr. Jindrich, we also investigated partial tertiary butyl ethers of CD. It is not possible to make these ethers by the alkylation of CDs in alkaline media -- the method used to prepare all other CD ethers. Acid catalyzed reaction of CDs with isobutylene had to be used instead. These ethers have high surface activity -- nearly as high as commercially used surfactants (Figure 4).

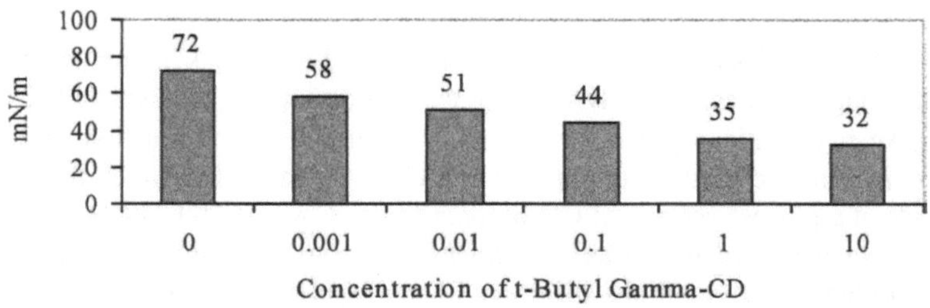

Figure 4. Surface Activity of Various Concentration of t-Butyl Gamma-CD

Furthermore, some of these ethers form inclusion complexes similar to other CDs. Solubilization of hydrocortisone in 5% (w/w) solutions of t-butyl gamma-CD was 8.2 mg/ml compared to 5.1 mg/ml for HPBCD.

5. Conclusion

Derivatives of CDs made using new chemistry differ enough from those in use to warrant thorough evaluation.

α-CYCLOALTRIN: CONFORMATION AND PROPERTIES IN THE SOLID-STATE AND AQUEOUS SOLUTION[1]

S. IMMEL, G. E. SCHMITT, AND F. W. LICHTENTHALER

Institute of Organic Chemistry, Darmstadt University of Technology, D-64287 Darmstadt, Germany

Abstract. Analysis of the conformational properties of α-cycloaltrin (α-CA) in aqueous solution by 800 MHz ^{1}H, 200 MHz ^{13}C NMR, and molecular dynamics (MD) simulations points towards a complex equilibrium of $^{4}C_1 \leftrightharpoons {}^{O}S_2 \leftrightharpoons {}^{1}C_4$ altropyranose units. Although the $^{3}J_{H\text{-}H}$ coupling constants do not reveal a preference for the alternating $^{4}C_1$ / $^{1}C_4$ or the *all*-$^{O}S_2$ conformation of α-CA, low-temperature ^{13}C NMR line-broadening indicates at least two different conformations of the altrose residues. Both α-CA geometries are stable during 600 ps MD simulations without conformational transitions, but constrained MDs forcing one altropyranose unit to vary along the $^{4}C_1 \rightarrow {}^{O}S_2 \rightarrow {}^{1}C_4$ reaction coordinate indicates cooperative conformational transitions $^{1}C_4 \rightarrow {}^{O}S_2 \leftrightharpoons {}^{3,O}B$ of neighboring units and statistical scrambling of the pyranose geometries in the macrocycle. In particular, the *all*-$^{O}S_2$ conformation of α-CA features a central cavity capable to form inclusion complexes, whereas alternate forms may have surface indentations only.

1. Introduction

In the course of our molecular modeling studies on the hydrophobic characteristics of the cyclodextrins[2], we have pointed out that inversion of the configuration at C-2 (→ D-mannose) leads to a significantly enhanced lipophilicity of the central cavities in the corresponding cyclomannins[3]. The opposite is to be expected for the C-3 epimers of the cyclodextrins, i.e. the as yet unknown cycloallins, although the axial 3-OH groups, now reaching into the cavities, are likely to introduce additional strain on the macrocycles and, hence, distortion. Even more intriguing are the synthetically readily accessible cycloaltrins[4,5], since two axial hydroxyl groups not only impose strain on the macrorings, but also render the monosaccharide units with considerable flexibility.

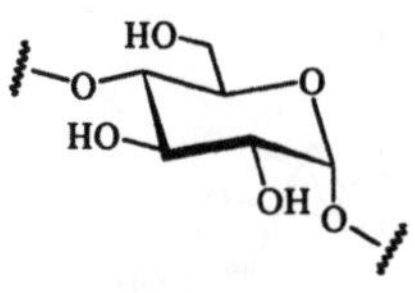

cyclodextrins

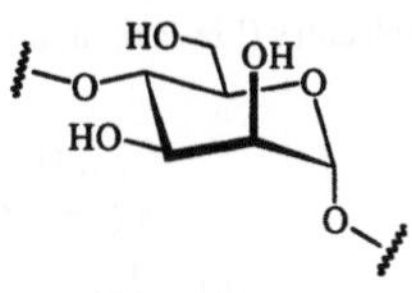

cyclomannins

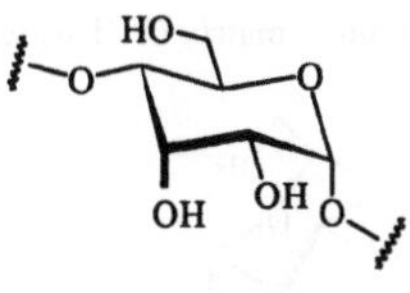

cycloallins

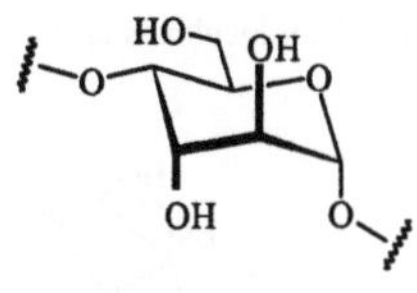

cycloaltrins

The very small energy difference for the $^{4}C_1$ (shown above) and $^{1}C_4$ conformation of α-D-altropyranose – computed values vary from +0.8[6] to -1.5 kJ/mol[7] – leads to a unique solid-state structure of α-cycloaltrin (**1**)[5], in which nearly perfect $^{4}C_1$ and $^{1}C_4$ chairs are arranged in an alternating sequence in a C_3 symmetrical conformation (cf. Fig. 1). Although this geometry lacks a "through-going" cavity, HTA calculations of **1** in vacuum yield an *all*-skew (twist-boat) $^{O}S_2$-geometry as the global energy-minimum structure, which features a cavity[5,8].

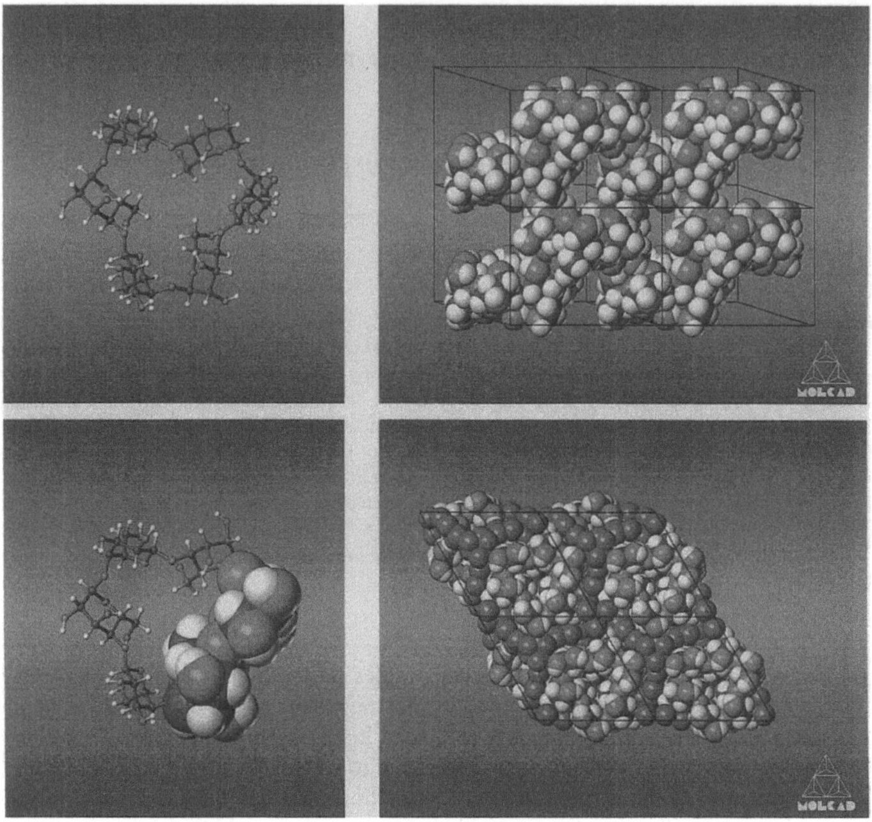

Figure 1. The α-cycloaltrin (**1**) solid-state structure features a rather compact conformation of 3-fold rotational symmetry with an alternating sequence of $^4C_1$ and $^1C_4$ altropyranose chairs (upper left); the constituent $^4C_1/^1C_4$-disaccharide units — of which one is represented in space filling form (lower right) — are of banana-type shape. In the crystal lattice, the molecules are stacked in transposed layers on top of each other (upper right side-view, water molecules were left of for clarity). The crystals are heavily hydrated, and each of the α-CA molecules is embedded into a matrix of 21 water molecules (lower right, top view).

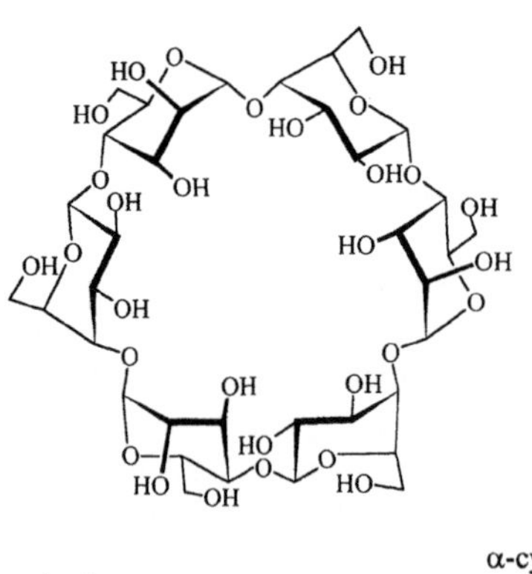

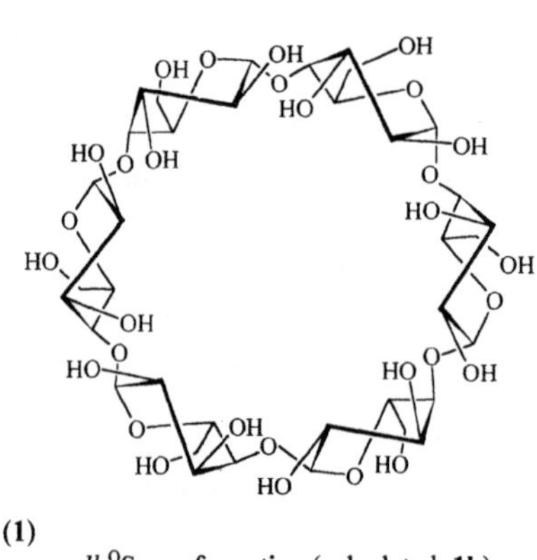

α-cycloaltrin (**1**)

$^4C_1/^1C_4$ conformation (solid-state, **1a**)

all-$^OS_2$ conformation (calculated, **1b**)

To get some insight into the conformation of α-cycloaltrin in aqueous solution we performed molecular dynamic simulations and temperature dependent NMR studies.

2. Methods

Temperature dependent ^{1}H (800 MHz) and ^{13}C (200 MHz) spectra were recorded at the "Large Scale Facility for Biomolecular NMR" in Frankfurt[9].

HTA[5] and the MD calculations were carried out using the CHARMM[10] force-field[11], comprising application of periodic boundaries (truncated octahedron filled with 606 (**1a**) and 611 (**1b**) water molecules, respectively), isothermal ($T = 300$ K), and isobaric conditions ($p = 1$ bar) during a simulation time of 600 ps (timestep 1 fs); analysis of the MD trajectories were accomplished with external computer programs[12].

3. Results and Discussion

Temperature Dependent NMR Analysis. — High-resolution ^{1}H and ^{13}C NMR spectroscopy of **1** (D_2O, 30 °C) displays only one set of signals for the six altropyranoses, indicating that, alternatively, fast conformational averaging between different pyranose ring forms takes place, or that – in contrast to the solid-state – a unique altrose geometry prevails in solution.

The ^{1}H NMR coupling constants are consistent with either scenario: as indicated in Table 1, the experimental values agree with an average between the $^4C_1$ and $^1C_4$ forms (geometry **1a**), but correlate also surprisingly well with the values calculated for the *all*-$^OS_2$ form **1b**.

Table 1. Comparison of the altropyranoid ring ^{1}H-^{1}H-coupling constants calculated for the $^4C_1$, $^1C_4$, and *all*-$^OS_2$ (skew) forms of α-cycloaltrin with those found (800 MHz, D_2O, 30 °C).

$^3J_{HH}$ [Hz][a]	$^4C_1$[b]	$^1C_4$[b]	$^OS_2$[c]	found	calc.[d]
$J_{1,2}$	2.2	8.0	4.0	4.73	4.9
$J_{2,3}$	2.7	10.2	10.2	8.19	8.3
$J_{3,4}$	3.4	2.4	4.5	3.71	3.5
$J_{4,5}$	9.7	1.4	6.4	5.32	5.5

[a] Calculation based on the generalized Haasnoot-equation[13]. [b] calc. for the pyranose conformations in **1a**. [c] calc. for **1b**. [d] least-squares fit for 26% $^4C_1$, 34% $^1C_4$, and 40% $^OS_2$ (cf. text)[14].

The presence of conformational equilibria in aqueous solutions of **1** is indicated by the temperature dependency of its ^{1}H NMR data: the value of $J_{2,3}$ decreases from 8.2 Hz at 30 °C to 7.6 and 7.1 Hz at 20 and 4 °C, respectively, thereby narrowing the well separated signal for 3-H. Although the decreasing value of $J_{2,3}$ with decreasing temperature points towards the $^4C_1$ altrose conformation to be the most stable form in the $^4C_1 \leftrightarrows {}^OS_2 \leftrightarrows {}^1C_4$ pyranose pseudorotational itinerary, as this form shows the lowest calculated coupling constant (cf. Table 1), the best fit of calculated vs. measured coupling constants was achieved for an approx. 1 : 1 : 1 mixture of $^4C_1$, $^OS_2$, and $^1C_4$ forms[14] (cf. Table 1).

More profound effects are observed for the ^{13}C signals for C-4 and C-5, which are substantially broadened upon lowering the temperature from 30 to 4 °C (cf. Fig. 2), providing further evidence for a dynamic equilibrium of various altrose geometries which starts to "freeze-out" at 4 °C.

However, the NMR data do neither provide a picture how different conformations are accessible within the "strait-jacket" of the α-CA macrocycle, nor do they answer the question how conformational transitions occur in **1** (e.g. **1a** ⇋ **1b**).

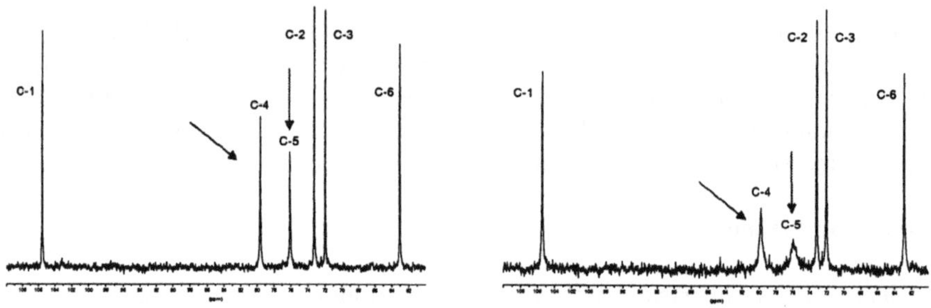

Figure 2. ^{13}C NMR spectra (200 MHz) of α-cycloaltrin (**1**) in D_2O at 32 °C (left) and 4 °C (right). As indicated by the arrows, the signals of C-4 and C-5 undergo significant broadening at the lower temperature.

$^4C_1$ / $^1C_4$ conformation **1a** *all-* $^OS_2$ conformation **1b** MD simulation box

Figure 3. Superimpositions of 60 α-cycloaltrin snapshot-geometries taken in 10 ps intervals from MD simulations of **1a** (left) and **1b** (center) display the flexibility of the macrocycles, although the pyranose ring conformations are rather rigid. On the right, a typical MD simulation box (truncated octahedron) filled with water is depicted.

Figure 4 (opposite page). Constrained MD simulations of α-CA display a conformational transition of the central $^4C_1$ altropyranose (gray shaded unit **B**) of a $^1C_4$-$^4C_1$-$^1C_4$-fragment (units **A-B-C**) in the macrocycle: as **B** is driven along the $^4C_1 \rightarrow {}^{3,O}B$ / $^OS_2 \rightarrow {}^1C_4$ reaction coordinate (formulas, left column), the conformation of the neighboring residue **A** varies very little with slight distortions $^1C_4 \rightarrow {}^3H_2$ only, but the altrose unit **C** cooperatively flips into the $^{3,O}B \leftrightharpoons {}^OS_2$ geometry (zoomed structures on the right). The unperturbed and cooperative conformational transitions of adjacent pyranose units point towards complex dynamic processes, in the course of which the altrose residues cannot be considered independent; each conformational change induces new strains in neighboring units (all α-CA snap-shot structures were taken from a 1300 ps MD perturbation, and constraints were applied to unit **B** only; for details see text).

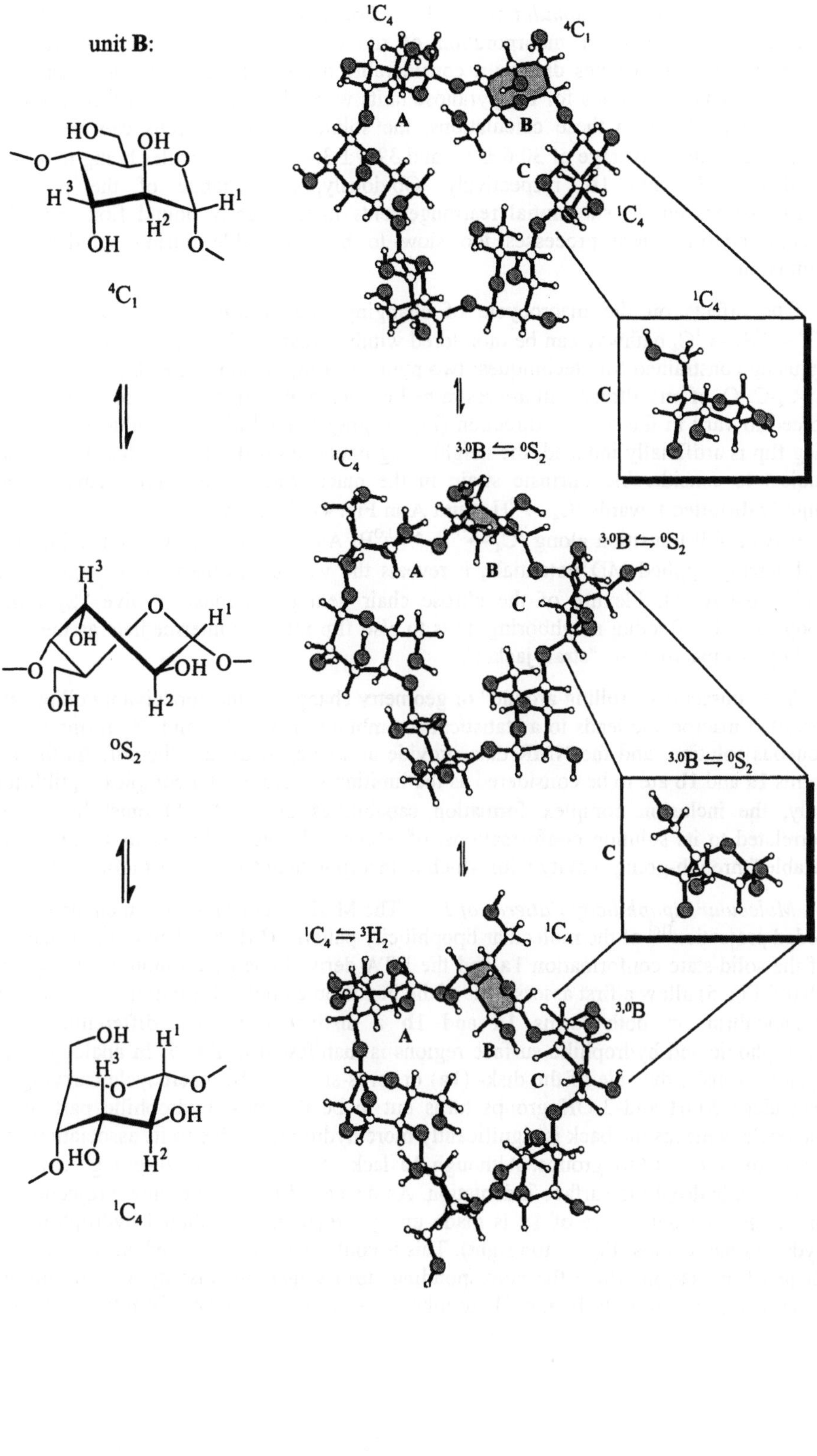

unit B:
HO
OH
O
H3
H1
H2
OH
O
4C1
H3
O
H1
OH
O
OH
OH
H2
0S2
OH
H1
H3
O
O
OH
O
HO
H2
1C4
1C4
4C1
A
B
C
1C4
1C4
C
3,0B ⇋ 0S2
1C4
A
B
C
3,0B ⇋ 0S2
3,0B ⇋ 0S2
C
1C4 ⇋ 3H2
1C4
A
B
C
3,0B

Molecular Dynamics Simulations of 1. — For both α-CA forms **1a** and **1b**[8] MD simulations with explicit incorporation of the (aqueous) solvent were performed. Although the macrocycles display a considerable degree of flexibility, no significant conformational transitions for the pyranose units were observed within a time frame of 600 ps (Fig. 3). From these calculations, the solute hydration shells can be readily identified, with an average of 30.6 ± 2.7 and 39.6 ± 3.4 water molecules being hydrogen bonded to **1a** and **1b**, respectively. Obviously, any change of the pyranose conformations entails substantial rearrangements in the tightly bound first hydration shell, rendering these processes too slow to be observable during standard MD simulations.

The effects on the macrocycle of changing one altrose unit in **1a** along the $^4C_1 \rightarrow {}^OS_2 \rightarrow {}^1C_4$ pathway can be monitored within a reasonable nanosecond time frame by using constrained MD techniques: two pyranose ring torsion angles O_5-C_1-C_2-C_3 and C_3-C_4-C_5-O_5 of one altrose unit are restrained in such a way, that the $^4C_1$ conformation is forced to vary in the desired direction (Fig. 4, gray shaded altrose residue **B**). As this ring flip is artificially induced, the neighboring unconstrained monosaccharide units are subject to considerable intrinsic strain in the macro-ring: as one unit remains only slightly distorted towards $^1C_4 \rightarrow {}^3H_2$ (unit **A** in Fig. 4), the second adjoining residue (**C**) displays a full transition along $^1C_4 \rightarrow {}^OS_2 \leftrightarrows {}^{3,O}B$. As this ring-flip was not induced by additionally applied MD potentials, it reveals the way conformational changes occur within α-CA (**1**). Flexure of the altrose chair geometries must involve $^OS_2$ forms, cooperatively affecting neighboring units to also flip into this intermediate as these are tied up in a macrocyclic "strait-jacket".

The consecutive "rolling around" of geometry changes in the conceivably elliptically distorted macrocycle leads to a statistical scrambling of $^4C_1$, $^1C_4$, and $^OS_2$ forms in **1** in aqueous solution, and the NMR data provide an averaged picture thereof. As the two forms **1a** and **1b** are to be considered as the limiting structures of a complex equilibrium only, the inclusion complex formation capabilities of α-CA (**1**) must be closely correlated to its solution conformations, of which only the *all*-$^OS_2$ form **1a** features a sizable "through-going" cavity with which to incorporate potential guest molecules.

Molecular Lipophilicity Patterns of 1. — The MOLCAD-program[15] mediated color-coded projection[16] of the molecular lipophilicity patterns (MLPs)[17] onto the surfaces[18] of the solid-state conformation **1a** and the HTA-derived energy-minimum *all*-$^OS_2$ form **1b** (cf. Fig. 5) allow a first assessment of the inclusion complex formation capabilities of α-cycloaltrin: in both forms **1a** and **1b** a distinct front/back differentiation of hydrophobic and hydrophilic surface regions is manifested in Fig. 5. In analogy to the cyclodextrins[2], the side of the disk- (**1a**) or torus-shaped (**1b**) macrocycle carrying the secondary 2-OH and 3-OH groups turns out to be the most hydrophilic part of the molecule, whereas the back is significantly more hydrophobic due to its association with the primary 6-CH_2OH groups. Although **1a** lacks a central "through-going" cavity, it features a hydrophilic surface indentation. As displayed in the side-view projections of the MLPs, the outer core of **1a** is made up by irregularly distributed hydrophilic and hydrophobic regions (Fig. 5, top right). This is contrasted by the cyclodextrin-like torus-shaped from **1b**, in which the corresponding surface qualities display a more uniform pattern (Fig. 5, bottom): the most hydrophobic surface areas extend from the 6-OH side well into the cavity. However, in relation to the cyclodextrins the *all*-$^{O}S_2$ form is characterized by an inverse tilt of the pyranose units in the macrocycle. Unlike α-CD[2], the cycloaltrin cavity is narrow at its hydrophilic end and wider at the hydrophobic surface. As the form **1b** of α-CA is the first cyclooligosaccharide of *"inverse conicity"* with an extended hydrophobic cavity, we expect — if this conformation is adopted — a high potential of this compound to form inclusion complexes in solution and the solid-state.

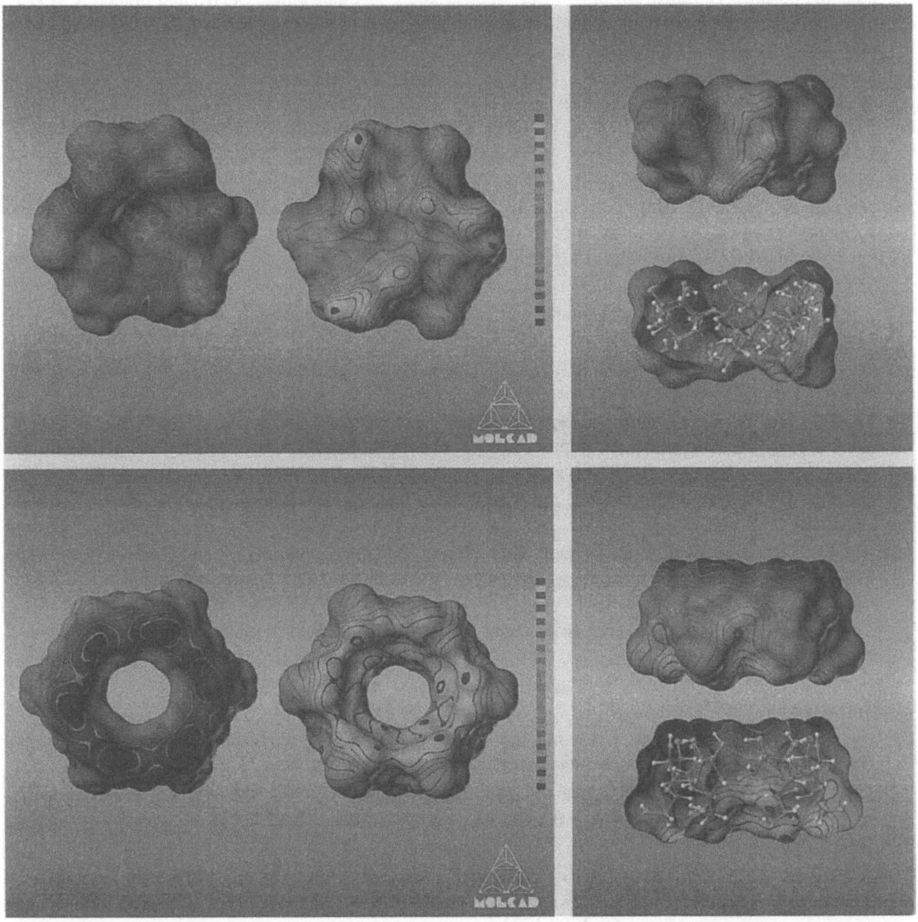

Figure 5. Molecular lipophilicity patterns (MLPs) of α-CA (**1**) projected onto the contact surfaces of the solid-state conformation (**1a**, top) and the *all*-$^{O}S_2$ form (**1b**, bottom). The color-code applied ranges from dark blue for the most hydrophilic surface areas to full yellow corresponding to the most hydrophobic regions. On the left each, the macrocycles are viewed from both sides perpendicular to the mean ring plane, exposing either the pronouncedly hydrophilic (blue) 2-OH/3-OH side of the torus or the opposite, hydrophobic rim made up by the primary 6-CH_2OH hydroxyls. The right graphics display the corresponding side-views in closed and bisected form each; the 2-OH and 3-OH groups are directed upwards, whereas the 6-OH point downwards.

References

[1] 'Molecular Modeling of Saccharides', Part 18. — Part 17: Gohlke, H.G., Immel, S., Lichtenthaler, F.W., and Schmitt, G.E. (1998). Molecular geometries of furanoid β(1→3)- and β(1→6)-linked cyclogalactins, in Torres Labandeira, J.J. (Ed.), *Proceedings 9th Internat. Symp. on Cyclodextrins*, Kluwer Acad. Publ., Dordrecht, NL.

[2] Lichtenthaler, F.W., and Immel, S. (1996). On the hydrophobic characteristics of cyclodextrins: computer-aided visualization of molecular lipophilicity patterns. *Liebigs Ann. Chem.*, 27-37.

[3] Lichtenthaler, F.W., and Immel, S. (1994). Cyclodextrins, cyclomannins, and cyclogalactins with five and six (1→4)-linked sugar units: a comparative assessment of their conformations and hydrophobicity potential profiles. *Tetrahedron: Asymmetry*, *5*, 2045-2060. — Lichtenthaler, F.W., and Immel, S. (1996). The lipophilicity patterns of cyclodextrins and non-glucose cyclooligosaccharides. *J. Inclusion Phenom. Mol. Recognit. Chem.*, *25*, 3-16.

[4] Fujita, K., Shimada, H., Ohta, K., Nogami, Y., Nasu, K., and Koga, T. (1995). β-Cycloaltrin: a cyclooligosaccharide consisting of seven α-(1→4)-linked altropyranoses. *Angew. Chem. Int. Ed. Engl.*, *34*, 1621-1622. - Nogami, Y., Fujita, K., Ohta, K., Nasu, K., Shimada, H., Shinohara, C., and Koga, T. (1996). A new synthetic strategy of cyclooligosaccharides: cyclodextrin-derived cycloaltrins made up from α-(1→4)-linked altropyranoses. *J. Inclusion Phenom. Mol. Recognit. Chem.*, *25*, 57-60.

[5] Nogami, Y., Nasu, K., Koga, T., Ohta, K., Fujita, K., Immel, S., Lindner, H.J., Schmitt, G.E., and Lichtenthaler, F.W. (1997). Synthesis, structure, and conformational features of α-cycloaltrin: a cyclooligosaccharide with alternating $^4C_1/^1C_4$ pyranoid chairs. *Angew. Chem. Int. Ed. Engl.*, *36*, 1899-1902.

[6] Angyal, S.J. (1968). The conformations and α-β-ratios of aldopyranoses in aqueous solution. *Aust. J. Chem.*, *21*, 2737-2746.

[7] Dowd, M.K., French, A.D., and Reilly, P.J. (1994). Modeling of aldopyranosyl ring puckering with MM3(92). *Carbohydr. Res.*, *264*, 1-19.

[8] The 3D-structures can be viewed at "http://caramel.oc.chemie.tu-darmstadt.de/immel/3Dstructures.html", MOLCAD graphics are available at "http://caramel.oc.chemie.tu-darmstadt.de/immel/molcad/gallery.html".

[9] We are most grateful to Prof. Dr. C. Griesinger and Dr. H. Schwalbe, for kindly determining the 800 (^1H) and 200 (^{13}C) MHz NMR data at the 'Large Scale Facility for Biomolecular NMR' of Frankfurt University.

[10] Brooks, B.R., Bruccoleri, R.E., Olafson, B.D., States, D.J., Swaminathan, S., and Karplus, M. (1983). CHARMM: a program for macromolecular energy, minimization, and dynamics calculations. *J. Comp. Chem.*, *4*, 187-217.

[11] Reiling, S., Schlenkrich, M., and Brickmann, J. (1996). Force field parameters for carbohydrates. *J. Comput. Chem.*, *17*, 450-468.

[12] Immel, S. (1997). *MolArch+ - MOLecular ARCHitecture Modeling Program*, Darmstadt University of Technology.

[13] Haasnoot, C.A.G., De Leeuw, F.A.A.M., and Altona, C. (1980). The relationship between proton-proton NMR coupling constants and substituent electronegativities. *Tetrahedron*, *36*, 2783-2792.

[14] Least-squares fitting of the J values for three conformations $^4C_1$, $^1C_4$, and $^OS_2$ was achieved with a root-mean-square deviations of $\sigma_J = 0.3$ Hz, whereas a two-state model $^4C_1 \leftrightharpoons {}^1C_4$ yielded $\sigma_J = 1.8$ Hz only.

[15] Brickmann, J. (1997). *MOLCAD - MOLecular Computer Aided Design*, Darmstadt University of Technology. — Brickmann, J. (1992). Molecular graphics: how to see a molecular scenario with the eyes of a molecule. *J. Chim. Phys.*, *89*, 1709-1721.

[16] Teschner, M., Henn, C., Vollhardt, H., Reiling, S., and Brickmann, J. (1994). Texture mapping: a new tool for molecular graphics. *J. Mol. Graphics*, *12*, 98-105.

[17] Heiden, W., Moeckel, G., and Brickmann, J. (1993). A new approach to analysis and display of local lipophilicity/hydrophilicity mapped on molecular surfaces. *J. Comput.-Aided Mol. Des.*, *7*, 503-514.

[18] Connolly, M.L. (1983). Analytical molecular surface calculation. *J. Appl. Cryst.*, *16*, 548-558.

SYMMETRICAL AND NON-SYMMETRICAL UREA DERIVATIVES OF β-CYCLODEXTRIN AND PIPERAZINE: APPLICATION IN CAPILLARY ELECTROPHORESIS

JÓZSEF KOVÁCS[1] ISTVÁN PINTÉR[1],, LÁSZLÓ JICSINSZKY[2], NATASCHA ROOS[2], ALAIN MARSURA[3]
[1] ProChem Ltd., H-1525 Budapest, P. O. Box: 17, Hungary
[2] CYCLOLAB R&D. Lab. Ltd., Budapest, H-1525 Budapest, P. O. Box: 435, Hungary
[3] Univ. H. Poincaré, Fac. de Pharmacie, F-54001 Nancy CEDEX, P. O. Box: 403, France

SUMMARY

Peracetylated 6-monoazido-6-monodeoxy-β-cyclodextrin (**1**) furnished – in a one-pot-reaction with triphenyl phosphine and carbon dioxide under strictly dry conditions – the isocyanate (**2**) [1]. Addition of 0.5 mol piperazine furnished a symmetrical urea derivative (**3a**) while the reaction with N-Methyl piperazine led to a non-symmetrical urea derivative (**4a**). The usual Zemplén deacetylation resulted in partially acetylated derivatives having good solubility in water.

Some of these basic cyclodextrin-ureas and their deacetylated derivatives (**3b**, **4b**) were applied in capillary electrophoretic chiral separation. Moderate chiral resolutions were obtained in case of mandelic acid and phenobarbital enantiomers, and good chiral separation was found in case of N-Methyl ephedrine racemate by **4b**. The partially acetylated, more lipophylic derivatives showed better selectivity than the naked urea derivatives.

Introduction

Sugar phosphinimines [2], obtained from azidosugars by application of the Staudinger reaction [3], proved to be versatile and appropriate synthons for various substituted aminosugars [4]. Symmetrical sugar carbodiimides were easily produced by treatment of phosphinimines with carbon dioxide in dry solvents [2, 4]. Intermediacy of the corresponding isocyanate should be supposed during the transformation, as was found in non-carbohydrate cases [5]. However, in the phosphinimine reactions, isocyanates of mono- or disaccharides have not been detected.

Recently, application of the phosphinimine reaction to acetylated 6-monoazido-6-monodeoxy-β-cyclodextrin (**1**) led to a new type of cyclodextrin dimer bonded by urea linkage [1, 6]. Formation of the urea was attributed to the presence of water closed into the cavity of β-cyclodextrin even after drying in solid form.

Experimental

Structure determination: IR spectra were recorded in KBr on *Nicolet 205*, ^{1}H- and ^{13}C-NMR in $CDCl_3$ (**3a**, **4a**) and DMSO-d_6 (**3b**, **4b**) on Varian *VXR-400* spectrometer. Some of the chemical shifts assignments are based on 2D-HETCOR experiments. Protons and carbons of the substituted CD-glucosyl residue were designated with H_a and C_a, and those of the unsubstituted units with H_b and C_b, respectively.

Capillary electrophoresis: 1 mM in 120 mM Britton-Robinson buffer, pH= 3.0, Preconditioned fused silica capillary 50 μm, 50/58.5 cm, 30 kV, 25 °C, 150 mbar*sec injection, λ=200 nm, ref.: 450 nm.

Synthesis: All materials used in syntheses were purchased from Fluka and used without further purification.

General synthetic method (for detailed description and characterization of **3a** and **3b** see [1] and [6]): The toluene solution of the azide **1** was dried by azeotropic distillation and, after saturation with dry CO_2, a solution of Ph_3P in dry toluene was added dropwise and the mixture was kept under CO_2 at room temperature for 20 h. The product precipitated with cyclohexane proved to be the isocyanate **2.**

Reaction of **2** with piperazine and N-methyl-piperazine resulted in urea derivatives **3a** and **4a**, respectively. Treatment of **3a** and **4a** with methanolic NH_4OH at 60 °C gave partially acetylated product (2.5-3 Ac in random distribution). Complete deacetylation was reached by MeOH/NaOH at the same temperature to give the crystalline **3b** and **4b**.

2: Yield: 92 %, white solid; **mp** 140-145 °C; $[\alpha]_D^{25}$= +122° ± 1° (c=1, $CHCl_3$). **IR**: 2267 (N=C=O), 1757 cm^{-1} (AcO). **^{1}H-NMR δ (ppm)**: 5.34-5.20 (m, 7H, H-3); 5.16-5.04 (m, 7H, H-1); 4.87-4.76 (m, 7H, H-2); 4.63-4.52 (m, 6H, H_b-6a); 4.32-4.02 (m, 13H, H-5, H_b-6b); 3.87-3.67 (m, 9H, H-4, H_a-6a,b); 2.15-2.04 (several s, 60H, CH_3CO). **^{13}C-NMR δ (ppm)**: 170.71-169.32 (CH_3CO); 125.03 (N=C=O); 96.94-96.53 (C-1); 77.36-76.43 (C-4); 71.06-69.40 (C-2,3,5); 62.68-62.33 (C_b-6); 43.58 (C_a-6); 20.82-20.71 (CH_3CO).

4a: Yield: 88 %, white solid; **mp** 152-162 °C; $[\alpha]_D^{25}$= +125° ± 0.5° (c=1.5, $CHCl_3$). **IR**: 1750 (AcO), 1655, 1520 cm^{-1} (NHCON). **^{1}H-NMR δ (ppm):** 5.38-5.24 (m, 7H, H_b-1,3); 5.21 (dd, 1H, H_a-3); 5.12-5.08 (m, 5H, H_b-1); 5.05 (br dd, 1H, NH); 5.01 (d, 1H, H_a-1); 4.88-4.76 (m, 6H, H_b-2); 4.75 (dd,1H, H_a-2); 4.66-4.50 (m, 5H, H_b-6a); 4.49-4.10 (m, 13H, H_b-5,6a,b); 4.04 (ddd, 1H, H_a-5); 3.80-3.66 (m, 7H, H_b-4, H_a-6a); 3.61 (dd, 1H, H_a-4); 3.55 (ddd, 1H, H_a-6b); 3.40(m, 4H, CON-CH_2); 2.39 (t, 4H, CH_3N-CH_2); 2.29 (s, 3H, NCH_3); 2.13-2.02 (several s, 60H, CH_3CO). **^{13}C-NMR δ (ppm):** 170.84-169.24 (CH_3CO); 157.51 (NHCON); 97.15 (C_a-1); 97.04-96.40 (C_b-1); 77.48-76.44 (C_b-4); 76.86 (C_a-4); 71.13 (C_a-5); 70.98-70.17 (C_b-2,3); 70.93 (C_a-3); 70.33 (C_a-2); 69.57-69.31 (C_b-5); 62.99-62.43 (C_b-6); 54.72 (CH_3N-CH_2); 46.11 (NCH_3); 43.84 (CON-CH_2); 41.35 (C_a-6); 20.89-20.73 (CH_3CO).

4b: Yield: 80 %, white solid;; **mp** > 330 °C; $[\alpha]_D$ +157° ± 1° (c=1, DMSO). **IR**: 1632, 1540 cm^{-1} (NHCON). **^{1}H-NMR (60 °C) δ(ppm)**: 6.06 (br t, 1H, NH); 5.63-5.45 (m, 14H, OH-2,3); 4.89-4.80 (m, 7H, H-1); 4.31-4.21 (m, 6H, OH-6); 3.79 (ddd, 1H, H_a-6a); 3.76-3.51 (m, 26H, H-3,5, H_b-6a,b); 3.46-3.10 (m, 14H, H-2,4); 3.25 (m, 4H, CON-CH_2); 2.97 (ddd, 1H, H_a-6b); 2.23 (t, 4H, MeN-CH_2); 2.17 (s, 3H, NCH_3). **^{13}C-NMR δ (ppm):** 157.50 (NHCON); 102.35-101.45(C-1);84.55-80.84(C-4); 73.50-70.58(C-2,3,5); 60.25-59.77 (C_b-6); 54.63 (MeN-CH_2); 45.98 (NCH_3); 43.30 (CON-CH_2); 41.76 (C_a-6).

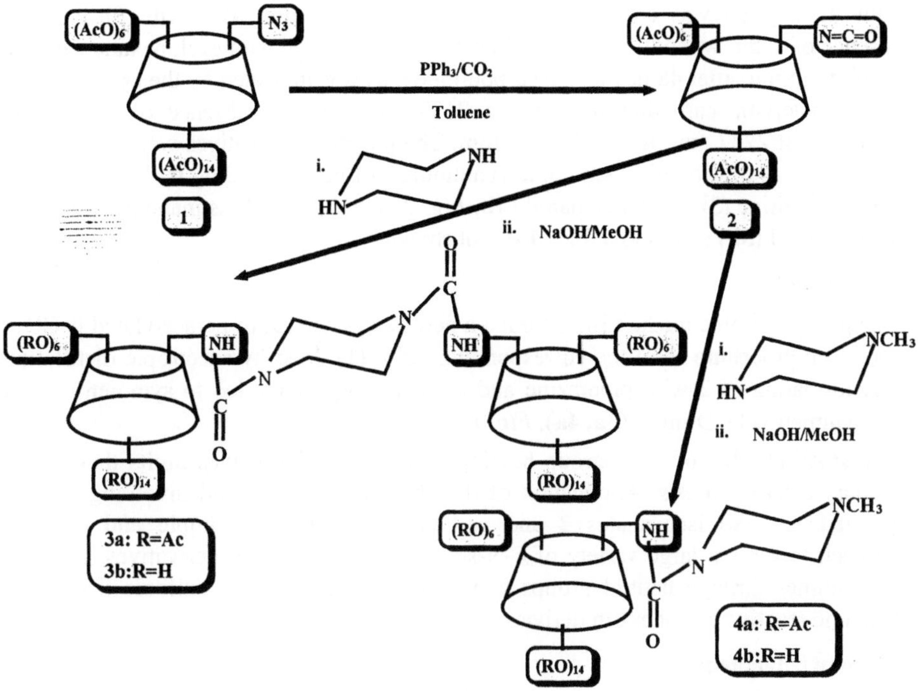

Fig. 1.: Preparation of β-cyclodextrin derivatives *via* isocyanate

Results

Formation of **2** instead of carbodiimide may be attributed to the decreased reactivity of both the isocyanate and the phosphinimine function linked to the bulky peracetyl β-cyclodextrin moiety. Consequently, the intermediate phosphinimine formed from **1** with Ph_3P reacts much faster with the small CO_2 molecule to give the isocyanate **2** than with the latter being in low concentration related to CO_2.

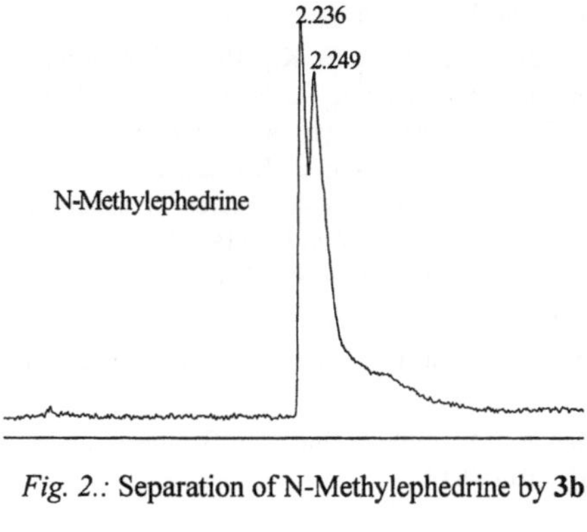

Fig. 2.: Separation of N-Methylephedrine by **3b**

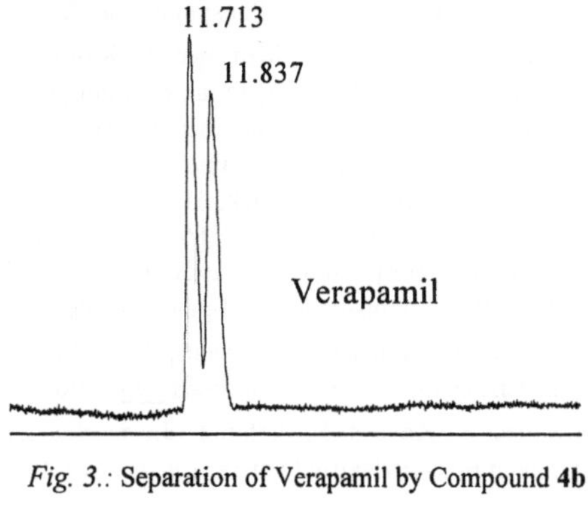

Fig. 3.: Separation of Verapamil by Compound **4b**

This assumption was corroborated by the one-pot reaction of **1** with Ph_3P in the presence of **2**, but without CO_2. The white solid, precipitated with cyclohexane from the reaction mixture after 48 h, was the carbodiimide, as was indicated by the appearance of the characteristic carbodiimide band at 2139 cm^{-1} and the absence of that of the isocyanate at 2267 cm^{-1} in the IR spectrum. The prepared derivatives were found to be useful in cases of difficultly separable enantiomers of important drug molecules like N-Methylephedrine (**3b**) and Verapamil (**4b**, *Fig. 2.* and *3.*) as well as mandelic acid (**3b**), Phenobarbital (**3b**) and Fluoxetine (**4b**) (not shown).

Conclusion

The phosphinimine method [6] afforded a new isocyanate (**2**) of peracetylated β-CD in a one-step procedure from the corresponding azide (**1**) dried by azeotropic distillation. Transformation of **2** with piperazine and N-methyl-piperazine led to symmetrical and non-symmetrical CD-ureas (**3a, 4a**), *Fig. 1.*

The small number of basic group bearing cyclodextrin derivatives limits their use in capillary electrophoresis. Application of the phosphinimine method involving reactive intermediates like isocyanates **2** and carbodiimides opens a simple and effective approach towards a large variety of 6-N-substituted cyclodextrin derivatives, from the basic amines through neutral groups to the acidic, carboxylic or sulfonic derivatives using one reactive cyclodextrin derivative.

Acknowledgement

This work was supported by the Hungarian Scientific Research Fund (OTKA T 14939 and T 23371). M. Kajtár-Peredy (Central Research Institute for Chemistry, Hungarian Academy of Sciences) is acknowledged for recording and evaluation of NMR spectra.

References:

[1] J. Kovács, P. Mészáros, I. Pintér, L. Jicsinszky, EUROCARB9, Utrecht (The Netherlands), 06. 07-06. 11. 1997., *Abstr. Book* p. A66

[2] Messmer, A., Pintér, I., Szegő, F., (1964), Acetylzucker-phosphinimine und –carbodiimide, *Angew. Chem.*, **76**, 227-228

[3] Staudinger, H., Meyer, J., (1919), Über neue organische Phosphorverbindungen III. Phosphinmethylenderivate und Phosphinimine, *Helv. Chim Acta*, **2**, 635-646

[4] a) Pintér, I., Kovács, J., Messmer, A., Kálmán, A., Tóth, G., Lindberg, B. K., (1979), Formation of phosphonioepimino salts of sugars: a novel neighbouring group participation, *Carbohydr. Res.*, **72**, 289-296; b) Kovács, J., Pintér, I., Messmer, A., Tóth, G., (1985), Unprotected sugar phosphinimines: a facile route to cyclic carbamates of amino sugars, *Carbohydr. Res.*, **141**, 57-65; c) Kovács, J., Pintér, I., Messmer, A., Tóth, G., Duddeck, H., (1987), A new route to cyclic urea derivatives of sugars via phosphinimines, *Carbohydr. Res.*, **166**, 101-111; d) Pintér, I., Kovács, J., Tóth, G., (1995), Synthesis of sugar ureas via phosphinimines, *Carbohydr. Res.*, **273**, 99-108; e) Mészáros, P., Kovács, J., Pintér, I., (1997), Synthesis of bis-cellobiosyl-guanidine derivatives, *Carbohydr. Lett.*, **2**, 355-361

[5] Molina, P., Alajarin, M., Arques, A., (1982), Convenient Improved Synthesis of Isocyanates and Isothiocyanates from Amines, *Synthesis* **1982** 596-597

[6] Sallas, F., Kovács, J., Pintér, I., Jicsinszky, L., Marsura, A., (1996), One Step Synthesis of New Urea-linked β-Cyclodextrin Dimers, *Tetrahedron Lett.*, **37**, 4011-4014

SULFATED CYCLODEXTRIN DERIVATIVES

ÁGNES MORVA[1], ÉVA FENYVESI[2], NATASCHA ROOS[2], BÉLA ZSADON[1], JÓZSEF SZEJTLI[2]

[1]*ELTE, Dept. Chem. Techn. and Environ. Chem., Budapest Pázmány P. sétány 2, H-1117, Hungary*
[2]*Cyclolab R&D Lab., Budapest Dombóvári út 5-7, H-1117, Hungary*

1. Introduction

Large number of sulfated derivatives of cyclodextrin (CD) monomers have been described in the literature [1], for preparation of water soluble CD polymer sulfates, however, only two references can be found [2,3]. Sulfated derivatives can be prepared by reacting the CD with chlorosulfonic acid or SO_3 complexes. The average degree of substitution (DS) of the product can be calculated based on sulfur content or from the results of MS, NMR, IR or titrimetric measurements.
The complex formation of βCD sulfate with Chlorpromazine was found to be supplemented by electrostatic interaction but hindered by the bulky substituents [4].
Sulfated CDs have various therapeutic effects, they show antiviral, anti-inflammatory and anti-angiogenic activity [5].
The CD derivatives are usually mixtures of isomers with different DS. The complex forming ability depends on the number and position of the substituents. The solubilizing effect of hydroxypropyl-βCD was found to decrease upon increasing the DS [6], while using sulfoalkylether derivatives it changed according to a maximum curve [7]. The effect of DS on the solubilizing potency of βCD sulfate has not been studied yet.
The DS has a large influence on the chiral separation by capillary electrophoresis (CE), too, where the CD derivatives act as chiral selectors. Chargeable CDs enable two separation modes, they can be used either in charged or in non-ionized form [8]. The DS of optimal resolution depends on the characteristics of the analyte [9].
In the present work sulfated derivatives of β-cyclodextrin and of water soluble β-cyclodextrin polymer with various DS were prepared. The effect of DS on the solubilizing properties and the enantiomeric separating potency of the products were studied.

2. Experimental

Preparation of the sulfated CD derivatives

Sulfated derivatives were prepared by reacting the βCD monomer (Wacker Chemie, Münich) and a βCD polymer crosslinked with butyleneglycol bis(epoxypropyl ether) (Cyclolab, Budapest) with $SO_3.NEt_3$ or $SO_3.DMF$ complexes (Fluka) in molar ratio 2.5-7.6 SO_3/CD at 25°C in DMF solvent, for 6 h. After 6 h an excess amount of water was added to the reaction mixture to decompose the reagent. After purifying the reaction mixture by dialysis the solution was passed through an ion exchange resin. The free acid was eluted with water, quantitatively collected and titrated with 0,1 N NaOH. Degree of substitution was calculated on the basis of the titration and expressed as the number of sulfate groups per cyclodextrin units. In case of the sulfated polymer the CD content obtained from iodometric titration after hydrolysis (62 %) was used for the calculation of DS.

Solubility studies

Excess amount of the drug (30-60 mg) was added to 2.5 ml pH 7 buffered 10 % CD solutions under stirring. Having stirred for 2 h at ambient temperature the suspensions were filtered and after proper dilution with 1:1 aqueous ethanol (Hydrocortisone, Furosemide) or 0.1 N HCl in 1:1 aqueous ethanol (Piroxicam, Terfenadine) the concentration of the drugs were measured UV spectrophotometrically.

Capillary electrophoresis

Capillary electrophoresis measurements were carried out on $HP^{3D}CE$ system. The conditions are the same as described in Ref. 8.

3. Results

3.1 SOLUBILIZING PROPERTIES OF CD POLYMER SULFATES (SβCDP)

SβCDPs were tested for their solubilizing properties as a function of DS with several poorly soluble drugs (Hydrocortisone, Piroxicam, Furosemide and Terfenadine).

SβCDPs form inclusion complexes with these drugs, thus improve their solubility. Studying the effect of DS on the solubilizing properties similar tendency can be observed with all guest molecules studied; the concentration of the drugs changed according to a maximum curve as a function of DS of SβCDP (Fig 1). The highest solubility was achieved with SβCDP of DS 3. The solubilizing effect of SβCDP with high DS was lower as a consequence of the steric hindrance by the sulfate groups, or the increased hydrophilicity of the CD rim.

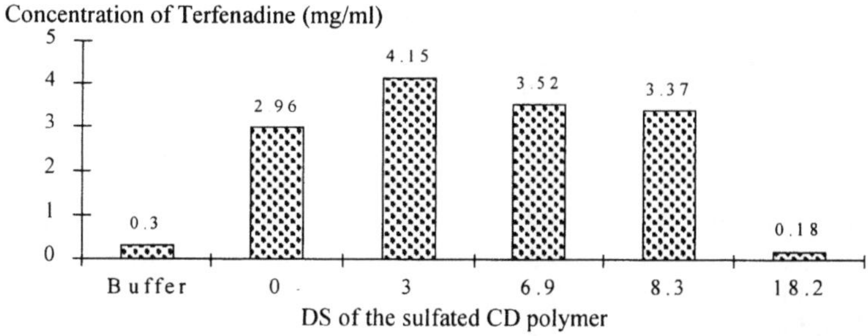

Figure 1. Concentration of Terfenadine as a function of DS of SβCDP at pH 7 in 10 % solutions

3.2 CAPILLARY ELECTROPHORESIS WITH SULFATED CD DERIVATIVES

3.2.1 Separating potency of the sulfated CD polymer (SβCDP)

SβCDPs with various DS were tested as chiral selector in comparison with the unsubstituted CD polymer (βCDP) for enantioseparation of basic drugs, Propranolol, N-Methylephedrine and Sympatomine at pH 9 in 120 mM Britton-Robinson buffer as basic electrolyte. The quality of the separation was characterised with the modified resolution factor (R') [8].

Enantiomeric separation of the compounds was not achieved applying the unsubstituted polymer as additive. Using sulfated derivatives, however, interaction between the chiral selector and the analytes were observable. In most cases the enantiomers were more or less separated due to the difference between the stability constants of the two enantiomer/selector complexes resulting in different electrophoretic mobility. The best results were obtained applying SβCDP of DS 16.1 (Table 1.).

Table 1. Enantioseparation of N-Methylephedrine with the unsubstituted and the sulfated CD polymer at pH 9

βCDP	*SβCDP DS 3*	*SβCDP DS 16.1*
R'= 0%	R'= 56%	R'= 99%

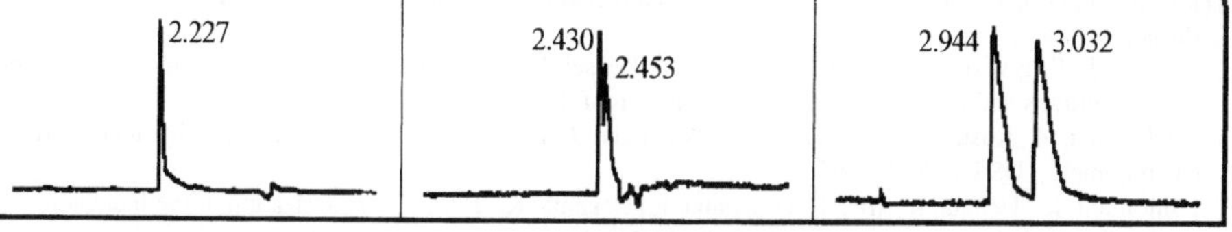

3.2.2 Separating potency of the sulfated CD monomer (SβCD)

The chiral recognition ability of SβCD for chargeable drugs as a function of DS and influence of pH of the electrolyte on the modified resolution factor were studied.

Good chiral separations were obtained with all SβCD types (DS 1.4, 4.2, 9.3) at high and low pH. The best results are listed in Table 2. Most of the enantiomers were separated using SβCD with higher DS, SβCD with DS 1.4 was acceptable as chiral selector only in certain cases (Fluoxetine, pH 5, R'= 86.8%). No direct influence of DS on the chiral resolution was found, indicating that prediction of the proper DS is difficult, the separation has to be optimised individually for each analyte.

Table 2. DS of SβCD and pH of the electrolyte used to achieve maximal R' applying 1% chiral selector in 120 mM BRB

Analyte	*DS of SβCD*	*pH*	*R`, %*
Terbutaline	4.2	5	331.0
Warfarin	4.2	5	92.5
Verapamil	4.2	5	40.0
Propranolol	4.2	11	33.0
Fluoxetine	9.3	9	118.0
Ketamine	9.3	5	17.1
Pindolol	9.3	5	89.0

4. Conclusion

βCD sulfates (SβCD) and βCD polymer sulfates (SβCDP) with various DS were prepared.

Solubilizing effect of SβCDP on poorly soluble drugs changes according to a maximum curve as a function of DS.

SβCDP was applied first as chiral selector in CE. It was shown to be a useful chiral selector with good enantioselectivity. Best separations were achieved with the SβCDP of DS 16.1.

Using SβCD as chiral selector it was observed that chiral resolution is strongly influenced by the DS of the CD and by the pH. The optimum conditions depend on the characteristic of the analyte, therefore the separation has to be optimised individually for each analyte. Further optimisations of the conditions can result in improved separation.

5. References

[1] Szejtli, J., Osa, T., 1996, Comprehensive Supramolecular Chemistry, Vol. 3., Cyclodextrins, Pergamon Press

[2] Szejtli, J., Fenyvesi, É., Zsadon, B., Szilasi, M., Décsei, L., 1985, Water soluble CD polymers substituted by ionic groups and process for the preparation thereof, US Pat. 4,535,152

[3] Uchiyama, H., Kasukawa, H., Takeda, E., Watanabe, J., Nanba, R., 1996, Sulfonated compounds of β-CD polymer..., US Pat. 5,512,665

[4] Shiontani, K., Uehata, K., Irie, T., Hirayama, F., Uekama, K., 1994, Characterization of the inclusion mode of β-CD sulfate..., *Chem. Pharm. Bull.*, 42(11), 2332.

[5] Szejtli, J., 1996, Sulfated CDs: Past, present and future, *Cyclodextrin News*, 10(14), 265.

[6] Matsuda, H. et all., 1993, Inclusion complexation of p-hydroxybenzoic acid esters with HPβCDs. On changes in solubility and antimicrobial activity, *Chem. Pharm. Bull.*, 41(8), 1448.

[7] Zia, V., Rajewski, R.A., Bornancini, E.R., Luna, E.A., Stella, V.J., 1997, Effect of alkyl chain length and DS on the complexation of sulfoalkyl ether β-CDs with steroids, *J. Pharm. Sci.*, 86(2), 220.

[8] Roos, N., Ganzler, K., Szemán, J., Fanali, S.,1997, Systematic approach to cost- and time-effective method development with a starter kit for chiral separation by CE, *J. Chromatogr. A*, 782, 257.

[9] Szemán, J., Ganzler, K., Salgó, A., Szejtli, J., 1996, Effect of the DS of CD derivatives on chiral separations by HPLC and CE, *J. Chromatogr. A*, 728, 423.

Acknowledgement

The work was partly supported by a grant of Hungarian Research Fund (OTKA T 019618). The authors are grateful to Ms. I. Megyeri and Ms. K. Lantos for their technical assistance.

THE MOLECULAR GEOMETRIES OF CYCLOFRUCTINS[1]

S. IMMEL, G. E. SCHMITT, AND F. W. LICHTENTHALER

Institute of Organic Chemistry, Darmstadt University of Technology, D-64287 Darmstadt, Germany

Abstract. Cyclofructins composed of six (CF_6, **1**) to ten (CF_{10}, **5**) β(1→2)-linked fructofuranose units (i.e. *cyclo*[D-Fru*f* β(1→2)]$_n$ with $n = 6 - 10$) were subjected to a detailed molecular modeling study. **1**, **2**, and **3** are disk-type shaped molecules without a cavity, whereas the nonamer (**4**) and decamer (**5**) adopt torus-like conformations that allow the formation of inclusion complexes similar to those of cyclodextrins.

1. Introduction

The cyclofructins[2] are a class of cyclooligosaccharides produced by the action of an extracellular enzyme of *Bacillus circulans* on the polysaccharide inulin; they are composed of six (**1**, CF_6), seven (**2**, CF_7), and eight (**3**, CF_8) β(1→2)-linked fructofuranose residues[3]. Although ring homologs with nine (**4**, CF_9) and ten (**5**, CF_{10}) fructose residues have as yet not been uncovered in enzymatic digests of inulin, the unspecificity of the *Bacillus circulans*-derived enzyme may lead to their formation and eventual isolation.

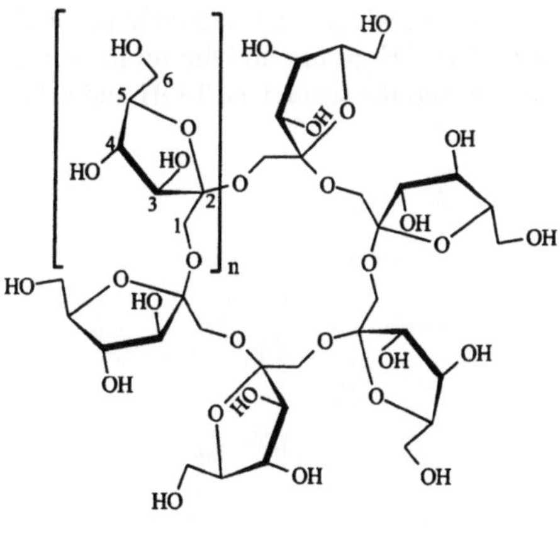

cyclofructins

1	*cyclo*[D-Fru*f* β(1→2)]$_6$	(CF_6)
2	*cyclo*[D-Fru*f* β(1→2)]$_7$	(CF_7)
3	*cyclo*[D-Fru*f* β(1→2)]$_8$	(CF_8)
4	*cyclo*[D-Fru*f* β(1→2)]$_9$	(CF_9)
5	*cyclo*[D-Fru*f* β(1→2)]$_{10}$	(CF_{10})

Experimental[4] and theoretical[5,6] studies have established the ability of the cyclofructins **1** - **3** to complex cations through electrostatic interactions with their characteristic crown ether skeletons. The question whether the cyclofructins of different ring sizes are able to form inclusion complexes similar to those formed by cyclodextrins[7] initiated our molecular modeling studies on **1** - **5** to be described in the sequel.

2. Methods

From the solid-state geometry of CF_6 trihydrate[8] (**1**) two different fructofuranosyl geometries of the asymmetric unit were assembled to symmetrical and asymmetrical macrocycles **1** - **5**, and for each of these starting geometries unrestricted "*Monte-Carlo*" (MC) and "*random-walk*" simulations[9] using adapted corner-flapping procedures[10] were initiated (PIMM91 force-field[11]). Out of the total of approx. 130 000 structures generated for each compound, the global energy-minimum structures[12] were used for the surface calculations and the molecular modelings utilizing the MOLCAD program[13].

MD simulations of the CFs in aqueous solution were carried out with the CHARMM[14] force-field[15] (timestep 1 fs, total simulation time ca. 750 ps for each CF) using periodic boundary conditions of the truncated octahedron (box size 32 Å, boxes filled with 606 - 558 water molecules for **1** - **5**).

3. Results and Discussion

Conformations of the Cyclofructins. — The global energy-minimum structures of the cyclofructins **1** - **5** obtained from Monte-Carlo and molecular dynamics simulations (vide supra) are shown in Fig. 1[12] together with their dotted contact surfaces[16] and the corresponding cross-section plots. Here, only the global shapes of the macrocycles are discussed; a detailed description of all used and calculated molecular geometry parameters will be published elsewhere.

The fructofuranosyl residues bear some degree of flexibility, preferring ${}^4T_3 \leftrightarrow E_3$ conformations with medium ring distortions. They are linked in a rigid, *spiro*-type fashion to the crown ether backbone of the macrocycles, which exhibit nearly perfectly planar *n*-polygons of the O-1 atoms. The primary 6-CH_2OH groups and the furanoid ring oxygens O-5 are located on one side of the molecule, and the secondary 3-OH and 4-OH groups on the other.

The most significant feature of the cyclofructins turned out to be a high preference for an alternating arrangement of the fructofuranoses being inclined "*inward*" (down) and "*outward*" (up) with respect to the mean plane of the macrocyclic backbone. This entails that half of the 3-OH groups point towards the center of the macrocycle ("*inward*" rotation), whereas the others are situated nearly perpendicular on the ring perimeter ("*outward*" tilting).

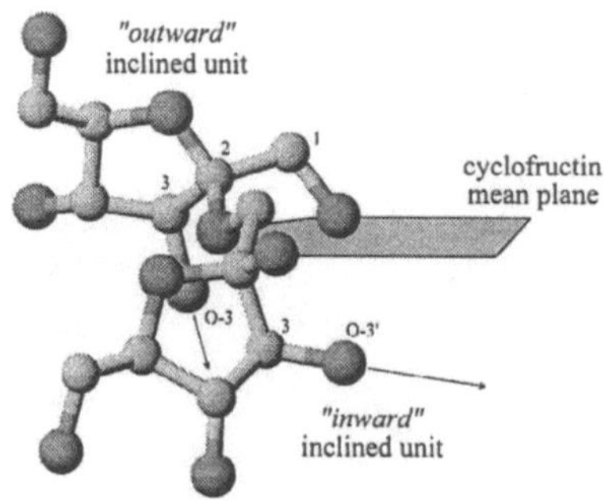

The even-membered cyclofructins **1**, **3**, and **5** prefer $C_{n/2}$ symmetrical structures, whereas for CF_7 (**2**) and CF_9 (**4**) the alternating alignment of the odd number of furanose residues is retained as largely as possible, but results in C_1 symmetry. Most notably, the computed structure for CF_6 (**1**) matches almost perfectly the geometry realized in the solid-state[8].

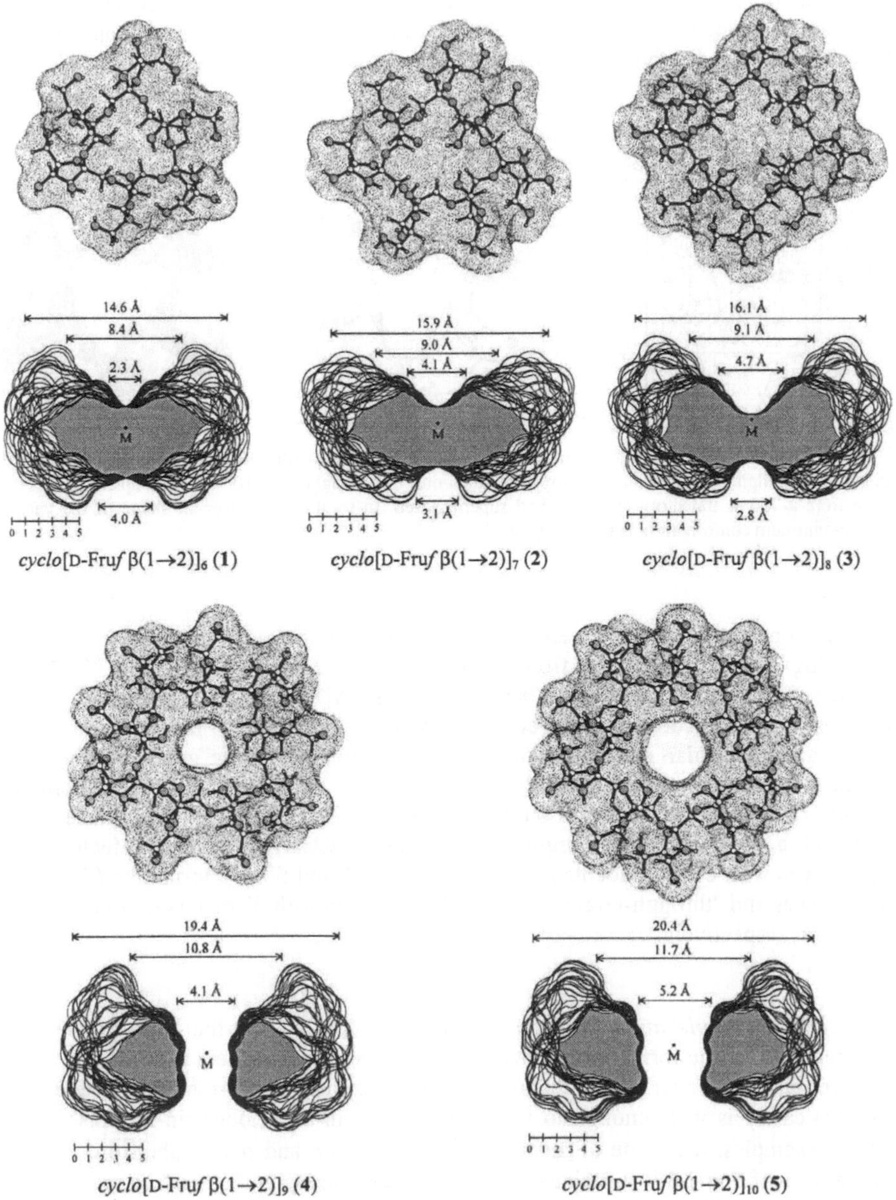

Figure 1. Ball-and-stick model representations of the global energy-minimum structures of the cyclofructins **1** - **5** with the contact surfaces superimposed in dotted form. Structures are shown perpendicular to the mean ring plane of the macrocycles; the 3-OH / 4-OH side of the furanoid rings points towards the viewer, and the 6-CH_2OH groups and the 5-O atoms away from him; oxygens are shaded. Surface cross-section contours (lower entries each) were obtained for successive 10° rotation steps around the geometrical center *M* and superimposed. In each case the 6-CH_2OH groups are located on the top, and the 3-OH / 4-OH groups on the bottom side of the cross-cut models; approximate molecular dimensions are given in Å.

In addition to the force-field calculations on the isolated molecules (vacuum conditions), molecular dynamics simulations on **1** - **5** (cf. Fig. 2) in aqueous solution provided evidence for the relevance of these conformations in water.

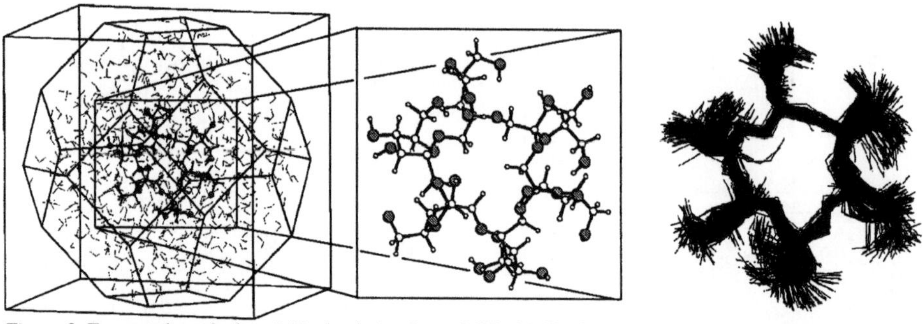

Figure 2. Truncated octahedron MD-simulation box of CF_6 (**1**) filled with 606 water molecules (left). In the middle, an exemplary solute conformation is zoomed out; on the right side, 100 solute snapshot geometries taken from a 750 ps trajectory are fitted and superimposed. In solution, the over-all shape of the vacuum energy-minimum conformation is largely retained.

The atomic distances between the O-1 crown ether oxygens diagonally across the macroring increase gradually from approximately 6.3 Å for CF_6 (**1**) to 10.2 Å in CF_{10} (**5**). In a more transparent way, the MOLCAD[13] program-generated contact surfaces[16] and the corresponding cross-section cuts[17] (cf. Fig. 1) display the effective shapes and molecular extensions of the CFs (approx. distances given in Fig. 1). In contrast to the nearly perfectly round-shaped cyclodextrins, the alternating "*inward / outward*" inclination of the fructofuranosyl residues renders the surface cuts of **1** - **5** with a deeply fissured, irregular over-all shape. Clearly, **1**, **2**, and **3** feature disk-type shapes with central indentations only, whereas **4** and **5** have torus-like (doughnut) appearances and "through-going" round-shaped cavities with diameters of approx. 4.1 Å and 5.2 Å, respectively.

Inclusion Complexation Potential of CF_9 and CF_{10}. — On the basis of the steric requirements for the formation of inclusion complexes, it is evident that CF_9 (**4**) is able to accommodate linear alkyl chains in *all-trans* zigzag conformations, whereas the CF_{10} (**5**) cavity is wide enough to fit aromatic rings in a cyclodextrin-like fashion. As typical examples, the mode of incorporation of β-alanin and *p*-aminobenzoic acid into the cavities of **4** and **5** is visualized in Fig. 3 by the corresponding surface cross-section cuts for the computer-generated global energy-minimum structures. Of the two possible guest-host alignments, the one shown in Fig. 3 is highly favored: in both complexes **6** and **7** the carboxylate groups of the guest molecules are located close to the 6-CH_2OH side of the macrocycles, whereas the amino functions are situated at the opposite side close to the 3-OH and 4-OH groups.

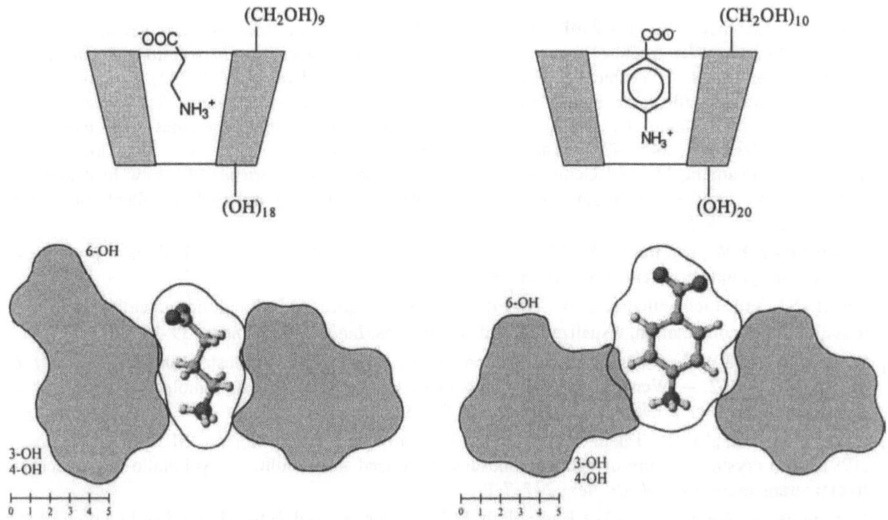

CF_9 · β-alanin (**6**) CF_{10} · p-aminobenzoic acid (**7**)

Figure 3. Surface cross section cuts through the global energy-minimum structures of the inclusion complex of CF_9 with β-aminopropionic acid (**6**, left) and the CF_{10} *p*-aminobenzoic acid adduct (**7**, right). The geometries depicted correspond to the energetically "*matched*" host-guest orientations; in both cases the host molecule was omitted for clarity.

Further insights into the guest complexation capabilities of the cyclofructins **4** and **5** are expected to emerge from generation of their molecular electrostatic potential profiles and lipophilicity patterns, which are surmised to provide a reliable assessment of the electrostatic interactions between the guest and host, as well as an indication to which extent hydrophobic effects contribute to the stability of the inclusion complexes; studies along this vein will be published in due course.

References

[1] 'Molecular Modeling of Saccharides', Part 16. — Part 15: Nogami, Y., Nasu, K., Koga, T., Ohta, K., Fujita, K., Immel, S., Lindner, H.J., Schmitt, G.E., and Lichtenthaler, F.W. (1997). Synthesis, structure, and conformational features of α-cycloaltrin: a cyclooligosaccharide with alternating $^4C_1/^1C_4$ pyranoid chairs. *Angew. Chem. Int. Ed. Engl.*, *36*, 1899-1902.

[2] Lichtenthaler, F.W., and Immel, S. (1994). Cyclodextrins, cyclomannins, and cyclogalactins with five and six (1→4)-linked sugar units: a comparative assessment of their conformations and hydrophobicity potential profiles. *Tetrahedron: Asymmetry*, *5*, 2045-2060.

[3] Kawamura, M., Uchiyama, T., Kuramoto, T., Tamura, Y., and Mizutani, K. (1989). Formation of a cycloinulo-oligosaccharide from inulin by an extracellular enzyme of *Bacillus Circulans* OKUMZ 31B. *Carbohydr. Res.*, *192*, 83-90. — Kawamura, M., and Uchiyama, T. (1994). Purification and some properties of cycloinulo-oligosaccharide fructotransferase from *Bacillus circulans* OKUMZ 31B. *Carbohydr. Res.*, *260*, 297-304.

[4] Takai, Y., Okumura, Y., Takahashi, S., Sawada, M., Kawamura, M., and Uchiyama, T. (1993). A permethylated cyclic fructo-oligosaccharide host that can bind cations in solution. *J. Chem. Soc., Chem. Comm.*, 53-54. — Yoshie, N., Hamada, H., Takada, S., and Inoue, Y. (1993). Complexation of cycloinulohexaose with some metal ions. *Chem. Lett.*, 353-356. — Uchiyama, T., Kawamura, M., Uragami, T., and Okuno, H. (1993). Complexing of cycloinulo-oligosaccharides with metal ions. *Carbohydr. Res.*, *241*, 245-248. — Takai, Y., Okumura, Y., Tanaka, T., Sawada, M., Takahashi, S., Shiro, M., Kawamura, M., and Uchiyama, T. (1994). Binding characteristics of a new host family of cyclic oligosaccharides from inulin: permethylated cycloinulohexaose and cycloinuloheptaose. *J. Org. Chem.*, *59*, 2967-2975.

[5] Lichtenthaler, F.W., and Immel, S. (1995). Computer simulation of chemical and biological properties of sucrose, the cyclodextrins and amylose. *Int. Sugar J.*, *97*, 12-22.

[6] Immel, S., and Lichtenthaler, F.W. (1996). The electrostatic and lipophilic potential profiles of α-cyclofructin: computation, visualization, and conclusions. *Liebigs Ann. Chem.*, 39-44.

[7] Saenger, W. (1980). Cyclodextrins inclusion compounds in research and industry. *Angew. Chem. Int. Ed. Engl.*, *19*, 344-362. — Wenz, G. (1994). Cyclodextrins as synthons for supramolecular structures and functional units. *Angew. Chem. Int. Ed. Engl.*, *33*, 803-822.

[8] Sawada, M., Tanaka, T., Takai, Y., Hanafusa, T., Taniguchi, T., Kawamura, M., and Uchiyama, T. (1991). The crystal structure of cycloinulohexaose produced from inulin by cycloinulo-oligosaccharide fructanotransferase. *Carbohydr. Res.*, *217*, 7-17.

[9] Metropolis, N., Rosenbluth, A.W., Rosenbluth, M.N., Teller, A., and Teller, E. (1953). Equation of state calculations by fast computing machines. *J. Chem. Phys.*, *21*, 1087-1092. — Van Gunsteren, W.F., and Berendsen, H.J.C. (1990). Molecular dynamics computer simulation. Method, application, and perspectives in chemistry. *Angew. Chem. Int. Ed. Engl.*, *29*, 992-1023.

[10] Goto, H., and Osawa, E. (1989). Corner flapping: a simple and fast algorithm for exhaustive generation of ring conformations. *J. Am. Chem. Soc.*, *111*, 8950-8951.

[11] Lindner, H.J., and Kroeker, M. (1997). PIMM91 - *Closed Shell PI-SCF-LCAO-MO-Molecular Mechanics Program*, Darmstadt University of Technology. — Smith, A.E., and Lindner, H.J. (1991). π-SCF-Molecular mechanics PIMM: formulation, parameters, applications. *J. Comput.-Aided Mol. Des.*, *5*, 235-262.

[12] The 3D-structures can be viewed at "http://caramel.oc.chemie.tu-darmstadt.de/immel/3Dstructures.html", MOLCAD graphics are available at "http://caramel.oc.chemie.tu-darmstadt.de/immel/molcad/gallery.html".

[13] Brickmann, J. (1997). *MOLCAD - MOLecular Computer Aided Design*, Darmstadt University of Technology. — Brickmann, J. (1992). Molecular graphics: how to see a molecular scenario with the eyes of a molecule. *J. Chim. Phys.*, *89*, 1709-1721.

[14] Brooks, B.R., Bruccoleri, R.E., Olafson, B.D., States, D.J., Swaminathan, S., and Karplus, M. (1983). CHARMM: a program for macromolecular energy, minimization, and dynamics calculations. *J. Comp. Chem.*, *4*, 187-217.

[15] Reiling, S., Schlenkrich, M., and Brickmann, J. (1996). Force field parameters for carbohydrates. *J. Comput. Chem.*, *17*, 450-468.

[16] Connolly, M.L. (1983). Analytical molecular surface calculation. *J. Appl. Cryst.*, *16*, 548-558.

[17] Immel, S. (1997). *MolArch$^+$ - MOLecular ARCHitecture Modeling Program*, Darmstadt University of Technology.

MOLECULAR GEOMETRIES OF FURANOID β(1→3)- AND β(1→6)-LINKED CYCLOGALACTINS[1]

H. GOHLKE, S. IMMEL, F. W. LICHTENTHALER, AND G. E. SCHMITT

Institute of Organic Chemistry, Darmstadt University of Technology, D-64287 Darmstadt, Germany

Abstract. The conformational properties of small cyclogalactins composed of β(1→3)- and β(1→6)-linked galactofuranose units, i.e *cyclo*[D-Gal*f* β(1→3)]$_n$, $n = 4, 5$ (**1** and **2**), and *cyclo*[D-Gal*f* β(1→6)]$_n$, $n = 3, 4$ (**3** and **4**), were investigated by means of adapted Monte-Carlo simulations. The flexibility of the macrocyclic backbone favors bend and asymmetrical conformations over round geometries. Visualization of the molecular surfaces of the global energy-minimum structures reveal the disk-type shapes of **1** - **4** without central cavities, but featuring distinct indentations close to the 2/3-O-groups, respectively. Molecular lipophilicity patterns prove these surface dents to be hydrophobic for **1** and **2**, whereas **3** and **4** display an inverse situation with a hydrophobic outer core structure.

1. Introduction

In analogy to cyclodextrins[2], cyclomannins[3], and cycloaltrins[4], the generic name *cyclogalactin* was proposed for cyclooligosaccharides composed of galactose units[3]. Although the former are all made up of α(1→4)-linkages, pyranoid (1→4)-galactose units in such a cyclic arrangement require β-type intersaccharidic bonds, and it was noted in an earlier molecular modeling study on *cyclo*[D-Gal*p* β(1→4)]$_6$ (i.e. α-cyclogalactin), that these compounds may be considered as "*inverso*"-cyclodextrins due to distinctly hydrophobic surface areas on the outside and hydrophilic central cavities[3]. Although the synthesis of pyranoid cyclogalactins has not been accomplished yet, cycloglycosylation of tritylated 1,2-*O*-(1-cyano)ethylidene-derivates of D-galactofuranose led to a number of their furanoid analogs[5], i.e. cyclogalactooligosaccharides with β(1→3)- (e.g. **1** and **2**), β(1→5)-, and β(1→6)-linked galactofuranose units (e.g. **3** and **4**).

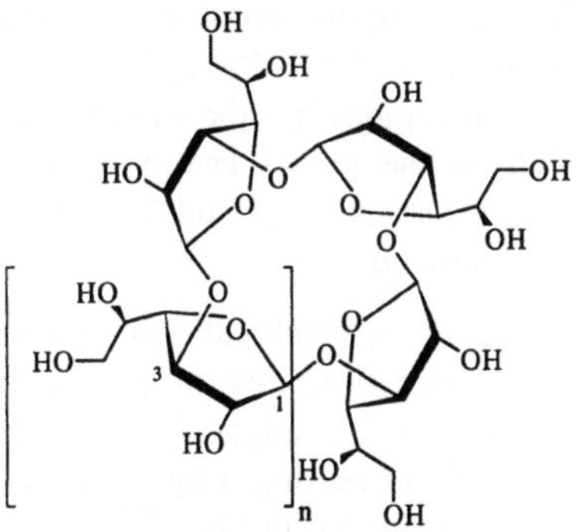

cyclo[D-Gal*f* β(1→3)]$_4$ **1**
cyclo[D-Gal*f* β(1→3)]$_5$ **2**

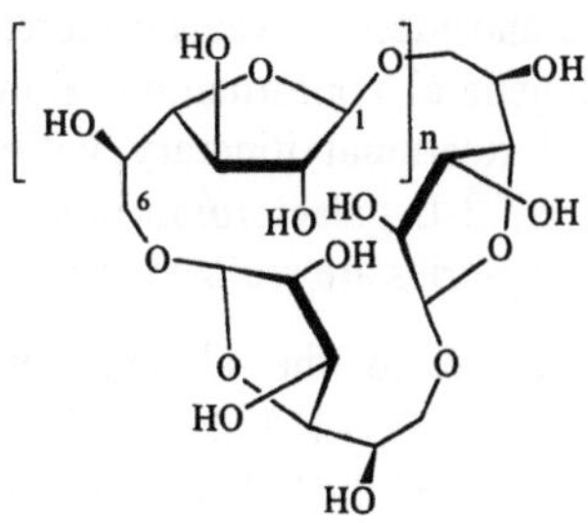

cyclo[D-Gal*f* β(1→6)]$_3$ **3**
cyclo[D-Gal*f* β(1→6)]$_4$ **4**

Due to the rather modest yields in their preparation, the capabilities of these compounds to form inclusion complexes could not be investigated. Their potential properties to function as hosts and their relationship to the cyclofructins which are equally composed of furanoid building blocks[6], induced us to perform a molecular modeling study on the β(1→3)- and β(1→6)-linked forms, whereby – due to the expected high conformational flexibility – we restricted our analysis to the tetra- and pentamer of *cyclo*[D-Gal*f* β(1→3)]$_n$ (**1** and **2**), as well as to the trimer (**3**) and tetramer (**4**) of the β(1→6)-type cyclooligosaccharides.

2. Methods

All calculations were carried out using the PIMM91 force-field[7] program with external conformational search algorithms[8]. A pre-optimized galactofuranose unit was assembled to various symmetrical and asymmetrical starting structures of the cyclogalactofuranosides **1** - **4**, which were subjected to conformational analysis using "*Monte-Carlo*" and "*random-walk*" techniques[9]. An adapted corner-flapping procedure[10] was used to effectively vary the ring geometries of the furanose units as well as of the macrocyclic backbone without breaking bonds in the different rings. Proper sampling of the conformational space was ensured by converging minimum energy structures and molecular parameters[11]. For the global energy-minimum structures the molecular contact surfaces[12], cross-section cuts[8], and lipophilicity patterns (MLPs) were computed; color-coded representations were generated using the MOLCAD[13-15] program.

3. Results and Discussion

Molecular Geometries of 1 - 4. — The conformational properties of the galactofuranose units are characterized by opposite tendencies of the anomeric substituent and the C5-C6-side-chain to adopt pseudo-axial and pseudo-equatorial arrangements, respectively. The adiabatic (PIMM91) energy potential surface of β-D-galactofuranose reveals a single, broad energy-minimum in the $^4T_3 \leftrightarrows {}^4E \leftrightarrows {}^4T_O \leftrightarrows E_O$ region for the furanose rings. Although the closely related $^1T_O \leftrightarrows {}^1E$ conformations seem to be preferred in the few solid-state structures of β-galactofuranose derivatives reported in the CCDF data file[16], an analysis of ^{1}H NMR $^3J_{H\text{-}H}$ coupling constants of various galactofuranosides[17] reveals preferred solution geometries in the $^4E \leftrightarrows {}^4T_O \leftrightarrows E_O \leftrightarrows {}^1T_O$ range. In general, conformations along these overlapping regions (i.e. the western part of the pseudorotational itinerary within 10 kJ/mol above the global energy-minimum) characterize β-D-galactofuranosides, and for the cyclogalactins ^1E (**1** and **2**) and $^4T_3$ (**3** and **4**) geometries are realized in the energy-minimum structures[11].

In contrast to the almost perfectly round-shaped, symmetrical cyclogalactins composed of β(1→4)-linked galactopyranoses[3], the high flexibility of the furanoid rings in **1** - **4**, and the β(1→6)-linkage in **3** and **4** with four torsion angles involved, leads to structures of lower symmetry to be favored with increasing ring size of the macrocycles. The energy-minimum geometries of **1** - **4** displayed in Fig. 1[11] show that the small ring homologs **1** and **3** retain symmetry as much as possible (C_4 and C_3, resp.), but **2** and **4** prefer C_1 (asymmetric) and C_2-type structures.

In the case of **3**, and to a lesser extent in **4**, the conformational symmetry is only retained through favorable intramolecular hydrogen bonding interactions of the 3-OH ··· O-6 and 5-OH ··· O-4'-type. In aqueous solution these effects are expected to become less important, hence, conceivable conformational equilibria are likely to be shifted towards asymmetric forms.

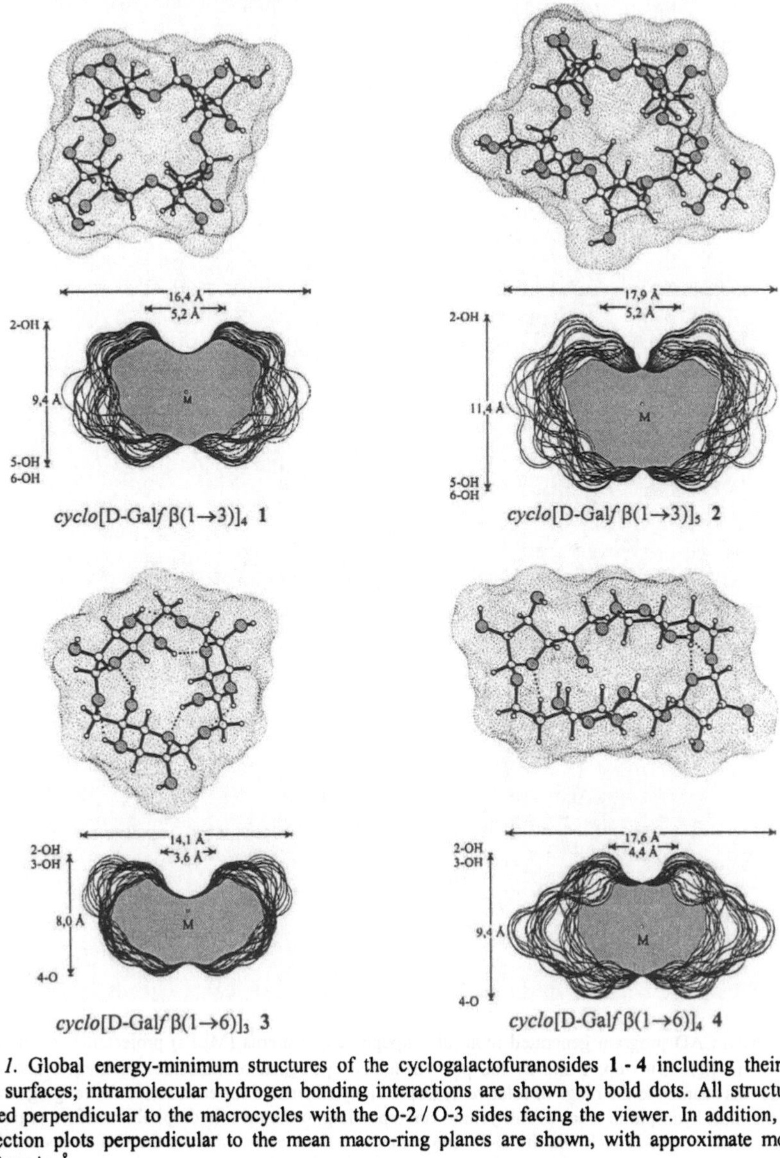

Figure 1. Global energy-minimum structures of the cyclogalactofuranosides **1** - **4** including their dotted contact surfaces; intramolecular hydrogen bonding interactions are shown by bold dots. All structures are displayed perpendicular to the macrocycles with the O-2 / O-3 sides facing the viewer. In addition, surface cross-section plots perpendicular to the mean macro-ring planes are shown, with approximate molecular dimensions in Å.

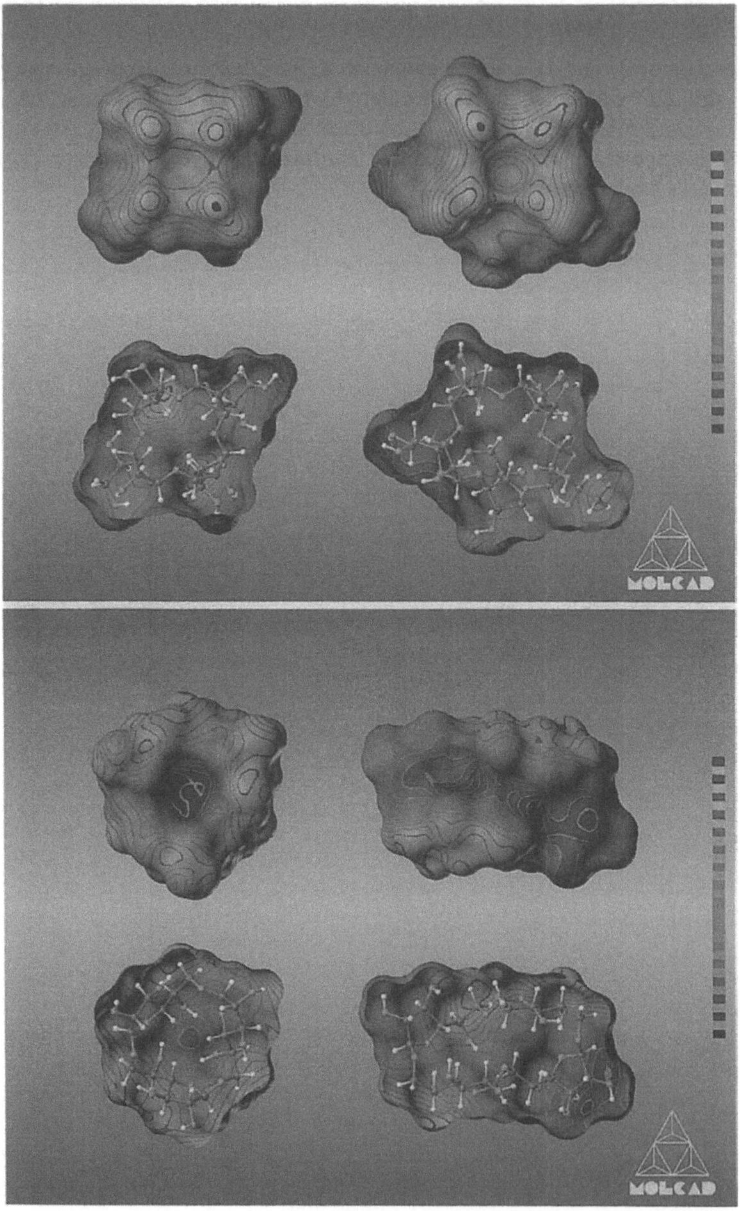

Figure 2. MOLCAD program generated molecular lipophilicity patterns (MLPs) projected onto the contact surfaces of the β(1→3)- (top: **1** and **2**) and β(1→6)-linked cyclogalactofuranosides (bottom: **3** and **4**). The color-code was adapted to the range of relative hydrophobicity calculated for each molecule, ranging from dark blue for the most hydrophilic areas to full yellow corresponding to the most hydrophobic surface regions. The orientation of all models is such that the O-2 / O-3 atoms face the viewer, whereas the O-4 atoms are directed to the back; the half-opened models on the bottom each visualize the molecular orientation as well as the surface properties of the back.

The molecular contact surfaces computed for **1 - 4** (cf. Fig. 1) as well as the corresponding surface cross-section cuts provide evidence for the disk-type shapes of these macrocycles, with small puckering amplitudes of less than 0.5 Å along their backbones. The surface models not only display the effective steric extensions of these molecules and their lack of central cavities to form inclusion complexes, but also allow to map their lipophilicity patterns in color-coded form, as has already been proved useful in earlier studies to assess the properties of cyclodextrins[2] and related compounds[3].

Molecular Lipophilicity Patterns (MLPs). — Beside the steric demands of **1 - 4**, their hydrophobic characteristics are of major interest to evaluate their properties. In Fig. 2, the corresponding MLPs were computed and mapped in color-coded form onto the contact surfaces of Fig. 1. Most remarkably, these graphics reveal a fundamental difference of the distribution of hydrophilic and hydrophobic surface areas for both classes of the β(1→3)- and β(1→6)-linked cyclogalactins. In the case of **1** and **2** (Fig. 2, top), the surface is significantly indented in the center without elaborating "through-going" central cavities. Due to their association with the O1-C2(H)-C3(H) fragments of the galactofuranose moieties, these surface dents are remarkably hydrophobic, whereas O-4 and the side chain 5-OH and 6-OH groups render the opposite molecular sides much more hydrophilic. Most notably, the outer rim of the disk-shaped molecules carrying the 2-OH hydroxyls is predominantly hydrophilic.

A different situation prevails for the β(1→6)-cyclogalactins **3** and **4** (Fig. 2, bottom): the type of intersaccharidic linkage, their increased flexibility, and the altered orientation of the furanoid rings in relation to the macrocycle cause an inverse alignment of hydrophilic and hydrophobic surface regions. The 5-OH groups pointing towards the central surface dents provide a hydrophilic environment that is augmented by the 3-OH groups, whereas on the back the close alignment of the O-4 ring atoms contributes to slightly less hydrophilic regions. The most hydrophobic surface areas of **3** and **4** are situated on the outer edge of the macrocycles, and are made up by the C1(H), C3(H), and C5(H)-C6(H_2) fragments.

In summation, this molecular modeling study of the furanoid cyclogalactins **1 - 4** provides reliable insights into their molecular architecture such, that they not only lack round-shaped geometries, but also central cavities with which to form inclusion complexes similar to those formed by cyclodextrins[18]. In contrast to the pyranoid cyclogalactohexaoside (α-cyclogalactin)[3], the intersaccharidic connectivity of **1 - 4**, the arrangement of their sugar units on the ring, as well as their molecular lipophilicity patterns may at most allow the formation of loose sandwich-type adducts.

References

[1] 'Molecular Modeling of Saccharides', Part 17. — Part 16: Immel, S., Schmitt, G.E., and Lichtenthaler, F.W. (1998). The molecular geometries of cyclofructins, in Torres Labandeira, J.J. (Ed.), *Proceedings 9th Internat. Symp. on Cyclodextrins*, Kluwer Acad. Publ., Dordrecht, NL.

[2] Lichtenthaler, F.W., and Immel, S. (1996). On the hydrophobic characteristics of cyclodextrins: computer-aided visualization of molecular lipophilicity patterns. *Liebigs Ann. Chem.*, 27-37.

[3] Lichtenthaler, F.W., and Immel, S. (1994). Cyclodextrins, cyclomannins, and cyclogalactins with five and six (1→4)-linked sugar units: a comparative assessment of their conformations and hydrophobicity potential profiles. *Tetrahedron: Asymmetry*, *5*, 2045-2060. — Lichtenthaler, F.W., and Immel, S. (1996).

The lipophilicity patterns of cyclodextrins and non-glucose cyclooligosaccharides. *J. Inclusion Phenom. Mol. Recognit. Chem.*, *25*, 3-16.

[4] Nogami, Y., Nasu, K., Koga, T., Ohta, K., Fujita, K., Immel, S., Lindner, H.J., Schmitt, G.E., and Lichtenthaler, F.W. (1997). Synthesis, structure, and conformational features of α-cycloaltrin: a cyclooligosaccharide with alternating $^4C_1/^1C_4$ pyranoid chairs. *Angew. Chem. Int. Ed. Engl.*, *36*, 1899-1902.

[5] Kochetkov, N.K., Nepogodiev, S.A., and Backinowsky, L.V. (1989). Synthesis of cyclo[(1-6)-β-galactofurano]-oligosaccharides. *Carbohydr. Res.*, *185*, C1-C3. — Backinowsky, L.V., Nepogodiev, S.A., and Kochetkov, N.K. (1990). Formation of cyclo-oligosaccharides by polycondensation of the 3- and 6-*O*-tritylated derivatives of 1,2-*O*-(1-cyanoethylidene)-α-D-galactofuranose. *Tetrahedron*, *46*, 139-150.

[6] Immel, S., and Lichtenthaler, F.W. (1996). The electrostatic and lipophilic potential profiles of α-cyclofructin: computation, visualization, and conclusions. *Liebigs Ann. Chem.*, 39-44.

[7] Lindner, H.J., and Kroeker, M. (1997). PIMM91 - *Closed Shell PI-SCF-LCAO-MO-Molecular Mechanics Program*, Darmstadt University of Technology. — Smith, A.E., and Lindner, H.J. (1991). π-SCF-Molecular Mechanics PIMM: formulation, parameters, applications. *J. Comput.-Aided Mol. Des.*, *5*, 235-262.

[8] Immel, S. (1997). *MolArch*$^+$ - *MOLecular ARCHitecture Modeling Program*, Darmstadt University of Technology.

[9] Metropolis, N., Rosenbluth, A.W., Rosenbluth, M.N., Teller, A., and Teller, E. (1953). Equation of state calculations by fast computing machines. *J. Chem. Phys.*, *21*, 1087-1092. — Van Gunsteren, W.F., and Berendsen, H.J.C. (1990). Molecular dynamics computer simulation. Method, application, and perspectives in chemistry. *Angew. Chem. Int. Ed. Engl.*, *29*, 992-1023.

[10] Goto, H., and Osawa, E. (1989). Corner flapping: a simple and fast algorithm for exhaustive generation of ring conformations. *J. Am. Chem. Soc.*, *111*, 8950-8951.

[11] The 3D-structures can be viewed at "http://caramel.oc.chemie.tu-darmstadt.de/immel/3Dstructures.html", MOLCAD graphics are available at "http://caramel.oc.chemie.tu-darmstadt.de/immel/molcad/gallery.html".

[12] Connolly, M.L. (1983). Analytical molecular surface calculation. *J. Appl. Cryst.*, *16*, 548-558.

[13] Brickmann, J. (1997). *MOLCAD - MOLecular Computer Aided Design*, Darmstadt University of Technology. — Brickmann, J. (1992). Molecular graphics: how to see a molecular scenario with the eyes of a molecule. *J. Chim. Phys.*, *89*, 1709-1721.

[14] Heiden, W., Moeckel, G., and Brickmann, J. (1993). A new approach to analysis and display of local lipophilicity/hydrophilicity mapped on molecular surfaces. *J. Comput.-Aided Mol. Des.*, *7*, 503-514.

[15] Teschner, M., Henn, C., Vollhardt, H., Reiling, S., and Brickmann, J. (1994). Texture mapping: a new tool for molecular graphics. *J. Mol. Graphics*, *12*, 98-105.

[16] *Cambridge Crystallographic Data File* (1995), Version 5.09. Refcodes: CIWVUZ, FACRAC, MACGAL, and VAJGIW.

[17] Zuurmond, H.M., van der Klein, P.A.M., Veeneman, G.H., and van Boom, J.H. (1990). A convenient iodonium-ion-assisted synthesis of an immunologically active tetrameric β(1→5)-linked D-galacto-furanoside from the extracellular polysaccharide of *Aspergillus* and *Penicillium* species. *Recl. Trav. Chim. Pays-Bas*, *109*, 437-441. — Veeneman, G.H., Notermans, S., Hoogerhout, P., and van Boom, J.H. (1989). Synthesis of an immunolocigally active component of the extracellular polysaccharide of *Aspergillus* and *Penicillium* species. *Recl. Trav. Chim. Pays-Bas*, *108*, 344-350. — Thiem, J., and Wessel, H.-P. (1983). Synthese des Tetrose-Bausteins der Chromomycinon-Seitenkette aus D-Galactose. *Liebigs Ann. Chem.*, 2173-2184.

[18] Lichtenthaler, F.W., and Immel, S. (1996). Towards understanding formation and stability of cyclodextrin inclusion complexes: computation and visualization of their molecular lipophilicity patterns. *Starch/Stärke*, *48*, 145-154.

SYNTHESIS AND CHARACTERIZATION OF PEPTIDO-CYCLODEXTRINS DEDICATED TO DRUG TARGETING

C. PÉAN [a], F. DJEDAÏNI-PILARD [a], C. CRÉMINON [b], A. WIJKHUISEN [c], J. GRASSI [b], P. GUENOT [d], Ph. JÉHAN [d], and B. PERLY [a]

a : DRECAM/SCM, CEA-Saclay, F-91191 Gif sur Yvette, (France)
b : DRM/SPI, CEA-Saclay, F-91191 Gif sur Yvette, (France)
c : UFR de Biologie, Université PARIS 7, F-75251, Paris, (France)
d : CRMPO, Université de Rennes 1, F-35042 Rennes, (France)

1. INTRODUCTION

Cyclodextrins (CD's) can be used as molecular carriers dedicated to drug targeting [(1)]. We report here on the grafting of CD's on the amino function(s) of neuropeptides, Substance P (SP), and of a shorter derivative, the SP4-11 fragment. The β- and γ-CD's rings were first functionalized with a carboxylic acid function by reacting succinic anhydride on 6-monoamino-6-deoxy-β- and γ-CD's, then coupled on the peptide backbone. Eight new peptidocyclodextrins were obtained and fully characterized by high resolution NMR and electrospray mass spectrometry[(2)], ESI-MS (high and low resolution). Preservation of the inclusion properties of CD's moieties of the peptido-cyclodextrins was also demonstrated for these compounds.

2. MATERIALS AND METHODS

β-CD and γ-CD were gifts from ROQUETTE and WACKER, respectively. All ^{1}H-NMR experiments were performed using a BRUKER DRX 500 spectrometer, at 298K in deuterium oxide. Mass spectra were obtained on MS/MS high resolution mass spectrometer ZabSpec-TOF (Micromass Company) with on orthogonal time of flight analyser. Electrospray (ESI) high resolution ionisation was used in positive and negative mode, with a mixture of H_2O/CH_3CN 50/50 as solvant. Peptides (SP and SP4-11) were synthesized with solid phase procedure, using FMOC amino-acids, and by using BOP/HOBT as coupling reagents. They were purified by HPLC and fully characterized by NMR and mass spectrometry.

3. SYNTHESIS

The active part of the neuropeptide Substance P is located at the C-terminus corresponding to the Phe7-XXX-Gly9-Leu10-Met11-NH_2 sequence. Therefore, a coupling reaction must involved the N-terminus extremity of the peptide and a carboxylic acid derivative of the CD's. These latters were obtained by reacting mono-6-amino-6-deoxy-CD derivatives [3] with one equivalent of succinic anhydride in DMF at room temperature.

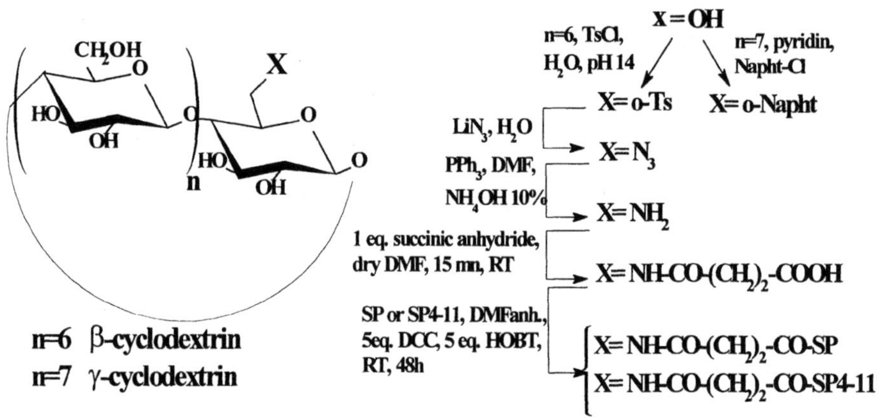

Fig. 1: Schematic synthesis of β- and γ-CD peptido derivatives

This reaction led to the CD-COOH adducts in quantitative yields, and introduced the expected carboxylic acid function on the CD core. Moreover, the succinic unit will play the role of spacer to avoid steric hindrance effects during the coupling reactions with peptides. These were performed under classical conditions (5 eq. dicyclohexyl-carbodiimide (DCC) (diisopropylcarbodiimide is also successful), 5 eq. hydroxybenzo-triazole (HOBT) per acid function, in dry DMF, at room temperature for 48 h), 2 molar eq. of CD's used for 1 eq. of peptide. After a precipitation step in aceton, the crude products were purified by HPLC under gradient conditions (delta-pack C_{18} column, H_2O/CH_3CN from 0 to 50% in 40 mn, on line UV (220nm) and light scattering detection (LSED)). Three major peaks were isolated (β- or γ-CD(Lys3)-SP (67%), β- or γ-CD(Arg1)-SP (11%) (fig. 2) and di-β- or di-γ-CD(Arg1,Lys3)-SP (22%)) when the grafting procedure involved SP, and only one (β- or γ-CD-SP 4-11) when the shorter SP4-11 fragment was used in the coupling process, since the latter has only one single primary amino reactive function.

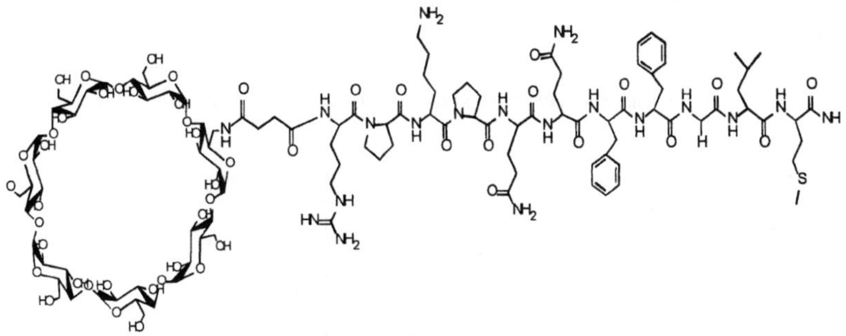

Fig. 2: Example of β-CD peptido derivative

4. STRUCTURAL ANALYSIS

All synthesized and isolated compounds were characterized by NMR (1D and 2D experiments) and ESI-MS. We performed TOCSY experiments (mixing time = 80 ms) to elucidate the primary sequence using the spectra of the native peptides as references. The grafting sites (Nα-Arg^1 or Nε-Lys^3) were demonstrated using NOESY experiments allowing a complete sequencing of the protons. The different ESI mass spectra exhibit two important ions $[M+H]^+$ and $[M+Na]^+$ in the mass range between 2000 and 3500 daltons and also dicharched ions $[M+2Na]^{++}$ and $[M+H+Na]^{++}$. All the formulae were established by high mass measurement with polyethyleneglycol as reference. NMR data further evidenced that, in the mono-SP-CD's derivatives, the remaining free amino group was trifluoroacetylated. Introduction of this protecting group occured during the coupling reaction between the free amino functions of the peptide and the TFA salts resulting from the HPLC purification step of the native peptide. The presence of this group was confirmed by mass spectrometry but was shown to be harmless in the activity and recognition processes. Moreover, TFA group can be remove by basic treatment.

5. INCLUSION PROPERTIES OF THE CD-SP ADDUCTS

To demonstrate the inclusion properties of the CD-SP adducts, we investigated the inclusion of indomethacin in the [Nε-(mono-6-amidosuccinylamido-6-deoxy-β-CD)-Lys3]-SP (further identified as β-CD(Lys3)-SP) by NMR. It is well established that indomethacin form a 1:1 inclusion complex [(4)] at slightly basic pH with β-CD. An equimolar mixture of β-CD(Lys3)-SP and indomethacin was prepared at pH 8. 2D-ROESY [(5)] (Rotating Frame Overhauser Effect SpectroscopY) experiment was carried out on this sample. Complex formation was demonstrated by presence of dipolar cross-peaks between the protons of β-CD(Lys3)-SP (H3 and H5, internal cavity protons of the CD) and those of the para-substituted aromatic ring of indomethacin (signals labelled with * on the Fig. 3). The NMR data is very similar to that obtained with the original complex performed with natural β-CD. It is therefore assumed that the presence of the peptide on primary hydroxyles face has a very weak impact on the formation of an inclusion complex with the present guest.

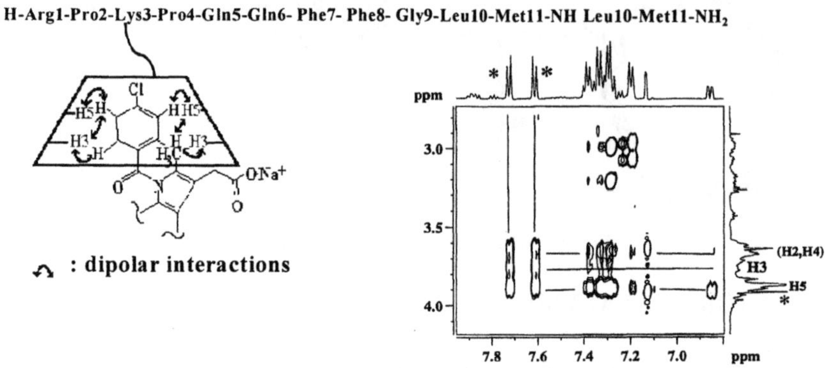

Fig. 3: Extension of the Roesy experiment performed between indomethacin and βCD(Lys3)-SP (right) and derived structural model for the inclusion complex in solution.

The observed data suggest that all peptido-CD compounds retain their inclusion properties as far as the secondary hydroxyles rim is involved.

6. CONCLUSION

With the aim to use cyclodextrins in the drug-targeting domain, we synthesized and fully characterized 8 new peptido-cyclodextrins such as $(CD's)_n$-Sub. P (or SP 4-11). The compounds were analyzed using ^{1}H-NMR and mass spectrometry. More, we demonstrated that the cyclodextrin moieties retained their inclusion properties by studying the β-CD(Lys3)-SP/indomethacin complex. The next step of this work will be the demonstration of the *in-vitro* recognition properties of the peptidocyclodextrins towards cells expressing the human NK1 receptor for Substance P.

This work was supported by the European Commission (DGXII) under the FAIR Programme CT95-0300

7. REFERENCES

(1) : F. Djedaini-Pilard, J. Désalos, B. Perly, *Tetrahedron. Lett.*, **34**, 2457-2460 (1993)
(2) : M. Yamashita, J.B. Fenn, *Phys. Chem.*, **88**, 4451 (1988)
(3) : E. S. Cotner, P. J. Smith, *J. Org. Chem.*, **63**, 1737-1739 (1998); F. Djedaini-Pilard, N. Azaroual-Bellanger, M. Gosnat, D. Vernet, B. Perly, *J. Chem. Soc. Perkin Trans.* 2, 723-730 (1995)
(4) : K-H. Fromming, J. Szejtli: *Cyclodextrins in pharmacy*, Topics in inclusion science, Kluwer Academic Publishers, Dordrecht
(5) : A. Bax, D.G. Davis, *J. Magn. Reson.* **63**, 207-213 (1985)

MONO-6-TOSYL-β-CYCLODEXTRIN: PREPARATION, HYDROLYSIS AND SELF-INCLUSION STUDIES IN AQUEOUS SOLUTION

F. DJEDAINI-PILARD [a], M. GOSNAT [a], S. STEINBRUCKNER [a],
J.P. DALBIEZ [a], G. CRINI [b], B. PERLY [a] AND A. GADELLE [c]
[a] DRECAM/SCM, CEA-Saclay, F-91191 Gif sur Yvette (France) [b] Centre de Spectrométrie, Université de Franche-Comté, F-25030 Besançon (France)
[c] DRFMC/SCIB, CEA-Grenoble, F-38054 Grenoble (France)

1. Introduction

The key step to attach functionnal groups on the primary hydroxyls of CD's is a selective sulfonation to afford a monotosyl derivative. The preparation method and substitution pattern of the popular sulfonated CD's are still unclear. Two major methods have been described in alkaline aqueous solutions[1-3] or organic solvents[4-5]. Using pyridine as solvent leads mainly to mono-6-tosyl-6-deoxy-β-cyclodextrin but the final product is contaminated with di- or tri-tosyl-β-CD derivatives. Conversely, in aqueous solution only mono-6-tosyl-6-deoxy-β-cyclodextrin is formed but with a relatively poor yield (less than 15%).

We show here that the kinetics of the reaction of tosylation in aqueous solution strongly depends upon the effective pH. Morever in alkaline aqueous solution, although the reaction is very fast and can yield up to 35% of the title compound, it is competing with hydrolysis of the mono-6-tosyl-6-deoxy-β-cyclodextrin **1**. A complete NMR study has demonstrated that this product is hydrolyzed in aqueous solution at pH > 6 and that acidification of the reaction medium can quench this process. Investigations of the structure of pure **1** in aqueous solution are presented showing that a strong intramolecular self-inclusion complex is formed. Dedicated two dimensional NMR experiments are used in conjunction with competition with external guests to evidence and estimate the strength of the auto-inclusion complex.

2. Materials and Methods

β-CD was obtained from Roquette Frères. *p*-toluene-sulfonyl chloride (TsCl) was recristallized from light petroleum prior to use. All NMR experiments were performed using Bruker DRX500 spectrometers operating at 500.13 MHz, after a freeze-drying step and redissolution in deuterium oxide (Euriso-Top, France). Mass spectrometry was performed on a SCIEX spectrometer with electrospray infusion mode.

Synthesis of Mono-6-tosyl-6-deoxy-β-cyclodextrin 1
β-CD (20g, 17.6 mmol) was dispersed in water. NaOH (8N, 7ml) was added resulting in a clear solution which was stirred vigorously and cooled at 5°C. 1.2 eq of TsCl (4.03g) dissolved in 10ml of CH_3CN was added dropwise. After neutralization using 10ml of 6N Hcl, the mixture was stirred at RT for 1 hour resulting in the formation of a copious white precipitate. The crude product was isolated by filtration and recristallized from boiling water affording pure **1** (6.80g, 29%),TLC (NH4OH-EtOH-BuOH 5 :5 :4) R_f= 0.34, *m/z*= 1290 $[M+H]^+$. δH (500MHz, d_6-DMSO): 7.72, 7.41 (4, ArH), 4.75 (1, grafted 1-H), 4.74 (1, 1-H), 4.82-4.81 (5, others 1-H), 4.33, 4.23 (2, grafted 6,6'-H), 2.41 (3, CH3).

3.Results and Discussion

3.1 KINETICS OF THE REACTION

The main results are displayed on Fig 1. It was evidenced that the yield of the reaction increased with the value of the pH. However the yield seems to reach a value around 15% independantly to the value of the basic pH.

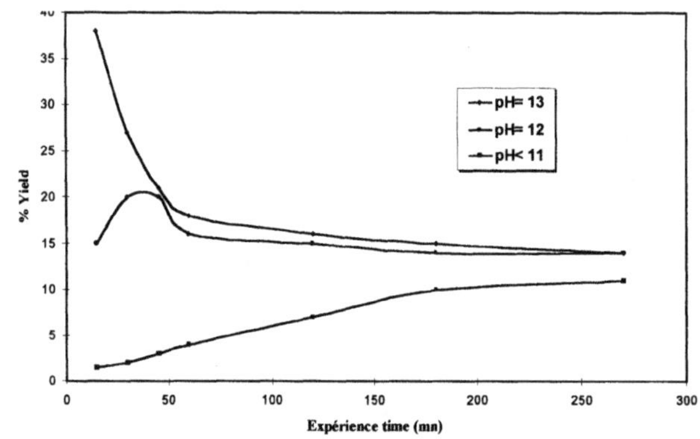

Figure 1: **Kinetics of reaction *vs* pH**. 3g of β-CD was dispersed in 25ml of water. After addition of appropriate buffering salts 3 eq of TsCl in solution on 1.5ml of CH3CN were added upon stirring at RT. At the indicated time, the solution was neutralized and the crude product was isolated by filtration and recristallized from boiling water.

This effect can be explained by a competition between the formation and the hydrolysis of the title compound as shown further. It should be noticed that highest yield achieved (40%) was obtained at pH=13, after 5 mn reaction and further quenching.

3.2 HYDROLYSIS OF MONO-6-TOSYL-6-DEOXY-β-CD **1**

The partial NMR spectrum in D_2O of pure **1** is displayed on Fig 2.a. The spectral region corresponding to anomeric protons is depicted since it is highly representative of the reduction of the seven-fold symmetry. On Fig 2b-d the presence of an additional signal due to the proton H1 of free β-CD is evidenced. The intensity of this peak increases with time reaching 55% of the total amount of **1** after 5 days at 35°C in water.

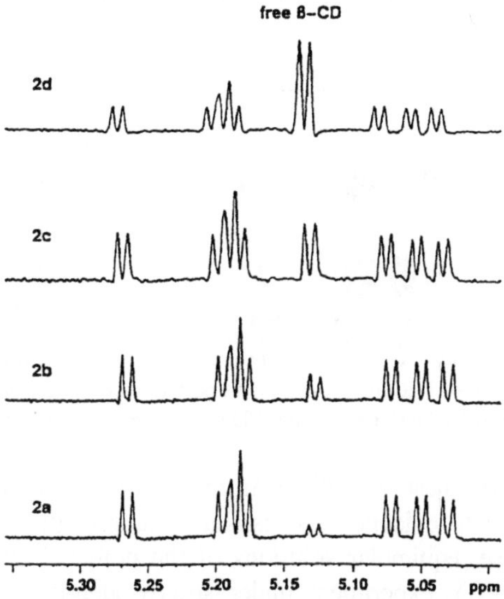

Figure 2: Partial 500 MHz NMR spectra of 1 (2.9 mmol, D_2O, 310K, spontaneous pH) at time zero: 2a, after 4H: 2b, after 24H: 2c and after 5 days: 2d.

Similarly, new signals corresponding to the aromatic protons of the free tosyl group can be observed. On the other hand, the same NMR experiments when performed in phosphate buffer (pH 5.16) have demonstrated that the signal of proton H1 of free β-CD did not appear indicating that the hydrolysis is inhibited at pH<6.
Finally, complete acetylation of all free hydroxyls of **1** fully suppresses the hydrolytic process even at pH > 6. This suggests that the primary hydroxyls are implicated in this process.

3.3 SELF-INCLUSION PHENOMENOM

Compound **1** displays a considerable spectral dispersion in D_2O. Conversely when the NMR spectrum is collected in d_6-DMSO as solvent, this spectral dispersion collapses and only weak inequivalences are retained (data not shown). This behaviour is the first clue to the possible inclusion of the aromatic ring in the cavity of the CD in aqueous solution as already described elsewhere[5].
A more convincing proof for the reality of self-inclusion is expected from direct observation of spatial proximities between protons of the tosyl and protons located in the internal cavity of the CD[5]. Fig 3 shows a partial contour plot of a T-ROESY[6] experiment performed at 500 MHz on **1** in D_2O. A large number of cross-peaks between the

aromatic protons of tosyl and protons of CD is observed. As almost all protons of the CD have been assigned previously, it is obvious that the aromatic protons experience interactions with protons 3, 5 located in the cavity of the CD. This observation fully supports the reality of the intramolecular self-inclusion complex for **1** as predicted by Toda and al[3].

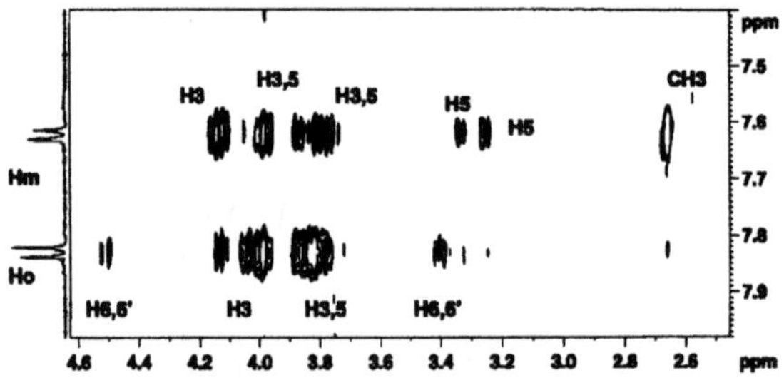

Figure 3: partial contour plot of T-ROESY experiment performed on **1** (2.9 mmol, D_2O, 310K, 500MHz, 300ms of spin lock). Horizontal scale: CD region; vertical scale: aromatic region.

Attempts to include external guest as Sodium Antraquinone-2-sulfonate (ASANa) in the cavity of **1** was performed to evidence and estimate the strength of the self-inclusion complex[7]. For this purpose, equimolar solutions of the potential host and of ASANa were subjected to T-ROESY experiment under strictly identical conditions and the results compared with those obtained in the absence of competing guest (data not shown). No cross-peaks between the protons of ASANa and those of the cavity of CD were observed clearly indicating that the added guest failed to enter the CD cavity. This observation implies that the self-inclusion complex is very strongly stabilized. This stabilization can implicate hydrogen bonds between primary hydroxyls of CD and sulfonyl group.

All these results provided an explanation for the exclusive of **1** in akaline solution, conversely to results obtained in pyridine. Morever the formation of a self-inclusion complex provides clues to a better understanding of the high sensitivity of **1** towards hydrolysis by involving the direct participation of primary hydroxyls of CD.

References

1 R.C. Petter, J.S. Salek, C.T. Sikorski, G. Kumaravel, F. Lin (1990) *J. Am. Chem. Soc*, 112, 3860-3868
2 S. Onozuka, M. Kojima, K. Hattori, F. Toda (1980) *Bull. Chem. Soc. Jpn*, 53 3221-3224
3 K. Takaheshi, K. Hattori, F. Toda (1984) *Tetrahedron Lett.*, 125, 3331-3334
4 L.D. Melton, K.N Slessor (1971) *Carbohydr Res*, 18, 29-37
5 F. Djedaïni-Pilard, N. Azaroual-Bellanger, M. Gosnat, D. Vernet and B. Perly (1995) *J. Chem. Soc. Perkin Trans II*, 723-730.
6 T.L. Hwang, A.J. Shaka (1992) *J. Am Chem. Soc*, 3157-3159
7 F. Djedaïni, B. Perly (1990) *Magn. Reson.Chem.*, 28, 372-374.

THIOUREA-BRIDGED β-CYCLODEXTRIN CONJUGATES

C. ORTIZ MELLET,[a] J.M. GARCIA FERNANDEZ,[b] J.M. BENITO,[b]
H. LAW,[c] K. CHMURSKI[c] AND J. DEFAYE[c]
[a] Departamento de Química Orgánica, Facultad de Química, Universidad de Sevilla, E-41071 Sevilla, Spain. [b] Instituto de Investigaciones Químicas, CSIC, Américo Vespucio s/n, E-41092 Sevilla, Spain. [c] CNRS (EP 811) and Université Joseph Fourier-Grenoble 1, Département de Pharmacochimie Moléculaire/Glucides, BP 138, F-38243 Meylan, France

Abstract

Saccharide as well as peptide antennae have been efficiently appended to the primary hydroxyl rim of the β-CD core through thiourea tethers. The synthetic strategy involves the coupling reaction of glycosyl or peptide isothiocyanates with amine funtionalized β-CDs and has been applied to the preparation of mono- as well as heptavalent derivatives. The new conjugates exhibited a dramatic increase in water solubility as compared to β-CD itself while retaining the inclusion properties towards the anticancer drug Taxotère®.

1. Introduction

Carbohydrates and peptides are known to mediate a number of important biological recognition phenomena at the cell membrane level. In many case, these interactions have a multivalent character, needing a local density and a precise spatial orientation of the active epitopes to be operative. Incorporation of peptide and oligosaccharide markers into cyclomaltooligosaccharides carriers (cyclodextrins, CDs) represents, thus, an attractive strategy for the design of site-specific drug delivery systems. The utility of this approach is highly dependent on the development of efficient synthetic routes to the corresponding CD-conjugates. Here we present a high yielding preparation of branched β-CDs incorporating both types of biomolecules at the primary (narrower) hydroxyl rim based on the coupling reaction of isothiocyanates and amines to give thiourea bridged adducts. Examples include mono- and hepta-antennary derivatives, the later ones bringing a biomimetic cluster-type arrangement. The use of alkyl and aryl spacers has been also explored, thus broadening the possibilities for control of the flexibility and spatial arrangement of the substituents.

2. Materials and Methods

2.1 MATERIALS

^{1}H and ^{13}CNMR (D_2O, 500 and 125.7 MHz, respectively) spectra were recorded at 343 K using a Bruker DRX 500 instrument. Mass spectra were performed in a High Resolution Fourier Transform Mass Spectrometer (FTMS) instrument equipped with an external electrospray ionisation (ESI) source, manufactured by IonSpec Corporation (Irvine, CA). The purity of all new compounds was checked by analytical HPLC (Bondapack C-18 column, 12:88 and 7:93 MeOH-water) and microanalysis (C, H, N).

2.2 METHODS

2.2.1 *General procedure for the preparation of 6^{I}-deoxy-6^{I}-[3-(methyl) or (phenyl) or (glycosyl) or (methoxycarbonylmethyl)thioureido]cyclomaltoheptaose.*
A solution of 6^{I}-amino-6^{I}-deoxycyclomaltoheptaose (160 mg, 0.142 mmol) and the corresponding isothiocyanate (0.156 mmol) in pyridine (2 mL) was stirred for 2 h at room temperature. The solvent was removed by distillation and further coevaporation with toluene, and the residue was purified by column chromatography (Silica Gel 0.04-0.06 mm, 3:1 MeCN/H_2O). The corresponding thioureido-β-CD conjugates **1-7** were obtained as white solids in 70-75% yield. Compounds **5-7** (0.1 mmol) were further dissolved in MeOH (20 mL) and the solution was adjusted to pH 8-9 by addition of freshly prepared 1 N methanolic NaOMe. After 5 min, a white precipitate appeared and the resulting suspension was stirred for 5 h at room temperature. Water (2 mL) was then added and the clear solution was stirred for 2 h, then neutralised with Amberlite IR-120 (H^+) ion exchange resin and further demineralised with Duolite MB 6113 (H^+,OH^-) mixed ion exchange resin. The fully unprotected monoantennary compounds **8-10** were obtained as white lyophilizates in 85-95% yield.

2.2.2. *General procedure for the preparation of heptakis{6-deoxy-6-[3-(methyl) or (phenyl) or (glycopyranosyl)]thioureido}cyclomaltoheptaose.*
A solution of heptakis(6-amino-6-deoxy)cyclomaltoheptaose as its heptahydrochloride (50 mg, 0.036 mmol) in water (6 mL) was adjusted to pH 8 by addition of saturated aq $NaHCO_3$. The corresponding isothiocyanate (0.276 mmol) in acetone (6 mL) was then added and the reaction mixture was stirred for 2 h. The solvent was removed by distillation and the residue was subjected to column chromatography (Silica Gel 0.04-0.06 mm, MeCN/H_2O, 5:1). Compounds **11** and **12** were obtained as white solids. The chromatographed reaction mixtures arising from the coupling reaction with glycosyl isothiocyanates contained partially deacetylated adducts (**13-15**), as seen from the corresponding ^{1}H and ^{13}C NMR spectra. After deacetylation and demineralisation as above, the corresponding fully unprotected heptaantennary glycoclusters **16-18** were obtained as white lyophilizates in 65-80% yield based on the starting heptahydrochloride.

3. Results and Discussion

In principle, the thiourea linker can be generated either from amine- or isothiocyanate-functionalised β-CD derivatives. Both functional groups can be incorporated onto the β-CD core following previously reported selective mono- and per-(C-6)-functionalisation methodologies as reported in [1-2] an references therein, with no need for hydroxyl protection. Initially, the coupling of β-CD-isothiocyanates with glycosylamines and amino acids was tried avoiding the use of protecting groups, an approach previously found successful for the attachment of non-anomeric amino sugars to CDs. Although reactions proceeded in pyridine with total consumption of the β-CD reagents, the formation of undesired side products seriously hampered the purification step, especially in the case of the hepataantennary glycoclusters. Nevertheless, the reverse strategy using readily available peracetylated glycosyl isothiocyanates [3] or amino acid methyl ester isothiocyanates and mono-(C-6^{I})-NH_2 β-CD was found to proceed in excellent yield (→**3-7**). Removal of the acetyl groups in **5-7** was accomplished by a mixed Zemplén-saponification methodology to give the fully unprotected compounds **8-10**.

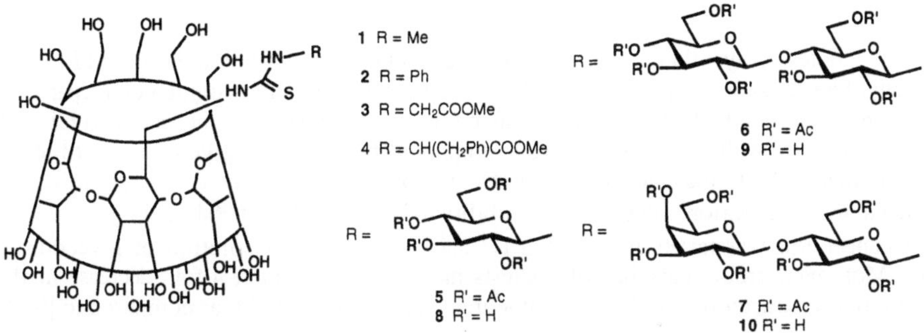

The above protocol failed, however, in providing the monofacially branched, heptaantennated glycoclusters. Reaction of per-(C-6)-NH_2 β-CD with glycosyl isothiocyanates in pyridine was not complete after 3 d at 50 °C using a high excess of the NCS-reagent. Moreover, extensive formation of bis(glycosyl)thiourea at the expenses of the glycosyl isothiocyanate took place under these conditions. Almost quantitative yields (→**13-15**) were obtained after 2 h in water-acetone at pH 8. Despite some deacetylation at this stage, subsequent base treatment afforded the target seven-fold symmetrical glycoclusters **16-18** in 65-80% overall yield.

It is noteworthy that purification of the thioureido-β-CD compounds, frequently the most problematic aspect in CD chemistry, can be achieved very efficiently with the partially acetylated derivatives **5-7** and **13-15** by a simple silica gel column chromatography using water-acetonitrile as eluent. After deacetylation, both the mono- and hepta-(glycosylthioureido) conjugates proved to be homogeneous by chromatographic (reversed phase HPLC), elemental analysis and NMR spectroscopy techniques. FAB (partially acetylated derivatives) and ES mass spectrometry (fully unprotected compounds), respectively, were well-suited techniques for the mass analysis.

An additional extension of this flexible synthetic concept was provided by the high-yielding coupling reaction of β-CD amines with methyl and phenyl isothiocyanates to give the corresponding methyl- (**1** and **11**) and phenyl-thioureido derivatives (**2** and **12**, respectively), thus covering the possibility for the use of either alkyl or aryl spacers.

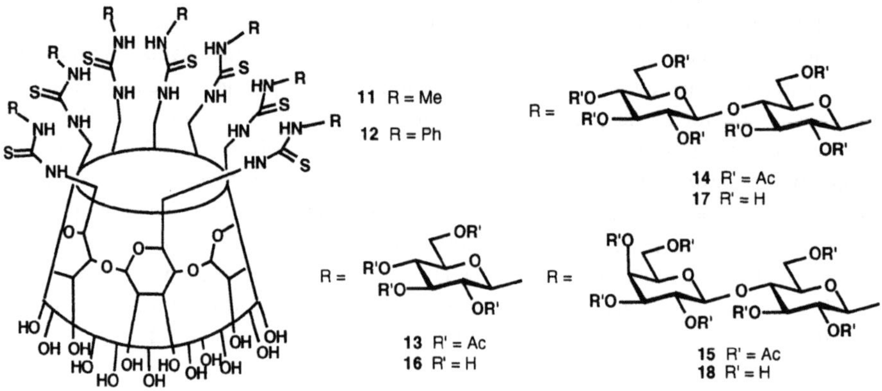

It is noteworthy that, except for the heptakis(methylthioureido) (**11**) and phenylthioureido derivatives (**2** and **12**), all the β-CD—thiourea conjugates prepared in this work exhibited a water solubility much higher as compared to the native β-CD (25 to 40-fold), even those incorporating hydrophobic peracetylated glycosyl substituents. Preliminary results indicate that the access to the internal cavity is not significantly altered by substitution. The water solubility of Taxotère® (0.004 g L^{-1}), a taxane anticancer drug was raised up to 1.5 g L^{-1} using a 50 mmolar solution of glycocluster **13**. Moreover, preliminary hemolysis tests have shown an about 4-fold decrease in the hemolytic character of β-CD bearing thioureido substituents at C-6 as compared to β-CD itself.

4. Acknowledgements

J. M. B. is grateful to the CSIC and Hoeft-Rademacher Ltd. Co. for a fellowship. This research was supported by the European Commission DG12 under the FAIR programme (contract no. FAIR-CT95-0300) and the Dirección General de Investigación Científica y Técnica (grant no. PB96-0820).

5. References

1. K. Chmurski, K., Defaye, J., (1997) An improved synthesis of 6-deoxyhalo cyclodextrins via halomethylenemorpholinium halides Vilsmeier-Haack t ype reagents, *Tetrahedron Lett.*, *38*, 7365-7368.

2. García Fernández, J.M., Ortiz Mellet, C., Maciejewski, S., Defaye, J. (1996) Thioureido β-cyclodextrins as molecular carriers, *Chem. Commun.*, 2741-2742.

3. García Fernández, J.M., Ortiz Mellet, C. (1996) The thiocarbonyl group in carbohydrate chemistry, *Sulfur Rep. 19*, 61-169.

ASSOCIATIONS OF AMPHIPHILIC DEGRADABLE POLYMERS WITH Β-CYCLODEXTRIN POLYMERS : PH - DEPENDENT NETWORK.

Catherine Amiel *, Laurence Moine*, Wyn Brown†, Estelle Renard*, Philippe Guérin *, Bernard Sébille*

** Laboratoire de Recherche sur les Polymères, UMR C7581 CNRS, 2-8 rue H. Dunant, 94320 Thiais ,France, † Departmentt of Physical Chemistry, University of Uppsala, Box 532, S-751 21 Uppsala, Sweden.*

1. Introduction

Within the framework of an extended study of macromolecular assemblies generated by host-guest interactions [1,2], this study has focused on assemblies of polyelectrolyte character. The system consists in the mixture of two different copolymers : a degradable copolyester bearing adamantyl pendant groups and a water soluble β-cyclodextrin / epichlorohydrin polymer (β-CD/EP). Inclusion of the hydrophobic groups inside the β-CD cavities can lead to an associating network above critical values of the polymers concentrations.

The adamantyl group is a highly symmetric tricyclic hydrocarbon which consits in three fused cyclohexane rings as shown in figure 1. Due to its perfect fit into β-cyclodextrin cavities, it is known to make very stable complexes with β-cyclodextrins [3]. Amphiphilic degradable copolyesters, poly(β-malic acid-co-ethyladamantyl β-malate) (PMLA-Ad), were synthesized by copolymerization of benzyl malolactonate, which is conducive after deprotection to malic acid units, and ethyladamantyl malolactonate [4] . These polymers are degradable by simple hydrolysis of main chain ester bonds. The hydrophobic pendant groups arranged in different proportions along the main chain (3 to 10%) allow the modification of the hydrophilic / hydrophobic balance. Presence of charged groups in the amphiphilic polymer increases its solubility and provides systems sensitive to ionic strength and pH changes. pKa of the copolymers are of the order of 4.

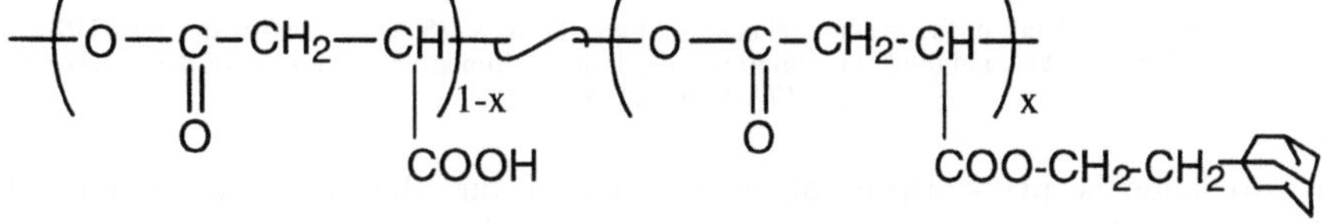

Figure 1: Poly (β-malic acid-co-ethyladamantyl β-malate)

Water-soluble β-CD polymers are prepared by polycondensation of β-CD with epichlorohydrin (β-CD/EP) under strongly alkaline conditions [5]. High molecular weight polymers (12 10^6 g/mol) were obtained by conducing the reaction close to the gelation threshold.

Associations between the host and guest polymers were studied by dynamic light scattering which is sensitive to apparent hydrodynamic radius of the species in the medium [6].

2. pH - dependent associations

Dynamic light scattering signals were recorded in mixtures copolyester/ β-CD/EP of given concentrations but increasing pH (addition of very small amounts of NaOH solution (2N)). At the initial pH values (2 - 2.5), there is no modification of the relaxation modes of the two components of the mixture. When the pH is increased, a slowing down of the relaxation process can be seen clearly in figure 1. There is a shift to longer relaxation times up to pH 4-5 followed by a decrease at higher pH values. The slower relaxation times are attributed to association between the hydrophobically modified copolyester and the β-CD copolymer, reflecting increasing hydrodynamic radius of the aggregates. Since association does not occur between the precursor copolyester (PMLA) and β-CD copolymer, it must involve the hydrophobic moieties (adamantyl groups) of the copolyester i.e. occurs through formation of inclusion complexes between adamantyl groups and the β-CD cavities.

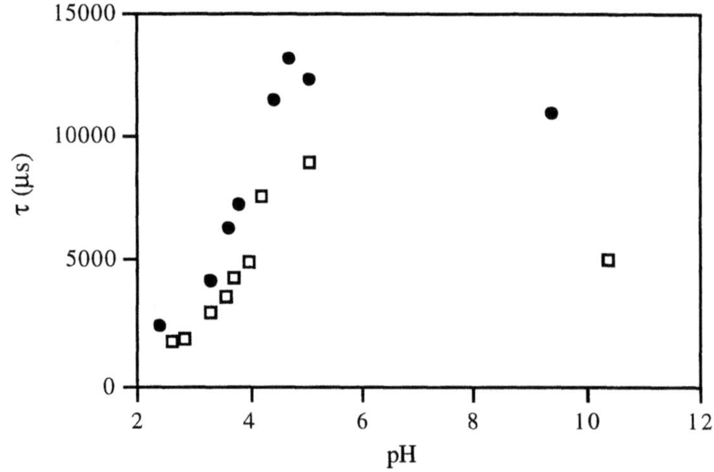

Figure 2: pH-dependence of the average relaxation times (DLS) at 25 °C for copolyester / β-CD/EP mixtures, weight ratio 40/60 and total concentration 20g/l. the hydrophobic molar ratio of the copolyesters are (□) 3.5% and (●) 7%.

The association is pH - dependent, as a result of the changes in the copolyester conformation . At the initial pH most of the carboxylic groups are protonated. The copolyesters probably adopt compact conformations with hydrophobic microdomains forming to limit contact with water [7,8]. As a result, the mobility and accessibility of the adamantyl groups is reduced and the probability of their inclusion in β-CD cavities is small. As the pH of the medium is increased, the carboxylic groups progressively dissociate. Electrostatic repulsions between these charged groups will oppose hydrophobic attraction between the pendant adamantyl groups. The hydrophobic microdomains should progressively diminish as the pH is increased and consequently the number of adamantyl group available for complexation should increase. The apparent hydrodynamic radius of the mixtures should increase monotonically as a function of the pH until the copolyester is completely charged (at pH~6) and then reach a limiting value. However, a slight decrease of the apparent hydrodynamic radius is observed above pH 4.5 - 5 . This effect is probably due to the added NaOH in order to increase the pH which screens the fixed charges as with an added simple electrolyte . The effect could also be attributed to a high charge density along the backbone of the copolyester which does not favor inclusion of the adamantyl groups into the cavities of β-CD [9].

3 - Influence of concentration

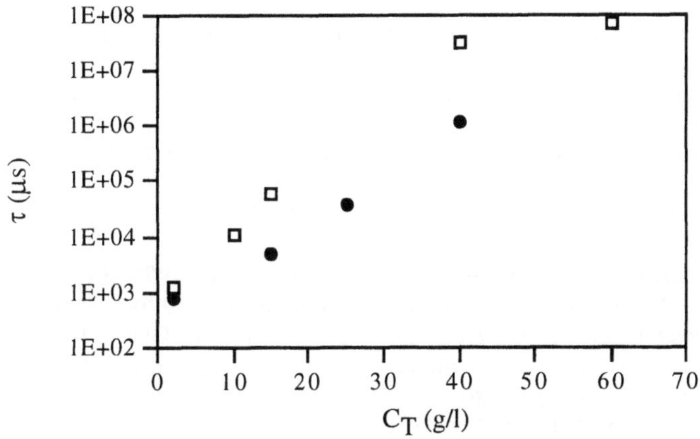

Figure 3: Concentration dependence of the average relaxation times (DLS) at 25 °C for copolyester / β-CD/EP mixtures, weight ratio 40/60 and pH 4. The hydrophobic molar ratio of the copolyesters are (□) 3.5% and (●) 7%.

Strong increases of the average relaxation times deduced from DLS measurements are observed when the total concentration is increased (figure 2). These results are well correlated to low shear viscosity data which show the same trend [10]. At low polymer concentrations, the proportion of adamantyl groups included in β-CD cavities is lower than the one expected from the stoichiometry of the mixture because of the finite value of the complexation constant. Moreover, probabilities to form interchains complexes are too low because chains are not overlapping. Above a critical concentration which correspond to the connectivity of the system, chains of both polymers can overlap and interchains complexes are favored. This leads to an associating network. Better associating properties are obtained for the higher hydrophobic molar ratio (7% compared to 3.5%). However, there should be optimal values of this parameter because there is a competition between auto-association of the hydrophobic moieties and host-guest interaction with cyclodextrins.

4. Conclusion

Mixtures of hydrophobically modified polyelectrolytes and β-cyclodextrin polymers have been studied by dynamic light scattering. At various concentrations, several diffusive relaxation modes are observed, which are identified with aggregates, single chains and sometimes gel modes. The original property of this polymer system is that the associative behavior seem to be controled by the net charge of the amphiphilic copolyester: at low pH (<3) no associations are observed, independantly of polymer concentrations , while large aggregates and viscosity enhancements are obtained when the pH is larger than the pK_a value of the amphiphilic copolyester (ca 4).

5. References

1. Amiel, C.; Sébille, B.,*Journal of Inclusion Phenomena and Molecular Recognition in Chemistry,* **1996**, *25*, 61.
2. Gosselet, N.M.; Borie, C.; Amiel, C.; Sébille, B., *Journal of Dispersion Science and Technology* (in press).
3. Eftink, M.R.; Andy, M.L.; Bystrom, K.; Perlmutter, H.D.; Kristol, D.S., *J. Am. Chem. Soc.,* **1989**, *111*, 6765
4. Moine, L.; Cammas, S.; Amiel, C.; Guérin, P.; Sébille, B., *Polymer,* **1997**, *38*, 3121
5. Renard, E.; Deratani, A.; Volet, G.; Sébille, B., *Eur. Polym. J.,* **1997**, *33*, 49
6. Nicolai, T., Brown, W., Hvidt, S., Heller, K., *Macromolecules*, **1990**, 23, 5088.
7. *Polymer in Aqueous Media : Performance Through Association,* Glass, J.E., Eds; Advances in Chemistry 223; American Chemical Society : Washington, DC, 1989
8. *Water-Soluble Polymers : Synthesis, Soluition Properties and Applications,* Shalaby, S.W.; Mc Cormick, C.L.; Butler, G.B., Eds; ACS Symposium Series 467; American Chemical Society : Washington, DC, 1991
9. *Cyclodextrins and their inclusion complexes,* Szejtli, J., Eds; Budapest : Akadémia Kiado, 1982
10. Moine, L.; Cammas, S.; Amiel, C.; Renard, E.; Sébille, B.; Guérin, P., *Macromolecular Symposia* , in press.

ORIGINAL ASSOCIATING SYSTEM BETWEEN HYDROPHOBICALLY MODIFIED N,N-DIMETHYLACRYLAMIDE HYDROXYETHYLMETHACRYLATE COPOLYMERS AND WATER SOLUBLE β-CYCLODEXTRINS POLYMERS

N. M. GOSSELET, F. BEUCLER and B. SEBILLE , *Laboratoire de Recherche sur les Polymères, CNRS, U.M.R. C7581, 2-8 rue Henri Dunant, 94320 Thiais, France*

1. Introduction

Hydrophobically modified polymers or associative polymers (AP) were used as thickeners in food or latex based paints, or as additives in enhanced oil recovery or water treatment [1-5]. They consist in hydrophylic backbones modified at the ends of the chain or along the chain , by hydrophobic groups introduced in small quantities , 1 to 5 mole %. Recently, we developed in our laboratory new systems of AP polymers consisting in a β-cyclodextrin epichlorhydrin polymer (βCD/EP) and a hydrophylic polymer PEO or N, N-dimethylacrylamide hydroxyethylmethacylate copolymers modified by adamantyl groups [6,7] which match the βCD cavities [8]. We will report here the main results obtained in viscosity, phase diagrams and fluorescence measurements.

2. Materials and methods

2.1 GEL PERMEATION CHROMATOGRAPHY

The molecular weight of the polymers was determined using gel permeation chromatography. TSK SW 3000 and SW 4000 columns were used for the analysis of βCD/EP polymers and copolymers respectively. The detectors were a refractometer RI 7 (Shodex) alone or coupled to a Light Scattering Minidawn (Wyatt) in some cases.

2.2 VISCOMETRY

The viscosity measurements were carried out with an Ubbelhode type viscometer at 25°C ± 0,5°C.

2.3 PHASE DIAGRAM

The concentration of βCD/EP in the liquid phase was determined with a polarimeter 141 (Perkin Elmer) under a mercury lamp (λ = 535 nm) after standardization. Refractive index measurements with a differential refractometer (R 401 Waters) connected to a small column (10 cm x 0,46 cm I. D.) of silica and a 1 mL loop permitted us to deduce the concentration of DMA/HEMA-Ad from the total value. In fact, the refractive index variations are additive when the two polymers are mixed in the concentration range 5.10^{-3} %< c < 0,2 %, w/w.

2.4 FLUORESCENCE

Stady-state fluorescence measurements were performed on air-equilibrated solutions at 25°C, using a SLM Aminco 8100 spectrofluorimeter equipped with a Xenon lamp. Pyrene, after recristallization in methanol, was used as probe in aqueous solution (5.10^{-7} mol.L^{-1}). All the measurements were carried out at an excitation wavelength of 335 nm.The pyrene emission spectra were registered and the value of I_1/I_3 at 373 nm and 384 nm respectively was determined.

2.5 MATERIALS

Different copolymers have been prepared by reaction of dimethylacrylamide (DMA) and hydroxyethylmethacrylate (HEMA) in the proportion 90/10 and modification with 1-adamantyl isocyanate [7]. The βCD/EP polymer was obtained as described previously [9] after a second ultrafiltration on a membrane which molecular weight cut off was 100 000.

3. Results and discussion

Table 1: Characteristics of the polymers

Polymer designation	M_n	M_w	I	Hydrophobe content (mol %)	βCD content (weight %)
DMA/HEMA1	170 000*	523 000	3.1		
DMA/HEMAAd 1				1.5***	
DMA/HEMA2	153 000*	515 000	3.4		
DMA/HEMAAd 2				0.1***	
DMA/HEMA3	24 700**	91 700	3.7		
DMA/HEMAAd 3				1♦	
DMA/HEMA4	31 200**	160 200	5.1		
DMA/HEMAAd 4				3♦	
βCD/EP1	157 000*	416 000	2.6		53♦
βCD/EP2	944 000*	1 950 000	2.1		59♦
βCD/EP3	> 100 000				44♦

* Absolute values
** Values in equivalent pullulane
*** Values determined from viscosity results
♦ Values determined from NMR results

Table 1 reports the characteristics of the synthesized polymers. The reduced viscosities of modified copolymers are lower than that of precursor copolymers at low concentrations, explained by intramolecular hydrophobic interactions. At concentrations higher than c_1, as the interactions become inter molecular ones, the reduced viscosities of modified copolymers are higher than that of precursor copolymers.The c* values have been determined from the 1/[η] values and are reported in Table 2. These c* values are as low as the molecular weight of the precursor or modified copolymer is higher. The variation of specific viscosity of mixtures of DMA/HEMA-Ad copolymers with βCD/EP polymers for a total constant concentration of 0,5 g/dL shows a significant enhancement and reaches a maximum for a well defined composition of the mixture depending upon the copolymer (molecular weight, hydrophobe content). These enhanced specific viscosities are attributed to the specific association of both polymers by complexation of adamantyl groups by the βCD cavities.The mixture of these DMA/HEMA-Ad and βCD/EP polymers can lead to very viscous solutions which look like " gels " for higher concentrations.

Phase diagrams were stated by mixing, at different ratios, solutions of DMA/HEMA-Ad and βCD/EP, both at 4%, w/w. Each polymer is present in both phases at concentrations almost neighbouring. These phase diagrams (fig.1) present a limited two-phases zone that expands as the hydrophobe content in the modified polymer increases. This phase behavior is of the same type as that of the mixtures of oppositely charged polyelectrolytes [10] or mixtures of strong hydrogen donor and acceptor [11] . In our system the two polymers develop strong attractive interactions due to the inclusion of the adamantyl groups in the βCD cavities.

Table 2: Values of critical concentrations

Polymer designation	c_1 (g/dL)	c^{*a} (g/dL)	c. a. c.(g/dL)
DMA/HEMA1		2.22	
DMA/HEMAAd1	2.9	3.12	4.3×10^{-2}
DMA/HEMAAd2		2.75	3.7×10^{-2}
DMA/HEMA3		2.7	
DMA/HEMAAd3		3.39	2.7×10^{-2}
DMA/HEMA4		2.48	
DMA/HEMAAd4	5.5	3.36	1.7×10^{-2}
βCD/EP1		16.95	
βCD/EP1		11.24	

c^{*a} : determined by viscosity
c. a. c.: determined by fluorescence

Fluorescence measurements in the presence of pyrene, pointed out the hydrophobic character of modified polymers, which differ according to their hydrophobic level. In fact, the variation of I_1/I_3 versus concentration (fig. 2) is characteristic of the change in environment of pyrene. First in water, pyrene, as the concentration of modified polymer increases, is then surrounded by hydrophobic groups. I_1/I_3 ratio decreases until it reaches a constant value. We so determined the critical agregation concentrations which are reported in Table 2. These measurements reflect the existence of hydrophobic microdomains. The c.a.c. values so determined are smaller than the c* values determined by viscosimetry.

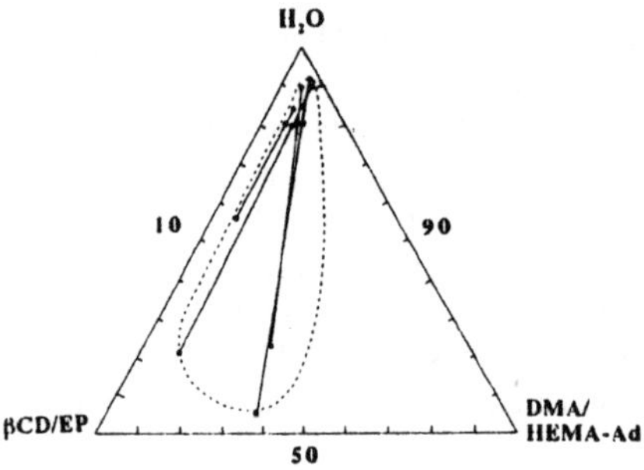

Figure 1: Phase diagram between polymer DMA/HEMA-Ad4 and βCD/EP3

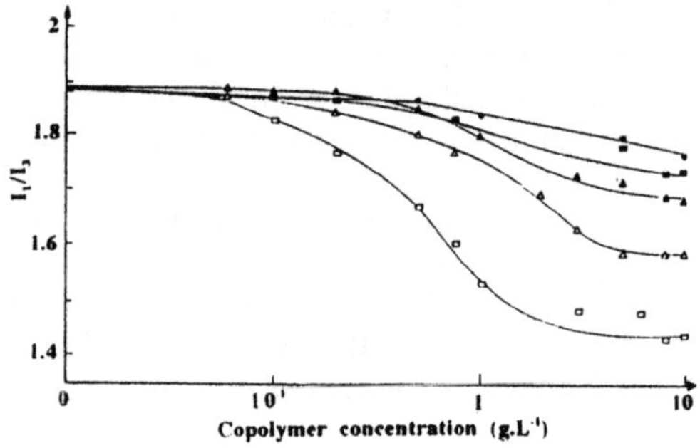

Figure 2: Plot of fluorescence ratio I_1/I_3 as a function of polymer concentration. (△) DMA/HEMA-Ad1; (▲) DMA/HEMA-Ad2; (■) DMA/HEMA-Ad3; (□) DMA/HEMA-Ad4; (●) DMA/HEMA3

4. Conclusion

We saw in this work the associative character of βCD/EP polymer in the presence of modified DMA/HEMA-Ad copolymer through the inclusion of hydrophobic adamantyl groups in the βCD cavities. Ternary diagrams confirm this associative character as they show a two-phase zone of which size increases as the hydrophobe content of modified polymer increases.

References

1. Glass, J. E. Ed.(1989): *Polymers in Aqueous Media*, Advances in Chemistry Series N°223, American Chemical Society: Washington DC.
2. Shaw, K. G. , Leipold, D. P., (1985) New cellulosic polymers for rheology control of latex paints, *J. Coat. Technol.*, 57, N° 727, 63-72.
3. Schaller, E. J. (1985) Rheology modifiers for water borne paints, *Surface Coatings Australia*, 22, 6-10, 12-13.
4. Constien, V. G. , Schlumberger D. (1986) Hydrophobically modified polyacrylamide gelling agents, *Polym. Prepr.(Am. Chem. Div. Polym. Chem.)* 27, 243-244.
5. Evani, S., Rose, G. D.. (1987) Water soluble hydrophobe association polymers, *Polym. Mater. Sci. Eng.*, 57, 477-481.
6. Amiel C., Sebille B.. (1996) New associating polymer' systems involving water soluble β-cyclodextrin polymers, *J. Inclusion Phenom. Mol. Recognit. Chem.*, 25, 61-67.
7. Gosselet N. M.. Borie C., Amiel C.. Sébille B., Aqueous two phase systems from cyclodextrin polymers and hydrophobically modified acrylic polymers, *J. Dispersion Sc. Techn. Festschrift* (in press).
8. Eftink M. R., Andy M. L., Bystrom K., Perlmutter H. D. and Kristol D. S. (1989) Cyclodextrin inclusion complexes: studies of the variation in the size of alicyclic guests, *J. Am. Chem. Soc.*, 111, 6765-6772.
9. Renard E., Deratani A., Volet G., Sébille B. (1997) Preparation and characterization of water soluble high molecular weight β-cyclodextrin epichlorhydrin polymers, *Eur. Pol. J.*, 33, 49-57.
10. Bundenberg de Jong H. G.(1949) : Complex Coacervation in Colloid Science, Kruyt H. ed., *Complex Colloid Systems*, Elsevier, New York, pp 335-432.
11. Dobry A;, Boyer-Kawenoki F. (1947) Phase separation in polymer solution, *J. Polym. Sc.*, 2, 90-100.

PRODUCTION OF CYCLOMALTONONAOSE (δ-CYCLODEXTRIN) BY VARIOUS CYCLODEXTRIN GLYCOSYLTRANSFERASES.

K.L. LARSEN, L. DUEDAHL-OLESEN, L.H. PEDERSEN AND W. ZIMMERMANN
Biotechnology Laboratory, Department of Civil Engineering, Aalborg University, Sohngaardsholmsvej 57, DK-9000 Aalborg, Denmark.

Abstract

The conversion of soluble starch to cyclomaltohexaose (α-CD), cyclomaltoheptaose (β-CD), cyclomaltooctaose (γ-CD) and cyclomaltononaose (δ-CD) by crude preparations of cyclodextrin glycosyltransferase (EC 2.4.1.19) (CGTases) from *Bacillus* spp. and bacterial isolates was studied. δ-CD was produced by all CGTases investigated in the range of 5 to 11.5 % of the total amount of α-, β-, γ-, and δ-CD produced, decreasing to 3 to 7 % after prolonged incubation. A CGTase from *Paenibacillus* sp. F8 was purified and characterised. Coupling experiments with *Paenibacillus* sp. F8 CGTase showed that δ-CD was more easily degraded than α-, β-, and γ-CD.

1. Introduction

The production of α-, β-, and γ-cyclodextrin (CD) from starch by cyclodextrin glycosyltransferases (CGTases) has been frequently reported. Large CD with more than 8 glucose units were first described by French and co-workers [1]. However, due to their low yields and difficulties in purification they have not been characterised until recently. CD with 9 to 17 α-(1→4)-linked D-glucose units, denoted δ-, ε-, ζ-, η-, θ-, ι-, κ-, λ-, and μ-CD, have been described [2-6]. Furthermore, recent studies have shown that CGTases may be capable of producing CD with several hundred α-(1→4)-linked D-glucose units [7]. Although the existence of larger CD has been known for many years, their production from starch has not been studied in any detail.

Employing a novel analysis method based on capillary electrophoresis with a dual electrolyte system for the detection of CD, a concomitant quantification of the production of α-, β-, γ-, and δ-CD by CGTases was possible [8]. In this paper, we present the results on the production and degradation of δ-CD by CGTases.

2. Materials and methods

Capillary electrophoretic analysis of α-, β-, γ-, and δ-CD was performed as described previously [8,9].

3. Results and discussion

Crude CGTase preparations from *Bacillus circulans* NRRL B-380, *B. macerans* NRRL B-3185, *B. macerans* NRRL B-3186, *Bacillus* sp. strain no. 169 (DSM No. 2518), *Bacillus* sp. strain no. 135 (DSM No. 2519), *Bacillus* sp. strain no. 13 (DSM No. 2523), *Bacillus* sp. strain no. 17-1 (DSM No. 2524) and four new isolates were analysed for their production of δ-CD. δ-CD was produced initially by all CGTases in a range of 5 to 11.5 % of the total amount of α-, β-, γ-, and δ-CD produced (Figure 1). After prolonged incubation, the ratio of δ-CD was reduced to 3 to 7 % of the total amount of CD.

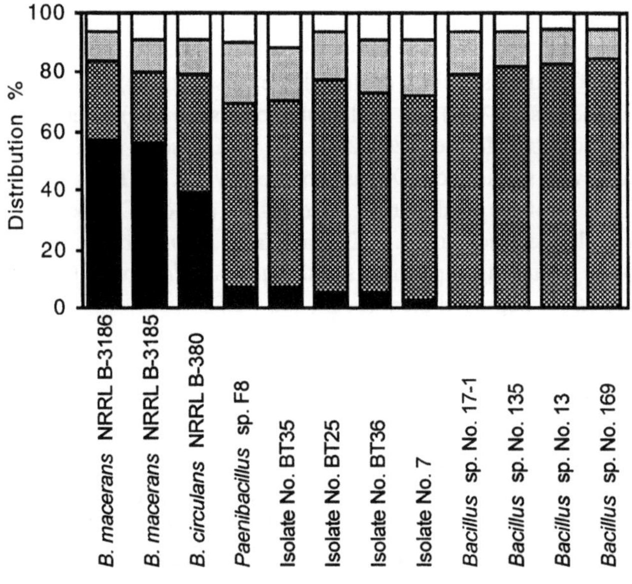

Figure 1: Distribution of α-, β-, γ-, and δ-CD produced initially by crude CGTase preparations from *Bacillus* spp. and bacterial isolates incubated with 5% soluble starch. Solid black bar: α-CD; dark gray bar: β-CD; light gray bar: γ-CD, white bar: δ-CD.

The CGTase from one of the new isolates, *Paenibacillus* sp. F8, was purified and characterised [9]. The molecular weight was estimated to be 72 kDa by SDS-PAGE. The pH optima of the enzyme were 7.5 for the cyclisation activity and 8.0 for the hydrolysis activity. The temperature optima for the cyclisation and hydrolysis activities were 50 °C and 60 °C, respectively, and Ca^{++} had a stabilising effect on the enzyme activity. Similar characteristics of other CGTases have been reported in the literature [9]. Although most of these CGTases have been characterised with respect to their catalytic activity towards the formation of α-, β-, and γ-CD, quantitative data on the production of large CD have not been reported. The ratio of the production of α-, β-, and γ-CD by *Paenibacillus* sp. F8 is similar to the values found in the literature for other CGTases [9].

The molar ratio of α-, β- γ- and δ-CD produced by the purified *Paenibacillus* sp. F8 CGTase, calculated from the initial slope of the production curve, was 0.09:1:0.25:0.14. However, this ratio changed during incubation over a 24 h period. The amount of β-CD produced remained almost constant during the production process and showed a small increase from 67 to 70% based on the molar ratio. The amount of γ- and δ-CD decreased from 18 to 17 % and 9 to 4 %, respectively, while the amount of α-CD increased from 6 to 9%.

Comparison of the coupling activity of the *Paenibacillus* sp. F8 CGTase on α-, β-, γ- and δ-CD showed that δ-CD was a much better substrate for coupling reactions compared to the smaller CD. The coupling rate with maltotetraose was higher than with maltose (Table 1). It is therefore likely that the increase in the ratio of α- and β-CD detected during prolonged incubation was caused by coupling of δ-CD and to some extent γ-CD. This may be the result of an higher flexibility of these molecules compared to α- and β-CD.

TABLE 1: Coupling activity of the purified CGTase (U/mg).

	Maltose	Maltotetraose
α-CD	0.4	0.2
β-CD	0.3	0.4
γ-CD	1.0	1.0
δ-CD	4.8	8.2

The structure of the substrate has to be taken into account in order to explain the production of δ-CD and larger CD. In its crystal structure, amylose can occur as double-stranded helices or, in the presence of organic compounds (e.g. fatty acids and long chain alcohols), as a single helix with 6 to 8 glucose molecules in one helical turn [10]. Amylose in solution has been described as an interrupted coil-like helical structure composed of helical and non-helical segments [11].

The production of small CD (α-, β- and γ-CD) by CGTases could be a consequence of the preferential helical structure of the amylose in solution, while δ-CD and larger CD could have been produced by CGTases as random "mistakes" due to the flexibility of the amylose chain. A site-directed mutagenesis study by Penninga et al. (1995) [12] has shown that an exchange of the highly conserved tyrosine 195 with leucine in the CGTase of *B. circulans* strain 251 resulted in an increased production of δ-CD as well as of other large CD. Tyrosine 195 is located at the center of the active site and the CD are formed around this residue [13]. This suggests that the CGTase lost its capability to control the formation of small CD by the amino acid exchange and the subsequent formation of larger CD was mainly influenced by the flexible structure of the substrate.

The presented results and the recent studies of Terada et al. (1997) [7] describe CGTases to show only little product specificity. Nevertheless, their main products are mainly α-, β-, and to some extent γ-CD. This may be explained by the low coupling activity with α-, β-, and γ-CD compared to δ-CD and probably also the larger CD.

Acknowledgement

Kim Lambertsen Larsen was supported by grant No. T.130-9 from the Danish Research Academy. Wacker Chemie Danmark Aps, Glostrup, Denmark is acknowledged for their donation of α-, β-, and γ-CD. Dr. H. Ueda, Hoshi University, Tokyo is acknowledged for a kind gift of δ-CD.

References

1. French, D. The Schardinger dextrins. In: Wolfrom, M.L. and Tipson, R.S: (eds.) *Advances in Carbohydrate Chemistry*, Vol. 12, Academic Press Inc., Publishers, New York, 189-260, 1957.
2. Fujiwara, T., Tanaka, N. and Kobayashi, S. (1990) Structure of δ-cyclodextrin 13.75 H_2O. *Chem. Lett.* 739-742.
3. Endo, T., Ueda, H., Kobayashi, S. and Nagai, T. (1995) Isolation, purification, and characterization of cyclomaltododecaose (η-cyclodextrin). *Carbohydr. Res.* 269, 369-373.
4. Ueda, H., Endo, T., Nagase, H., Kobayashi, S., and Nagai, T. (1996) Isolation, purification, and characterization of cyclomaltodecaose (ε-cyclodextrin). In: Szejtli, J. and Szente, L. (Eds.) *Proceedings of the Eighth International Symposium on Cyclodextrins.* Kluwer Academic Publishers, Dordrecht. pp. 17-20.
5. Endo, T., Nagase, H., Ueda, H., Kobayashi, S., Nagai, T., (1997) Isolation, purification, and characterization of cyclomaltodecaose (ε-cyclodextrin), cyclomaltoundecaose (ζ-cyclodextrin) and cyclomaltotridecaose (θ-cyclodextrin). *Chem. Pharm. Bull.* 45, 532-536.
6. Endo, T., Nagase, H., Ueda, H., Shigihara, A., Kobayashi, S., Nagai, T. (1997) Isolation, purification, and characterization of cyclomaltotetradecaose (ι-cyclodextrin), cyclomaltopentadecaose (κ-cyclodextrin), cyclomaltohexadecaose (λ-cyclodextrin) and cyclomaltoheptadecaose (μ-cyclodextrin) *Chem. Pharm. Bull.* 45, 1856-1859.
7. Terada, Y., Yanase, M., Takata, H., Takaha, T. and Okada, S. (1997) Cyclodextrins are not the major cyclic α-1,4-glucans produced by the initial action of cyclodextrin glucanotransferase on amylose. *J. Biol. Chem.* 272, 15729-15733.
8. Larsen, K.L. and Zimmermann, W. (1998) Capillary electrophoretic separation of α-, β-, γ-, and δ-cyclodextrins using a dual electrolyte system. *J. Chromatogr.*, In press
9. Larsen, K.L. (1998) *Production and analysis of large cyclodextrins.* Ph.D.-thesis. Aalborg University.
10. Kubik, S., Höller, O., Steinert, A., Tolksdorf, M., Vand der Leek, Y., and Wulff, G. (1996) Molecular inclusion within polymeric carbohydrate matrices. In: van Bekkum, H., Röber, H. and Voragen, A.G.J. (Eds.) *Carbohydrates as Organic Raw Materials* III, VCH, Weinheim, pp 169-187.
11. Szejtli, J. (1991) Helical and cyclic structures in starch chemistry. In: Friedman, R.B. (Ed.) *Biotechnology of Amylodextrin Oligosaccharides*, ACS Symposium Series 458, American Chemical Society, Washington DC, pp 2-10.
12. Penninga, D., Strokopytov, B., Rozeboom, H.J., Lawson, C.L., Dijkstra, B.W., Bergsma, J., and Dijkhuizen, L. (1995) Site-directed mutations in tyrosine 195 of cyclodextrin glycosyltransferase from *Bacillus circulans* strain 251 affect activity and product specificity. *Biochemistry* 34, 3368-3376.
13. Schmidt, A.K., Cottaz, S., Drigues, H., and Schulz, G.E. (1998) Structure of cyclodextrin glucosyltransferase complexed with a derivative of its main product β-cyclodextrin. *Biochemistry* 37, 5909-5915.

CAPILLARY ELECTROPHORETIC INVESTIGATION OF THE INCLUSION COMPLEX FORMING PROPERTIES OF LARGE CYCLODEXTRINS

K.L. LARSEN[a], T. ENDO[b], H. UEDA[b] AND W. ZIMMERMANN[a]

[a]Biotechnology Laboratory, Department of Civil Engineering, Aalborg University. Sohngaardsholmsvej 57, DK-9000 Aalborg, Denmark

[b]Department of Physical Chemistry, Faculty of Pharmaceutical Sciences, Hoshi university, 4,41 Ebara 2-chome Shinagawa-ku, Tokyo 142, Japan

Abstract

By use of capillary electrophoresis (CE), the binding constants of inclusion complexes formed with cyclomaltononaose (δ-CD), cyclomaltodecaose (ε-CD), cyclomaltoundecaose (ζ-CD), cyclomaltododecaose (η-CD) and cyclomaltotridecaose (θ-CD) and various anions were determined and compared to the corresponding binding constants of maltoheptaose and α-, β- and γ-cyclodextrin (CD). All of the large CD were able to form complexes. Dependent on the type of guest molecule, δ- and ε-CD were the weakest complex formers. The complex forming ability of ε-, ζ-, η-, and θ-CD increased with increasing size of the ring structure and θ-CD displayed inclusion complex formation constants comparable to γ-CD.

1. Introduction

Enzymatic degradation of starch by the extracellular microbial enzyme cyclodextrin glycosyltransferase (CGTase) (E.C 2.4.1.19) yields a mixture of cyclic and linear oligosaccharides. Prolonged incubation times leads to the accumulation of α-, β-, and γ-CD, whereas only a small amount of larger CD is present [1]. Only recently, large CD ranging form 9 to 17 (1→4)-linked α-D-glucose units, denoted δ-, ε-, ζ-, η-, θ-, ι-, κ-, λ-, and μ-CD, respectively, have been purified and characterised [2-6]. The complex forming capacity of the larger CD has been assumed to be negligible due to an expected unsuitable cavity size and increased flexibility of the macrocycles [7].

The large interest in cyclodextrins is based on their ability to form inclusion complexes with a wide range of molecules [7]. Several hypotheses have been proposed to explain the mechanism of inclusion complex formation. These include a relief from conformational strain, release of cavity-bound water, hydrophobic interaction, dipole-

dipole and hydrogen-bonding interactions as well as induction and dispersion forces [7-9]. It is believed that hydrophobic interactions, dipole-dipole interactions and dispersion forces are the major factors involved in inclusion complex formation [9]. However, the size of the guest molecule and the size and flexibility of the CD cavity is also considered to be important in the formation of strong inclusion complexes, since too wide and flexible CD cavities would be unable to hold small guest molecules and to prevent a dissociation of the complex. On the other hand, a guest molecule which is too large will not be able to enter the CD cavity and thus a strong complex will not be formed [7-9].
In the present study, we have measured the binding constants between CD ranging from α-CD to θ-CD and several anions using CE and compared the complex forming properties of the larger CD with maltoheptaose and α-, β-, and γ-CD.

2. Materials and methods

Materials. Substituted benzoic acids and 1-adamantane carboxylic acid were obtained from Aldrich, Steinheim, Germany. Benzoic acid and ibuprofen were obtained from Sigma Chemical CO., St. Louis, MO, USA. 3-phenyl propionic acid and Na_2HPO_4 were obtained from E. Merck, Darmstadt, Germany. All chemicals were of analytical grade. Pharmaceutical grade α-, β-, and γ-CD were a gift from Wacker Chemie. The large CD were purified as described previously [3].

Electrophoresis. CE was performed as described previously [11]. The inclusion complex formation constant was calculated from the effective mobilities of the individual CD separated in two series of at least 7 different concentrations ranging from 10 to 200 mM aromatic anion and calculated using a y-reciprocal plotting method with least squares linear regression [10].

3. Results and discussion

A linear relationship between the concentration of the anion and the concentration of the anion divided by the effective mobility was obtained for all combinations of CD and anions in the y-reciprocal plots which supports the assumption that 1:1 complexes were predominating. Inclusion complex formation constants could be determined for most combinations of CD and anions (Table 1). Most of the inclusion complex formation constants were low. However, aromatic anions such as benzoate derivatives, are known to display inclusion complex formation constants 10-100 times lower than neutral molecules [9]. A comparison of the obtained inclusion complex formation constants with previously published data showed that comparable results were found using the CE method [11]. The larger CD showed overall lower binding capacities. δ-CD displayed the lowest complex formation capacity with substituted benzoates, whereas ε-CD was the poorest complexant with 4-*tert.*-butyl benzoic acid, adamantane carboxylate and the ibuprofen anion. Surprisingly, ε-, ζ-, η-, and θ-CD exhibited an increasing binding

Table 1: Inclusion complex formation constants of the 1:1 complexes between cyclodextrins and various anions measured by capillary electrophoresis at 25°C.

	Inclusion complex formation constant (M^{-1})								
Compound	Maltoheptaose	α-CD	β-CD	γ-CD	δ-CD	ε-CD	ζ-CD	η-CD	θ-CD
Benzoate	N.D.	16	23	3	3	3	5	4	5
2-Methyl benzoate	5	13	13	7	6	6	5	6	7
3-Methyl benzoate	N.D.	26	40	6	3	5	6	7	8
4-Methyl benzoate	N.D.	36	66	8	2	4	6	6	7
2,4-Dimethyl benzoate	4	45	42	8	3	4	5	7	6
2,5-Dimethyl benzoate	5	41	27	6	4	5	5	6	6
3,5-Dimethyl benzoate	5	39	9	7	2	5	7	8	8
3,5-Dimethoxy benzoate	7	47	63	10	8	9	10	9	12
Salicylate	7	11	65	13	9	8	8	9	10
3-Phenyl propionate	6	35	79	7	2	3	5	4	6
4-*tert.*-Butyl benzoate	13	51	382	74	47	3	9	15	25
Ibuprofen anion	18	56	>2500[a]	67	27	2	12	29	39
1-Adamantane carboxylate	N.D.	114	501	42	8	N.D.	4	4	8

N.D.: Not determined
a) Too high to be accurately determined

ability with an increasing number of glucose units in the molecules and θ-CD obtained inclusion complex formation constants comparable to γ-CD and in some cases even to α-CD. The linear β-CD analogue maltoheptaose displayed complex formation constants which were generally a factor of 10 lower than those of its cyclic counterpart, β-CD (Table 1). Only in the case of the ibuprofen anion and 4-*tert.*-butyl benzoic acid did the complex formation constant of maltoheptaose exceed those of ε-, and ζ-CD.
The results strongly suggest that the cavities of the large CD are much smaller than what would be expected if they had a planar structure like α-, β-, and γ-CD [8]. Indeed, the crystal structures which have been obtained for δ- [2] and ε- [4], as well as ^{13}C-NMR measurements [6] strongly suggest that the large CD have a distorted cyclic structure in solution. Furthermore, the results indicate that increased flexibility as a consequence of the increased number of glucose molecules in the macrocycle can also lead to an increased binding, which was observed for the larger CD and for maltoheptaose. This effect could be due to an increased ability of the larger CD to adapt to the geometry of the included molecule.

Acknowledgement

Kim Lambertsen Larsen was supported by grant No. T.130-9 from the Danish Research Academy. Wacker Chemie Danmark ApS, Glostrup, Denmark is acknowlegded for their donation of α-, β-, and γ-cyclodextrin.

References

1. Terada, Y., Yanase, M., Takata, H., Takaha, T., and Okada, S. (1997) Cyclodextrins are not the major cyclic α-1,4-glucans produced by the initial action of cyclodextrin glucanotransferase on amylose. *J. Biol. Chem.* 272, 15729-15733.
2. Fujiwara, T., Tanaka, N., and Kobayashi, S. (1990) Structure of δ-cyclodextrin 13.75 H_2O. *Chem. Lett.*, 739-742.
3. Endo, T., Ueda, H., Kobayashi, S., and Nagai, T. (1995) Isolation, purification and characterization of cyclomaltododecaose (η-cyclodextrin). *Carbohydr. Res.*, 269, 369-373.
4. Ueda, H., Endo, T., Nagase, H., Kobayashi, S., and Nagai, T. (1996) Isolation, purification, and characterization of cyclomaltodecaose (ε-cyclodextrin). In Szejtli, J. and Szente, L. (Eds.) *Proceedings of the Eighth International Symposium on Cyclodextrins*. Kluwer Academic Publishers, Dordrecht. pp. 17-20.
5. Endo, T., Nagase, H., Ueda, H., Kobayashi, S., and Nagai, T. (1997) Isolation, purification, and characterization of cyclomaltodecaose (ε-cyclodextrin), cyclomaltoundecaose (ζ-cyclodextrin) and cyclomaltotridecaose (θ-cyclodextrin). *Chem. Pharm. Bull.*, 45, 532-536.
6. Endo, T., Nagase, H., Ueda, H., Shigihara, A., Kobayashi, S., and Nagai, T. (1997) Isolation, purification, and characterization of cyclomaltotetradacaose (ι-cyclodextrin), cyclomaltopentadecaose (κ-cyclodextrin), cyclomaltohexadecaose (λ-cyclodextrin) and cyclomaltoheptadecaose (μ-cyclodextrin). *Chem. Pharm. Bull.*, 45, 1856-1859.
7. Szejtli, J. (1988) *Cyclodextrin Technology*, Kluwer Academic Publishers, Dordrecht.
8. Frömming, K.-H. and Szejtli, J. (1994) *Cyclodextrins in pharmacy*. Kluwer Academic Publishers. Dordrecht.
9. Connors, K.A. (1997) The stability of cyclodextrin complexes in solution. *Chem. Rev.*, 97, 1325-1357.
10. Rundlett, K.L. and Armstrong, D.W. (1996) Examination of the origin, variation, and proper use of expressions for the estimation of association constants by capillary electrophoresis. *J. Chromatogr. A*, 721, 173-186.
11. Larsen, K.L., Endo, T., Ueda, H., and Zimmermann, W. (1998) Inclusion complex formation constants of α-, β-, γ-, δ-, ε-, ζ-, η-, and θ-cyclodextrins determined with capillary electrophoresis. *Carbohydr. Res.*, In press.

NEW SORBENTS CONTAINING β-CYCLODEXTRIN : SYNTHESIS, CHARACTERISATION, AND SORPTION PROPERTIES.

L. JANUS[a], G. CRINI[a], M. MORCELLET[a] G. TORRI[b], A. CAMBIAGHI[b], A. NAGGI[b] and C. VECCHI[c].

a : Laboratoire de Chimie Macromoléculaire, UA CNRS 351, Université des Sciences et Technologies de Lille, 59655, Villeneuve d'Asq cedex, France

b : Istituto di Chimica et Biochimica « G ; Ronzoni », 81 via G. Colombo, 20133 Milano, Italy

c : Stazione Sperimentale per i combustibili, 3 viale De Gasperi, 20097 Milano, Italy.

1. Introduction

There has been considerable interest in the preparation, properties and applications of insoluble polymer and copolymers containing βcyclodextrin (βCD). These materials which form water insoluble crosslinked polymers are obtained using polyfunctional crosslinkers such as aldehydes, ketones, isocyanates or epoxides. For example, the utility of βCD containing polymers obtained by condensation between βCD and epichlorohydrin has been recognized in many applications : catalysis, chromatography (M. Morcellet and al 1995a, 1995b, 1996), pharmaceutical and food industries, and in the clean-up of industrial waste effluents. These operations use the property of βCD to form inclusion complex with various compounds, especially aromatics.

The purpose of this work is to prepare a macroporous resin able to get rid of pollutants as pentachlorophenol (PCP) or polychlorobiphenyl (PCB). A new cyclodextrin derivative monomer called 2-hydroxy-3-methacryloyloxy- βcyclodextrin (β W7 MAHP) :

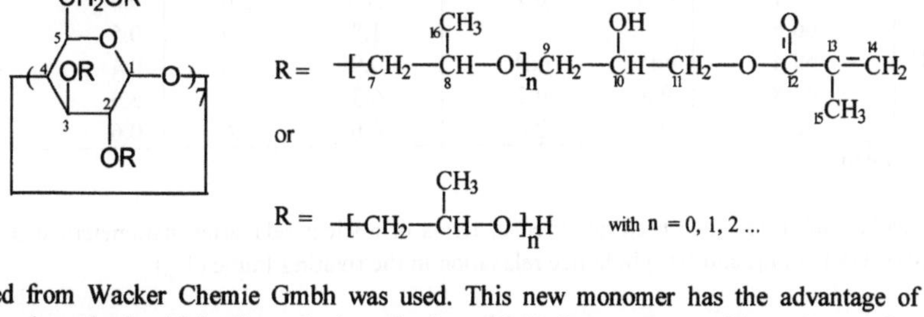

obtained from Wacker Chemie Gmbh was used. This new monomer has the advantage of being readily polymerized, which allows the introduction of βCD into a system without the usual methods of chemical crosslinking of βCD.

2. Results and discussion

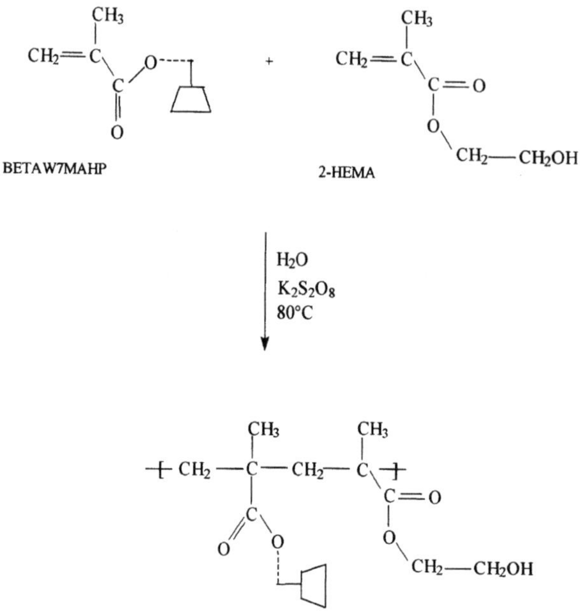

Scheme 1. Copolymerization of ßW7MAHP 1

We describe here the synthesis of βCD based crosslinked homopolymers and copolymers obtained by simple radical polymerization, or by different ways of suspension polymerization :
These polymers and copolymers of βW7MAHP and HEMA were characterized using IR spectroscopy and solid state ^{13}C NMR techniques (CP/MAS).

Table 1 NMR features of copolymers containing 253 (copo 3) and 24 μmole of CD/g

Peaks	Chemical shifts	pBETAW7MAHP		Copo 3		Copo 5	
	ppm	T1	T1ρ	T1	T1ρ	T1	T1ρ
12	178.1	nd*	nd*	12.7	3.6	11	5.2
1	101.5	25.5	0.7	15.7	3.5	5.9	nd*
4	80.6	21.4	0.6	12.5	3.1	8.1	4.5
2,3,5	72.1	17.4	0.7	11.1	3.6	9.6	5.3
CH2O	66.6			1.8	3.9	0.5	5.3
6,CH2OH	59.4	nd*	nd*	0.9	3.6	0.4	4.8
14'	44.8	9.5	0.7	6.3	4.1	5.9	5.5
15,16	18.8	0.4	0.6	0.6	3.6	0.6	5.3

* Not determined

Their molecular mobilities have been analyzed in terms of nuclear relaxation parameters such as ^{13}C spin lattice relaxation (T_1) and ^1H spin lattice relaxation in the rotating frame ($T_{1\rho}$) .

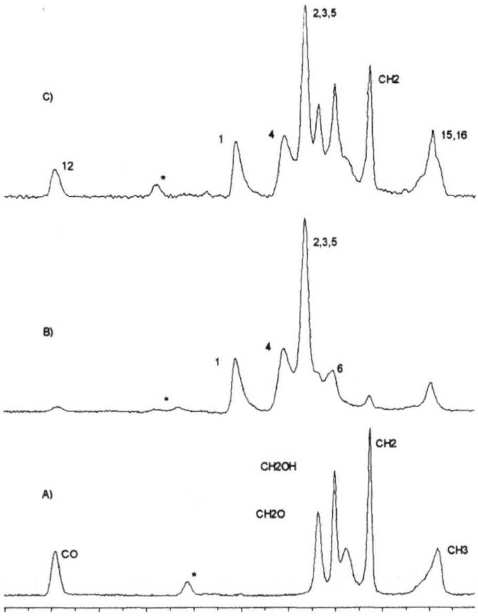

Figure 2. Solid state ^{13}C NMR spectra of a HEMA/ βW7MAHP copolymer, p βW7MAHP and pHEMA (from top to bottom)

The T_1 values of the copolymers showed that the βCD trapped inside the materials seemed to change its mobility irrespective of the amount of comonomer . The $T_{1\rho}$ values were equivalent, thereby indicating the homogeneous nature of the samples .The sorption of tertiobutyl benzoic acid (TBBA), was then studied. The capacity for TBBA varies between 393 and 449 μmol /g for resins obtained by suspension polymerization. For the copolymer prepared by simple radical polymerization the maximum sorption capacity is about 140 mg of TBBA per g of copolymer :

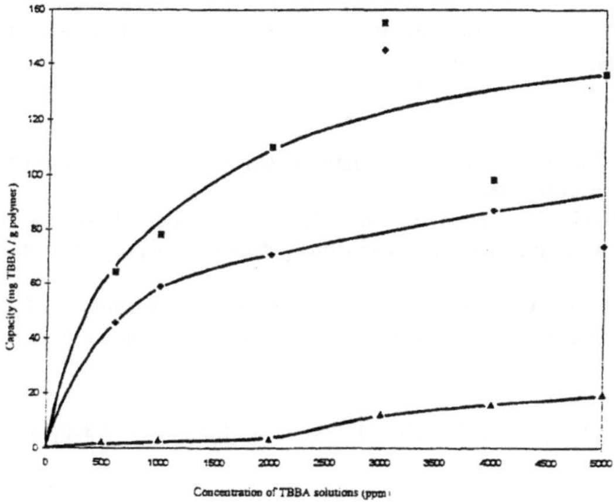

Figure 3 : Sorption isotherms for copolymer 3, pβW7MAHP and pHEMA (from top to bottom)

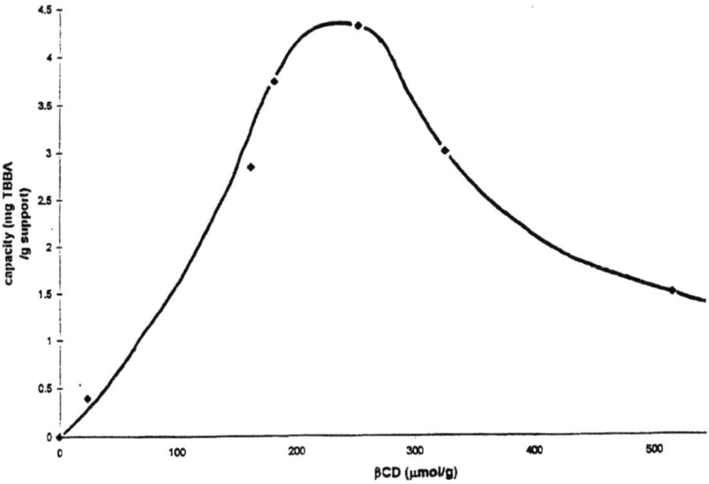

Figure 4 : Sorption capacity vs CD content in the copolymers

At this step, we can summarize the result as following :

- The βW7 MAHP monomer can be polymerized with others monomers.
- To obtain macroporous resins, the organic suspension polymerization method seems to be the better way (maximum of incorporation of βW7 MAHP).
- Nevertheless, the resins obtained have still a gel structure, not yet a macroporous structure.
- The adsorption capacity versus TBBA of the two type of resins (organic and aqueous suspension) seems to be about the same, and high (70 to 80 mg TBBA/g polymer) although the quantity of βCD contained into the two resins is very different. This mean that the adsorption is not only due to the βCD.
- Actually, it does not seem, anyway, that it should be necessary to have a great quantity of βCD into the resin to have a good adsorption capacity :

References :

Crini G., Lekchiri Y. and Morcellet M. (1995) Separation of structural isomers on cyclodextrin-polymers coated on silica beads *Chromatographia* **40**, 296-302

Crini G., Torri G., Lekchiri Y., Martel B., Janus L. and Morcellet M. (1995) High performance liquid chromatography of structural isomers using a cyclodextrin - poly(allylamine) coated silica column *Chromatographia* **41**, 424-430

Crini G. and Morcellet M. (1996) High performance liquid chromatography of structural isomers using cyclodextrin-poly(vinylamine) coated silica columns : 2 - Retention behavior and separation. *J.Chromatogr. Sci.*, **34**, 485-494

Aknowledgments : Thanks to the European Community for his financial support through program FAIR CT 95300

PREPARATION AND CHARACTERIZATION OF SORBENTS CONTAINING βCD DERIVATIVES COATED OR GRAFTED ONTO SILICA GEL

T.N.T PHAN, M. BACQUET, M. MORCELLET *Laboratoire de Chimie Macromoléculaire, UA CNRS 351, Université des Sciences et Technologies de Lille, 59655, Villeneuve d'Ascq cedex, France*

1. Introduction

We have previously reported the synthesis and applications of βCD grafted polymer coated onto silica gels (1).This paper presents preparation and characterization of new βCD/silica composite supports used for water pollutants decontamination. Two βCD derivatives are coated or chemically bonded onto silica. A detailed characterization is undertaken to find the optimum conditions of synthesis.

2.Materials and methods

2.1 MATERIALS

Silica, type Davisil, with an irregular shape, an average diameter porosity of 150Å and a specific area of 330 m^2/g is chosen for its low price and its convenient porosity.
βCD derivatives are a kind gift from Wacker Chemie GmbH. The first one called W_7MAHP has the following structure :

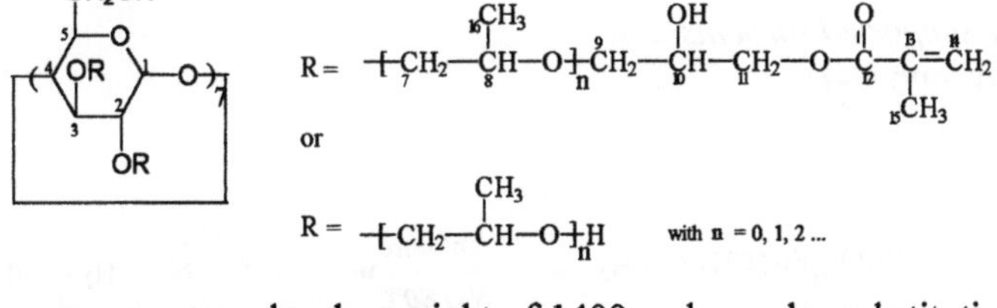

Wacker gives an average molecular weight of 1400 and a molar substitution degree by 2-hydroxy-3-methacryloyloxy-propyl of 0.3-0.5 per anhydroglucose.
The detailed structure of the second one, called W7MCT, is seen on scheme 2. For this molecule the average molecular weight is 1560 and the molar substitution degree by chlorotriazinyl is 0.3-0.5 per anhydroglucose.

2.2 METHODS

W_7 derivatives were quantified by Thermogravimetric analysis and were revealed on each bead by Raman microspectrophotometry.

2.2.1 W_7MAHP supported sorbents

- *Coating way*

Radical homopolymerization of W_7MAHP monomer in water, at a temperature of 80°C, with $K_2S_2O_8$ as initiator was undertaken.

- *Grafting way (2)*

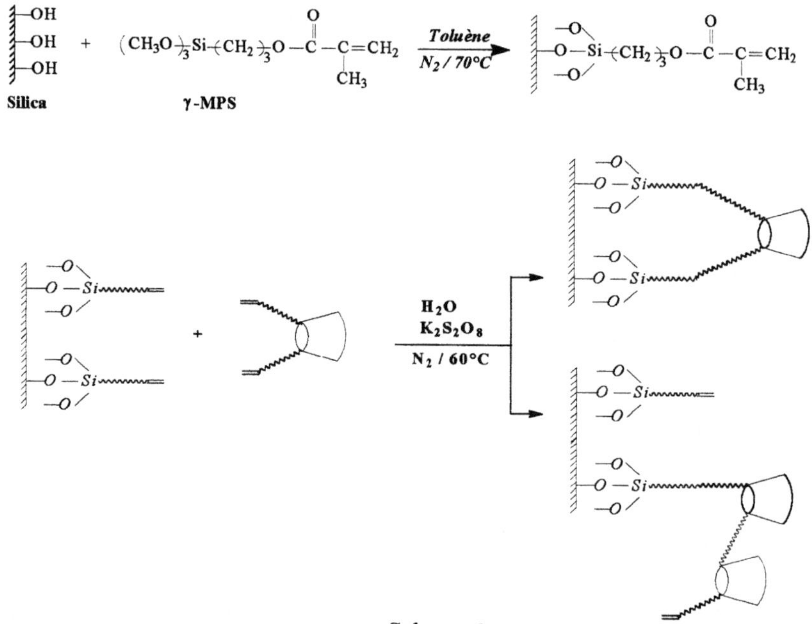

Scheme 1

2.2.2 W_7MCT supported sorbents

- *Grafting way only(3-4)*

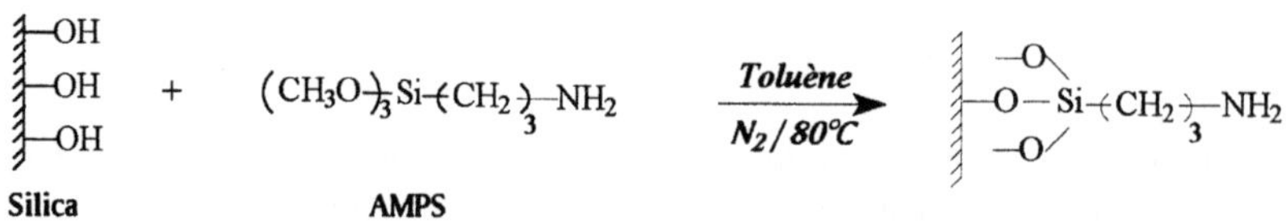

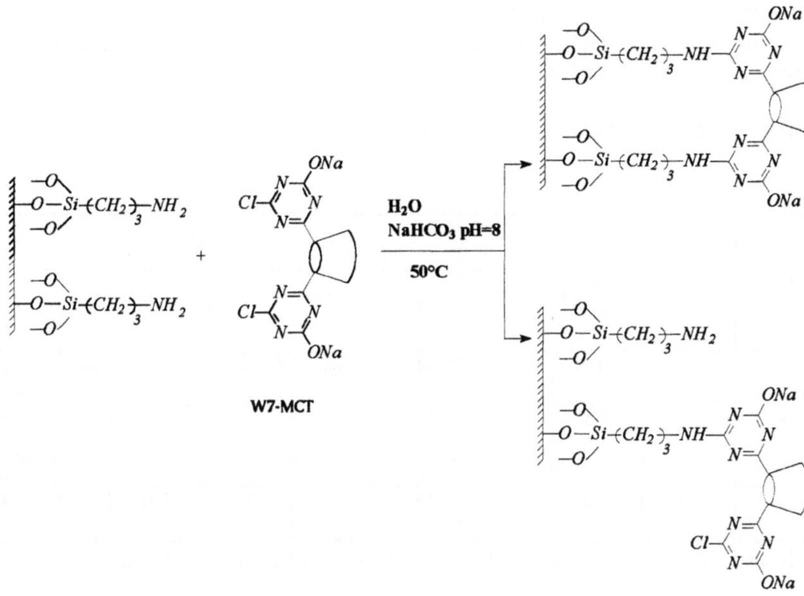

Scheme 2

3. Results and discussion

3.1 COATING WITH W7MAHP

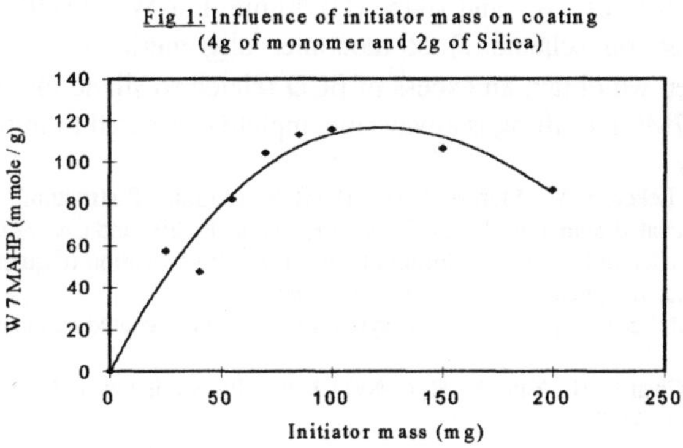

For low concentration of initiator, polymerization has a low yield and for high concentration we synthesize oligomers which don't coat efficiently. So a maximum of coating is obtained for 100 mg of initiator.

3.2 GRAFTING WITH W7MAHP AND WITH W7MCT

Sample First step	G_1D_1			G_1H_1	G_1J_1				G_1K_1
Reac.Time (H)	5				20				15
γMPS µmole/g	76,1			42,3	270				163
Sample Second step	G_1D_2A	G_1D_2B	G_1D_2C	G_1H_2	G_1J_2A	G_1J_2B	G_1J_2C	G_1K_2A	G_1K_2B
W7MAHP = N γMPS	10	7,5	10	20	0,5	1	2	5,2	5,2
Reac.Time (H)	4	4	15	4	24	4	4	4	20
W_7MAHP µmole/g	110	104	137	98	26	59	107	120	142

Table1 : Characteristics of grafted W_7MAHP sorbents (G1 series)

Sample (First step)	**G_4A_1**			**G_4B_1**
AMPS µmole/g	256			274
Sample (Second step)	**G_4A_2A**	**G_4A_2B**	**G_4A_2C**	**G_4B_2**
W7MCT = N AMPS	5	10	10	2
Reac.Time (H)	24	24	48	48
W7MCTµmole/g	14,2	57	56,5	27,5

Table 2: Characteristics of grafted W_7MCT sorbents (G4 series)

The methacryloyl or amino silane groups grafting (first step) depends on the drying of silica, the reaction time. The W_7 derivatives grafting (second step) depends *strongly on molar ratio between W7 and silane (N) * time. For W7MAHP grafting, a side reaction (as described on scheme 1) of monomer oligomerization initiated from one silane occurs. Then we obtain an excess of βCD related to silane initially present (for high N values). W7MCT grafting is always incomplete compared to initial aminosilane groups.

References

1. Crini G., Lekchiri Y., Morcellet M. (1995) Separation of structural isomers using cyclodextrin polymers coated onto silica beads *Chromatographia*, 40, 296 and *Chromatographia*, 41, 424
2. Gorski D., Klemm E., Fink P., Horhold H. H. (1988) Investigation of quantitative SiOH determination by the silane treatment. *Colloid Interface Science* 126
3. Huhn G., Muller H., Eppert G. (1992) synthesis of polymer- coated cation exchangers *React. Polymers,* 170, 109
4. Lin C. E., Chen C. H., Yang M. H. (1989) Chemically bonded cyclodextrin Silica Stationary phases *J. Chromatog. Sci*, 27, 665.

Thanks to the European Community for its financial support through the FAIR program CT 95 0300

COMPARATIVE CATION CHELATING PROPERTIES OF PER (3,6-ANHYDRO),- AND PER (3,6-ANHYDRO 2-OMe) α-CYCLODEXTRINS

J.C. DEBOUZY[1], F.FAUVELLE[1], A. GADELLE[2], V. DABOUIS[1], A. PERRIN[1], B. BRASME[1], A. PEINEQUIN[1], B.PERLY[3].

1 CRSSA/Biophysics, 24 Av. des maquis du Grésivaudan, 38702-La Tronche cedex, France.

2 CEA Grenoble, BP85X, 38000 Grenoble, France.

3 CEA Saclay, 91191 Gif/Yvette cedex, France.

1. Abstract.

The cation chelating properties of per(3,6 anhydro) α-cyclodextrin, [A36] and of per(3,6 anhydro, 2-O methyl) α-cyclodextrin, [A36M] were investigated by mass and NMR spectroscopy. A36 forms 1: 1 complexes with lead (K=2500 M^{-1}), and also with strontium and potassium with a fast exchange rate kinetics. However, the formation of A36-Pb complex results in a dramatic enhancement of the haemolytic properties. Permethylation at the position 2 (A36M) confers an extreme affinity for Ba^{2+}, Pb^{2+}, Sr^{2+} and Ca^{2+} following a slow rate exchange process and a 1: 1 stoichiometry. Besides, a weak 1:1 A36M-K complex is also found, with a fast exchange rate. By opposition with A36, A36M complexes showed no haemolytic properties.

2. Introduction.

In terms of public health, the removal of toxic cations from both environment and biological systems is still of great interest. Hence, significant levels of heavy metals are clearly identified[1], resulting from current domestic and industrial waste and also as consequences of major accidents (e.g. Chernobyl 1989, or Andalucia 1998). A solution for biological and physical treatment by chelation supposes both high affinity and selectivity of the chelating reagent, and also the absence of toxicity to be bioavailable. These points were investigated for per(3,6 anhydro) α-cyclodextrin, [A36] and per(3,6 anhydro, 2-0 methyl) α-cyclodextrin, [A36M]: Mass and NMR spectroscopies were used for physical aspects of cation complexation, and absorption method to measure haemolytic activity of A36, A36M and their complexes.

3. Material and methods,

A36 and A36M were synthesized following the method of Gadelle and Coll[2]. Proton numbering is given on the figure 1.

Metal nitrate salts were from Sigma (La Verpillere, France) and were used without further purification. ^{1}H-NMR spectra were recorded on a AM400 Brucker spectrometer (9.4T) using 32K acquisition points, a spectral width of 400 Hz, and a presaturation of water resonance Mass spectra were recorded on a quadrupole instrument (Quattro II, Micromass, Altricham,UK) equipped with an ESI. The solutions (H_20/MeOH 1/1 V/V) of metal salts and cyclodextrins were introduced at a rate of 5 µL/min with a syringue pump. Ions were detected by scanning the first quadrupole over the range m/z 260-1260 in 3 s. All data were carried out in the continuum mode. At least 50 scans were averaged to obtain representative spectra. Haemolysis tests were performed on blood from healthy volunteers as describes elsewhere using absorption spectroscopy at 540 nm on a Shimazu MS-2100 spectrometer.

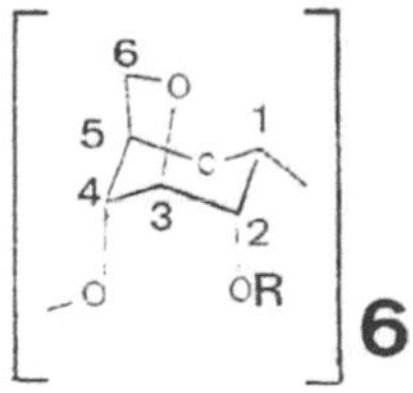

Figure 1: Proton nomenclature; R=H (A36), R=OCH_3 (A36M)

4. Results.

4. 1. Cation complexation

Among the cations of interest, a first screening was operated for both cyclodextrin derivatives, using Thin Layer Chromatography on ion loaded plates (not shown). The following mass spectroscopy and NMR study only reports the results for selected ions.

4.1.1. Cation complexation by A36.

NMR Spectra similar as those recorded on A36 in the presence of various amounts of $Pb(NO_3)_2$ (figure 2A) are typical of a complex formation, with a fast exchange rate. This evolution was also found for Sr^{2+}, K^+, Cs^+ and NH_4^+, allowing Job-plots constructions (see figure 2B for Pb^{2+} that provided in all cases a 1: 1 stoichiometry. The apparent affinity constant k(M^{-1}) was also calculated using the Benesi-Hildebrand method (NH_4^+, Cs^+, K^+) or the SIMPLEX algorithm (program COMPLEX[5]) for Pb^{2+} and Sr^{2+}, leading to the step scale affinity Pb^{2+} (2500 M^{-1}) > Si^{2+} (400 M^{-1}) > K^+ (130 M^{-1}) > Cs^+ (80 M^{-1}) > NH_4^+ (40 M^{-1})

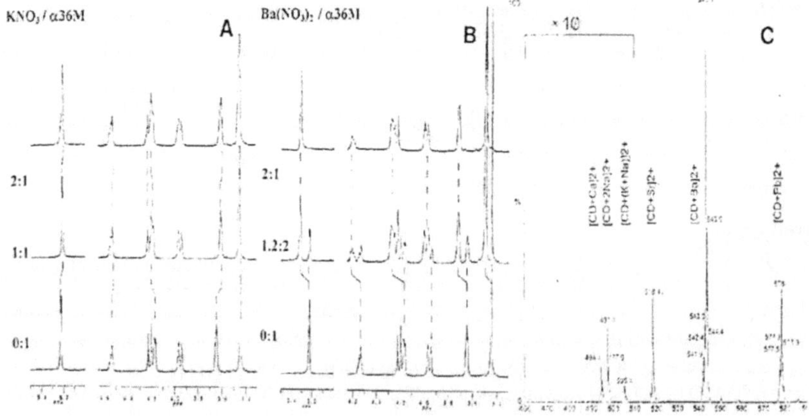

Figure 2: A/^{1}H-NMR spectra of A36 in the presence of lead nitrate at various molar ratios R =[A36]/([A36]+[Pb]) with a total concentration of 5mM. B/ Job plot construction derived from A/ with chemical shifts of H1(♦), H2(□), H5(+)and H6 (●) protons. C/ ES-MS spectrum recorded on an equimolar mixture of A36, $Pb(NO_3)_2$, $Sr(NO_3)_2$, and KNO_3 (1.6mM).

Mass study (Electrospray) . Direct competition conditions between lead, strontium and potassium were realized by injecting an equimolar solution of their nitrate salts (1.6 mM) in the presence of A36. The resulting ES spectrum displayed on figure 2C was found in great agreement with the NMR datas, i.e. an affinity in the range Pb^{2+}>Sr^{2+} >K^{+}, considered the relative intensities of the cations/A36 completes identified.

4.1.2. Cation complexation by A36M.

NMR. Cation complexation by A36M was found very dependant on the class of ion:

- Most of alcalis gave very weak associations with A36M following a fast exchange kinetic, with a 1: 1 stoichiometry and apparent association constants less than 100 M^{-1} (except K^{+}, k= 190 M^{-1}), as displayed on figure 3A; and Cs^{+}, that was not complexed by A36M
- conversely strong complexations with slow exchange process were found for some alcaline earths (Ba^{2+}, Ca^{2+} and Sr^{2+}) and lead: hence, on a ^{1}H-NMR spectrum recorded on A36M in the presence of those ions, new resonances of complexed A36M are clearly identified (see figure 3B), which develope in intensity at the expense of the resonances of free A36M. The respective contribution of the two species at different molar ratios allowed to determinate a 1: 1 stoichiometry. These datas, and also direct competition with well-known chelating reagents (EDTA ...) were used in conjunction with ESI mass spectroscopy to give an estimation of the apparent affinity constants.
- no argument for cation complexation was found for the other divalent ions such as Mg^{2+}, Hg^{2+}, Co^{2+}.

Mass spectroscopy. The ES-MS spectrum recorded on equimolar mixture of nitrate salts of Pb^{2+}, Ba^{2+}, Ca^{2+}, Sr^{2+}, K^{+}+ and of A36M (1.6mM) is shown on the figure 3C. Hence, the relative repartition of A36M7ions completes was found fully in agreement with NNM results, providing a final step scale of affinity :

$$Ba^{2+}(\log k = 6) > Pb^{2+}(\log k = 5.4) > Ca^{2+} (\log k = 4.4) > Sr^{2+}(\log k = 3.8) >>> K^{+} (\log k = 2.3)$$

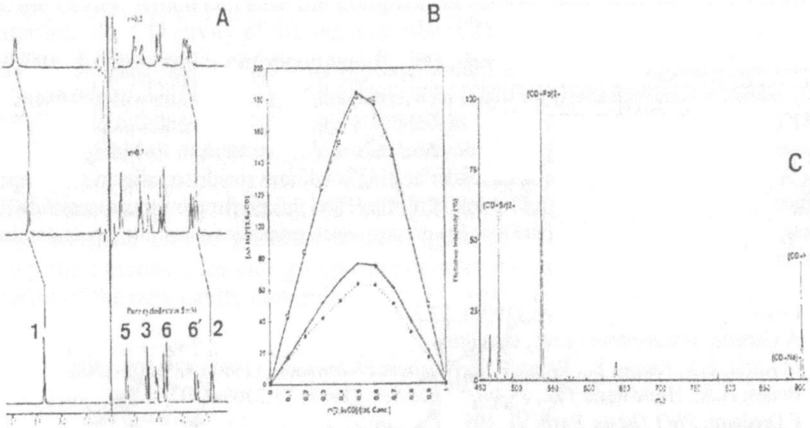

Figure 3: ^{1}H-NMR spectra of A36M in the presence of various molar ratios (Ion/A36M) of A) KNO_3, total concentration [A36M]+[K^{+}], of 5 mM, B) $Ba(NO_3)_2$, [A36M] = 5 mM. C) ES-MS spectrum of an equimolar mixture of A36M(1.6 mM) and $Ba(NO_3)_2$, $Pb(NO_3)_2$, $Sr(NO_3)_2$, $Ca(NO_3)_2$, KNO_3.

4.2. HAEMOLYSIS.

Figure 4 shows the haemolysis curves built with natural α-cyclodextrin (ACD), A36, A36M and their complexes with lead. It is noteworthy that both peranhydro cyclodextrins are significantly less haemolytic than

ACD (DH_{50} = 10 M°) since in the range of solubility (less than 50 nM), it was not possible to reach 50% haemolysis. However the consequences of a complex formation (with Pb shown here, but also true for Sr) on the haemolytic activity are evident: Thus, the concentration of A36/Pb sufficient to induce 50% haemolysis (i.e. DH_{50}) was found less than 3mm. A previous study could evidence neither direct membrane damage or pore formation, while electronic microscopy photographs were evocative of a neutral cotransport [7]. Conversely, the formation of A36M/Pb complex only induced a limited increase of haemolytic properties (DH_{50} around 40nM).

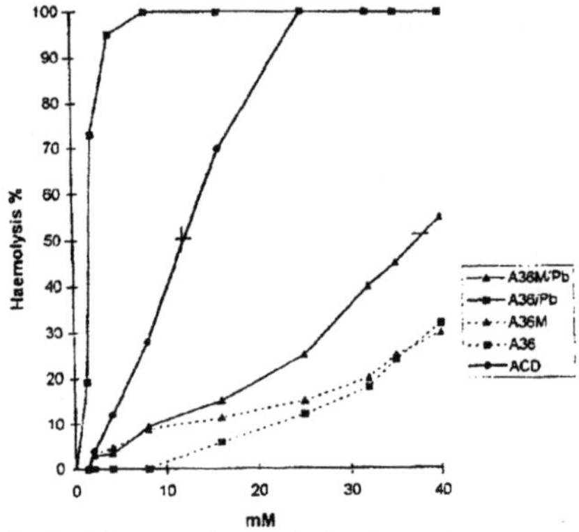

Figure 4: Haemolysis curves obtained for natural (α-cyclodextrin (ACD), A36, A36M and their lead complex.

5.Conclusions.

A36 and A36M were synthetized to chelate toxic cations, especially for biological uses. Under this point of vie, A36M clearly exhibits a better selectivity for toxics such as baryum, lead or strontium with relative respect to alcalis, even if calcium is also chelated by A36M. Besides, A36 appears less efficient, except for lead, but the intrinsic haemolytic properties of the A36/Pb complex leads is a severe limit to its biological use. Since A36M/Pb or A36M/Sr complexes remain stable under heating conditions (no dissociation occurs under 85°), the introduction of A36M in agarose gel (2.8%) was performed, and this preparation was successfully used to decontaminate wounds polluted with lead (and also Sr) on rats. Such properties constitute a potential hope for nuclear external decontamination.

1 Linqvist O., *J. Power Sources,* (1995), 57:3-7.

2 A.Gadelle, *Tetrahedron. Lett.*, submitted.

3 Macarák E.J., Kumor K., Weisz P.B., *Biochem.Pharmacol.*, (1991) 42:1502-1508.

4 Benesi H.A., Hildebrand J.H., *J. Am. Chem. Soc.* (1949) 71:2703-2707

5 F.Djedaini, *PhD Thésis,* Paris VI, 1993.

6 P.B.Weisz, Greene M.I., *Biochem. Pharm.* (1994), 42 (7): 1502-1503

7 J.C.Debouzy, F.Fauvelle.A.Gadelle, C.Baudin, M.Richard, B.Perly, F.Chouteau, J.Joets, J.J.Tazz, D.Daveloose, *Bull. Chim. Farm.* (1998) under press.

SYNTHESIS OF CYCLODEXTRIN GLYCEROL ETHERS AND INVESTIGATION OF THEIR BINDING PROPERTIES.

MÁR MÁSSON[1], JOSEF PITHA[2] and THORSTEINN LOFTSSON[1]
[1]Department of Pharmacy, University of Iceland, PO Box 7210, IS-127 Reykjavik, Iceland. [2]417 Anglesea Street, Baltimore, MD 21224 U.S.A.

1. Introduction

Cyclodextrin (CD) derivatives introduced for pharmaceutical applications are usually present as mixtures of compounds, differing in both the number and location of subtituded hydroxyl groups. These mixtures generally lead to more highly soluble complexes than are obtained with a single parent CD [1]. Introduction of side groups can affect the ability to form inclusion complexes in two ways. The side group can extend the cavity, which can ease the complex formation. Alternatively, the substituent can enter into the CD cavity of the same or other CD molecule and reduce its ability to encapsulate hydrophobic compounds. It has been shown that, in the case of hydroxypropyl-β-CD (HPβCD), both effects are present and the stability constant (K_c) of complexation is either greater or smaller than that of the parent βCD depending on the degree of substitution and the position of the substituents [2].
In the present work, a new class of CD derivatives, glycCD, was investigated. The glycCDs were made by forming $-CH_2CH(CH_2OH)-$ bridges between the 2 and 3 hydroxyl oxygens on the glucosyl units (Figure 1). This extends the cavity in a relatively rigid manner, the side group can not enter the cavity, and the extension keeps the interior of the new cavity non-polar.

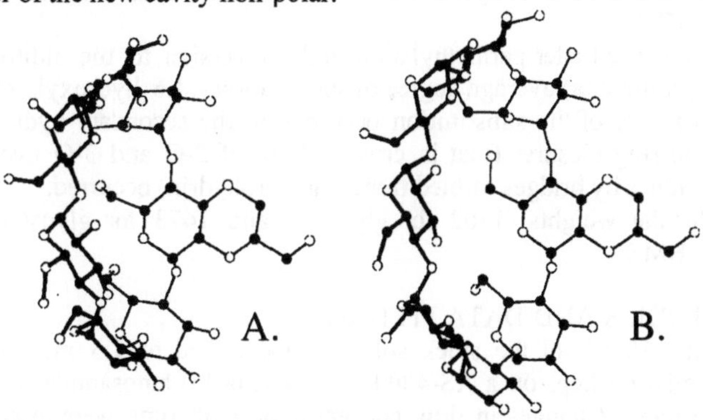

Figure 1. Figure showing the proposed structure of mono-glycCD derivative. A: βCD derivative, B: γCD derivative

Entering guest compounds are accommodated in a cavity which is larger and of different shape from that of the parent CD. Therefore, the K_c value for large and odd shape

molecules should increase. Nevertheless, not all bridges formed in the reaction are closed and some simple monoethers of glycerol are formed as well.
The binding properties of these compounds were studied through its solubilization and degradation rate investigations.

2. Materials and Methods

2.1 MATERIALS

2-Hydroxypropyl-β-CD (DS = 0.6) and γCD were kindly donated by Wacker-Chemie (Germany); Encapsin® (i.e. hydroxypropyl-β-CD with DS = 0.42) was purchased from American Maize-Products (Indiana, USA); βCD was obtained from Nihon Shokuhin Kako Co. Ltd. (Tokyo, Japan). The drug compounds were obtained from the following suppliers: acetazolamide from Icelandic Pharmaceuticals (Reykjavik), indomethacin and 6-ethoxyzolamide from Sigma Chemical Company, and hydrocortisone from Norsk Medisinaldepot (Oslo, Norway). Chlorambucil was supplied by the courtesy of Wellcome Foundation Ltd. (UK). All other chemicals were commercially available chemicals of reagent or analytical grade.

2.2 SYNTHESIS.

Previously, we have synthesized monosubstituted glycβCD through intramolecular cyclization of 2-O-(2',3'-epoxypropyl)-β-CD performed in anhydrous trifluoroacetic acid [3]. In the present work, mixtures of the glycCDs were obtained through condensation of parent CDs with epichlorohydrin in mildly alkaline media.
glycβCD: a suspension of βCD (15 g), calcium hydroxide (3.48 g) and sodium borohydride (0.15 g) in water (800 mL) was stirred and heated until reflux occurred; then epichlorohydrin (6.6 mL) was added drop-wise. The reflux was continued for 75 minutes and then 200 ml of the water was distilled off at atmospheric pressure. The suspension was then cooled to room temperature, filtered and the basic solution neutralized by hydrochloric acid. The solution was then dialyzed against water and evaporated in vacuo to dryness, leaving 15.59 g of white amorphous material. Calcium salts had to be removed from this material by the addition of an excess of sodium carbonate solution and another dialysis. The same procedure as above was applied to γCD to form glycγCD.
The product was analyzed after permethylation and conversion to the alditol acetates with the following results: (a) average degree of substitution - 4.6 hydroxyls of gamma-CD substituted, (b) 97% of the substitution occurred on the secondary hydroxyls, (c) 55% of all possible ring closures (that is cross-linking of 2-O and 3-O hydroxyls of glucopyranosyl residues by bridges formed from epichlorohydrin) occurred.
The average molecular weights (1202 for glycβCD and 1473 for glycγCD) were determined by FABMS.

2.3. KINETIC STUDIES AND DATA FITTING

In the experiments, 10 μL of the stock solution was added to 1.5 mL of the CD solution, and the mixture kept on a AS-4000 (Merck-Hitachi) autosampler temperature controlled sample rack. Changes in drug concentration with time were monitored by HPLC. In indomethacin degradation studies, a 10 mM Na_2CO_3/HCl (pH 10 at 40 °C) buffer was used; a 10 mM NaH_2PO_4/NaOH (pH 7.5 at 30 °C) buffer was used for the chlorambucil studies. The overall degradation constant (k_{obs}) depends on the degradation constant for the drug in solution (k_o), the degradation rate constant for the drug within the complex (k_c), the stability constant of the drug-CD complex (K_c) and the total CD concentration $[CD]_t$ as described by Equation 1. [4]

$$k_{obs} = \frac{k_o + k_c\ K_c[CD]_t}{1 + K_c[CD]_t} \qquad (1)$$

The data were fitted to the equation using non-liner fitting of the Kaleidagraph program (Synergy Software, USA).

2.4 SOLUBILITY STUDIES.

CD solutions (1%) were made in H_2O or in case of ethoxyzolamide in 10 mM aqueous NaH_2PO_4/NaOH (pH 6.0) solution.

Three mL of the solutions were added to vials containing approximately 10 mg of the drug; three vials were measured at each CD concentration. The vials were then autoclaved for 20 minutes, cooled to room temperature, equilibrated for up to 5 days, and the suspensions filtered through a 0.45 µm cellulose acetate membrane filter and the drug concentration of drug in the solution (filtrate) was measured by HPLC.

3 Results and Discussion

At conditions where basicity was minimal, but high enough to allow the reaction to proceed, epichlorohydrin formed the non-cyclic and cyclic glycerol ethers of CDs. Characteristically, at these conditions, more than half of the ethers formed were cyclic; that is, the secondary hydroxyls of anhydroglucose residues became connected by $CH_2CH(CH_2OH)$- bridges, forming CDs with extended cavities as shown in Figure 1.

The K_c values and the degradation rate (k_c/k_o) ratios were obtained from non-linear fitting of the data for degradation in various CD concentrations.\ Complexation of chlorambucil was larger with βCD, which was followed by the CD glycerol ethers, which were more powerful complexation agents than the corresponding HPβCD and γCD (Table 1). Indomethacin has a larger aromatic core than chlorambucil and K_c with glycβCD was 26% or 180 % larger than with the corresponding βCD and HPβCD respectively. The k_c/k_o ratios were similar for all the chlorambucil CD inclusion complexes, except for glycγCD where it was significantly smaller. Indomethacin was better protected in the glycγCD complex than in the γCD complex, but less protection was observed for glycβCD than for the two other βCDs tested.

Acetazolamide, ethoxyzolamide and hydrocortisone form 1:1 complexes with CDs at low CD concentrations [5,6]. Thus, the complexation efficacy ($K_c[D]_o$) was calculated from the solubilization of the drug at a 1% CD concentration.

The complexation with glycγCD increased relative to γCD with acetazolamide and hydrocortisone. The complexation with glycβCD also improved relative to βCD and hydroxypropylβCD (Encapsin®) in the case of acetazolamide and hydrocortisone. (Table 2).

The general observation was that complexation with glycCD was improved relative to other CDs, especially with glycγCD, where the extended shape of the cavity would lead to better complexation. In the case of glycβCD, the complexation increased

Table 1. K_c and the k_c/k_o ratio as determined in the degradation rate study.

Cyclodextrin	$K_C \times 10^{-3}$ M^{-1}	k_c / k_o
Indomethacin (40 °C; pH 10),		
$k_o = (8.53 \pm 0.36) \times 10^{-3}$ min^{-1}		
βCD	0.31 ± 0.03	0.13
γCD	0.27 ± 0.04	0.65
HPβCD	0.69 ± 0.06	0.09
glycβCD	0.87 ± 0.09	0.22
glycγCD	0.22 ± 0.06	0.43
Chlorambucil (30 °C; pH 7.5)		
$k_o = (1.95 \pm 0.03) \times 10^{-3}$ min^{-1}		
βCD	5.34 ± 0.11	0.071
γCD	0.35 ± 0.01	0.056
HPβCD	3.84 ± 0.09	0.059
glycβCD	4.10 ± 0.08	0.061
glycγCD	0.38 ± 0.01	0.029

Table 2. Drug solubility in 1% cyclodextrin solutions.

Solution	CD conc. ($M \times 10^{-3}$)	Drug conc. (M)	K_c[CD]
Acetazolamide			
Water		2.97×10^{-4}	
glycβCD	8.32	4.77×10^{-4}	0.0221
glycγCD	6.79	3.55×10^{-4}	0.0086
Encapsin	7.65	4.50×10^{-4}	0.0204
βCD	8.80	4.41×10^{-4}	0.0130
γCD	7.71	3.35×10^{-4}	0.0050
Ethoxyzolamid (ph 6.0)			
Buffer		0.43×10^{-5}	
glycβCD	8.32	2.69×10^{-5}	0.0027
glycγCD	6.79	0.73×10^{-5}	0.0004
Encapsin	7.65	2.97×10^{-5}	0.0033
βCD	8.80	2.36×10^{-5}	0.0022
γCD	7.71	0.76×10^{-5}	0.0004
Hydrocortisone			
Water		1.34×10^{-4}	
glycβCD	8.32	7.74×10^{-4}	0.083
glycγCD	6.79	8.17×10^{-4}	0.112
Encapsin	7.65	5.81×10^{-4}	0.062
βCD	8.80	8.00×10^{-4}	0.082
γCD	7.71	6.59×10^{-4}	0.073

with larger molecules like indomethacin and hydrocortisone. Changing the shape of the cavity had little influence on k_c which suggests that other factors, like polarity, are responsible for the reduction in k_c seen with some other CD derivatives.

4. References

1. Loftsson, T. and Brewster, M. E. (1996) Pharmaceutical Applications of cyclodextrins. 1. Drug solubilization and stabilization. *Journal of Pharmaceutical Sciences* , 85, 1017-1025
2. Brewster, M. E., Simpkins, J. W., Hora, M. S., Stern, W. C. and Bodor, N. (1989) The Potential use of cyclodextrins in parenteral formulations. *Journal of Parenteral Science & Technology* , 43, 231-240
3. Jindrich, J., Harata, K., Lindberg, B., Pitha, J. and Seffers, P. (1997) Synthesis of 2, 3-O-(propane-1,2-diyl) -and 2,3-O-(3-hydroxypropane-1,2-diyl)-cyclomaltoheptaose. *Carbohydrate Research* , 300,
4. Loftsson, T. (1995) Effects of cyclodextrins on the chemical stability of drugs in aqueous solutions. *Drug Stability* , 1, 22-33
5. Loftsson, T., Fridriksdottir, H., Ingvarsdottir, G., Jonsdottir, B. and Sigurdardottir, A. M. (1994) The influence of 2-hydroxypropyl-beta-cyclodextrin on diffusion rates and and transdermal delivery of hydrocortisone. *Drug Development and Industrial Pharmacy* , 20, 1699-1708
6. Loftsson, T., Fridriksdottir, H., Thorisdottir, S., Stefansson, E., Sigurdardottir, A. M., Gudmundsson, O. and Sigthorsson, T. (1994) 2-Hydroxypropyl-ß-cyclodextrin in topical carbonic anhydrase inhibitor formulations. *European Journal of Pharmaceutical Sciences* , 1, 175-180

CONFORMATIONAL EQUILIBRIA OF α-CYCLOALTRIN IN SOLUTIONS

K. FUJITA,* K. OHTA,* Y. NOGAMI, K. NASU, T. SHIRATANI, M. SUDO, and T. KOGA, **Faculty of Pharmaceutical Sciences, Nagasaki University, 1-14 Bunkyo-machi, Nagasaki 852-8521, Japan and Daiichi College of Pharmaceutical Sciences, Tamagawa-cho, Fukuoka 815-8511, Japan*

We reported the practical synthetic method of cycloaltrins (CAs, **1a,b**), in which we used cyclodextrins (CDs) as the starting materials and converted all the constituent glucose units into altrose units (Scheme 1) [1, 2]. X-ray analysis of α-CA revealed that the altrose units

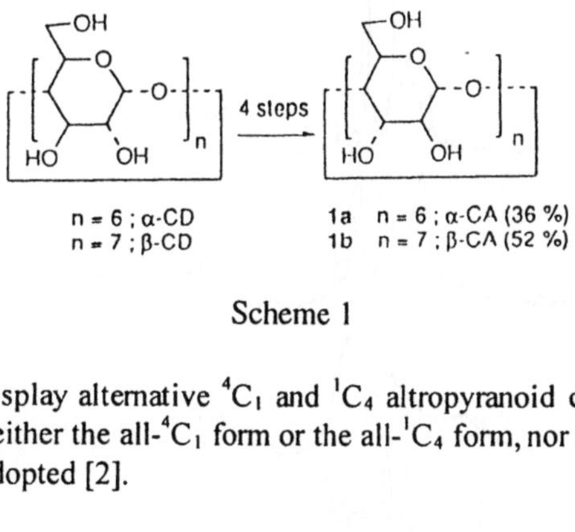

Scheme 1

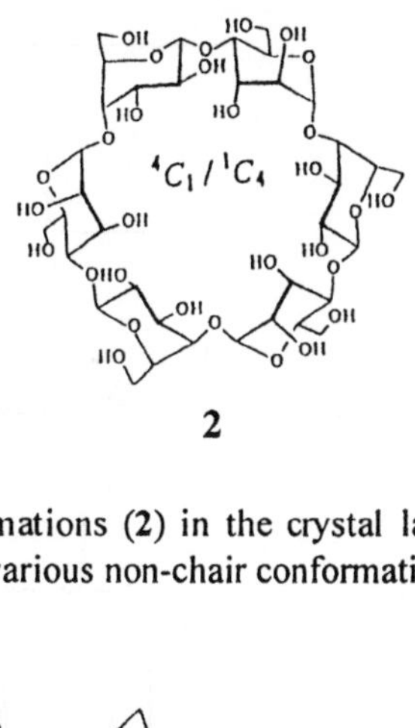

2

display alternative $^4C_1$ and $^1C_4$ altropyranoid chair conformations (**2**) in the crystal lattice; neither the all-$^4C_1$ form or the all-$^1C_4$ form, nor one of the various non-chair conformations is adopted [2].

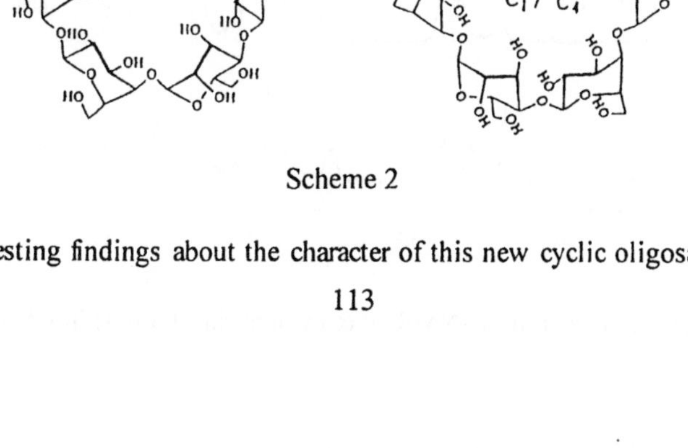

Scheme 2

We report interesting findings about the character of this new cyclic oligosaccharide. From dynamic NMR study of *a*-CA we did find out new facts that the molecule also adopts continuous alternative $^4C_1$ / $^1C_4$ conformational array in solution, and surprisingly, the conformation of altropyranoid chairs exchange each other dynamically in the solutions (Scheme 2). Thus, *a*-CA is a mobile molecule offering a marked contrast with rigid CD molecule.

In ^{1}H-NMR spectrum of *a*-CA (Fig. 1), coupling constant of H-1 and H-2 was observed as 4.04 Hz in CD_3OD at +20 °C. The fact that the value is almost equivalent with the mean value of $J_{H\text{-}1,H\text{-}2}$ estimated for two altropyranoid chair forms ($^4C_1$ and $^1C_4$) suggests that the

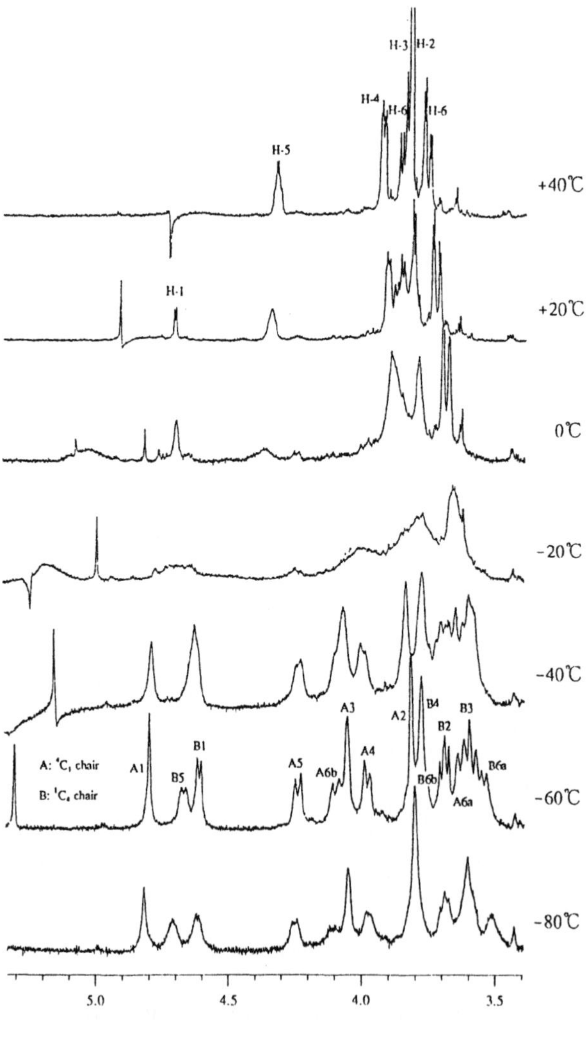

Figure 1. Dynamic ^{1}H-NMR of α-cycloaltrin (0.5 mg) in CD_3OD

non-chair twist boat conformation of altroside is predominant, or otherwise, there may be an rapid equilibrium between $^4C_1$ and $^1C_4$ conformer in each altropyranoid ring, in special case, between all-$^4C_1$ and all-$^1C_4$ conformers in solution.

It is therefore necessary first to freeze out the exchange by lowering the temperature. So, we tried to get temperature dependent ^{1}H-NMR data of *a*-CA (Fig. 1). Lowering the temperature, strongly exchange-broadened spectra were disclosed. All proton signals showed broadening, and along with further temperature down, all became sharp, again. At -60 °C, the signals clearly exhibited the presence of a set of two altropyranoid rings in the solution, and the conformations of both $^4C_1$ and $^1C_4$ chairs were assigned on the basis of H-H COSY spectrum measured in CD_3OD-D_2O (9 : 1) at - 50 °C. The coalescence temperature of the signals was about -15 °C on 500 MHz NMR equipment. These data ruled out the possibility of equilibrium between $^4C_1$ and $^1C_4$ conformer in each altropyranoid chairs, that is random conformational changes of **5a**.

Since there remains a possibility of the equilibrium between all-$^4C_1$ and all-$^1C_4$, low temperature NOE experiments were carried out. Irradiation at H-1 of $^4C_1$ conformer afforded obvious NOE signal on H-4 of $^1C_4$ conformer, and reverse irradiation at H-4 of $^1C_4$ gave NOE on H-1 of $^4C_1$, certainly. Thus the equilibrium existing in the solution is not between all-$^4C_1$ and all-$^1C_4$, but is a type of degenerated rearrangement depicted in Scheme 2. Dynamic ^{13}C-NMR experiment also supports this consideration. By the use of band-shape analysis of C-1 signals, two-site exchange program, we obtained the exchange rate constant (k_a). From these k_a values, activation energy of the molecular movement is estimated as 8.5 kcal/mol. This value corresponds to an activation energy for conformational change of one pyranoid ring. In conclusion, *a*-CA is a dynamic molecule having "jelly-fish"-like, but methodical motions in the solution at normal temperature and interestingly, the first conformational change of one pyranoid ring enforces the change on the other pyranoid rings and controls the kind of each change.

References

1. Fujita, K., Shimada, H., Ohta, K., Nogami, Y., Nasu, K. and Koga T. (1995) β-Cycloaltrin: A Cyclooligosaccharide Constituted of Seven α(1→4)-linked Altropyranoses, *Angew. Chem. Int. Ed. Engl.*, 34, 1621.
2. Nogami, Y., Nasu., Koga, T., Ohta, K., Fujita., Immel, S., Lindner, H. J., Schmitt, G. E., Lichtenthaler, F. W. (1997) Synthesis, Structure and Conformational Features of α-Cycloaltrin: a Cyclooligosaccharide with Alternating $^4C_1$/$^1C_4$ Pyranoid Chairs, *Angew. Chem. Int. Ed. Engl.*, 36, 1899-1902.

THE SYNTHESIS OF OLIGOSACCHARIDE-BRANCHED CYCLODEXTRINS

Toshiyuki INAZU, Takashi YAMANOI, Katsuji HANEDA, Mamoru MIZUNO, Keisuke MATSUDA†, Tomohiro YAMAZAKI†, Masahito TAKEUCHI†, Hisayo TSURUI† and Kenjiro HATTORI†

The Noguchi Institute, 1-8-1, Kaga, Itabashi-ku, Tokyo 173-0003, Japan
†Faculty of Engineering, Tokyo Institute of Polytechnics, Atsugi, Kanagawa 243-0297, Japan

1. Introduction

Cyclodextrins (CDs) can be used in various drugs as an auxiliary additive such as a carrier, diluent, and solubilizer for tablet ingredients.[1-3] The syntheses of natural oligosaccharide-branched CDs, which showed potential binding to lectin, have been studied for the development of a drug carrier for targeting drug delivery systems. We have already reported the chemo-enzymatic syntheses of three natural oligosaccharide-branched CDs by the transglycosylation of Fmoc-Asn(GlcNAc)-NH-β-CD (**4**) with

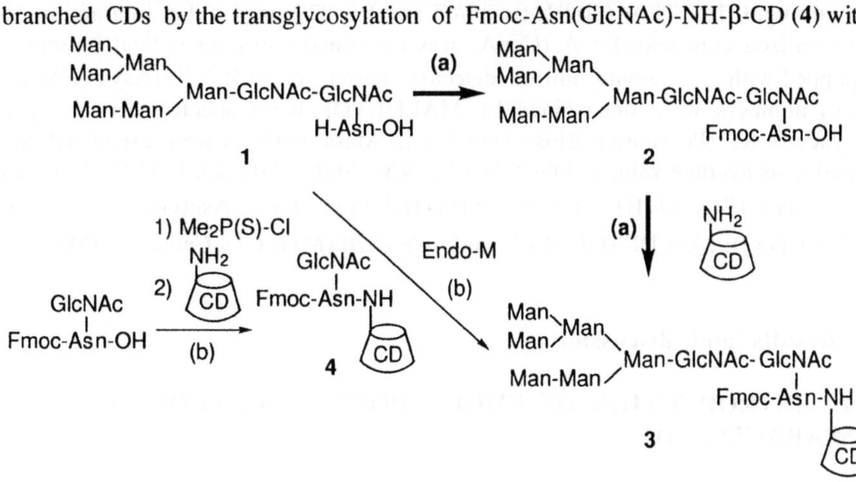

Fig 1. The present syntheses (a) and previous syntheses (b) of oligosaccharide- branched CD.

glycosyl asparagines prepared from ovalbumin[4] and human transferrin[5] using *endo*-β-*N*-acetylglucosaminidase of *Mucor hiemalis* (Endo-M).[6] A large-scale synthesis of these CDs was required for the evaluation of these CDs. However, it is difficult to synthesize a large quantity of these CDs by this method, because of the availability limitations of Endo-M. In this paper, the large-scale chemical synthesis of high-mannose type oligosaccharide-branched CD (**3**) using glycosylasparagine, H-Asn($GlcNAc_2Man_6$)-OH (**1**)[5] as the starting material, which was used as a transglycosylation substance of Endo-M, was described.

2. Experimental

2.1. MATERIALS

Glycosylasparagine, H-Asn($GlcNAc_2Man_6$)-OH (**1**) and 6-*mono*-amino-6-deoxy-β-cyclodextrin (H_2N-β-CD) were prepared according to the literature.[5,7] Fmoc-OSu (9-Fluorenylmethyloxycarbonyl-oxysuccinimide) was used as the protecting reagent. The Amberlite XAD-2 (nonionic polymeric adsorbent) is commercially available (Organo Co.). PyBOP (Benzotriazole-1-yl-oxy-tris-pyrroridino-phosphonium hexafluorophosphate) or HBTU (2-(1H-Benzotriazole-1-yl)-1,1,3,3-tetramethyluronium hexafluorophosphate)-HOBt (1-Hydroxybenzotriazole) as the condensing agents was used.

2.2. METHOD

Reversed-phase HPLC was performed on an Inertsil ODS-3 column (20 x 250mm) (GL Sciences Inc.) using a linear gradient of acetonitrile containing 0.1% TFA/H_2O containing 0.1% TFA (10/91 to 35/65, v/v). The molecular interaction to immobilized concanavalin A (ConA) was examined using an optical biosensor equipped with a resonant mirror detector based on SPR,[8-10] IAsys (FAST). Mass numbers were determined by MALDI-TOF mass spectrometry using a Voyager® RP (PerSeptive Biosystems Inc.). Mass numbers were calculated and reported as average values. Fmoc-Asn($GlcNAc_2Man_6$)-OH (**2**): MALDI-TOF MS Found: *m/z* 1735 $(M+H)^+$. Calcd for $(M+H)^+$ 1734. Fmoc-Asn($GlcNAc_2Man_6$)-NH-β-CD (**3**): MALDI-TOF MS Found: *m/z* 2870 $(M+Na)^+$. Calcd for $(M+Na)^+$ 2871.

3. Results and discussion

3.1. INTRODUCTION OF FMOC GROUP TO GLYCOSYL-ASPARAGINE (1)

Natural oligosaccharides were planned as the starting materials for the synthesis of natural oligosaccharide-branched CDs. However, it is not very easy to use natural oligosaccharides of this level of organic synthesis because of the problems with isolations of the oligosaccharide related products in the reaction mixture. It is known that Amberlite XAD-2 is useful for the removal and recovery of aromatic compounds. If the Fmoc group could be introduced into the glycosylasparagines, the Fmoc group would have function as the marker of XAD-2 and the natural oligosaccharide-related products having an Fmoc group could be easily purified from the reaction mixtures by solid-phase extraction using XAD-2. The synthesis of Fmoc-Asn($GlcNAc_2Man_6$)-OH (**2**) was then examined. The corresponding **2** was prepared by the reaction of **1** with Fmoc-OSu in dioxane-water in the presence of sodium hydrogencarbonate, and was isolated in 88% yield by adsorbing on XAD-2 followed by eluting with methanol. Product **2** was obtained ≥150 mg / 30 eggs.

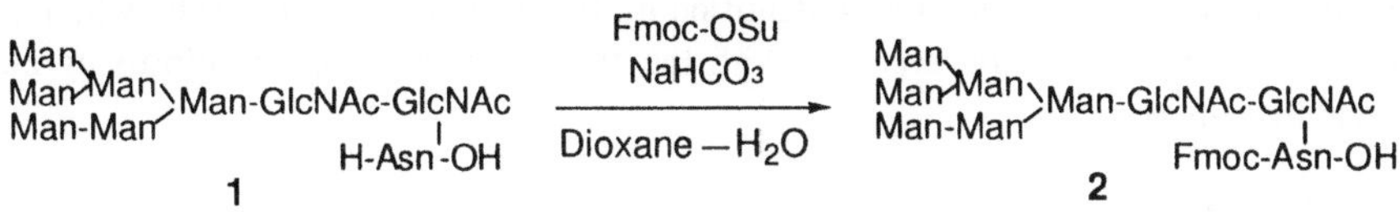

Fig 2. The synthesis of **2** by the reaction of **1** with Fmoc-OSu.
Reaction conditions; **1**:Fmoc-OSu:$NaHCO_3$=1:1.2:3, r.t., overnight, dioxane/water=1/1.

3.2. SYNTHESIS OF NATURAL OLIGOSACCHARIDE-BRANCHED CD (3)

The condensation of **2** with H_2N-β-CD to synthesize the desired Fmoc-Asn($GlcNAc_2Man_6$)-NH-β-CD (**3**) was investigated. The condensation of **2** with H_2N-β-CD in DMF using several condensing agents gave **3** in low yields. This condensation was influenced by the poor solubility of **2** in DMF. However, **2** was dissolved in NMP (*N*-methyl-2-pyrrolidinone). Several condensing agents were adopted for the reaction. As a result, it was found that the reaction using PyBOP or HBTU-HOBt as condensing agents in the presence of *N,N*-diisopropylethylamine (DIEA) in NMP gave product **3** in 74% and 54% yields, respectively.
The oligosaccharide-branched CD (**3**) was isolated by preparative HPLC and identified by its retention time and a MALDI-TOF MS spectra compared with the authentic

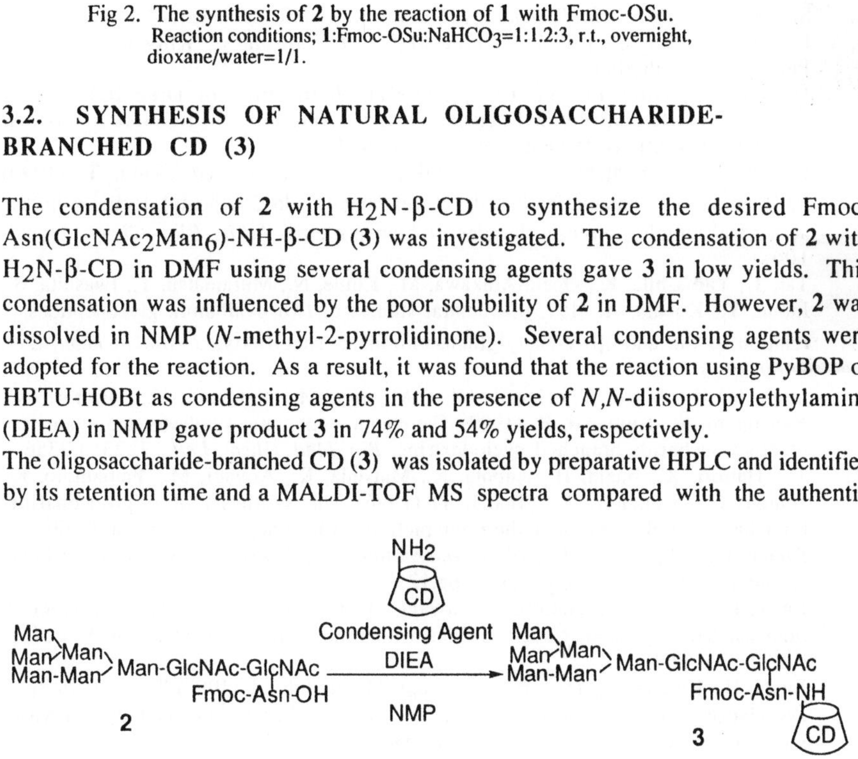

Fig 3. The synthesis of **3** by the reaction of **2** with H_2N-β-CD.
Reaction conditions; **3**:DIEA:condensing agent:H_2N-β-CD=1:1:1:1, r.t., overnight

sample made by the chemo-enzymatic method.[6] The association constant of **3** with ConA using a biosensor coincided with the previous value[6] of ConA with the oligosaccharide-CD obtained by the transglycosylation.

4. Conclusion

A large-scale chemical synthesis of the high-mannose type oligosaccharide-branched CD (**3**) was achieved by a two-step reaction using glycosylasparagine, H-Asn($GlcNAc_2Man_6$)-OH (**1**) and H_2N-β-CD as the starting materials. Fmoc-Asn($GlcNAc_2Man_6$)-OH (**2**) was easily isolated by the adsorbent XAD-2. Also, the condensation of **2** with H_2N-β-CD was influenced by the solubility of **2**. The condensation using PyBOP in NMP in the presence of DIEA gave the corresponding **3** in high yield. The product **3** was identified by HPLC and a MALDI-TOF MS, and assessed for its association constant with the saccharide-binding protein using an optical biosensor.

5. References

1. Szejtli, J. : *Cyclodextrin Technology*, Kluwer Academic Publishers, Netherlands.
2. Froming, K. H., Szejtli, J.: *Cyclodextrin in Pharmacy*, Kluwer Academic Publishers,, Netherlands.
3. Uekama, K., Hirayama, F., Irie, T. (1991) Modification of Drug Releases by Cyclodextrin Derivatives, in Duchene, D. (ed.), *New Trends in Cyclodextrins and Derivatives*, Editions de Sante, Paris, pp. 409-446.
4. Kadowaki, S., Yamamoto, K., Fujisaki, M., Izumi, K., Tochikura, T. (1990) Purification and characterization of a novel fungal endo-β-*N*-acetylglucosaminidase acting on complex oligosaccharides of glycoproteins, *Agric. Biol. Chem.*, 54, 97-106.
5. Tai, T., Yamashita, K., Ogata-Arakawa, M., Koide, N., Muramatsu, T., Iwashita, S., Inoue, Y., Kobata, A. (1975) Structural studies of two ovalbumin glycopeptides in relation to the endo-β-*N*-acetylglucosaminidase specificity, *J. Biol. Chem.*, 250, 8569-8575.
6. Matsuda, K., Inazu, T., Haneda, K. , Mizuno, M., Yamanoi, T., Hattori, K., Yamamoto, K., Kumagai, H. (1997) The chemo-enzymatic synthesis and evaluation of oligosaccharide-branched cyclodextrins, *Bio. Med. Chem. Lett.*, 7, 2353-2356.
7. T., Hattori, K., Imata, H., Kubota, K., Matsuda, K., Aoyagi, M., Yamamoto, K., Jindoh, C., Yamanoi, T., Inazu, T. (1996) The synthesis of oligosaccharide-branched cyclodextrins and their interaction with concanavalin A, in Szejtli, J., Szente, L., (ed.), *Proceeding of 8th International Cyclodextrins Symposium*, Kluwer Academic Publishers, Netherlands, pp.121-124.
8. Imata, H., Kubota, K., Hattori, K., Aoyagi, M., Jindoh, C. (1997) The interaction of oligosaccharide-branched cyclodextrins with immobilized concanavalin A, *Polym. J.*, 29, 563-567.
9. Imata, H., Kubota, K., Hattori, K., Aoyagi, M., Jindoh, C. (1997) The specificity of association between concanavalin A and oligosaccharide-branched cyclodextrins with an optical biosensor, *Bio. Med. Chem. Lett.*, 7, 109-112.
10. Shuster, S. C., Swanson, R. V., Alex, L. A., Bourret, R. B., Simon, M. I. (1993) Assembly and function of a quaternary signal transduction complex monitored by surface plasmon resonance, *Nature*, 365, 343-347.

CRYSTALLIZATION OF PARTIALLY O-ALKYLATED β-CYCLODEXTRINS BY INCLUDING POLYMERS

K. SUEHIRO, T. SUGINO, and M. KURAMORI
Saga University. Faculty of Science and Engineering. Department of Chemistry and Applied Chemistry. Saga. 840-8502 Japan

Introduction

β-cyclodextrin (β-CD) has poor solubility in water and organic solvents. However, methylation of the hydroxyl groups significantly increases its solubility. 2,6-O-dimethyl- and 2,3,6-O-trimethyl-β-CD show better solubility and in particular, partially O-methylated β-CD (PM-β-CD) is soluble in most of common organic solvents [1]. However, little attention has focused on the inclusion ability of PM-β-CD. We attempted to prepare the complexes of PM-β-CD and partially O-ethylated β-CD (PE-β-CD) with polyethers, $[-O(CH_2)_m-]_n$ (m=2-4) and $[-OCH_2CH(CH_3)-]_n$, and polyesters, $[-O(CH_2)_mCO-]_n$ (m=2,4,5).

Experimental

PM-β-CD was kindly supplied by Mercian Co., Ltd. The degree of methylation of the hydroxyl groups was 70% on average. PE-β-CD was prepared by treating β-CD with diethyl sulfate in 40% (w/v) aqueous NaOH at 0-15℃.
The preparation of the complexes was tried at room temperature by slow evaporation of the solvent after dissolving the CDs with polymers (ca. 10:1 in weight) in appropriate solvents. The formation of the complexes was identified by the X-ray diffraction patterns of the evaporated residues.

Results and Discussion

1. Inclusion Complexes with Polyethers

Figure 1 shows the X-ray diffraction diagrams of β-CD, PM-β-CD, and PE-β-CD in the uncomplexed state. In contrast to unmodified β-CD, the guest-free PM- and PE-β-CD are amorphous due to their irregular chemical structures based on the incomplete alkylation of the hydroxyl groups. When the solvent was evaporated from an ethanolic solution of PM-

β-CD and poly(ethylene glycol) (PEG, m=2)(MW=1,000) mixture, a crystalline complex between PM-β-CD and PEG was obtained, although unmodified β-CD does not form a complex with PEG of any molecular weight [2]. PEG bis(3,5-dinitrobenzoate) did not form a complex with PM-β-CD, because the dinitrophenyl groups bonding to both ends of the PEG chain are bulky in comparison with the diameter of the cavity of PM-β-CD and they cannot pass through it. That is, crystallization of PM-β-CD is caused by including the PEG chain into the cavity and being aligned. Similar crystalline complexes were obtained with tri(ethylene glycol), higher oligomers, and high molecular weight poly(ethylene oxide), but ethylene glycol and di(ethylene glycol) were too short to form complexes.
PM-β-CD also formed crystalline complexes with poly(propylene glycol)(PPG) analogs whose molecular weight is greater than 200 (n≒3). Other polyethers, such as poly(oxacyclobutane)(POCB, m=3) and poly(tetrahydrofuran)(PTHF, m=4), also formed complexes with PM-β-CD. The X-ray diffraction patterns of these complexes appear as shown in Fig. 2.
PE-β-CD also formed crystalline complexes with PEG and other polyethers, and gave similar X-ray diffraction patterns to each other (Fig. 3). The cavity of β-CD is too large to fit the PEG chain but the alkyl groups of PM- and PE-β-CD may expand the depth of the hydrophobic space and also make the bores narrower to fit the PEG chain.

2. Inclusion Complexes with Polyesters

Neither PM-β-CD nor PE-β-CD formed complexes with poly(β-propiolactone)(PPL, m=2), similar to β-CD. Though a cross section of the PPL chain is slightly larger than that of PEG and appreciably smaller than that of PPG, PPL was not included into the cavities of these partially O-alkylated β-CDs. This may be due to PPL having a carbonyl group that is rigid compared to the polyethers. Poly(δ-varelolactone)(PVL, m=4) and poly(ε-caprolactone) (PCL, m=5), which have a longer methylene chain and are more flexible, formed complexes with both PM-β-CD and PE-β-CD. Their X-ray diffraction patterns are similar to those of the complexes with previously mentioned polyethers (Figs. 4 and 5). These results are summarized in Table 1.

3. Crystal Structures of the Inclusion Complexes

The X-ray diffraction patterns of the PM- and PE-β-CD complexes could be tentatively indexed based on the hexagonal unit cells: a=15.38 and c=9.99Å for the PM-β-CD complex, and a=15.64 and c=11.10Å for the PE-β-CD complex. Stacking on top of each other, these β-CD molecules form columnar channel structure and the guest polymers are present in the channels. If the CDs are alternately stacked in a head-to-head and tail-to-tail fashion, the repeat unit along the c-axis must be double. In any case, the height of the β-CDs increased from 7.8Å in β-CD to 11.1Å in PE-β-CD by partial ethylation of the hydroxyl groups.

Conclusion

The ability of β-CD to include polymers is increased by its partial O-methylation and O-ethylation. The polymer chains successively passing through the cavities of PM- or PE-β-CD, which are amorphous in the guest-free state, cause these CD molecules to align, inducing crystallization. So far as we know, this would be the first example of the inclusion crystallization of amorphous host compounds. Polyesters, which are rigid compared to polyethers, are much more difficult to be included into the cavities of these β-CDs to form complexes rather than polyethers.

References

1. Yamamoto, K., Matsuda, A., Tsuchiyama, Y., Sato, M., Yagi, Y. and Ishikura, T. (Sanraku Co. Ltd.) (1988) Partially methylated cyclodextrin and its manufacturing, Jpn. Pat. 63, 41,505.
2. Harada, A. and Kamachi, M. (1990) Complex formation between poly(ethylene glycol) and α-cyclodextrin, *Macromolecules*, *23*, 2821.

TABLE 1. Preparation of Inclusion Complexes between β-CD and polymers

Polymer		β-CD	PM-β-CD	PE-β-CD
Polyether	$[-O(CH_2)_m-]_n$			
PEG	(m=2)	−	+[a]	+
POCB	(m=3)	+	+	
PTHF	(m=4)	+	+	+
PPG	$[-OCH_2CH(CH_3)-]_n$	+	+[b]	+
Polyester	$[-O(CH_2)_mCO-]_n$			
PPL	(m=2)	−	−	−
PVL	(m=4)	+	+	+
PCL	(m=5)	+	+	+

[a] n≧3

[b] MW≧200

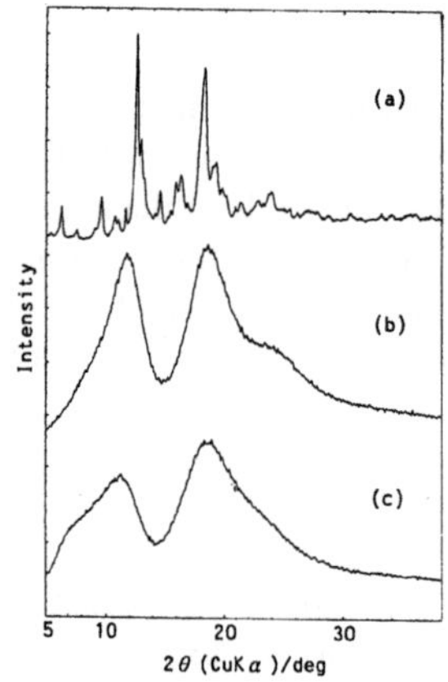

Figure 1. XRD diagrams of β-CDs in the uncomplexed state: (a) β-CD, (b)PM- β-CD, and (c)PE- β-CD.

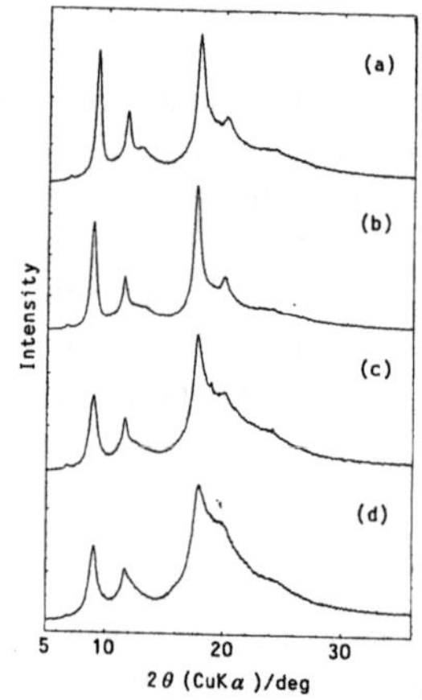

Figure 2. XRD diagrams of PM- β-CD complexes with (a)PEG, (b)PPG, (c)POCB, and (d)PTHF.

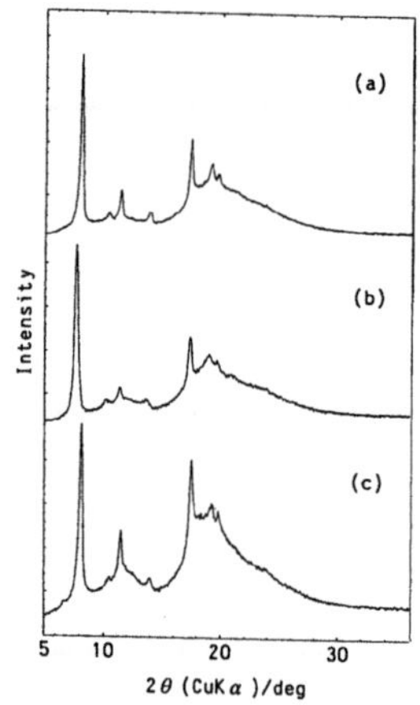

Figure 3. XRD diagrams of PE- β-CD complexes with (a)PEG, (b)PPG, and (c)PTHF.

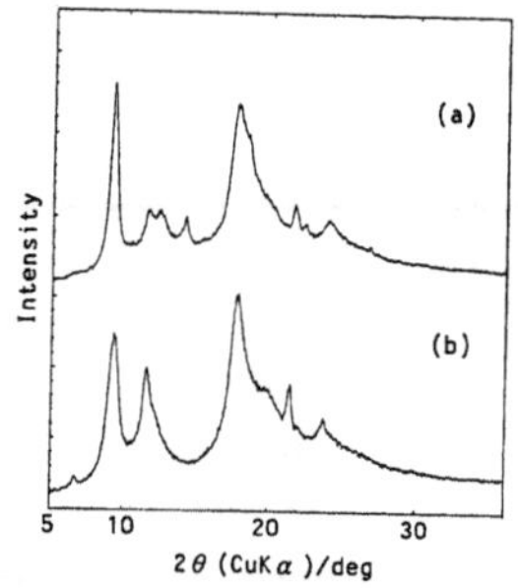

Figure 4. XRD diagrams of PM- β-CD complexes with (a)PVL and (b)PCL.

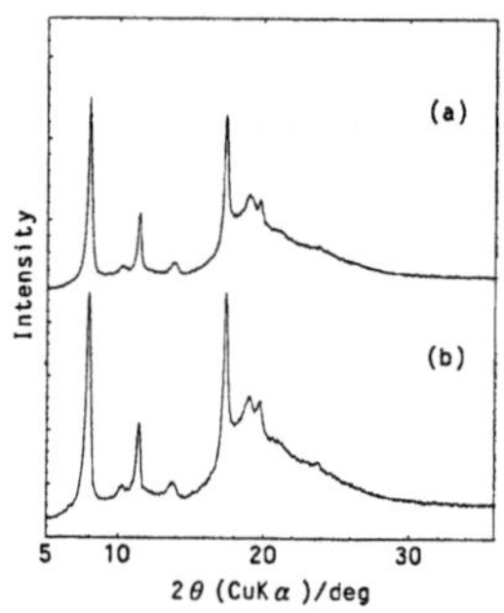

Figure 5. XRD diagrams of PE- β-CD complexes with (a)PVL and (b)PCL.

SYNTHESIS OF PER (2,3 -DI- O- HEXANOYL)- β- CYCLODEXTRIN AND CHARACTERIZATION OF AMPHIPHILIC β--CYCLODEXTRIN NANOPARTICLES

E. MEMISOGLU*, D. CHARON**, D. DUCHENE***, A. A. HINCAL*

**Hacettepe University, Faculty of Pharmacy, Dept. of Pharmaceutical Technology, 06100 Ankara-TURKEY*

***Université Paris-Sud, Faculty of Pharmacy, URA CNRS 1116, Laboratoire d'Endotoxines, 92296 Chatenay-Malabry-FRANCE*

****Université Paris-Sud, Faculty of Pharmacy, URA CNRS 1218, Centre d'Etudes Pharmaceutique, 92296 Chatenay-Malabry-FRANCE*

Introduction:

Studies on the synthesis of amphiphilic β-cyclodextrins to possibly obtain self-assemblying supramolecular aggregates have been carried out particularly in the last decade by many researchers by grafting alkyl chains of variable length (C2-C18) on the primary or secondary faces. In order to modify the cyclodextrin molecule, the reaction type most frequently used is electrophilic attack at the hydroxyl groups. However, it is evident that the selective and efficient modification of cyclodextrins is very complicated due to the difference of reactivity of hydroxyl groups in different positions, electronic reasons and steric factors of the torus. The large number of hydroxyl groups in the molecule (21 for β-cyclodextrin) results in statistical problems of selective polysubstitution (1-3).

Some of these amphiphilic cyclodextrin derivatives are able to form nanoparticles such as nanospheres and nanocapsules but the complete structure of these derivatives have not been fully enlightened. Usually, these derivatives were not selectively substituted but rather they were the mixtures of the intended derivative and other derivatives substituted on different positions.

In this study, the objective was to obtain a homogenous and pure modified cyclodextrin with C6 alkyl chains on secondary hydroxyl groups via ester bonds that could form nanospheres with or without the presence of surface active agents. The characterization of the obtained nanoparticles were carried out by measurement of particle size and imaging the nanoparticles.

Material and Methods:

β-cyclodextrin used in the synthesis procedure was Kleptose® (Lot EΦ 221, Code 341 001 106 R) and Hexanoic anhydride (Aldrich Chemical Co., USA) was the main reagent in this process. In the preparation of nanoparticles, the modified β-cyclodextrin was used in addition to Pluronic PE® F 68 (ICI Surfactants, UK), which was a kind gift of Chemsol-Istanbul, Turkey.

The synthesis was largely based on the previously reported works of Hirayama et al (4) and Takeo et al (5). To obtain a selectively acylated β-cyclodextrin, the primary hydroxyl groups on position 6 were blocked with *tert*-butyldimethylsilyl chloride. This step was carried out in pyridine at 0°C for 3 hours and was kept at room temperature overnight. Grafting of alkyl chain on positions 2 and 3 on the secondary face was realized by the treatment of silylated β-cyclodextrin with hexanoic anhydride in the presence of dimethylaminopyrine in anhydrous pyridine for 2 days at 70°C. The silyl groups on position 6 were eliminated with tetrabutylammonium fluoride. The final product was assayed with TLC and H-NMR and further purification was realized by repeated column chromatography.

Nanoparticles of amphiphilic β-cyclodextrin synthetized were obtained using the nanoprecipitation method previously described by Skiba et al (6). The method consists of dropwise addition of organic phase to the aqueous phase with constant stirring rate at 300 rpm. Different ratios of amphiphilic β-cyclodextrin/surfactant were essayed indicated in Table 1 and the resulting suspension was then filtered with a 0.45 μm membrane filter (RC Minisart®-Sartorius) and concentrated to the desired volume under vacuum.

Obtained particles of different formulations were characterized by measurement of mean particle size and polydispersity index in Coulter Nanosizer N4MD. The results of these mesaurements are seen in Table 1.

Photomicrographs of the obtained nanoparticles were taken with Scanning Electron Microscopy (Hacettepe Univ, Fac. of Medicine, Dept. of Anatomy) after staining with gold.

Results and Discussion

The purity of the obtained per(2,3-di-O-hexanoyl)-β-cyclodextrin was evaluated by H-NMR Spectroscopy 500 μHz seen in Figure 1 displays the structure of the intended product.

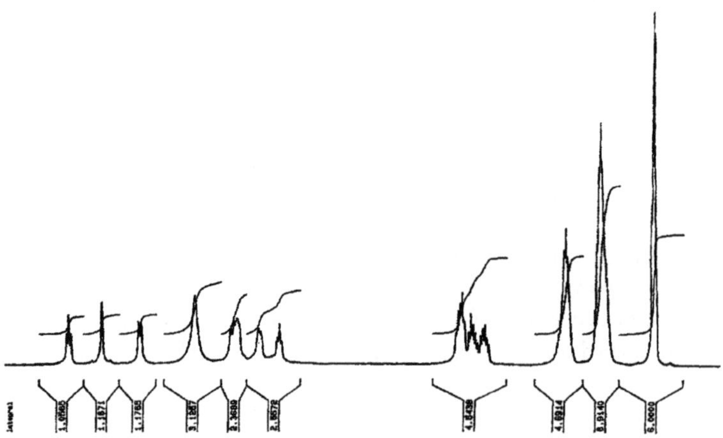

Figure 1-H-NMR spectrum of amphiphilic β-cyclodextrin

Mass spectrum of the product has also been taken which displays the purity and homogeneity of the modified β-cyclodextrin in Figure 2.

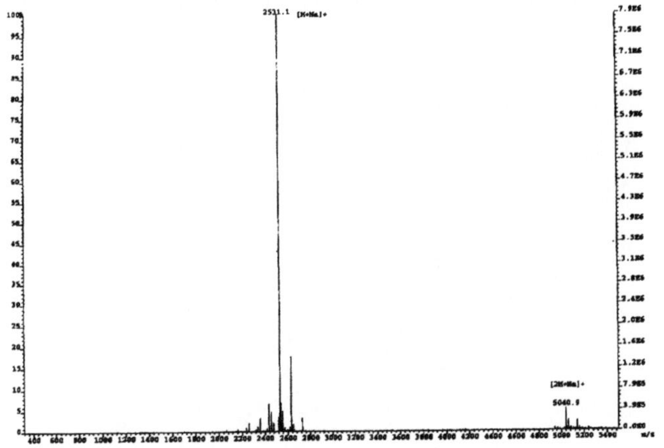

Figure 2-Mass spectrum of per(2,3-di-O-hexanoyl)-β-cyclodextrin

These data assure that the product obtained in this synthesis is of high purity and that selective polysubstitution is eliminated which will facilitate both the illumination of the structure of the new modified β-cyclodextrin and the characteristics of the supramolecular assemblies obtained by these cyclodextrin derivatives. To obtain homogeneous and selectively substituted amphiphilic β-cyclodextrin, the reaction conditions must be optimized to a great extent. Any source of hydratation may yield with a polysubstituted or impure product in this 3-step synthesis procedure. On the other hand, usage of hexanoic anhydride as an alkylating agent was the main difference from the previously reported derivatives of amphiphilic cyclodextrins. the purity of the product was further enlightened assuring the reproducibility of the product in the industrial administration.

Results of particle size measurements of amphiphilic β-cyclodextrin nanoparticles are shown in Table 1.

TABLE 1- Mean particle size and polydispersity index of different formulations of amphiphilic β-cyclodextrin nanoparticles. (n=3)

β--Cyclodextrin/ surfactant ratio	**1:4**	**1:1**	**1:0.5**	**No surfactant**
Mean Particle size (polydispersity index)	175±80.1 (1.633)	222.2±78.7 (0.114)	283.1±119.9 (0.580)	1229.4±572.6 (1.615)

As seen in table, increase in surfactant amount leads to an expected decrease in particle size. On the other hand, the nanosphere suspension prepared without surfactant displayed a bimodal particle size distribution suggesting the formation of unstable nanospheres that have a tendency to aggregate without the presence of surface active agents.

Photomicrographs taken with Scanning Electron Microscopy display the presence of nanoparticles in the shape of smooth surfaced spheres obtained with surface active agent Pluronic PE® F68 as can be seen in Figure 3.

Taking into account that there rests a vast amount of research to be carried out in order to fully illuminate the physicochemical characteristics, in vitro and in vivo behaviour of these nanoparticles, they may be considered as promising agents for targeted drug delivery, in enhancing the bioavailability of poorly soluble drugs or drugs with serious side effects which can be masked by entrapment in cyclodextrin.

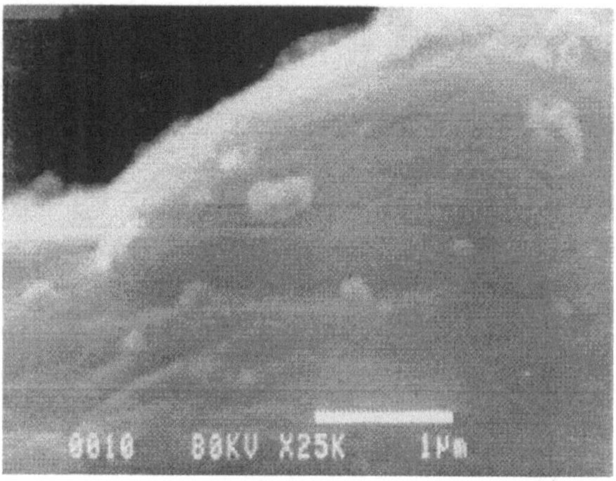

Figure 3-- SEM photomicrograph of amphiphilic β-cyclodextrin nanoparticles

Acknowledgement

The authors wish to thank the **CIES** of French Foreign Affaires Ministry and **TUBITAK** Turkish Scientific and Technical Council of Research for their financial support in this work, Chemsol, Istanbul-Turkey for the kind supply of Pluronic PE F 68 and Prof. Dr. Denis Wouessi Djewe for his scientific cooperation.

References

1. Wenz, G., (1994) «Cyclodextrins as building blocks for supramolecular structures and functional units » *Angew. Chem. Int. Ed. Engl.*, **33**, 803-822
2. Wouessidjewe, D., Skiba, M., Leroy-Lechat, F., Puisieux, F., Duchene, D., (1996) « A new concept in drug delivery based on skirt-shaped cyclodextrin aggregates. Present state and future prospects », *S.T.P. Pharma. Sci.*, 6(1), 21-28
3. Croft, A.P., Bartsch, R.A., (1983), « Synthesis of chemically modified cyclodextrins » *Tetrahedron,* 39(9), 1417-1474
4. Hirayama, F., Yamanaka, M., Uekama, K., (1995), « Characterization of peracylated β-cyclodextrins with different chain length as a sustained release carrier for water soluble drugs », *Chem. Pharm. Bull.*, 43 (1), 130-136
5. Takeo, K., Mitoh, H, Uemura, K. (1989), « Selective chemical modification of cyclomaltooligosaccharides via tert. butyldimethyl silylation », *Carbohydrate Res.*, 187, 203-221
6. Skiba, M., Wouessidjewe, D., Puisieux, F., Duchene, D., Gulik, A., (1996) «Characterization of amphiphilic β-cyclodextrin nanospheres », *Int. J. Pharm.*, 142, 121-124

THIAMINE-APPENDED CYCLODEXTRIN DIMER AS A LIGASE MODEL

H. IKEDA, Y. HORIMOTO, M. NAKATA, and A. UENO

Department of Bioengineering, Faculty of Bioscience and Biotechnology, Tokyo Institute of Technology, 4259 Nagatsuta-cho, Midori-ku, Yokohama 226-8501, Japan

1. Introduction

Cyclodextrins (CD's) have been extensively studied as enzyme models and as molecular receptors due to their abilities to bind various hydrophobic compounds into their hydrophobic cavities with substrate selectivity [1, 2]. The most successful enzyme model in our studies is imidazole-appended 2,6-dimethyl-β-cyclodextrin [3]. This model caused about 1000-fold acceleration of the hydrolysis of *p*-nitrophenyl acetate, and k_{cat} for this reaction of this model is 2.67×10^{-2} s^{-1}, which is over twice as much as that of α-chymotrypsin. Recently we found an artificial enzyme using a CD dimer has cooperative binding ability and allosteric regulation ability for hydrolysis reactions [4]. As the acyl chain length of the substrate, *p*-nitrophenylalkanoate was longer, the transition state of the hydrolysis reaction was more stabilized by the CD dimer. This new artificial enzyme shows the homotropic allosteric effect with the Hill constant of 1.8 for the hydrolysis reaction of *p*-nitrophenyl methoxyethoxyethoxyacetate as the substrate. So, CD dimers are expected to be also effective building blocks for enzyme models that catalyze synthetic reactions such as a carbon-carbon bond formation reaction. Thiamine pyrophosphate participates as the essential cofactor in numerous enzymatic

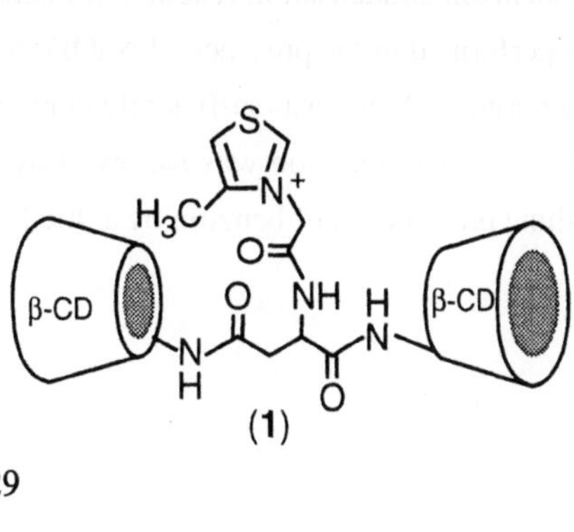

(1)

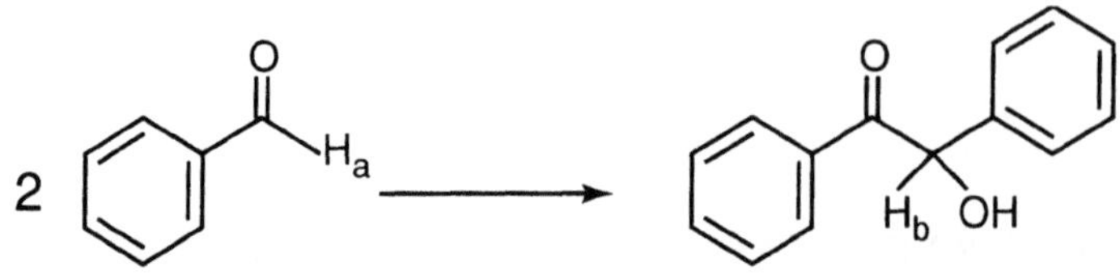

Scheme 1 Benzoin Condensation

reactions involving formation and breaking of carbon-carbon bonds, which are also important for synthetic organic reactions. In this symposium, we present the synthesis and characterization of the thiazolium-appended β-cyclodextrin dimer (**1**) for the benzoin condensation reaction (Scheme 1).

2. Materials and Methods

2.1. MATERIALS

β-Cyclodextrin was kindly gifted by Nihon Shokuhin Kako Co., Ltd. All chemicals were reagent grade and were used without further purification unless otherwise noted. Deuterium oxide and dimethyl sulfoxide-d_6 with an isotopic purity of 99.95%, were purchased from the Merck Co. The synthesis details of the thiazolium-appended β-cyclodextrin dimer will be reported elsewhere.

2.2. METHODS

1D and 2D NMR spectra were recorded in D_2O at 25 °C on Varian VXR-500S and UNITY plus-400 spectrometers operating at 499.843 MHz and 399.973 MHz, respectively for ^1H. The benzoin condensation reaction of benzaldehyde (400 mM) catalyzed by **1** (3.8 mM) was performed in the presence of NaOD (8 mM) in D_2O / DMSO-d_6 (6/4 (v/v)) at 60 °C under argon. Argon was sufficiently bubbled into the reaction solution, before incubation at 60 °C. The reaction was followed by the appearance of the ^1H resonance for the methine proton (H_b) of benzoin by a 400 MHz NMR spectrometer.

3. Results and Discussion

3.1. SYNTHESIS

The thiazolium-appended β-cyclodextrin dimer (**1**) was prepared by the condensation of 6-amino-6-deoxy-β-cyclodextrin and *N-tert*-butyloxycarbonylaspartic acid with dicyclohexylcarbodiimide (DCC) and 1-hydroxybenzotriazole (HOBt) followed by the condensation with 3-(4-methylthiazolio)acetic acid after the deprotection of the amino group of the aspartate unit. The product was purified with CM Sephadex C-25. The product was identified by elemental analysis, some kinds of NMR spectra including 2D NMR, and mass spectrum.

3.2. BENZOIN CONDENSATION

The catalytic activity of the enzyme model (**1**) was studied with measuring the rate for the benzoin condensation reaction (Scheme 1). The benzoin condensation reaction of benzaldehyde (400 mM) catalyzed by **1** (3.8 mM) was performed in the presence of NaOD (8 mM) in D_2O/DMSO-d_6 (6/4 (v/v)) at 60 °C under argon. The reaction was followed by the appearance of the ^1H resonance for the methine proton (H_b) of the product, benzoin by a 400 MHz NMR spectrometer. Figure 1 shows the time course of the ^1H resonance for H_b of benzoin. The yields of the product were quantified from the integral of ^1H NMR spectra. The dimer model (**1**) showed turnover reactivity for the benzoin condensation

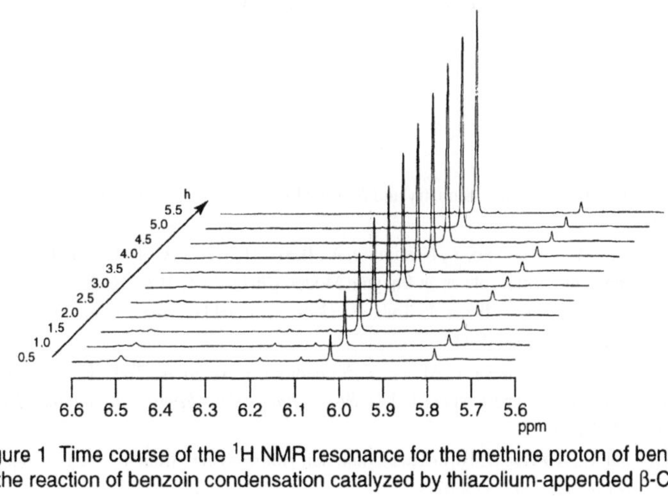

Figure 1 Time course of the ^1H NMR resonance for the methine proton of benzoin in the reaction of benzoin condensation catalyzed by thiazolium-appended β-CD (1)

reaction of benzaldehyde as shown in Figure 2. It is reported that only one cavity of β-CD is not enough for the benzoin condensation reaction [5]. Our results suggest that the β-CD dimer is efficient for the condensation reaction of two substrates. The kinetic date can be fitted by the first order rate treatment in the benzaldehyde concentration, and the first order rate constant is 0.264 h^{-1}.

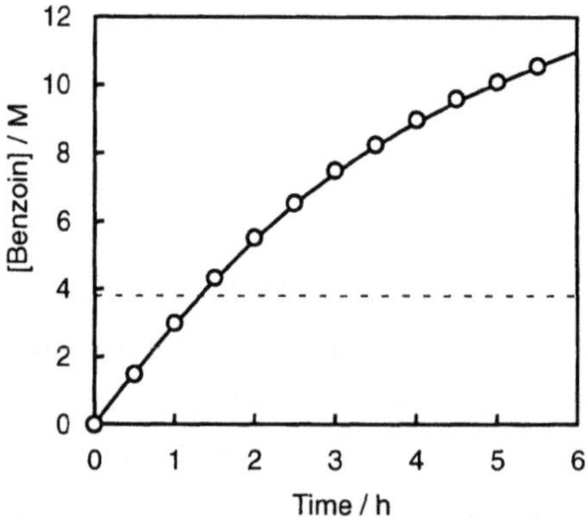

Figure 2. Benzoin condensation catalyzed by thiazolium-appended β-cyclodextrin dimer (1); [cat] = 3.8 mM, [benzaldehyde]= 400 mM, [NaOD] = 8 mM in DMSO-d_6 / D_2O (v/v) = 6 / 4, at 60 °C.

4. Conclusion

The thiazolium-appended β-CD dimer was synthesized as a ligase model. This model effectively catalyzed the benzoin condensation reaction of benzaldehyde.

Acknowledgments

We are grateful to Nihon Shokuhin Kako Co. LTD. for providing us with cyclodextrins

References

1. Breslow, R. (1995) Biomimetic Chemistry and Artificial Enzymes: Catalysis by Design, *Acc. Chem. Res.*, *28*, 146-153.
2. Akiike, T., Nagano, Y., Yamamoto, Y., Nakamura, A., Ikeda, H., Ueno, A., Toda, F. (1994) Coupled Cyclodextrin Appending Imidazole as an Enzyme Model, *Chem. Lett.*, 1089-1092.
3. Ikeda, H., Kojin, R., Yoon, C-j., Ikeda, T., Toda, F. (1989) Artificial Hydrolase Using Modified Dimethyl-β-Cyclodextrin, *J. Incl. Phenom.*, *7*, 117.
4. Ikeda, H., Nishikawa, S., Takaoka, J., Akiike, Y., Yamamoto, Y., Ueno, A., Toda, F., (1996) Cyclodextrin Homo- and Hetero-Dimers as Enzyme Models, *Proceedings of the Eighth International Symposium on Cyclodextrins*, 277-280.
5. Breslow, R., Kool, E., A γ-Cyclodextrin Thiazolium Salt Holoenzyme Mimic for the Benzoin Condensation (1988) *Tetrahedron Lett.*, *29*, 1635-1638.

SUBSTITUTED CYCLODEXTRINS WITH AMINO ACIDS AS HYDROLYSIS ENZYME MODELS

Woo-Sang Lee, Hisakazu Mihara and Akihiko Ueno
Department of Bioengineering, Faculty of Bioscience and Biotechnology, Tokyo Institute of Technology, 4259 Nagatsuta-cho, Midori-ku, Yokohama, 226-8501, Japan

Abstract
Modified cyclodextrins bearing a 4-residue peptide, Asp-Gly-Pro-His, were synthesized as hydrolysis enzyme models. N-terminal of peptide part was substituted with acetyl in the β–CyD and 2-naphthalene acetyl in the γ–CyD. Substituted naphthalene moiety play role as a spacer in the γ–CyD. The activity of synthesized enzymes was measured with *p*-nitrophenyl ester, Boc-L, D-alanine *p*-nitrophenyl esters as substrates.

Keywords: Enzyme modeling, Ester hydrolysis, Modified cyclodextrins, Naphthalene

1. Introduction

Cyclodexrins (CyDs) have property to bind substrates into their cavity utilizing hydrophobic interaction in aqueous solution. This property has been used to construct enzyme models. α- and β-CyDs have been widely utilized for various purposes because of their remarkable property of forming inclusion complexes with a variety of small molecules. However, γ-CyD has not been frequently used as enzyme model due to its large cavity size.

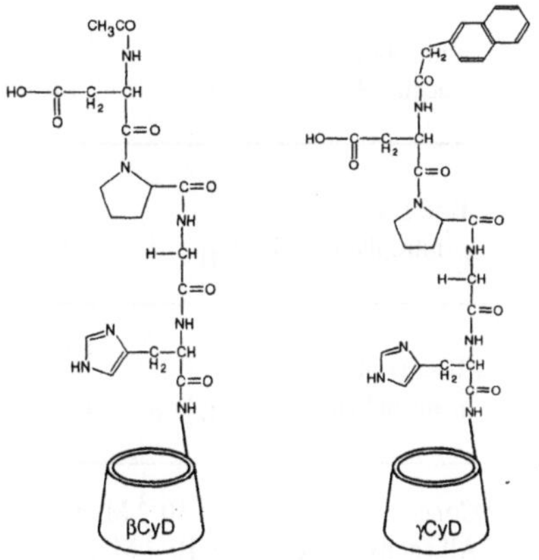

I II

Fig.1 The structures of synthetic enzymes

We have prepared β-CyD and γ-CyD derivatives I and II, both containing 4-residue peptide (asp-pro-gly-his), with acetyl and with 2-naphthalene acetyl, respectively, at their N-terminals. The peptide residues were expected to take a turn structure. In order to fix this turn structure, we introduced naphthalene moiety in the γ-CyD derivative as a spacer so as to bind guest species tightly by narrowing the γ-CyD cavity. This complexation behavior may be regarded as an induced-fit type of complexation .

2. Material and Methods

Compound I and II were synthesized by coupling of β-and γ-CyD-6-NH_2 and the peptides. The peptide were synthesized by Fmoc method using wang resin, and then purified by reverse phase HPLC and identified by FAB-MS and ^{1}H-NMR.

Enzyme activity was investigated in 200 mM phosphate buffer (pH7.2) at 25°C. The concentration of I and II was 2.5 x 10^{-5} M and the absorbance at 400 nm was monitored.

3. Results and Discussion

The ratio of the catalyzed and anticatalyst reaction rates of I to 295,145 and 36 for p-

Table 1 Hydrolysis of Esters Catalyzed by Modified Cyclodextrins at 25°C

Substrate	Catalyst[c]	Kun(s^{-1}) x 10^{-5}	Kcat(s^{-1}) x 10^{-3}	Km(s^{-1}) x 10^{-3}	Kcat/Km	Krel[d]
p-nitrophenly acetate[a]	I	1.08	3.19	16.3	0.2	295
	II	1.08	0.68	2.7	0.25	63
Boc-L-ala-*p*-nitrophenyl ester[b]	I	2.97	4.3	4.2	1.03	145
	II	2.82	0.87	1.78	0.49	31
Boc-D-ala-*p*-nitrophenyl ester[b]	I	3.46	1.24	1.54	0.81	36
	II	3.26	0.50	1.2	0.41	15

[a] Concentration was 2.5 x 10^{-3} M in acetonitrile. [b] Concentration was 5.0 x 10^{-4} M in acetonitrile.
[c] All catalysts were 2.5 x 10^{-5} M in 200 mM phosphate buffer, pH 7.2.
[d] Krel = Kcat/Kun.

nitrophenyl acetate, Boc-ala(L)-p-nitrophenyl ester and Boc-ala(L)-p-nitrophenyl ester, respectively. On the other hand, the catalytic enhancement of II was 63 , 31 and 15 for p-nitrophenyl acetate, Boc-ala(L)-p-nitrophenyl ester and Boc-ala(L)-p-nitrophenyl ester. Kinetic parameters are shown in Table 1.

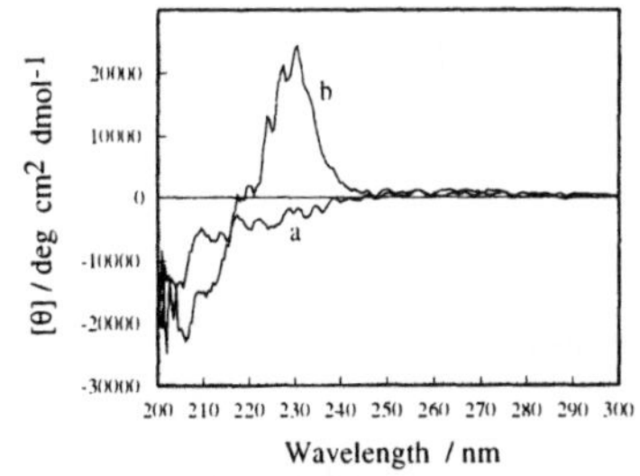

Fig.2 Circular Dichroism of AcDPGH-βCyD(a) and NpDPGH-γCyD(b) in 200 mM phosphate buffer pH 7.2. Concentration of synthetic enzyme were 2.5 x 10^{-5}M.

Compound I was revealed *ca*. 5 fold higher K_{rel} value for L-form of Boc-ala-*p*-nitrophenyl ester than that of D-form. As shown in Table 1, substrate binding of the II slightly increased in comparison with that of I. This result suggests that naphthalene ring is included together with substrate in the cavity of γ–CyD. That is, II undergoes induced-fit type complexation. But, the reaction rate (Kcat) of II are smaller than those of I. Thus enhancement of reaction (K_{rel}) for II is smaller. This results suggests that introduced naphthalene ring of II increase

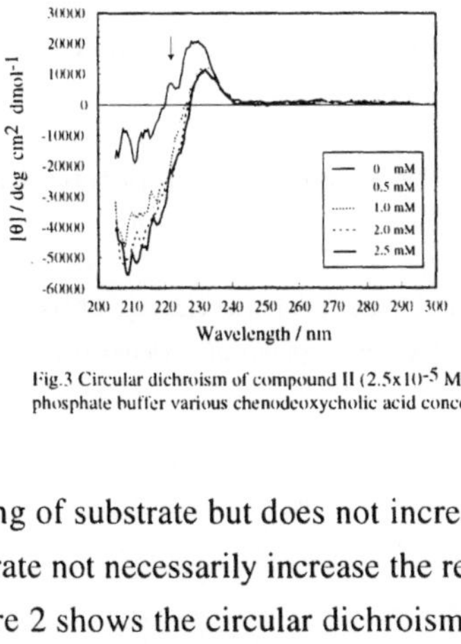

Fig.3 Circular dichroism of compound II (2.5x10^{-5} M) in phosphate buffer various chenodeoxycholic acid concentrations

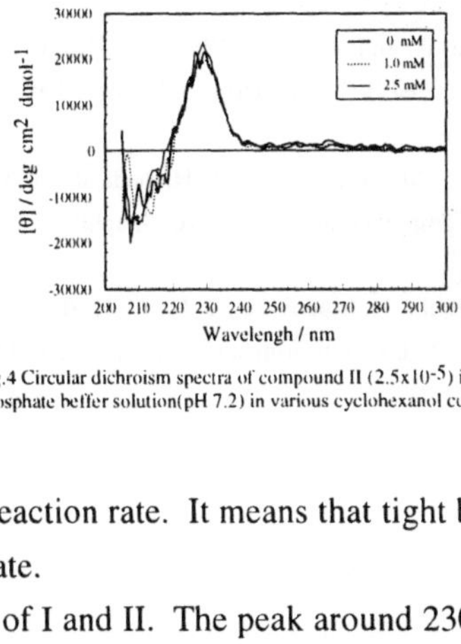

Fig.4 Circular dichroism spectra of compound II (2.5x10^{-5}) in phosphate beffer solution(pH 7.2) in various cyclohexanol concentration.

binding of substrate but does not increase the reaction rate. It means that tight binding of substrate not necessarily increase the reaction rate.

Figure 2 shows the circular dichroism spectra of I and II. The peak around 230 nm of II appears in the ^1B_b transition region of naphthalene ring. From the circular dichroism, we can not obtain any information about the structure of the peptide part. The intensity of positive peak of ^1B_b transition of naphthalene ring diminishes with increasing guest concentration. Figure 3 shows the variation of the circular dichroism pattern of II in the case of chenodeoxycholic acid as a guest. This result suggests that naphthalene ring is excluded from the cavity in complexation with the guest. But, when the small size guest (cyclohexanol) was used, almost no change was induced upon addition of cyclohexanol in the circular dichroism spectra (Figure 4). This result indicates that the location of the naphthalene ring of II is not affected by cyclohexanol. That is, naphthalene ring may contribute to the catalytic reaction as a spacer in the cavity of II.

4. Conclusion

Synthesized enzyme models, compounds I and II enhance the catalytic activity of ester hydrolysis reaction.

The naphthalene ring in II, which was introduced in order to fix structure of peptide part strengthened substrate binding in ester hydrolysis reaction but not increased reaction rate. Substituted naphthalene ring was bound together with substrate in the cavity of γ–CyD. Naphthalene ring play a role as a spacer in the II.

Reference

[1]. K. Hamasaki and A. Ueno (1995) Significant Enanthioselectivity in Alanine Ester Hydolysis Catalyzed by Imidazole attached β-Cyclodextrins. *Chemistry Letters* 859-860

[2]. A. Ueno, F. Moriwaki, T. Osa, F. Hamada and K. Murai (1986) Fluorescence and Circular Dichoism Studies on Host-guest Complexation Bearing two 2-naphthyl moieties, *Bull. Chem. Soc. Jpn.*, 59, 465-470

[3]. K. Hamasaki, H. Mochizuki, T. Ikeda, H. Ikeda, A. Ueno and F. Toda (1994) The cleavage of aryl esters by modified γ-cyclodextrin bearing a histamine moiety at C3 position, *The 7th International cyclodextrins symposium* 342-345

SYNTHESIS AND MOLECULAR RECOGNITION ABILITIES OF MODIFIED γ-CYCLODEXTRINS BEARING TWO PYRENE MOIETIES.

T. Aoyagi, H. Ikeda and A. Ueno
Tokyo Institute of Technology. Faculty of Bioscience and Biotechnology. Department of Bioengineering. 4259 Nagatsuta-cho, Midori-ku, Yokohama, 226-8501, Japan

Abstract
In this study, a series of bis-pyrene modified γ-cyclodextrins ($Py_2\gamma$-CDs) were synthesized as new chemosensors for molecules. Because γ-CD has eight glucose units, we have four isomers of $Py_2\gamma$-CDs which are different only in the mutual position of pyrene-modified glucose units. We measured fluorescence, absorption and induced circular dichroism (ICD) spectra under the conditions of various guest concentrations and investigated guest induced conformational changes and the molecular recognition abilities of $Py_2\gamma$-CDs.

1.Introduction

Recently, many chromophore-modified hosts have been prepared as chemosensors for detecting ions or molecules[1]. Among many of these systems, spectroscopic variations occurring associated with guest binding have been utilized for the sensing purpose. We have prepared a variety of fluorescent CDs, which have one or two fluorophore units, as fluorescent chemosensors for detecting molecules[2]. Pyrene is well known to exhibit excimer emission and used in some of chemosensors for detecting metal cations or molecules. In particular, because γ-CD has a large cavity enough to include two pyrene units, γ-CD derivatives having one or two pyrene moieties have been shown to be very effective as such sensor systems. In the course of constructing fluorescent chemosensors, we have recently prepared novel γ-CD derivatives **1-4** (Chart I). These derivatives have almost the same structure with two pyrene units linked to the primary hydroxyl group. However only the position, to which pyrene moieties are linked, are different each other. Because γ-CD consists of eight glucose units, there are four isomers (AB, AC, AD and AE). Each have two pyrene moieties and one methylene unit as a flexible linker. Their abilities as fluorescent chemosensors and guest induced conformational changes are described in this paper.

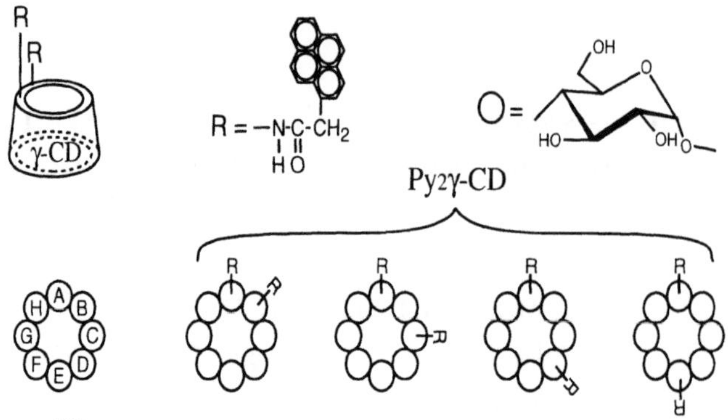

γ-CD 1(AB) 2(AC) 3(AD) 4(AE)

Chart I

2. **Experimental**

Each of **1-4** was prepared by reaction of the corresponding diamino-γ-CD[3] with 1-pyreneacetic acid with dicyclohexylcarbodiimide in DMF. The purification was performed by HPLC and identified by ^{1}H-NMR and TOFMS spectra. The hosts concentrations were 0.15 and 20 μM to measure fluorescence and ICD spectra, respectively. The binding constants of **1-4** for several guests were estimated by monitoring the guest-induced changes of the excimer emission intensities (490 nm) using a nonliner least squares fitting procedure.

3. **Result and Discussion**

3.1 CONFORMATION ANALYSIS OF **1-4**

Fig.1 shows a fluorescence spectrum of **3** in a 20% DMSO aqueous solution. **3** shows a small normal emission and an extremely large broad excimer emission. It is a very natural result because two pyrene units exist very near in the same CD, and when these two hydrophobic chromophores are included in the cavity they may take a face-to-face position. **1**,**2** and **4** also show almost the same spectra as **3**. Fig.2 shows a ICD spectrum of **3**, and we could observe cotton ef-

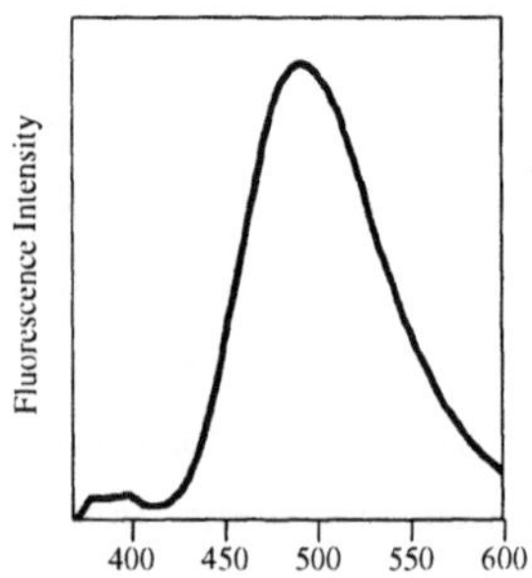

Fig.1 The fluorescence spectrum of 3 (0.15 μM) in a 20% DMSO aqueous solution at 25°C. Ex.=344 nm

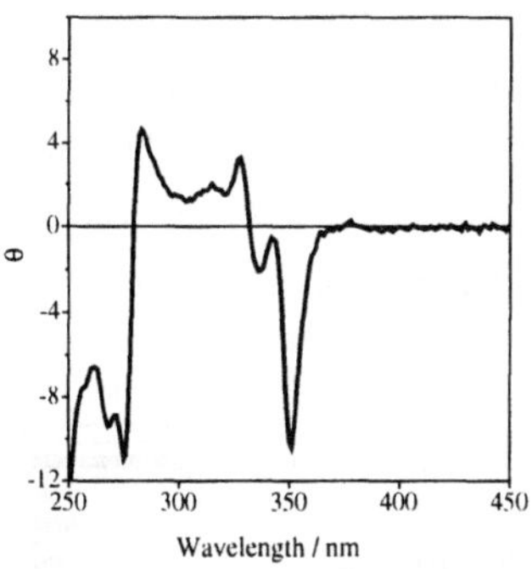

Fig.2 The fluorescence spectrum of **3** (20 μM) in a 20% DMSO aqueous solution at 25°C.

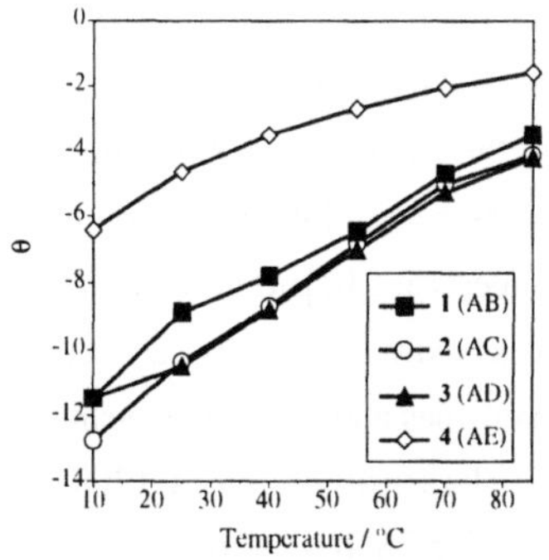

Fig.3 The temperature dependence of the θ values at 351 nm of **1-4**.

fects in the pyrene absorption regions. To investigate the conformation of **1-4**, we measured the ICD spectra at various temperatures and the intensities of cotton effects of **1-4** at 351 nm were plotted to the temperature at Fig.3. When the temperature is low, because the mobilities of two pyrene units are very mild, the modified chromophores are inside the cavity and the restricted conformation shows the strong ICD. On the other hand, when the temperature is high, the mobilities of pyrene moieties are strong enough to be excluded from the CD cavity, and the weak ICD is observed owing to the free movement of chromophores. We also considered the effect of the organic solvent on ICD spectra, and the result were almost same as that of temperature experiments. Increasing the percentage of organic solvent causes the decreasing of the cotton effects because in the organic solvent the hydrophobic cavity becomes null.

3.2 FLUORESCENT SENSOR USING EXCIMER EMISSION INTENSITIES.

As described above **1-4** shows remarkably strong excimer emission and the guest induced conformational changes. Fig.4 shows the fluorescence spectra of **3** without and with various concentrations of lithocholic acid in a 20% DMSO aqueous solution at 25°C. The peak of excimer emission around 490 nm decreases with adding the guest, while the normal emission peaks around 380 and 400 nm increase, which means that the formation is

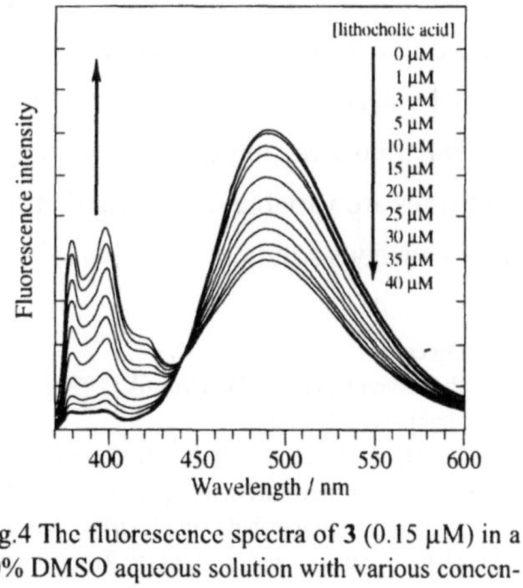

Fig.4 The fluorescence spectra of **3** (0.15 μM) in a 20% DMSO aqueous solution with various concentrations of lithocholic acid at 25°C. Ex.=344 nm.

depressed upon guest addition and two pyrenes are separated by the guest included. The value of $\Delta I/I^0$ was used as the measure of the sensitivity of **1-4** and, where $\Delta I=I^0-I$, and I^0 and I are the excimer emission intensities in the absence and presence of guests, respectively, measured at 490 nm. The sensitivities are summarized in Fig.5. The concentrations of guests are 10 μM and 100 μM for cholic acid derivatives and others, respectively. The values below zero means the decrease of the intensities and the above means the increase. The sensitivities for C12OH are larger than for C8OH, indicating that the size of cavity is good for C12OH and C8OH is too small for the cavity. Unfortunately there are not so large differences in the sensitivities between two borneol isomers. When five biologically important steroidal compounds were used as guests, marked differences were observed in spite of the similar structure they have. For all isomers, LCA is most sensitive, while CA is less sensitive. Only **1** show the increase of the excimer emission intensities with some steroids. This observation suggests that the polality of the guest molecules is an important factor for the sensitivities.

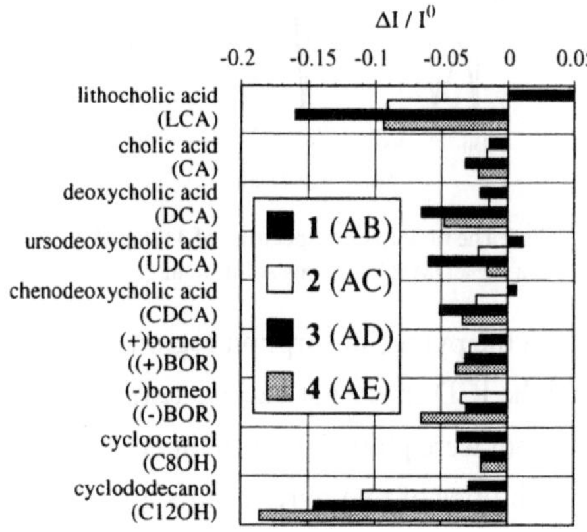

Fig.5 The sensitivity parameters ($\Delta I/I^0$) of **1-4** for various guests.

4.**Conclusion**

Two pyrene units existing on the same CD framework form an excimer inside the cavity and exhibit strong excimer emission. These chromophores are excluded from the CD cavity by heating or adding organic solutions. Excimer formation is depressed not only by temperature raising but also by guest addition. The effects to the excimer emission intensities are variable indicating that these systems are useful in detecting organic species in aqueous solution.

References

[1] J. P. Desvergne and A. W. Czarnik (1977): *Chemosensors of Ion and Molecule Recognition*, Kluew Academic Publishers, Dorddrecht.

[2]H. Ikeda, M. Nakamura, N. Ise, N. Oguma, A. Nakamura, T. Ikeda, F. Toda,A. Ueno (1996) Fluorescence cyclodextrins for molecule sensing: Fluorescent Properties, NMR characterization, and inclusion phenomena of *N*-dansylleucine-modified cyclodextrins, *J. Am. Chem. Soc.*, 118, 10980-10988.

[3]A. Ueno, I. Suzuki, and T. Osa (1989) Associatiion dimers, excimers, and inclusion complexes of pyrene-appended γ-cyclodextrins, *J. Am. Chem. Soc.*, 111, 6391-6397.

MOLECULAR "LEGO" COMPOSED OF MONOSUBSTITUTED CYCLODEXTRINS

KEIKO TAKAHASHI*, MASAHIKO KITSUTA AND KOUICHI IMOTANI, *Department of Industrial Chemistry, Faculty of Engineering, Tokyo Institute of Polytechnics, 1583 Iiyama Atsugi, Kanagawa 243-0239 JAPAN*

1.Introduction

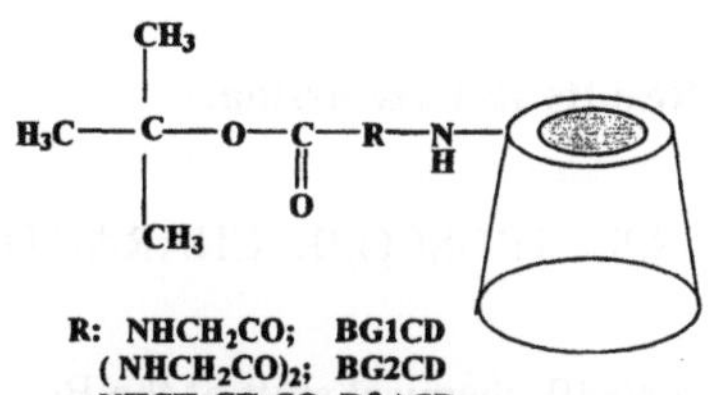

Figure 1. Structure of the modified CDs

The development of supramolecular chemistry involves more complete control over molecules and supramolecules. To compose a more organized system using cyclodextrins(CD), "flexible" modified CDs are more suitable than "rigid" modified CDs. We have already prepared some "flexible" modified CDs which have an aromatic amino acid and observed the location change of the modified group by 1D and 2D NMR measurements[1]. The formation and release of the self-inclusion complex should reflect the cooperative interactions around the hydrophobic cavity, Van der Waals force, hydrogen bonding, *etc.*; therefore, it is important to clarify the stability of the self-inclusion complex with various modified CDs and their molecular recognition ability. In order to quantitatively discuss the above interactions, we have synthesized dozens of CDs having phenyl groups in a carbobenzoxy group(Z) and determined the correlation between the structure of the arm and molecular recognition[2]. When an aliphatic hydrophobic group, such as the *tert*-butoxycarbony group (Boc group) instead of the aromatic group was introduced to the CD, the "self- guest" group is not included in its cavity, rather than in the other CD cavity. We now report a new molecular assembly like "Lego" due to the formation of an intermolecular inclusion complex.

2.Experimental

2.1 PREPARATION OF MODIFIED CDS.

N'-tert-butoxycarbonylamino acid-binding CDs(BCD; Fig. 1), such as *6N-(N'-tert*-butoxycarbonyglycyl)amino-6-deoxy-β-CD(BG1CD), *6N-(N'-tert*-butoxycarbonylglycylglycyl) amino-6-deoxy-β-CD(BG2CD), *6N-(N'-tert*-butoxycarbonyl-β-alanyl)amino-6-deoxy-β-CD (BβA1CD), *6N-(N'-tert*-butoxycarbonylphenylalanyl)amino-6-deoxy-β-CD(BPCD) and *6N-(N'-tert*-butoxycarbonylphenylalanylglycyl)amino-6-deoxy-β-CD(BPGCD) were prepared from 6-amino-6-deoxy-β-CD and *N-tert*-butoxycarbonylamino acid in DMF with *N,N*-dicyclohexylcarbodiimide(DCC) according to the same previously reported method [1].

2.2 MEASUREMENTS.

NMR spectra were taken using a JEOL Lambda 500 spectrometer in D_2O with acetone as the internal reference(2.100 ppm) at room temperature. The association constants(K) between the modified CD and β-CD were estimated using the NMR titration method[3].

3. Results and discussion.

3.1 SPECTROSCOPIC CHARACTERIZATION OF BCD SERIES.

The NMR chemical shift of the Boc groups of all the BCDs was observed at a lower field than that without the CD groups. The C1 proton was observed in a single shift, which is different from the previously reported "self-inclusion" complex type modified CD. Other CD resonances of BG1CD and BPCD were observed as broad peaks. The low-field shift was released by the addition of cyclohexanol. The broaden resonance of BG1CD and BPCD also narrowed in the presence of cyclohexanol. The induced circular dichroism spectra of BG1CD, BG2CD and BβACD due to the amide bonds were observed around 215 nm and also disappeared by adding cyclohexanol. These results suggest that the Boc groups of BCD are included in the CD cavity.

3.2 BIMODAL COMPLEX FORMATION BETWEEN BCD AND β-CD.

The modified groups at the CD stay inside and outside the cavity to maintain the equilibrium state. If the equilibrium leans to the outside type, the modified group should be included in the other CD(Scheme 1). Fig. 2 shows that the chemical shift of the C3 proton on the CD ring depended on the concentration of the modified CD. In the absence of BCDs, chemical shifts of C3 proton were observed at 3.840 ppm. The addition of BG1CD shifted the C3 proton to saturate at a high concentration as far as 3.82 ppm. When BPCD or BβACD was added to the β-CD solution, a high-field shift was also observed. On the basis of the NMR titration method assuming a 1:1 complex formation, the association constants β-CD and BG1CD, BPCD and BβACD were calculated as 82, 98 and 8 M^{-1}, respectively. The chemical shift due to the C5 proton, which locates near the narrow side of cavity, slightly changed. When BCDs were added to the α-CD solution instead of the β-CD, no change was observed. This suggested that the Boc groups of BG1CD and BPCD are included in the wider rim of the cavity of β-CD. The broaden resonance of the BG1CD and the BPCD narrowed in the presence of β-CD. Moreover, the water solubilities of both BG1CD and BPCD increased in the presence of β-CD. These results suggest that the Boc group of BG1CD is included in the other BG1CD like "Lego blocks", and BPCD also forms a similar complex. The Boc group at the BβACD is included in its own cavity and forms an intramolecular complex.

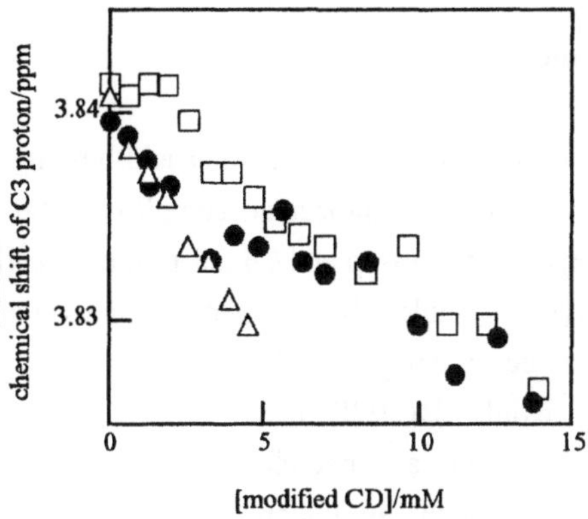

Figure 2. Dependence of the chemical shift of the C3 proton on the β-CD ring on the concentration of BG1CD(●), BβACD(□) and BPCD(△).

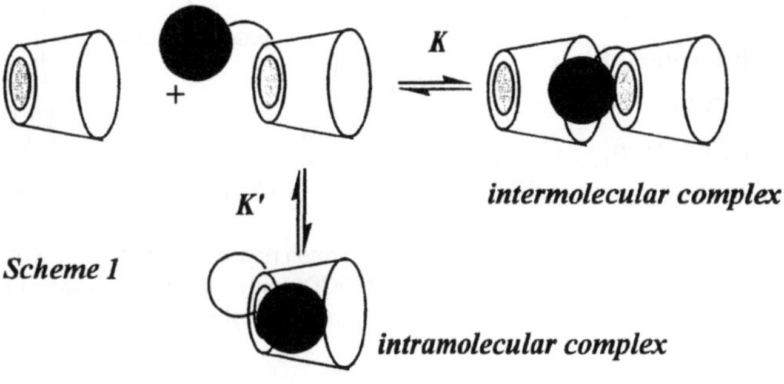

Scheme 1

3.3 ENVIRONMENT CONSISTING OF MOLECULAR LEGOS

The location and environment of the phenyl group could be easily determined by using the characteristic chemical shifts of the phenyl protons[1]. As shown in Fig. 3, the phenyl group of BPCD is affected by the CD cavity while that of the BPGCD is not. When adding cyclohexanol, the phenyl proton resonance changed. The fluorescence intensity of BPCD in an aqueous solution was stronger than that of the solution that contained the same concentration of chromophore. The addition of cyclohexanol to the solution decreased the intensity. The edge effect, which is known as the evidence that the chromophore stays around the "edge" between the hydrophobic and hydrophilic areas, was not observed. The phenyl group of BPCD should be embeded in the hydrophobic environment. The hydrophobic environment seem to be composed of many CDs.

4. Conclusion

N'-tert-butoxycarbonylmonoamino acid-binding β-CDs, such as BG1CD and BPCD, prefer to form an "intermolecular complex". The length of an arm between the *N'-tert*-butoxycarbonyl group and C6 position on the glucose ring was as long as -NH-C-CO-NH- which was suitable. The modified CDs having a longer arm form intramolecular rather than intermolecular complexes. In other words, BG1CD and BPCD form a "pseudo polymer" with inclusion phenomenon.

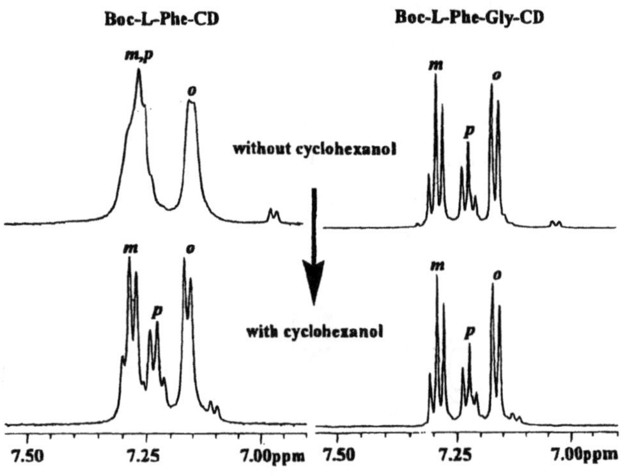

Figure 3. Phenyl proton region of 500 MHz ^{1}H-NMR spectra of BPCD(left) and BPGCD(right) in the absence(upper) and presenceof cyclohexanol(bottom) in D_2O.

Acknowledgment

The present work was supported by a Grant-in-Aid No.09680579 from the Ministry of Education, Science, Sport and Culture.

References

1.(a)Takahashi,K.; Ohtsuka,Y.;Hattori,K., *Chem.Lett.*, **1990**, 2227 (b)Saka,W.; Inoue,Y.;Yamamoto,Y.; Chûjô,R.; Takahashi,K.; Hattori,K., *Bull.Chem.Soc.Jpn.*,**1990**, *63*, 3175 (c Takahashi,K., *Bull.Chem.Soc.Jpn.*, **1993**, *66*, 550 (d) Takahashi,K.,*Polymer J.*,**1996**,*28*, 458.

2.Takahashi,K.(1997), *Proceedings of 15th CD Symposium in Japan*

3.Wilcox,C.S. (1991) Design, Synthesis, and Evaluation of an Efficancious Functional Group Dyad. Methods and Limitations in the Use of NMR for Measuring Host-Guest Interactios, in Schneider,H-J and Dürr,H. (Ed.), *Frontiers in Supramolecular Organic Chemistry and Photochemistry*,VCH,Weinheim pp123-143.

STRUCTURAL ANALYSES OF SUGAR BRANCHED CYCLODEXTRIN DERIVATIVES BY THE POST-SOURCE DECAY FRAGMENT METHOD IN MALDI-TOF MASS SPECTROMETRY

TOHRU YAMAGAKI,[a,b] YASUKO ISHIZUKA,[a] SHIN-ICHIROU KAWABATA,[c] AND HIROSHI NAKANISHI[a*]

[a]*National Institute of Bioscience and Human-Technology, 1-1, Higashi, Tsukuba, Ibaraki, 305-8566, Japan*
[b]*Department of Biochemistry, Saitama University, Urawa, Saitama, 338, Japan*
[c]*Shimadzu Corporation, 380-1, Horiyamashita, Hadano, Kanagawa, 259-13, Japan*

Post-source decay fragment analyses by MALDI-TOFMS were applied to the sugar-branched β-cyclodextrin (CD) derivatives. In the fragment spectra, all possible fragment ions larger than trisaccharide produced by the cleavage of glycosidic linkage were clearly observed. The detailed ion intensity analyses enable us to distinguish the ions produced from branched and CD moieties because the ion intensities were influenced by the structure. The relative intensities of the ions produced from β-CD moiety have the characteristically cleavaged pattern which was common in the branched β-CDs.

Post-source decay fragment analysis by matrix-assisted laser desorption/ionization time-of-flight mass spectrometry (MALDI-PSD fragment analysis) provides sequential information in biomolecules[1-5] such as peptide, nucleic acid, and oligosaccharide. In neutral branched oligosaccharides such as xyloglucan, the structural analyses were already performed by the MALDI-PSD fragment method.[6,7] The branched cyclodextrins (CDs) with the same mass number were distinguished by the MALDI-PSD fragment analysis in our previous reports.[8,9] In this paper, we study how to relate the intensities of the fragment ions to the length of branches in CD derivatives.

Cyclomaltoheptasaccharide (β-CD; **1**), 6-O-glucosyl-β-CD (G1-β-CD; **2**), and 6-O-maltosyl-β-CD (G2-β-CD, **3**) consist of α-D-glucopyranose residues (see structure) and were commercially available.

MALDI-TOFMS measurements were made by using a KOMPACT MALDI IV mass spectrometer (Shimadzu Corporation, Japan). Operation conditions were as follows: nitrogen laser (337 nm), acceleration energy of 20 kV, reflectron mode, positive-ion detected. In MALDI-PSD fragment measurements, the laser power was adjusted to about 40 mJ. 2,5-Dihydroxybenzonic acid (DHBA) was used as the matrix.[10,11] 0.5% NaCl solution was added to assist the formation of sodium adduct.

All MALDI-PSD fragment ions were observed as a sodium adduct.

In the MALDI-PSD fragment spectrum of **1**, four fragment ions (A through D) were observed with the 162 amu interval corresponding to the loss of anhydroglucose (Fig. 1-a). These results indicate that the cleavage occurs at the glycosidic linkage in the MALDI-PSD fragmentation. The relative intensity of ion B was higher than ion A, and that of ion C was higher than that of ion D. These trends in the fragment spectrum patterns were highly reproduced.

In the spectrum of **2** (Fig. 1-b), five fragment ions were observed.

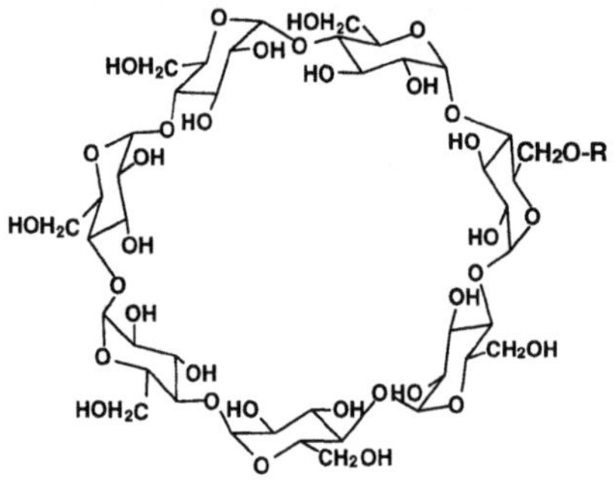

1: β-cyclodextrin (β-CD)
R =-H

2: Glucosyl-β-CD (G1-β-CD)
R =

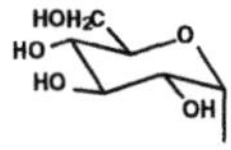

3: Maltosyl-β-CD (G2-β-CD)
R =

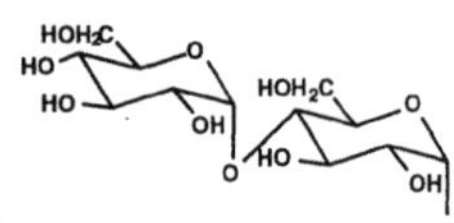

Each ion interval was 162 amu, and ions A' through D' in **2** had the same mass number in **1**. Ion E produced by the loss of one anhydroglucose had a much higher intensity than ions A' through D'. Comparing the fragment spectra between **1** and **2**, the relative intensity pattern among ions A' through D' was almost common.

In the case of **3** (Figure 1-c), six fragment ions were observed. Each ion interval was also 162 amu, and four ions (A" through D") in **3** had the same mass number like in **1**. Ions E' and F had higher intensity than ions A" through D", which were produced by the loss of one and two anhydroglucose, respectively. The relative intensity of ion F was higher than that of ion E. Comparing the fragment spectra between **2** and **3**, the relative intensity pattern among ions A" through D" was common. These results suggested that ions A', A" through D', and D" were produced from the β-CD moiety as structurally common moiety in **2** and **3**. Therefore, it can be reasonably considered that ion E in **2** was produced by the loss of the branched glucose residue. In **3**, ions E' and F were produced by the loss of the branched glucosyl and maltosyl group. These results indicate that the relative intensity of the fragment ions produced from the branch was much higher than that from the CD moiety.

4: GUG-β-CD

R =

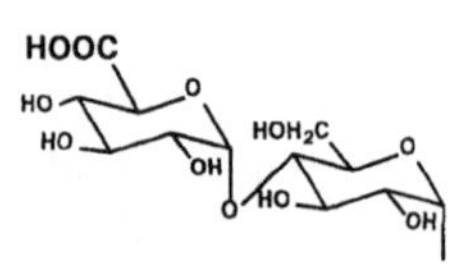

Above described results were confirmed by the same analysis of GUG-β-CD which has a glucuronic acid residue at the branch as shown in **4**. In the PSD fragment spectrum of GUG-β-CD, the fragment ions produced from the branch were distinguished from those of the CD moiety due to the different mass number of the glucuronic acid.

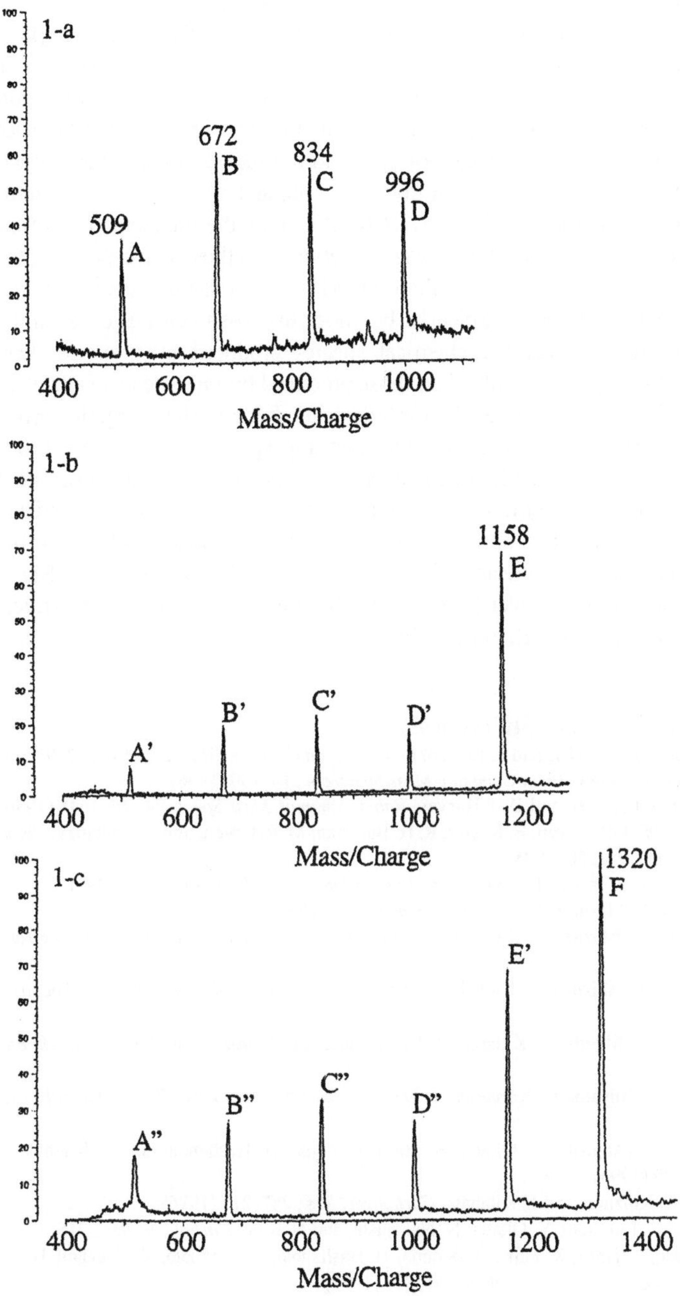

Figure 1. The MALDI-PSD fragment spectra of **1** (1-a), **2** (1-b), and **3** (1-c).

These detailed results will be soon published somewhere.

Ions E, E', and F in the PSD fragment spectra of **2** and **3** were produced by one-site cleavage of glycosidic linkage and their relative intensities were higher than those of ions A', A" through D', and D". Ions A', A" through D', and D" were produced by two-site cleavage at least, and their relative intensities were lower than those of ions E, E', and F. Thus, one-site cleavage of glycosidic linkage occurs with much higher probability than two-site cleavage at glycosidic linkage in MALDI-PSD fragmentation.

In the previous studies of FAB-MS,[12-17] only the fragment ions corresponding to ion E in **2** and ions E' and F in **3** were observed (not those corresponding to ions A', A" through D', and D"). Thus, in the FAB-MS spectra, the fragment ions produced by one-site cleavage were observed clearly, but those of two-site cleavage were not detected. These results suggest that the chemical species produced by two-site cleavage has considerably lower producibility than those produced by one-site cleavage of glycosidic bonds in FAB-MS spectrometry for cyclodextrins, because the energetic barrier of two-site cleavage was so high that the corresponding species was scarcely produced. In contrast, in the MALDI-PSD fragment analyses of the branched CDs, the fragment information about the CD moiety was obtained as well as those of branched sugar chain moiety, even requiring two-site cleavage at glycosidic linkage. These similar results were confirmed in xyloglucan oligosaccharides.[6,7] Thus, MALDI-PSD fragment analyses are advantageous over FAB-MS for the sugar-branched CD derivatives, and the method gave us detailed structural information.

References

1 R. Kaufmann, *J. Biotechnol.*, **41**, 155 (1995).
2 R. Kaufmann, B. Spengler, and F. Lutzenkirchen, *Rapid Commun. Mass Spectrom.*, **7**, 902 (1993).
3 G. Talbo and M. Mann, *Rapid Commun. Mass Spectrom.*, **10**, 100 (1996).
4 B. Kuster, T. J. P. Naven and D. J. Harvey, *Rapid Commun. Mass Spectrom.*, **10**, 1645 (1996).
5 D. J. Harvey, T. J. P. Naven, B. Kuster, R. H. Bateman, M. R. Green, and G. Critchley, *Rapid Commun. Mass Spectrom.*, **9**, 1556 (1995).
6 T. Yamagaki, Y. Mitsuishi, H. Nakanishi, *Biosci. Biotech. Biochem.*, **61**, 1411 (1997).
7 T. Yamagaki, Y. Mitsuishi, H. Nakanishi, *Chem. Lett.*, **1998**, 57.
8 T. Yamagaki, Y. Ishizuka, S. Kawabata, and H. Nakanishi, *Rapid Commun. Mass Spectrom.*, **10**, 1887 (1996).
9 T. Yamagaki, Y. Ishizuka, S. Kawabata, and H. Nakanishi, *Rapid Commun. Mass Spectrom.*, **11**, 527 (1997).
10 T. Yamagaki, M. Maeda, K. Kanazawa, Y. Ishizuka, and H. Nakanishi, *Biosci. Biotech. Biochem.*, **60**, 1222 (1996).
11 Y. Kimura, S. Kajiyama, J. Kanaeda, T. Izukawa, and M. Yonekura, *Biosci. Biotech. Biochem.*, **60**, 2099 (1996).
12 K. Koizumi, T. Tanimoto, Y. Okada, K. Hara, K. Fujita, H. Hashimoto, and S. Kitahata, *Carbohydr. Res.*, **278**, 129 (1995).
13 Y. Okada, K. Koizumi, and S. Kitahata, *Carbohydr. Res.*, **287**, 213 (1996).
14 K. Koizumi, T. Tanimoto, Y. Okada, N. Nakanishi, and N. Kato, *Carbohydr. Res.*, **215**, 127 (1991).
15 K. Hamayasu, K. Hara, K. Fujita, Y. Kondo, H. Hashimoto, T. Tanimoto, K. Koizumi, H. Nakano, and S. Kitahata, *Biosci. Biotech. Biochem.*, **61**, 825 (1997).
16 K. Hara, K. Fujita, N. Kuwahara, T. Tanimoto, H. Hashimoto, K. Koizumi, and S. Kitahata, *Biosci. Biotech. Biochem.*, **58**, 652 (1994).
17 T. Oguma, T. Kurokawa, T. Nagata, M. O-Kameyama, S. Kitao, K. Tobe, and M. Kobayashi, *J. Appl. Glycosci.*, **44**, 37 (1997).

THREE DIMENSIONAL STRUCTURE OF BRANCHED α-CYCLO-DEXTRINS IN SOLUTION

YASUKO ISHIZUKA†, TADASHI NEMOTO, MASAKO FUJIWARA*, KEN-ICHI FUJITA*, and HIROSHI NAKANISHI

National Institute of Bioscience and Human-Technology, Higashi 1-1, Tsukuba, Ibaraki, 305-8566, Japan and

JEOL DATUM Ltd., Nakagami-cho 1156, Akishima, Tokyo 196, Japan*

^{1}H NMR spectra of glucosyl-α-CD, maltosyl-α-CD, maltotriosyl-α-CD and panosyl-α-CD were measured analysing NOE correlations. Three dimensional structure of these substituted cyclodextrins were effectively simulated using DADAS method with distance constraints estimated by the strength of NOE correlation peaks.

1. Introduction

Substitution of cyclodextrin (CD) with oligosaccharides increases the solubility of the cyclodextrins in aqueous solution. The inclusion interaction of these cyclodextrins is expected to be strongly affected by the existence of the branch saccharides. Particularly, the position of the branch moiety relative to the core cyclodextrin must be important to the inclusion phenomena. In this report, aiming to know the three dimensional structures of branched cyclodextrins, ^{1}H NMR spectra were analysed, NOE correlational peaks were detected, and further the three dimensional structures were simulated by DADAS method, in which distance constraints evaluated from the strength of the NOE correlational peaks were used.

Four substituted cyclodextrins for our experiments are constituted with only α-D-glucoses. For the assignments of the signals, glucosyl pyranoses of branch moiety are definited to be A, B and C from the neighbor to the core cyclodextrin, and core moiety are definited to be 1D, 2D, 3D, ..,6D. The branch saccharide bonds to the 6-methylene carbon in 1D glucose of the CD ring. Maltotriosyl-α-CD and panosyl-α-CD are structural analogous isomers each other, in which the difference is in the non-reducing terminal glucosyl bond; the former has an α1-4 bondage and the latter α1-6 bondage.

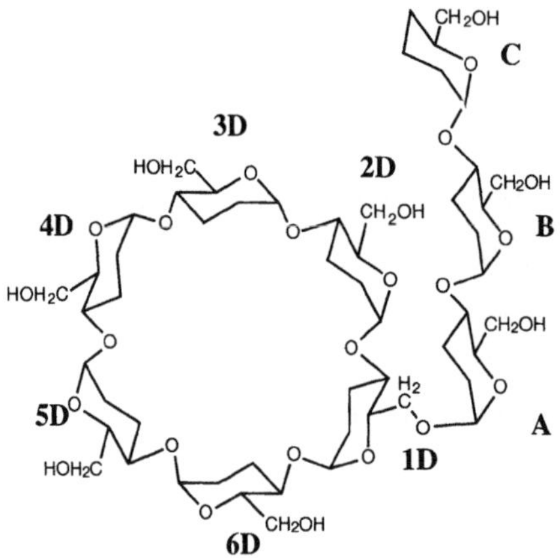

Maltotriosyl-α

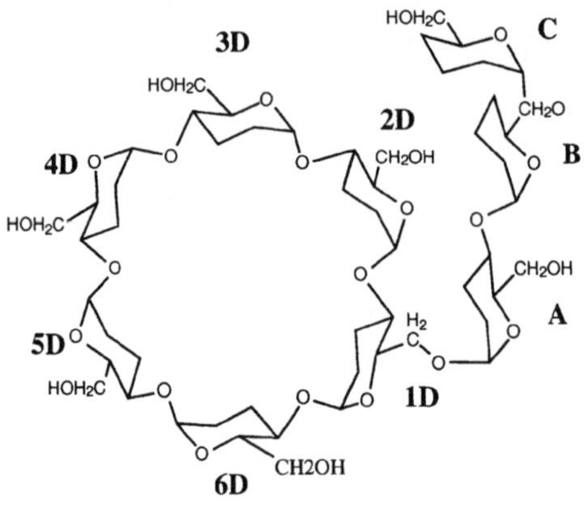

Panosyl-α

2. Experiment

Glucosyl-α-CD, maltosyl-α-CD, maltotriosyl-α-CD and panosyl-α-CD were used as substituted cyclodextrins. 1D and 2D TOCSY, DQF-COSY, NOESY and HSQC spectra were recorded to assign the ^{1}H signals with 750MHz and 500Mhz NMR spectrometers (Bruker DMX750 and DMX500) at 298K in aqueous solution. External sodium-2,2-dimethyl-2-silapentonate-d4 (DMSP) was used as a reference. NOE correlational peaks were detected by 2D NOESY and ROESY spectra.

3. Results and Discussions

3.1 ASSIGNMENT OF ^{1}H SIGNALS

Anomeric signal (D1) of the glucoses in glucosyl-α-CD was devided to three doublets with different integral strength and with considerably small difference in chemical shift. One doublet in the lowest magnetic field has one proton integral strength, the second has also one proton integral strength. The third doublet has four proton integral strength in the highest magnetic field. 1D6 proton signals were expected to shift to the lower field for the substitution by a glucosyl residue at 6 carbon. 1D6 proton signals clearly showed TOCSY correlation with the central D1 signal among three doublets. So, the central D1 signal was assigned as 1D1. Also, the doublet in the lowest magnetic field were assigned as 2D1, because the effect of the substitution was expected to be stronger with 2D1 proton than with 6D1 proton. These NMR signal features of D1 protons were similarly observed with other three substituted cyclodextrins. Table 1 shows the ^{1}H chemical shifts of maltotriosyl-α-CD as the results of various kind of 2D NMR analysis.

Table 1. Chemical Shifts of Maltotriosyl-α-CD (ppm from DMSP)

	C	B	A	1D	2D	3D -6D
1	5.375	5.305	4.904	5.048	5.059	5.043, 5.054
2	3.573	3.614	3.578	3.658	3.616	3.625
3	3.681	3.941	3.996	4.008	3.989	3.982
4	3.405	3.649	3.644	3.610	3.553	3.584
5	3.706	3.848	3.852	3.832	3.852	3.832
6	3.752	3.795	3.798	3.905-3.950	3.905-3.960	3.905-3.970
6	3.812	3.850	3.870	4.015	3.810-3.866	3.810-3.867

The assignments of branch moiety were in good agreement with previously reported data.[1] The other three substituted cyclodextrins were similarly analysed, all ^{1}H chemical shifts were successfully obtained.

3.2 DETECTED NOE

Interresidual NOE correlational peaks were detected between all two protons at the position of O-glucosyl-bond for the four substituted cyclodextrins. Particularly strong NOEs were observed between D1 and D4 in the neighboring residues. NOEs in medium degree were also detected between D1 and D6 in the neighboring residues. Characterisitically in the case of maltotriosyl-α-CD, medium NOE was detected between A1 and 1D5, and in the case of panosyl-α-CD, between D1 and B6. In the case of glucosyl-α-CD, medium NOE was detected between 2D1 and A5. Several weak NOEs were also detected between core protons and branch protons.

3.3 THREE DIMENSIONAL STRUCTURE SIMULATED WITH DADAS METHOD

Three dimensional structures for the saccharides were simulated using DADAS90 software. DADAS method was originally proposed for proteins [2] in which only dihedral angles of amide bonding were used as variables in the simulation. For the simulations of the substituted cyclodextrins, distance constraint and van der Waals Radii of every atoms are important factors. Distance constraint of two protons with strong NOE was selected to be 3.0Å or 3.5Å and those with medium NOEs to be 4.0Å or 4.5Å.

Two calculation groups were obtained as possible structure for panosyl-α-CD. One result is shown in the Fig. 1. In this case branch saccharides are clearly outside of the cyclodextrin ring, and do not cover the entrance of the ring cavity. In the other possible calculation, branch part of panosyl-α-CD does not cover the ring cavity. However, branch part of maltotriosyl-α-CD covers one site of entrance of hydrophobic inner space in the CD ring. Thus, there are large difference in the three dimensional structure between maltotriosyl-α-CD and panosyl-α-CD. These simulation results give very useful informations about inclusion phenomena.

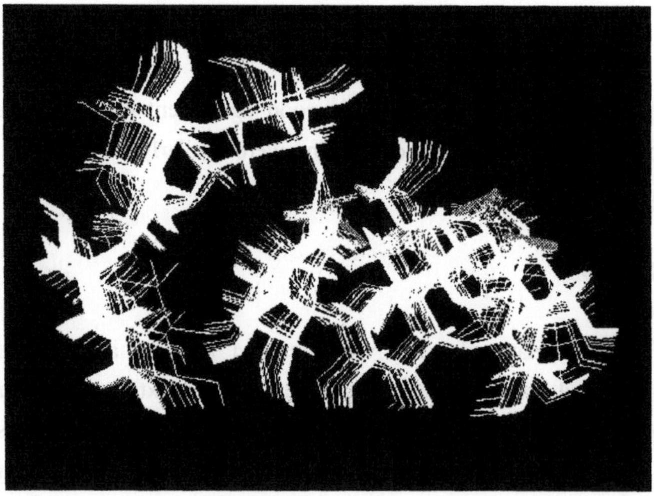

Fig. 1 Panosyl-α-CD, 100 structures converged from 100 initially random structures

4. References

[1] Yasuko Ishizuka, Hiroshi Nakanishi, Takanori Shiraishi, Yasuo Kogure and Shoichi Kobayashi, J. Carbohydr. Chem. 10(4), 583-592 (1991)

[2] W. Braun and N. Go, J. Mol. Biol., 186, 611-626,1885

SYNTHESIS AND ENHANCED CHEMILUMINESCENCE OF NEW MONO-CYCLOMALTOOLIGOSACCHARIDE-BOUND 6-PHENYLIMIDAZO[1,2-*a*]PYRAZIN-3(7*H*)-ONES

KATSUNORI TERANISHI, SAORI TANABE, ATSUKO KOMODA, MAKOTO HISAMATSU, TETSUYA YAMADA
Faculty of Bioresources, Mie University, Kamihama, Tsu, Mie 514, Japan

1. Introduction

Aequorin is a photoprotein found in the jellyfish, *Aequorea victoria*, that generates blue light (460 nm) with high light-emitting efficiency (ϕ = 0.2) upon treating this luminescence system with calcium ions [1-3]. Aequorin is known to consist of apoaequorin, molecular oxygen, and 2-(*p*-hydroxybenzyl)-6-(*p*-hydroxyphenyl)-8-benzylimidazo[1,2-*a*]pyrazin-3(7*H*)-one (*Oplophorus* luciferin, coelenterazine) [4,5]. The chemiluminescent reaction of coelenterazine and its analogues such as MCLA without apoaequorin in aqueous buffer solutions in the presence of active oxygen has been being used for analyses, whereas its light-emitting efficiency is extremely low [6-8]. Many studies on the enhancement of chemiluminescence have dealt with the regulated medium, such as using a micellar surfactant or cyclomaltooligosaccharide (cyclodextrin) [9-13]. It has been shown that cyclodextrin enhances the intensity of chemiluminescence of MCLA in aqueous solvents by Toya [12]. That study showed that very large amounts of cyclodextrin are needed for the enhancement. For analysis by chemiluminescence, the analytical system should not affect the subject except to promote luminescence. It is noteworthy to construct a novel chemiluminescent system which promotes greater chemiluminescent activity. In this communication, we report a first example of the synthesis of light-producing compounds, in which MCLA, a chemiluminescent chromophore, is covalently bound to one cyclodextrin molecule, and show that the chemiluminescence is effectively enhanced in an aqueous solvent.

2. Materials and Methods

The HPLC analysis was done with a JASCO Gulliver system and an MD-910 three-dimensional UV-VIS detector. A Nacalai Tesque Cosmosil 5C18-MS column (4.6 mm x 150 mm) was used. Preparative chromatography for cyclodextrin compounds was conducted with a Fuji Silysia Chromatorex-ODS DM1020T. Analytical TLC was performed on Merck Kieselgel 60 F254 precoated, glass-packed plates of 0.25 mm layer thickness. Elution was by 6:3:1 CH_3CN/H_2O/concd. aq. NH_3, and the compounds were visualized under a UV lamp and with a *p*-anisaldehyde-HOAc-H_2SO_4-EtOH solution. Structure of synthesized compounds was determined by ^{1}H- and ^{13}C-NMR spectroscopy. ^{1}H- and ^{13}C-NMR data were recorded with a JEOL JNM-A500 spectrometer. ^{1}H- and

^{13}C-NMR chemical shifts were assigned on the basis of ^{1}H-^{1}H COSY, DEPT ^{13}C-NMR, and ^{1}H-^{13}C COSY data. FAB mass spectra (positive) were measured with a JEOL DX-303 instrument, using glycerol as a matrix, and SIMS data were measured with a Hitachi M-80B instrument. Elemental analyses were performed with a Yanaco CHNCORDER MT-3 instrument. All chemicals not otherwise mentioned were purchased from Tokyo Kasei Kogyo Co., Ltd. (Tokyo, Japan) or Aldrich Chemical Co. (St. Louis, MO., U.S.A.) in chemically pure grade and were used as such. Cyclodextrins were dried for 12 h at 140 °C under vacuum. Mono-6-amino-6-deoxy-cyclodextrins [14] were prepared according to the published method. 2-Amino-5-(*p*-methoxyphenyl)pyrazine was prepared using the method of Kishi [15].

Chemiluminescent intensity time curves were obtained using a ALOKA LUMINESCENCE READER BLR-301, and chemiluminescent spectra and fluorescent spectra were obtained using a JASCO FP 750 spectrofluorometer. Circular dichroism spectra were obtained using a JASCO J-720M spectrometer. Chemiluminescent intensity time curves were obtained as follows: 40 μL of 0.25 mM cyclodextrin-bound MCLA compounds dissolved in distilled water was added to 0.96 mL of 30 mM phosphate buffer (pH 8.3) at 30 °C. The reaction mixture was placed in the photometer and chemiluminescent intensity time curve was obtained at 30 °C. The chemiluminescent quantum efficiencies (Φ) were determined as relative values to that of luminol as a standard ($\Phi = 0.013$) [16]. Yields of the amide compounds generated by the chemiluminescent reaction were obtained follows: HPLC system was employed for analyzing the luminescence-spent products and the yields of amides were determined by comparison with the corresponding synthesized amides.

3. Results and Discussion

We have found a convenient regioselective mono-2-*O*-tosylation of cyclodextrins at 32-36% yields using 1-(*p*-toluenesulfonyl)imidazole and molecular sieves in *N,N*-dimethylformamide at 20 °C. In this sulfonation no 3-*O*-sulfonates or 6-*O*-sulfonates was given and decomposition of the 2-*O*-sulfonates was not observed. The 2-*O*-sulfonates were changed to 3-amino-3-deoxycyclodextrins by the treatment with aqueous NH_3. Then, the 3-aminocyclodextrins and MCLA-COOH, which was prepared from 2-amino-5-(*p*-methoxyphenyl)pyrazine and 2-ketoglutaric acid in EtOH in the presence of H_2 and Pd-c followed by treatment with acetic anhydride in THF, were coupled to afford cyclodextrin-bound MCLA compounds, **3-α-1**, **3-β-1**, and **3-γ-1**, at 54-64% yields. Moreover, a useful mono-2-*O*-aminoalkylation has been developed. Reactions of cyclodextrins with NaH and *N*-(bromoalkyl)phthalimides having more than three alkyl carbons such as *N*-(3-bromopropyl)phthalimide in DMSO gave mono-2-*O*-(phthalimidoalkyl)cyclodextrins, whereas *N*-(2-bromoethyl)phthalimide afforded 6-*O*-(2-phthalimidoethyl)cyclodextrins. Mono-2-*O*-(3-phthalimidopropyl)cyclodextrins were changed to mono-2-*O*-(3-aminopropyl)cyclodextrins followed by condensation with MCLA-COOH to cyclodextrin-bound MCLA compounds, **2-α-5**, **2-β-5**, and **2-γ-5**. Synthesis of 6-*C*-MCLA cyclodextrin analogues, **6-α-1**, **6-β-1**, and **6-γ-1**, was achieved by the formation of amide bond between MCLA-COOH and known mono-6-amino-6-deoxycyclodextrins. In order to investigate the influence of distance between MCLA and the cyclodextrins, glycine(s) also was set between MCLA and the cyclodextrins to give **3-γ-4**, **6-γ-4**, and **6-γ-7**.

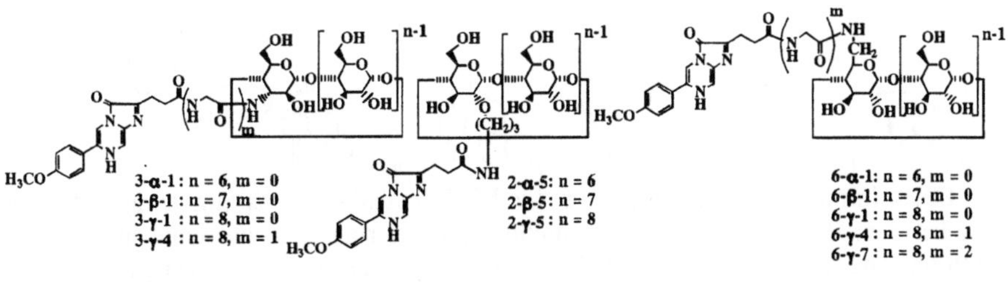

TABLE 1. Chemiluminescent Properties of Cyclodextrin-bound MCLA Compounds

Compound	Φ	Φ_R	$\Phi_S \times \Phi_F$	Compound	Φ	Φ_R	$\Phi_S \times \Phi_F$
MCLA	0.00048	0.27	0.0018	**2-γ-5**	0.0044	0.14	0.031
2-α-5	0.0015	0.16	0.0096	**3-γ-1**	0.021	0.41	0.052
3-α-1	0.00062	0.14	0.0045	**3-γ-4**	0.0096	0.36	0.027
6-α-1	0.00059	0.39	0.0015	**6-γ-1**	0.0096	0.33	0.029
2-β-5	0.0048	0.27	0.018	**6-γ-4**	0.010	0.24	0.042
3-β-1	0.0036	0.41	0.0087	**6-γ-7**	0.0058	-	-
6-β-1	0.0054	0.49	0.011	**5**	0.082	0.33	0.25

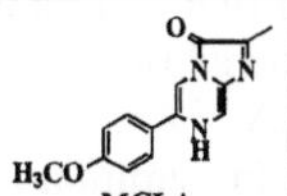

MCLA

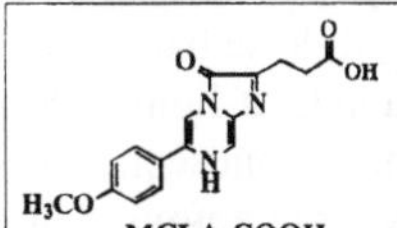

MCLA-COOH

The chemiluminescent reaction consists of many reaction steps such as binding with oxygen molecule, generation of intermediate(s) having high energy, formation of excited singlet state, and emission of light, and efficiencies in these reaction steps determine the quantum yield (Φ) of chemiluminescence. Generally, the chemiluminescent quantum efficiency (Φ) is given by the following equation; $\Phi = \Phi_R \times \Phi_S \times \Phi_F$ (Φ_R: formation yield of emitter, Φ_S: quantum efficiency of excited singlet state formation of the emitter, Φ_F: quantum efficiency of fluorescence of the emitter). Light-emitting intensity of the oxygen-induced chemiluminescence was tested in phosphate buffer (pH 8.3, 0.03M). The intensity was remarkably dependent on the kind of bound cyclodextrin, the binding site of chromophore to cyclodextrin, and the length of spacer between the chromophore and cyclodextrin as shown in Table 1. When MCLA was covalently attached to C3-carbon of cyclodextrin with short spacer, namely **3-γ-1**, its light-emitting intensity showed an up to 44-fold enhancement ($\Phi = 0.021$) than that of MCLA. Φ_R and $\Phi_S \times \Phi_F$ of **3-γ-1** were 0.41 and 0.052, respectively, as shown in Table 1, and its $\Phi_S \times \Phi_F$,, which was 29-fold as against MCLA, was highest in the cyclodextrin-bound MCLA compounds.

$\Phi_S \times \Phi_F$ is dependent on the molecular state of the light-producing chromophore, amide compound, at generating singlet excited state, staying on the excited sate, and generating light. Induced circular dichroism spectra of all the cyclodextrin-bound MCLA compounds exhibited too small bands. It is possible that MCLA chromophores are not included in cyclodextrins, because the MCLA seems not to be sufficiently hydrophobic. Induced circular dichroism spectrum of an amide compound given from the chemiluminescent reaction of **3-γ-1** showed strong positive bands in contrast with those of other compounds. This suggests that the light-emitting chromophore of **3-γ-1** is included in the γ-cyclodextrin cavity. In the chemiluminescent reaction of **3-γ-1**, after **3-γ-1** reacts with oxygen molecule, 5-(*p*-methoxyphenyl)pyrazine moiety can approach to the γ-cyclodextrin cavity to give the highest $\Phi_S \times \Phi_F$.

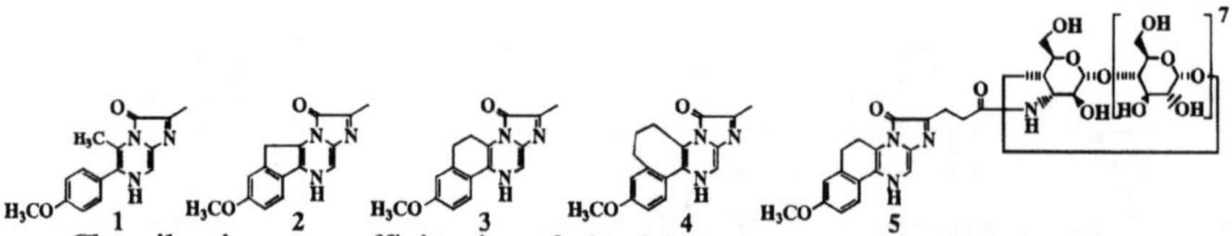

Chemiluminescent efficiencies of the MCLA analogues having a rigid dihedral angle between the pyrazine and the *p*-methoxyphenyl ring were measured in the same condition. The decrease of the dihedral angle has some enhancement effects; the five and six-membered ring derivatives, **2** and **3**, are superior to the 5-methyl derivative **1** and seven-ring derivative **4** which have larger dihedral angle. Cyclodextrin-bound six-membered ring MCLA compound **5** gave 170-fold enhancement (Φ = 0.082) as against MCLA.

In summary, the light-emitting efficiency of the cyclodextrin-bound MCLA compound in which γ-cyclodextrin was covalently attached with short spacer showed the high enhancement. This study indicats that the strategy involving covalently attaching a light-producing chromophore onto a cyclodextrin for the enhancement of chemiluminescence is more efficient than the use of an aqueous solution containing very large amounts of cyclodextrin.

References

1. Shimomura, O., Johnson, F. H., and Saiga, Y. (1962) Extraction, purification and properties of aequorin, a bioluminescent protein from the luminous hydromedusan, Aequorea, *J. Cell. Comp. Physiol.* 59, 223-240.
2. Shimomura, O., and Johnson, F. H. (1969) Properties of the bioluminescent protein aequorin, *Biochem.* 8, 3991-3997.
3. Shimomura, O., and Johnson, F. H. (1970) Caluciun binding, quantum yield, and emitting molecule in aequorin bioluminescence, *Nature.* 227, 1356-1357.
4. Shimomura, O., and Johnson, F. H. (1972) Structure of the light-emitting moety of aequorin, *Biochem.* 11, 1602-1608.
5. Shimomura, O., Johnson, F. H., and Morise, H. (1974) Mechanism of the luminescent intramolecular reaction of aequorin, *Biochem.* 13, 3278-3286.
6. Sawada, H.,and Nakayama, M., (1989) Detection of linoleic acid hydroperoxide by use of chemiluminescence emitted by cypridina luciferin analogs, *Yukagaku,* 38, 103-105.
7. Goto, T., and Takagi. T. (1980) Chemiluminescence of a cypridina luciferin analogue, 2-methyl-6-phenyl-3,7-dihydroimidazo[1,2-a]pyrazin-3-one, in the presence of the xanthine-xanthine oxidase system, *Bull. Chem. Soc. Jpn.* 53, 833-834.
8. Toya, Y., Kayano, T., Sato, K., and Goto, T., (1992) Synthesis and chemiluminescence properties of 6-(4-methoxyphenyl)-2-methylimidazo[1,2-a]-pyrazin-3(7H)-one and 2-methyl-6-(2-naphthyl)-imidazo[1,2-a]pyrazin-3(7H)-one, *Bull. Chem. Soc. Jpn.*, 65, 2475-2479.
9. Grayeski, M. L. and Woolf, E. J. (1985) Effect of cyclodextrin solutons on chemiluminescence of 10,10'-dimethyl-9,9'biacridinium nitrate, *J. Lumin.*, 33, 115-121.
10. Woolf, E. J. and Grayeski, M. L. (1987) Effect of cyclodextrin solutions on aqueous peroxyoxalate chemiluminescence, *J. Lumin.*, 39, 19-27.
11. Karatani, H. (1986) Effects of cyclodextrin on enhancement for chemiluminescence of the luminol related compound, *Chem. Lett.*, 377-380.
12. Toya, Y., (1992) *Nippon Nogeikagaku Kaishi*, 66, 742-747.
13. Mitani, M., Sasaki, S., Koinuma, Y., Toya, Y., Kosugi, M., (1995) Enhancement effect of 2,6-*O*-dimethyl-β-cyclodextrin on the chemiluminescent detection of β-D-galactosidase using a cypridina luciferin analog, *Anal. Sci.*, 11, 1013-1015.
14. Hamasaki, K., Ikeda, H., Nakamura, A., Ueno, A., Toda, F., Suzuki, I., and Osa, T., (1993) Fluorescent sensors of molecular recognition. modified cyclodextrins capable of exhibiting guest-responsive twisted intramolecular charge transfer fluorescence, *J. Am. Chem. Soc.*, 115, 5035-5040.
15. Kishi, Y., Tanino, H., and Goto. T., (1972) The structure confirmation of the light-emitting moiety of bioluminescet jellyfish aequorea, *Tetrahedron Lett.*, 2747-2748.
16. Ferguson, J. F. III and Seliger, H. H., (1966) ,The use of luminol as a standard of photon emission, in Johnson, F. H. and Haneda, Y (ed.), *Bioluminescence in progress*, Princeton university press, pp.35-43.

THERMAL BEHAVIOUR AND PHASE TRANSITIONS OF CYCLOMALTONONAOSE (δ-CYCLODEXTRIN)

G. P. BETTINETTI[1], H. UEDA[2], M. SORRENTI[1], A. NEGRI[1]
[1]Dipartimento di Chimica Farmaceutica, Università di Pavia, Viale Taramelli 12, I-27100 PV, Italy
[2]Department of Physical Chemistry, Hoshi University, 4-41, Ebara 2-chome, Shinagawa-ku, Tokyo 142-8501, Japan

Abstract

Thermal properties and phase transitions of δ-cyclodextrin were investigated using differential scanning calorimetry, thermogravimetry, thermomicroscopy and X-ray powder diffractometry. The occurrence of crystalline δCD·13.75 H_2O and δCD·7 H_2O hydrates and amorphous δCD was demonstrated.

1. Introduction

δ-Cyclodextrin (δCD) (stereoisomer of 5,10,15,20,25,30,35,40,45-nonakis (hydroxymethyl)-2,4,7,9,12, 14,17,19,22,24,27,29,32,34,37,39,42,44-octadecaoxadecacyclo[41.2.2.$2^{3,6}$.$2^{8,11}$.$2^{13,16}$.$2^{18,21}$.$2^{23,26}$.$2^{28,31}$.$2^{33,36}$.$2^{38,41}$] triexacontane-46,47,48,49,50,51,52,53,54,55,56,57,58,59,60,61,62,63-octadecol)

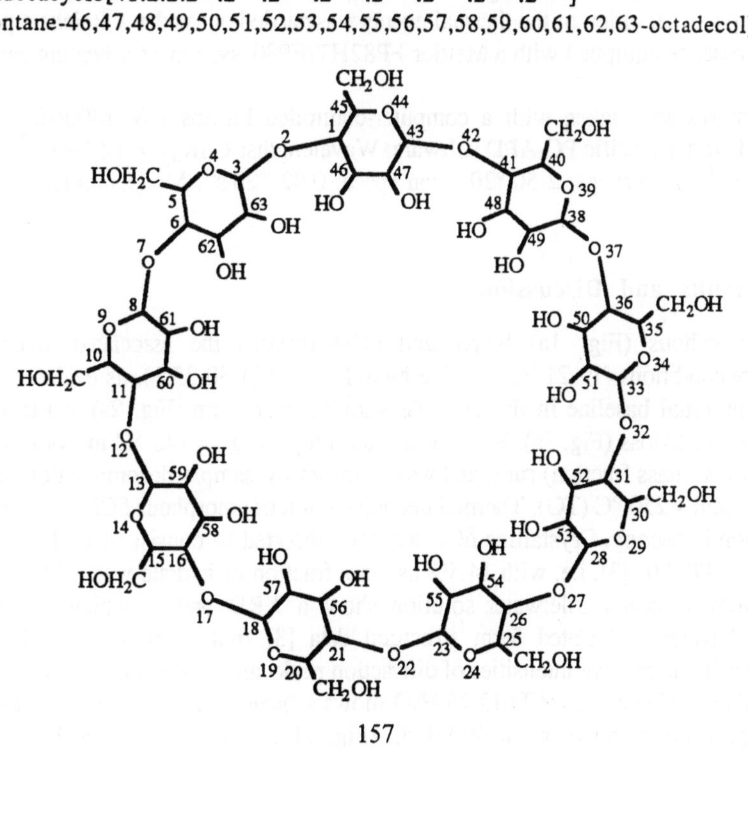

is the smaller homologue of the family of large-ring cyclomaltoses including also ε-, ζ-, η- and θ-CD which has been isolated from CD powder [1,2]. The X-ray crystal structure [3] and some physicochemical properties of δCD such as aqueous solubility [4] and complexation ability [4,5] have already been assessed. The purpose of this study was to characterize the solid state of δCD by means of differential scanning calorimetry (DSC), thermogravimetry (TG), thermomicroscopy and X-ray powder diffractometry (XRD).

2. Materials and Methods

2.1. MATERIALS

Lyophilized δCD isolated from CD powder [1,2] was used. Recrystallizations were carried out by spontaneous evaporation at room temperature of a 1:1 (by volume) acetonitrile:water solution containing ≈ 20 mg δCD per mL.

2.2. METHODS

Temperature and enthalpy values were measured with a METTLER STAR[e] system equipped with a DSC821[e] Module on 3-5 mg samples in open Al pan, heating rate 10 $K{\cdot}min^{-1}$, temperature range 30-300 °C, static air atmosphere.
Thermogravimetry was carried out with a Mettler TA 4000 apparatus equipped with a TG 50 cell on 6-10 mg (Mettler M3 microbalance) samples in open alumina crucibles between 30-300 °C, at the heating rate of 10 $K{\cdot}min^{-1}$ under static air atmosphere.
Microscopic observation of the thermal events was carried out under a Reichert polarized light microscope equipped with a Mettler FP82HT/FP80 system at a heating rate of 10 $K{\cdot}min^{-1}$.
XRD patterns were taken with a computer-controlled Philips PW 1800/10 apparatus equipped with a specific PC-APD software. Wavelengths: $CuK_{\alpha,1}$ = 1.54060 Å, $CuK_{\alpha,2}$ = 1.54439 Å. Scan range: 2-50 °2θ. Scan speed: 0.02 $°2\theta{\cdot}s^{-1}$. Monochromator: graphite crystal.

3. Results and Discussion

X-ray amorphous (Fig. 1a) lyophilized δCD released the associated water under laboratory conditions (19-21 °C, relative humidity (R.H.) 30-35%), as evidenced by the lack of the initial baseline in the DSC dehydration endotherm (Fig. 2a) and in the first TG step of mass loss (Fig. 3a). Water loss was complete at ≈ 140 °C in both DSC and TG (11.6% as mass fraction) runs, and was followed by sample decomposition at ≈ 265 °C (DSC) and ≈ 250 °C (TG). Thermal decomposition of amorphous δCD was confirmed by thermomicroscopy. Crystalline δCD has been reported to consist of $C_{54}H_{90}O_{45}{\cdot}13.75\ H_2O$ (F.W. 1707.0) [3], i.e. with 14.5% as mass fraction of hydrate water. δCD crystals grown from an acetonitrile:water solution show a XRD pattern which matches the theoretical pattern calculated form structural data [3] (not shown) (Fig. 1b). Some differences in the relative intensities of diffraction peaks are attributable to isoorientation effects. The DSC curve of $\delta CD{\cdot}13.75\ H_2O$ shows a broad dehydration endotherm with a distinct peak emerging between 40-90 °C (Fig. 2b), which is reflected by the first-derivative TG curve with an associated water loss of 14.3% as mass fraction (Fig. 3b). Loss of hydrate water was complete at ≈ 140 °C and an associated change of crystals from translucent to opaque followed by melting with decomposition at ≈ 275 °C can be seen by thermomicroscopy. Rehydration of the sample scanned by DSC to 180 °C under high R.H. conditions (≥95% R.H., room temperature) gave a product with the same XRD pattern as the initial δCD hydrate (Fig. 1b). Rehydration under ambient

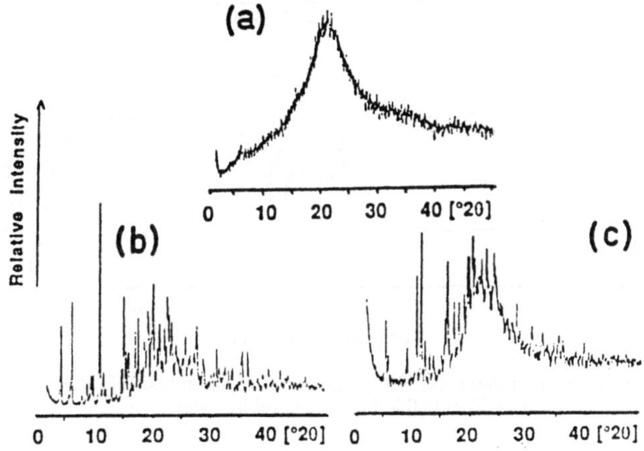

Figure 1. XRD spectra of amorphous δCD (a) and crystalline δCD·13.75 H_2O (b), δCD·7 H_2O (c) hydrates.

R.H. (≈ 32%) and temperature conditions gave a new δCD hydrated phase (Fig. 1c). The thermal properties, i.e. a triplet of DSC dehydration endotherms (Fig. 2c) and the corresponding first-derivative TG peaks (Fig. 3c), were markedly different from those of the starting δCD hydrate. The TG mass loss associated with the triplet of endotherms (7.8% as mass fraction) corresponded to 7 molecules of crystal water per $C_{54}H_{90}O_{45}$ molecule.

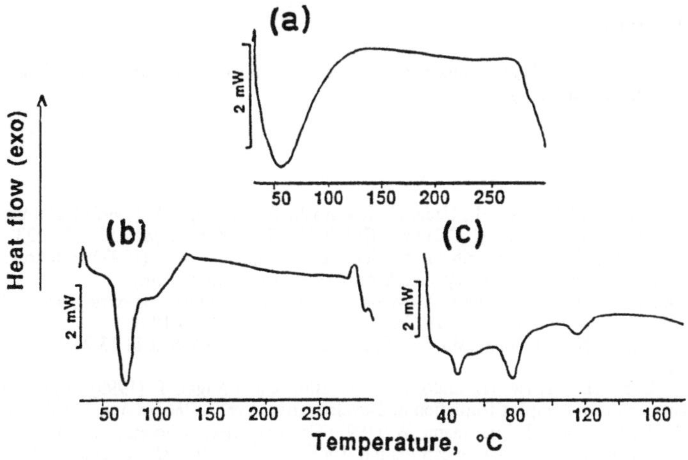

Figure 2. DSC curves of amorphous δCD (a) and crystalline δCD·13.75 H_2O (b), δCD·7 H_2O (c) hydrates.

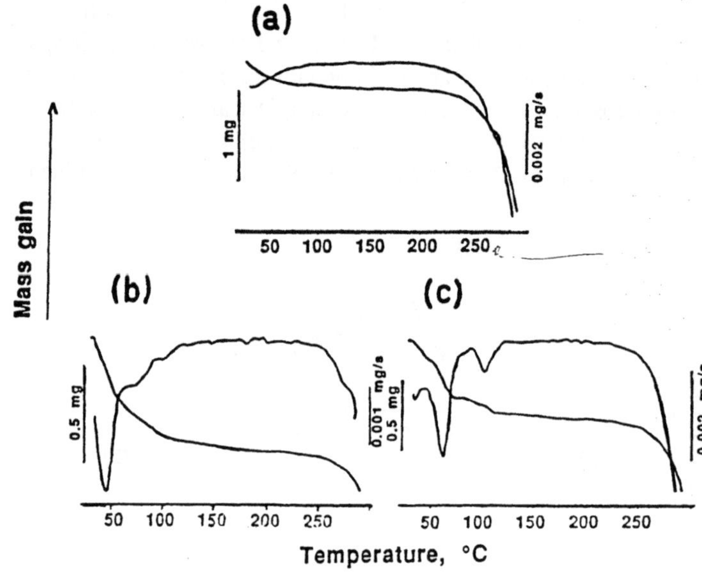

Figure 3. TG (lower curves) and first-derivative TG (DTG) curves of amorphous δCD (a) and crystalline δCD·13.75 H_2O (b), δCD·7 H_2O (c) hydrates.

4. Conclusion

The thermal behaviour of δCD (water loss followed by melting with decomposition at temperatures > 265 °C) resembles that of native α-, β, and γ- cyclodextrins. Differences in the dehydration endotherms and X-ray powder diffraction patterns make it possible to discriminate between amorphous and crystalline δCD and between δCD·13.75 H_2O and δCD·7 H_2O.

Acknowledgements

Work supported by a grant from Istituto Superiore di Sanità - Roma (Progetto proprietà fisico-chimiche dei medicamenti e loro sicurezza d'uso).

References

1. Wakamiya, A., Endo, T., Nagase, H., Ueda, H., Kobayashi, S. and Nagai,T. (1997) Isolation and purification of cyclomaltononaose (δ-CD) from CELDEX SG-30, *Yakuzaigaku* 57, 220-223.
2. Endo, T., Nagase, H., Ueda, H., Shigihara, A., Kobayashi, S. and Nagai, T. (1997) Isolation, purification, and characterization of cyclomaltotetradecaose (ι-cyclodextrin), cyclomaltopentadecaose (κ-cyclodextrin), cyclomaltohexadecaose (λ-cyclodextrin), and cyclomaltoheptadecaose (μ-cyclodextrin), *Chem. Pharm. Bull.* 45, 1856-1859.
3. Fujiwara, T., Tanaka, N. and Kobayashi, S. (1990) Structure of δ-cyclodextrin 13.75H_2O, *Chem. Lett.* 739-742.
4. Miyazawa, I., Ueda, H., Nagase, H., Endo, T., Kobayashi, S. and Nagai, T. (1995) Physicochemical properties and inclusion complex formation of δ-cyclodextrin, *Eur. J. Pharm. Sci.* 3, 153-162.
5. Larsen, K. L., Ueda, H. and Zimmermann, W. (1997) Capillary electrophoretic evaluation of large cyclodextrins as hosts for molecular encapsulation, *8th Eur. Congr. Biotechnol.*, Budapest (HU).

TEXTILE FINISHING WITH MCT-β-CYCLODEXTRIN

J.-P. MOLDENHAUER, H. REUSCHER
Wacker-Chemie GmbH, Johannes-Heß-Str. 24,
D-84489 Burghausen, Germany
Wacker Biochem Corp., 3301 Sutton Road,
Adrian, MI 49221-9397, USA

Abstract

Monochlorotriazinyl-β-cyclodextrin (BETA W7 MCT) is a reactive cyclodextrin derivative that can be covalently fixed to nucleophilic substrates by a condensation reaction. This new type of surface modification means a permanent transfer of cyclodextrin properties to the treated materials.
An important application field of BETA W7 MCT is the textile finishing process. Analogues to reactive dyes the MCT-cyclodextrin can be fixed to the fabric by well known methods and with common equipment. Cellulosic fibres are most suitable for the MCT-CD modification. Detailed fixation parameters have been worked out for cotton, but also mixed fibre materials like cotton / polyurethan or cotton / polyamide can be finished with BETA W7 MCT in good yields. MCT-CD modified textiles can either absorb unpleasant odour by complexation of sweat components or release included guest substances like fragrances or pharmaceuticals over a long period of time.

1. Introduction

Monochlorotriazinyl-β-cyclodextrin (BETA W7 MCT) was presented in 1996 as the first reactive CD derivative for permanent surface modification manufactured on an industrial scale [1-3].

Fig.1: Molecular formula of BETA W7 MCT

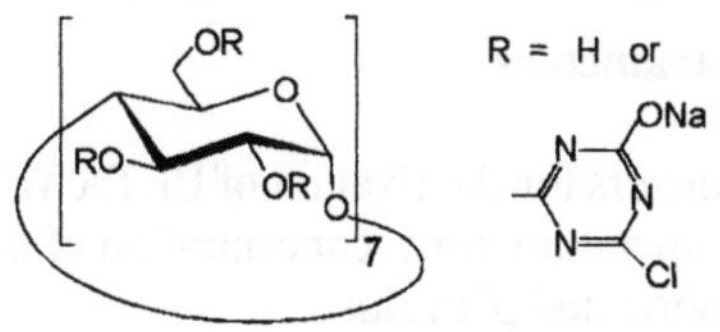

The favourable CD properties like odour masking and slow release of guest compounds can be transferred to different substrates by a finishing process well known from the dyeing technology. Cellulosic materials are especially suitable for the modification with BETA W7 MCT. The investigation of application methods revealed a successful fixation of BETA W7 MCT to cotton textiles either by a Foulard process with dry heat or contact heat or by a printing process [4].
Details of the dry heat fixation process on cotton have been worked out in this study.

2. Materials and Methods

Materials: BETA W7 MCT is a product of Wacker-Chemie GmbH, Munich, Germany. The cotton fabric to be treated with BETA W7 MCT is a product of Testfabrics, Inc., Middlesex, NJ, USA.

Methods: BETA W7 MCT was applied to cotton samples by a process comprising dipping in aqueous solution of BETA W7 MCT, squeezing (solution uptake 80 – 100% of dry fabric), drying, heating and rinsing. The amount of fixed CD (in % by weight) was determined by exact weighing (gravimetric method) or was calculated via the increase of the nitrogen content measured by elemental analysis of the treated and untreated fabric.

3. Results and Discussion

3.1 Fixation process

Analogues to many reactive dyes BETA W7 MCT is fixed to cotton by a nucleophilic substitution reaction of the fibre's Hydroxy-groups at the Chlorotriazine ring, induced by elevated temperature.

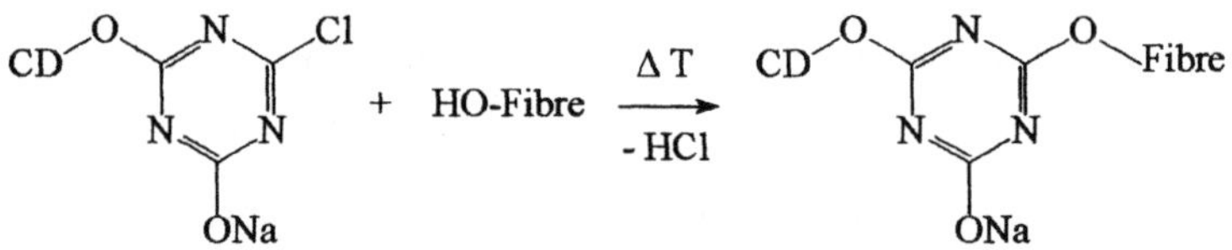

The fixation process can be run either under alkaline conditions (pH 10-11) by adding Na_2CO_3 or NaOH to the BETA W7 MCT solution or under acidic conditions (pH 5), achieved e.g. by the addition of HCl. The highest fixation yield is usually reached by the alkaline procedure (s. 3.2), whereas the acidic process is preferred for white fabric due to minimized yellowing during the CD modification.

3.2 Fixation parameters

Five main parameters for the fixation of BETA W7 MCT to cotton have been investigated: temperature, time, concentration of the BETA W7 MCT solution, moisture content of the fabric and pH value.

3.2.1 Fixation temperature / Moisture content

Pieces of cotton were finished at different temperatures between 90 and 170°C for 5 minutes under alkaline conditions, comparing the MCT finish reached with samples pre-dried at room temperature before fixation and with samples immediately finished after squeezing.

Fig. 2: Fixation temperature

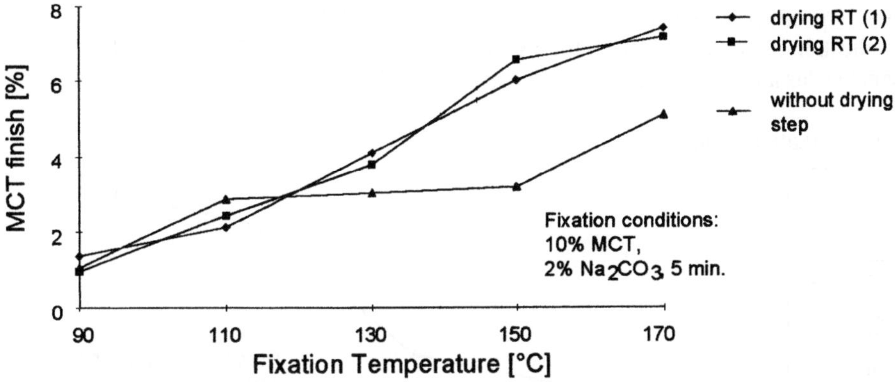

The MCT finish of pre-dried samples increases with the fixation temperature in a linear way from 1.5 to about 7% by weight (Fig. 2). When finishing cotton samples at 130 – 170°C without pre-drying the amount of fixed CD is significantly lower due to partial hydrolysis of the Chlorotriazin moiety. A similar effect is observed with the acidic procedure.

3.2.2 Fixation time / pH value

Pieces of cotton were exposed after pre-drying to dry heat of 150°C for 1 - 15 minutes under alkaline and acidic conditions in order to evaluate the influence of the pH value on the fixation process.

Fig. 3: Fixation time

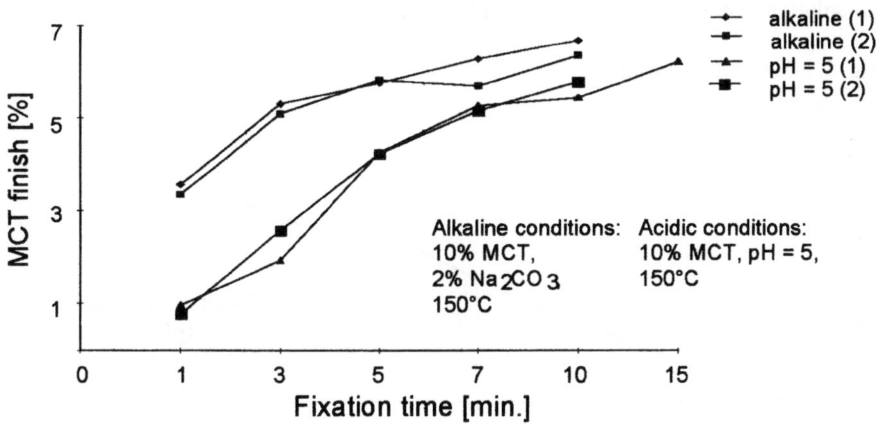

With very short times of heat exposure the acidic fixation is less effective than the alkaline one. However, a comparable MCT finish can be reached either by an extension of the fixation time (Fig. 3) or by an increase of the fixation temperature.

3.2.3 Fixation yield

A very good fixation yield of 80 – 85% is achieved with BETA W7 MCT by the application of dry heat (150°C) for 5 minutes under alkaline conditions (Fig. 4). The same yields are observed for the acidic process at 170°C.

Fig. 4: Fixation yield

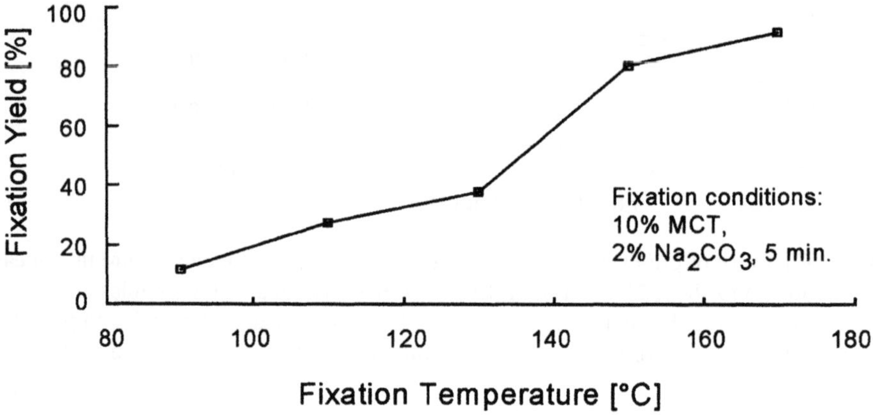

3.2.4 Concentration of BETA W7 MCT

The amount of fixed CD on the fabric at certain fixation conditions (temperature, time, pH) can be controlled by the concentration of BETA W7 MCT in the dipping bath. A linear relation is observed between the MCT concentration and the MCT finish to be reached (Fig. 5). However, the fixation yield is independent from the concentration of the solution, it only depends on the fixation conditions.

Fig. 5: MCT concentration

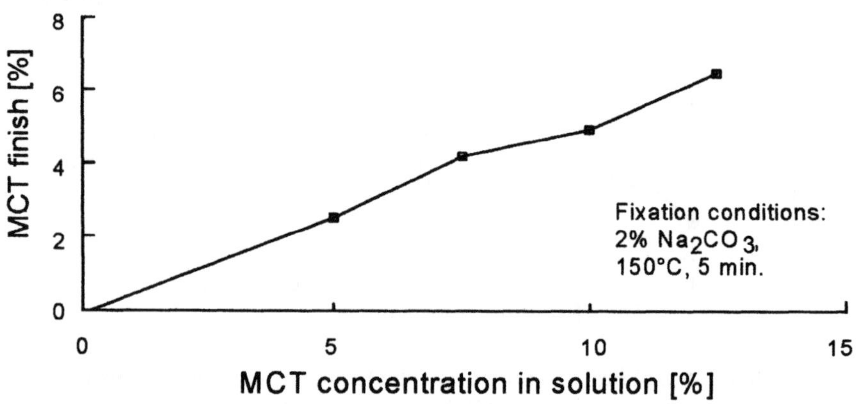

Fixation yield: 80 - 85%

3.3 Fastness of the MCT-β-CD finish against washing

Cotton T-shirts and socks with a finish of about 5% of BETA W7 MCT were washed ten times with washing powder at 60°C in a common washing machine and the MCT finish was analysed after 1, 5 and 10 washing steps. No significant decrease of the amount of fixed BETA W7 MCT by this procedure was detected (Fig. 6).

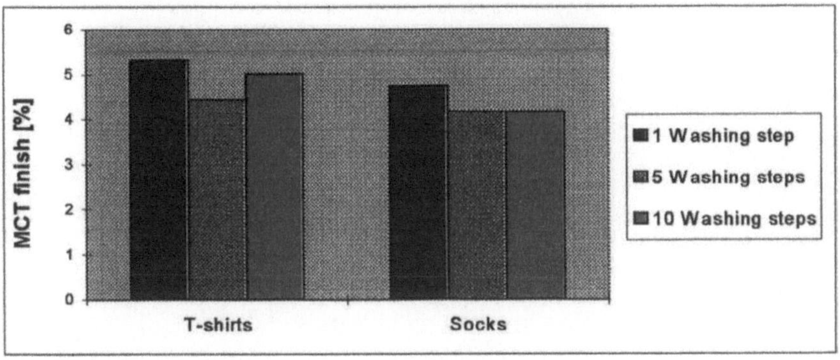

Fig. 6: Fastness of the MCT finish against washing

3.4 Application of actives to MCT-β-CD finished textiles

Two main procedures for the application of actives, e.g. fragrances, to textiles finished with BETA W7 MCT have been worked out. On a laboratory scale the moist CD modified fabric is stored in an atmosphere of the active at 60°C for some hours (vapour method). For technical applications the active is sprayed onto the CD finished fabric as a diluted aqueous solution or suspension. After drying at low temperature a controlled release of the active can be achieved by re-wetting the textile.

4. Conclusion

Cotton textiles can be finished with BETA W7 MCT in good yields according to common technical procedures transferring the useful CD properties permanently to the fabric. The finish is fast against washing and the CD cavity is still accessible for guest compounds, which can be applied easily.

References

[1] H. Reuscher, R. Hirsenkorn, BETA W7 MCT – New ways in surface modification, Proceedings of the 8th Intern. Symp. on Cyclodextrins, Budapest 1996, p. 553 – 558.

[2] Consortium für elektrochemische Industrie GmbH, Cyclodextrinderivate mit mindestens einem stickstoffhaltigen Heterocyclus, ihre Herstellung und Verwendung, Offenlegungsschrift DE 44 29 229 A1 (1996).

[3] Consortium für elektrochemische Industrie GmbH, Cyclodextrin derivatives having at least one nitrogen-containing heterocycle, their preparation and use, US 5728823.

[4] U. Denter, H.-J. Buschmann, D. Knittel, E. Schollmeyer, Textilveredlung 32 (1997), Nr. 1/2, p. 33-39.

ISOLATION, PURIFICATION, AND CHARACTERIZATION OF CYCLOMALTOOCTADECAOSE (ν-CD) AND CYCLOMALTONONADECAOSE (ξ-CD)

H. UEDA, T. ENDO, H. NAGASE, S. KOBAYASHI* and T. NAGAI
Hoshi University, 4-41, Ebara 2-chome, Shinagawa-ku, Tokyo 142-8501, Japan.
**National Food and Research Institute, Ministry of Agriculture, Forestry and Fisheries, Tsukuba, Ibaraki 305-8642, Japan.*

ABSTRACT

Although it was previously found that cyclodextrin glucanotransferase (CGTase) produces cyclodextrins (CDs) composed of six to seventeen glucopyranose units, the existence of CDs which have more than eighteen glucopyranose units has still not been confirmed. Cyclomaltooctadecaose (ν-CD) and cyclomaltononadecaose (ξ-CD) are new large-ring CDs (LR-CDs) composed of eighteen and nineteen glucopyranose units, respectively. ν- and ξ-CD were purified from the commercially available CD powder produced by CGTase, by a combination of HPLC and column chromatography. The molecular weights of ν- and ξ-CD were determined by FAB-MS, and their cyclic structures were identified by ^{1}H-NMR and ^{13}C-NMR. The ^{13}C-NMR chemical shifts of ν- and ξ-CD were elucidated and compared with those of already confirmed LR-CDs , and their structures were predicted from the results.

KEY WORDS

Cyclomaltooctadecaose (ν-CD), Cyclomaltononadecaose (ξ-CD),
Large-ring Cyclodextrin (LR-CD), Isolation, Purification, Characterization

1. INTRODUCTION

For several years, we have focused on the large-ring cyclodextrins (LR-CDs) composed of more than nine glucopyranose units. We have already isolated and purified nine kinds of LR-CDs, which are δ-, ε-, ζ-, η-, θ-, ι-, κ-, λ- and μ-CD, from a commercially available CD powder produced by CGTase [1 - 4]. Furthermore, crystallographic analyses of ε- and ι-CD, and the complex-forming ability of several

LR-CDs for some drugs were also reported [4 - 6]. However, LR-CDs composed of more than eighteen glucopyranose units have not been studied in detail, although there has been a report concerning glucoamylase-resistant large molecules in amyloses synthesized with CGTase [7]. Here, we describe the isolation, purification, and characterization of cyclomaltooctadecaose (ν-CD, composed of eighteen glucopyranose units) and cyclomaltononadecaose (ξ-CD, composed of nineteen units). The ^{13}C-NMR chemical shifts of ν- and ξ-CD were also elucidated and compared with those of other LR-CDs, and their structures are predicted from these results.

2. MATERIALS AND METHODS

2.1 MATERIALS

CD powder (DEXY PEARL K-50), β-amylase, and glucoamylase were purchased from Ensuiko Sugar Refining Co. (Yokohama, Japan), Tokyo Kasei Kogyo Co. (Tokyo, Japan) and Seikagaku Kogyo Co. (Tokyo), respectively. Pullulanase [Promozym 200L™] was a gift from Novo Nordisk Bioindustry Co. (Chiba, Japan). All other chemicals were from reliable commercial sources and were used without further purification. Milli-Q water was used as purified water in all preparation and purification steps.

2.2 PURIFICATION OF ν-CD AND ξ-CD

The preparation of LR-CD mixture was described in detail previously [4]. Purification was achieved by a combination of column chromatography and HPLC (Fig. 1).

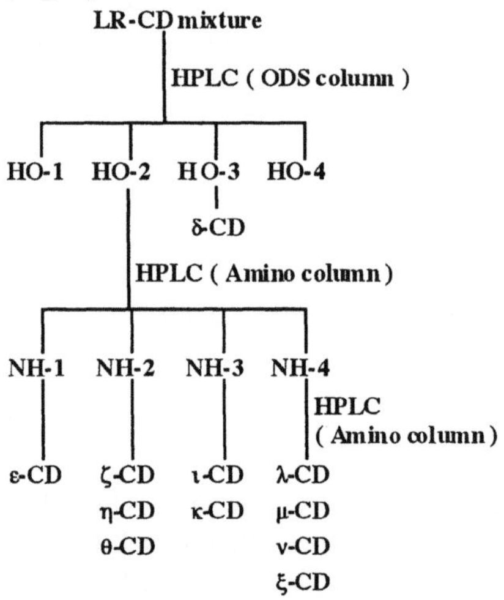

Figure 1 **Purification of ν- and ξ-CD**

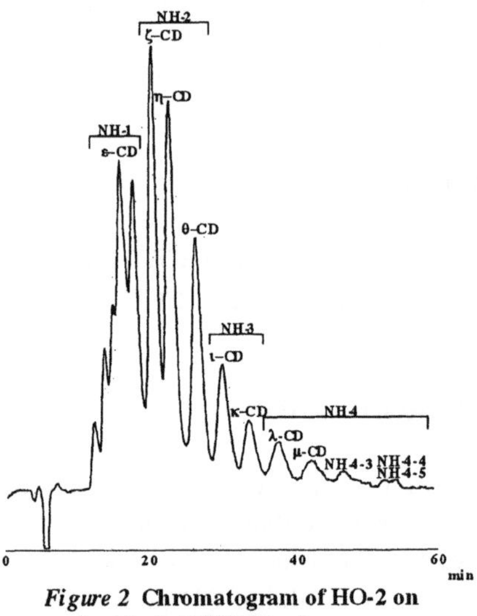

Figure 2 **Chromatogram of HO-2 on an Asahipak NH2P-50 column**

2.3 IDENTIFICATION OF ν-CD AND ξ-CD

2.3.1. *HPLC*

ν-CD and ξ-CD were identified by HPLC using the following two columns;
(1) column, Senshu Pak ODS-1251-SS; eluent, methanol-water (5 : 95); flow rate, 0.8 ml/min; column temperature, 25°C, and, (2) column, Asahipak NH2P-50; eluent, acetonitrile-water (55 : 45); flow rate, 0.7 ml/min; column temperature, 20°C.

2.3.2. *Fast Atom Bombardment Mass Spectrometry (FAB-MS)*

FAB-MS spectra were measured in a positive ion mode with an SX-102A mass spectrometer (JEOL) using Magic Bullet as a matrix. The acceleration voltages were 7 kV.

2.3.3. *Nuclear Magnetic Resonance (NMR)*

^{1}H-NMR, ^{13}C-NMR and two-dimensional ^{1}H-^{13}C correlation (H, C COSY) NMR spectra were taken on a JNM-LA500 spectrometer (500 MHz ^{1}H, JEOL) at 50°C. The samples were dissolved in 99.8% deuterium oxide. Chemical shifts are reported in δ-units (ppm) downfield from the signal of external tetramethylsilane.

3. RESULT AND DISCUSSION

Fig. 2 shows the chromatogram of HO-2 by HPLC on an analytical Asahipak NH2P-50 column. The results suggested that some components with larger retention times than μ-CD in HO-2 would be larger CD, because generally the elution sequence with this column and acetonitrile-water system follows the order of number of glucopyranose units. Then, NH-4-3 (ν-CD) and NH-4-4 (ξ-CD) were purified by HPLC on an analytical Asahipak NH2P-50 column. Their yields were about 18 mg (1.6×10^{-4}%) for ν-CD and about 10 mg (0.9×10^{-4}%) for ξ-CD, and both purities were above 98% (Fig. 3).

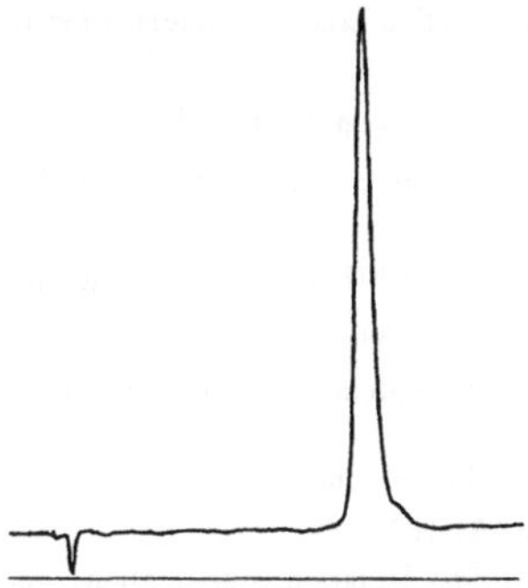

Figure 3 **Chromatogram of ν-CD on an Asahipak NH2P-50 column**

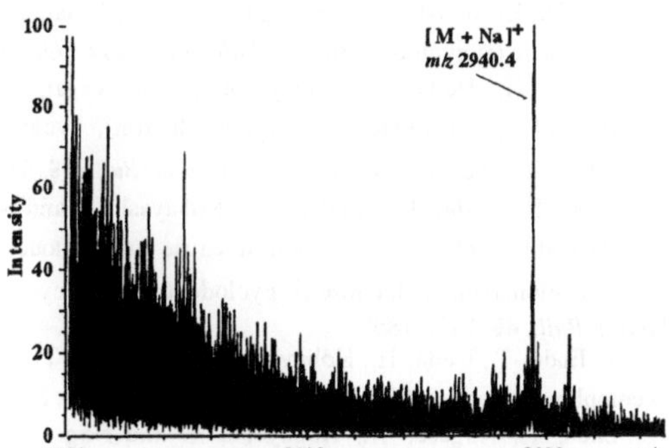

Figure 4 **FAB-MS spectrum of ν-CD**

Table 1. ^{13}C-NMR Chemical Shifts of LR-CDs

Carbon	δ value (ppm)										
	δ–CD	ε–CD	ζ–CD	η–CD	θ–CD	ι–CD	κ–CD	λ–CD	μ–CD	ν–CD	ξ–CD
1	100.96	99.74	99.83	100.17	100.34	100.43	100.29	100.08	100.07	100.48	100.62
2	73.08	72.65	72.56	72.53	72.51	72.56	72.59	72.54	72.48	72.76	72.76
3	73.74	73.68	73.66	73.67	73.64	73.69	73.76	73.82	73.88	74.23	74.28
4	79.26	78.03	78.31	78.87	78.97	78.89	78.51	78.09	78.00	78.37	78.46
5	72.34	71.78	71.74	71.82	71.90	71.97	72.00	71.97	71.96	72.29	72.34
6	61.29	61.49	61.53	61.54	61.43	61.39	61.40	61.42	61.44	61.72	61.72

ppm downfield from external tetramethylsilane at 50 °C in D_2O soln.

Fig. 4 shows the FAB-MS spectrum of ν-CD. Each FAB-MS spectrum gave the following: m/z peak 2940.4 $[M+Na]^+$ from ν-CD and m/z peak 3102.6 $[M+Na]^+$ from ξ-CD. These findings were in agreement with each calculated molecular weight. Finally, the spectra of ν- and ξ-CD each showed six clear and distinct single peaks attributable to equivalent glucopyranose units in the ^{13}C-NMR spectrum. Chemical shifts of each carbon with ν- and ξ-CD were similar to those that had been attributed to the cyclic structures of other LR-CDs except for δ-CD (Table 1).

4. CONCLUSION

We isolated and purified ν- and ξ-CD from commercially available CD powder produced by CGTase, and their identities were confirmed by HPLC, FAB-MS and NMR. These results indicated that CGTase could produce larger CDs than μ-CD similarly to potato D-enzyme. The ^{13}C-NMR data further suggested that structures of ν- and ξ-CD were similar to those of other LR-CDs except for δ-CD.

REFERENCES

1. Endo, T., Ueda, H., Kobayashi, S., and Nagai, T. (1995) Isolation, purification, and characterization of cyclomaltododecaose (η-cyclodextrin), *Carbohy. Res.*, **269**, 369-373.
2. Endo, T., Nagase, H., Ueda, H., Kobayashi, S. and Nagai, T. (1997) Isolation, purification, and characterization of cyclomaltodecaose (ε-cyclodextrin), cyclomaltoundecaose (ζ-cyclodextrin) and cyclomaltotridecaose (θ-cyclodextrin), *Chem. Pharm. Bull.*, **45**, 532-536.
3. Endo, T., Nagase, H., Ueda, H., Shigihara, A., Kobayashi, S. and Nagai, T. (1997) Isolation, purification, and characterization of cyclomaltotetradecaose (ι-cyclodextrin), cyclomaltopentadecaose (κ-cyclodextrin), cyclomaltohexadecaose (λ-cyclodextrin), and cyclomaltoheptadecaose (μ-cyclodextrin), *Chem. Pharm. Bull.*, **45**, 1856-1859.
4. Miyazawa, I., Endo, T., Ueda, H., Kobayashi, S. and Nagai, T. (1995) Physicochemical properties and inclusion complex formation of δ-cyclodextrin, *Eur. J. Pharm. Sci.*, **3**, 153-162.
5. Ueda, H., Endo, T., Nagase, H., Kobayashi, S. and Nagai, T. (1996) Isolation, purification, and characterization of cyclomaltodecaose (ε-CD), *J. Inclusion Phenom. Mol. Recognit. Chem.*, 25, 17-20.
6. Harata, K., Endo, T., Ueda, H. and Nagai, T., *Supramol. Chem.*, in press.
7. Terada, Y., Yanase, M., Takata, H., Takaha, T. and Okada, S. (1997) Cyclodextrins are not the major cyclic α-1,4-glucans produced by the initial action of cyclodextrin glucanotransferase on amylose, *J. Biol. Chem.*, **272**, 15729-15733.

SUPRAMOLECULAR ASSEMBLY OF POLYCHARGED β-CYCLODEXTRIN DERIVATIVES: FORMATION OF HETERODIMERS BETWEEN A POLYAMINO- AND A POLYSULFONATO-β-CYCLODEXTRIN

PASCALE SCHWINTÉ, ANN HOLOHAN, RAPHAEL DARCY* AND FRANCIS O'KEEFFE
Laboratory for Carbohydrate and Molecular Recognition Chemistry, Department of Chemistry, National University of Ireland, University College, Dublin 4, Ireland

1. Introduction

Supramolecular assemblies involving cyclomalto-oligosaccharides (cyclodextrins) [1] have mostly made use of the host-molecular properties of these compounds [2]. In contrast, self-organisation of cyclodextrins without guests, leading to Langmuir layers [3], liquid crystals [4], or micelles [5], has been brought about by their conversion to glycolipids through covalent attachment of hydrophobic chains to the (externally) hydrophilic cyclodextrin as headgroup.

The concept of cyclodextrins as templates for the ordering of multiple large atomic groupings has not been tested to date. Arrays on the cyclodextrin would be circular, in multiples of six, seven or eight for the most common cyclodextrins, with potential for high-multiple branching [6]. The reported dimerisation of charged cyclodextrins in a complex with lead [7], and of an oppositely charged primary-amino with a thiocarboxy cyclodextrin [8], leads to the conclusion that electrostatic forces could be used to array the templated molecules on the host's surface.

Here we show that extended structures on the rims of two β-cyclodextrins (**1**) and (**2**) can be brought to interact and form a circular array, formed by the strong electrostatic interaction between the seven hydroxyethylamino cationic groups on one torus, and the seven aromatic anionic groups on the other.

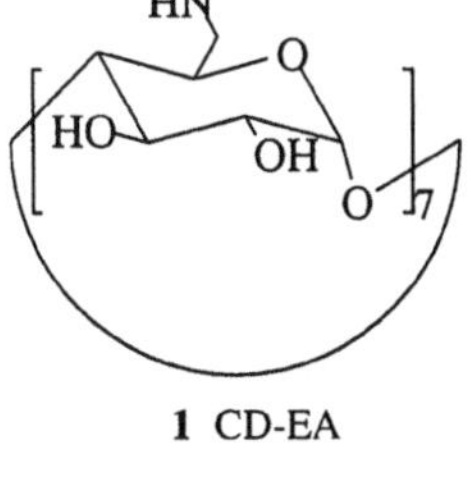

1 CD-EA

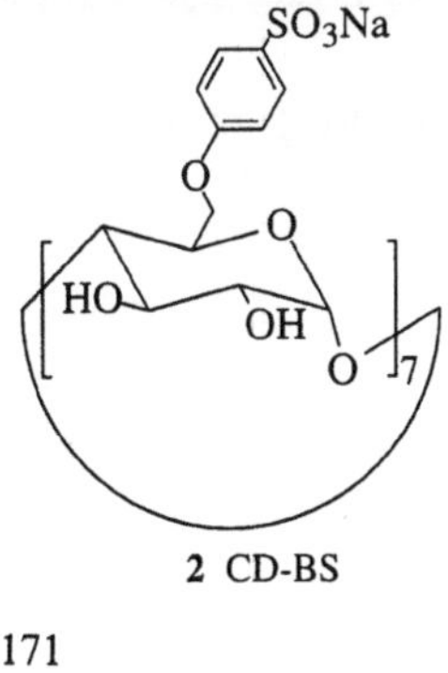

2 CD-BS

2. Materials and methods

Heptakis(6-hydroxyethylamino-6-deoxy)-β-cyclodextrin (**1**) (CD-EA) was synthesised as previously described [6]. Heptakis(6-sulfonatophenyloxy-6-deoxy)-β-cyclodextrin (**2**) (CD-BS) was synthesised by reaction of 4-hydroxyphenylsulfonic acid disodium salt with heptakis(6-bromo-6-deoxy)-β-cyclodextrin under conditions similar to those used by us already to glycosylate cyclodextrin [9]. The structure was confirmed by NMR spectroscopy (500 MHz), and by FABMS.

Potentiometric titrations of **1** in the presence of an equimolar quantity of **2**, were conducted in water at 25°C to detect the formation of possible complexes and determine their stability. The procedure has previously been described by us [10]. Potentiometric data were treated with the programme Hyperquad [11]. The protonation constants of CD-EA have been measured previously [10], and full dissociation of CD-BS was confirmed.

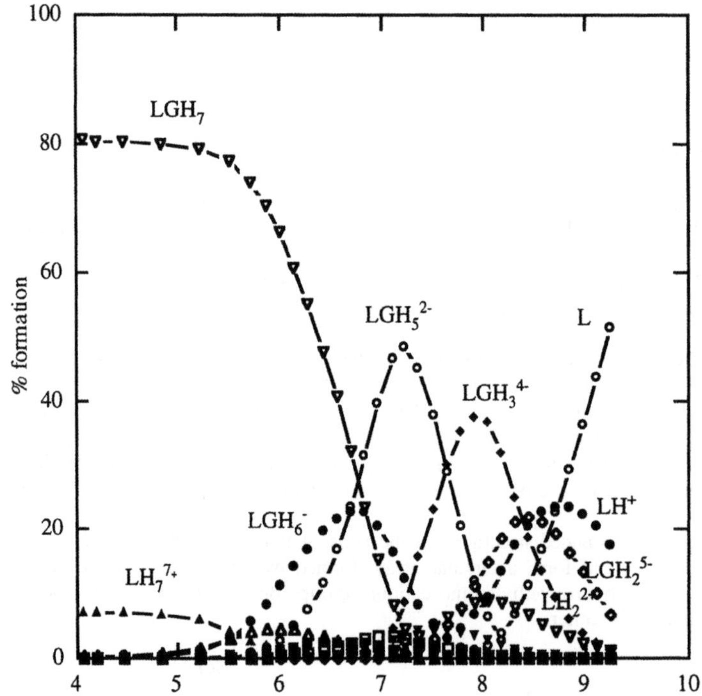

FIGURE 1: Distribution curves of the heterodimers versus pH

3. Results and Discussion

The experimental titration curve was compared to the simulated titration curve of a mixture of the compounds assuming no interaction between them. A significant deviation of the experimental curve in the pH region 5-9 was observed, supporting the concept of ionic interaction between the two CD's. In order to obtain a fit of the experimental points, new equilibria describing the interaction of protonated forms of CD-EA with CD-BS were introduced, assuming that a 1:1 stoichiometry would be the most probable. A satisfactory fit was found after the introduction of five heterodimers of the general formula $(EA)(BS)H_i$, i=2,3,5,6,7. The distribution curves of the dimers versus pH are shown in Figure 1. The percentage of formation generally decreases as the global charge of the species is increased.

The logarithms of the overall (β) and stepwise (K) stability constants of these heterodimers are given in the Table. It should be mentioned that the existence of other species is not excluded but could not be detected.

The stability of the dimers increases with the protonation degree of CD-EA. The most stable dimer, as expected, is formed between the fully protonated CD-EA (LH_7^{7+}) and the heptasulfonato CD-BS, consistent with maximal electrostatic interaction. The resulting species is neutral and, because of the similar circular distribution of the charges on each CD, may reasonably be considered as a face-to-face dimer.

TABLE: Logarithms of the overall (β) and stepwise (K) stability constants of the complexes of CD-EA (L) with CD-BS (G) (I = 0.1M; 25°C). Charges are omitted for simplicity

$L + G + 2H \rightleftharpoons LGH_2$	21.1(3)
$L + G + 3H \rightleftharpoons LGH_3$	29.5(2)
$L + G + 5H \rightleftharpoons LGH_5$	44.9(3)
$L + G + 6H \rightleftharpoons LGH_6$	51.8(6)
$L + G + 7H \rightleftharpoons LGH_7$	58.6(4)
$LH_2 + G \rightleftharpoons LGH_2$	4.4(3)
$LH_3 + G \rightleftharpoons LGH_3$	5.3(2)
$LH_5 + G \rightleftharpoons LGH_5$	7.0(3)
$LH_6 + G \rightleftharpoons LGH_6$	7.2(6)
$LH_7 + G \rightleftharpoons LGH_7$	8.5(4)

4. Conclusions

The stability of the neutral heterodimer, although high (log K = 8.5), is lower than that reported for the heptazwitterionic heterodimer formed between βCD derivatives bearing respectively seven primary amino groups and seven thiocarboxylate groups (log K = 10.25) [8]. There is obviously an unfavourable effect due to the interacting charges on one cyclodextrin being on the aromatic-group extension rather than on the more fixed circle of the cyclodextrin face. In contrast, we have shown already that the presence of multiple hydrophobic chains on the cavity influences complexation by cyclodextrins, since they form an extension of the cavity on the more hydrophobic side [4]. It is possible that the hydroxyethyl and aromatic groupings in the present modified cyclodextrins also aid the polar interactions by providing a less aqueous environment.

The formation constants of these dimers show that a very stable assembly is formed between the fully deprotonated sulfonato derivative and the fully protonated amino derivative, which constitutes the assembly of multiple extended atomic groupings on cyclodextrin as template.

5. Acknowledgments

We gratefully acknowledge the award of a Marie Curie Fellowship (to P. S.) by the European Commission, and of research studentships (to A. H. and F. O'K.) by Forbairt, the Irish Agency for Science and Innovation.

References

[1] Szejtli, J. and Osa, T. (eds.), *Comprehensive Supramolecular Chemistry*, Pergamon Press, Oxford, vol.3, 1996

[2] Wenz, G., *Angew. Chem., Int. Ed. Engl.*, **33**, 803 (1994)

[3] Kawabata, Y., Matsumoto, M., Tanaka, M., Takahashi, H., Irinatsu,Y., Tamura, S., Tagaki, W., Nakahari, H. and Fukuda, K., *Chem. Lett.*, 1933, (1986); Parrot-Lopez, H., Ling., C.-C., Zhang, P., Baszkin, A., Albrecht, G., de Rango, C. and Coleman, A.W., *J. Am. Chem. Soc.*, **114**, 5479 (1992)

[4] Ling, C.-C., Darcy, R. and Risse, W., *J. Chem. Soc., Chem. Commun.*, 438 (1993)

[5] Dey, J., Schwinté, P., Darcy, R., Ling, C.-C., Sicoli, F. and Ahern, C., *J. Chem. Soc. Perkin Trans. 2.*, in pre

[6] Ahern, C., Darcy, R., O'Keeffe, F. and Schwinté, P., *J. Incl. Phenom. Mol. Recog. Chem.*, **25**, 43 (1996)

[7] Klufers, P., Schumacher, J., *Angew. Chem. Int. Ed. Engl.*, **33**, 1863 (1994)

[8] Hamelin, B., Jullien, L., Guillo, F., Lehn, J.-M., Jardy, A., De Robertis, L. and Driguez, H., *J. Phys. Chem.*, **99**, 17877 (1995)

[9] Defaye, J., Gadelle, A., Coste-Sarguet, A., Darcy, R., McCarthy, K. and Lynam, N., in Duchêne, D. (ed.), *Proceedings Fifth Internat. Symposium on Cyclodextrins*, Editions de Santé, Paris, 1990, p. 184

[10] Schwinté, P., Darcy, R. and O'Keeffe, F., *J. Chem. Soc. Perkin Trans. 2*, 805 (1998)

[11] Gans, P., Sabatini, A. and Vacca, A., *Talanta*, **43**, 1739 (1996)

INSOLUBLE POLYMERS WITH HIGH AMOUNTS OF βCD: CHARACTERIZATION AND ADSORPTION CAPACITY

Sabrina Bertini[a], Gregorio Crini[a], Anna-Maria Naggi[a], Roberta Suardi[b], Giangiacomo Torri[a], Carmen Vecchi[b] , L. Janus[c], B. Martel[c] and M. Morcellet[c].

a) Istituto Scientifico di Chimica e Biochimica "G.Ronzoni" Vía G. Colombo 81, 20133 Milano, Italy

b) *Stazione Spetimentale per i Combustibili, Viale De Gasperi 3, 20097 San Donato Milanese, Italy*

c) Laboratoire de Chemie Macromoléculaire, UA CNRS 351, Université des Sciences et Technologies de Lille, 59655 Villeneuve d'Ascq., France

ABSTRACT.

New insoluble gels with a high content of βCD have been synthesized and their adsorption capacity compared to those of known products.

The sorption capacities of these products toward some aromatic water pollutants have been tested and compared with those of traditional gels.

The compounds tested show a greater affinity for the products with high amounts of βCD than for traditional gels. The increase of the sorption capacity is a function of the experimental conditions used for the polymerization.

INTRODUCTION

Cyclodextrins polymerized with epichlorohydrin (Epi) form well known insoluble polymer gels (1-3) used in many fields such as the pharmaceutical and food industries (4-6). These materials, able to form inclusion complexes, have also a potential use in waste water treatment (7).

To increase the sorption capacity of the gels, new products with a high content of βCD have been synthesized.

Samples with 60%-80 % weight of βCD have been obtained with a polymerization carried out in presence of low quantities of water.

The sorption capacities of these products toward some water pollutants, like β-naphtol (BN) and p-nitrophenol (pNP), have been tested both in batch and in open and continuous column and compared with those of traditional gels (20% weight of βCD)

EXPERIMENTAL

Materials

The βCD supplied by Janssen Chimica was recrystallized from water and dried overnight under vacuum at 110 °C.

The gel A was obtained with the synthetic procedure already describes (8).

The new gets were obtained with low quantities of NaOH, low temperatures and short reaction times.

In Table 1 are summarized the experimental conditions of reaction

TABLE 1

Sample	V NAOH 23%(wlw)	Reaction Time (h)	Reaction Temperature °C	%CD (w/w
A	100 ml	24	80	20
B	4 ml	5	50	57
C	4 ml	2	50	60
D	3 ml	5	50	70
E	3 ml	24	22	80

Batch Tests
15 mg. of the gel were mixed with 5 ml of an acqueous solution of pollutant (5.10^{-4} M) and stirred for a fixed time. After centrifugation the residual concentration of the organic solute was detected by UV measurements and the sorption capacity calculated.

Column Tests
Two tanks, one for the solvent and the other for the sample, a peristaltic pump, a two-way collector, a 1 ml column and a fraction collector were used for the open column tests. The column was filled with 50 mg of gel and eluted, at the beginning, only with water and then with a water solution of the aromatic compound (5.10^{-4} M).
For the continuous column tests, a fixed volume of the water solution of the pollutant (5.10^{-4} M) was eluted trough the column many times (more than 500 minutes).
The sorption capacity was calculated by UV measurements of the eluted solutions at fixed times.

RESULTS
An excess of βCD with respect to the pollutant was used for the batch tests.
In Fig.1 are reported the % of the complexed BN and pNP obtained with all the gels tested in aqueous solution.

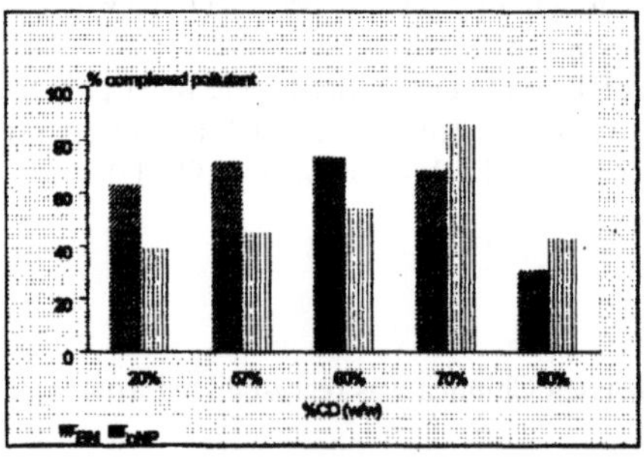

FIGURE 1

To evaluate the influence of the pH on the sorption capacity of the gels, the tests have also been carried out in buffer solution at pH 11.

The results are reported in Fig.2.

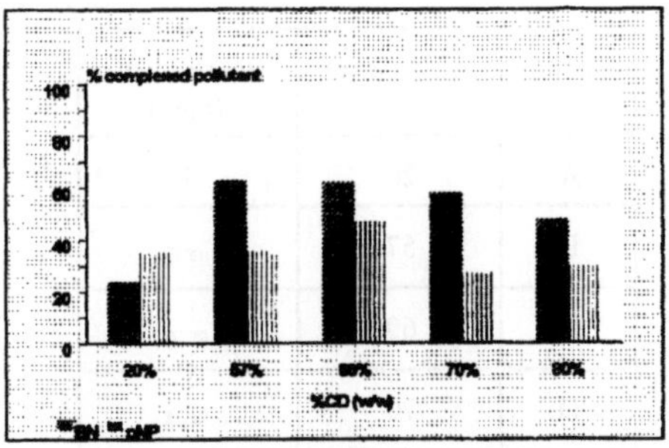

FIGURE 2

A general decrease of the sorption is observed at pH 11. This reflects the fact that the inclusion completes with βCD and aromatics are less stable at pH 11 than in water (9). The results obtained at pH 11 are more in agreement with the association constants between βCD and the guests than those obtained in water.

The new gels exhibit higher sorption capacity toward the aromatic compounds tested than gel A and this property is better pointed out at pH 11.

The low adsorption capacity of gel E could be attributed to a high and different reticulation induced by the great Epi/CD molar ratio (1/60 instead of 1/20) used in the reaction mixture. This supports the view that the sorption capacity of the gels is not only a function of the CD content but also of the structure of the material.

To better simulate real conditions, some tests have been done in open and continuous columns. The results are shown in Table 2 and 3.

TABLE 2 Open system

Sample	%CD (w/w)	% CD complexed
A	20	>100
B	57	>100
C	60	>100
D	70	87
E	80	65

TABLE 3 Continuous system

Sample	%CD (w/w)	% CD complexed
A	20	>>100
B	57	>100
C	60	>100
D	70	98
E	80	93

An adsorption higher than expected is observed for all he gels when an excess of BN is used and part of the adsorbed is not removed by washing with water whatever the quantity of water used. These results confirm the view that some interactions take place between Epi and the BN.

When an excess of βCD is used in a continuous column system, the BN is almost completely adsorbed by the gels (Table 4).

TABLE 4

Sample	%CD (w/w)	BN residual (ppm)
A	20	1
B	57	≅4
C	60	≅ 4
D	70	≅ 4
E	80	≅ 6

CONCLUSIONS

By using particular conditions of reaction, it is possible to obtain gels with a very high amount of βCD.
The increase of the βCD content increases the sorption capacity of the gels toward the aromatic compounds tested.
It was found that sometimes, besides the complexation of the guest, a chemical physical interaction between the material and the pollutant takes place. This kind of interactions change the adsorption capacity of the products which becomes not only a function of the quantity of βCD and the value of the association constant of the inclusion complex, but also of the chemical structure. In this case the availability of the βCD becomes more important than its quantity.

ACKNOWLEDGEMENTS
We gratefully acknowledge the financial support of the European Union through the FAIR program no CT 950300.

References

1] J. Solms and R. H. Egli, Helv. Chim. Acta, **48**,1225-1228,1965
2] J. L. Hoffman, J. Macromol. Sci. Chem., 1147-1157, **A7** 1973
3] M.Komiyama, I.Sugiura and H Hirai, Polym. J., **17**,1225-1227,1985
4] E. Fenyvesy, J. Incl. Phenom., **6**, 537-545,1988
5] C. S. Su and C. P. Yang, J. Sci. Food Agric., **54**, 635-643, 1991
6] P. L. Irwin, G. King and K. B. Hicks, Carbohydr. Res., **282**, 65-79,1996
7] G. Crini, S. Bertini, , G.Torri, A. Naggi, D. Sforzini, C. Vecchi, L. Janus, Y. Lekchiri and M. Morcellet, accepted by J. Applied Polym. Sci.
8] G. Crini, Y. Lekchiri and M. Morcellet, Chromatographia, **40**,296-302, 1995
9] J. Szejtii, "Cyclodextrins and their Inclusion Complexes", Akademiai Kíado, Budapest, 1982

Thanks to the European Community for its financial support through the FAIR program CT 95 0300

INFLUENCE OF pH ON THE STABILITY OF THE INCLUSION COMPLEX OF METHYL ORANGE AND β-CYCLODEXTRIN

J. CARRAZANA GARCÍA, W. AL-SOUFI, M. NOVO RODRÍGUEZ, J. VÁZQUEZ TATO

Departamento de Química Física, Facultad de Ciencias de Lugo, Universidad de Santiago, E-27002 Lugo, Spain.

1. Introduction

Methyl Orange (MO) has often been used as a model guest for physicochemical studies of cyclodextrins (CDs) inclusion complexes. This may be due to the fact that the stability constants of MO with CDs can be determined by visible absorption spectroscopy, a technique which is available in most laboratories. Moreover, this indicator also allows the determination of stability constants of CDs with other guests by means of the competition method [1]. Nevertheless, the published values for the stability constants of MO with the native CDs differ significantly depending on the experimental conditions used and on the care taken in the data treatment. Thus, for the complex formation of MO with β-cyclodextrin (β-CD), values of 2.8 x 10^3 mol^{-1} dm^3 [2], 3.8 x 10^3 mol^{-1} dm^3 [3] and 4.9 x 10^3 mol^{-1} dm^3 [4] for the stability constant of the MO/CD complex at pH 7 and 25ºC have been reported. In more basic media (around pH 10) other authors found MO to bind to β-CD with stoichiometries 1:1 and 1:2, reporting values of 3 x 10^3 mol^{-1} dm^3 and 0.6 x 10^3 mol^{-1} dm^3 for the first and second association constants at 25ºC, respectively [5,6]. In acid media much lower values of the complex stability constant have been reported, as 2.7 x 10^2 mol^{-1} dm^3 [7] and 3.52 x 10^2 mol^{-1} dm^3 [8]. The first of these values is for the apparent stability constant, since the existence of two complexes of 1:1 stoichiometry, the ammonium and the azonium inclusion complexes, is proposed in that work.

These disagreements show the complex nature of the system MO/CD and lead to the question whether and under which conditions MO is a good model guest for the study of CDs inclusion complexes. In order to answer this question we have performed a systematic study of the inclusion-complex formation between MO and β-CD as a function of β-CD concentration, pH and absorption wavelength.

2. Materials and methods.

β-Cyclodextrin (Roquette) was dried to constant weight and Methyl Orange (Panreac) was used without further purification. The measuring solutions were prepared using the step-by-step dilution-extraction method [9]. A phosphate buffer (pH 7.0, 0.05 mol dm^{-3}) was used to prepare the solutions at constant pH 7.0, whereas in the solutions at pH 1.0

no buffer was needed and the proton concentration was adjusted by adding HCl to the stock solutions. The concentration of MO was 2.2×10^{-5} mol dm^{-3} in all these solutions and the β-CD concentration was varied between 6×10^{-5} and 9×10^{-3} mol dm^{-3}. For the measurements with varying pH, the MO concentration was 4.4×10^{-5} mol dm^{-3} and the concentration of β-CD was not higher than 5×10^{-3} mol dm^{-3}. No buffer was used in these solutions. In all measuring solutions the β-CD concentration was at least ten times higher than the MO concentration. Absorption spectra were measured with a Varian Cary 1E spectrophotometer thermostated by a water bath at 20ºC.

3. Results and discussion.

The absorption spectra of MO with different concentrations of β-CD at pH 7.0 were measured. A blue shift of the absorption band of MO (λ_{max} = 465 nm) is observed when increasing the β-CD concentration due to the inclusion complex formation. Isosbestic points are observed up to a β-CD concentration of about 1.5×10^{-3} mol dm^{-3}, but at higher concentrations the spectra shift further in blue and do not go through the isosbestic points anymore. Furthermore, the non-linear analysis of the experimental data on the basis of the formation of a single 1:1 complex leads to different values of the stability constant at different wavelengths. Moreover, this simple model does not explain the experimental data at wavelengths below 480 nm. These facts lead to the conclusion that two inclusion complexes MO/β-CD with stoichiometries 1:1 and 1:2 are formed. In order to determine the corresponding stability constants, a global non-linear analysis has been performed in which a function based on the model of formation of these two complexes was fitted to the experimental data of absorbance vs. β-CD concentration at two different wavelengths, 430 nm and 500 nm. The values calculated in this way were $K_{B1} = (3.5 \pm 0.1) \times 10^3$ mol^{-1} dm^3 and $K_{B2} = 32 \pm 8$ mol^{-1} dm^3, for the formation constants of the 1:1 and the 1:2 complexes, respectively. These values have been used to determine the pure absorption spectra of the two inclusion complexes. The resulting spectrum for the 1:2 complex shows a slight shift to lower wavelengths (λ_{max} = 450 nm) with respect to the absorption spectra of MO and the 1:1 complex, and at wavelengths above 500 nm it has molar extinction coefficients very similar to those of the 1:1 complex. Since most of studies at pH 7 have used wavelengths in this range for the data treatment and did not reach high enough concentrations of β-CD, it is explainable that the 1:2 complex was not observed [2-4].

The existence of the two inclusion complexes of MO with β-CD has only been reported in more basic media [5,6]. In these studies a slightly lower value of K_{B1} and a much higher value of K_{B2} (see introduction) were obtained in comparison to our values at pH 7. The small disagreement between the two K_{B1} values may be due to the effect of the β-CD acid-base equilibria at basic pH values, since the pK_a of β-CD is 12.20 at 25ºC [1]. This effect has already been observed in our preliminary measurements of the MO/β-CD system at pH 12, whose analysis yield a significantly lower value of K_{B1} at this pH value than at pH 7. Concerning the great difference between the values of K_{B2}, it may be due to the range of wavelengths used for the data analysis (490-530 nm), where the two complexes can hardly be distinguished.

Taking into account the pK_a value of MO ($pK_a = 3.53$ at 20°C) another series of measurements has been performed at pH 1.0, where the MO is totally protonated. Under these conditions the MO absorption spectrum shifts to the red ($\lambda_{max} = 507$ nm) with respect to that at pH 7. When adding β-CD, no spectral shift is observed, but the absorbance decreases as the β-CD concentration increases. An isosbestic point is observed at all β-CD concentrations, suggesting the formation of only one complex MO/β-CD of stoichiometry 1:1 under these conditions. The non-linear analysis under this assumption of the experimental data absorbance/β-CD concentration at different wavelengths yields the same value of $K_{A1} = 305 \pm 10$ mol^{-1} dm^3. This result supports the hypothesis that only a 1:1 complex is formed between the protonated MO and β-CD at pH 1 and is in agreement with the literature [7,8]. This seems to be reasonable since the protonation of the tertiary nitrogen in MO leads to a positive charge which makes it more difficult for this side of the molecule to enter into the β-CD.

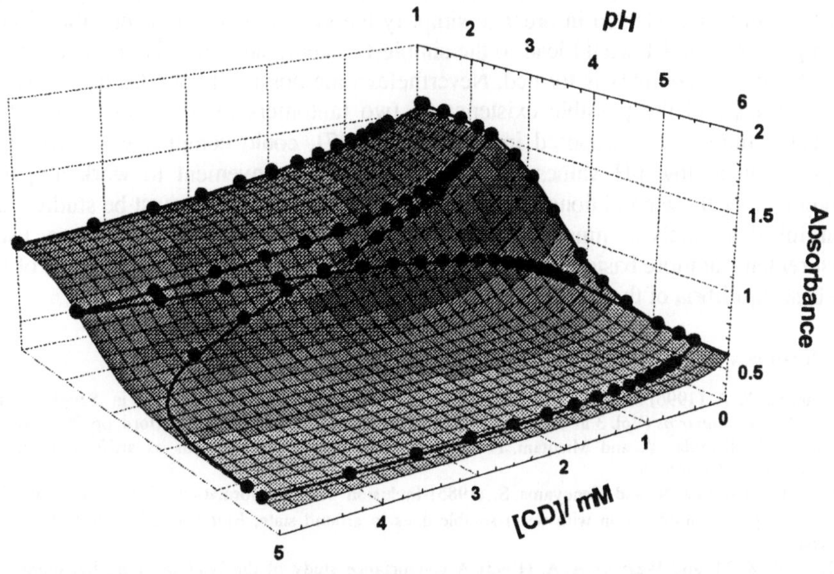

Fig.1. Absorbances at 500 nm of the MO/β-CD system versus β-CD concentration and pH. Filled circles are the experimental data and lines are fitted curves using the model proposed.

Until now we have studied the MO/β-CD system at such pH values that the MO is either in its acid form (pH 1) or in its basic form (pH 7), but the system becomes more complicated if a pH value in between is chosen. Under such acidity conditions both the acid and the basic forms of MO are present and can form complexes with the β-CD. The concentrations of the different species depend then on pH and β-CD concentration, and their contribution to the absorbance is different at the different wavelengths. This complicated system can be simplified if the interference of the 1:2 complex formed by the basic form of MO is avoided by using adequate experimental conditions (i.e., low β-CD concentrations and wavelengths for data analysis longer than 490 nm). Under such conditions a model which takes into account the acid-base equilibrium of MO (acidity constant: K_a), the formation of a 1:1 complex between the acid form of MO and β-CD (stability constant: K_{A1}) and the formation of a second 1:1 complex between the basic form of MO and β-CD (stability constant: K_{B1}) can satisfactorily explain the experimental data. This model was tested with a set of experimental data at 500 nm obtained from absorption spectra measured with varying β-CD concentrations and pH values (Fig. 1). On the basis of the proposed equilibria, an equation was deduced which relates the three variables (absorbance, β-CD concentration and proton concentration) through seven parameters: absorbances of the four pure species, K_a, K_{A1} and K_{B1}. Fixing the parameter K_a with the value determined in an independent experiment, the other six parameters could be fitted to the experimental data. The values obtained for both stability constants were in agreement with those obtained before from the series at constant pH values.

In view of all these results, we conclude that MO, as a model guest for the study of the formation of CDs inclusion complexes, must be used with care. The experimental conditions must be selected in order to simplify the system as much as possible. Thus, working at about pH 1 would lead to the simplest system, since MO is in its acid form and only one 1:1 complex is formed. Nevertheless, the possibility of CDs hydrolysis at high acidities and the possible existence of two tautomers of MO which form two different complexes as proposed in the literature [7], could complicate the study of complexation at low pH values. It is therefore more convenient to work at pH 7, although under these conditions a large range of CD concentration must be studied and a careful data analysis must be done in which different wavelengths are used. Measurements at more basic pH values are not recommended due to the existence of the acid-base equilibria of the CDs.

4. References

1. Connors, K. A. (1996) Measurement of cyclodextrin complex stability constants, in Szejtly, J. and Osa, T., *Cyclodextrins* (Vol. 3 in *Compreh. Supramol. Chem.*), Elsevier Science, Oxford, pp. 205-241.
2. Tabushi, I., Kuroda, Y. and Mizutani, T. (1984) Functionalized cyclodextrins as artificial receptors, *Tetrahedron*, 40, 545-552.
3. Hirai, H., Toshima, N. and Uenoyama, S. (1985) Inclusion complex formation of γ-cyclodextrin. One host-two guest complexation with water-soluble dyes in ground state, *Bull. Chem. Soc. Jpn.*, 58, 1156-1164.
4. Tawarah, K. M. and Wazwaz, A. A. (1993) A conductance study of the binding of methyl orange, o-methyl red and p-methyl red anions by β-cyclodextrin in water, *Ber. Bunsenges. Phys. Chem.*, 97, 727-731.
5. Fujita, K., Ejima, S. and Imoto, T. (1984) Fully collaborative guest binding by a double cyclodextrin host, *J. Chem. Soc., Chem. Commun.*, 1277-8.
6. Schneider, H.-J. and Xiao, F. (1992) Binding and catalysis with a metal-induced temary complex of a ethylenediamine-sustituted cyclodextrin, *J. Chem. Soc., Perkin Trans. 2*, 387-391.
7. Tawarah, K. M. (1992) A thermodynamic study of the inclusion processes of α- and β-cyclodextrins with the acid forms of methyl orange and methyl yellow, *J. Incl. Fen. Mol. Recogn.*, 14, 195-204.
8. Wang, A. S. and Matsui, Y. (1994) Solvent isotope effect on the complexation of cyclodextrins in aqueous solutions, *Bull. Chem. Soc. Jpn.*, 67, 2917-2920.
9. Jover, A., Meijide, F., Mosquera, V. and Vázquez Tato, J. (1990) A step-by-step dilution-extraction method for laboratory experiments, J. Chem. Edu., 67, 530-532.

AN IMPROVED SYNTHESIS OF PER(6-DEOXYHALO) CYCLODEXTRINS USING *N*-HALOSUCCINIMIDES — TRIPHENYLPHOSPHINE IN DIMETHYLFORMAMIDE

KAZIMIERZ CHMURSKI AND JACQUES DEFAYE
CNRS (EP 811) and Université Joseph Fourier - Grenoble 1, Département de Pharmacochimie Moléculaire / Glucides, BP 138, F-38243 Meylan, France

Abstract

Per(6-deoxy-6-halo) cyclodextrins (bromo, chloro or iodo) have been prepared in excellent yield and high selectivity by treatment of the native cyclodextrins with the corresponding *N*-halosuccinimide and triphenylphosphine in *N,N*-dimethylformamide.

1. Introduction

Per(6-deoxyhalo)cyclomaltooligosaccharides (cyclodextrins, CD) constitute an important and versatile class of compounds which has already been widely exploited for the preparation of numerous useful cyclodextrin derivatives arising from either intra- [1] or intermolecular displacement reactions [2]. A number of methods for the per(C-6) halogenation of cyclodextrins have been already reported involving, as a general concept, reaction of the native CD with a Vilsmeier dimethylimidoyl halide reagent in *N,N*-dimethylformamide. In the pioneer report of Takeo et al. [3], the reagent was generated in situ from methanesulfonyl bromide and DMF. Improved results have been reported with the Evans' et al. [4] methanesulfonyl chloride in DMF reagent, alone [5] or in the presence of imidazole [6]. In situ formation of the Vilsmeier iodide or bromide from triphenylphosphine and iodine or bromine in DMF was a further convenient improvement for the preparation of per(6-deoxyiodo) and per(6-bromodeoxy) α- and β-CD [1, 7] although prior isolation of the Vilsmeier bromide was subsequently claimed to be preferred for the preparation of **1-3** [8]. The more stable and easier to handle bromo- or chloromethylenemorpholinium bromide and chloride were quite recently proposed [9] for the preparation of per(6-bromodeoxy) **1-3** and per(6-chlorodeoxy) **4-6** cyclodextrins. We now report an even more general and high yielding methodology for the preparation of per(6-bromodeoxy)- **1-3**, per(6-chlorodeoxy)- **4-6**, and per(6-iododeoxy)cyclomaltooligosaccharides **7-9** based on the use of commercially available *N*-halosuccinimides (bromo-, chloro-, or iodo-) in the presence of triphenylphosphine, a reagent previously introduced by Hanessian et al. [10] for the halogenation of monosaccharides and nucleosides.

2. Results and Discussion

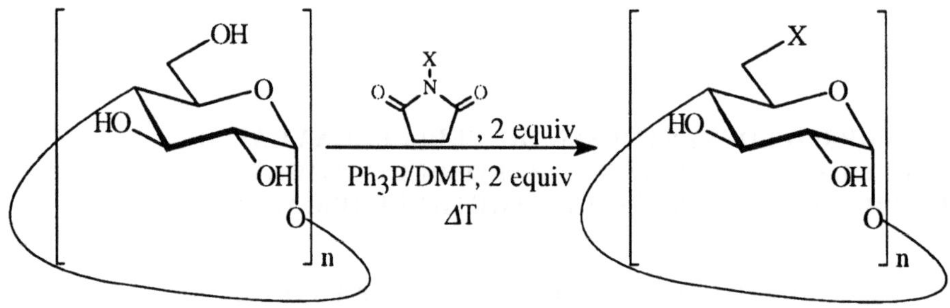

	n=6	n=7	n=8
CD	α	β	γ

	n=6	n=7	n=8
X = Br	**1**	**2**	**3**
X = Cl	**4**	**5**	**6**
X = I	**7**	**8**	**9**

Reaction of cyclomaltooligosaccharides (-hexaose, -heptaose, or -octaose) in DMF with 2 equiv of each *N*-bromo- or *N*-chlorosuccinimide and triphenylphosphine at 70-80 °C for 3-4 h, followed by the addition of methanol in order to destroy the excess of reagent and subsequent alkalinisation to pH 9 of the reaction mixture, yielded the corresponding per(6-bromo) or per(6-chloro) cyclodextrins as almost pure amorphous powders by pouring the solution into water, and washing out the residual triphenylphosphine oxide with methanol from the recovered precipitate.

Table I. Yields and physical data for per(6-deoxyhalo) cyclodextrins **1-9**

CD	yield (%)	mp (°C)	$[\alpha]_D$ (°) (*c* 1, DMF)	FAB (glycerol) *m/z* $[M + H]^+$	Lit.	mp (°C)	$[\alpha]_D$ (°)
1	85	194	+92.1	1350.4	[7]	222	+124 (*c* 1.5, DMF)
2	96	202	+84.7	1576.5	[7]	214	+78 (*c* 1.8, DMF)
3	88	198	+105.5	1800.6	[2]	223	+111.6 (*c* 0.2, Me_2SO)
4	76	256	+112.4	1104.5 $[M + Na]^{+a)}$	[5]	-	-
5	93	240	+113.8	1285.0 $[M + Na]^{+b)}$	[5][6]	> 265	-
6	93	222	+134.8	1445.1	[5]	-	-
7	81	231	+64.7 (*c* 0.5, DMF)	1654.4 $[M + Na]^+$	[1][7]	227	+95 (*c* 1.3, DMF)
8	89	218	+66.3 (*c* 0.5, DMF)	1925.9 $[M + Na]^+$,	[1] [7]	235 224	+66.1 (*c* 1.1, DMF) +79.5 (*c* 1, DMF)
9	81	213	+75.3 (*c* 1.1, Me_2SO)	2198.7 $[M + Na]^+$	[12]	231	+72.8 (*c* 1.1, Me_2SO)

a) NBA matrix
b) glycerol/Me_2SO matrix.

Physical data for per(6-bromo-6-deoxy) α-, β-, and γ-CD **1-3** agree with lit. (Table 1). For the chlorodeoxy derivatives **4-6**, for which such data have been scarcely reported, ^1H and ^{13}C NMR spectra displayed only one set of signals for C-1—C-6 as required for the

six, seven, and eight fold symmetries expected for uniformly substituted compounds. The FABMS, which showed for halo derivatives **1-6** prominent molecular ions, confirmed the purity of all compounds (Table 1).
When, however, the same protocol as above was applied to the preparation of per(6-deoxy-6-iodo) CD **7-9**, using *N*-iodosuccinimide as halide donor, the FABMS displayed minor signals at 110 mu or multiple of 110 mu below the expected molecular ion indicating some uncomplete substitution. Kuszmann and coll. [11] have already pointed out that two pathways involving either a dimethylformamidium halide or a phosphonium halide are both operative in the *N*-halosuccinimide—triphenylphosphine reaction scheme with alcohols. Enhanced susceptibility towards hydrolysis of the formamidinium iodide salts might explain the uncomplete substitution with the *N*-iodosuccinimide reagent. In fact, when the solution of α-, β-, and γ-CD in DMF was dried by azeotropic distillation with toluene prior to the reagent addition, pure per(6-deoxy-6-iodo) α-, β-, and γ-cyclodextrins were obtained in high yield (Table 1).

3. Material and Methods

3.1 MATERIALS

α-, β-, and γ-cyclodextrins were pharma grade from Wacker Chemie (München), dried at 110 °C (1.33 10^{-4} kPa) for 48 h prior use. DMF was freshly distilled over calcium hydride and stored over molecular sieves. TLC was performed on E. Merck plastic sheets coated with Silica Gel 60 F_{254}, using 7:7:5:4 EtOAc — 2-propanol — 25% aq NH_4Cl — water as eluent. The spots were visualised in UV light and by dipping into 15% aq H_2SO_4 containing 2% of ammonium dimolybdate and 1% of cerium(IV) sulfate followed by heating at 110 °C. FAB mass spectra were recorded with an AMD604 Intectra spectrometer.

3.2 METHODS

3.2.1 General procedure for the preparation of per(6-bromo-6-deoxy) and per(6-chloro-6-deoxy) cyclodextrins **1-6**.
To a solution of dried cyclomaltooligosaccharide (1 mmol) in DMF (40 mL), the corresponding *N*-halosuccinimide (12-16 mmol, 2 equiv/OH-6) and triphenylphosphine (12-16 mmol, 2 equiv/OH-6) were added at ambient temperature. The reaction mixture was protected by a drying tube ($CaCl_2$) and heated for 3-4 h at 70-80 °C. After completion (TLC), MeOH was added at room temperature and stirring was continued for 30 min. The reaction mixture was then cooled to ~15 °C and the pH was adjusted to 9 with 3 M MeONa in MeOH, while stirring for further 30 min. It was then poured into stirred ice-water (600 mL) resulting in the formation of a precipitate which was filtered (fritted glass no. 3), washed with MeOH, and dried. Yields and physical data with reference to lit. are collected in Table 1.

3.2.2 Per(6-deoxy-6-iodo) cyclomaltooligosaccharide **7-9**.
To a solution of dried cyclomaltooligosaccharide (1 mmol) in DMF (40 mL), toluene (3 x 10 mL) was added and evaporated using a rotatory evaporator. To the resulting solution, *N*-iodosuccinimide (18-24 mmol, 3 equiv/OH-6) and triphenylphosphine (18-

24 mmol, 3 equiv/OH-6) were added with stirring. The reaction mixture was heated at 90 °C for 20 h while protected by a drying tube. Isolation of the products followed the above procedure.

4. References

1. Gadelle, A., Defaye, J. (1991) Selective halogenation at primary positions of cyclomaltooligosaccharides and a synthesis of per-3,6-anhydro cyclomaltooligosaccharides, *Angew. Chem. Int. Ed. Engl.*, 30, 78-80; *Angew. Chem.*, 103 (1991) 94-95.
2. Ling, C-C., Darcy, R. and Rise, W. (1993) Cyclodextrin liquid crystals: synthesis and self-organisation of amphiphilic thio-beta-cyclodextrins, *J. Chem. Soc., Chem. Commun.*, 438-440; Parrot-Lopez, H., Ling, C-C., Zang, P., Baszkin, A., Albrecht, G., de Rango, C. and Coleman, A.W., (1992) Self-assembling system of the amphiphilic cationic per-6-amino-beta-cyclodextrin, *J. Am. Chem. Soc.*, 114, 5479-5480; Chmurski, K., Coleman, A.W. and Jurczak, J. (1996) Direct synthesis of amphiphilic α-, β, and γ-cyclodextrins, *J. Carbohydr. Chem.*, 15, 787-796.
3. Takeo, K., Sumimoto, T. and Kuge, T. (1974) An improved synthesis of 6-deoxy-analogues of cyclodextrins and amylose, *Stärke*, 26, 111-118.
4. Evans, M.E., Long, L., Jr and Parrish, F.W. (1968) Reaction of carbohydrates with methylsulfonyl chloride in *N,N*-dimethylformamide. Preparation of some methyl 6-chloro-6-deoxyglycosides, *J. Org. Chem.*, 33, 1074-1076.
5. Guillo, F., Hamelin, B., Jullien, L., Canceill, J., Lehn, J-M., De Robertis, L. and Driguez, H. (1995) Synthesis of symmetrical cyclodextrin derivatives bearing multiple charges, *Bull. Soc. Chim. Fr.*, 132, 857-866.
6. Khan, A.R., D'Souza, V.T. (1994) Synthesis of 6-deoxychlorocyclodextrin via Vilsmeier-Haack-type complexes, *J. Org. Chem.*, 59, 7492-7495.
7. Baer, H.H., Vargas Berenguel, A., Shu, Y.Y., Defaye, J., Gadelle, A. and Santoyo Gonzáles, F. (1992) Improved preparation of hexakis(6-deoxy)cyclomaltohexaose and heptakis(6-deoxy)cyclomaltoheptaose, *Carbohydr. Res.*, 228, 307-314.
8. Vizitiu, D., Walkinshaw, C.S., Gorin, B.I. and Thatcher, G.R.J. (1997) Synthesis of monofacially functionalized cyclodextrins bearing amino pendent groups, *J. Org. Chem.*, 62, 8760-8766.
9. Chmurski, K., Defaye, J. (1997) An improved synthesis of 6-deoxyhalo cyclodextrins via halomethylenemorpholinium halides Vilsmeier-Haack type reagents, *Tetrahedron Lett.*, 38, 7365-7368.
10. Hanessian, S., Ponpipom, M.M. and Lavallee, P. (1972) Procedures for the direct replacement of primary hydroxyl groups in carbohydrates by halogen, *Carbohydr. Res.*, 24, 45-56.
11. Hodosi, G., Podányi, B. and Kuszmann, J. (1992) The mechanism of the hydroxyl→halogen exchange reaction in the presence of triphenylphosphine, *N*-bromosuccinimide, and *N,N*-dimethylformamide: application of a new Vilsmeier-type reagent in carbohydrate chemistry, *Carbohydr. Res.*, 230, 327-342.
12. García Fernández, J.M., Ortiz Mellet, C., Jiménez Blanco, J.L., Fuentes Mota, J., Gadelle, A., Coste-Sarguet, A. and Defaye, J. (1995) Isothiocyanates and cyclic thiocarbamates of α,α'-trehalose, sucrose, and cyclomaltooligosaccharides, *Carbohydr. Res.*, 268, 57-71.

Chapter 3

CYCLODEXTRINS IN PHARMACY AND BIOTECHNOLOGY

ANTI-ASTHMATIC-CYCLODEXTRIN INCLUSION COMPLEX FOR PULMONARY DELIVERY

JOSE LEITE PINTO, HELENA MARIA CABRAL MARQUES
UCTF
Faculdade de Farmácia, Universidade de Lisboa, Av. Forças Armadas, 1600 LISBOA, Portugal

1. Introduction

Dry Powder Inhalers (DPIs) are devices used for pulmonary administration of drugs. They have known an increasing importance, namely in asthma treatment, in substitution of MDIs due to the ban of CFC propellants, which destroy the ozone layer (1).

The efficiency of the pulmonary administration of a drug is restricted by the anatomic configuration of the lung tree, the aiways becoming progressively narrower as the drug approaches the action site (lower respiratory tract - bronchioli). In order to successfully achieve that site, two major factors have to be considered in the pharmaceutical formulation design:

- The powder formulation - which must deliver the drug in the lower respiratory tract.
- The delivery device - which must be designed to generate an aerosolised cloud of the powder mixture.

Both factors are intimately related since the efficiency determined for a given drug formulation-Inhaler device combination cannot be expected to be reproducible if one of these factors is changed. DPIs formulations are typically composed by powder mixtures of the active substance and an excipient, usually lactose.

Cyclodextrins (CYD) can be used in order to suitably modify the biopharmaceutical properties of drugs (2). In the present work a new approach to the formulation of powders for inhalation is proposed, by using an anti-asthmatic drug-CYD complex.
After isolation of this complex in the solid state, a powder formulation was developed with the following purposes: to facilitate the preparation of homogeneous powder mixtures for DPIs; to increase the stability of the drug and the respiratory fraction.

In order to determine the deposition of the emitted dose, an inhalation device - the Micro-haler™ and the Twin Impinger apparatus proposed in the European Pharmacopoeia were used. A comparative study of different powder mixtures is presented: the drug-CYD complex, the physical mixture of the drug and the CYD and the drug alone.

2. Materials

HP-B-CYD, supplied by Janssen; micronised drug, batch n° 23; Purified water; Lactose, supplied by V. Reis, Ltd.; Alcohol, commercial grade; Methanol P.A.; Metallic sieve of 180 μm mesh aperture; gelatin capsules, size n° 3.

3. Methods

Phase solubility studies: phase solubility studies were performed according to Higuchi and Connors (3) and at least a week was necessary to reach the equilibrium. UV spectrophotometry was used as the analytical method (r=0.999; n=4). By determining the linear regression from the straight line portion of the curve obtained by plotting drug molar concentrations against CYD molar concentrations, the equilibrium constant (Ks) was calculated. This constant measures the strength of the complex obtained.

Preparation of the complex: drug in excess was added to a 0.05M CYD aqueous solution and placed in a water bath at 37°C, with constant stirring, for a week. After equilibrium, the solution was filtered through a 0.45 micron filter and 15 ml aliquots were distributed in glass vials for freeze-drying to obtain the solid complex.

Evidence of complex formation: three different instrumental methods of analysis were used in liophylised samples:
Differential Scanning Calorimetry (DSC); all measurements were performed using a Mettler DSC 25 apparatus.
X-Ray Diffractometry; the X-ray powder diffraction patterns were performed using a «Miniflex» desktop X-ray diffractometer.
Infra-Red Spectrophotometry (IR); the IR spectra were performed using a Pye-Unicam apparatus.

Preparation of powder formulations: the complex containing 0.1 mg of drug per 88.3 mg of powder) was passed through a 180 μm mesh operture. For the physical mixture, 5.9 mg of the drug were mixed with 4.994 g of CYD (same proportion as that of the complex) as follows: the powders were first passed through a metallic sieve of 180 μm mesh aperture and then mixed, by rolling for 10 minutes, in a glass vial. The powder mixture was passed through the same 180 μm sieve.

Testing of the capsules in the «Twin impinger»: the two powders above and the micronised drug alone, were filled in gelatin capsules, which were then tested for determination of the deposition of the emitted dose through an inhalation device, the Micro-haler™, using the Twin Impinger apparatus proposed in the European Pharmacopoeia. The Micro-haler™ was adjusted to the Twin Impinger Throat by means of a mouthpiece adapter. 7 ml and 30 ml of methanol PA were transferred to the Upper impingement chamber and Lower impingement chamber, respectively, and the system was operated for 5 seconds at 60 l/min vacuum. After operation, the apparatus was dismantled and the pieces corresponding to each of the four compartments (A - Device+Capsule ; B - Throat ; C - Upper and D - Lower chamber) were washed with the same solvent, the washings being used to perform the required dilution volumes.

4. Results and Discussion

Phase solubility studies: according to Higuchi and Connors (3), the strength of the drug-CYD complex can be estimated from the phase-solubility curve. This is obtained by plotting drug molar concentrations against CYD molar concentrations (Fig 1).

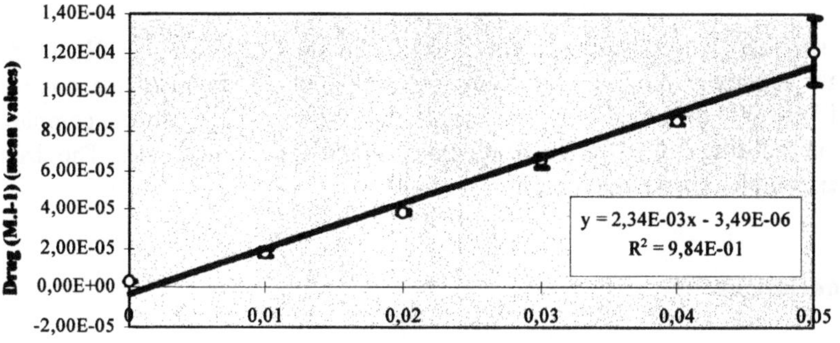

Fig 1. Phase solubility Drug-HP-B-CYD diagram

The strength of the complex formed, as measured by the solubility constant was Ks = 805 M-1 (R = 0.9920) and is within the range reported in the literature (4), thereby allowing to conclude that a stable complex anti-asthmatic drug-CYD was formed. A A_L type curve was obtained showing that a soluble complex was formed.

Proof of complex formation: the DSC thermograms of the drug-CYD complex show that the endothermic peak corresponding to the melting temperature of the drug has disappeared in the drug-CYD preparation obtained by freeze drying, while it is present in the physical mixture of the two compounds, suggesting the complex formation.
The X-ray diffraction pattern of the drug is not comparable to those of the two drug-CYD preparations complex showing that a new solid phase was constituted.
The Infra-Red studies show a spectral change in the region between 1600 - 1760 cm^{-1} (carbonyl-stretching bands), suggesting the inclusion of these groups in the CYD.

Twin Impinger test: the results obtained from the Twin Impinger for the three formulations tested are shown in Fig 2.

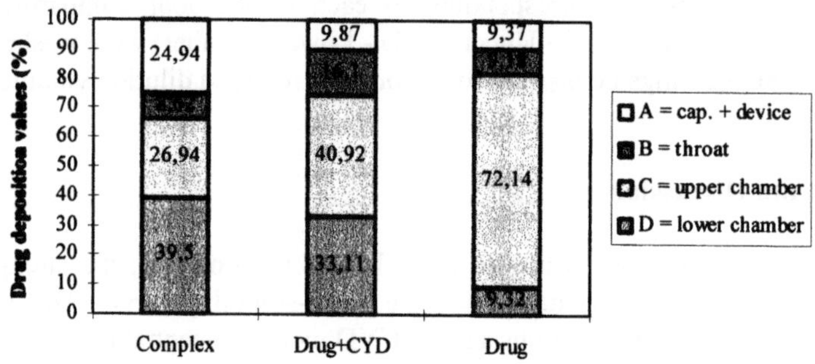

Fig 2. Twin Impinger drug deposition

The *in vitro* aerosol deposition results obtained with the Micro-haler™ DPI and the Twin Impinger apparatus show a significant respirable fraction for both the complex and drug-HP-B-CYD mixtures used. The results are in the following order: complex alone > micronised drug + CYD physical mixture > micronised drug alone. The drug-CYD complex sample was not micronised.

5. Conclusions

A complex of drug-HP-B-CYD was formed as shown by Differential Scanning Calorimetry, X-Ray spectroscopy and Infra-Red absorption studies.

From the above results it is apparent that the use of anti-asthmatic drug-CYD complexes show a potential therapeutic interest for pulmonary administration.

6. References

(1) Ogden, J., Rogerson, C. and Smith, I., Scrip Mag., June, 56 (1996)
(2) Cabral Marques, H.M., Rev. Port. Farm., Vol. XLIV, 77-83 (1994)
(3) Higuchi, T. and Connors, K., Adv. Anal. Chem. Instrum. Vol. 4, 117-212 (1965)
(4) Uekama, K., Fujinaga, F., Hirayama, F., Otagiri, M. and Yamasaki, M., Int. J. Pharm., 10, 1-15 (1982)

7. Acknowledgements

The authors express their appreciation to the Rigaku International Corporation, Japan, for performing the X-Ray analysis of the samples.

CYCLODEXTRIN-CONTAINING MEMBRANES. SYNTHESIS AND SEPARATION PROPERTIES IN PERTRACTION

P. BALDET-DUPY AND A. DERATANI
Laboratoire des Matériaux et Procédés Membranaires
UMR 5635 CNRS-ENSCM-Université Montpellier 2,
2, place Eugène Bataillon, 34095 Montpellier cedex, France

Abstract

Polymeric membranes were prepared by crosslinking polyvinylalcohol (PVA) and cyclodextrins (CDs) with epichlorohydrin or hexamethylenediisocyanate. Immobilized CDs were shown to enhance the permeation rate of toluene in a water/methanol pertraction system depending on the CD content (0-50 wt%) and the cavity size (α-, β-, γ-CD). Further increases were observed by replacing α-CD with an α-CD polyrotaxane. This effect was interpreted by the cooperativity of CD cavities in the toluene permeation across channels formed inside the membrane.

1. Introduction

Advantage can be taken of the molecular recognition by CDs to perform extractions of valuable compounds or organic pollutants from aqueous solutions. Several techniques for removal and recovery by complex formation have been proposed such as selective precipitation, liquid-liquid extraction and sorption onto CD-immobilized polymer network (see for examples [1-3]). We now report on a continuous liquid-liquid extraction technique based on a membrane process using CD-containing films.

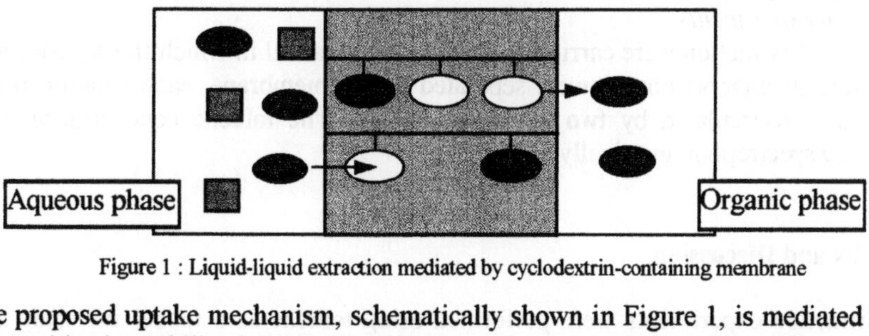

Figure 1 : Liquid-liquid extraction mediated by cyclodextrin-containing membrane

The proposed uptake mechanism, schematically shown in Figure 1, is mediated by the CD cavities immobilized on the membrane matrix. It consists of three different steps : complex formation at the upstream side of the membrane (aqueous phase), diffusion from a complexing site to another inside the membrane, and complex dissociation at the downstream side (organic phase). It is expected that molecules having a high affinity for CDs can be transfered to the downstream side more rapidly than others, providing that their diffusivites into the membrane polymer are not too low. From this assumption, selective extraction of compounds from aqueous solutions can be carried out using this membrane process so-called pertraction, the driving force being the differential of solvation energy between the aqueous and the organic phases.

2. Experimental

Materials

PVA (degree of polymerisation 1600, degree of hydrolysis 99.5 mol.-%) from Fluka, epichlorohydrin and hexamethylenediisocyanate from Sigma-Aldrich were used as purchased. CDs were gifts from Orsan (α-, β-) and from Wacker (γ-). All other materials and solvents obtained from Sigma-Aldrich were of analytical grade. 18 MΩ MilliQ water was used for the preparation of aqueous solutions.

Preparations of membranes

- *method A : supported membranes* - A 4 wt.-% PVA solution in DMSO (5 g) was added to the desired amount of a 5wt.-% CD solution in 2M NaOH under stirring. 0.33 cm^{-3} of epichlorohydrin were added when the mixture was homogeneous. The supported film was then cast onto an ultrafiltration membrane (Molecular/Por® type C, MWCO 10,000; Spectrum Medical Industries Inc.) using a doctor blade technique with a gap of 100 μm and dried at 60°C for 3h30. The resulting thickness of the dense layer was about 20μm.

- *method B : free standing membranes* - The desired amount of CD was dissolved into a 4 wt.-% PVA solution in DMSO (10 g). 0.027 cm^{-3} of 1.5 wt.-% hexamethylenediiso-cyanate solution in DMSO were then added and the free standing was immediately cast on a glass plate according to the same work-up as above described. The film was removed from the glass plate by immersion into warm water.

The membranes obtained by both methods were kept in water which was exchanged several times to remove all residual reactants. The CD content was determined by quantitative FT-IR measurements. The absence of defects (crack, pin hole) was checked by gas permeation.

Pertraction experiments

Liquid/liquid extraction were carried out at 25°C using a cell in which the aqueous and the organic (methanol) phases were separated by the membrane, each solution being continuously recirculated by two peristaltic pumps. The toluene concentration was determined spectrophotometrically at 254 nm.

3. Results and Discussion

The pertraction experiments were performed using toluene as a model molecule and methanol as the organic solvent. It should be noted that the organic phase can be miscible in water in this process since the membrane acts as a separator. Liquid/liquid extractions can be achieved by to two procedures : (a) the feed solution contains a given starting amount of compound and (b) the concentration of the feed solution is kept constant. Results of permeation kinetics were plotted in Figure 2 for both cases. As shown, the permeation rate remained at a constant value in procedure (b) equal to the initial rate in procedure (a). This measure is proportionnal to the membrane efficiency.

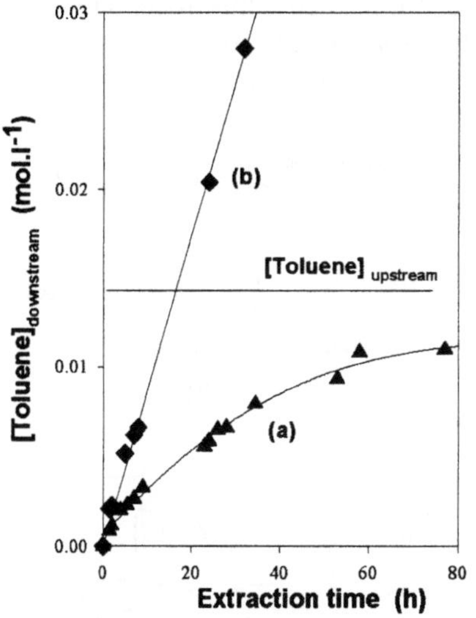

Fig.2 : Kinetics of toluene extraction by procedure (a) ▲ and (b) ◆ (see text)

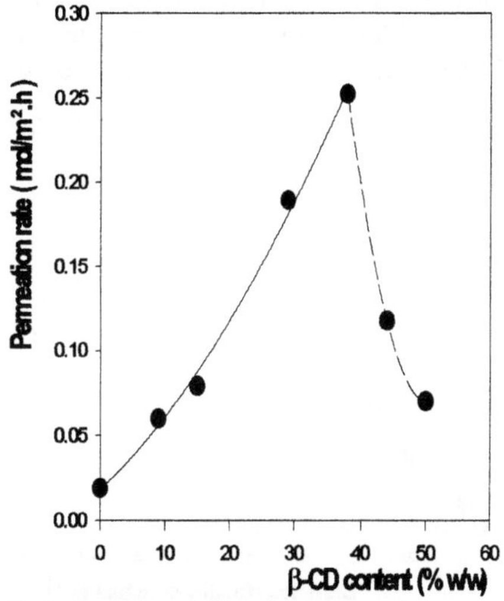

Fig.3 : Dependance of the toluene permeation rate on the membrane β-CD content

The facilitated transport of toluene is linked to the presence of CD cavities inside the membrane as shown in Figure 3. The strong enhancement of the pertraction rate observed by increasing the β-CD content demonstrates that the complex binding to CD can accelerate the transfer of molecules to the downstream side of the membrane. The small amount of toluene extracted when using the PVA membrane (β-CD content 0 %) is related to the diffusivity through the polymer matrix. However, as can be seen , the permeation rate decreased dramatically beyond a content value of 38 wt.-%. The formation of crystallized β-CD microdomains, as evidenced by X-ray diffraction, can account for this observation (Fig. 4). The same phenomena were also observed for α- and γ-CD content values higher than 50 and 40 wt.- %, respectively.

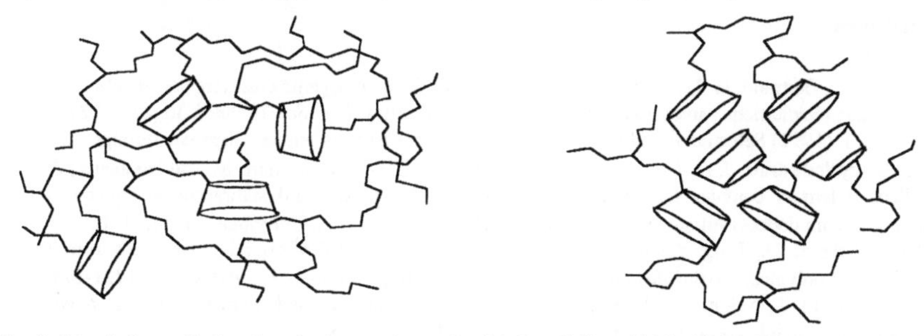

Fig. 4 : Morphology of microdomains in membranes having low (left) and high (right) CD content (see text).

The binding affinity of toluene depends on the size of the CD cavity thereby affects the membrane efficiency. Figure 5 displays the toluene permeation rate for membranes containing the same molar fraction of α-, β- and γ- CD (33, 38 and 41 wt.-%) as a function of the stability constant of the inclusion complex in solution. An excellent correlation is observed, the better efficiency being obtained for β-CD based films. These results seem to indicate that the efficiency of CD-containing membranes can be evaluated from the binding constant values in solution. Actually, this assumption was not verified for all the hydrocarbons tested (results not shown). Further studies will be undertaken to understand the permeation mechanism through these membranes.

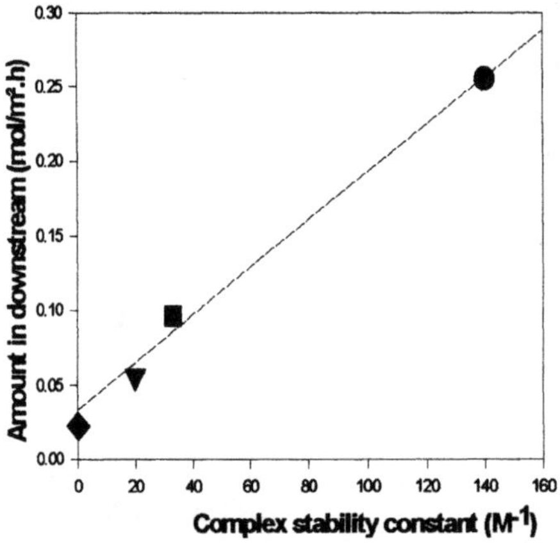

Fig; 5 : Toluene pertraction rate for PVA (◆), α- (■), β- (●) and γ- (▼) CD-containing membranes

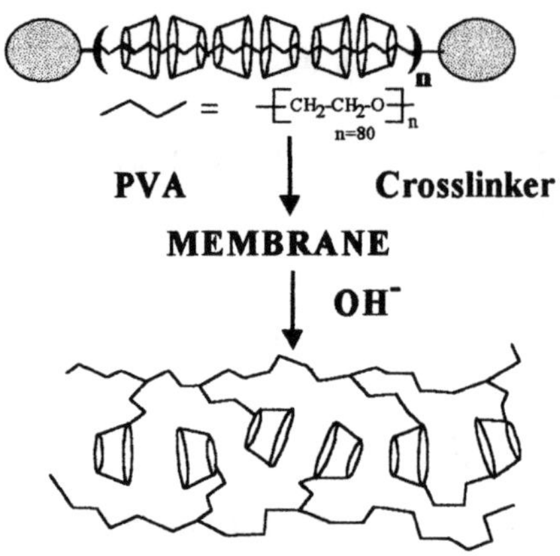

Fig. 6 : Preparation of α-CD polyrotaxane membranes

Finally, in order to improve the performances of pertraction process, membranes were prepared from α-CD polyrotaxane prepared according to the procedure described by Harrada [4]. The films obtained were then treated with a strong alkali to remove the end-stoppers and the linear chain (Fig. 6). The idea is to form channels inside the membrane in which cooperativity can occur between the attached CDs. The efficiency of the toluene pertraction was doubled by using polyrotaxane instead of native α-CD as a starting material. We are now working to determine the selectivity of this process in liquid/liquid extractions of complex mixtures.

References

[1] Zhou, E. Y., Bertrand, G. L., Armstrong, D. W. (1995) Effects of organic cosolvents on enantio-enrichments via CD-based precipitations : an examination of production efficiency, *Sep. Sci. Technol.* 30, 2259-2276.

[2] Uemasu, I. (1992) Selective liquid-liquid extraction of xylene isomers and ethylbenzene through inclusion by branched α-cyclodextrins, *J. Incl. Phenom.* 13, 1-7; Andreaus, J., Draxler,J., Marr, R., and Hermetter, A. (1997) The effect of ternary complex formation on the partitioning of pyrene and anthracene in aqueous solutions containing sodium dodecyl sulfate and methylated γ-cyclodextrin, *J. Colloid Interface Sci.* 193, 8-16.

[3] Warner-Schmid, D., Tang, Y.and Armstrong, D. W. (1994) Removal of organic compounds from water via adsorption onto polymethylhydrosiloxane pentenyl-β–cyclodextrin, *J. Liquid Chromatogr.* 17, 1721-1735.

[4] Harada, A., Li, J., Nakamitsu, T., Kamachi, M. (1993), Preparation and characterization of polyrotaxanes containing many threaded α-cyclodextrins, *J. Org. Chem.* 58, 7524-7528

EFFICACY AND SAFETY OF CYCLODEXTRINS IN NASAL DRUG DELIVERY

F.W.H.M. MERKUS, P.H.M. VAN DER KUY, E. MARTTIN, S.G. ROMEIJN, J.C. VERHOEF,
Division of Pharmaceutical Technology and Biopharmaceutics, LACDR, Leiden University, PO Box 9502, 2300 RA Leiden, the Netherlands

1. Introduction

The intranasal application of tobacco snuff, cocaine, various psychotropic and hallucinogenic agents has been known for a long time. It is therefore surprising that only in the past two decades intranasal administration of systemic drugs has attracted much attention. The nasal route circumvents the first-pass elimination associated with oral drug delivery. Furthermore, nasal drug delivery is an attractive alternative to the injection therapy, it is easily accessible and suitable for self administration.
Despite numerous references in the recent literature on nasal drug delivery, the list of compounds that are currently on the market or investigated in patients or volunteers is limited. Examples are desmopressin, vasopressin, oxytocin, buserelin, nafarelin, calcitonin, insulin, glucagon, human growth hormone, butorphanol, dihydroergotamine, vit B 12, metoclopramide, midazolam, nicotine, steroid hormones, scopolamine, sumatriptan.
The majority of the investigations published so far demonstrate, mainly in animal experiments, the large potential of nasal drug delivery, but only a few authors realise that large interspecies differences exist in the nasal absorption of drugs. In human subjects the potential for nasal drug formulations is limited to drugs which are active in a low dose and possess a sufficient aqueous solubility. Many lipophilic drugs are poorly soluble in water and large hydrophilic drugs like peptides and proteins show an insufficient nasal absorption. Cyclodextrins, especially methylated ß-cyclodextrins, have proven to be excellent solubilizers and absorption enhancers in nasal drug delivery.

2. Cyclodextrins as excipients in nasal drug formulations

A pharmaceutical excipient used as solubilizer and absorption promoter in nasal drug delivery should be potent in a very low concentration, but inert from a pharmacological-toxicological point of view. This means that the selected excipient should (a) have no local or systemic effect, (b) exert no damage to the mucosal integrity, (c) show no severe ciliostatic effect, (d) enhance the drug permeation through the nasal epithelium in a transient and reversible way and (e) should be nonirritating and nonallergenic. Also, the chemical and pharmaceutical quality of the selected cyclodextrin are important issues. For instance, methylated ß-cyclodextrins are available in various qualities. It is possible to prepare a pure 2,6 dimethyl ß-cyclodextrin (DMßCD; degree of substitution 2.0), but selective methylation of the 2- and 4 -OH group requires expensive solvents and a production process causing environmental pollution. Commercially available products consist of about 75% dimethylated ß-cyclodextrin (of which 65% is 2,6 dimethylated). Randomly methylated ß-cyclodextrin (RAMEB; degree of substitution 1.8) consists of about 50% dimethylated ß-cyclodextrin (of which about 25% is 2,6 dimethylated), but the production is much cheaper, whereas the chemical and pharmaceutical properties are similar. An additional advantage is that the randomly methylated product is amorphous and avoids the risk of an in vivo crystallization. The (di)methylated ß-cyclodextrins are extremely water soluble.

2.1 LIPOPHILIC DRUGS

The nasal administration of the female steroid hormones estradiol and progesterone has been studied in animals and humans [1-4]. Nasal administration of estradiol makes it possible to decrease the dose administered compared to oral administration, circumventing high blood levels of the metabolite estrone and thus providing a physiological estrone/estradiol ratio [4]. Estradiol was administered with DMßCD to rats and rabbits, resulting in mean absolute bioavailabilities of 94.6% and 67.2%, respectively [1]. Also in oophorectomized women estradiol and DMßCD were administered nasally, giving a rapid absorption of estradiol [3]. During a 6-month trial estradiol replacement therapy was achieved in oophorectomized postmenopausal women without side effects [3]. A combination of progesterone and estradiol with DMßCD was administered in rats and humans, resulting in nasal absorption comparable to the separate administration of both steroids [2, 4].

The lipophilic antiviral drug pirodavir was given intranasally to humans, with 10% hydroxypropyl-ß-cyclodextrin as solubilizer [5]. Frequent intranasal sprays (6 times daily) were effective in preventing the development of clinical colds following experimentally induced rhinovirus infection. However, irritating effects of the formulation on the nasal mucosa were observed, such as nasal dryness and blood in the mucus. These side effects were attributed to the viscosity of the administration vehicle, and the high frequency of administration [5].

Another example is dihydroergotamine (DHE), an effective antimigraine drug. In a number of countries DHE is on the market as a nasal preparation (e.g. Migranal, Diergo). This nasal formulation contains 4 mg/ml DHE, glucose (5%) and caffeine (1%). The spray is available in an ampule which has to be opened when the migraine attack occurs, then provided with a spraying device, and subsequently 4 puffs (2 puffs in each nostril) of 0.125 ml can be administered to achieve a dose of 2 mg DHE. This large volume of 0.500 ml that has to be administered and the fact that the open ampule is only stable for 24 hours are serious disadvantages.

New nasal DHE formulations have been developed by combining DHE with a cyclodextrin to enhance the concentration and improve the stability. Liquid and powder formulations were prepared, containing dihydroergotamine mesylate (DHEM) in combination with the cyclodextrin derivative RAMEB. In rabbits, liquid and powder formulations were compared with the currently available product and it turned out that it was possible to prepare a stable nasal formulation with a pharmacokinetic profile in rabbits similar to the product on the market [6].

In a randomized cross-over study, 5 different preparations of DHEM (with at least 1 week interval) were administered to 9 healthy human subjects [7]. Blood samples were taken at t=0 and during 8 hours after drug administration. The preparations and doses administered were: [A] DHEM i.m. 0.5 mg (Dihydergot, in which the drug is dissolved in an ethanol-glycerol-water solution); [B] DHEM nasal 2 mg as Diergo nasal spray, which means 1 puff of 0.125 ml in each nostril, repeated after 1 minute, thus in total 4 puffs; [C] DHEM nasal 2 mg as liquid (DHEM 10 mg, RAMEB 20 mg, mannitol 50 mg, water 1 g), which means 1 puff of 0.100 ml in each nostril; [D] DHEM nasal 2 mg as powder (DHEM 2 mg, RAMEB 4 mg, lactose 4 mg), which means about 5 mg powder in each nostril; and [E] DHEM 2 mg oral as solution. No serious adverse effects were reported by the volunteers. No statistically significant difference in Tmax, Cmax, AUC and absorption rate could be found between the three nasal applications, indicating that the two DHEM/RAMEB formulations have pharmacokinetic properties which are comparable to the currently available product. The preference of the volunteers was clearly in favour of the liquid DHEM/RAMEB nasal spray, compared to the Diergo ampule-spray because (i) a much less complicated handling of the spray and (ii) reduction of the number of puffs from 4 to 2 [7]. The better stability of the novel formulations is an additional pharmacoeconomical advantage.

2.2 HYDROPHILIC DRUGS

Nasal absorption in human subjects of hydrophilic drugs, e.g. peptides and proteins, is rather low and decreases with an increasing molecular size. In numerous animal studies, it has been demonstrated that cyclodextrins, particularly α-cyclodextrin and the methylated ß-cyclodextrins, are efficient absorption enhancers. However, large interspecies differences have been observed between rats, rabbits, other animals and human subjects in nasal absorption of peptides and proteins using cyclodextrins as absorption promoters [8, 9]. Sometimes these differences are so large, that the clinical relevance of a lot of these animal studies is questionable. In the rat-in-situ-perfusion-technique, the model itself deviates so much from the clinical reality that the results obtained by this technique are meaningless (Fig. 1) [10].

In rabbits a twofold increase in bioavailability of Org 2766 was obtained with 5% DMßCD, whereas in rats the increase in bioavailability was about fivefold [11]. The intranasal administration of calcitonin with DMßCD resulted in a 12% reduction in calcium levels in rats, but in rabbits the hypocalcemic effect of intranasal calcitonin was lower (9.5%) [12]. For leuprolide administered as drops, using α-cyclodextrin as absorption enhancer, a bioavailability of 35% was achieved in rats versus 4% in humans [13]. Nasal insulin absorption with DMßCD has been extensively investigated. The

Perfusion in Rats ➡ Extrapolation to Man

	Rats	Man	Clinical Reality	Deviation
Nasal volume	0.4 ml	20 ml		
Perfusion volume	5 ml	250 ml	0.1 ml	**2,500 x** ↓
Perfusion time	2 hr	2 hr	15 min	**8 x** ↓
				20,000 x

Figure 1. Rat-in-situ-perfusion extrapolated to men [9]

Species / Formulation Effects

insulin bioavailability

species	insulin + DMβCD solution	insulin + DMβCD powder
rat	100 %	100 % (expected)
rabbit	0 %	13 %
man	0 %	5 %

Figure 2. Species differences in nasal absorption [14-17]

intranasal bioavailability of a liquid insulin formulation was 100% in rats [14], but 0% in rabbits and man [15, 16].

However, by using a powder formulation of insulin and DMßCD an insulin bioavailability could be achieved of 3.4% in healthy volunteers and 5.1% in diabetes mellitus patients (Fig. 2) [17]. Nevertheless, the amount of insulin absorbed is very low and the absorption is too variable to warrant the large investment needed to bring such formulation to the market place. Obviously the molecular size of insulin, or aggregates of insulin, is the limiting factor in the nasal absorption in human subjects. The second, and more general conclusion is, that results in rats are not predictive for a possible absorption in man.

3. Safety of cyclodextrins in nasal drug delivery

New excipients in nasal formulations should be without local and systemic toxicity. Adverse effects on the nasal epithelial tissue and the mucociliary clearance should be avoided. The function of the mucociliary clearance is to protect the nose and the lower airways from damage by inhaled noxious substances. Several in vitro and in vivo models which investigate the effects of substances on the mucociliary system are known. In vitro ciliary beat frequency measurements have shown to be a good indicator for the effects of substances on nasal tissue morphology. With this in vitro model the potential of a substance to inhibit or arrest ciliary beating is determined. The effects of methylated ß-cyclodextrins on ciliary beat frequency demonstrate that 2% RAMEB and 2% DMßCD had similar cilio-inhibitory effects to those of physiological saline. The effects of both cyclodextrins were smaller than those of the preservative benzalkonium chloride (0.01%) [9, 18].

Ciliary beat frequency data should be interpreted carefully, because the effects of nasal formulations in vitro are more pronounced than their effects in vivo. The cilia are protected by the mucus layer and the nasal formulations are diluted by the mucus in vivo, whereas in vitro the cilia are in direct contact with the investigated substances. Furthermore, in vivo the respiratory nasal epithelium can be expected to recover from damage. Consequently, it is not possible to make predictions regarding the effects of chronic use of a formulation on mucociliary clearance in vivo based solely on the effects on ciliary beat frequency in vitro. The acute histological effects of 2% RAMEB and 2% DMßCD are minor, and quite similar to those of physiological saline. Benzalkonium chloride at 0.01% caused more changes of the nasal epithelium than the methylated ß-cyclodextrins [19]. Systemic toxicity after nasal administration of methylated ß-cyclodextrins is not expected, because very low doses are administered and only very small amounts are absorbed [9].

REFERENCES

1. Hermens, W.A.J.J., Deurloo, M.J.M., Romeijn, S.G., Verhoef, J.C. and Merkus, F.W.H.M. (1990) Nasal absorption enhancement of 17-ß-oestradiol by dimethyl-ß-cyclodextrin in rabbits and rats, *Pharm. Res.* 7, 500-503.

2. Schipper, N.G.M., Hermens, W.A.J.J., Romeijn, S.G., Verhoef, J. and Merkus, F.W.H.M. (1990) Nasal absorption of 17-beta-estradiol and progesterone of a dimethyl-ß-cyclodextrin inclusion formulation in rats, *Int. J. Pharm.* 64, 61-66.
3. Hermens, W.A.J.J., Belder, C.W.J., Merkus, J.M.W.M., Hooymans, P.M., Verhoef, J and Merkus, F.W.H.M. (1991) Intranasal estradiol administration to oophorectomized women, *Eur. J. Obs. Gynecol. Reprod. Biol.* 40, 35-41.
4. Hermens, W.A.J.J., Belder, C.W.J., Merkus, J.M.W.M., Hooymans, P.M., Verhoef, J. and Merkus, F.W.H.M. (1992) Intranasal administration of estradiol in combination with progesterone to oophorectomized women: a pilot study, *Eur. J. Obs. Gynecol. Reprod. Biol.* 43, 65-70.
5. Hayden, F.G., Andries, K. and Jansen, P.A.J. (1992) Safety and efficacy of intranasal pirodavir (R77975) in experimental rhenovirus infection, *Antimicrob. Agents Chemother.* 36, 727-732.
6. Marttin, E., Romeijn, S.G., Verhoef, J.C. and Merkus, F.W.H.M. (1997) Nasal absorption of dihydroergotamine from liquid and powder formulations in rabbits, *J. Pharm. Sci.* 86, 802-807.
7. Van der Kuy, P.H.M. Lohman, J.J.H.M., Hooymans, P.M., Ter Berg, J.W.M. and Merkus, F.W.H.M. (1998) Abstract in *Brit. J. Clin. Pharmacol.*, in press.
8. Merkus, F.W.H.M., Schipper, N.G.M., Hermens, W.A.J.J., Romeijn, S.G. and Verhoef, J.C. (1993) Absorption enhancers in nasal drug delivery: efficacy and safety, *J. Control. Rel.* 24, 201-208.
9. Marttin, E., Verhoef, J.C. and Merkus, F.W.H.M. (1998) Efficacy, safety and mechanism of cyclodextrins as absorption enhancers in nasal delivery of peptide and protein drugs, *J. Drug Target.*, in press.
10. Marttin, E., Schipper, N.G.M., Verhoef, J.C. and Merkus, F.W.H.M. (1998) Nasal mucociliary clearance as a factor in nasal drug delivery, *Adv. Drug Del. Rev.* 29, 13-38.
11. Schipper, N.G.M., Verhoef, J.C., De Lannoy, L.M., Romeijn, S.G., Brakkee, J.H., Wiegant, V.M., Gispen, W.H. and Merkus, F.W.H.M. (1993) Nasal administration of an ACTH(4-9) peptide analog with dimethyl-ß-cyclodextrin as an absorption enhancer: pharmacokinetics and dynamics, *Brit. J. Pharmacol.* 10, 335-340.
12. Schipper, N.G.M., Verhoef, J.C., Romeijn, S.G. and Merkus, F.W.H.M. (1995) Methylated ß-cyclodextrins are able to improve the nasal absorption of salmon calcitonin, *Calcif. Tissue Int.* 56, 280-282.
13. Adjei, A., Sundberg, D., Miller, J. and Chun, A. (1992) Bioavailability of leuprolide acetate following nasal and inhalation delivery to rats and healthy humans, *Pharm. Res.* 9, 244-249.
14. Merkus, F.W.H.M., Verhoef, J., Romeijn, S.G. and Schipper, N.G.M. (1991) Absorption enhancing effect of cyclodextrins on intranasally administered insulin in rats, *Pharm. Res.* 8, 588-592.
15. Merkus, F.W.H.M., Verhoef, J., Romeijn, S.G. and Schipper, N.G.M. (1991) Interspecies differences in the nasal absorption of insulin, *Pharm. Res.* 8, 1343.
16. Schipper, N.G.M., Romeijn, S.G., Verhoef, J.C. and Merkus, F.W.H.M. (1993) Nasal insulin delivery with dimethyl-ß-cyclodextrin as an absorption enhancer in rabbits: powder more effective than liquid formulations, *Pharm. Res.* 10, 682-686.
17. Merkus, F.W.H.M., Schipper, N.G.M. and Verhoef, J.C. (1996) The influence of absorption enhancers on the intranasal insulin absorption in normal and diabetic subjects, *J. Control. Rel.* 41, 69-75.
18. Romeijn, S.G., Verhoef, J.C., Marttin, E. and Merkus, F.W.H.M. (1996) The effect of nasal drug formulations on ciliary beating in vitro, *Int. J. Pharm.* 135, 137-145.
19. Marttin, E., Verhoef, J.C., Romeijn, S.G., Zwart, P. and Merkus, F.W.H.M. (1996) Acute histopathological effects of benzalkonium chloride and absorption enhancers on rat nasal epithelium in vivo, *Int. J. Pharm.* 141, 151-160.

A PRELIMINARY STUDY OF A β-CYCLODEXTRIN/SALBUTAMOL COMPLEX FOR POSSIBLE USE IN A DRY POWDER INHALER

A.M. REIS[1,2], H.M. CABRAL MARQUES[1], I. W. KELLAWAY[2]
[1]*Faculdade de Farmácia da Univ. de Lisboa, U.C.T.F.*
Avenida das Forças Armadas, 1600 Lisboa, Portugal
[2]*The Welsh School of Pharmacy, Cardiff University,*
Cardiff, CF1 3XF, Wales, U.K.

1. Introduction

Among the many potential pharmaceutical applications of cyclodextrins, their ability to prolong a biological effect can be a useful characteristic. Sustained pulmonary release of drugs that require multiple administrations per day, such as salbutamol (a β_2-selective adrenoreceptor agonist used in Asthma therapy), would be advantageous.

β-CYD is composed of 7 glucose units surrounding a 0.6-0.8 nm rigid cavity. Previous studies indicate the formation of soluble complexes between salbutamol and β-CYD with a molar ratio of 1:1 and a Ks = 66-69 M^{-1}.[1]

Lung delivery could be considered effective and efficient compared to other routes of administration (nasal and oral) since there is almost quantitative absorption - i. e. 100% bioavailability of the fraction of drug that deposits in the lower airways - due to its large surface area, low enzymatic activity and avoidance of hepatic first-pass effect. The extent of drug absorption from the lung depends on deposition and distribution of the aerosolised drug in airways. [2]

Therefore, the main aerosol characteristics affecting bioavailability of inhaled drugs are: Particle size and shape - particles greater than 10 µm deposit predominantly in the upper airways (throat and trachea) by impaction, particles smaller than 3 µm generally deposit in the lower airways (alveoli and acini) by sedimentation, particles below 0.5 µm are exhaled without deposition in the lung; Moisture and interparticle attraction; Velocity of aerosolised drug particles that results in impaction losses in aerosol devices and in the nasopharynx thus decreasing pulmonary absorption. [2,3]

The aim of this preliminary study was to characterise the CYD/salbutamol inclusion complexes in order to achieve an efficient Dry Powder Inhalation formulation.

2. Materials and Methods

2.1 MATERIALS

β-CYD was obtained from Chinoin, Budapest, Hungary, and from Berck Ltd., Basingstocke, Hants, U.K; Micronised salbutamol sulphate was obtained from Glaxo, U.K.; Chloroform was obtained from Fisher Scientific, UK Limited, Loughborough, U. K.; Polyethylene glycol 300 was obtained from BDH Organics, BDH Limited Poole, U.K.; Sorbitan trioleate (SPAN 85) was obtained from Sigma Chemical Co., St. Louis, USA; Hydrochloric acid 2 M was obtained from Fisons, Fisher Scientific Equipment, Loughborough, U.K.; 0.2 μm pore size nitro-cellulose filters, Whatman; Glass microfiber filter, 90 mm circle GF/A, Whatman.

2.2 METHODS

Preparation of a β-CYD/salbutamol inclusion complex: Three mixed 100 ml aqueous solutions were prepared by adding: 0.3590 g of micronised salbutamol (1.5×10^{-3} mol), 1.7025 g of β--CYD (1.5×10^{-3} mol) and recently distilled water. These solutions were sonicated for 10 min, protected from light and placed in a stirring water bath for approximately 24 hours (180 strokes) at 20°C as the complex formation is not significantly affected by temperature [(1)]. An 80 ml portion of each of the mixed solutions of salbutamol/β-CYD was filtered and freeze-dried in 500 ml round bottom flasks for approximately 48 hours (Super Modulyo Edwards high vacuum from Edwards, UK). In order to obtain a smaller particle size the complex samples were micronised on a Jet Mill (Glen Creston Ltd., Stanmore, Middlesex, UK). The same methodology was applied to three 250 ml -CYD aqueous solutions (C1=1.5x10-2 mol/L; C2=1.25x10-2 mol/L; C3=1.0x10-2 mol/L)

Particle sizing of β-CYD/salbutamol complex and β-CYD on a Malvern 2600 Particle Sizer:
A small amount of each sample (10 mg) was dispersed in 10 ml of a chloroform solution and SPAN 85 at 0.5% and then sonicated for 5 min. The dispersing solution was filtered twice through a 0.2 μm pore size Whatman nitro-cellulose filter membrane.

Confirming β-CYD/salbutamol complex presence in the micronised powder: In order to confirm the existence of a true complex, in the solid state, the thermal behaviour of the obtained micronised complex and the drug were studied by Differential Scanning Calorimetry - DSC (Differential Scanning Calorimeter, Perkin Elmer, DSC 7, Norwalk, Connecticut, USA) [(1,4)]. Indium was the reference material use for calibration purposes. Both the samples and the calibration compound were weighed using a Sartorius MC5 balance and sealed in standard Aluminium pans used for non-volatile samples (Aluminium's melting point = 660°C; kit no. 0219-0041). Each sample was scanned at 5°C min^{-1} in the range 50-250°C.

Preliminary aerodynamic particle size analysis: Gelatine capsules were filled with approximately 30.0 mg of β-CYD/salbutamol complex (weighed on a Sartorius MC5 balance). These capsules were then placed in a Rotahaler™ device (Allen & Hanburrys, Glaxo Group, England, U. K.).
An eight stage Andersen Mark II Cascade Impactor (1 ACFM Non-viable Ambient Particle Sizing Sampler, Graseby Andersen, USA) was used in order to obtain an aerodynamic particle

size distribution of the micronised complex samples. No preseparator was used because previous results from the Malvern Particle Sizer showed the absence of particles above 4.5 µm. A solution of 1% PEG 300/Acetone was chosen as a coating solution as PEG 300 is soluble in 1M HCl and has minimal UV absorption capacity at 276 nm. The flow rate was adjusted at 28.3 l min-1 (Singer 802 ATM-115, American Meter Division). An absolute glass microfiber filter was placed below the 6th stage to collect particles less than 0.5 µm. The aerosol was discharged from the Rotahaler manually only once at a rate of 28.3 l min^{-1} for 9 seconds assuming an average of 4000 ml lung capacity after a maximal inspiratory effort. (Total Lung Capacity (TLC) = 4200-6400 ml) [(2)]. The throat piece and each stage (1-7), as well as the filter, were washed with 10 ml of 1M HCl and the solutions were analysed by UV (Ultrospec II 4050, LKB Biochrom, Cambridge, UK) according to the method described in BP 1993. It was confirmed that β-CYD does not show absorption at 276 nm.[(1)]

3. Results and Discussion

The particle size of the micronised complex samples (Fig. 1) were as follows:

Table 1
Particle size analysis of micronised β-CYD/salbutamol complex

Sample No.	D[v, 0.5] (µm)	D[v, 0.9] (µm)	Max. Diam. (µm)	Min. Diam. (µm)
1	2.63	3.23	4,50	1,75
2	2.63	3.38	4,84	1,63
3	2.60	3.37	4,50	1,63

Quantitative size analysis of the freeze-dried and micronised β-CYD/salbutamol complex revealed a monomodal size distribution. Optical microscopy showed mainly rounded irregular forms. The particle size range was suitable to use on a Cascade Impactor as a means of obtaining the aerodynamic size distribution.

The thermogram of micronised salbutamol shows an endothermic peak at 159.4°C. The β-CYD/ salbutamol complex thermogram (Fig. 2) does not shows any peak until the higher temperature of 190.7°C. The disappearance of salbutamol endothermic peak may be attributed to the formation of an inclusion complex.

Figure 1: Complex Particle size distribution

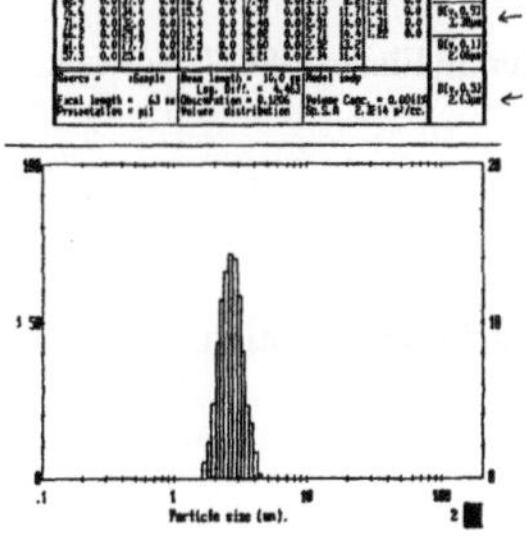

Figure 2: Micronised Complex Thermogram

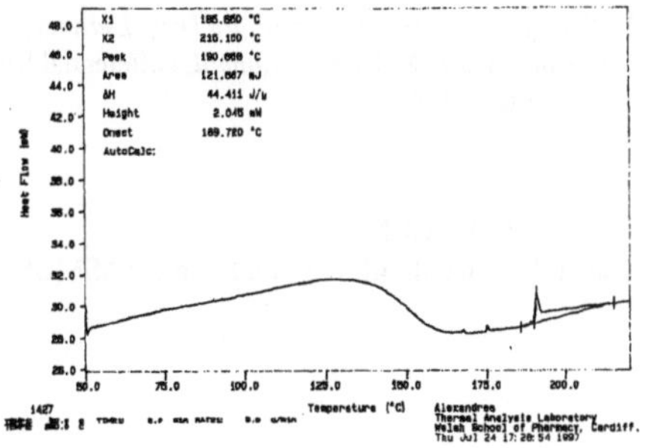

The results of aerodynamic particle size analysis and data interpretation are shown in Table 2.

TABLE 2

Stage	Diluton	Absorbance	Weight mg	% retention	Cumultive	ECD μm
Device			28,28	95,21	5,03	
Throat	10	0,179	0,15	0,51	4,50	
0	10	0,045	0,04	0,14	4,35	9
1	10	0,469	0,39	1,31	2,98	5,8
2	10	0,212	0,18	0,60	2,35	4,7
3	10	0,168	0,14	0,48	1,85	3,3
4	10	0,141	0,12	0,40	1,43	2,1
5	10	0,139	0,12	0,40	1,01	1,1
6	10	0,114	0,10	0,33	0,67	0,7
7	10	0,105	0,09	0,30	0,35	0,4
F	10	0,116	0,10	0,33		
Total in C. I.	4,2817 mg					
FPF	0,0831					
MMAD	3,8 μm					

The particle size analysis was obtain using a multistage stage Andersen Mark II Cascade Impactor. Micronised complex samples were delivered using a Rotahaler™ device on a total amount of about 5,0 mg of salbutamol and 24,5 mg of excipient per capsule. The limited number of experiments did not allow any conclusions to be drawn although a M.M.A.D. of 3,8 μm was determined. However, only a small fraction of the powder (4,7%) was aerosolized and alternative strategies are required to ensure a higher emitted dose.

4. Conclusions

Particle size distribution results clearly showed a size range with potential for alveolar deposition, i.e., within the respirable fraction (<5 μm) [(2)]. The low percentage of aerosolised powder may have occurred due to humidity exposure, agglomeration caused by electrostatic activity and/or lack of a carrier, such as lactose.

Further research is therefore required to establish whether CYDs inclusion complexes are useful as a new generation of DPI formulation.

References

1. MARQUES, H. M. C. (1991) *Salbutamol - Cyclodextrins Inclusion Complexes for Pulmonary Drug Delivery*, Ph.D., UWCC
2. HICKEY, A. J. (1996) *Inhalation Aerosols - Physical and biological basis for therapy*, Marcel Dekker, inc., USA
3. BYRON, P.R. (1990) *Respiratory Drug Delivery*, CRC Press, USA
4. *Perkin Elmer DSC 7 Users Manual*, Differential Scanning Calorimeter, Perkin Elmer, DSC 7, Norwalk, Connecticut, USA

ACKNOWLEDGEMENTS:

The authors would like to thank the ERASMUS Program for the financial contribution.

SOLUBILITY OF A ISOXAZOLYL-NAPHTHOQUINONE BY COMPLEXATION WITH HYDROXYPROPYL-ß-CYCLODEXTRIN

MARCELA LINARES, MARIA DE BERTORELLO, MARCELA LONGHI,*
Departamento de Farmacia, Facultad de Ciencias Químicas, Universidad Nacional de Córdoba, Agencia Postal 4, C.C. 61, 5000 Córdoba, Argentina.
FAX: 54-51-33-4163.
* *Corresponding author. E-mail: mrlcor@dqo.fcq.unc.edu.ar*

1. Introduction

2-Hydroxy-N-(4-methyl-5-isoxazolyl)-1,4-naphthoquinone-4-imine (**1**) has been studied as a potential synthetic antibacterial and trypanosidal agent due to the well-known biological properties of a structurally related compound (Bogdanov et al., 1993; Albesa et al., 1995; Schwarcz et al., 1990). As **1** is very poorly soluble in water weak acid, the study of its biological properties turned many times difficult.

Cyclodextrins are cyclic oligosaccharides which are known to form inclusion complexes with many lipophilic drugs, changing their physicochemical and biopharmaceutical properties (Loftsson and Brewster, 1996), with the result of an increased aqueous solubility and rate of dissolution (Loftsson et al., 1990). The hydroxyalkylated cyclodextrins appear to be more suitable for the solubilization of drugs than the non-substituted parent cyclodextrin (CD), because of their greater aqueous solubility and lack of toxicity (Backensfeld et al., 1991).

In this study, the possibility of inclusion complexes formation between **1** and hydroxypropyl-ß-cyclodextrin (HP-ß-CD) was studied with the aim of improving water solubility characteristics.

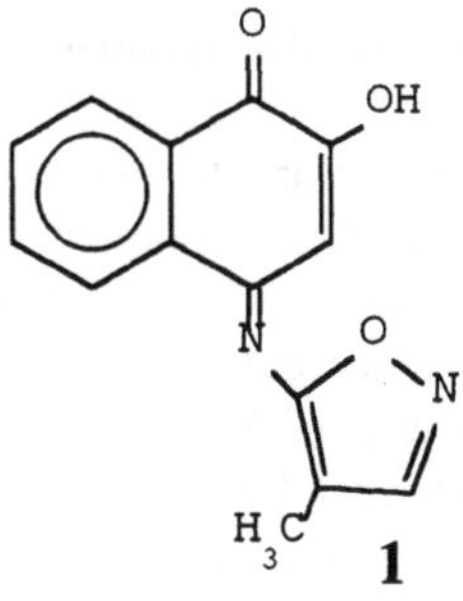

1

2. Materials and Methods

2.1 MATERIALS

The synthesis and identification procedures for **1** have been described previously (Fernández et al., 1982). HP-ß-CD (mw 1326-1400, degree of molar substitution, 7.0) was a gift from CERESTAR USA, INC. (Hammond, IN). All other materials and solvents were of analytical reagent grade.

2.2 BUFFERS

McIlvaine buffers (pH 3.00-8.00) were prepared according to the literature (Elving et al., 1956). A KH_2PO_4/Na_2HPO_4 buffer was used at pH 7.40. The water used for the buffers was generated by a Millipore Milli-Q Water purification system.

2.3 SOLUBILITY STUDIES

Phase solubility studies were carried out as described by Higuchi and Connors (1965). Excess amounts of **1** (≈ 5.00 mg) were added to the buffer solutions at different pH values which varied from 3.50 to 8.00, containing different CD concentrations (from 0.0 to 53.0 % (w/v)). The suspensions formed were sonicated in an ultrasonic bath for 1 h and then placed in a 25.0 ± 0.1°C constant-temperature water bath. After equilibration up to 72 h, an aliquot was filtered through a 0.45 μm membrane filter (Micron Separations Inc., USA), the equilibrium pH of each solution was measured (ORION SA520 pH-meter), suitably diluted, and analyzed by UV.

2.4 ANALYTICAL METHODS

The quantitative determinations of **1** were made spectrophotometrically (Shimadzu UV 260 UV/visible spectrophotometer).

3. Results and discussion

The phase solubility diagrams of **1** in aqueous HP-ß-CD solutions and various pH values at 25° C are shown in Fig. 1. At pH values below 6.00, the curves obtained were of A_L-type, however at pH values above 6.00 the solubility curves showed negative curvature and could be classified as those of A_N-type. In order to investigate the A_N-type behaviour, the solubility of **1** in buffer at pH 6.52 was compared with that in distilled water (pH 6.48). The obtained results are shown in Fig. 2. An A_L diagram was obtained for **1** and HP-ß-CD in water whereas in the buffer the solubility curve was of the A_N-type. Although the slopes of the initial parts of the curves were very

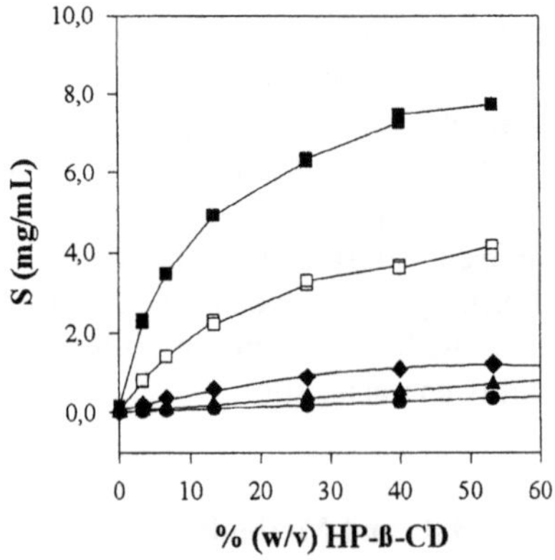

Figure 1. Phase solubility diagrams of **1** at 25°C at different pH values. Key: (●) pH 3.53; (▲) pH 5.05; (◆) pH 6.55; (□) pH 7.41; (■) pH 8.03.

similar, when the concentration of HP-ß-CD was raised the solubility of **1** in water was higher than the solubility in the buffer. The pKa for **1** is 5.54 (Longhi, 1989), i.e., the drug is almost fully ionized at pHs higher than 6.00. Hence, if for water the isotherm was linear, the negative curvature in the buffer could be associated with the salt formation. This interpretation is well supported by the study carried out using two buffers of different composition at the same pH. The results are shown in Fig. 3. A_N isotherms were obtained for **1** and HP-ß-CD both in the citrate and in the phosphate buffer, pH 7.41, albeit the solubility and the stability constant ($K_{1:1}$) were found to be higher in citrate buffer.

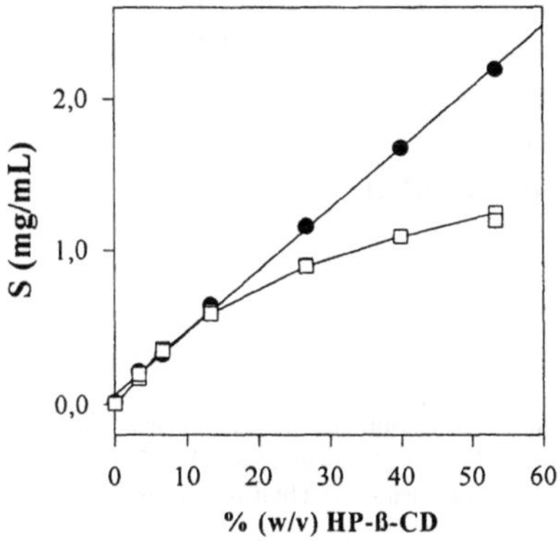

Figure 2. Phase solubility diagrams of **1** in water (●) and in buffer pH 6.52 (□).

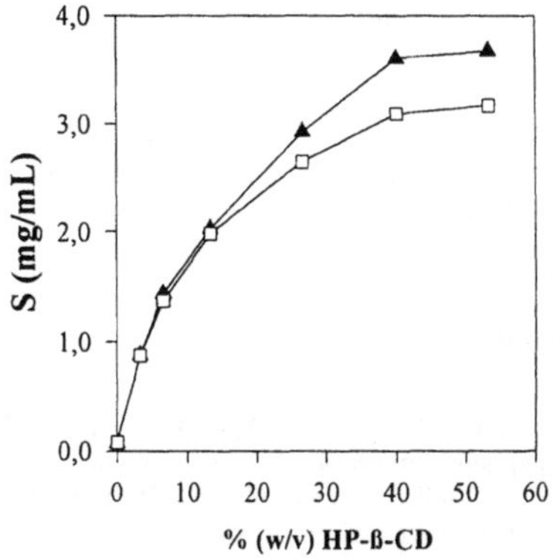

Figure 3. Phase solubility diagrams of **1** at pH 7.41: citrate buffer (▲) and phosphate buffer (□).

The advantage of the combined use of pH adjustment and complexation with HP-ß-CD is clearly demonstrated in Fig. 4, which shows the solubility profiles of **1** in the absence and presence of 53.3 % HP-ß-CD. As can be seen, the solubility of **1** underwent a 5,500-fold rise at pH 8.03 using 53.5% HP-ß-CD. Recently we reported about the convenience of pH increases for improving the solubility of another isoxazolyl-naphthoquinone-imine (Linares et al., 1997).

The apparent stability constants $K_{1:1}$ were calculated from the solubility data using the formula:

$$K_{1:1} = \text{slope} / [S_0 (1 - \text{slope})] \qquad (1)$$

where S_0 (the intercept) denotes the solubility of the substrate in the absence of HP-ß-CD and slope is the slope of the linear part of all the solubility curves (Higuchi and Connors, 1965). The intrinsic solubilities (S_0), the type of diagram and the apparent complex constants ($K_{1:1}$) are summarized in Table 1. Since the stability constants of the **1** complexes are larger in the less ionized form, it is obvius that the degree of dissociation has a decisive influence on the complexability of ionic guest molecules.

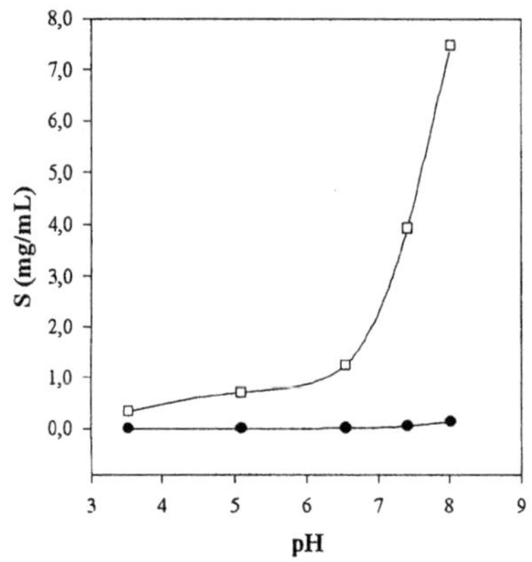

Figure 4. Solubility of **1** vs. pH in aqueous solutions. Key: (●) 0.00 % HP-ß-CD; (□) 53.3 % HP-ß-CD.

TABLE 1. Data of the phase solubility curves of **1** with HP-ß-CD at different pH values.

pH	**S_0 (M)**	**$K_{1:1}$ (M^{-1})**	**Type of curve**
3.52[a]	3.2×10^{-4}	5.1×10^{3}	A_L
5.05[a]	2.1×10^{-3}	2.4×10^{3}	A_L
6.48[b]	2.4×10^{-2}	4.3×10^{2}	A_L
6.55[a]	7.8×10^{-3}	1.5×10^{3}	A_N
7.41[a]	5.3×10^{-2}	9.3×10^{2}	A_N
7.41[c]	7.8×10^{-2}	5.2×10^{2}	A_N
8.03[a]	1.4×10^{-1}	9.0×10^{2}	A_N

[a] Citrate buffer
[b] Water
[c] Phosphate buffer

4. References

Albesa I., Bogdanov P., Eraso A., Sperandeo N., and de Bertorello M.M. (1995) Antibiotic Activity of Isoxazolyl-naphthoquinone Imines on Mice Infected with Staphylococcus aureus, *J. Applied Bacteriol.*, 78, 373-377.

Backensfeld, T., Müller, B., and Kolter K. (1991) Interaction of NSA with cyclodextrins and hydroxypropyl-cyclodextrin derivatives. *Int. J. Pharm.*, 74, 85-93.

Bogdanov P., Albesa I., Sperandeo N., and de Bertorello M.M. (1993) Actividad Antibacteriana in vitro de Isoxazolilnaftoquinonas, *Rev. Arg. Microbiol.*, 25, 119-128.

Elving, P., Markowitz, J., and Rosenthal, I. (1956) Preparation of buffers systems of constant ionic strength, *Anal. Chem.* 28, 1179-1180.

Fernandez, A., de Bertorello, M.M., and Manzo, R. (1982) Síntesis y propiedades espectroscópicas de 1,2-naftoquinona-4- aminoisoxazoles. *Anal. Asoc. Quim. Argentina.* 70, 49-55.

Higuchi, T.; Connors, K. (1965) Phase solubility techniques, in Reilly, C. (de.), *Advances in Analytical Chemistry and Instrumentation*, Wiley Interscience, New York, pp. 117-212.

Loftsson, T., Olafsdóttir, B. and Fridriksdóttir, H. (1990) Comparative study on inclusion complexation acetylsalicylic acid, cholecalciferol and melphalan with ß- and γ-cyclodextrin and some of their derivatives, *Acta Pharm. Nord.*, 2, 303-312.

Loftsson, T. and Brewster, M.E. (1996) Pharmaceutical applications of cyclodextrins. 1. Drug solubilization and stabilization, *J. Pharm. Sci.*, 85, 1017-1025.

Linares, M., de Bertorello, M. and Longhi, M. (1997) Effect of Hydroxypropyl-β-cyclodextrin on the Solubility of a Naphthoquinone-imine, *Int. J. Pharm.*, 159, 13-18.

Longhi, M. (1989) *Síntesis y Estabilidad de Nuevos Derivados de Isoxazolil-naftoquinonas*, Ph.D. dissertation, National University of Córdoba.

Schwarcz M., Goijman S., Molina M., and Stoppani A. (1990) Effects of Isoxazolyl-naphthoquinoneimines on Growth and Oxygen Radical Production in Trypanosoma cruzi and Crithidia fasciculata, *Experientia*, 46, 502-505.

Acknowledgements

The authors thank the Consejo Nacional de Investigaciones Científicas y Técnicas (CONICET), the Secretaría de Ciencia y Técnica de la Universidad Nacional de Córdoba (SECyT), and the Consejo de Investigaciones Científicas y Tecnológicas de la Provincia de Córdoba (CONICOR), for financial support, and the CERESTAR USA, INC. (Hammond, IN) for their gift of the 2-hydroxypropyl-β-cyclodextrin.

A COMPARATIVE STUDY OF THE DISSOLUTION PROPERTIES OF PIROXICAM-β-CYCLODEXTRIN INCLUSION COMPLEXES PREPARED BY DIFFERENT METHODS.

T. VAN HEES, B. EVRARD, G. PIEL and L. DELATTRE.
Laboratoire de Technologie Pharmaceutique, Institut de Pharmacie, CHU tour 4, Bâtiment B36, avenue de l'Hôpital 1, 4000 Liège 1, Belgium.

1. Introduction.

Many drugs possess poor water solubility and their bioavailabilities are limited by their dissolution rates. Several approaches are known to increase the dissolution rate and hence the bioavailability of such drugs including micronization, formation of solid dispersions, solvates, adsorbates or complexes. [1]

CHF 1194 is an inclusion complex of β-cyclodextrin (β-CD) with piroxicam (PIR), an NSAID, marketed in many countries as Brexin® or Cycladol®. This inclusion complex shows an improved wettability and water solubility as compared to plain piroxicam. [2]

The usual preparation methods of cyclodextrin inclusion complexes include kneading, grinding, freeze-drying, slow evaporation, spray-drying and coprecipitation. [3] But as for many non-polar drugs, the poor aqueous solubility of piroxicam restricts the possible preparation methods. Giordano and co-workers first studied the interaction of SCF with drug-cyclodextrin inclusion compound. [4]

The aim of this work was to take benefit from the dissolving power of supercritical carbon dioxide (S.C. CO_2) in order to include PIR into β-CD. In the second part of this study, the aqueous solubility of the inclusion complex was compared with plain piroxicam and physical mixture. We also compared this potential new way of inclusion with more classical techniques.

2. Materials and methods.

2.1. MATERIALS.

Piroxicam was purchased from Certa (Braine-l'Alleud, Belgium) and β-cyclodextrin from Cavitron (Neuilly-sur-Seine, France). CO_2 was of quality 5.0 from Air Product.

2.2. FORMATION OF INCLUSION COMPLEXES.

Inclusion experiments with supercritical CO_2 were carried out in a Suprex SF Extractor Autoprep 44. Freeze-drying was performed in a LAB CONCO lyoph-look 6 and spray-drying in a NIRO Atomizer "mobile minor".

2.3. THERMAL ANALYSIS.

A Mettler TC11 TA Processor DSC apparatus was used for the thermal analysis of the inclusion complexes. Samples (10-15 mg) were scanned between 30 and 230°C under a nitrogen gas stream at the heating rate of 10 °C/min.

2.4. DISSOLUTION RATE AND INTRINSIC DISSOLUTION RATE.

Dissolution tests were performed on a USP XXIII N°2 dissolution apparatus (SOTAX AT7) connected with a Hewlett Packard 8452A Diode Array Spectrophotometer. 500 ml of solution at pH 1.2 (diluted HCl) and at pH 6.8 (phosphate buffer) were used as a dissolution medium at 37.0 ± 0.1°C. The stirring rate was 50 rpm. Every two minutes, samples were removed and analyzed for piroxicam content by U.V. spectrometry at 334 nm (pH 1.2) or 354 nm (pH 6.8).

3. Results and discussion.

3.1. INCLUSION OF PIROXICAM INTO β-CD BY MEANS OF S.C. CO_2.

A series of experiments at different temperature and pressure of S.C. CO2 was designed. Table 1 gives a summary of the tested conditions.

T° (°C)	Pressure (bar)	Density of CO_2 (g/cm^3)	% of inclusion after 3 hours
50	500	0.965	7.10
100	400	0.761	6.70
150	150	0.238	99.61
150	350	0.560	98.40
150	450	0.655	94.50

Table 1 - Density of S.C. CO_2 and percentage of inclusion after 3 hours at the different conditions of temperature and pressure.

The ability of S.C. CO_2 to form a piroxicam-β-cyclodextrin inclusion complex was demonstrated by thermal analysis. In figure 1, typical DSC curves of both physical mixture and inclusion compound are represented.

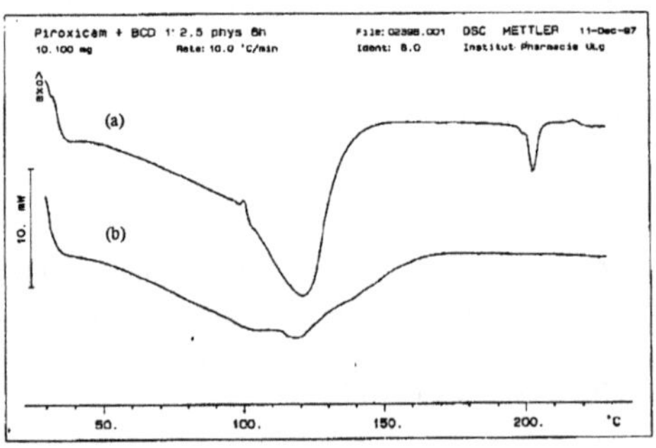

fig. 1 - Typical DSC curves (a) : physical mixture of piroxicam and β-cyclodextrin separately treated by CO_2; (b) : product obtained after 6 hours at 150°C and 150 bar.

3.2. DISSOLUTION TESTS.

Complex I (supercritical condition) was obtained by keeping approximately 6.5g of a physical mixture of piroxicam-β-cyclodextrin 1:2.5 (mol-mol) in a 10ml cartridge for 6 hours at 150°C and 150 bar of CO_2. Complex II (freeze-drying) and complex III (spray-drying) were isolated from an aqueous solution of piroxicam-β-cyclodextrin 1:2.5 (mol-mol), adjusted to a pH value of 10.0 with NH_4OH.

4g of complex (corresponding to ± 400mg of piroxicam) were placed in the dissolution medium for 30 minutes. Figure 2 shows the profiles of dissolution in the medium at pH 1.2

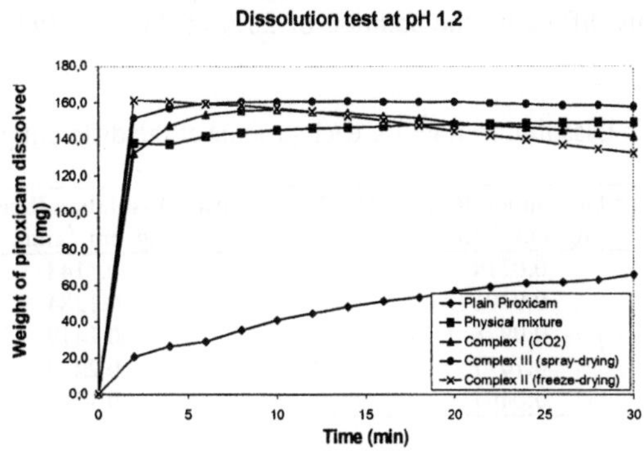

Fig. 2 – Dissolution test at pH 1.2.

Figure 3 shows the profiles of dissolution in the buffer at pH 6.8.

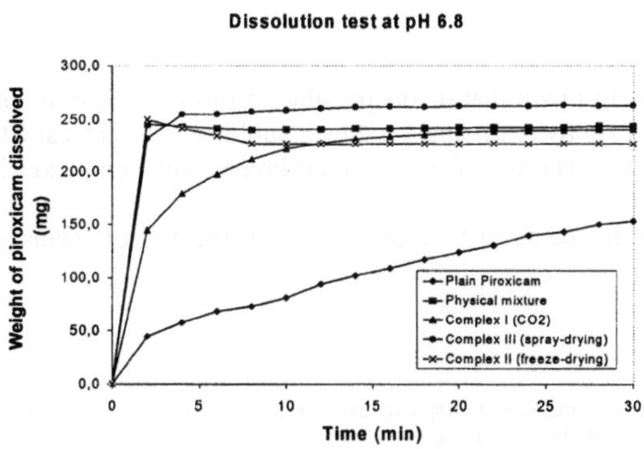

Fig 3 – Dissolution test at pH 6.8.

From these tests, we can notice that Complex I gives a lower dissolution rate than Complex II or III at pH 1.2 and even lower than the physical mixture at pH 6.8. This result can be explained by the differences in the particle size characteristics of the products. In order to suppress the influence of this parameter, the intrinsic dissolution rate has been determined.

3.3. DETERMINATION OF THE INTRINSIC DISSOLUTION RATE BY THE DISC METHOD.

500mg of the complex were compressed into a disc using a 1.3 cm diameter die in a IR press (Perkin-Elmer) at a force of 10 tons during 1 minute. The disc was then placed in a stainless-steel holder, which was previously lined with a silicone gel, so as to permit only one face to be exposed to the dissolution medium.[5] The holder (thickness 0.7cm, inner diameter 1.40cm and outer diameter 2.0cm) was then introduced in the dissolution apparatus. The area in contact with the dissolution medium is now 1.327 cm^2 for each preparation method.

Disc of pure drug could not be prepared because of severe capping occurring after compression. In order to overcome this difficulty, the addition of 20% of Avicel® PH 102 in the powder blend was necessary.

Table 2 gives the Intrinsic Dissolution Rate for each product at the different pH values.

Product	Intrinsic Dissolution Rate at pH 1.2 ($mg. cm.^{-2} min.^{-1}$)	Intrinsic Dissolution Rate at pH 6.8 ($mg. cm.^{-2} min.^{-1}$)
Plain piroxicam	0.0214	0.1148
Physical mixture	0.0701	0.1284
Complex I	0.0978	0.1419
Complex III	0.1027	0.2270
Complex II	0.0407	0.1378

From the results shown in table 2, it is obvious that both complexes I and III give the highest rates of dissolution. We can also observe that the physical mixture dissolves less rapidly than the inclusion complexes. The poor result obtained with the freeze-dried product (complex II) can be explained by the formation of air bubbles at the surface of the disc.

4. Conclusions.

The goal of that work was to prove that it was possible to form a inclusion complex of a non-polar drug with a cyclodextrin by rendering the drug soluble in Supercritical CO_2. The thermal analysis and the dissolution tests didn't show a real difference with complexes isolated by more classical methods.
Supercritical carbon dioxide seems to be a new potential method of inclusion for non-polar drugs.

5. References.

[1] Sencar-Bosic P., Srcic S., Knez Z. and Kerc J., Improvement of nifedipine dissolution characteristics using supercritical CO_2. *Int. J. Pharm.*, 148 (1997) 123-130.
[2] Acerbi D., Poli G., Rondelli I. and Ventura P., A pilot phamacokinetic study after single oral administration of a sachet formulation of piroxicam-β-cyclodextrin inclusion complex versus a liotablet formulation of plain piroxicam in healthy volunteers. *Proc. 8th Int. Symposium on Cyclodextrins.*
[3] Lin S.Y. and Kao Y.H., Solid particulates of drug-β-cyclodextrin inclusions complexes directly prepared by spray-drying technique. *Int. J. Pharm.*, 56 (1989) 249-259.
[4] Giordano F., Rillosi M., Bettinetti G.P., Gazzaniga A., Majewski W. and Perrut M., Interaction of supercritical fluids with drug/cyclodextrin inclusion compounds and physical mixtures. *Proc. 8th Int. Symposium on Cyclodextrins*, (1996) 193-196.
[5] Oth M.P. and Moës A.J., In vitro release studies of naproxen from non-aqueous suspensions using naproxen/polyvinylpyrrolidone coevaporate. *Acta Pharm. Technol.*;33(3), 131-135 (1987)

A COMPARATIVE PHARMACOKINETIC STUDY OF INTRAVENOUS SOLUTIONS OF MICONAZOLE WITH OR WITHOUT CYCLODEXTRINS.

G. PIEL, T. VAN HEES, B. EVRARD and L. DELATTRE

Laboratoire de Technologie Pharmaceutique, Institut de Pharmacie, CHU tour 4, Bâtiment B36, avenue de l'Hopital 1, 4000 Liège, Belgium

1. Introduction

The objective of this work was to compare the intravenous pharmacokinetic of miconazole in sheep after its administration in a polyoxyl 35 castor oil/lactic acid mixture, a 100 mM HP-βCD - 50 mM lactic acid solution and a 50 mM SBE_7-βCD - 50 mM lactic acid solution.

Miconazole is an antimycotic drug practically insoluble in water and is consequently formulated with a non-ionic surfactant, Cremophor® (polyoxyl 35 castor oil), for parenteral administration. Polyoxyl 35 castor oil is associated with several side effects most notably an allergic reaction comparable with anaphylactic shock [1].

Cyclodextrins (CDs) have been chosen to develop a new parenteral solution containing miconazole without polyoxyl 35 castor oil in order to avoid the anaphylactic reactions involved to this excipient.

It has been shown that acids and CDs have a synergistic effect on the aqueous solubility of miconazole which allows to solubilize more than 10 mg of miconazole per ml (which is the concentration of the marketed solution) [2].

The aim is to demonstrate that a parenterally safe CD solution (HP-βCD or SBE_7-βCD solution) would have a minimal effect on the pharmacokinetic of miconazole compared to the marketed surfactant formulation (Daktarin IV®).

2. Materials and methods

2.1 MATERIALS

Miconazole and HP-βCD were obtained from Janssen Pharmaceutica (Beerse, Belgium). SBE_7-βCD was kindly supplied by Cydex (Lawrence, Kansas, USA).

2.2 METHODS

2.2.1 Preparation of the IV solution.

The miconazole-HP-βCD solution is prepared by dissolving miconazole (10 mg/ml) in a solution containing HP-βCD 100 mM, lactic acid 50 mM, NaCl (4.47 mg/ml) and water for injection. The obtained solution is sterilized by filtration and diluted 5 times with NaCl 0.9% before administration.
The miconazole-SBE_7-βCD is prepared by the same way. The solution contains miconazole (10 mg/ml), SBE_7-βCD 50 mM and lactic acid 50 mM.
The Daktarin IV® solution is also diluted 5 times before administration.

2.2.2 Animal experimental protocol.

6 sheep (males and females) weighing between 47 and 75 kg were used. IV administration of 4 mg/kg of miconazole was completed within 5 minutes. Blood samples were taken from the jugular vein before administration and after 5 minutes (just after the end of the perfusion), 10 15, 30, 45, 60, 90 minutes, 2, 3, 4, 5, 6, 8, 10, 12, 15, 18 and 24 hours.
The blood samples were centrifuged and the plasma was stored at - 20°C until assayed.

2.2.3 Extraction and analysis of the sample.

The method described by Hosobuto was slightly modified.
250 µl of acetonitrile was added to 250 µl of plasma or plasma standard to precipitate the proteins. The tube was vortexed for 30 sec., kept standing for 5 minutes at room temperature and then centrifuged at 3000 rpm for 5 minutes. 50 µl of the supernatant were injected in the HPLC system.
The plasma standard for calibration curve contains 0.5, 1.0, 1.5 and 2.0 µg of miconazole per ml of drug-free plasma.
HPLC was performed using a Lachrom Merck Hitachi system consisting of a L-7100 pump, a L-7400 UV detector operating at 230 nm, a L-7200 autosampler and a D-2500 chromato integrator.
Elution of each 50 µl sample was accomplished on a Lichrocart column (125 x 4 mm i.d.) prepared with an octylsilane (C8) phase Lichrospher 60 RP - select B 5 µm (Merck) maintained at 25°C, using a mobile phase consisting of a mixture of 0.05 M Na acetate pH 7.2 : acetonitrile (30:70).
The method was validated and showed good linearity, reproducibility and accuracy.

2.2.4 Pharmacokinetic analysis.

The pharmacokinetic of miconazole from the surfactant mixture and the CDs solutions were considered as a two-compartment model, fitting to Eq. 1:

$$C = A.e^{-\alpha t} + B.e^{-\beta t} \quad (1)$$

where C is the plasma concentration of miconazole in µg/ml and t is the time in min. A, B, α and β were calculated. α and β are respectively the distribution and the elimination rate constants and A and B are the intercepts or the extrapolated concentrations at the origin. The apparent distribution and elimination half-lives were calculated from the values of α and β.

3. Results and discussion.

Figure 1 shows the mean miconazole plasma concentration vs. time for the 3 dosage forms. From this figure, it seems that there are no differences between the three dosage forms.

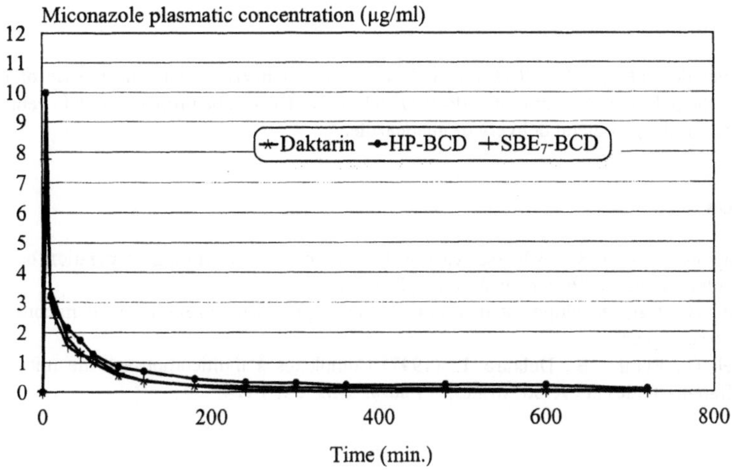

Figure 1: Miconazole plasmatic concentrations versus time for the Daktarin IV solution (*), the HP-βCD IV solution (•) and the SBE_7-βCD solution (+)

Miconazole plasma concentration vs. time profiles from the surfactant, HP-βCD and SBE_7-βCD solutions can be defined by Eq. 2, 3, and 4 respectively:

$$C = 83.94\ e^{-0.148t} + 2.56\ e^{-0.012t} \quad (2)$$
$$C = 104.47\ e^{-0.508t} + 2.66\ e^{-0.011t} \quad (3)$$
$$C = 91.2\ e^{-0.497t} + 2.79\ e^{-0.013t} \quad (4)$$

Some pharmacokinetic parameters are listed in table 1. It can be seen that the different pharmacokinetic parameters obtained are very close. The half-life of distribution is very short (less than 2.4 minutes) showing that miconazole is very rapidly distributed in the organism. The mean values for the half-life of miconazole from the three solutions are not significantly different ($p > 0.05$). The AUC $_{0\text{-}240}$ values which were calculated by linear trapezoidal rule were not significantly different. The clearance value and the half-life of distribution were also calculated and no significant difference was found between the three solutions for both parameters.

Table 1 - Average miconazole pharmacokinetic parameters for the three different dosage forms after IV administration (4 mg/kg) to sheep (n=6).

Dosage form	T ½ α (min)	T ½ β (min)	AUC $_{0\text{-}240\ min}$ (µg .ml^{-1} . min)	Cl (ml . min^{-1} . kg^{-1})
DAK	2.33 ± 0.69	55.60 ± 12.12	247.03 ± 85.89	16.54 ± 7.09
HP	1.42 ± 0.42	57.72 ± 22.48	240.33 ± 82.08	11.11 ± 4.26
SBE	1.42 ± 0.48	51.80 ± 11.25	239.47 ± 64.75	12.78 ± 4.54

HP = miconazole-HP-βCD solution
SBE = miconazole-SBE_7-βCD solution
DAK = Daktarin® IV solution

4. Conclusions

We can conclude that both HP-βCD and SBE_7-βCD do not interfere with the release of miconazole compared to the polyoxyl-35 castor oil. HP-βCD and SBE_7-βCD can be proposed as safe vehicles instead of surfactants for the parenteral delivery of miconazole.

5. References

[1] Hopkins, C.S. (1988) Adverse reaction to a cremophor containing preparation of intravenous vitamin. *Intensive Ther. Clin. Monit.* 9, 254-255

[2] Piel, G. et al. Development of a non-surfactant parenteral formulation of miconazole. *Int. J. Pharm.* in press.

[3] Piel, G., Evrard, B., Delattre, L. (1997) Complexes à multicomposants de miconazole avec différents acides et cyclodextrines, *J. Pharm. Belg.* 52, 3 124

DEVELOPMENT OF A SUSTAINED RELEASE DOSAGE FORM CONTAINING A DICLOFENAC-CYCLODEXTRIN INCLUSION COMPLEX.

G. PIEL, B. EVRARD, T. VAN HEES, C. FERNANDEZ DEL POZO and L. DELATTRE

Laboratory of Pharmaceutical Technology, Institute of Pharmacy, CHU - Tour 4 - Bât. B 36, 1 Avenue de l'Hôpital, 4000 Liège 1, Belgium

1. Introduction

Diclofenac is a non steroidal anti-inflammatory drug with distinct anti-inflammatory, analgesic and antipyretic properties. Diclofenac has a very poor aqueous solubility (± 1 μg/ml). Its Na salt has an aqueous solubility of ± 18 μg/ml but the solubility in an acidic medium is less than 1 μg/ml.
This low solubility in an acidic medium may give rise to problems of dissolution and absorption with a low and erratic bioavailability.

The aim of our study is to develop a diclofenac sustained release dosage form with a diclofenac Na-cyclodextrin (CD) inclusion complex which should exhibit an increased solubility in an acidic medium, such as the gastric juice.

2. Materials and methods

2.1 MATERIALS

Diclofenac was obtained from Ludeco (Bruxelles, Belgium).
β-CD, γ-CD and HP-β-CD were obtained from CNI (Neuilly sur Seine, France), Wacker Chemie GmbH (Munchen, Germany) and Janssen Pharmaceutica (Beerse, Belgium) respectively.

2.2 METHODS

2.2.1 Phase solubility studies
Solubility studies were performed as described by Higuchi and Connors. Excess amounts of diclofenac were added to various concentrations of β-CD, γ-CD or HP-β-CD in 25 ml purified water, pH 1 and pH 3 phosphate buffer solutions. The suspensions were shaken in a water bath at 25°C and when the equilibrium was reached (approximately 24 hours), an aliquot was filtered through a 0.45 μm filter and assayed for diclofenac content by UV spectrophotometry.

2.2.2 Preparation of the complexes
Different complexes were prepared: precise quantities of diclofenac and of cyclodextrin (to respect the molar ratio 1:1, 1:2 or 2:1 and to obtain approximately 15g of complex) were dissolved in

purified water (approximately 500 ml). Agitation was maintained for 15 minutes and then the solution was spray-dried (Niro Atomizer Mobile Minor, DK) using the following conditions: air pressure: 3 bars, inlet temperature: 140°C, outlet temperature: 75°C, flow rate: 12.5 ml/min.

2.2.3 Dissolution kinetics

Dissolution kinetics were realized with the paddle method (rotating speed 100 rpm) in 1000 ml of a pH 5 phosphate buffer solution (0.05M, μ=0.2) containing 0.05% of polysorbate 20.
A quantity of complex equal to 12.5 mg of diclofenac was introduced in a hard gelatin capsule and the kinetics of dissolution were measured over 2 hours.

2.2.4 Preparation of the sustained release dosage form

Granules of diclofenac Na-βCD (1:1) were prepared with 15% of Cutina® HR by a melt granulation process. Only the granulometric fraction between 400-800 μm was used.

3. Results and discussion

Cyclodextrins (β-CD, γ-CD and HP-βCD) are able to increase the aqueous solubility of both diclofenac and diclofenac Na in water and in an acidic buffer solution (pH 1 and 3). The HP-βCD gives an A type diagram in all conditions while the β-CD and γ-CD give a B type profile. In a pH 1 buffer solution, a 200 mM HP-βCD solution increases the solubility of diclofenac Na around 500 times (Figure 1).

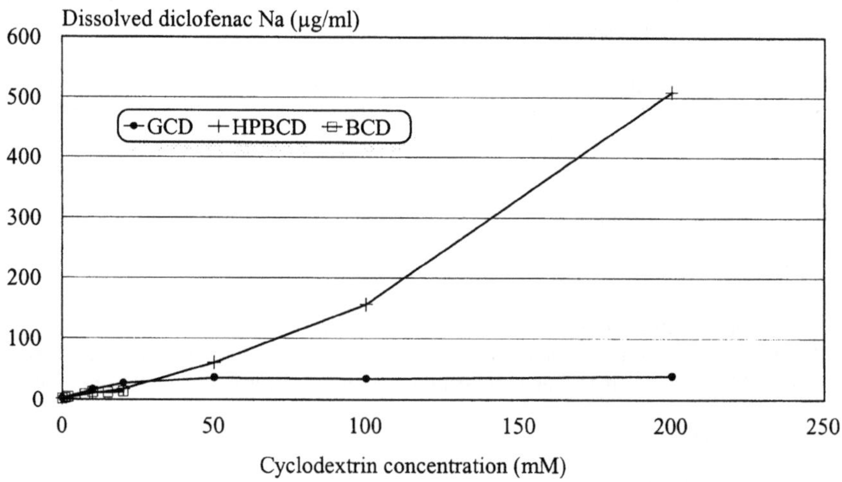

Figure 1: Phase solubility diagram of the diclofenac Na salt in a pH 1 phosphate buffer

The diclofenac Na-CD (β, γ and HP-βCD) inclusion complexes were isolated by spray-drying in different ratios (1:2, 1:1 and 2:1).
The kinetics of dissolution of these complexes were studied in a pH 5 phosphate buffer solution containing 0.05% polysorbate 20. Compared to their corresponding physical mixture, the best

results were obtained with the diclofenac Na-βCD 1:1 complex and with the diclofenac Na-HP-βCD 1:2 complex (Figure 2) which are completely dissolved within 5-10 minutes while their corresponding physical mixtures are not completely dissolved even after 2 hours.

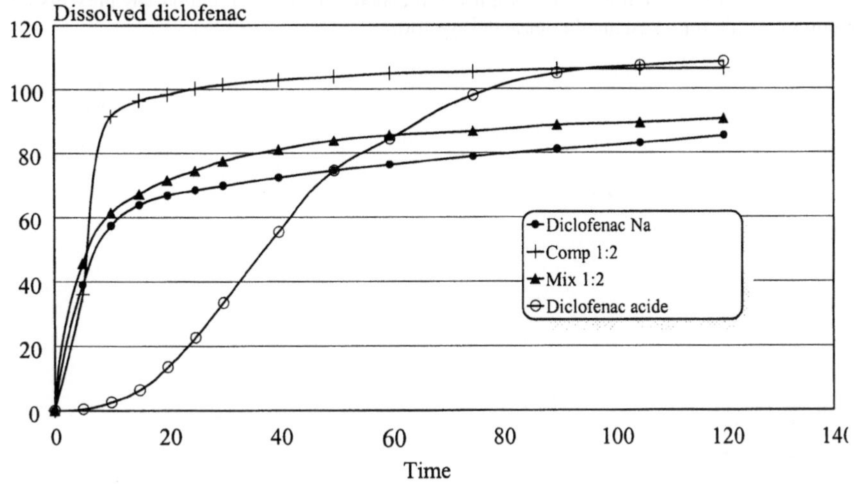

Figure 2: Kinetics of dissolution of capsules containing 12.5 mg of diclofenac as diclofenac, diclofenac Na, diclofenac Na-HP-βCD (1:2) complex (comp 1:2)and its corresponding physical mixture (Mix 1:2)

The sustained release dosage form was developed from the diclofenac Na-βCD 1:1 complex by a melt granulation process with 15 % of Cutina HR®. This figure shows the dissolution profile of the 400-800 μm fraction of the granulate in a pH 5 phosphate buffer solution containing 0.05% of polysorbate 20. We can observe that diclofenac is completely released within 20 hours.

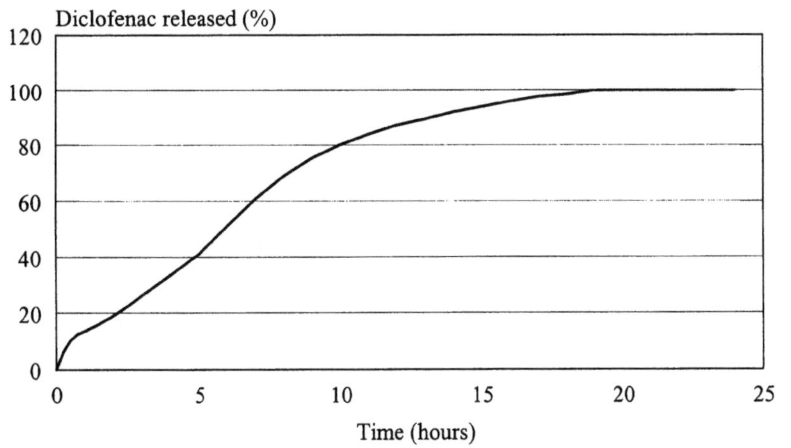

Figure 3: Dissolution profile at pH 5 of granules containing 12.5 mg of diclofenac as diclofenac Na-βCD 1:1 inclusion complex with 15% of Cutina HR®.

4. Conclusion

We can conclude that the complex diclofenac Na-βCD (1:1) presents good solubility characteristics even in a pH 5 buffer solution. This complex granulated with 15 % of cutina HR presents a good dissolution profile for a sustained release dosage form.

INFLUENCE OF CYCLODEXTRINS ON THE SOLUBILITY AND THE PHARMACOKINETICS OF ALBENDAZOLE.

B. EVRARD[1], P. CHIAP[2], G. PIEL[1], T. VAN HEES[1] , F. GHALMI[3], B. LOSSON[3] and L. DELATTRE[1]

Laboratoire de Technologie Pharmaceutique[1], Laboratoire d'Analyse des Médicaments, Institut de Pharmacie, CHU tour 4, Bâtiment B36, avenue de l'Hôpital 1, 4000 Liège, Belgium

Laboratoire de Parasitologie et Pathologie des Maladies Parasitaires[3], Ulg, bd de Colonster, 20, 4000 Liège, Belgique

1. Introduction

Albendazole (ABZ) is a benzimidazole derivative with a broad spectrum of activity against human and animal helminthe parasites. The ABZ therapy is very important in systemic cestode infection specially in inoperable or disseminated cases of hydatosis [1]. The main problem encountered with benzimidazoles is their low and erratic bioavailability owing to their low aqueous solubility.

Several authors reported an improvement of the aqueous solubility of ABZ by using cyclodextrins (CD's) (Bassani et al. ,1996; Kata and Schauer, 1991; Evrard et al., 1998).

The aim of this study is the in vivo evaluation of an oral solution of ABZ-CD in comparison with the suspension Valbazen® (Pfizer Animal Health).

2. Materials and methods.

2.1. MATERIALS.

Albendazole (USP 23)was supplied by INDIS (Belgium). Hydroxypropyl-β-CD was purchased from JANSSEN BIOTECH (Belgium). SBE-7-β-CD was kindly supplied by Cydex (USA). All other materials were of analytical grade.

2.2. METHODS.

2.2.1. Phase solubility studies.

Solubility studies were performed as described by Higuchi and Connors [4]. After cooling at 25°C an aliquot was filtered through a 0.45 μm membrane filter and assayed for ABZ content by a HPLC method. 20 μl samples were injected on a Lichrocart column (125 x 4 mm i.d.) prepared with an octylsilane (C8) phase Lichrospher 60 RP Select B - 5 μm. (Merck) and maintained at 25°C. The mobile phase consisted of a 70:30 (v:v) mixture of methanol HPLC grade and a pH 3.5 acetate buffer 0.05 M. The flow rate was 1.0 ml/min.

The analytes were monitored photometrically at 295 nm. The method was successfully validated in the range of ABZ concentrations of 5 and 25 μg/ml.

2.2.2. Bioanalysis method

A fully automated HPLC method was developed for the determination of ABZ and its metabolites in plasma. This method was based on the use of dialysis as purification step, followed by enrichment of its dialysate on a precolumn and subsequent liquid chromatographic analysis.

Chromatographic conditions

The dialysis block contained a cellulose acetate membrane and the trace enrichment precolumn was packed with C18 phase.

The mobile phase was a mixture of acetonitril and pH 6.0 phosphate buffer (20:80; v/v). ABZ and its metabolites were separed on a C8 column (Lichrospher® RP Select B, 5 μm, Merck) and monitored spectrophotometrically t 295 nm.

2.2.3. In vivo study design.

Six healthy cross-bred sheep were used as experimental animals. The sheep were weighed on the day of drug administration and were ranging from 28 to 62 kg of body weight. During the tests, the animals were maintained in individual lodges and were fed and watered ad libitum.

The ABZ-CD solution was prepared by dissolving ABZ (1.5 mg/ml) in a solution containing HP-β-CD 200 mM, Citric acid 0.05 M and water. The ABZ suspension was Valbazen® 1.9% (Pfizer Animal Health).

The drug was given at a dose regimen of 5 mg/kg of body weight. Blood samples were taken from the jugular vein before administration and after 0.25, 0.5, 1, 1.5, 2, 3, 4, 6, 8, 10, 12, 24, 28, 32, 48, 52, 56, 72, 76 and 80 hours.

The blood samples were centrifuged and the plasma was stored at –20°C until assayed.

3. Results and discussion.

3.1. FORMULATION OF THE SOLUTION.

3.1.1. Phase solubility diagrams .

Phase solubility diagrams of ABZ were established at 25 and 37°C in presence of HP-β and SBE-7-CD, respectively. The intrinsic solubility of ABZ in water is very low (0.16 μg/ml). Table 1 summarizes the stability constants (K_c) of the complexes at the two temperatures.

Table 1 : K_c of the complexes

	25°C	37°C
ABZ-HPβCD	12290 M^{-1}	4129 M^{-1}
ABZ-SEB7βCD	30127 M^{-1}	4971 M^{-1}

The values of K_c obtained suggest that in purified water the cavity of both HP-β-CD and SBE-7-β-CD accommodates very well the molecular portion involved in the inclusion.

A 200 mM HP-β-CD solution increases the aqueous solubility of ABZ about 2000 times at 25°C while a 200 mM SEB-7-β-CD solution increases it about 3400 times at the same temperature.

A 400 mM HP-β-CD concentration allows to dissolve 2.4 mg of ABZ per ml.

3.1.2. Influence of multicomponent systems.

Several multicomponent systems with HP-β-CD or SBE-7-β-CD and gluconic, citric or lactic acid were also evaluated (figure 1). We observed a similar synergistic effect between acids and both HP-β-CD and SBE-7-β-CD allowing an important increase of the solubility (20 000 times) of ABZ in water.

For the in vivo evaluation, the ABZ-CD solution was prepared by dissolving ABZ (1.5 mg/ml) in a solution containing HP-β-CD 200 mM, citric acid 0.05M and water.

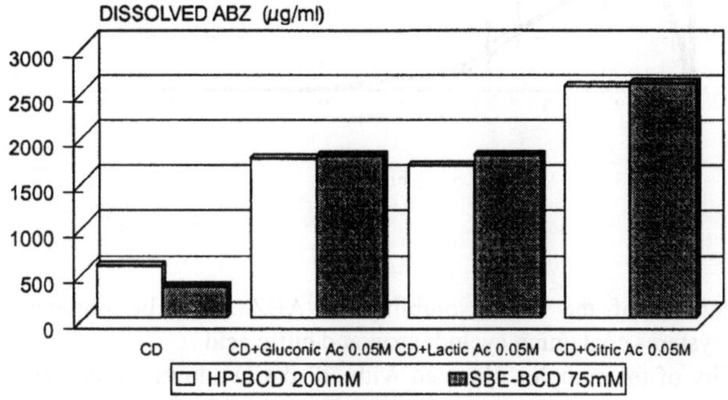

Figure 1

3.2. PHARMACOKINETIC EVALUATION.

During this evaluation, unmodified ABZ was never detected so that the main two metabolites [ABZ:SO (active metabolite) and $ABZSO_2$] were evaluated. The main results obtained for ABZ:SO are shown in table 2 and in figure 2.

Table 2 : Pharmacokinetic parameters of ABZSO after oral administration of ABZ in a solution or in a suspension.

	Solution	Suspension
T_{max} (h)	4 (2 – 4)	8 (8 – 12)
C_{max} (µg/ml)	2.1 (1.7-2.7)	1.1 (0.9-1.3)
$AUC_{0\rightarrow 24h}$ (µg.h.ml^{-1})	28.6 (27.1 - 30.1)	19.5 (14.8 – 25.6)

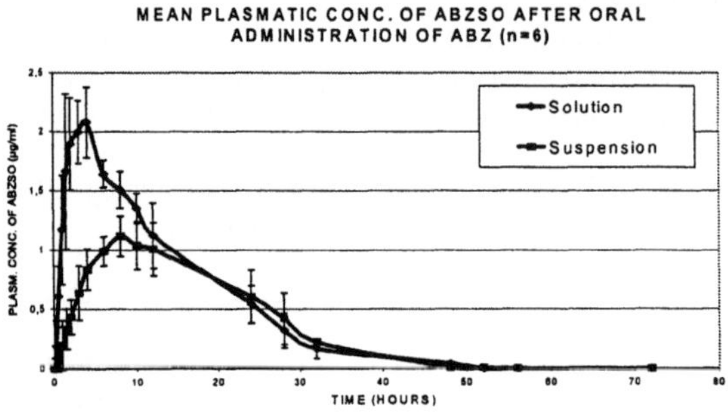

Figure 2

4. Conclusions

A noticeable increase of the water solubility of ABZ could be obtained by the use of multicomponent systems containing cyclodextrin and citric acid .
The good solubility of the complex obtained with HP-β-CD allows the preparation of a solution which induces a better and reproducible bioavailability of ABZ given by the oral route; the value of T_{max} is reduced while the C_{max} is increased twofold.

5. Bibliography

[1] Bassani V. L. et al., Proc of the 8th Intern. Symp. On Cyclodextrins, 321-324 (1996).
[2] Kata M. and Schauer M., Acta Pharm. Hungarica, 61, 23-31 (1991).
[3] Evrard et al., Proc. of the 2nd World Meeting on Pharm., Biopharm., Pharm. Tech., Paris,1998.
[4] Higuchi T. and Connors K.A., Advances in Analytical Chemistry and Instrumentation, 4, 117-212, (1965).

6. Acknowledgment

This work is supported by the Ministry of the Walloon Region of Belgium in application of a F.I.R.S.T. Program.

EVALUATION OF SPIRONOLACTONE BIOAVAILABILITY FROM SOLUTIONS OF β-CYCLODEXTRIN DERIVATIVES IN RATS

A. M. KAUKONEN, H. LENNERNÄS* AND J-P MANNERMAA,
*Pharmaceutical Technology Division, Department of Pharmacy, P.O.Box 56, FIN-00014 University of Helsinki, Finland, *Division of Biopharmaceutics and Pharmacokinetics, Department of Pharmacy, University of Uppsala, Sweden*

1. Introduction

Spironolactone (SP), a specific aldosterone antagonist, is used as a potassium sparing diuretic to reduce pulmonary oedema in premature infants suffering from bronchopulmonary dysplasia. In neonates oral medication is given through a nasogastric tube making liquid formulations preferable. Lacking suitable liquid preparations, powder papers prepared from commercial tablets are presently used, often causing obstruction of nasogastric tubes and loss of a variable part of the dose. Water-soluble derivatives of β-cyclodextrin have therefore been considered for the solubilization of SP in order to formulate a safe liquid preparation, which would also provide more concistent delivery of SP (Tötterman et al. 1997, Kaukonen et al. 1997).
The aim of this study was to evaluate the possible need to adjust SP dosage in prospective clinical studies. Oral absorption of SP in rats from the potential formulation containing sulfobutyl ether β-cyclodextrin (Captisol^{TM}) (SBE7βCD) and dimethyl-β-cyclodextrin (DMβCD) solutions was compared to that from SP-containing powder papers (reference preparation). DMβCD solutions were included in the study to evaluate the possible effects of SP degradation in SBE7βCD solutions (Kaukonen et al. 1997). SP in SBE7βCD solution was administered intravenously to assess the extent of intestinal absorption from the different formulations and also provided new information on the pharmacokinetics of SP in the rat.

2. Materials and methods

2.1. CHEMICALS

Spironolactone (SP) was donated by Orion Pharmaceuticals (Turku, Finland). Degradation products and metabolites of SP, 7α-thiospirolactone (TSP), 7α-thiomethylspirolactone (TMSP) and 6β-hydroxy-7α-thiomethylspirolactone (OH-TMSP) and canrenone (CAN) were donated by Searle (Skokie, IL, USA). Sulfobutyl ether β-cyclodextrin (SBE7βCD), DS 7 (Captisol^{TM}) was donated by CyDex, Inc. (Overland Park, KS, USA). Heptakis 2,6-di-*O*-methyl-β-cyclodextrin (DMβCD),

DS 14, was purchased from Sigma Chemicals (St. Louis, MO, USA).

2.2. ANIMALS AND TREATMENT

Adult male Wistar rats (611 ± 65 g, age 36 ± 5 weeks) were used. The rats were deprived of food for 20-24 hours before drug administration. Food was provided 4 hours after drug administration, while tap water was freely available throughout the experiment. A catheter was inserted into the carotid artery of anesthetized rats the day before the experiment. Rats administered intravenous doses also had their jugular vein catheterized in a similar fashion. The heparinized catheters were passed under the skin and exteriorized at the back of the neck. A polypropene "hat" was sown onto the neck to shield the catheters from the rats.

Serial blood samples were withdrawn before dosing (control) and up to 12 or 14 hours postdose of intravenous or oral administration, respectively. The samples were withdrawn into vials for serum collection, centrifuged and the serum stored at –20°C. After the last sample was collected the rats were immediately killed with carbon dioxide.

2.3. PREPARATION OF FORMULATIONS

SP-containing powder papers (5 mg per 100 mg of powder) were used as the oral reference formulation. Prior to administration, distilled water (1 ml per 100 mg of powder) was added to the powder and the mixture thoroughly shaken.

Oral solutions at 5 mg/ml of SP were prepared in autoclaved 48 mM solutions of DMβCD and SBE7βCD (molar ratio of SP vs. CD 1:4). The temperature of the SBE7βCD solution was 6°C when SP was added and the solutions were kept on ice during solubilization to minimize degradation of SP (Kaukonen et al 1997). DMβCD solutions were prepared at room temperature. Solutions for intravenous dosing were prepared as above in 48 mM solutions of SBE7βCD.

2.4. HPLC-ANALYSIS OF SERUM SAMPLES

SP and the metabolites TSP, TMSP, OH-TMSP and CAN were determined from serum samples according to an HPLC-method developed for this purpose (Kaukonen et al. 1998).

2.5. CALCULATION OF PHARMACOKINETIC PARAMETERS

Serum AUC-values for SP and its metabolites were calculated by the trapezoidal method. The elimination coefficients and half-lives ($t_{½}$) were determined by linear regression analysis of the linear terminal phase of the serum concentration time curve (iv – rats). The fraction absorbed, with reference to the main active metabolites TMSP and CAN (Overdiek & Merkus 1987), was calculated from the oral AUC-values against the dose-corrected means of intravenously treated rats. Relative oral bioavailabilities of SP from DMβCD and SBE7βCD solutions compared to powder papers were calculated from $AUC_{0\text{-}14}$ values of TMSP.

3. Results

3.1. INTRAVENOUS ADMINISTRATION

SP, TSP, TMSP and CAN were determined in rat serum samples following intravenous administration of SP at 20 mg/kg. The half-lives for SP, TMSP and CAN were 0.72 ± 0.17, 1.5 ± 0.3 and 2.2 ± 0.3 h, respectively. The half-life for TSP seems to be very short, 7.9 ± 2.3 min, but this should be taken as an approximation due to the low number of datapoints during the elimination phase. According to C_{max} values, TMSP (C_{max} 2105 ± 64 ng ml^{-1}) was the major serum metabolite in all the rats, seconded by CAN (C_{max} 780 ± 100 ng ml^{-1}). In rats the $AUC_{0-\infty}$ value of CAN was higher than that of TMSP, thus differing from results in man (Overdiek & Merkus 1987).

3.2. ORAL BIOAVAILABILITY OF SPIRONOLACTONE

TMSP and CAN were determined in serum samples of all rats following oral administration of spironolactone at 50 mg/kg as powder papers or cyclodextrin solutions. Samples collected 0.5-2 hours postdose contained small amounts of SP and/or TSP in two, three and four of the rats administered SP in powder papers, DMβCD and SBE7βCD solutions, respectively. Administration of SP in SBE7βCD and DMβCD solutions produced higher values of C_{max} and AUC_{0-14} for both TMSP (FIGURE 1) and CAN in comparison to the administration of powder papers.

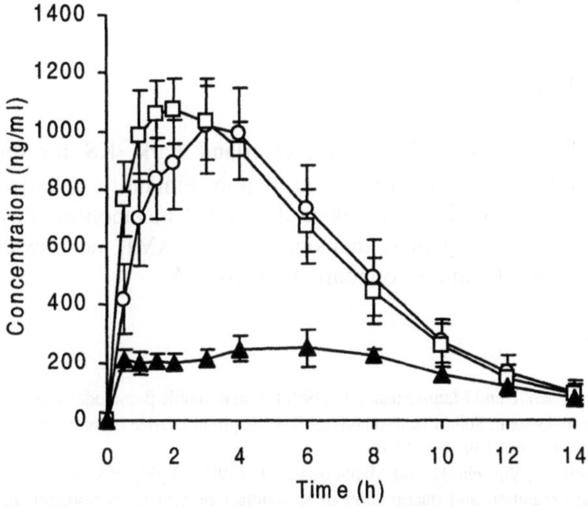

Figure 4. Plots of serum concentrations of 7α-thiomethylspirolactone vs. time in rats (mean ± SEM, n = 5) after oral administration of 50 mg/kg of spironolactone. Powder papers (▲); Solutions of 48 mM SBE7βCD (□); Solutions of 48 mM DMβCD (O).

The oral bioavailability of SP, determined with respect to AUC values of TMSP (F_{metab}), was significantly higher from DMβCD and SBE7βCD solutions than from powder papers (81.3 ± 28.8% and 82.8 ± 28.6% vs. 27.5 ± 9.3%). Respective values for

CAN were 67.7 ± 15.8%, 73.7 ± 34.0% and 37.6 ± 9.0%. The relative bioavailabilities of SP evaluated from the AUC values of TMSP further demonstrated the enhanced bioavailability from formulations containing DMβCD (296 ± 105%) and SBE7βCD (302 ± 104%) compared to the powder papers.

In the presence of SBE7βCD SP degrades through deacetylation producing TSP (Kaukonen et al. 1997), which is the metabolic precursor to both TMSP and CAN (Overdiek & Merkus 1987). The deacetylation of SP in the SBE7βCD-formulation did not seem to affect the bioavailability of SP, as evidenced by the similar relative bioavailabilities of both cyclodextrin formulations. Oral administration of solid SP-HPβCD complexes, likely to contain most of the dose as deacetylated SP (Kaukonen et al 1997), has been reported to produce AUC values of canrenone 3.6 times higher than from spironolactone alone in dogs (Soliman et al. 1997). This suggests that TSP could indeed be credited with therapeutic effects upon oral administration.

4. Conclusions

According to the high degree of improved oral bioavailability, a reduced dosage of spironolactone would be recommendable when clinical studies are pursued in premature infants. These and previous results indicate that that SBE7βCD could be a safe and suitable excipient for the solubilization of SP in paediatric formulations, provided absorption characteristics and pharmacokinetics of SP are confirmed to be similar in infants and adults.

5. Acknowledgements

We gratefully acknowledge CyDex, Inc., Overland Park, KS for the donation of SBE7βCD (Captisol^{TM}). Orion Corporation, Espoo, Finland is acknowledged for the donation of SP and Searle & Co, Skokie, IL for SP metabolites. Pirjo Aaltonen is thanked for assistance during anaesthetising of rats. AMK acknowledges financial support from the Finnish Cultural Foundation and NorFA.

6. References

Kaukonen, A.M., Kilpeläinen, I. and Mannermaa, J-P. (1997) Water-soluble β-cyclodextrins in paediatric oral solutions of spironolactone: Solubilization and stability of spironolactone in solutions of β-cyclodextrin derivatives, *Int. J. Pharm.* 159, 159-170

Kaukonen, A.M., Vuorela P., Vuorela, H. and Mannermaa, J-P. (1998) High-performance liquid chromatography methods for the separation and quantitation of spironolactone and its degradation products in aqueous formulations and of its metabolites in rat serum, *J. Chromatogr. A.* 797, 271-281

Overdiek, H. W.P.M., Merkus, F.W.H.M. (1987) The metabolism and pharmacokinetics of spironolactone in man, *Drug Metab. Drug Inter.* 5, 273-302

Soliman, O. A. E., Kimura, K., Hirayama, F., Uekama, K., El-Sabbagh, H. M., El-Gawad, A. E-G. A., Hashim, F. M. (1997) Amorphous spironolactone-hydroxypropylated cyclodextrin complex with superior dissolution and oral bioavailability, *Int. J. Pharm.* 149, 73-83

Tötterman, A.M., Schipper, N.G.M., Thompson, D. and Mannermaa, J-P. (1997) Intestinal safety of water-soluble β-cyclodextrins in paediatric oral solutions of spironolactone: Effects on human intestinal epithelial Caco-2 cells, *J. Pharm. Pharmacol.* 49, 43-48

CHEMOSENSORS OF MODIFIED CYCLODEXTRINS FOR DETECTING MOLECULES

AKIHIKO UENO, SACHIKO MATSUMURA, TAKUYA KANAI, HISAKAZU MIHARA

Department of Bioengineering, Faculty of Bioscience and Biotechnology, Tokyo Institute of Technology, 4259 Nagatsuta-cho, Midori-ku, Yokohama 226 Japan

1. Introduction

Cyclodextrins (CDs) with an appending chromophoric moiety usually form self-inclusion complexes, and the appending moiety is excluded from the cavity to water environment associated with guest accommodation (Fig. 1). This induced-fit behavior changes the environment around the chromophoric moiety from hydrophobic cyclodextrin cavity to polar water solution, and consequently fluorescence intensity changes upon addition of the guest species. In the case of dansyl fluorophore as the appending moiety, the fluorescence intensity decreases upon guest addition because the dansyl fluorescence is strong when included in the hydrophobic CD cavity whereas weak in bulk water solution. On this basis, we have prepared various dansyl-modified CDs as chemosensors for detecting molecules [1]. On the other hand, we have prepared CDs bearing two fluorophores [2]. When the fluorophore is naphthalene or pyrene, we observed guest-induced variations in the excimer emission. All these studies have been performed with simple derivatives in which one or two chromophores are almost directly connected with CD unit. Since the orientation and distance between CD and fluorophore are not easy to be controlled as desired because of the flexibility of the systems, we have attempted to connect both CD and fluorophore to α-helix peptide. Here we report some results performed with CD-fluorophore-α-helix peptide systems.

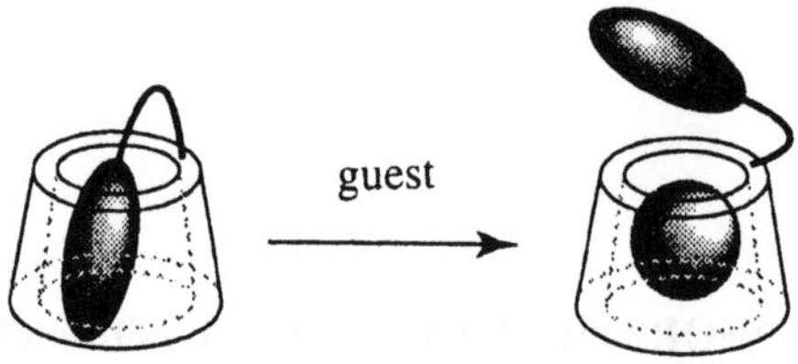

Fig. 1 Schematic representation for induced-fit type of complexation

2. α-Helix Peptides Bearing Dansyl and β-Cyclodextrin Units

2.1 DESIGNED PEPTIDES

Peptides, which contain three pairs of glutamic acid and lysine and consist of totally 17 amino acid, were prepared. β-CD and dansyl (DNS) units are connected at Glu(4) and lys(8) or lys(4) and Glu(8), each abbreviated as EK3 and EK3R. These peptide have three amino acids between the modified amino acids, and DNS and β-CD are separated by one turn of the helix. On the other hand, the peptides with six amino acids between the modified amino acids, EK6 and EK6R were prepared as the samples in which DNS and β-CD are separated by two turns of the helix.

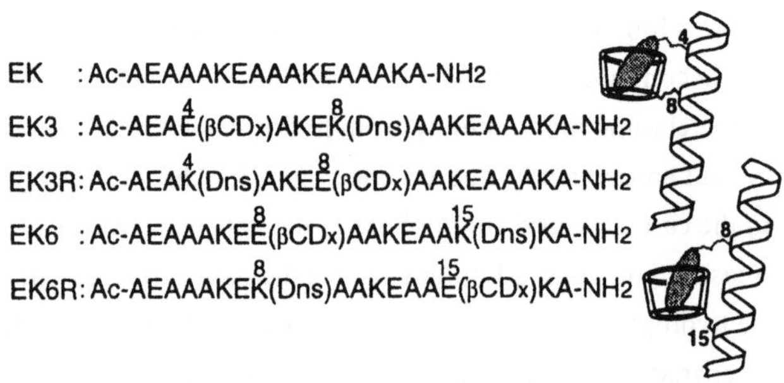

Fig. 2 Designed structures of peptides with β-CD and dansyl moiety.

2.2 FLUORESCENCE SPECTRA

Circular dichroism spectra of all these peptide samples exhibits troughs around 222 nm and 207 nm, which are typical of α-helical conformation. On the other hand, fluorescence spectra of these peptides exhibit a peak above 520 nm region, and the peak intensities of EK3R and EK6R are much stronger than those of EK3 and EK6. The peak intensities of these peptides decreases upon addition of 1-adamantanol as a guest (Fig. 3).

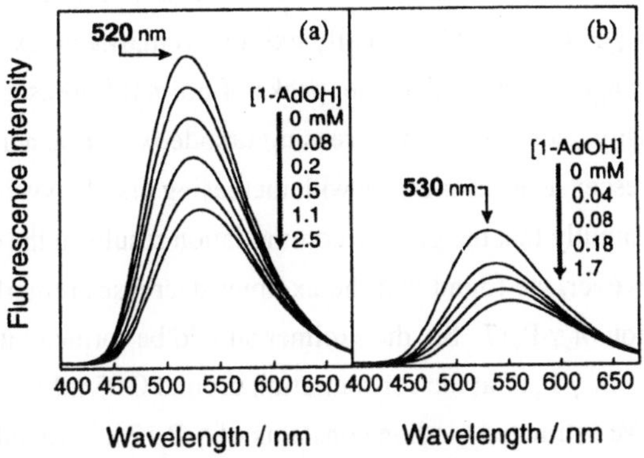

Fig. 3 Fluorescence spectra of peptides with various concentrations of 1-adamantanol. [Peptide]=10 μM. (a) EK3R, λex=338 nm. (b) EK6, λex=340 nm.

We obtained the binding constants from the analysis of the guest-induced fluorescence variations. The binding constants for 1-adamantanol are 4900, 2200, 19000, and 13000 M^{-1} for EK3, EK3R, EK6, and EK6R, respectively. In the case of (-)-menthol as the guest, the values are 410, 410, 2900, and 1300 M^{-1} for EK3, EK3R, EK6, and EK6R, respectively. Including the case of other guests, EK6 and EK6R give much larger binding constants than EK3, EK3R. The larger binding constants of EK6 and EK6R may be explained by loose binding and easy accommodation of the guest species.

3. α-Helix Peptide Bearing Two Pyrene Moieties and γ-Cyclodextrin

3.1 TWO GUEST INCLUSION IN THE CAVITY OF γ-CYCLODEXTRIN

Pyrene is a large aromatic compound and is known to form excimer when two pyrene rings interact with each other in a face-to-face manner. Although it is too large to be included in the β-CD cavity, pyrene and γ-cyclodextrin form a 2:2 complex in which two pyrenes form an excimer. On the other hand, Ueno et al. prepared γ-CD derivatives with two pyrene moieties, and found that the pyrene moieties form excimers intramolecularly [2]. Since the pyrene-modified γ-CDs change the excimer intensity, they were used as chemosensors for detecting guest molecules. In this study, we prepared the following peptide (γ-$P_2$17), which has one γ-CD and two pyrene moieties at opposite sides of γ-CD.

γ-$P_2$17: Py γ-CD Py
Ac-AEAAKKEAEAKEKAAKA-NH2

This peptide is expected to form an excimer in the side chain of α-helix peptide. The fluorescence spectra of γ-$P_2$17 (10 μM) actually exhibits remarkable excimer emission around 500 nm with stronger intensity than the peaks of normal fluorescence below 450 nm. The excimer intensity decreases with increasing ursodeoxycholic acid while the intensities of normal fluorescence peaks increase with increasing ursodeoxycholic acid. This phenomenon seems apparently that the guest accommodation results in the decrease in the excimer emission. However, we found that the excimer decreases remarkably with decreasing the concentration of γ-$P_2$17. So, the excimer should be formed intermolecularly. The fluorescence spectra of γ-$P_2$17 at the concentration around 0.05 mM exhibit predominant monomer peaks. We estimated binding constants of γ-$P_2$17 and found that the binding constants are markedly large. It is well known that hydrophobic capping of CD cavity enhances the binding strength. In the complex of γ-$P_2$17, two pyrene moieties act as hydrophobic cap and floor, extremely enhancing the guest binding.

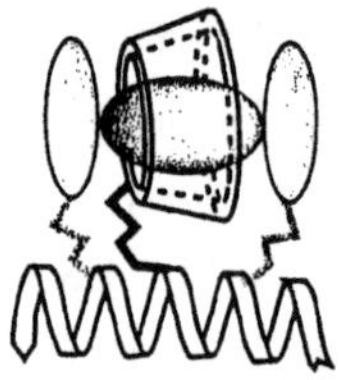

Fig 4. Hydrophobic cap and floor effect of cyclodextrin

4. Conclusion

α-Helix peptide can be used as an template to arrange chromophores and CD at desired positions for construction of effective chemosensors.

5. References

1. Ikeda, H., Nakamura, M., Ise, N., Oguma, M., Nakamura, A., Ikeda, T., Toda, F. and Ueno, A. (1996) Fluorescent cyclodextrins for molecule sensing. fluorescent properties, NMR characterization and inclusion phenomena of N-dansyl-modified cyclodextrins, J. Am. Chem. Soc. 118, 10980-10988.
2. Suzuki, I., Ohkubo, M., Ueno, A., Osa, T. (1992) Detection of organic compounds by dual fluorescence of bis(1-pyrenecarbonyl)-γ-cyclodextrins, Chem. Lett., 267-272.

MOLECULARLY IMPRINTED CYCLODEXTRIN POLYMERS AS ARTIFICIAL RECEPTORS -THE REQUISITES FOR REMARKABLE IMPRINTING-

H. ASANUMA, M. SHIBATA, T. HISHIYA, and M. KOMIYAMA*

Department of Chemistry and Biotechnology, Graduate School of Engineering, The University of Tokyo, Hongo, Tokyo 113-8656 Japan.

ABSTRACT

Polymeric receptors for cholesterol were synthesized by crosslinking cyclodextrin (β-CyD) with hexamethylene diisocyanate or toluene 2,4diisocyanate in, dimethyl sulfoxide (DMSO) in the presence of cholesterol as the template. Non-imprinted β-CyD polymers were much poorer in the cholesterol adsorption. Use of DMSO as the crosslinking solvent is necessary for the imprinting, since β-CyD molecules can satisfactorily form inclusion completes with cholesterol in this solvent and thus their mutual conformation in the polymer is regulated appropriately for cholesterol binding.

1. INTRODUCTION

Molecularly imprinting technique is one of the most promising methods to synthesize tailor-made artificial polymeric receptors.[1,2] This technique is based on the polymerization of functional monomers in the presence of the target molecule as the template, so that the positions and orientations of the functional residues of the monomers are immobilized in the polymer complementarily to the target molecule. We have successfully applied this technique to the crosslinkage of cyclodextrins in the presence of template molecules.[3] In the present paper, cholesterol is chosen as templates and the requisites for the preparation of efficient artificial receptors are investigated in detail. The choices of the reaction solvent and crosslinking agent are crucially important.

2. EXPERIMENTAL

2.1 Preparation of imprinted polymer

The desired amounts of β-CyD and cholesterol were dissolved in dry DMSO or DMSO-pyridine mixed solvent (DMSO/pyridine=1/1). Then the diisocyanate agents were added ([diisocyanate]/[β-CyD] = 4.2), and the mixtures were kept at 65 °C for a few hours. The diisocyanate agents used are hexamethylene diisocyanate (HMDI) and toluene 2,4-diisocyanate (TDI). The obtained polymer was sufficiently washed with hot water, THF, and hot ethanol. Nonimprinted β-CyD polymers were similarly prepared in the absence of cholesterol template.

2.2 Guest adsorption

The polymeric receptor (0.5 mmol of β-CyD residue) was incubated in 11 ml of water/THF mixture (6/5 in volume) containing 0.05 mmol of guest. After being stirred by magnetic stirrer at 25 °C, the mixture was centrifuged and the amount of guest in the liquid phase was analyzed by gas chromatography or HPLC.

3. RESULTS AND DISCUSSION

3.1 Selective recognition of cholesterol by the imprinted CyD polymer.

The β-CyD-immobilized polymer crosslinked with TDI (TDI-Imp), prepared in the presence of cholesterol (β-CyD/cholesterol = 3/1) in DMSO, efficiently adsorbed cholesterol (see Table 1). However, the corresponding non-imprinted β-CyD polymer (TDI-Non) adsorbed only 19 % of cholesterol. Similarly, imprinted β-CyD polymer crosslinked with HMDI (HMDI-Imp), instead of TDI, adsorbed cholesterol, whereas non-imprinted polymer (HMDI-Non) did not adsorb it to a measurable extent. Thus, the "imprinting effect" for the regulation of mutual conformation of β-CyD residues is unequivocal.

Table 1. Adsorption of cholesterol from the water/THF mixture by crosslinked β-CyD polymers

Guest	Adsorbing activity [a]		Imprinting[b] effect
	TDI-Imp	TDI-Non	
Cholesterol	0.70	0.19	3.7
Testosterone	0.30	0.23	1.3
Stilbestrol	0.27	0.14	1.9
Phenol	0.16	0.08	2.0

a. Molar ratio of the adsorbed guest to the initially fed guest
b. The ratio of the adsorption activity of TDI-Imp to that of TDI-Non

In contrast, the adsorbing activity of the imprinted polymer on testosterone or stilbestrol was smaller than that on cholesterol. The adsorbing activity on phenol, which is known to form 1:1 complex with β-CyD, was far smaller. The maximum imprinting efficiency (defined as the ratio of the adsorption activity of imprinted β-CyD to that of non-imprinted one) was observed for cholesterol adsorption by using β-CyD 100 polymer crosslinked in the presence of cholesterol as the template (see Table 1). Thus, the selectivity for cholesterol adsorption against other guests was enormously improved by using the molecularly imprinting technique.

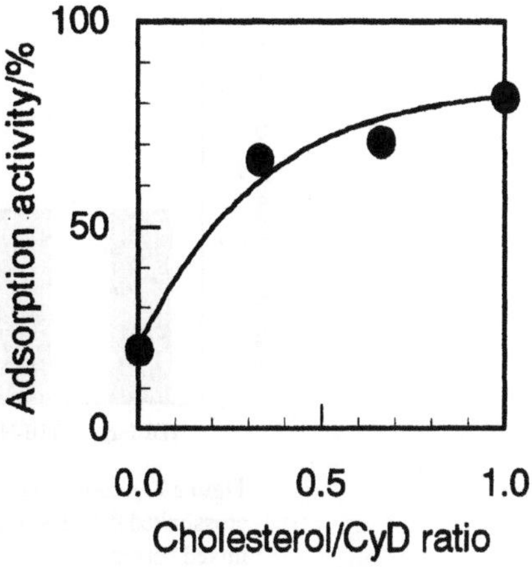

Figure 1. Effect of the amount of the template cholesterol on the adsorption activity

Interestingly, the adsorption activity of the imprinted polymer increased with increasing amount of cholesterol, even when [template]/[β-CyD] ratio in the polymerization mixture was greater than 1/3 (see **Figure 1:** note that β-CyD and cholesterol form a 3:1 complex in solid).[4] Thus, direct polymerization of the 3:1 complex for the imprinting is unlikely. Rather, β-CyD molecules should be connected each other in a stepwise manner to the conformationally regulated polymers.

3.2. Requirement of the Use of DMSO as the Crosslinking Solvent

The use of DMSO as polymerization solvent is essential for the imprinting, since β-CyD-cholesterol inclusion complex can be satisfactorily formed there as confirmed by the ^{1}H-NMR measurement. When 30 *mM* of β-CyD was added to DMSO solution of cholesterol (2 mM at 65°C (the temperature used for the crosslinking), the 18-methyl protons of cholesterol shifted toward the higher magnetic field (by 0.05 ppm) with respect to the value in the absence of β-CyD. The complex formation between β-CyD and cholesterol under the crosslinking conditions is conclusive. The binding constant for β-CyD-cholesterol complex formation was estimated to be 550 M^{-1} at 65°C from NMR-titration experiment. This argument is further confirmed by the fact that no imprinting effect was observed when DMSO/pyridine mixed solvent (DMSO/Pyridine=1/1) in which CyDs can not form inclusión complex effectively was used as solvent instead of DMSO (see **Figure 2**). Formation of well-defined complexes should be responsible for the present efficient and successful imprinting.

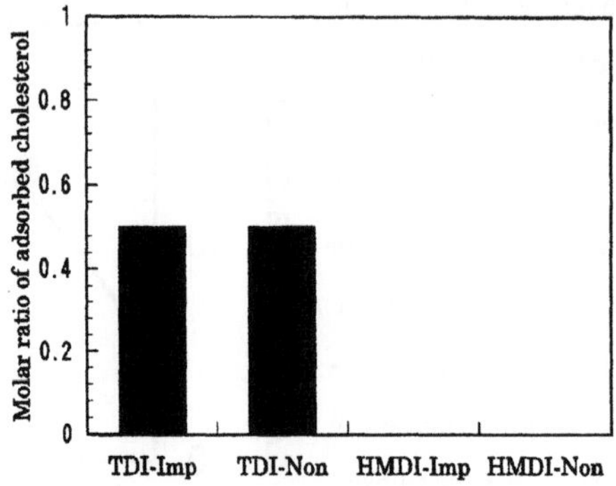

Figure 2. Adsorption of cholesterol by the crosslinked β-CyDs prepared in DMSO/pyridine mixed solvent.

In conclusion polymeric receptors for cholesterol were synthesized by crosslinking CyDs with the diisocyanates in DMSO in the presence of cholesterol as a template. Artificial receptors for steroids other than cholesterol, as well as those for amino acid derivatives, could be also synthesized by applying this technique. The present finding indicates that artificial polymeric receptors for varieties of large guest molecules can be prepared by using the molecular imprinting technique for the immobilization of CyD.

ACKNOWLEDGMENTS. This work was partially supported by a Grant-in Aid for Scientific Research from The Ministry of Education, Science, and Culture, Japan.

REFERENCES

1) G. Wulff, *Angew. Chem. Int. Ed. Engl.*, **34**,1812 (1995).

2) K.Mosbach et al., *Jam.Chem.Soc*, **117**, 5853 (1988).

3) H. Asanuma, M. Kakazu, M. Shibata, T. Hishiya, and M. Komlyama, *Chem.Commun.,1997,1971.*

4) P.Claudy et al., J. Thermal Anal, **37**, 2497 (1991).

NEW DEVELOPMENTS IN IMPROVEMENT OF DRUG SOLUBILITY AND AVAILABILITY BY CYCLODEXTRINS

D. DUCHENE, G. PONCHEL and D. WOUESSIDJEWE

Physico-chimie, Pharmacotechnie, Biopharmacie, URA CNRS 1218,
Faculté de Pharmacie, Université Paris-Sud,
Rue Jean Baptiste Clément, 92290 Châtenay Malabry, France

ABSTRACT

Cyclodextrins and their water-soluble derivatives are classicaly proposed to increase the solubility of poorly water-soluble drugs. Some examples are given and the influence of parameters and additives is highlighted. Two possibilities of using cyclodextrins in the preparation of nanoparticles for drug targeting are presented, one consisting in increasing the loading capacity of poly(isobutylcyanoacrylate) nanoparticles by cyclodextrins, the other being the use of amphiphilic cyclodextrins to prepare nanoparticles. Both types of nanoparticle present a very fast release of the poorly water-soluble drug that they contain.

1. INTRODUCTION

One of the major problems of drug formulation consists in poor water-solubility of the drug, resulting in its low bioavailability. Cyclodextrins represents a means to overcome this drawback. From the industrial standpoint, the choice of a cyclodextrin depends on its not only on its existence and cost on the market, but also on its safety, recognition by health authorities, and simplicity of use. However, solubility and bioavailablity problems are such that many researchs are carried out in order to propose new water-soluble cyclodextrin derivatives or new possibilities of utilization of cyclodextrins. These different routes are reviewed here.

2. SOLUBILIZATION BY CYCLODEXTRINS OR WATER-SOLUBLE DERIVATIVES

2.1. Water-soluble cyclodextrins and derivatives

Among the three natural cyclodextrins, β-cyclodextrin which is the less water-soluble product (18.5 g/l) was, for years, the most frequently studied product. This is the consequence of the too small cavity size of α-cyclodextrin, preventing its use for inclusion of most of the therapeutic molecules, and, on the other hand, of the difficulty in the preparation of γ-cyclodextrin resulting in a cost incompatible with its industrial use. For these reasons, β-cyclodextrin was intensively studied, leading to its inscription at the US Pharmacopoea. Its safety after oral administration was demonstrated [1], when its tolerance by parenteral route was discussed because of possible hemolytic effect and nephrotoxicity which is not observed with more water-soluble α- or γ-cyclodextrins [2].

In order to have a cyclodextrin more water-soluble than β-cyclodextrin, and usable for parenteral administration, many derivatives were synthetized.

Methyl derivatives, oppositely to what could be expected, present a high water solubility, especially the dimethyl derivative (570 g/l for dimethyl β-cyclodextrin). However, these products have an exothermic dissolution process, resulting in a decrease in solubility with an increase in temperature. Furthermore, they are much more hemoltyic than the parent cyclodextrins. Their use is not recommended for parenteral administration, but they display interesting properties for dermal administration [3, 4].

Hydroxylpropyl derivatives, are highly water-soluble (> 500 g/l). This is not only the consequence of their chemical nature, but also of their amorphous nature. Like methyl derivatives, they can be interestingly used in dermal products [3, 4]. However, their main advantage relies in the fact that they have a low hemolytic effect and thus are especially recommended for parenteral administration [5, 6]. However, due to industrial protection, they are not, up to now, easily available industrial use.

For this reason, a new product was recently proposed. It is a sulphobutylether of β-cyclodextrin randomly substituted, but with an average degree of substitution of 7 [7]. It is claimed as not hemolytic, and is especially intended for parenteral use.

Other water-soluble derivatives were synthetized which have to be mentioned, even if they are the subject of only few works in the pharmaceutical field, and, for some of them, not marketed. Among these products are the branched cyclodextrins (glycosyl and maltosyl) [8, 9], the hydroxyethyl derivatives, very similar to the hydroxypropyl derivatives [10], and more recently the polyoxyethylene derivatives [11].

2.2. Some examples of utilization

Solubilization obtained by the use of cyclodextrins depends not only on the nature of the cyclodextrin employed, but also on the type of association between the active drug and the cyclodextrin.

2.2.1. Influence of the cyclodextrin nature

The increase in solubility obtained in the presence of a cyclodextrin depends on the cyclodextrin water-solubility, but also on the stability constant of the inclusion compound simultaneously obtained, a low stability constant resulting in the dissociation of the inclusion compound and reprecipitation of the free drug. For example solubility of 1 mmole of nimesulide was assessed in 65 ml of water at 25°C for 5 days in the presence 1mmole of a series of cyclodextrins and derivatives (Table 1) [12]. Because of a better adaptability of the active drug to the β-cyclodextrin cyclodextrin cavity than to that of γ-cyclodextrin, the best increase in solubility were observed in the presence of the β derivatives and especially the permethylated β-cyclodextrin (PM β-cyclodextrin), followed by Captisol (SEB7 β-cyclodextrin) and a polyethylene oxide β-cyclodextrin (PEO β-cyclodextrin).

Table 1

Increase in solubility of numesulide in the presence of cyclodextrins

Product		Solubility (mg/l)	Increase in solubility
nimesulide alone		4.32	
—	+ β-cyclodextrin	62.95	x 14.5
—	+ γ-cyclodextrin	7.91	x 2
—	+ HP β-cyclodextrin	52.16	x 12
—	+ SEB7 β-cyclodextrin	83.33	x 19
—	+ PM β-cyclodextrin	106.12	x 24.5
—	+ PEO β-cyclodextrin	79.74	x 18.5
—	+ PEO γ-cyclodextrin	32.73	x 7.5

2.2.2. Influence of the association method

A simple physical mixture of the active drug and the cyclodextrin very often results in an increase in water-solubility. For exemple β or γ-cyclodextrin associated to progesterone lead to an increase in solubility depending on the solubility of the cyclodextrin used [13].

Amorphization of such a physical mixture, either by freeze-drying or co-grinding, results in a dramatic increase in water-solubility. For example, the simple physical mixture of tenoxicam with β-cyclodextrin results in a negligeable increase in water-solubility, when their freeze-drying or co-grinding lead to 80% dissolution in less than 10 minutes, opposite to some 10-12% in the same time for the free drug or its physical mixture [14].

Whatever the interest of the previous forms, their major drawback is their physical instability by possible recrystallization in the presence of some humidity. This phenomenon can be prevented by the use of a drug/cyclodextrin inclusion compound, which is very often used to increase drug water-solubility [15].

2.2.3. Influence of additives

The increase in water-solubility by the complexation method is very often limited by the stability constant of the inclusion. It has been proposed to increase the affinity of the drug for the cyclodextrin cavity by increasing the water structure with hydrotropic agents. However the results are very disputable and, for example, instead of an increase in indomethacin water-solubility, by β-cyclodextrin, in the presence of either sucrose or mannitol, it is a decrease which is observed [16].

In liquid formulation, low stability constant of the dissolved inclusion compound can result in a progressive reprecipitation of the free active drug following theinclusion dissociation. It was demonstrated, not only, that this drawback can prevented be prevented by association to a polymer, but also that this polymer can increase dramaticaly the drug solubility in the presence of a cyclodextrin [17]. A very interesting result was obtained for tretinoin associated either to β-cyclodextrin, dimethyl β-cyclodextrin, or hydroxypropyl β-cyclodextrin in the presence of polyvinylpyrrolidone [18]. This could allow the formulation of stable hydrogels with the poorly water-soluble tretinoin.

3. NANOPARTICLES AND CYCLODEXTRINS

Nowadays, nanoparticles represent powerfull tools not only in parenteral targeting, but also for improving drug bioavailability through the gastrointestinal tract. However, for nanoparticles prepared by nanoprecipitation in an aqueous medium, it is almost impossible to obtain a high loading of the nanoparticles with poorly water-soluble drugs.

3.1. Cyclodextrins in nanoparticles

In the past few years, several authors incorporated cyclodextrins in microparticles in order to increase the encapsulation of drugs [19, 20] or to modulate the release of the incorporated drug [19, 20, 21, 22, 23]. However, none of them worked on nanoparticles despite their interest for drug administration and bioavailability. In order to increase the drug loading of poly(isobutylcyanoacrylate) particles we studied the influence of a series of cyclodextrins [24].

3.1.1 Incorporation of cyclodextrins to nanoparticles

Poly(isobutycyanoacrylate) nanoparticles were prepared by anionic polymerization of isobutylcyanoacrylate in 0.01 M hydrochloric acid containing 1% of poloxamer 188 and in the presence of a series of cyclodextrins and derivatives. Nanoparticle size, zeta potential and cyclodextrin content were influenced by the nature of the cyclodextrin (Table 2) [25].

Table 2
Influence of cyclodextrin nature on poly(isobutylcyanoacrylate) nanoparticle characteristics

Cyclodextrin (5mg/ml)	Size (nm) ±SD	Zeta potential ± SD	Cyclodextrin content (mg CD/g particles)
β-CD	369 ±7	- 24.7	360
γ-CD	286 ±9	- 22.9	240
HP β-CD	103 ±6	- 8.6	247
HP γ-CD	87 ±3	- 2.6	220

Loading of hydroxypropyl β-cyclodextrin nanoparticles was studied with a series of steroids. Nanoparticles were prepared, as previously, by anionic poplymerization in the presence of poloxamer 188, and addition of either steroid/hydroxypropyl β-cyclodextrin complexes (10.0 mg hydroxypropyl β-cyclodextrin per ml of solution) or steroid solutions. The increase in drug loading varied from 5.5 to 130 folds for megestrol acetate and prednisolone respectively (Table 3) [24].

Table 3
Drug loading of poly(isobutylcyanoacrylate) nanoparticles in the presence or not of hydroxypropyl β-cyclodextrin

Steroid	CD content (mg/g)		Drug loading (µmol/g)		Loading
	poloxamer 1%	poloxamer 1% + HPβCD	poloxamer 1%	poloxamer 1% + HPβCD	increase (fold number)
hydrocortisone	-	180	6,04	42.21	6.98
prednisolone	-	210	0,33	43.00	129.17
spironolactone	-	230	18,36	127.23	6.93
testosterone	-	180	7.87	67.60	8.59
megestrol acetate	-	220	0.65	3.64	5.59
danazol	-	280	1.01	33.19	32.94
progesterone	-	242	2.51	69.60	27.70

DSC analysis of nanoparticles loaded with progesterone showed that the drug is in amorphous state in the particles and probably molecularly dispersed.

The release of progesterone in ABB medium (pH 8.4) occurs very rapidely (50% in 1.30 h, for 70 nm diameter particles) but is never complete and is limited to 60% is this example. The release is faster with small particles than with large ones and increases in the presence of PEG 400, but remains limited. The total release is obtained only in the presence of esterase in the dissolution medium. The very rapid release of hydroxypropyl β-cyclodextrin itself suggests that progesterone loading of nanoparticles occurs in at least two different manners: free progesterone in molecular state is entrapped inside the nanoparticle cores and inclusion of progesterone in hydroxypropyl β-cyclodextrin is adsorbed at the particle surface.

3.2. Amphiphilic cyclodextrin nanoparticles

Among the various amphiphilic cyclodextrins described in the literature, skirt-shapped cyclodextrins have been shown to present the remarquable ability to lead to the formation of nanoparticles [26]. These amphiphilic cyclodextrins [27, 28] are obtained by esterification of secondary hydroxyl groups by acyl chloride of variable length.

Working on either β- [29] or γ-cyclodextrin [30] with C_6 chain length substitutions, we prepared nanoparticles by the nanoprecipitation method [29, 30] or the emulsion solvent evaporation method [31].

These skirt-shaped cyclodextrins have surfactive properties allowing the preparation of blank nanospheres by nanoprecipitation method without the presence of surfactant. Their loading by hydrophobic drugs can be carried out either on blanck nanoparticles or during the nanosphere preparation. In this latter case, the presence of surfactant (pluronic F68) is recommanded. Preparation of nanospheres by emulsion solvent evaporation method in the presence of surfactant does not result in so well monodispersed nanoparticles than by nanoprecipiation and the yield is lower than in this latter method.

Nanospheres of the amphiphilic γ-cyclodextrin were loaded with progesterone, testosterone or hydrocortisone as model drugs (Table 4) [12]. It appears that the loading capacity increases with an increase in the stability constant of the drug and the parent γ-cyclodextrin, and with a decrease in water-solubility of the drug.

DSC analysis, as well as XR diffractometry, showed that progesterone is molecularly dispersed in the nanoparticles [30]. The immediate release obtained of progesterone indicates that the drug is probably concentrated at the nanoparticle surface [30]. This technique shows that it is possible to deliver rapidly high amounts of water-insoluble drugs, allowing their faster bioavailability.

Due to the different amphiphilic cyclodextrins which can be potentially used to prepare nanoparticles: different parent cyclodextrin, different chain length, different susbtitution degree, different localization of substitution, amphiphilic cyclodextrins can be a powerfull tool for varying the loading capacity of nanoparticles and the drug release profile.

Table 4

Loading capacity of amphiphilic γ-cyclodextrin (C_6 chains) with a series of steroids

Drug	Loading capacity (mg/g)	Water-solubility (mg/l)	Kγ
progesterone	60-80	3.3	24.000
testosterone	20-30	23.0	16.500
hydrocortisone	< 15	326.0	2.240

4. CONCLUSION

By their diversity, cyclodextrins represent powerfull tool to improve the apparent water-solubility of drugs. The choice of the cyclodextrin to use depends on the administration route and the pharmaceutical form looked at: solid dosage form for oral administration, hydrogel for local application, nanoparticles for drug targeting.

REFERENCES

[1] β-cyclodextrin orale

[2] β-cyclodextrin hémolytique

[3] Vollmer, U., Müller, B.W., Mesens, J., Wilfert, B., Peters, T., In vivo skin pharmacokinetics of liarozole: Percutaneous absorption studies with different formulations of cyclodextrin derivatives in rats. *Int. J. Pharm.*, ***99***, *51-58 (1993).*

[4] Vollmer, U., Müller, B.W., Peeters, J., Mesens, J., Wilfert, B., Peters, T., A study of the percutaneous absorption-enhancing effects of cyclodextrin derivatives in rats. *J. Pharm. Pharmacol.*, ***46***, *19-22 (1994).*

[5] HPβ-cyclodextrin parenteral

[6] HPβ-cyclodextrin parenteral

[7] Thompson, D.O., Cyclodextrins–Enabling excipients: their present and future in pharmaceuticals. *Crit. Rev. Ther. Drug Carrier Syst.*, ***14***, *1-104 (1997).*

[8] Koizumi, K., Okada ,Y., Kubota, Y., Utamara, T., Inclusion complexes of poorly water-soluble drugs with glycosyl cyclodextrins. *Chem. Pharm. Bull.*, ***35***, *3413-3418 (1987).*

[9] Yamamoto, M., Yoshida, A., Hirayama, F., Uekama, K., Some physicochemical properties of branched β-cyclodextrins and their inclusion characteristics. *Int. J. Pharm.*, ***49***, *163-171 (1989).*

[10] Yoshida, A., Arima, K., Uekama, K., Pitha, J., Pharmaceutical evaluation of hydroxyalkyl ethers of β-cyclodextrins. *Int. J. Pharm.*, ***46***, *217-222 (1988).*

[11] POE-CD

[12] Duchêne, D., Wouessidjewe, D., *Solubilization of drugs via cyclodextrins.* In Proceedings of the International Pharmaceutical Applications of Cyclodextrins Conference, 29 June-2 July 1997, Lawrence.

[13] Lin, S.-Z., Skiba, M., Wouessidjewe, D., Agnus, B., Duchêne, D., *Inclusion complexes of progesterone and its analogue X with cyclodextrins.* In Minutes of the 6th International Symposium on Cyclodextrins, (Ed. A.R. Hedges), Editions de Santé, Paris, 1992, 460-464.

[14] Senel, S., Cakoglu, Ö., Sumnu, M., Duchêne, D., Hıncal, A., Preparation and investigation of tenoxicam/β-cyclodextrin complex. *J. Inclusion Phenom.*, ***14***, *171-179 (1992)*.

[15] Lin, S.-Z., Wouessidjewe, D., Poelman, M.-C., Duchêne, D., Indomethacin and cyclodextrin complexes. *Int. J. Pharm.*, ***69***, *211-219 (1994)*.

[16] Pioger, E., Wouessidjewe, D., Duchêne, D., Bogdanova, S., Effects of some hydrotropic agents on the formation of indomethacin/β-cyclodextrin inclusion compounds. *J. Incl. Phenom.*, ***30***, *151-161, (1998)*.

[17] Loftsson, T., Fridriksdóttir, H., Sigurdardóttir, A.M., The effect of water-soluble polymers on drug/cyclodextrin complexation. *Int. J. Pharm.*, ***110***, *169-177 (1994)*.

[18] Montassier, P., Duchêne, D., Poelman, M.-C., Inclusion complexes of tretinoin with cyclodextrins. *Int. J. Pharm.*, ***153***, *199-209 (1997)*.

[19] Loftsson, T., Kristmunsdóttir, T., Ingvarsdóttir, K., Olafsdóttir, B.J., Baldvinsdóttir, J., Preparation and physical evaluation of microcapsules of hydrophilic drug-cyclodextrin complexes. *J. Control. Rel.*, ***9***, *375-382 (1992)*.

[20] Loftsson, T., Brewster, M., Pharmaceutical applications of cyclodextrins. 1. Drug solubilization and stabilization. *J. Pharm. Sci.*, ***85***, *1017-1025 (1996)*.

[21] Filipovic-Grcic, J., Becirevic-Lacan, M., Skalko, N., Jalsenjak, I., Chitosan microspheres of nifedipine and nifedipine-cyclodextrin inclusion complexes. *Int. J. Pharm.*, ***135***, *183-190 (1996)*.

[22] Sutinen, R., Kekki, M., Paronen, P., Urtti, A., *Effect of cyclodextrin on drug release from pH-controlled silicone microspheres*. In Formulation of Poorly Available Drugs for Oral Administration, (Ed. D. Duchêne), Editions de Santé, Paris 1996, 261-264.

[23] Utsuki, T., Brem, H., Pitha, J., Loftsson, T., Kristmundsdóttir ,T., Tyler, B.M., Olivi, A., Potentiation of anticancer effects of microencapsulated carboplatin by hydroxypropyl α-cyclodextrin. *J. Control. Rel.*, ***40***, *251-260 (1996)*.

[24] Silveira, A.M., Formulation et caractérisation de nanoparticules combinées de poly(cyanoacrylate d'isobutyle) et de cyclodextrines destinées à l'administration de principes actifs faiblement solubles dans l'eau. Thesis N°562, Université Paris-Sud, 1998.

[25] Silveira, A.M., Ponchel, G., Puisieux, F., Duchêne, D., Combined poly(isobutylcyanoacrylate) and cyclodextrin nanoparticles for enhancing the encapsulation of lipophilic drugs. *Pharm. Res., in print.*

[26] Wouessidjewe, D., Skiba, M., Leroy-Lechat, F., Lemos-Senna, E., Puisieux, F., Duchêne, D., A new concpt in drug delivery based on "skirt-shaped cyclodextrin aggregates'. Present state and future prospects. *S.T.P. Pharma Sci.*, ***6***, *21-28 (1996)*.

[27] Zhang, P., Ling, C.C., Coleman, A.W., Parot-Lopez, H., Galons, H., Formation of amphiphilic cyclodextrins via hydrophobic esterification at the secondary hydroxyl face. *Tetrahedron Lett.*, ***32***, *2769-2770 (1991)*.

[28] Memisoglu, E., Charon, D., Duchêne, D., Hıncal, A.A., Synthesis of per(2,3-di-*O*-hexanoil)-β-cyclodextrin and characterization of amphiphilic β-cyclodextrin nanoparticles. Proccedings of the 9th International Cyclodextrin Conference, Santiago de Compostela, 31 May-3 June 1998.

[29] Skiba, M., Duchêne, D., Puisieux, F., Wouessidjewe, D., Development of a new colloidal drug acrrier from chemically-modified cyclodextrins: nanospheres, and influence of physicochemical and technological factors on particle size. *Int. J. Pharm.*, ***129***, *113-121 (1996)*.

[30] Lemos-Senna, E., Wouessidjwe, D., Lesieur, S., Puisieux, F., Couarraze, G., Duchêne, D., Evaluation of the hydrophobic drug loading characteristics in nanoprecipitated amphiphilic cyclodextrin nanospheres, *Pharm. Dev. Technol.*, ***3***, *1-10 (1998)*.

[31] Lemos-Senna, E., Wouessidjewe, D., Lesieur, S., Duchêne, D., Preparation of amphiphilic cyclodextrin nanospheres using the emulsion solvent evaporation method. Influence of the surfactant on preparation and hydrophobic drug loading. *Int. J. Pharm., in print.*

DESIGN AND EVALUATION OF COLON-SPECIFIC DRUG DELIVERY SYSTEM BASED ON CYCLODEXTRIN CONJUGATES

F. HIRAYAMA, K. MINAMI, and K. UEKAMA
Faculty of Pharmaceutical Sciences, Kumamoto University, 5-1, Oe-honmachi, Kumamoto 862-0973, Japan

1. Introduction

α-1,4 glucosidic bonds of cyclodextrins (CyDs) are barely capable of being hydrolyzed and only slightly absorbed in passage through the stomach and small intestine; however they are fermented by colonic microflora into small saccharides and thus absorbed in the large intestine. This kind of biodegradable property of CyDs is particularly useful as a colon-targeting carrier [1], and thus CyD conjugates may serve as a source of site-specific delivery of drugs to colon. Therefore, we prepared six conjugates of the antiinflammatory drug, 4-biphenylylacetic acid (BPAA) coupled to α-, β- and γ-CyDs through an ester or amide linkage (Fig. 1), and investigated the *in-vitro* and *in-vivo* release behavior of BPAA from the prodrugs, oral absorption behavior, and antiinflammatory effect in rat model to evaluate them as a colon-specific drug delivery system.

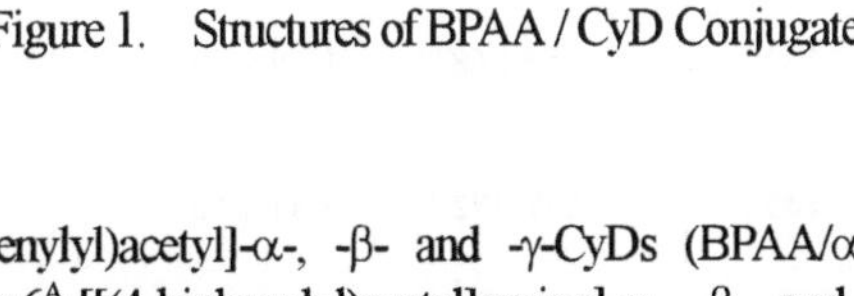

X= -NH- : Amide type
-O- : Ester type

Figure 1. Structures of BPAA / CyD Conjugates

2. Experimental

Materials: 6^A-*O*-[(4-biphenylyl)acetyl]-α-, -β- and -γ-CyDs (BPAA/α-, β- and γ-CyD ester conjugates) and 6^A-deoxy-6^A-[[(4-biphenylyl)acetyl]amino]-α-, -β-, and -γ-CyDs (BPAA/α-, β-, and γ-CyD amide conjugates) were prepared according to the method reported in previous papers [2, 3]. The NMR and mass spectroscopic results indicated that the BPAA moiety was introduced onto one of the primary hydroxyl groups of CyDs and the substitution degree was unity. The BPAA/β-CyD complex in a molar ratio of 1:1 was prepared by the kneading method as reported [3].

Hydrolysis in Rat Intestinal Contents: Stomach and other contents of male Wistar rats, 400-500 g, were diluted to 20%w/v with isotonic acetate buffer (pH4.4) and phosphate buffer (pH7.4), respectively, and the dispersions of contents were filtered through a gauze to remove large particles.

The conjugates solution (5.0mL, $2.0x10^{-5}$M in 2.0%v/v DMF/isotonic buffer) was added to the filtrate (5.0mL) in air-tight vessels and incubated at 37°C. At appropriate intervals, the reaction solutions were determined for BPAA and intact conjugates by HPLC.

In-Vivo Absorption Studies: The conjugates, the BPAA/β-CyD complex or BPAA alone (equivalent to 10mg BPAA/5mL water per kg of rat body weight) were orally administered to male Wistar rats weighing about 200g. Blood samples were taken periodically from the jugular vein, and was assayed for BPAA and intact conjugates by HPLC.

Antiinflammatory Response [4]: A 1%w/v carrageenan (Nacalai Tesque, Tokyo, Japan) in saline (0.1mL) was injected subcutaneously into the right hind paw at various times after the conjugates had been administered orally (equivalent to 10mg of BPAA/5mL of water per kg of rat body weight). Paw volume was periodically monitored using a water plethysmometer after the carrageenan injection.

3. Results and Discussion

3.1. CHARACTERIZATION OF CyD CONJUGATES

Table I summarizes some physicochemical properties of 6 conjugates prepared. The solubility of the conjugates in water at 25°C decreased in the order of the α-CyD conjugate > the γ-CyD conjugate >> the β-CyD conjugate. The solubility of β-CyD conjugates was increased by the addition of parent β-CyD that is well fitted to the BPAA moiety or the guest molecules that are fitted to the β-CyD cavity, suggesting that the low solubility of the β-CyD conjugates may be

Table I. Some Physicochemical Properties of BPAA/CyD Conjugates

Compound	Molecular weight	Melting point (°C)	Solubility a) (M)	Solubility ratio (conjugate / BPAA)
BPAA	212	164-165	1.26×10^{-4}	1
α-CyD ester conjugate	1167	255 b)	1.18×10^{-2}	94
β-CyD ester conjugate	1329	258-268 b)	1.29×10^{-5}	0.10
γ-CyD ester conjugate	1492	277-282 b)	4.34×10^{-4}	3.4
α-CyD amide conjugate	1166	249-256 b)	1.28×10^{-2}	102
β-CyD amide conjugate	1328	258-263 b)	1.42×10^{-5}	0.11
γ-CyD amide conjugate	1491	279-280 b)	1.19×10^{-3}	9.4

a) In water at 25 °C. b) Decomposition.

ascribed to the self-interaction of the conjugates. The amide conjugates were stable in aqueous solution (half-life in 0.1 M NaOH at 60°C > 12h), whereas the ester conjugates were hydrolyzed at moderate rates resulting in V-shaped rate-pH profiles (half-life at pH8.7 and 37°C: 8h).

3.2. *IN-VITRO* DRUG RELEASE BEHAVIOR

In rat cecal and colonic contents, the α- and γ-CyD ester conjugates released more than 95% of BPAA within 1-2h, and the β-CyD ester conjugate released about 77% of the drug within 24h. The amide conjugates did not release BPAA in the cecal content, but gave BPAA/maltose or BPAA/triose conjugates linked through an amide bond. On the other hand, these conjugates were stable in the contents of rat stomachs and small intestines, in intestinal or liver homogenates, and in rat blood. These *in-vitro* release studies indicated that the conjugates are firstly subject to the ring opening to give the maltose and triose conjugates, and the ester bond is then hydrolyzed to give BPAA.

3.3. *IN VIVO* DRUG RELEASE BEHAVIOR

The CyD conjugates were stable in rat stomach and small intestine and negligibly absorbed from these tracts. Most of drug had been moved to the cecum and colon 2-3 h after the oral administration. The α- and γ-CyD amide conjugates were hydrolyzed to the maltose conjugate in the cecum and colon, and these conjugates were negligibly absorbed. On the other hand, the α- and γ-CyD ester conjugates produced BPAA in the cecum and colon, and BPAA appeared in the blood after 3-6h. Both β-CyD amide and ester conjugates released only small or negligible amounts of the maltose conjugates or BPAA in the cecum and colon within 24 h, probably due to the low solubility in the biological media.

The serum levels of BPAA increased about 3h after oral administration of the α- and γ-CyD ester conjugates to rats, accompanying a marked increase in the serum levels, whereas the β-CyD ester and the amide conjugates resulted in little increase of the serum levels (Fig. 2).

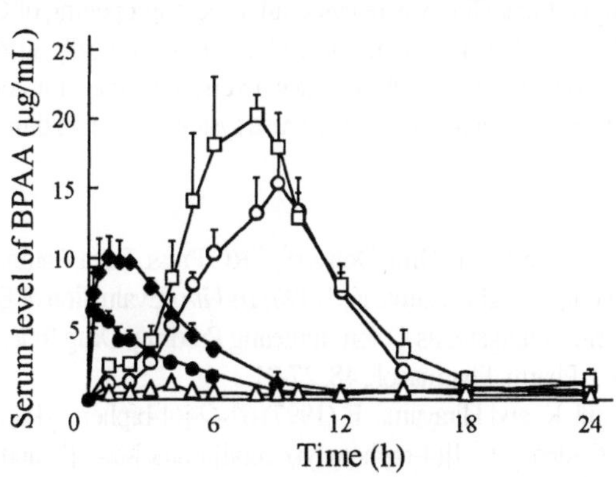

Figure 2. Serum levels of BPAA after oral administration of α- (○), β- (△), γ-CyD ester conjugates (□), BPAA (●), or BPAA/β-CyD complex (◆) (equivalent to 10mg/kg BPAA) to rats. Each point represents the means ± S.E. of 3 experiments.

3.4. THE ANTIINFLAMMATORY RESPONSES
The antiinflammatory effect of the γ-CyD ester conjugate and β-CyD complex was evaluated using the model of carrageenan-induced acute edema in rat paw. In the case of β-CyD complex, a rapid antiinflammatory response was observed, compared to BPAA alone, because the drug was mainly absorbed from the small intestine after a fast dissolution of the complex. In sharp contrast, the γ-CyD ester conjugate needed a fairly long lag time to exhibit the drug activity, because BPAA was produced after it had reached the cecum and colon.

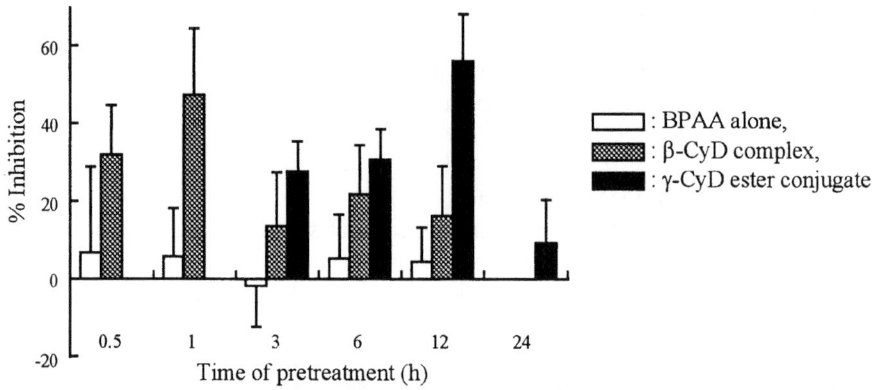

Figure 3. Antiinflammatory responses of BPAA, β-CyD complex and γ-CyD ester conjugate administered orally to carrageenan-induced edema in rat paw.
Each data point represents the mean ± S.E. of 7-8 experiments.

4. Conclusion

The present results suggest that BPAA is released after the ring opening of CyDs followed by the ester hydrolysis, and the BPAA activation takes place site-specifically in the cecum and colon. Therefore, the present conjugate approach provides a versatile means for construction of not only colon-specific delivery system but also delayed-release system of certain drugs.

5. References

[1] Friend, D.R.:Oral Colon-Specific Drug Delivery, CRC Press, Boca Raton, 1992.
[2] Hirayama, F., Minami, K. and Uekama, K. (1996) *In-Vitro* Evaluation of Biphenylylacetic Acid-β-Cyclodextrin Conjugates as Colon-Targeting Prodrug: Drug Release Behaviour in Rat Biological Media, J. Pharm. Pharmacol., **48**, 27-31.
[3] Uekama, K., Minami, K. and Hirayama, F. (1997) 6^A-*O*-[(4-biphenylyl)acetyl]-α-, -β- and -γ-cyclodextrins and 6^A-deoxy-6^A-[[(4-biphenylyl)acetyl]amino]-α-, -β-, and -γ-cyclodextrins: Potential Prodrugs for Colon-specific Delivery, J. Med. Chem., **40**, 2755-2761.
[4] Minami, K., Hirayama, F. and Uekama, K. (1998) Colon-specific Drug Delivery Based on Cyclodextrin Prodrug: Release Behavior of Biphenylylacetic Acid from Its Cyclodextrin Conjugates in Rat Intestinal Tracts after Oral Administration, J. Pharm. Sci., in press.

THE EFFECT OF 2-HYDROXYPROPYL-β-CYCLODEXTRIN ON THE SOLUBILITY, STABILITY AND BRAIN TARGETING OF CHEMICAL DELIVERY SYSTEMS FOR NEUROPEPTIDES

N. BODOR, W-M. WU AND J. WU
Center for Drug Discovery, University of Florida,
P.O. Box 100497 JHMHC, Gainesville, FL 32610, USA

1. Introduction

A general method for brain targeting of neuropeptides such as enkephalins, kyotorphin and various TRH (thyrotropin releasing hormone) analogs, based on their molecular packaging and subsequent, sequential metabolism, was recently described (Bodor et al. 1992, Bodor 1997, Prokai-Tatrai et al., 1996). Some salient aspects of this approach were recently reviewed (Bodor and Buchwald 1998). The main objective of the method is to specifically target neuropeptides to the brain, which otherwise are unaccessible due to their hydrophilic nature, as well as, being excellent substrates of peptidases found in the blood-brain barrier (BBB). Brain delivery, and particularly brain targeting of neuropeptides is very important, as of the some 100 neuropeptides isolated, most, if not all are constituents of the brain and their hyper or hypo production is involved in various CNS diseases. The successful brain targeting of various neuropeptides was achieved by multiple chemical modifications of the target peptides in order to produce lipophilic constructs, which are not recognized by peptidases as peptides or peptide derivatives. But converting a peptide into a lipophilic form would only allow the brain delivery of this disguised peptide construct, which however, would not result in effective brain delivery of the peptide. In order to produce effective targeting, we have introduced a number of other modifications, such as, the use of the earlier described dihydrotrigonellyl « trigonellyl redox brain targeting system, which is linked to the N-terminal of the peptide through strategically selected spacer functions. In addition, when necessary the C-terminal is further modified in order to allow specific enzymes to produce the active terminal, such as the case of some TRH analogs. The redox targetor system provides a lock-in of the main precursor for a sustained release of the peptide after the strategically planned enzymatic reactions take place. In the same time, the rest of the body is spared from the peptide as the complex CDS is destroyed by various enzymes, primarily in the liver, before the active peptide can be released. The actual brain targeting, the presence of the predicted intermediates and high and sustained pharmacological activity for various "packaged" peptides was demonstrated before. One of the general problems related to these packaged peptide constructs is their intrinsic high lipophilicity. In these constructs in which the peptides represent a perturbation, in order to avoid premature peptidase deactivation, large lipophilic functions on both the C and N-terminals are used. The result is that the original chemical delivery systems (CDS) for these peptides are highly lipophilic, in general their log *P* is over 6 and consequently they have negligible solubility in water. In addition, due to presence of the 1,4-dihydrotrigonellyl function, these molecules are sensitive to acid catalyzed water addition, as well as, chemical oxidative conversion to the corresponding quaternary pyridinium, trigonellyl salts. We have recently developed a molecular based approach to estimate partition properties for organic solutes (Bodor and Buchwald, 1997), including non-zwitterionic peptides and their derivatives (Buchwald and Bodor, 1998). This method allows us to estimate with excellent accuracy the partition and solubility properties of all potential lipophilic constructs, as well as, their predicted enzymatic intermediates and byproducts. However, the administration of the first, fully packaged form of the peptide is always a problem. Water miscible organic solvents, which would solubilize these molecules, such as DMSO, cannot be used since these highly lipophilic molecules would immediately precipitate upon injection. We have previously found that 2-hydroxypropyl-β-cyclodextrin (HPBCD) is very effective in solubilizing and stabilizing protein drugs (Brewster et al., 1991). Accordingly, we were expecting that the brain targeted peptide derivatives could also benefit from HPBCD. Accordingly, the aim of the present study was to provide a way to convert these highly lipophilic constructs into water soluble forms, and to provide chemical stability for the various sensitive functions. The class of compounds that we are focusing on involves three different types of TRH (Glp-His-Pro-NH_2) analogs, in which histidine was replaced by leucine (Leu) or norvaline (Nva), respectively, and the packaging was accomplished by opening the pyroglutamate ring to glutamine to allow the coupling at the N-terminal with Dhtr-Spacer, while the active prolineamide was generated from the additional glycine on the C-terminal, which is oxidatively converted to the prolineamide by peptidyl glycine a-amidating monooxygenase (PAM). In one alternative, the C-terminal proline amide was replaced by pipecolinic acid, which does not require the amide form for activity. All three analogs contained a cholesterol as the ester function, a dihydrotrigonellyl targetor and two L-prolines as the spacer, which provides timely and sustained release of the active peptide.

2. Results

The solubilities of the packaged peptides in HPBCD solutions are listed in Table 1. All three molecules show an about 15 mg/ml solubility in 50% HPBCD.

TABLE 1. Solubilities of brain-targeted TRH analog CDSs in HPBCD solution at pH 6.5

Compound	HPBCD (%)	Solubility (mg/ml)
Dhtr-ProPro-GlnLeuPro-Gly-O-cholesterol [Leu2]-TRH-CDS (CDS-**1**)	0	< Detection limit*
	10	2.56
	20	4.66
	30	6.38
	40	9.70
	50	13.80
Dhtr-ProPro-GlnNvaPro-Gly-O-cholestero l [Nva2]-TRH-CDS (CDS-**2**)	50	14.10
Dhtr-ProPro-GlnNvaPip-O-cholesterol [Leu2,Pip3]-TRH-CDS (CDS-**3**)	50	15.97

*Detection limit was 0.05 mg/ml. Dhtr: 1,4-dihydrotrigonellyl

The phase solubility diagram shows a 1:1 complex up to 30% cyclodextrin, but at 40% and 50% the two:one complex is dominating. Since none of the amino acids in these peptides contain a phenyl or similar rings, the complexation most likely occurs at the molecule ends, as indicated on Fig. 1.

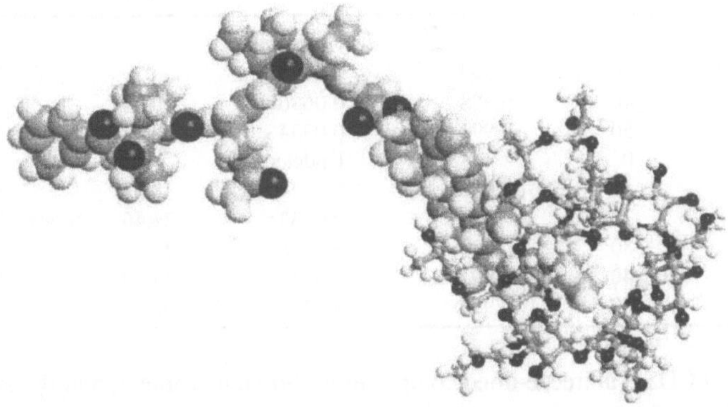

FIGURE 1. A possible complexation between [Leu^2]-TRH-CDS and HPBCD

Given the expected high activities of these neuropeptides, the achieved solubility should be enough to be practical for application. The second important issue relates to stability. The water addition as depicted on Fig. 2, produces a 6-hydroxy-tetrahydrotrigonellyl derivative, a process which is irreversible and as such highly undesirable.

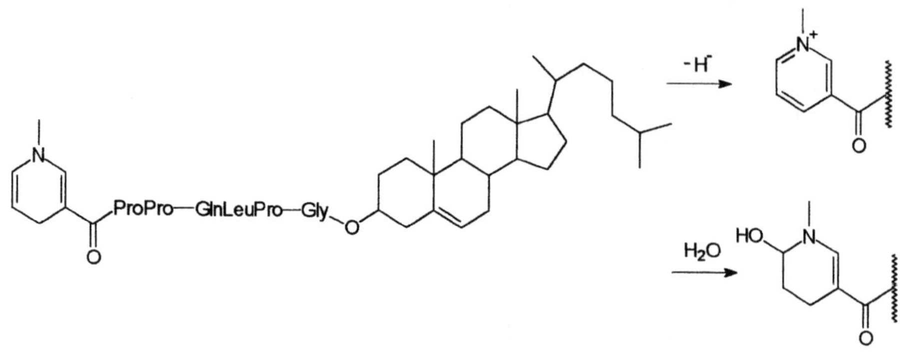

FIGURE 2. Oxidation and water addition reactions of [Leu^2]-TRH-CDS.

As shown on Table 2, the CDS-1, shows limited aqueous stability at pH-s near neutral, with a t_{90} of about 4½ hours in 40-50% HPBCD at pH 7.4. While this is sufficient for use as a freshly reconstituted injectable, it is clear that these compounds cannot be formulated in aqueous solutions.

TABLE 2. Stabilities of [Leu^2]-TRH-CDS in HPBCD solution, pH 6.5 and 7.4, at 25°C.

HPBCD, % (w/w)		*r*	*k*, hr^{-1}	$t_{1/2}$, hr	t_{90}, hr
pH 6.5	0	-	Undetectable		-
	30	0.999	0.0764	9.07	1.37
	40	0.998	0.0646	10.73	1.63
	50	0.994	0.0334	20.72	3.14
pH 7.4	0	-	Undetectable		-
	10	0.998	0.0374	18.51	2.81
	20	0.998	0.0283	24.46	3.71
	30	0.998	0.0258	26.82	4.06
	40	0.997	0.0237	29.30	4.44
	50	0.999	0.0231	30.06	4.55

The amount of CDS-s in freeze-dried complexes is shown in Table 3, and the stability of the freeze-dried complexes is shown in Table 4.

TABLE 3. Amount of CDSs in freeze-dried CDS / HPBCD complex.

Compounds	Content, mg/g
Dhtr-ProPro-GlnLeuPro-Gly-O-cholesterol [Leu2]-TRH-CDS (CDS-1)	26.22
Dhtr-ProPro-GlnNvaPro-Gly-O-cholesterol [Nva2]-TRH-CDS (CDS-2)	26.79
Dhtr-ProPro-GlnNvaPip-O-cholesterol [Leu2,Pip3]-TRH-CDS (CDS-3)	30.34

TABLE 4. Stability of freeze-dried [Leu^2]-TRH-CDS / HPBCD complex

	r	k, d^{-1}	$t_{1/2}$, d	t_{90}, d
Argon protected				
-15°C	No detectable change observed in 3 month			
4°C	0.998	0.00166	417	63
25°C	0.917	0.00666	104	16
Air exposed				
-15°C	No detectable change observed in 3 month			
4°C	0.962	0.00152	457	69
25°C	0.930	0.00759	91	14

At lower temperatures, such as 4°C, the stability is significantly increased to t90 over two months. But in order to provide expected shelf life of 1-2 years, the complexes should be kept below freezing. In order to demonstrate the usefulness of the TRH-CDS-s as cyclodextrin complexes, pharmacological studies were undertaken. One generally accepted method to demonstrate the CNS-activity of TRH and its analogs is to determine their effect on the barbiturate induced sleeping time, by their effect on the cholinergic system. Swiss Webster mice were used, and various doses of the CDS-2, the norvaline analog, which was not previously studied was injected intravenously. Ten minutes after the i.v. injection, the animals received an intraperitoneal injection of sodium pentobarbital at a dose of 60 mg/kg. The sleeping time was then recorded as the time elapsed from the onset of loss of the writing reflex, until the reflex was regained. Groups of 8-9 animals were used, including the control group which received only the cyclodextrin solution. The dose response curve is shown on Fig. 3. The packaged TRH analog is highly potent, maximum effect was achieved at doses above 2 mg/kg. This highly significant, about 50% reduction in the sleeping time is shown on Fig. 4.

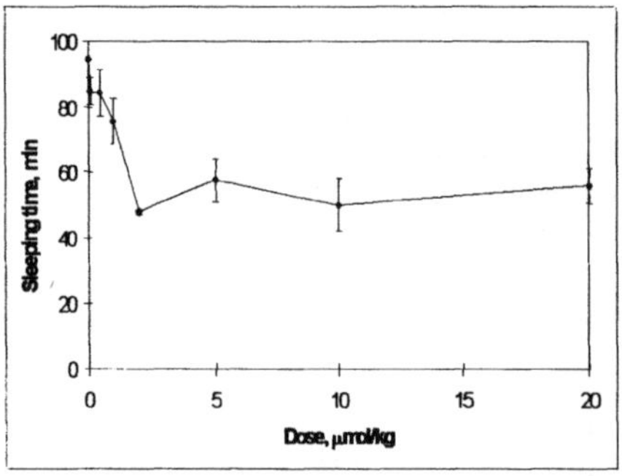

FIGURE 3. Dose response after i.v. administration of [Nva^2]-TRH-CDS in mice.

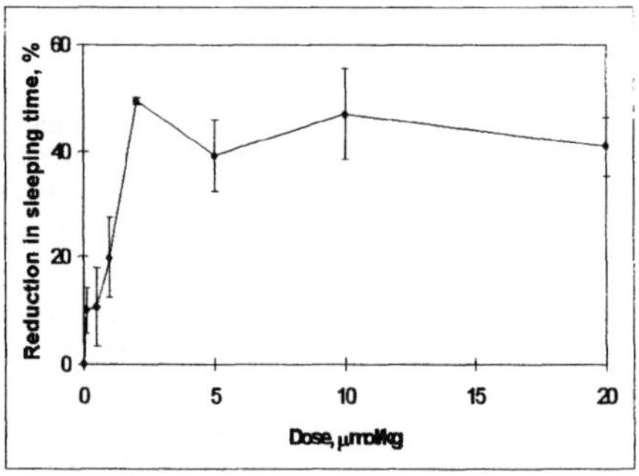

FIGURE 4. Reduction in sleeping time after i.v. administration of [Nva^2]-TRH-CDS in mice.

3. Conclusions

It was demonstrated that both aqueous solubility and chemical stability (both in solution and in solid form) of the various CDS-s of TRH analogs can be significantly improved by complexation with HPBCD. The stabilizing effect of HPBCD is significantly greater in the solid, freeze-dried complex form, which at temperatures below freezing demonstrate sufficient shelf life stability to be potential useful pharmaceutical preparations. The reconstituted freeze-dried forms show high pharmacological activity at low doses, demonstrating the success of the concept of brain targeting of neuropeptides, benefiting from formulation, using cyclodextrins.

4. References

Bodor, N., Prokai, L., Wu, W., Farag, H., Jonnalagadda, S., Kawamura, M. and Simpkins, J. (1992) A strategy for delivering peptides into the central nervous system by sequential metabolism, Science, 257, 1698-1700.

Bodor, N. and Buchwald, P. (1998) All in the mind, Chemistry in Britain, 34(1), 36-40.

Bodor, N. (1997) Brain targeting of drugs and neuropeptides by retrometabolic design approaches, S.T.P. Pharma sciences Editions De Sante, 7(1), 43-52.

Prokai-Tatrai, K., Prokai, L., and Bodor, N. (1996) Brain-targeted delivery of a leu-enkephalin analog by retrometabolic design, Journal of Medicinal Chemistry, 39, 4775-4782.

Buchwald, P. and Bodor, N. (1997) Molecular size-based model to describe simple organic liquids, Journal of Physical Chemistry B, 101, 3404-3412.

Bodor, N. and Buchwald, P. (1998) Octanol-water partition coefficient for non-zwitterionic peptides: predictive power of a molecular size based model, Proteins, Structure, Function, and Genetics, 30, 86-99.

Brewster, M., Hora, M., Simpkins, J., and Bodor, N. (1991) Use of 2-hydroxypropyl-β-cyclodexdrin as a solubilizing and stabilizing excipient for protein drugs, Pharmaceutical Research, 8(6), 792-795.

Prokai, L., Ouyang, X., Wu, W-M. and Bodor, N. (1994) Chemical delivery system to transport a pyroglutamyl peptide amide to the central nervous system, Journal of the American Chemical Society, 116(6), 2643-2644.

ENHANCED COMPLEXATION EFFICACY OF CYCLODEXTRINS

T. LOFTSSON, M. MÁSSON AND J. F. SIGURJÓNSDÓTTIR
Department of Pharmacy, University of Iceland, PO Box 7210, IS-127 Reykjavik, Iceland

1. Introduction

For various reasons, including drug availability, formulation bulk and toxicological considerations, it is important to use as little cyclodextrin (CD) as possible in pharmaceutical formulations. The complexation efficacy of CDs is generally rather low and, consequently, relatively large amount of CD has to be used to solubilize relatively small amount of drug. Thus, it is important to develop methods which enhance the complexation efficacy [1]. Furthermore, the aqueous solubility of the parent CDs and their complexes is low which limits their usage in pharmaceutical formulations.
If one drug molecule forms a complex with one CD molecule then the complexation efficiency will be equal to the intrinsic solubility of the drug (S_0) times the stability constant of the drug-CD complex (K_C). Thus, increased complexation efficiency can be obtained by either increasing S_0 or by increasing K_C or by increasing both simultaneously (Figure 1). Addition of organic solvents, such as ethanol, to the

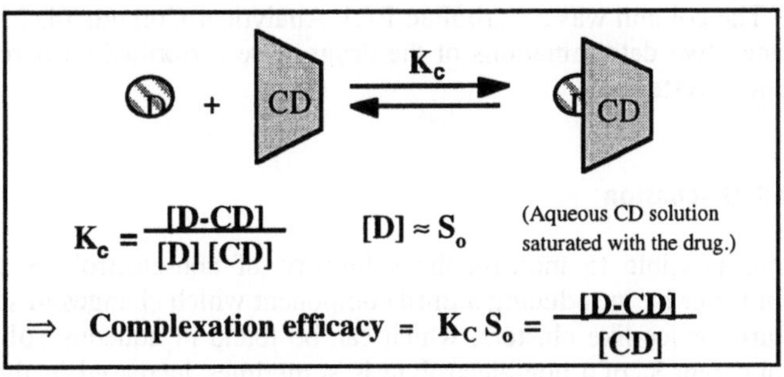

FIGURE 1. The complexation efficacy of 1:1 drug-CD complexes.

aqueous complexation media increases S_0 but at the same time they decrease the apparent K_C by competing with the drug molecule for a space in the CD cavity [2]. Thus, addition of organic solvents frequently decreases the efficiency. Ionisation of an ionizable drug molecule increases S_0 and although some decrease in K_C is generally observed, the increase in S_0 is frequently more than enough to compensated for the decrease [3]. Addition of water-soluble polymers to the aqueous complexation media increases the apparent K_C and, thus, increased efficiency is observed [4]. The purpose of this study was to investigate methods which enhance the complexation efficacy and which increased the aqueous solubility of the parent CDs, especially βCD.

2. Materials and Methods

2.1 MATERIALS

β-Cyclodextrin (βCD) was obtained from Celedex (Japan), 2-hydroxypropyl-β-cyclodextrin or Encapsin HPB (HPβCD) from Janssen Biotech (Belgium) and hydroxypropyl methylcellulose 4000 (HPMC) from Mecobenzon (Denmark). All other chemicals and drugs used were of pharmaceutical or special analytical grade.

2.2 SOLUBILITY STUDIES

An excess amount of the drug to be tested was added to aqueous CD solution, or aqueous solution containing CD and a water-soluble polymer, or aqueous buffer solutions containing the same. The suspension formed was heated in an autoclave in a sealed container (120-140°C for 20-40 min). After equilibration at room temperature (22-23°C) for 4 to 7 days, the suspension was filtered through a 0.45 μm membrane filter (Nylon Acrodisc from Gelman, USA), diluted with methanol-water (7:3) solution, and analysed by HPLC. The apparent stability constants (K_C) of the drug-CD (1:1) complexes were determined from the phase-solubility diagrams according to the method of Higuchi and Connors [5].

2.3 QUANTITATIVE DETERMINATIONS

The quantitative determination of βCD was performed on high-performance liquid chromatographic (HPLC) system equibbed with a PAD-2 pulsed amperometric detector from Dionex (USA) with a gold working electrode and a silver-silver chloride reference electrode [6]. The column was a CarboPac PA1 Analytical Column (4x250 mm) from Dionex. Quantitative determinations of the drugs were performed on a reversed-phase HPLC component system [4].

3. Results and Discussion

It is sometimes possible to increase the solubility of non-electrolytes (increase the apparent S_O) in water by introducing a third component which changes in some way the ordered structure, or ice-like clusters, which can be found in aqueous solutions. For example, the apparent S_O of a non-electrolyte is sometimes increased in the presence of carbohydrates such as glucose. This has been called sugaring-in effect [7]. This phenomenon is usually observed at relatively high sugar concentrations. It is possible that similar phenomenon could be observed at high CD concentrations which then should increase the complexation efficacy. Thus, we investigated the effect of increased HPβCD concentration on the complexation of several drugs (Figure 2). However, no enhancement could be observed. No sugaring-in effect was observed.

Unionised drugs usually form more stable CD complexes (i.e. they have larger K_C value) than their ionic counterparts. However, it is sometimes possible to enhance CD solubilization of ionizable drugs by appropriate pH adjustments (i.e. by increasing the value of S_0 through ionisation of the drug molecule). For example, at room temperature (25 °C) the solubility of phenytoin (which is a weak acid with pKa of 8.1) in water is

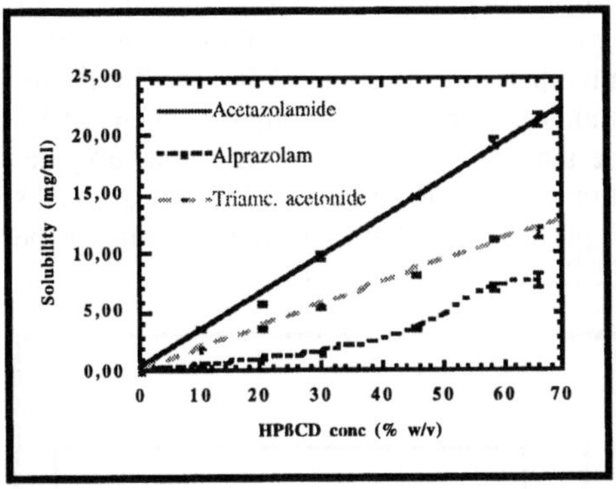

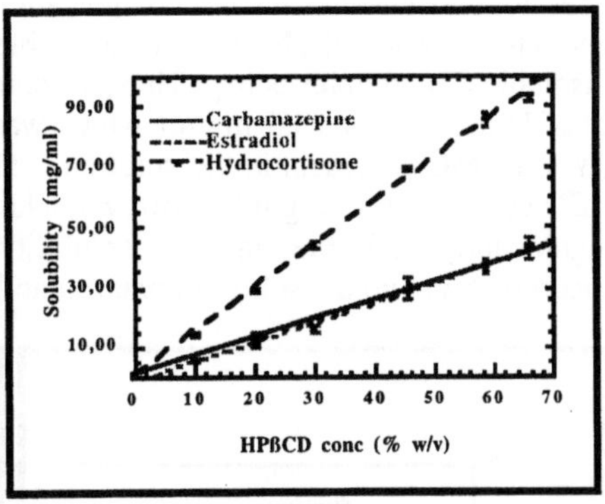

FIGURE 2. The phase-solubility diagrams of acetazolamide, alprazolam, triamcinolone acetonide, carbamazepine, 17β-estradiol and hydrocortisone in aqueous HPβCD solutions at 22-23°C.

only about 18 μg/ml at pH below 5, 22 μg/ml at pH 7.5 and about 0.5 mg/ml at pH 10. Addition of 5% (w/v) HPβCD to the aqueous solutions increases the solubility of phenytoin to about 0.7 mg/ml at pH below 5, 1.1 mg/ml at pH 7.5 and 15 mg/ml at pH 10 (Figure 3). Thus, it is possible to obtain much higher total solubility by adding CD to the aqueous solution and at the same time increasing ionisation of the drug by increasing the pH of the media (i.e. the increase in S_0 does more than to compensate for the decrease in K_C, with regard to solubility).

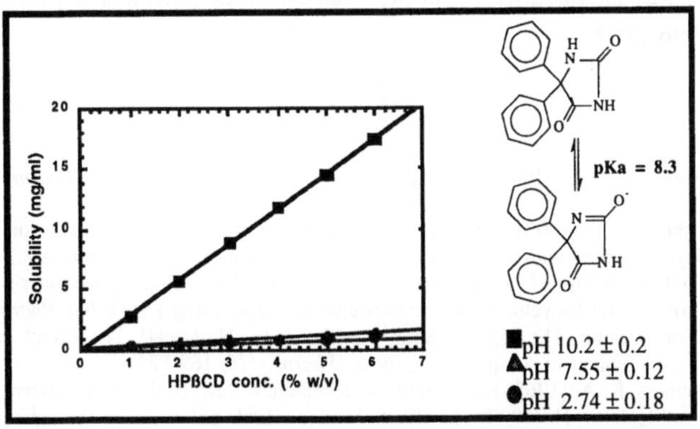

FIGURE 3. The effect of phenytoin ionization on the HPβCD solubilization of phenytoin in aqueous buffer solutions (25°C).

It has been shown that various pharmaceutical polymers, such as water-soluble cellulose derivatives and other rheological agents, can form complexes with CDs and that such complexes possess physicochemical properties distinct from those of individual CD molecules [4, 8]. In aqueous solutions, water-soluble polymers increase the solubilizing effect of CDs on various hydrophobic drugs by increasing the apparent stability constants (K_C) of the drug-CD complexes. Addition of water-soluble polymer does not only enhance the CD complexation of water-soluble and water-insoluble drugs and other compounds, but they are also able to enhance the aqueous solubility of water-insoluble CD complexes [6]. Thus, at room temperature (approx. 23 °C) the solubility of βCD in water is 18.6 mg/ml which was increases to about 21 mg/ml when 0.10% (w/v) HPMC is present in the solution. The aqueous solubility of the carbamazepine-βCD complex is 32 mg/ml. However, addition of a small amount of HPMC improves significantly both the aqueous solubility of the complex and the carbamazepine incorporation into the solid carbamazepine-βCD complex (Figure 4).

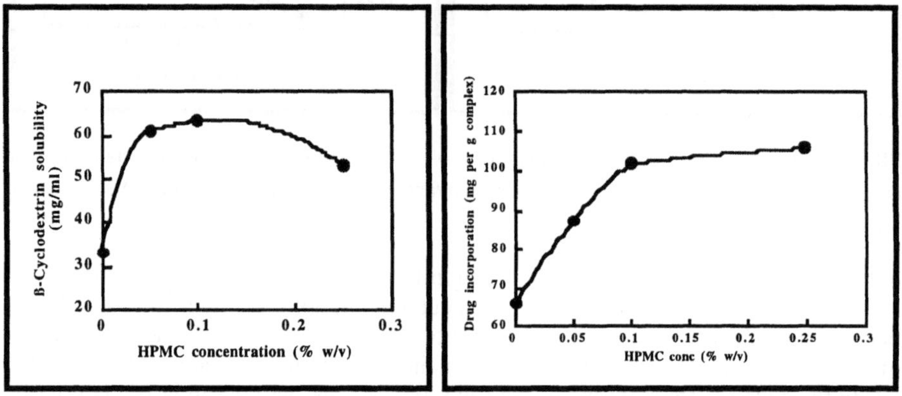

FIGURE 4. The effect of HPMC concentration on (A) the solubility of βCD in an aqueous solutions which was saturated with both βCD and carbamazepine, and (B) the carbamazepine incorporation into βCD.

4. References

1 Loftsson, T.; Brewster, M. E. (1996) Cyclodextrins as pharmaceutical excipients, *Pharm. Tech. Eur., 9(4),* 26-34.
2. Pitha, J.; Hoshino, T. (1992) Effects of ethanol on formation of inclusion complexes of hydroxypropylcyclodextrins with testosterone or with methyl orange, *Int. J. Pharm., 80,* 243-252.
3. Loftsson, T.; Gudmundsdottir, T. K.; Fridriksdottir, H. (1996) The influence of water-soluble polymers and pH on hydroxypropyl-β-cyclodextrin complexation of drugs, *Drug Devel. Ind. Pharm., 22,* 401-405.
4. Loftsson, T.; Fridriksdottir, H.; Sigurdardottir, A. M.; Ueda, H. (1994) The effect of water-soluble polymers on drug-cyclodextrin complexation, *Int. J. Pharm., 110,* 169-177.
5. Higuchi, T.; Connors, K. A. (1965) Phase-solubility techniques, *Adv. Anal. Chem. Instrum., 4,* 117-212.
6. Loftsson, T.; Fridriksdottir, H. (1998) The effect of water-soluble polymers on the aqueous solubility and complexing abilities of β-cyclodextrin, *Int.J.Pharm., 163,* 115-121.
7. James, K. C.: *Solubility and related properties,* Marcel Dekker, New York 1986.
8. Ganzerli, G.; Santvliet, L. v.; Verschuren, E.; Ludwig, A. (1996) Influence of beta-cyclodextrin and various polysaccharides on the solubility of fluorescein acid and on the rheological and mucoadhesive properties of ophthalmic solutions, *Pharmazie, 51,* 357-362.

COADMINISTRATION OF A WATER-SOLUBLE POLYMER INCREASES THE USEFULNESS OF CYCLODEXTRINS IN SOLID ORAL DOSAGE FORMS

JOUKO SAVOLAINEN[1], KRISTIINA JÄRVINEN[2], HANNU TAIPALE[1], PEKKA JARHO[1], THORSTEINN LOFTSSON[3], TOMI JÄRVINEN[1]

[1]Department of Pharmaceutical Chemistry, [2]Department of Pharmaceutics, University of Kuopio, P.O. Box 1627, FIN - 70211, Kuopio, Finland. [3]Department of Pharmacy, University of Iceland, P.O. Box 7210, IS-127 Reykjavik, Iceland.

Abstract

The aim of this study was to investigate the effect of cyclodextrins (β-CD, HP-β-CD and $(SBE)_{7m}$-β-CD), and coadministration of water-soluble polymer (hydroxypropyl-methylcellulose; HPMC) and cyclodextrins (CDs) on aqueous solubility and oral bioavailability of glibenclamide (GBA). GBA was administered orally and intravenously to dogs. GBA showed A_L-type diagram for β-CD and A_P-type diagrams for both β-CD derivatives studied. HPMC (0.05 %) enhanced the solubilizing effect of CDs but did not affect the type of phase-solubility diagram. Orally administered GBA and its physical mixture with HP-β-CD showed poor absolute bioavailability while GBA/β-CD, GBA/HP-β-CD and GBA/$(SBE)_{7m}$-β-CD enhanced significantly the absolute bioavailability of GBA. Orally administered GBA/β-CD/HPMC and GBA/$(SBE)_{7m}$-β-CD/HPMC showed similar absolute bioavailability compared to formulations not containing HPMC even though 80 % (in the case of $(SBE)_{7m}$-β-CD) or 40 % (in the case of β-CD) less CD was used. In conclusion, the oral bioavailability of GBA was significantly increased by CD-complexation and the amount of CD needed was significantly reduced by co-administration of HPMC.

1. Introduction

The use of cyclodextrins (CDs) in solid oral dosage forms is limited to low-dose drugs with large stability constants due to mass limitations of oral dosage units. The solubilizing effect of CDs in aqueous solutions has been increased by addition of water-soluble polymers [1-3] which might be an useful strategy to decrease the amount of CD needed in oral dosage forms and therefor to increase the pharmaceutical usefulness of CDs in solid oral dosage forms. The aim of this study was to determine the effects of β-CD and it's derivatives, $(SBE)_{7m}$-β-CD and HP-β-CD, and coadministration of CDs and water-soluble polymer (HPMC) on the oral bioavailability of glibenclamide.

2. Materials and methods

2.1 MATERIALS

Glibenclamide was purchased from Research Biochemicals (Natick, USA). Sulfobutyl ether β-CD sodium salt ($(SBE)_{7m}$-β-CD); Captisol™) was kindly supplied by CyDex, Inc. (Kansas City, USA). Hydroxypropyl-β-cyclodextrin (HP-β-CD; Encapsin®) was obtained from Janssen Biotech N.V (Belgium). β-CD was kindly supplied by Wacker-Chemie (München, Germany). Hydroxypropyl-methylcellulose (HPMC; 4000 cP) was purchased from Sigma (Steinheim, Germany).

2.2 PHASE-SOLUBILITY STUDIES

Effects of CDs on the solubility of glibenclamide, with (at pH 7.4) and without (at pH 3.0 and 7.4) HPMC, were studied using the phase solubility method. In the presence of HPMC the suspensions were sonicated at + 70 °C in an ultrasonic bath for 3 h prior to equilibration.

2.3 PREPARATION OF DOSAGE FORMS

Solid complexes of glibenclamide with CDs (with and without HPMC) were prepared by dissolving the maximum amount of glibenclamide in CD solutions. The HPMC containing solution was sonicated 3 h (+70 °C). All solutions mentioned above were freeze-dried. Glibenclamide was used as received when CD-free glibenclamide capsule and physical mixture of glibenclamide and HP-β-CD were prepared. Oral formulations were administrated in hard gelatin capsules. The solution for i.v. administration was prepared by dissolving glibenclamide into HP-β-CD solution

2.4 IN VIVO ABSORPTION STUDIES

The oral capsules were administered to four beagle dogs for a randomized crossover desing. I.v. solution was injected directly into the cephalic vein of conscious dogs in i.v. study. All formulations were equal to 3.0 mg of glibenclamide. Blood samples were withdrawn and the plasma was withdrawn and stored at -20 °C until analyzed.

Glibenclamide and diazepam (internal standard) were extracted from plasma and analyzed by HPLC using fluorescence detection.

3. Results and discussion

3.1 SOLUBILITY STUDIES

The phase solubility-diagrams for $(SBE)_{7m}$-β-CD and HP-β-CD are A_p -type at both pH-values used . In the case of β-CD only 1:1-complexes were observed (Figure 2). At pH 3.0, HP-β-CD and $(SBE)_{7m}$-β-CD had similar solubility enhancement ability but at pH 7.4 neutral HP-β-CD increased the solubility of glibenclamide more effectively than $(SBE)_{7m}$-β-CD (Figure 1).

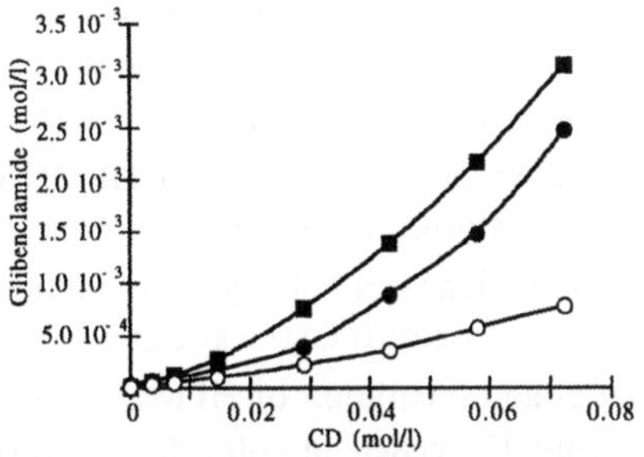

Figure 1. Phase-solubility diagrams for glibenclamide in the presence of $(SBE)_{7m}$-β-CD with (•) and without (o) HPMC and in the presence of HP-β-CD (■) at pH 7.4.

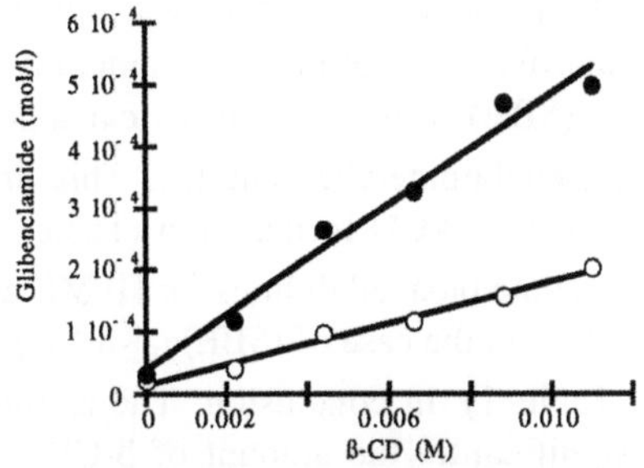

Figure 2. Phase-solubility diagrams for glibenclamide in the presence of β-CD with (•) or without (o) HPMC at pH 7.4.

Addition of HPMC (0.05 %) enhanced thesolubility of glibenclamide about 2.5-fold at pH 7.4. The solubilizing effect of CD and HPMC together was synergistic which is in good agreenment with an earlier study [1]. Water-soluble polymers have been reported to increase the stability constants of drug/CD complexes, which result in enhanced solubility of drugs [1, 4]. Although the addition of HPMC enhanced the solubilizing effect of CDs it did not change the type of phase-solubility diagrams (Figures 1 and 2).

3.2 *IN VIVO* STUDY

Plain glibenclamide and physical mixture of glibenclamide and HP-β-CD showed a poor absolute bioavailability (Table 1). Compared to plain glibenclamide, substantially higher C_{max} values and bioavailability were achieved by administering glibenclamide/β-CD, glibenclamide/$(SBE)_{7m}$-β-CD or glibenclamide/HP-β-CD inclusion complexes (Table 1).

Table 1. Pharmacokinetic parameters of glibenclamide (GBA) in plasma after oral administration to beagle dogs (mean ± sem, n=4).

Treatment	C_{max} (ng/ml)	Absolute Bioavailability (F, %)[d]	$F_{formulation}$ / $F_{plain\ GBA}$	Cyclodextrin needed (mg)
Crystal GBA	83.9 ± 4.7	14.7 ± 3.4	1	-
Physical mixt.	117.4 ± 19.1	14.8 ± 2.6	1	200
GBA/β-CD	499.9 ± 84.0[a,b]	72.4 ± 4.2[a,b]	4.9	300
GBA/β-CD /HPMC	433.5 ± 22.6[a,b]	67.3 ± 6.6[a,b]	4.6	120
GBA/HP-β-CD	610.0 ± 80.4[a,b]	84.3 ± 4.2[a-c]	5.7	200
GBA/SBE7-β-CD	569.0 ± 50.8[a,b]	80.1 ± 5.1[a,b]	5.4	1200
GBA/SBE7-β-CD/HPMC	638.3 ± 32.0[a-c]	90.5 ± 4.8[a-c]	6.2	250

[a]Significantly different from the value for the capsules containing crystalline GBA. [b]Significantly different from the value for the capsules containing physical mixture. [c]Significantly different from the value for the capsules containing GBA/β-CD ($p<0.05$ by ANOVA, Fisher´s PLSD test).

The present study indicates that orally administered glibenclamide/CD-complexes improve the poor oral bioavailability of glibenclamide. However, especially in the case of $(SBE)_{7m}$-β-CD, unpractical amount of CD was needed to formulate the solid glibenclamide/CD-complex. This drawback was overcome by the use of HPMC in $(SBE)_{7m}$-β-CD formulation (Table 1). Compared to HPMC-free formulations similar absolute bioavailabilities for HPMC containing formulations were obtained even though 80 % (in the case of $(SBE)_{7m}$-β-CD), and 40 % (in the case of β-CD) less CD was used (Table 1). In conclusion: The results show that coadministration of HPMC reduced significantly the amount of β-CD and $(SBE)_{7m}$-β-CD needed in solid dosage form without reducing the bioavailability of glibenclamide achieved by CD-complexation.

Acknowledgments

The authors gratefully acknowledge the financial support of The Academy of Finland, the Technology Development Centre (Finland) and Finnish Cultural Foundation. The authors are greatful to CyDex, Inc. for supplying $(SBE)_{7m}$-β-CD and to Wacker-Chemie for supplying β-CD.

References

1. T. Loftsson, H. Fridriksdóttir, A. M. Sigurdadóttir and H. Ueda. The effect of water-soluble polymers on drug-cyclodextrin complexation. *Int. J. Pharm.* **110**: 169-177 (1994).
2. T. Loftsson, H. Fridriksdóttir and T. K. Gudmundsdóttir. The effect of water-soluble polymers on aqueous solubility of drugs. *Int. J. Pharm.* **127**: 293-296 (1996).
3. H. Fridriksdóttir, T. Loftsson and E. Stefánsson. Formulation and testing of methazolamide cyclodextrin eye drop solutions. *J. Contr. Rel.* **44**: 95-99 (1997).
4. T. Loftsson, T. K. Gudmundsdóttir and H. Fridriksdóttir. The Influence of Water-soluble Polymers and pH on Hydroxypropyl-ß-Cyclodextrin Complexation of Drugs. *Drug Dev. Ind. Pharm.* **22**: 401-405 (1996).

EVALUATION OF DEGRADATION STUDIES PERFORMED IN AQUEOUS CYCLODEXTRIN SOLUTIONS

MÁR MÁSSON AND THORSTEINN LOFTSSON.
Department of Pharmacy, University of Iceland, PO Box 7210, IS-127 Reykjavik, Iceland

1. Introduction

Cyclodextrins (CDs) have unique properties as pharmaceutical excipients as they can enhance both the solubility and the stability of drugs in aqueous solutions, through complexation.
Sufficient solubility is a prerequisite to develop an aqueous dosage form, but sufficient stability is also required. The general requirement is that the drug should maintain more than 90% of it's activity for more than 2 years.
The most common reactions of drugs in aqueous solutions are, hydrolysis, oxidation and photoreactions. It has been reported that CDs can protect drugs from oxidation and photoreactions. However CDs have most often been used to prevent hydrolysis, which is a very common degradation degradation pathway in aqueous solutions.
CDs can catalyse the hydrolysis of drugs [1] but it is much more common that inclusion in the CD cavity will lead to decreased reaction rates. More than 10 fold increase in drug stability is common and more than 1000 fold increase in stability has been reported [2].
Investigation of aqueous stability of drugs is an integral part the development of aqueous dosage forms. In CD solutions the stability can be defined through three parameters; k_o the intrinsic degradation rate constant for the free drug in the aqueous solution, k_c the degradation rate constant for the drug included in the CD cavity and K_c the stability constants of the drug-CD complex. All these parameters can be determined from a degradation rate study where the degradation rate is measured at different CD concentrations. These parameter also provide information about the physical nature of the complex and the transition state in the degradation reaction.
In the present work we evaluate the methods used to determine the three constants and discuss factors limiting the accuracy of such determinations.

2. Materials and Methods

The degradation studies were done by placing the CD (or buffer) solution in vial on a heat controlled rack in a Merck-Hitachi AS-4000 Intelligent autosampler and adding small volume of drug stock solution. The degradation reaction was followed by

withdrawing samples with a fixed interval and injecting them into an HPLC system. Relative concentrations were followed by measuring the relative peak heights or peak areas for the characteristic peak of each compound. The HPLC conditions for indomethacin and chlorambucil have been reported elsewhere [3].
The data was fitted linear and non-linear equations using the Kaleidagraph software (Synergy Software, USA).

3 Results and Discussion

3.1 THEORY

The hydrolysis reactions usually involve two components: the drug molecule (D) and a second component which can be water molecule, hydroxide ion or hydronium ion. The concentration of the latter is usually constant during the reaction and therefore these reactions can be treated as pseudo-first order reactions and described by Equation 1. In CD solutions the observed degradation rate for the total drug concentration ($[D]_T$) is the weighted average of k_o and k_c. The observed degradation rate is therefore dependent on the relative concentration of free drug ([D]) and drug in complex ([D•CD]) (Equation 1).

$$-\frac{d[D]_T}{dt} = k_{obs}[D]_T = k_o[D] + k_c[D \bullet CD] \quad (1) \qquad \ln\left(\frac{[D]_T}{[D]_{T_o}}\right) = -k_{obs}t \quad (2)$$

By integration of (1), Equation 2 is obtained. The k_{obs} value can then be obtained from graphical representation as slope of the line $\ln([D]_T)$ vs. t.
The stability constant for the complex formation is described by Equation 3. By combining equations (1) and (3), k_{obs} can expressed in terms of k_o, k_c, K_c and the CD concentration [CD] (Equation 4).

$$K_c = \frac{[D \bullet CD]}{[D][CD]} \quad (3) \qquad k_{obs} = \frac{k_o + k_c K_c [CD]}{1 + K_c[CD]} \quad (4)$$

Equation 5 can be derived form Equation 4. This equation allows k_{obs} data for different CD concentrations to be fitted by linear regression [4].

$$\frac{k_o}{k_o - k_{obs}} = \frac{k_o}{k_o - k_c} + \frac{k_o}{K_c(k_o - k_c)} \frac{1}{[CD]} \quad (5)$$

3.2. EXPERIMENTAL RESULTS.

Most researchers have used linear regression based on equation 5 to determine the k_c and K_c values. This method offers the advantage that data for 1:1 complexes will follow a straight line. Figure 1. shows an example of this method applied to degradation study for indomethacin in γ-CD solutions.
Although this method is widely used it has the disadvantage of being based on subtraction and therefore experimental error will by magnified as the k_{obs}/k_o ratio approaches 1 (Table 1.)

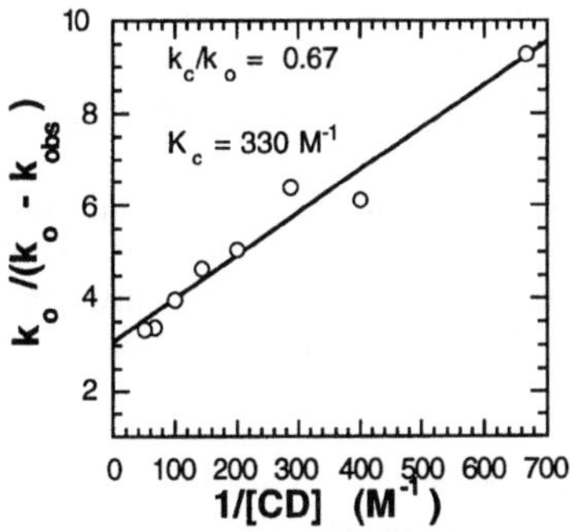

Figure 1. Linear regression of degradation rate data. Degradation of indomethacin in pH 10, 10 mM $Na_2HCO_3/NaCO_3$ buffer solution, 40 °C.

Table 1. Magnification of the error in relation to the k_o/k_c ratio. (Shown for fixed k_{obs} errors 1% and 3%).

	Percentage error		Percentage error	
k_o/k_{obs}	k_{obs}	k_o-k_{obs}	k_{obs}	k_o-k_{obs}
0.25	1%	1.3%	3%	4%
0.50	1%	2%	3%	6%
0.75	1%	4%	3%	12%
0.90	1%	10%	3%	30%

Alternatively Equation 4 can be fitted directly by non-linear regression. In this case the data is plotted directly and experimental error is not affected. Figures 2 and 3 show that it is easy to asses form the graphs if appropriate range for the data has been selected.

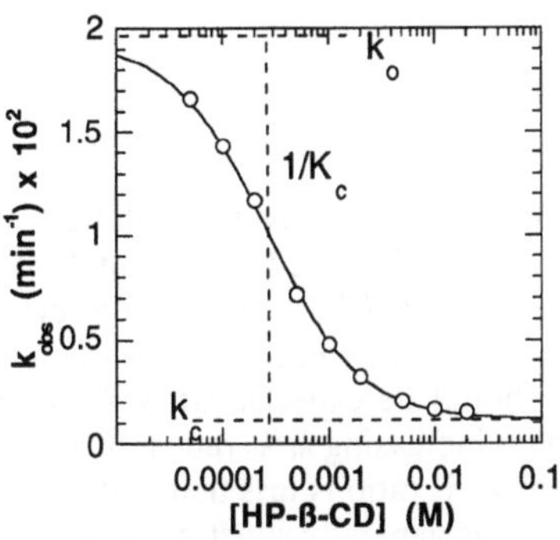

Figure 2. Non-linear regression of degradation rate data for chlorambucil. Degradation in a 10 mM NaH_2PO_4/NaOH buffer, pH 7.5, 30 °C

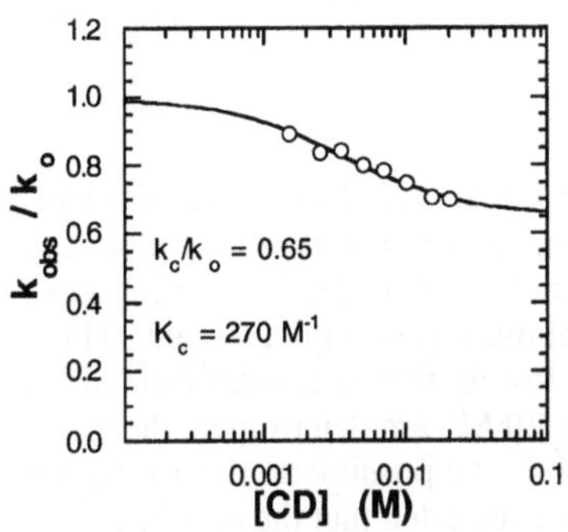

Figure 3. Non-linear regression of degradation rate data for indomethacin. Same data as in figure 1.

Figure 3 shows data for chlorambucil. The CD concentrations are well distributed from high concentration were most of the drug is bound in complex to low concentration were most of the drug is free in the solution. The K_c value can therefore be determined with high accuracy (2%). Additional benefit of non-linear regression method is that the k_c and K_c values can be obtained directly from the graph.

Figure 4 show a non-linear regression of the data displayed in Figure 2. The relatively large k_c/k_o ratio leads to relatively large error in the linear regression of the data. Figure 4. shows that greater accuracy would have been obtained if the degradation rates would have been determined at higher and lower CD concentration.

The fitting of data, in the linear and non-linear case, is based on the approximation that $[CD] \approx [CD]_T$, ie that the CD concentration is almost the same as the total CD concentration. This approximation is justified when [CD]/[D•CD] is large and this is the case when $[CD]_T >> [D]_T$. From figures 2 and 3 it can be understood that the highest accuracy in determination of K_c is achieved if the degradation study is carried out for a series of CD concentrations, from high concentrations where most of the drug is bound in a complex to low concentrations where most of the drug is free in solution. Calculations show that to have 80% of the drug free in solution the CD concentration has to be $1/(5K_c)$. If K_c is to be determined with more than 5% accuracy then the [D•CD]/[CD] ratio should be less than 4 at this concentration. Figure 5. shows that in order determine K_c in the range 1000 to 10,000 M^{-1} $[D]_T$ should be 10 μM or less.

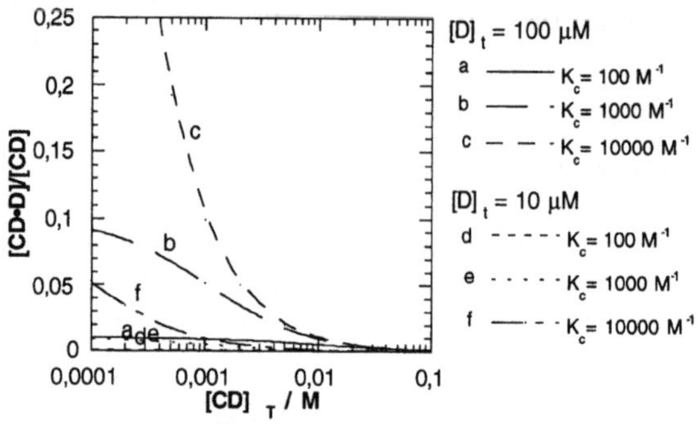

Figure 4. Calculated [D•CD]/[CD] ratios

CDs are relatively large molecules and therefore an 0.1 M β-CD derivative solution will contain approximately 90% water. Less water content should lead to lower hydrolysis rates and therefore the degradation rates may change, at high CD concentration, even if the complexation is not affected. This concentration effect introduces an error when low K_c or low k_c values are determined. Low accuracy should be expected when K_c values below 20 M^{-1} are determined, the error increasing with increasing k_c/k_o ratios. In practice it will not be possible to determine the k_c value if the k_c/k_o ratio is less than 0.01.

It was concluded that the non-linear regression method is generally superior to the linear regression method. It is recommended that degradation rate studies should be carried out with CD concentrations in the range 0.1-0.0001 M at 10 μM drug concentration.

4. References.

1. Loftsson, T. and Ólafsdóttir, B. J. (1991) Cyclodextrin-accelerated degradation of β-lactam antibiotics in aqueous solutions. *International Journal of Pharmaceutics*, 67, 5-7
2. Jarho, P., Urtti, A., Jarvinen, K., Pate, D. W. and Jarvinen, T. (1996) Hydroxypropyl-β-cyclodextrin increases aqueous solubility and stability of anandamide. *Life Sciences*, 58, 181-185
3. Masson, M., Loftsson, T., Jonsdottir, S., Fridriksdottir, H. and Petersen, D. S. (1998) Complexation and stabiliztion of ionic drug compounds with ionic and non-ionic cyclodextrins. *International Journal of Pharmaceutics*, 164, 45-55
4. Loftsson, T. (1995) Effects of cyclodextrins on the chemical stability of drugs in aqueous solutions. *Drug Stability*, 1, 22-33

THE EFFECT OF CELLULOSE ETHERS ON INDOMETHACIN/CYCLODEXTRIN COMPLEXATION

Lj. Tasić, J.Milić, M. Primorac, M.Stupar and S.Simović

Faculty of Pharmacy, Pharmaceutical Technology Department, Belgrade University, 11221 Belgrade, V.Stepe 450, p.o.box 146, Yugoslavia

1.INTRODUCTION

Cyclodextrins (CD) are water-soluble, hydrophobic torus-shaped cyclic oligosacharides that can accommodate in their cavities water-insoluble drugs to form water-soluble inclusion complexes. Because of these properties CD have been used to enhance drug solubility, to stabilize formulations, and to increase the bioavailability of drugs (1). Natural and chemically modified CD are biocompatible polymers and are regarded as relatively safe for therapeutic applications (2). Particullary safe profile as parenteral and ophtalmic drug delivery systems were evaluated on drug/hydroxypropyl-β-cyclodextrin (HPβCD) complexes (3). A many publication dealing with effect of different vehicle additives commonly used in drug formulation such as polymers, non-ionic surfactants, buffer salts, preservatives, which could be reduced the CD abilities of drug complexation (3,4).

Indomethacin (INDO) is known non-steroidal anti-inflamatory drug, which have been well established on the oral route of administration. Topical (dermal or ophtalmic) application of INDO formulation leads to poor bioavailability, because of its poorly water solubility and UV light unstability. For these reason, INDO has been a candidate for inclusion into either β- or hydroxipropyl-β-cyclodextrin (β-CD, HP-β-CD) (5).

The purpose of this study was to investigate the effects of two cellulose ethers (hydroxypropylmethyl cellulose HPMC and hydroxyethyl cellulose HEC) on the HP-β-CD complexation of INDO. Evaluation was carried out by the solubility curve, the stability constants of complex, pH, viscosity and computer modeling.

2. MATERIALS AND METHODS

2.1. Materials

The following materials were used: INDO (Dolder, Basel, Switcherland); HP-β-CD(MS 0.6, Cyclolab, Budapest, Hungary), HPMC (grade 3000 cps, Fluca) and HEC (Hercules, Germany). All other chemicals were commercially available products of special reagent grade.

2.2. Methods

2.2.1.*Solubility studies*

An excess amount of the drug to be tested was addead to aqueous solution containing both a polymer and cyclodextrin. The polymer concentration was 0.1 % and 0.2 % in range of CD concentration from 0-10% (w/v). The suspension formed was sonicated in an ultrasonic bath (KARLKOLB,Dreieich, Germany) for 2 h and then heated in an autoclave (Sutjeska, Belgrade, Yu) in a sealed container to 120 ^{0} C for 20 min. After equilibration at room temperature (20±2 ^{0} C) for at 3 days, the suspension was filtered through a 0.45 μm membrane filter (CST, Belgrade, Yu) diluted with methanol and analysed by an UV-VIS spectrophotometry method (318 nm) (GBS 914, Dordrecht Australia). The stability constant (Kc) of the INDO/ HP-β-CD was calculated from the slope of the phase-solubility diagrams according to method of Higuchi and Connors (6).

2.2.2. *Viscosity measurements*

The viscosity of the samples at room temperature was examined at rotary viscometer Rheothest 2.1 (VEB MLW Prufgerate-Werk,Germany) with cylindrical measuring device; HP-β-CD and polymers were dissolved in water and the solution heated in a sealed container (120 ^{0} C for 20 min).After equilibrium at room temperature (24 h) and appropriate adjustments of the viscometer the viscosity of the solutions was determined with shear rate of 131.2 s^{-1} .

2.2.3. *pH measurements*

The pH of solutions was determined at room temperature with a pH-meter Hanna instrument 8417 (Mauricius) at room temperature.

2.2.4. *Computer graphics*

All molecules were constructed and minimized using the HyperChem program version 5.0; the geometry optimization was doing according to Single point and SemiEmpirical MNDO calculation; convergence limit 0.1. Using the numerous data of energies and gradient we are has possibility to obtain the best conformation of INDO/ HP-β-CD complexes.

3. RESULTS AND DISCUSSION

The most common cyclodextrin derivatives presently used in drug formulation are the hydroxypropyl derivatives of β-cyclodextrin. In our study we used HPβCD MS 0.6 according to experiences that their complexing abilities increases when the MS decreases. The effect of two polymers in conc. 0.1% and 0.2% in presence increasing concentration of HPβCD (0-10% w/v) on the solubility of INDO were present in Fig.1. Loftsson reported that the optimum amount of the polymer in the aqueous cyclodextrin solutions appeared to be between 0.05 and 0.25% (w/v) polymer concentration (4). Our results was shown that the lower concentration (0.1%) of HPMC has better solubilizing effect than higher (0.2%). Contrary, the HEC decrease the solubility of INDO regarding the corresponding INDO/HPβCD solution without polymer. The HPMC increased the efficiency of the complexation INDO in HPβCD solution by increasing Kc (65.7 mol $^{-1}$ and 66.5 mol $^{-1}$, regarding 51.7 mol $^{-1}$). The results of pH and viscosity of our samples are present in Table I. The data of pH were almost the same at all solution. Results of visosity is realtively similar between all samples, and agree with some data of Loftsson report (4). But, the slightly decrease of viscosity we have been remark with the increased concentration of HPβCD. If we compared the two used polymer (HPMC and HEC) the mentioned phenomenon was remarkable with HPMC polymer. As well known the HPβCD even et high aqueous concentration (25% w/v) have low viscosity. The water soluble polymers can form of water-soluble drug-polymer complexes. The polymers mainly interact with drug molecules via electrostatic bonds, i.e., ion-to-ion, ion-to-dipole and dipole-to-dipole bonds, but other types of forces, such as van der Walls forces and hydrogen bridges, frequently participate in complex formation (7).

Table 1 Viscosity and pH of aqueous HPBCD solutions with HPMC and HEC polymers at room temperature

Aqueous solution	*Viscosity (mPa s)*	*pH*
0.1% HPMC and 2% HPβCD	2.41	5.67
0.1% HPMC and 6% HPβCD	1.97	5.45
0.1% HPMC and 10%HPβCD	1.86	5.45
0.2% HPMC and 2% HPβCD	2.28	5.73
0.2% HPMC and 6% HPβCD	2.19	5.65
0.2% HPMC and 10%HPβCD	2.15	5.94
0.1% HEC and 2% HPβCD	1.69	5.51
0.1% HEC and 6% HPβCD	1.65	5.56
0.1% HEC and 10%HPβCD	1.76	5.58
0.2% HEC and 2% HPβCD	2.30	5.70
0.2% HEC and 6% HPβCD	1.84	5.85
0.2% HEC and 10%HPβCD	1.80	5.81

Fig 1.Solubility curves of Indomethacin in HP βCD solutions in the presence of different polymers

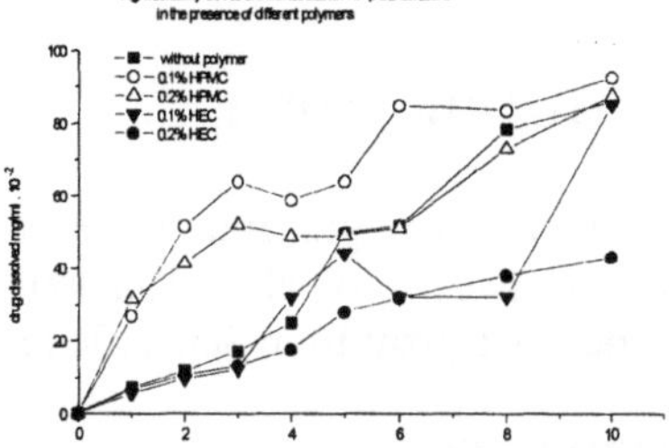

The solubilization enhancement of drug in HPβCD aqueous solution with polymers is more than simply additive, it is synergistic. This synergistic phenomenon have been discussed in Hladon et al. and Ganzerly et al. study and has a different explanation (8,9). The postulated explanation based on formation of thermodinamically stable ternary complexes, seems reasonable (10).The more physicochemical methods must bee used for the further investigation of this phenomenon. Since that we were used the computer modeling method for the simulation of complex formation INDO/ HPβCD (Fig.2). According to results which we are obtained the INDO molecule is fitted into HPβCD cavity on 4-Chlorobenzoyl ring. The further modelings with presence of polymer (HPMC) are now under the investigation.

4. CONCLUSION

The experimental results show that the polymers (HPMC and HEC) used in aqueous solution of HPβCD has influence on solubilisation of INDO. The preferable concentration of polymer is 0.1% w/v, and the HPMC is powerful, regarding to HEC. The some decrease in viscosity of the observed solution attended on synergistic phenomenon in this system. With additional further investigation we could have more detail for the explanation of such phenomenon.

INDOMETACIN-HPBCD

TOTAL CHARGE DENSITY

ELECTROSTATIC POTENTIAL

```
HyperChem log start -- Wed May 13 10:41:24 1998.
Geometry optimization, MolecularMechanics, molecule = (untitled).
mmplus
PolakRibiere optimizer
Default parameters being used for torsions...
Default parameters being used for stretches...
Default parameters being used for bends...
```

ACKNOWLEDGEMENT

The authors are gradeful to prof.dr. Mirjana Vojinovic -Miloradov, Chemical Institute,University of Novi Sad (Yugoslavia) for computer modeling.This work was supported by a grant from Scientific fondation of Republic Serbia.

REFERENCES

1. Loftsson, T. and Brewster,M.E.: Pharmaceutical application of Cyclodextrins.1. Drug solubilization and stabilization, *J.Pharm.Sci.* 85, 1017-1025, 1996.
2. Irie, t. and Uekama, K.: Pharmaceutical application of Cyclodextrins. 3. Toxicological issues and safety evaluation, *J.Pharm.Sci.* 86, 147-162, 1997.
3. Kraus, C., Mehnert,W. and Fromming, H.-H., Interactions of β-cyclodextrin with Solutol HS 15 and their influence on diazepam solubilization, *Pharm.Ztg.Wiss.*, 136/4, 11-15, 1991.
4. Loftsson, T., Frioriksdottir, H., Thorisdottir, S. and Ueda, H., The effect of water-soluble polymers on drug-cyclodextrin complexation, *Int. J. Pharm.,* 110 , 169-177, 1994.
5. Loftsson, T., Olafsdottir, B.J., Frioriksdottir, H., and Jonsdottir, S., Cyclodextrin complexation of NSAIDs: physicochemical characteristics. *Eur.J.Pharm.Sci.,* 1 , 95-101, 1993.
6. Higuchi,T. and Connors,K.A., Phase-solubility techniques. Adv.Anal.Chem.Instrum., 4, 117-212,1965.
7. Racz,I., Drug Formulation, Wiley, Budapest, 1989, pp.212-242.
8. Hladon, T. and Cwiertnia, B.: Physical and chemical interactions between cellulose ethers and β-cyclodextrins, *Pharmazie*, 49, 497-450, 1994.
9. Ganzerli, G., van Santvliet, L., Verschuren, E. and Ludwig, A.: Influnce of β-cyclodextrin and various polysaccharides on the solubiliy of fluorescein and on the rheological and mucoadhesive properties of ophtalmic solutions, *Pharmazie*, 51, 357-362, 1996.
10. Brewster, E.M., Anderson, R.W., Lotsson,T., Huang, M.J., Bodor,N. and Pop.E.: Preparation, characterization, and anesthetic properties of 2-hydroxypropyl-β-cyclodextrin complexes of Pregnanolone and Pregnenolone in rat and mouse, *J.Pharm.Sci.*, 84, 1154-1159, 1995.

PHYSICO-CHEMICAL CHARACTERIZATION AND PHARMACOLOGICAL PROPERTIES OF DIFFERENT CYCLODEXTRIN COMPLEXES OF A PLATELET ACTIVATING FACTOR RECEPTOR ANTAGONIST: SR 27417A.

Th. Breul*, J.C. Gautier*, J.M. Herbert**, R. Ghirlando#, B. Perly##, Ph. Saudemon*.

*Sanofi Recherche 371 rue du Pr Blayac 34184 Montpellier, France.
**Sanofi Recherche 195 route d'Espagne 31036 Toulouse, France.
INSERM U.414 75 rue de la cardonille 34093 Montpellier, France.
DSM/DRECAM C.E.A. Saclay, 91191 Gif/Yvette, France.
Present address: LMB-NIDDK-NIH, Bethesda, MD 20892-0540, USA.

INTRODUCTION

SR 27417A: [N-(2-dimethylamino ethyl)-N-(3-pyridinyl methyl)[4-(2,4,6-triisopropylphenyl) thiazol-2-yl]amine] fumarate salt is a highly potent, competitive and selective antagonist of the binding of platelet activating factor (PAF) to its receptor in rabbit platelets, and potently inhibits PAF-induced aggregation of human platelets in vitro (1).
The low aqueous solubility (0.4 mg/ml) and poor chemical stability of SR 27417A when solubilized in aqueous media have led to study its molecular encapsulation in cyclodextrins, in order to improve both solubility and stability thereby enabling its parenteral administration.

PHASE SOLUBILITY STUDIES

Phase solubility studies of SR 27417A with different cyclodextrins were determined according to the method of Higuchi and Connors (2) and led to the selection of 2-hydroxypropyl-β-cyclodextrin as the most efficient pharmaceutically approved injectable solubilizer. Its phase solubility diagram is presented in Fig. 1. The two portions of linear increase for SR 27417A solubility as a function of 2-hydroxypropyl-β-cyclodextrin concentration suggests the formation of a 1:1 stoichiometric ratio soluble complex exhibiting two apparent stability constants : Ks1 and Ks2. These stability constants were calculated from the linear portions of the phase solubility diagram according to equation Ks = slope / So (1-slope) where So is SR 27417A solubility in water. According to this equation Ks1 was found to be 1320 M-1 and Ks2 was found to be 159 M^{-1}.

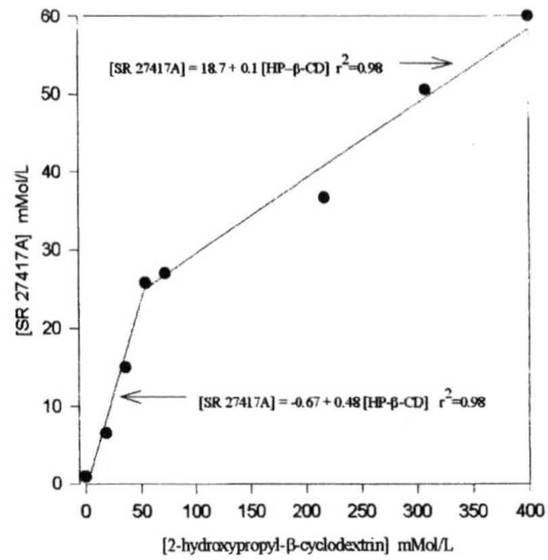

Fig 1: **Phase solubility diagram of SR 27417A vs 2-hydroxypropyl-β-cyclodextrin in aqueous medium at 25°C**

NMR STUDIES

Changes in the chemical shifts Dd of SR 27417A and H3 - H5 β-cyclodextrin protons were measured with NMR. Two-dimensional ROESY ^{1}H NMR studies were used to determine the location of binding interactions between SR 27417A and β-cyclodextrin, and therefore, inclusion geometry and complex stoichiometry. Comparison of the partial proton NMR spectra of β-cyclodextrin and SR 27417A- β-cyclodextrin complex at 500 Mhz and 310 °K in deuterium oxide have been determined.

These data led to the structural model of SR 27417A- β-cyclodextrin complex detailing binding sites and overall inclusion geometry.

ISOTHERMAL TITRATION MICROCALORIMETRY (ITC) STUDIES

ITC studies of equilibrium association constants for the formation of SR 27417A and β-cyclodextrin or 2-hydroxypropyl-β-cyclodextrin confirmed the 1:1 encapsulation complex stoichiometries. The equilibrium association constants for the formation of the complex with each cyclodextrin were found to be identical Ka = 230 ± 40 M-1 at 25°C. Ka value thus obtained seems to be a combination of Ks1 and Ks2 values obtained by phase solubility technique. Temperature-dependent molar enthalpy and molar free energy changes associated with encapsulation processes in aqueous medium are presented in Fig.2. Even though the temperature dependent molar enthalpies for the association with the two cyclodextrins were different, the corresponding molar free energy and heat capacity changes were identical.

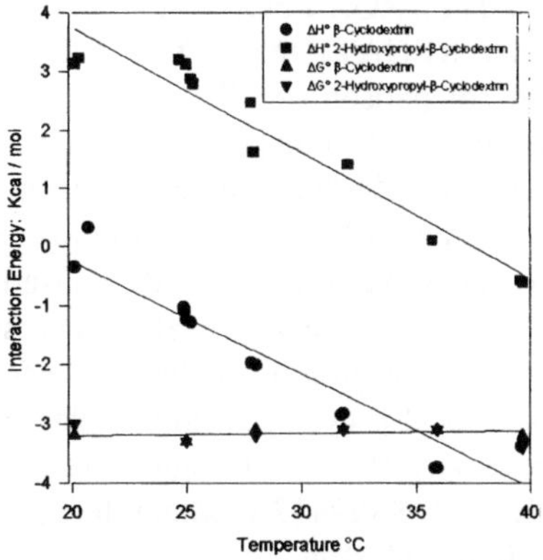

Fig.2 : SR-27417A - Cyclodextrins interaction molar enthalpy and molar free energy variations as a function of temperature

THERMOGRAVIMETRIC (TGA) STABILITY STUDY

Poor stability of SR27417A - 2-hydroxypropyl-β-cyclodextrin complexes in aqueous solution led to the formulation of a freeze-dried dosage form whose temperature-dependent stability improvement was characterized by thermogravimetric (TGA) technique at 10°C/min. between 25°C and 500°C. On set degradation temperatures of freeze-dried SR27417A-2-hydroxypropyl-β-cyclodextrin complex are presented in Fig 3 as a function of the encapsulation complex molar ratio R = [2-hydroxypropyl-β-cyclodextrin] / [SR 27417A]. The degradation temperature increase was found to be a linear function of the encapsulation molar ratio showing that this cyclodextrin complexed formulation of SR 27417A had an improved overall temperature dependent stability.

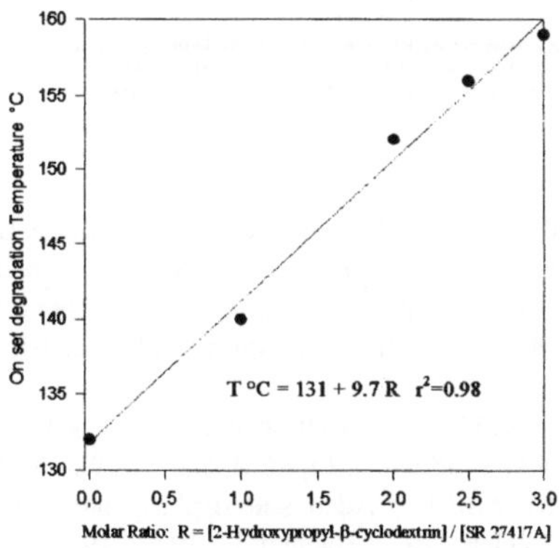

Fig 3: Thermogravimetric determined on set degradation temperature of freeze-dried SR 27417A - 2-hydroxypropyl-β-cyclodextrin complex vs [2-hydroxypropyl-β-cyclodextrin] / [SR 27417A] molar ratio

EFFECT OF SR 27417A-2-HYDROXYPROPYL-β-CYCLODEXTRIN ON PAF-INDUCED PLATELET AGGREGATION EX VIVO

A placebo-controlled randomized comparison of SR 27417A solubilized in saline and SR27417A - 2-hydroxypropyl-β-cyclodextrin complex was performed. Single doses (10 μg/kg) were administered i.v. to male New-Zealand rabbits. A treatment group of 6 rabbits was assigned to each dose level. An ex vivo PAF-induced platelet aggregation test was used to evaluate the PAF antagonist activity (3). Washed rabbit platelets were prepared according to the method described by Ardlie et al. (4). Platelet aggregation was assessed optically with a 2-channel aggregometer, using a modification of the technique described by Born (5). When administered as a 2-hydroxypropyl-β-cyclodextrin complex, SR 27417A inhibited PAF-induced platelet aggregation ex vivo (Fig.4). No significant difference was observed between the antiaggregating effect of SR27417A administered in saline solution or complexed with 2-hydroxypropyl-β-cyclodextrin.

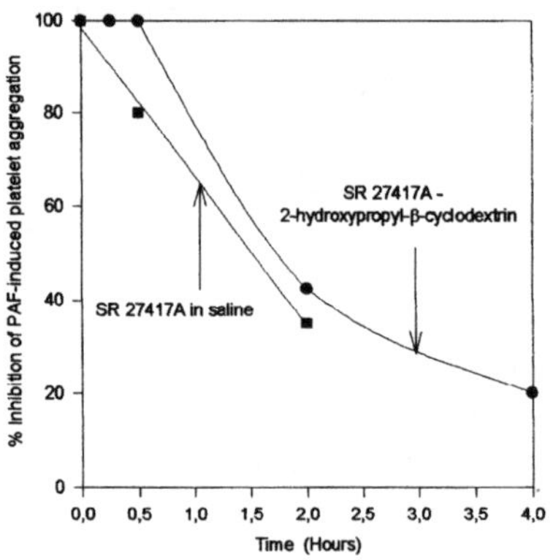

Fig. 4: Time-course of inhibition of PAF-induced platelet aggregation after a single dose of SR 27417A (10 μg/kg) administered per i.v. at T=0 in saline or complexed with 2-hydroxypropyl-β-cyclodextrin

REFERENCES

(1) Herbert J M; Bernat A; Valette G; Gigo V; Lale A; Laplace M C; Lespy L; Savi P; Maffrand J P; Le Fur G: Biochemical and pharmacological activities of SR 27417, a highly potent, long acting platelet-activating factor receptor antagonist. J. Pharmacol. Exp. Ther., 1991, 259, 1, 44-51

(2) Higuchi T.; Connors K.A. : Phase solubility techniques. Adv. Analyt. Chem. Instrum. 1965, 4, 117-212

(3) Herbert J M; Laplace M C; Maffrand J P: Ex vivo effects of SR 27417, a novel PAF antagonist on rabbit platelet aggregation and (3H)-PAF binding J. Lipid Mediat., 1992, 5, 1, 1-12

(4) Ardlie N.G., Packman M.A., Mustard J.F.: Adenosine triphosphate-induced platelet aggregation in suspensions of washed rabbit platelets. Br. J. Haematol., 1970, 19, 7-17

(5) Born G.V. Aggregation of blood platelets by adenosine diphosphate and its reversal. Nature (London) 1962, 194, 927-929.

MUTUAL SOLUBILITY ENHANCEMENT BY BINARY AND MULTICOMPONENT COMPLEXATION OF CLOMIPHENE AND TAMOXIFEN

A.GERLÓCZY[1], M.VIKMON[1], J.SZEJTLI[1], E.REDENTI[2], P.VENTURA[2]
[1]*Cyclolab, Cyclodextrin R.&D.Lab. Ltd. Budapest, H-1525, P.O.Box 435, Hungary*
[2]*Chiesi Farmaceutici S.p.A. 43100 Parma, Via Palermo 26/A., Italy*

1. Introduction

Clomiphene (CL) and Tamoxifen (TM) (Fig.1.) are non-steroidal, synthetic, structurally related triphenylethylene derivatives bearing N,N-dialkylamino moiety.

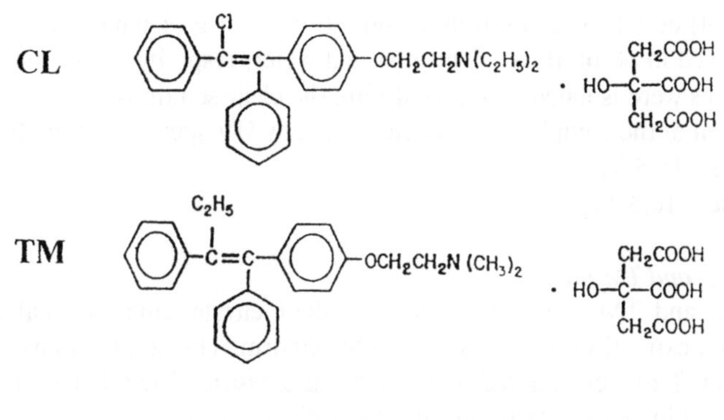

***Fig.1.:** Chemical structure of Clomiphene and Tamoxifen*

As monocitrate salts, these drugs are used for the treatment of female infertility and estrogen receptor positive human breast cancer. Single oral dose of CL is 20 or 50 mg, while for TM citrate is 10 or 20 mg.

The solubilities of the CL and TM bases are approx. 0.04 and 0.03 mg/ml, respectively. Salt formation by citric acid, however, increased the aqueous solubility to 6.2 mg/ml for CL and 0.3 mg/ml for TM [1.].

Multicomponent complexation of base type drugs e. g. terfenadine, ketoconazole in the presence of salt-forming agent with CDs or CD derivatives results in the formation of freely soluble multicomponent complexes if the components are used simultaneously in an appropriate molar ratio [2.].

In this study we compared the attainable mutual solubility enhancement - for the guest and the host - by binary βCD complexation of commercial CL and TM citrate salts and by multicomponent complexation using the base form of the drugs and performing the salt formation by citric acid and complexation with βCD simultaneously in the preferred molar ratio.

2. Experimental

2.1. MATERIALS

Clomiphene citrate and Tamoxifen citrate were provided by Chiesi Farmaceutici S.p.A., Parma, Italy
βCD was supplied by Wacker Chemie, Munich, Germany
All other materials were of analytical grade.

2.2. METHODS

2.2.1. Preparation of CL and TM citrate/βCD binary complexes
1.2 g CL citrate (2mmol) or 1.13 g TM citrate (2 mmol) and 5.4 g (4 mmol) or 6.75 g βCD (2.5 mmol) were dissolved in 60 or 100-105 ml distilled water, resp., by ultrasonication for 5 minutes. Solid complexes were isolated by freeze drying the clear solutions.
Active ingredient content of the complexes was determined by UV spectrophotometry.
CL citrate/βCD complex: 19.5 %
TM citrate/βCD complex: 16.5 %

2.2.2. Preparation of CL and TM bases
Free base forms of CL and TM were prepared from their citrate salts by alkalizing their aqueous suspensions and extracting the bases with chlorophorm. The organic phase was dried on Na_2SO_4 siccum, filtered, and evaporated under reduced pressure. The oily CL base and the solid TM base were stored in an exsiccator above P_2O_5 until constant weight.

2.2.3. Preparation of CL/citric acid/βCD multicomponent complexes of 1:2.5:1 and 1:2.5:2 molar ratio
1:2.5:1 M complex: 1 g CL base (2.5 mmol) 3.3 g βCD (2.5 mmol water content 14%), approx. 7 ml 1 M citric acid and 10 ml distilled water was ultrasonicated to obtain a clear solution and the solution was immediately freeze dried. The active ingredient content of the solid complex : 17.9 %,
1:2.5:2 M complex: 0.6 g CL base (1.5 mmol) 3.9 g βCD (3 mmol water content 14%), approx. 3.75 ml 1 M citric acid and 10 ml distilled water was ultrasonicated to obtain a clear solution and the solution was immediately freeze dried. The active ingredient content of the solid complex : 11.5 %,

2.2.4. Preparation of TM/citric acid/βCD multicomponent complex of 1:2:2 molar ratio
0.74 g TM base (2mmol), 5.2 g βCD (4 mmol), 4 ml 1 M citric acid and 50 ml distilled water were ultrasonicated for about five minutes to obtain a clear solution and the solution was immediately freeze-dried. The active ingredient content of the complex: 12.2%.

3. Results and Discussion

Salt formation with organic or inorganic acids is a common method which is very often used to increase the solubility of base-type drugs, but sometimes even the solubility of the salts is not satisfactory.
The mutual solubility enhancement of the commercial CL and TM citrate salts by binary complexation and that of the corresponding base forms by multicomponent complexation by βCD and citric acid are summarized in Table I.
Solubility values given in the table mean the calculated concentration of solutions to be freeze-dried for the preparation of solid complexes.

TABLE 1. Maximum attainable aqueous solubility of CL and TM citrate and βCD by binary and multicomponent complexation

	Aqueous solubility (mg/ml)			
	CL citrate	βCD	TM citrate	βCD
Citrate salts	**6.2** (0.04*)	-	**0.3** (0.03*)	-
Binary complex citrate salt/βCD	**20** (1:2 M)	**75**	**11** (1:2.5 M)	**~60**
Multicomponent complex base/citric acid/βCD	**100** (1:2.5:1 M)	**170**	**20** (1:2.2 M)	**>80**
	65 (1:2.5:2 M)	**240**		

*Solubility of the base form
Molar ratio of the components are in brackets.

Taking CL citrate:βCD in 1:2 molar ratio results in freely soluble complex with 20 mg/ml drug and 75 mg/ml βCD concentration which means approx. 3 and 4 fold solubility enhancement for CL citrate and βCD, resp. By multicomponent complexation, 1 mol CL base could be completely and quickly dissolved in the presence of 1 mol βCD and 2.5 mol citric acid resulting in a highly supersaturated solution with 100 mg/ml CL citrate and approx. 170 mg/ml βCD concentration in solution, which means 16 and approx. 9-10 fold solubility enhancement for CL citrate and βCD, resp. When, - similarly to the CL:βCD ratio in the binary system - , 2 mol βCD is applied in the multicomponent system, the amount of citric acid cannot be reduced i.e. 2.5 mol citric acid is necessary to obtain a clear solution with decreased CL citrate and increased βCD concentration compared to that obtained in 1:2.5:1 multicomponent system.

Taking TM citrate:βCD in 1:2.5 molar ratio easily results in the formation of freely soluble complex with 11 mg/ml TM citrate and approx. 60 mg/ml βCD concentration in solution, which means about 36 and more than 3 fold solubility enhancement for TM citrate and βCD.
By multicomponent complexation 1 mol TM base could be easily dissolved in the presence of 2 mol βCD and 2 mol citric acid resulting in a highly supersaturated solution approx. 20 mg/ml TM citrate and more than 80 mg/ml βCD concentration in solution, which means 65 and more than 4 fold solubility enhancement for TM citrate and βCD, resp.
The optimal molar ratios in binary and multicomponent systems for both drugs are different, which points to formation of different types of complexes when the pre-prepared citrate salts or the base forms were used, in the latter case salt formation taking place simultaneously with complexation.
By this solubilization method, the low aqueous solubility of βCD - which hinders its application for complex preparation from homogeneous solutions of large volume by removing the water content either by freeze- or spray drying - can be overcome.
The solid complexes can easily be reconstituted with distilled water to the original volume.

4. Conclusion

Multicomponent complexation of CL and TM base with citric acid and βCD results in an extremely high mutual solubility enhancement for both the host and guest.
The solubility increase for the bases was higher than that observed either in the presence of acid or βCD alone and it was also significantly higher than the solubility enhancement with binary complexation of prepared CL and TM citrate salts.

5. References

1. AHFS Drug Information® 97 on Clomiphene and Tamoxifen citrates, pp 2854 and 861
2. Chiesi, P., Ventura, P., Pasini, M., Szejtli, J., and Vikmon, M. PCT Patent Application, WO94/16733 29.01.93

Acknowledgement: Thank is due to Ms K.Dobo and K.Gera for their contribution in the experimental work.

PREPARATION AND CHARACTERIZATION OF PIROXICAM ALKALI-SALT γ-CYCLODEXTRIN COMPLEXES

M. VIKMON,. I. KOLBE, J. SZEJTLI, *CYCLOLAB Cyclodextrin Research and Development Laboratory Ltd., Budapest, Hungary*

E. REDENTI, P. VENTURA, *CHIESI Farmaceutici S.p.A., Parma, Italy*

ABSTRACT

Piroxicam sodium-, potassium- and ammonium salts form complexes with γCD by precipitation from even highly alkaline solution, giving stoichiometric compounds in crystalline state with good yield.
The stoichiometry of the complexes corresponds to 1:1 molar ratio of Piroxicam to γCD.
Powder X-ray diffractometry proved the complex formation in solid state, the diffraction patterns of the complexes are clearly distinct from that of the superposition of the components.
The aqueous solubility of different cation containing complexes is comparable, 1.2-1.4 mg/ml dissolved Piroxicam can be measured.

1. Introduction

Piroxicam is a widely used non-steroidal antiinflammatory drug. It is poorly soluble in water and biological fluids at physiological pH values.
βCD complexation increased the wettability, solubility and dissolution rate of the drug. Piroxicam βCD complex is characterized by improved pharmacokinetic properties like increased rate of absorption and better gastric tolerability than the standard formulation of Piroxicam. βCD complexed Piroxicam has been a successfully marketed drugs since years.
Piroxicam, being a very weak acid-type drug, can be dissolved by salt formation in aqueous solution of sodium-, potassium and ammonium hydroxide, taking in stoichiometric ratio.
Evidences of complex interaction of Piroxicam sodium salt with βCD in solution was established by ^{1}H-NMR spectroscopy [1]. Preparation of Piroxicam sodium βCD complex of different molar ratio by freezed-drying has been reported [2].
In this study we describe the preparation of Piroxicam sodium potassium and ammonium salt/γCD complexes by precipitation, their characterization by powder X-ray diffractometry and comparison of the solubility properties of solid complexes.

2. Experimental

2.1. MATERIALS

Piroxicam was provided by Chiesi Farmaceutici S.p.A., Parma, Italy
γ-cyclodextrin was supplied by Wacker Chemie, Munich, Germany
All other materials were of analytical grade.

2.2. PREPARATION OF PIROXICAM SODIUM/γCD COMPLEX

1 g Piroxicam (0.30 mmol) and 3.9 g γCD (0.30 mmol) were suspended in 10 ml of distilled water. 7-8 ml of approx. 0.5 N sodium hydroxide solution was added to solubilize the drug obtaining a yellow solution of pH above 10. Contrary to the βCD complex preparation a clear solution could be obtain even above pH 12, by alkalising Piroxicam in the presence of γCD. Reducing the strong alkalinity of the solution a fine light-yellow complex is precipitated around pH 10.
Further reducing the pH below 8 under stirring, the precipitated complex were let to stand overnight. Thereafter the solid complex was filtered and dried in an exsiccator under vacuo in the presence of P_2O_5. The dried complex were homogenised.

Yield: 4.1 g
Piroxicam content: 20.2 % (calculated on the dry substance)
Piroxicam potassium and ammonium salt/γCD complexes can be prepared on similar way, but using potassium or ammonium hydroxid for solubilization.

2.3. DETERMINATION OF PIROXICAM CONTENT OF COMPLEXES BY UV-SPECTROPHOTOMETRY

20 mg of Piroxicam alkali salt/γCD complex was accurately weighed and dissolved in 25.0 ml of 0.1 N HCl in 50% ethanol. After appropriate dilution absorbances of the solutions were measured at λ_{max}=333 ± 1 nm. For calculation $E^{1\%}_{1cm}$ = 680 was used for Piroxicam substance.

2.4. X-RAY DIFFRACTOMETRY OF SOLID COMPLEXES

Powder X-ray diffraction patterns of the product were recorded on a Philips diffractometer Typ. PW 3710, using Cu-Kα irradiation. In the diffractograms the relative intensity of reflexion peaks were recorded in the function of diffraction angles 2 θ°.

2.5. SOLUBILITY TEST IN DISTILLED WATER

100 mg of complex samples were equilibrated in 2.5 ml of distilled water by ultrasonication for 15 minutes after filtration. Piroxicam content of the filtrates was determined by UV-spectrophotometry.

3. Results and discussion

Piroxicam sodium-, potassium- and ammonium salts form complexes with γCD by precipitation from even highly alkaline solution, giving stoichiometric compounds in crystalline state with good yields. Piroxicam sodium/βCD complex preparation is also possible by precipitation with lower yield. Its crystal structure has been recently determined [3].

Piroxicam content of complexes fairly corresponds to 1:1 molar ratio of Piroxicam salt to γCD.
X-ray diffractometry proved the complex formation in solid state. X-ray diffractograms of alkali and ammonium salt complexes as well as those of Piroxicam and γCD are shown on Fig. 1. Different complexes showed very similar crystalline structure.
The diffraction patterns of the complexes are clearly distinct from that of the superposition of the components.

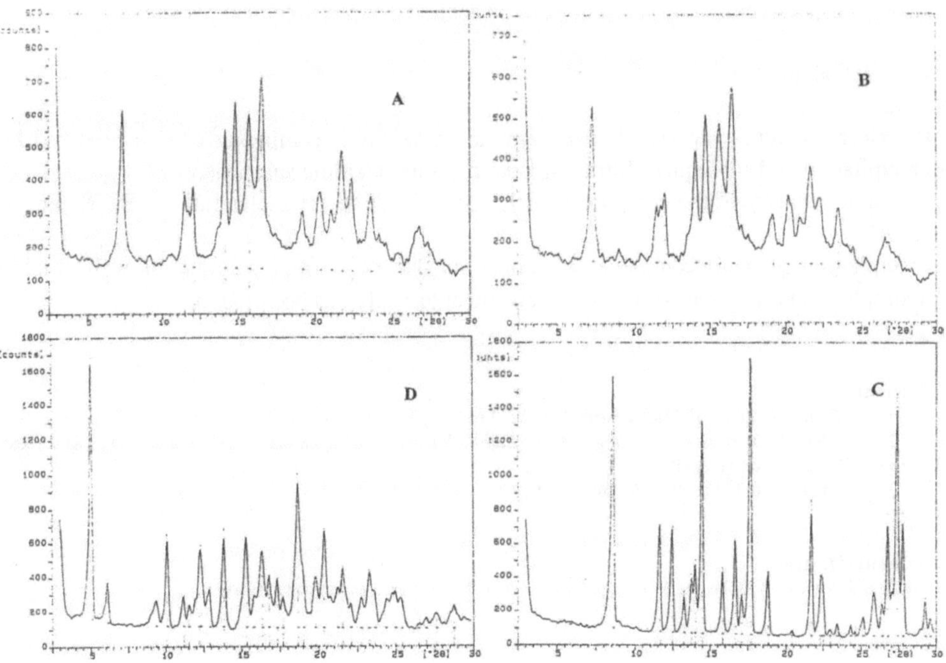

Figure 1. Powder X-ray diffractograms of *(A) Piroxicam sodium/γCD*
(B) Piroxicam potassium /γCD
(C) Piroxicam substance
(D) γCD

Analysis of the different complexes are summarized in Table 1.

TABLE 1.: Composition of Piroxicam alkali and ammonium salt γCD complexes

	Na-salt	K-salt	NH_4-salt
Piroxicam content*	20.20%	18.60%	19.90%
Theoretical Piroxicam content (in 1:1 M ratio):	20.12%	19.92%	20.15%

*related to the dry substance

Aqueous solubility of different complexes and pH of the obtained solution were as follows:

Complex	Dissolved (mg/ml)	Piroxicam	pH of the solution
Piroxicam sodium salt/γCD	2.2		7.3
Piroxicam potassium salt/γCD	1.5		7.1
Piroxicam ammonium salt/γCD	1.3		7.2

As it is seen, a remarkable solubility can be achieved at physiological pH value.

4. Conclusion

Piroxicam sodium-, potassium- and ammonium salt γCD complexes can be prepared by precipitation and subsequent filtration from common alkaline solutions.
The stoichiometry of the complexes fairly corresponds to 1:1 molar ratio of Piroxicam to γCD.
Precipitation of solid complexes in stoichiometric ratio from solution of well soluble hydrophilic guests, even with γCD is new, no anteriority has been found.

References

1. Fronza, G. et al. (1992) *J. Pharm. Sci., 81,* 1162-1165
2. Glass, B.D. et al. *Proceedings of the Eighth International Symposium on Cyclodextrins,* Budapest, Hungary, (1996) March 31-April 1.
3. Villa, A. et al., Results to be published

Acknowledgement
Thank is due to Ms. K.Dobo and M.Balogh for their contribution in the experimental work.

DESIGN AND EVALUATION OF A POROSITY-CONTROLLED OSMOTIC PUMP TABLET FOR CHLORPROMAZINE USING (SBE)7m-β-CD.

K. OKIMOTO[1], M. MIYAKE[1], O. AOKI[1], N. OHNISHI[1], T. IRIE[2], K. UEKAMA[2], R. A. RAJEWSKI[3] and V. J. STELLA[3]

[1] *Technological Development Laboratories, Fujisawa Pharmaceutical Co., Ltd.*, [2] *Faculty of Pharmaceutical Science, Kumamoto University*, [3] *Center for Drug Delivery Research, The University of Kansas*

1. INTRODUCTION

We have developed a porosity-controlled osmotic pump tablet (OPT) for prednisolone, a poorly water soluble and neutral drug, using sulfobutyl ether β-cyclodextrin, (SBE)7m-β-CD, which acts as a solubilizer and an osmotic agent. OPT showed a pH-independent release profile, controlling the *in-vivo* absorption rate by adjusting the *in-vitro* release rate (Okimoto et al., [1, 2]). Chlorpromazine (CLP) is a weak basic drug, which shows a pH-dependent solubility, and is poorly soluble at pH 6.8. In this study, (SBE)7m-β-CD was used for designing OPT from which CLP is released in a pH-independent manner. Moreover, factors responsible for the CLP release rate from OPT were discussed.

2. SYSTEM OF OPT

An OPT is one coated by a semipermeable membrane containing leachable materials (Zentner et al. [3]). A suspension composed of micronized lactose/cellulose acetate (CA-390-10)/triethyl citrate (TEC) in ethanol /methylene chloride was coated onto a core tablet composed of CLP and an osmotic pump agent, (SBE)7m-β-CD, HP-β-CD or sugar mixture (lactose : fructose = 1:1 as a weight ratio). When the OPT is placed into water, CLP is released from the OPT by hydrostatic pressure through pores created by the dissolution of micronized lactose incorporated into the membrane. The hydrostatic pressure is created by the osmotic

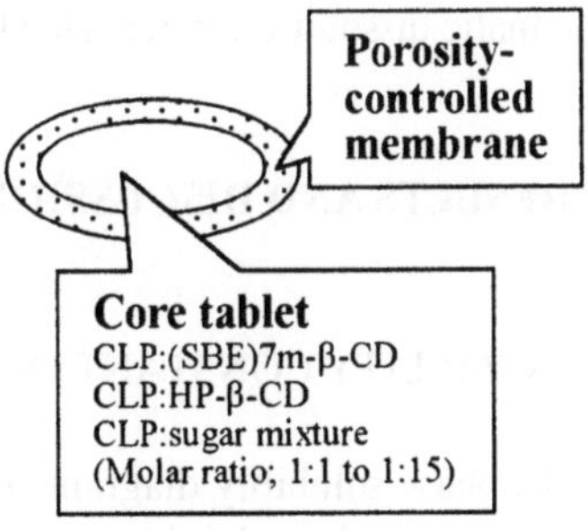

Scheme 1. The cross-section of OPT

agent, after water is imbibed across the semipermeable membrane.

3. METHODS

3.1 PREPARATION OF OPT

A kneaded powder of CLP and the respective osmotic agents at a molar ratio of 1:1 to 1:15 was prepared by vacuum drying for 12 hours at 40°C after mixing the compositions with minimal water using a mortar and pestle. Core tablets for the OPT were prepared by using an eccentric tableting machine (Okada Seikou Co.) using 6 to 9 mmϕ punches with the kneaded powder.

The components of membrane for OPT are shown in Table 1. A Flow Coater Mini® (Floint Co.) was used for the film coating onto the core tablet.

Table 1. Composition of OPT membrane

Ingredient	Formulation (composed ratio)*							
	#1	#2	#3	#4	#5	#6	#7	#8
CA-390-10	1	1	1	1	1	1	1	1
TEC	0.5	0.5	0.5	0.5	0.5	0.5	0	1
Micronized lactose								
450 mesh pass	0.5	1	2	3			1	1
350 mesh pass					1			
200 mesh pass						1		

* The total concentration of ingredients in film solution for all formulations was 10.6% (w/w).

3.2 RELEASE STUDIES

The release of CLP from uncoated cores or OPT (containing the equivalent to 10 mg CLP) was evaluated by using the Japanese Pharmacopoeia (JP) XIII dissolution test, paddle method (50 rpm, 37°C). The dissolution medium (900 mL) was the JP first fluid (pH 1.2) or the JP second fluid (pH 6.8). The CLP release was monitored by an automatic dissolution tester (Hewlett 8451A Diode Array Spectrophotometer).

4. RESULTS AND DISCUSSION

4.1 STABILITY CONSTANT OF CLP WITH β-CD DERIVATIVES

In the phase solubility diagrams for CLP with (SBE)7m-β-CD and HP-β-CD in varying pH solutions, the solubility of CLP linearly increased with increasing concentration of both β-CD derivatives up to 0.05M, suggesting a 1:1 complexation. Stability constants (K_1) of the neutral CLP and that (K_2) of positively charged CLP with β-CDs calculated by the method reported by Okimoto et al. [4] were as follows; K_1 = 73,100 M^{-1}, K_2 =

32,100 M^{-1} for (SBE)7m-β-CD, and K_1 = 44,600 M^{-1}, K_2 = 7010 M^{-1} for HP-β-CD.

4.2 FACTORS RESPONSIBLE FOR CLP RELAESE RATE FROM OPT

It is important for the design of OPT to choose an appropriate membrane component. Factors responsible for CLP release rate from OPT were investigated using various membrane components (see Table 1) with the core tablet of CLP:(SBE)7m-β-CD at a 1:1 molar ratio. Since CLP shows a pH-dependent solubility, the JP first fluid (pH 1.2), in which the solubility of CLP is higher than in the JP second fluid (pH6.8), was used as the dissolution medium to evaluate its controlled-release. As shown in Table 2, the CLP release rate increased with increasing amounts of micronized lactose (the release rate; #1 < #2 < #3 < #4), and TEC (the release rate; #5 < #2 < #8) and decreasing the lactose size (the release rate; #6 < #5 < #2) in the membrane. Of the formulations tested, the formulation #2 was chosen as the optimal membrane composition for the OPT of CLP, because of the reasonable release rate as an oral sustained-release tablet. Furthermore, the release rate of CLP from the formulation #2 was controlled by changing the membrane thickness as shown in Table 3.

Table 2. CLP release rates from OPT prepared varying with the membrane component (membrane thickness 0.25 mm)

	Formulation							
	#1	#2	#3	#4	#5	#6	#7	#8
Release rate (%/hr)	13	20	40	60	11	2	85	5

Table 3. CLP release rates from OPT (#2) varying with the membrane thickness

	Membrane thickness			
	0.1 mm	0.13 mm	0.25 mm	0.33 mm
Release rate (%/hr)	62	37	20	13

4.3 DESIGN OF OPT FOR CLP

The dissolution rate of CLP from the core tablet composed of CLP: (SBE)7m-β-CD at a molar ratio of 1:1 was not so pH-dependent, and CLP was completely dissolved in dissolution media of pH 1.2 and pH 6.8 as shown in Fig. 1. However, the CLP release rate from OPT prepared by the core tablet depended on pH of the dissolution media, the CLP release rate at pH 1.2 being much greater than that observed at pH 6.8 (see Fig.1). This pH-dependency of CLP release rates from OPT decreased with increasing molar ratios of (SBE) 7m-β-CD in the core tablets, and was completely disappeared over 1:10 (CLP:(SBE)7m-β-CD), as shown in Fig. 2. On the other hand, the CLP release rates from OPT prepared using core tablets, containing CLP:HP-β-CD or the sugar mixture at a molar ratio of 1:10, were pH-dependent (see Fig.2).

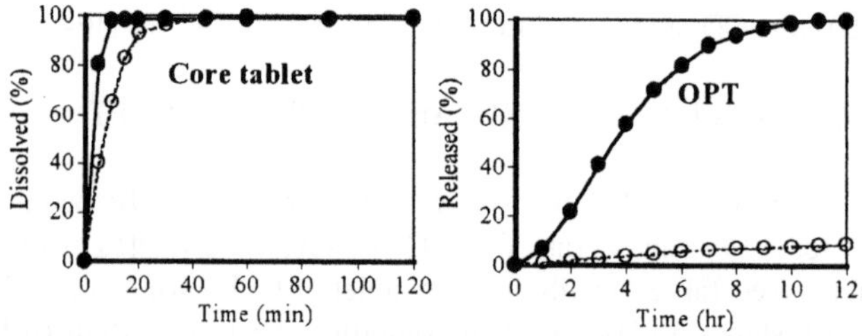

Figure 1. Influence of pH on CLP release rate from core tablet composed of CLP: $(SBE)_{7m}$-β-CD = 1:1 and OPT prepared using the core tablet. In this case, the membrane thickness of OPT was 0.25 mm. ●; pH 1.2, ○; pH 6.8

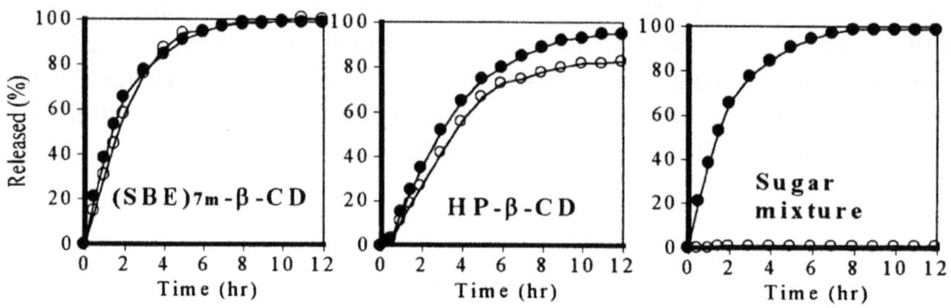

Figure 2. Comparison of CLP release rates from OPT (membrane thickness; 0.25 mm) prepared using core tablet composed of CLP: $(SBE)_{7m}$-β-CD = 1:10 with those containing HP-β-CD or the sugar mixture at a molar ratio of 1:10. ●; pH 1.2, ○; pH 6.8

In conclusion, the present results suggest that (SBE)7m-β-CD can serve as both a solubilizer and as an osmotic agent for designing the OPT of CLP with a pH-independent release.

5. REFERENCES

1. Okimoto, K., Rajewski, R. A. and Stella, V. J. (1998) Release of testosterone from an osmotic pump tablet utilizing (SBE)7m-β-CD as both a solubilizing and an osmotic pump agent. Submitted.

2. Okimoto, K., Miyake, M., Ohnishi, N., Rajewski, R. A., Stella, V. J., Irie, T. and Uekama, K. (1998) Design and evaluation of an osmotic pump tablet (OPT) for prednisolone, a poorly water soluble drug, using (SBE)7m-β-CD, *Pharm. Res.* in press.

3. Zentner, G. M., Rork, G. S., and Himmelstein, K. J. (1985) Osmotic flow through controlled porosity films: An approach to delivery of water soluble compounds. *J. Control. Rel.* 2, 217-229.

4. Okimoto, K., Rajewski, R. A., Uekama, K., Jana, J. A. and Stella, V. J. (1996) The interaction of charged and uncharged drugs with neutral (HP-β-CD) and anionically charged (SBE-β-CD) β-cyclodextrins. *Pharm. Res.* 13, 256-264.

COMPARISON OF THE SOLUBILIZING EFFECT OF ETHYL CARBONATE OF γ-CYCLODEXTRIN TO OTHER CYCLODEXTRIN DERIVATIVES

E. FENYVESI[1], J. SZEJTLI[1], F. TROTTA[2], E. REDENTI[3], P. VENTURA[3]

[1]Cyclolab R&D Lab., Budapest, Dombóvári út 5-7, H-1117 Hungary

[2]Dip. Chim. Inorg., Univ. di Torino, via Pietro Giuria 7, I-10125 Torino, Itally

[3]Chiesi Farmaceutici S.p.A., via Palermo 26/A, Parma, Italy

1. Introduction

As none of the CD derivatives was found to be a universal solubilizer, the search for new derivatives with solubilizing activity is still in progress. To find the best solubilizer for various poorly soluble compounds is one of the main target of the CD-research. Ethyloxycarbonyl γ-cyclodextrin, a new acyclic carbonate derivative of γ-cyclodextrin can be a member of the arsenal together with the methyl, hydroxypropyl, acetyl derivatives.

A selective reaction was found for the preparation of acyclic carbonates of CDs [1]. The product is a mixture considering the degree and position of the substitution, as usual in case of CD derivatives. It can be characterized by an average degree of substitution (DS) which depends on the reaction conditions. The solubility of the product is the function of the alkyl chain length: the ethyloxycarbonyl βCD is about 30 times more soluble than the parent βCD, while the propyloxycarbonyl derivative is less soluble. The solubility also depends on the DS: the products with low DS have higher solubility.

These experiences obtained studying the alkyloxycarbonyl βCD derivatives were utilized in the preparation of γCD derivatives. Ethyloxycarbonyl γ-cyclodextrin (γCD-OCOO-CH_2CH_3, ECγCD) (DS ~ 5) was synthesized and characterized by FT-IR and ESI-MS. This derivative being well soluble in water (more than 50 % solubility) was studied as a solubilizing agent for various drugs. Its solubilizing effect was compared with that of some other γCD derivatives (acetyl, hydroxypropyl) and βCD derivatives of good drug-solubilizing capacity (random methyl and hydroxypropyl) as well as with γCD.

2. Experimental

2.1. SYNTHESIS OF ECγCD

The synthesis was carried out with minor modification of the reported method for βCD carbonates [1]:

36.4 g (0.0281 moles) of anhydrous γCD were dissolved in 320 mL of anhydrous pyridine. The solution was thermostated at 80 °C with an oil bath and 19.7 g (0.141 moles) of previously activated ethyl alcohol with carbonyldiimidazole were added under vigorous magnetic stirring. The solution became clear in one hour and was allowed to react for further two hours.

Once the reaction occurred, the solution was concentrated to small volume under vacuum below 40 °C. A large excess of dichloromethane was added to the viscous oil obtained and a white precipitate was recovered by filtration.

In order to remove the residual pyridine the precipitate was Soxleth extracted with dichloromethane. The compound obtained showed no TLC spot of the parent γCD.

2.2. DETERMINATION OF DEGREE AND PATTERN OF SUBSTITUTION

The IR spectra were obtained on a Perkin Elmer 1710 FT-IR spectrophotometer. The electrospray-ionization mass spectrometry (ESI-MS) was carried out using a Finnigan SSQ 7000 single quadrupole instrument. The sample was dissolved in water/acetonitrile 50/50 (v/v).

2.3. SOLUBILITY STUDIES

Excess amount of the drug (usually 20 mg, but 60 mg in case of the acidic drug/ECγCD systems) was added to 2.5 mL aqueous cyclodextrin solutions under stirring. Having stirred for 2 h at ambient temperature the suspensions were filtered and the filtrates measured spectrophotometrically after proper dilution with 1:1 aqueous ethanol or 0.1 N HCl in 1:1 aqueous ethanol.

The following cyclodextrins were used:

ethylcarbonate γ-cyclodextrin (ECγCD), DS ~ 4
acetyl γ-cyclodextrin (AcγCD), Wacker Chemie, Munich, DS = 7.2 W8 A 0.9 Lot No. 88C001
hydroxypropyl γ-cyclodextrin (HPγCD), Wacker Chemie, Munich, Lot No. 110238
γ-cyclodextrin (γCD), Wacker Chemie, Munich, Lot No. 80P063
hydroxypropyl β-cyclodextrin(HPBCD), 8911390 (Chinoin, Budapest)
randomly methylated β-cyclodextrin (RAMEB), Wacker Chemie, Munich, Beta W7 M 1.8, Lot No. 71B006

3. Results

The average degree of substitution (DS) calculated by FT-IR spectroscopy is about 5. The analysis by electrospray-ionization mass spectrometry (ESI-MS) in the positive mode yields predominantly the $[M+Na]^+$ ions at m/z 1607 (M.w. 1584, DS = 4), m/z 1679 (M.w. 1656, DS = 5), m/z 1751 (M.w. 1728, DS = 6), m/z 1823 (M.w. 1800, DS = 7), m/z 1895 (M.w. 1872, DS = 8).

In the comparative solubility studies γCD, and two γCD derivatives (acetyl and hydroxypropyl) were involved together with two commercially available βCD derivatives of high solubilizing capacity (methyl and hydroxypropyl) The results are summarized in Table I.

Table I Solubility of drugs in various cyclodextrin solutions of 10 % CD content

	Water	ECγCD (pH=7.8)	γCD (pH=5.3)	AcγCD (pH=3.6)	HPγCD (pH=4.8)	HPBCD (pH=6.0)	RAMEB (pH=4.4)
Furosemide	0.06	**16**	0.16	0.35	1.9	0.7	0.94
Piroxicam	0.02	**3.1**	0.07	1.0	0.19	0.6	1.3
Chlorothiazide	0.2	**7.7**	0.8	2.5	1.5	0.7	4.0
Ibuprofen	0.08	15.6	0.65	2.6	n.d.	9.5	17.0
Beclomethasone	0.002	2.3	0.1	3.6	1.0	0.2	2.1
Testosterone	0.03	11.1	2.4	14.0	8.5	15.0	24.3
Hydrocortisone	0.4	16.1	4.5	16.9	9.9	18.5	17.8
Flunisolide	0.05	9.0	0.7	11.1	2.4	1.9	6.8
Budesonide	0.03	0.57	0.1	3.8	n.d.	1.5	3.9
Ketoconazole	0.001	1.3	0.25	2.3	n.d.	5.3	7.9
Terfenadine	0.01	1.5	1.2	0.86	n.d.	1.2	4.5
Domperidone	0.005	0.09	0.12	0.4	n.d.	0.1	0.47
Amphotericine B	0.003	0.3	0.4	0.3	0.2	0.02	0.03

n.d. = not determined

Solubility isotherms were measured at 25 °C for various model drugs using 0 to 10 % ECγCD concentration. A_L or A_N type isotherms were obtained even for those drugs which show B_S type isotherm with the parent γCD (e.g. Beclomethasone, Testosterone, Flunisolide, etc.).

ECγCD was superior to the other γ-cyclodextrins in case of the acidic drugs. It surpasses even RAMEB. The effect can be partly attributed to the higher pH of these solutions (Fig. 1).

The underivatized cyclodextrin forms insoluble complex with most of the steroids (B_s type isomers). All the studied derivatives increase the solubility with increasing cyclodextrin concentration resulting in A_L or A_N isotherms. RAMEB is the best solubilizer for these types of drugs. The performance of ECγCD is comparable or slightly inferior to that of AcγCD.

Amphotericine B forms water-soluble complex with γ-cyclodextrin. The solubilizing effect of the derivatives is lower than that of the parent γ-cyclodextrin.

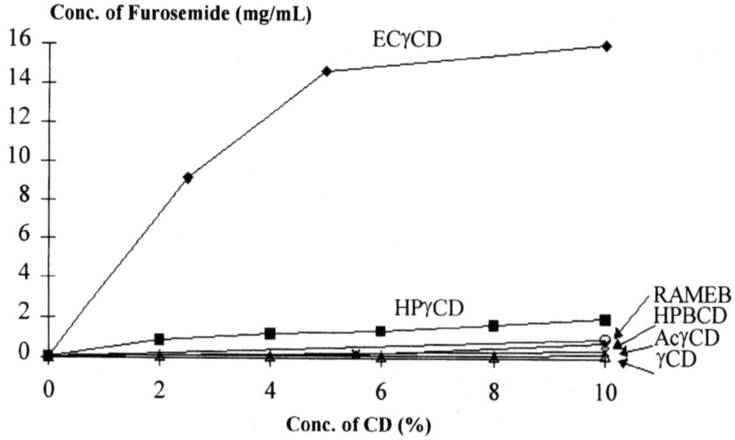

Figure 1. Solubility of Furosemide in aqueous ECγCD (◆), HPγCD (■), RAMEB (O), HPβCD (+), AcγCD (X) and γCD (Δ) solutions

4. Summary

A reproducible method has been developed for the preparation of ethyloxycarbonyl γ-cyclodextrin (ECγCD). The degree and pattern of substitution was measured by FT-IR and ESI-MS.

The ECγCD shows extremely high solubility enhancement for acidic drugs (piroxicam, furosemide, chlorothiazide), partly in consequence of the higher pH of its solutions. For steroids the solubilizing power is slightly lower than that of the AcγCD and RAMEB, but much higher than that of HPγCD or γCD. Amphotericine B dissolves the best in γ-cyclodextrin solutions, any of the γ-cyclodextrin derivatives can not surpass its effect.

The ECγCD is a potential solubilizer of the poorly soluble substances, especially those of acidic character.

Acknowledgment

The technical assistance of I. Megyeri is greatly acknowledged.

5. References

1. Trotta, F., Moraglio, G., Marzona, M., Maritano, S. (1993) Acyclic carbonates of β-cyclodextrins, *Gazzetta Chimica Italiana* 123, 559-562

IMPROVED SOLUBILITY AND ORAL BIOAVAILABILITY OF CYCLOSPORIN A BY HYDROPHILIC CYCLODEXTRIN COMPLEXATION

K. MIYAKE, T. IRIE, F. HIRAYAMA and K. UEKAMA
Faculty of Pharmaceutical Sciences, Kumamoto University, 5-1, Oe-honmachi, Kumamoto 862-0973, Japan

1. Introduction

Cyclosporin A (CsA) is a potent immunosuppressant used primarily in the prevention of allograft rejection after organ transplantation and the treatment of autoimmune diseases. CsA exhibits a low therapeutic index and a poor oral bioavailability with large intra- and inter-individual variations, because of the limited solubility, the influence of foods and bile flow, the poor intestinal membrane permeability, the intestinal and hepatic first-pass metabolism and P-glycoprotein counter-transport processes [1]. The present study addresses how complexation of CsA with hydrophilic cyclodextrins (CyDs) impacts upon the intestinal absorption and the lymphatic transport of the cyclic and hydrophobic peptide in rats, with emphasis on inter-individual variations in its intestinal absorption.

2. Materials and Methods

2.1 MATERIALS

CsA was donated by Shiseido Co. Ltd (Yokohama, Japan). Natural CyDs, 2-hydroxypropyl-CyDs (HP-CyDs), 2,6-di-*O*-methyl-CyDs (DM-CyDs) and 2,3,6-tri-*O*-methyl-CyDs (TM-CyDs) were donated by Nihon Shokuhin Kako Co. Ltd (Tokyo, Japan). Maltosyl-CyDs (G_2-CyDs) were donated by Ensuiko Sugar Refining Co. Ltd (Yokohama, Japan). Other chemicals and solvents were of analytical reagent grade, and deionized double-distilled water was used throughout the study.

2.2 METHODS

Solubility studies: A constant and excess amount of CsA was added to aqueous solution of various CyDs at different concentrations and shaken for 4 days at 25°C. An aliquot was filtered with the cellulose acetate membrane. The concentration of CsA in the filtrate was determined by HPLC. The stability constants of the complexes of CsA with CyDs ($K_{1:n}$), assuming that 1:n higher-order complexes occur in a stepwise reaction, were calculated from the ascending curvatures of the solubility diagrams according to the optimization technique described [2].

In vitro CsA metabolism studies: Male Wistar rats, 250-300 g, were fasted overnight prior to the studies. Treatment regimens consisted of Dexamethasone, a P-450IIIA inducer (80 mg/kg/day) suspended in 1 mL of olive oil was administered intraperitoneally to rats for 2 days. The final dose of the P-450IIIA inducer was administered 16-18 h prior to phlebotomy lethality and laparotomy. Microsomes were prepared from epithelial cell of rat small intestine. After the incubation of CsA with the microsomal suspension, the amounts of intact drug and its metabolites were assayed by HPLC [3].

In situ permeability studies: Male Wistar rats, 250-270 g, were fasted for overnight, and was anesthetized with 25% ethyl carbamate in saline (6 mL/kg). After laparotomy, about 10 cm segment of the jejunum was ligated at both ends to form small intestinal sac. One milliliter of CsA solution (20 μg/mL) or suspension (200 μg/mL) was injected into the intestinal sac. All mesenteric venous blood draining the loop was collected. The intestinal sac was removed after 1h, and the amounts of CsA into the sac and blood concentration of CsA and its metabolites were determined by HPLC [3].

In vivo absorption studies: Male Wistar rats, 200-250 g, were fasted for overnight, and CsA suspension (10 mg/kg) with or without hydrophilic CyDs were administered orally to the rats. At appropriate intervals, the rats were anaesthetized with an intraperitoneal injection of sodium pentobarbital (30 mg/kg) and a heparinized polyethylene canula was implanted into the thoracic lymph duct. The lymph sample (100~500 μL) was collected for 15 min after cannulation into the thoracic lymph duct, and at the same time, the blood sample (1 mL) was collected from the heart. The plasma and lymph concentrations of CsA were determined by HPLC.

3. Results and Discussion

3.1 Interaction of CsA with CyDs

Figure 1 shows the phase solubility diagrams of CsA-CyDs systems in water at 25°C. The hydrophilic CyDs increased the solubility of CsA in water with a positive deviation from linearity, forming higher-order complexes. The solubilizing ability of hydrophilic CyDs for CsA increased in order of HP-γ-CyD ≈ γ-CyD < HP-β-CyD ≈ G_2-β-CyD < β-CyD < HP-α-CyD < G_2-α-CyD ≈ α-CyD << DM-β-CyD ≈ DM-α-CyD, suggesting that the smallest cavity of α-CyDs is the most appropriate partner for CsA. The weaker interaction of HP-CyDs and G_2-CyDs compared with the parent CyDs may be due to the steric hindrance of thier substituents, while the greater ability of DM-CyDs may be attributable to an increase in hydrophobic space of the CyD cavity and to the surface activity. The ascending curves in Figure 1 were analyzed according to the higher-order complexation ($K_{1:n}$), and the results are summarized in Table I. Insight into the inclusion mode of CsA with DM-CyDs was gained by employing proton nuclear magnetic resonance (^{1}H-NMR) spectroscopy. Upon binding to DM-α-CyD, the proton signals of amino acid residues at the position of 1, 4 and 9 in CsA molecule were largely shifted to downfield. By contrast, DM-β-CyD showed a biphasic effect on the chemical shift displacement of those amino acid residues; i.e. the upfield shifts at relative low concentrations and the downfield shifts at higher concentrations. These results indicate that the CsA molecule is too bulky to be wholly included in the cavity of DM-CyDs, and DM-α- and DM-β-CyD interact with the cyclic peptide in a different manner.

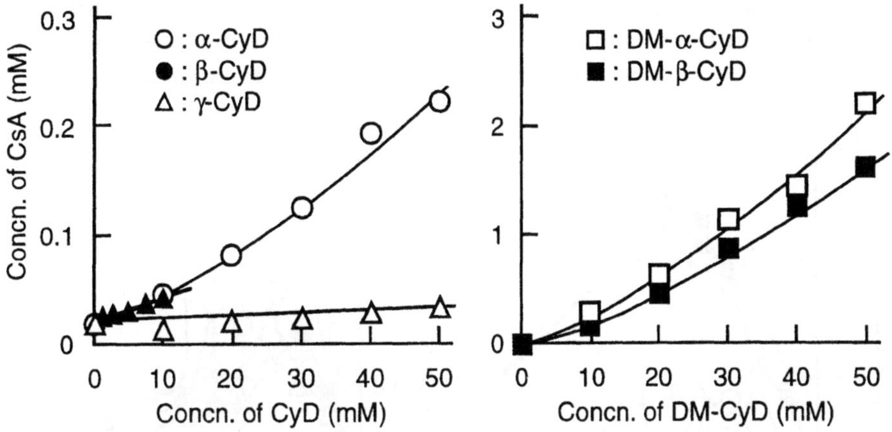

Figure 1. Phase Solubility Diagrams of CsA-CyD Systems in Water at 25°C

Table I. Stability Constants (M^{-1}) of Complexes of CsA with CyDs in Water at 25°C

System	α-CyDs		β-CyDs		γ-CyDs	
	$K_{1:1}$	$K_{1:2}$	$K_{1:1}$	$K_{1:2}$	$K_{1:1}$	$K_{1:2}$
Natural CyDs	150	130	120	–	10	6
DM-CyDs	1060	15	1050	21		
TM-CyDs	80	150	20	–		
HP-CyDs	100	7	20	28	2	37
G_2-CyDs	100	20	40	14		

3.2 Effects of CyDs on metabolism and membrane permeability of CsA

When incubated with rat intestinal microsomes, CsA was metabolized mainly to two metabolites, 9γ-hydroxy form (M1) and 1η-hydroxy form (M17). DM-α- and DM-β-CyDs inhibited the conversion of CsA by the microsomal enzymes including cytochrom P-450IIIA4, the latter being more effective. When CsA in solution or in suspension was injected into rat small intestinal sac, DM-β-CyD increased the cumulative amount of CsA transferred into the mesenteric venous blood about 2-fold. In addition, DM-β-CyD decreased the molar ratio of M1 to CsA, probably through restricting the formation of a catalytic complex of CsA with enzymes or enzymatic saturation.

3.3 *In vivo* Absorption Behavior

Figure 2 shows the plasma and lymph levels of CsA after oral administration of CsA suspensions (equivalent to 10 mg/kg CsA) with or without CyDs to rats. DM-α- and DM-β-CyDs enhanced the extent of oral bioavailability of CsA about 5-fold, reaching ~25 % of that of the intravenous administration in rats. On the other hand, both DM-CyDs did not affect the lymphatic transfer of CsA. It is noteworthy that DM-CyDs decreased the inter-individual variability in plasma levels of CsA after oral administration, as expressed by the decrease in the coefficient of variation for plasma from 63 % to 22 %. These results suggested that the increased bioavailability and the decreased variability may be due to the enhanced solubility and membrane permeability, and the inhibited drug metabolism.

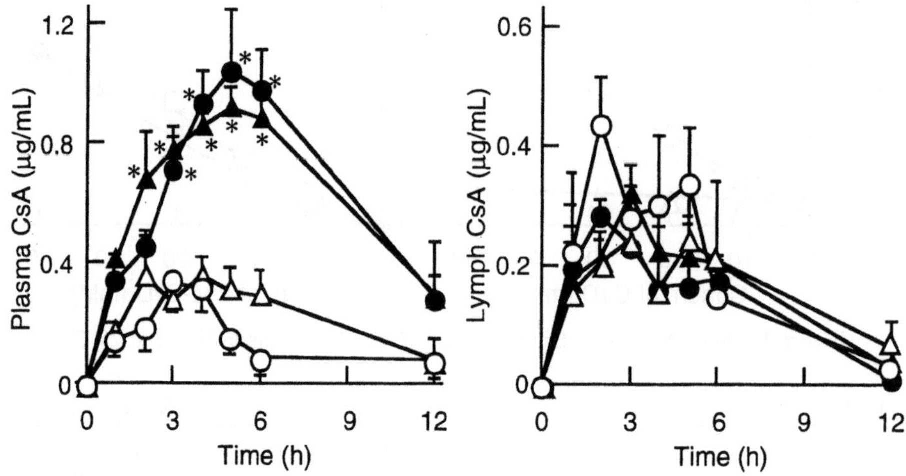

Figure 2. Plasma and Lymph CsA Levels after Oral Administration of CsA (10 mg/kg) Suspensions with or without CyDs (molar ratio of 1:10 (CsA:CyD)) to Rats
O: CsA alone, Δ: with α-CyD, ▲: with DM-α-CyD, ●: with DM-β-CyD.

4. Conclusion

The present results suggest that DM-CyDs are particularly useful in improving the intestinal absorption of CsA and decreasing its inter-individual variability in oral bioavailability.

Acknowledgements

This work is partly supported by the Sasagawa Scientific Research Grant from The Japan Science Society.

5. References

[1] Lampen, A., Christians, U., Bader, A., Hackbarth, I. and Sewing, K.F. (1996) Drug interaction and interindividual variability of ciclosporin metabolism in the small intestine. *Pharmacology*, 52, 159-168.

[2] Higuchi, T., and Connors, K.A. (1965) Phase solubility techniques. *Advan. Anal. Chem. Instr.* 4, 117-212.

[3] Khoschsorur, G., Semmelrock, H.J., Rödl, S., Auer, T., Petek, W., Iberer, F. and Tscheliessnigg, K.H. (1997) Rapid, sensitive high-performance liquid chromatographic method for the determination of cyclosporin A and its metabolites M1, M17 and M 21. *J. Chromatogr.*, 690, 367-372.

EFFECT OF AMORPHOUS β-CYCLODEXTRINS ON CRYSTALLIZATION AND POLYMORPHIC TRANSITION OF TOLBUTAMIDE IN SOLID STATE

K. KIMURA, F. HIRAYAMA and K. UEKAMA
Faculty of Pharmaceutical Sciences, Kumamoto University, 5-1 Oe-honmachi, Kumamoto 862-0973, Japan

1. Introduction

It is important to control crystallization and polymorphic transition of solid drugs, since crystal modifications affect various pharmaceutical properties, such as stability, solubility, dissolution rate and oral bioavailability of drugs. Amorphous cyclodextrins (CyDs) such as 2-hydroxypropyl-β-cyclodextrin (HP-β-CyD) are useful for the control of solid properties of poorly water-soluble drugs, because they can convert crystalline drugs to amorphous complexes which are usually water-soluble [1]. In this study, the crystallization and polymorphic transition behavior of tolbutamide (TB), a hypoglycemic agent, in the absence and presence of HP-β-CyD was investigated by powder X-ray diffractometry and differential scanning calorimetry (DSC), and compared with those in the presence of polyvinylpyrrolidone (PVP) or hydroxypropylcellulose-M (HPC-M). Furthermore, dissolution and *in-vivo* absorption studies with TB polymorphs and its HP-β-CyD complex were carried out. TB was chosen as a model drug because it has several polymorphic forms such as Form A (stable form), B and C (metastable forms).

2. Experimental

Preparation of TB Polymorphs and Amorphous TB Powders: The polymorphs of TB (Forms A and B) were prepared by recrystallization from benzene and ethanol, respectively, at room temperature, according to the method of Simons [2]. Form C of TB was obtained by standing Form D, a new polymorph of TB prepared in this study, at 60°C, 75% relative humidity (R.H.). Forms A, B and C were confirmed to be identical to those reported by Traue [3]. Form D and amorphous TB powder were prepared by spray-drying method. Solid complex of TB with HP-β-CyD (degree of substitution= 4.8) and solid dispersion of TB with PVP (K-30) or HPC-M were obtained by dissolving them in a mixed solvent of ethanol/dichloromethane (1.2/1 %v/v), and spray-drying.

Ageing Studies: The test powders were placed in desiccators at 75% R.H. (saturated solution of sodium chloride) and then stored in incubators at constant temperatures [4]. The crystallization and polymorphic transition rates of TB were monitored by powder X-ray diffractometry.

Dissolution Studies: The dissolution test was performed according to the dispersed amount method in Japanese Pharmacopoeia XIII (JP XIII) second fluid (pH 6.8) and stirred at 91 rpm at 37°C. At appropriate intervals, an aliquot was withdrawn with a cotton-pluged pipette and analyzed spectrophotometrically for TB at 230 nm.

Absorption Studies: The absorption studies were carried out using male beagle dogs, fasted for 24 hours before drug administration. The sample was wrapped with a wafer and administered orally using a catheter, together with water. Blood samples were withdrawn from the cephalic vein by heparinized syringe. Plasma levels of TB and glucose were determined by HPLC and Glucose CII Test Wako (Wako Pure Chemical Ind. LTD., Tokyo, Japan), respectively.

3. Results and Discussion

3.1. Crystallization and Polymorphic Transition Behavior of TB

A new metastable form of TB, Form D, was obtained when TB was spray-dried from a mixed solvent of ethanol/dichloromethane. Form D showed two strong diffraction peaks at 2θ = 10.6° and 18.9° in the powder X-ray diffractogram, as shown in Fig. 1. This diffraction pattern was apparently different from those of Forms A, B and C reported (2θ = 8.7° and 12.1°, 19.6° for Form A, 2θ = 11.2°, 15.4°, 18.2° for Form B, 2θ= 10.3°, 11.3°, 19.6° for Form C). Form D was easily converted to stable form of TB, Form A, through the metastable intermediate, Form C, when it was stored at 60°C, 75% R.H. (Fig. 1). The conversion of Form D to Form A was completed in 3 days under the above storage condition.

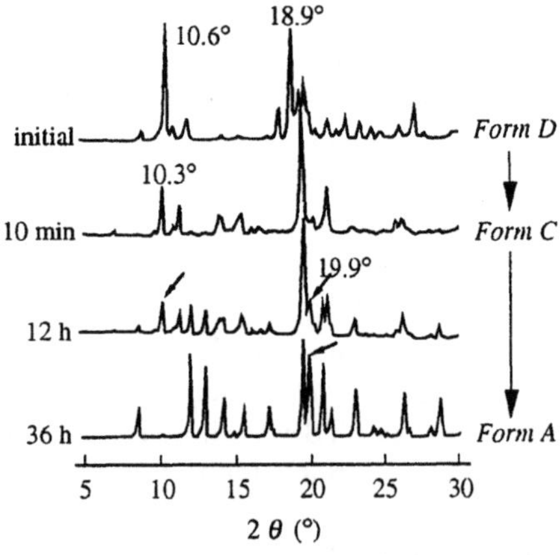

Figure 1. Changes in Powder X-ray Diffraction Pattern of Form D, during Storage at 60°C, 75% R.H.

An amorphous complex of TB with HP-β-CyD was obtained when they were spray-dried from a mixed solvent of ethanol/dichloromethane. The crystallization and polymorphic transition of TB were significantly inhibited by the complexation with HP-β-CyD, as shown in Fig. 2. For example, it took about 18 hours and 3 months to appear the intermediate, Form C, and the stable form, Form A, respectively, and the complete conversion to From A took over 4 months. On the other hand, the amorphous TB in PVP and HPC-M despersions was easily converted to Form A within several days (Fig. 2).

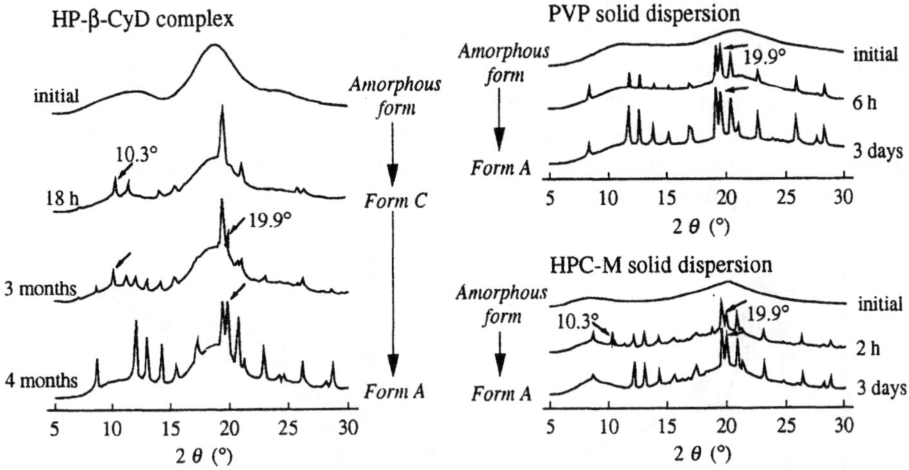

Figure 2. Changes in Powder X-ray Diffraction Patterns of TB Polymorphs in HP- β-CyD Complex and in PVP and HPC-M Solid Dispersions, during Storage at 60°C, 75% R.H.

3.2. Dissolution Behavior

The dissolution rate of TB increased in the order of Form A ≤ Form B < Form C ≤ Form D < HP-β-CyD complex. The dissolution rate of TB from the PVP or HPC-M dispersion was much faster than those of TB polymorphs, but decreased significantly after the storage because of the polymorphic transition to Form A having the slow dissolution property. On the other hand, the HP-β-CyD complex maintained the superior dissolution even after the storage, because HP-β-CyD prevented the polymorphic transition to Form A (Fig. 3).

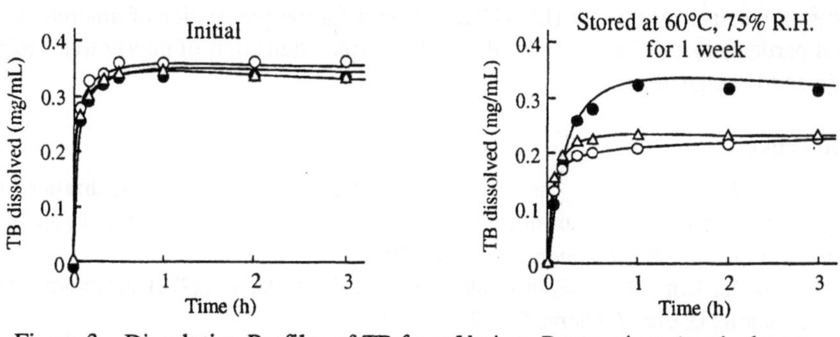

Figure 3. Dissolution Profiles of TB from Various Preparations (equivalent to 100 mg TB) in JP XIII Second Fluid at 37°C, Measured by Dispersed Amount Method at 91 rpm

●: HP-β-CyD complex, O: PVP solid dispersion, Δ: HPC-M solid dispersion.
Each point represents the mean of 2-3 experiments.

3.3. *In-Vivo* Absorption Behavior

Figure 4 shows plasma TB and glucose levels following the oral administration of TB-HP-β-CyD complexes immedately after the spray-drying or after storage for 1 week at 60°C, 75% R.H. The *in-vitro* dissolution characteristics were clearly reflected in the *in-vivo* absorption behavior of TB, *i.e.*, no appreciable ageing effect was observed for the plasma TB and glucose levels after the storage, which may be due to the prevention of the crystallization and polymorphic transition of TB in the HP-β-CyD matrix.

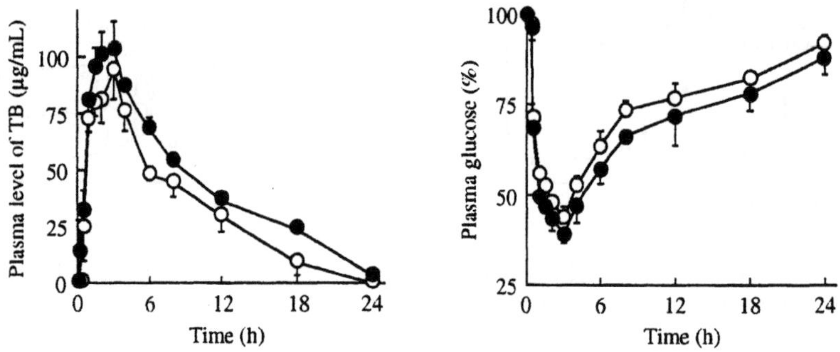

Figure 4. Effect of Storage on Plasma Levels of TB (Left) and Glucose Levels (Right) after Oral Administration of TB/HP- β-CyD Complex (equivalent to 100 mg/body TB) to Dogs

●: initial, O: after storage for 1 week at 60°C, 75% R.H.

Each point represents the mean ± S.E. of 3-4 dogs.

4. Conclusion

The present results suggest that HP-β-CyD is useful for the preparation of amorphous TB, and will particularly provide for controlling polymorphic transition of poorly water-soluble drugs in solid dosage forms.

5. References

[1] Hirayama, F., Usami, M., Kimura, K. and Uekama, K. (1997) Crystallization and polymorphic transition behavior of chloramphenicol palmitate in 2-hydroxypropyl-β-cyclodextrin matrix, *Eur. J. Pharm. Sci.*, **5**, 23-30.

[2] Simons, D. L., Ranz, R. J., Gyanchandani, N. D. and Picotte, P. (1972) Polymorphism in pharmaceuticals II, *Can. J. Pharm. Sci.*, **7**, 121-123.

[3] Traue, J., H. Kala, M. Köhler, U. Wenzel, A. Wiegeleben, B. Förster, P. Pollandt, K. Pintye-Hódi, P. Szabó-Révész und B. Selmeczi (1987) Untersuchungen zur polymorphie von arzneistoffen in pulvern und tabletten, *Pharmazie*, **4 2**, 240-241.

[4] Nyqvist, H. (1983) Saturated salt solutions for maintaining specified relative humidities, *Int. J. Pharm.*, **4**, 47-48.

A NEW METHOD FOR DETERMINATION OF STABILITY CONSTANT OF CYCLODEXTRIN COMPLEXES BY MEMBRANE PERMEATION TECHNIQUE

Naomi Ono, Fumitoshi Hirayama and Kaneto Uekama

Faculty of Pharmaceutical Sciences, Kumamoto University, 5-1 Oe-honmachi, Kumamoto 862-0973, Japan

1. Introduction

It is important to determine the stability constant ($K_{m:n}$ defined by Eq. 1) of cyclodextrin (CyD) complexes with great accuracy, because this parameter is useful for the estimation of changes in physicochemical and biopharmaceutical properties of the guest molecule included in the CyD cavity:

$$Kc = \frac{[\text{complex}]}{[\text{CyD}]_m \cdot [\text{guest}]_n} \qquad (1)$$

where [complex], [CyD] and [guest] are molar concentrations of the complex and free CyD and guest molecules, respectively, and m and n are binding mole numbers of CyD and guest in the complex, respectively. There are many methods for determination of the stability constant of CyD complexes, using various techniques such as solubility, potentiometry, kinetic and spectroscopic methods [1]. All these methods are based principally on titrating certain physical or chemical properties of the guest with CyD, and analyzing the CyD concentration dependencies by means of Benesi-Hildebrand equation, Lineweaver-Burk equation or their modified forms. Therefore, the change in physicochemical property of the guest molecule has to be measured as a function of CyD concentration. In this study, we developed a simple and reliable method for the determination of stability constant of CyD complexes by analyzing permeation rate of the guest in the presence of CyD through a cellophane membrane.

2. Experimental

Materials: Phenacetin (**1**) was obtained from Sigma-Aldrich Co. *o*-, *m*- and *p*-Toluic acids (**2**, **3** and **4**) were purchased from Wako Pure Chemical Co. *o*-Chlorobenzoic acid (**5**), *m*- and *p*-Bromobenzoic acids (**7** and **8**) and *o*- and *m*-Iodobenzoic acids (**9** and **10**) were obtained from Nacalai Tesque. *p*-Chlorobenzoic acid (**6**) was obtained from Tokyo Kasei Kogyo. α-, β-, and γ-CyDs were donated by Nihon Shokuhin Kako Co. All other materials and solvents were of analytical reagent grade, and deionized double-distilled water was used.

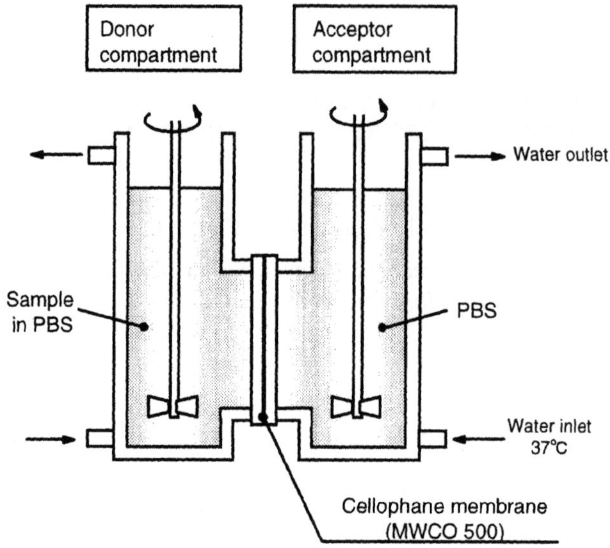

Figure 1. Apparatus Used for Cellophane Membrane Permeability Study

Solubility Study: Excess amounts of guest compounds were added to aqueous solution containing various concentrations of CyDs and the mixtures were shaken in isotonic phosphate buffer (PBS, pH 7.4) at 37°C. After equilibration (approximately 5 days), an aliquot was centrifuged and filtered by using Advantec® DISMIC-13CP (0.20 μm). A portion of the sample was diluted and analyzed spectrophotometrically. The apparent 1:1 stability constant (Kc) was calculated from the initial linear portion of the phase solubility diagrams according to Eq. 2 [2].

$$Kc = \frac{\text{slope}}{\text{intercept} \cdot (1 - \text{slope})} \quad (2)$$

In Vitro Permeation Study: The permeation studies were carrierd out using the apparatus shown in Figure 1. The donor and acceptor phases (55 mL of PBS) were separated by a cellulose ester membrane (Spectra/Por® MWCO500) through which the guest molecules having molecular weights smaller than about 500 are permeable whereas CyDs are not permeable. The temperature of both phases was kept at 37°C, and the guests and CyDs in both phases were determined by HPLC.

3. Results and Discussion

The permeation process of a guest in the presence of CyD through a cellophane membrane was depicted in Figure 2, where only free guest molecule can transfer into the acceptor phase. Because CyDs were not detected in the acceptor phase (< 0.1%) under the experimental conditions, the complex formation occurs only in the donor phase. The Kc is defined by Eq. 3 assuming the 1:1 stoichiometry, and the total concentration, $[D]_0$, of the guest is expressed as Eq. 4, where $[D]_f$ and $[D]_A$ are concentrations of free guest in the donor and guest in the acceptor, respectively. Then, $[D]_f$ is written as Eq. 5.

$$Kc = \frac{[D]_0 - [D]_f - [D]_A}{[D]_f \cdot [CyD]_f} \quad (3)$$

$$[D]_0 = [D]_A + [D]_f + [complex] \quad (4)$$

$$[D]_f = \frac{[D]_0 - [D]_A}{1 + Kc \cdot [CyD]_f} \quad (5)$$

The permeation rate equation is expressed as Eq. 6 where k is permeation rate constant. Substitution of Eq. 5 into Eq. 6 yeilds Eq. 7 which is integrated into Eq. 8.

$$\frac{d[D]_A}{dt} = k \cdot [D]_f - k \cdot [D]_A \quad (6)$$

$$\frac{d[D]_A}{dt} = k \cdot \frac{[D]_0 - (2 + Kc \cdot [CyD]_f) \cdot [D]_A}{1 + Kc \cdot [CyD]_f} \quad (7)$$

$$[D]_A = \frac{[D]_0}{2 + Kc \cdot [CyD]_f} \cdot (1 - e^{-A \cdot t}) \quad (8)$$

$$\text{where; } A = \frac{k \cdot (2 + Kc \cdot [CyD]_f)}{1 + Kc \cdot [CyD]_f}$$

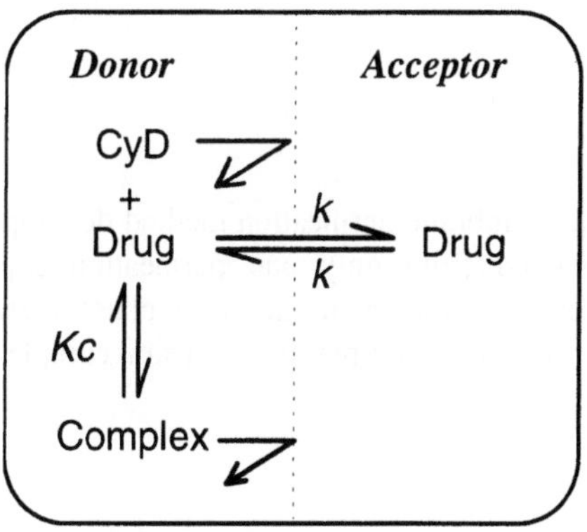

Figure 2. Schematic Representation for Permeation of Drug with CyD through a Cellophane Membrane (MWCO 500)

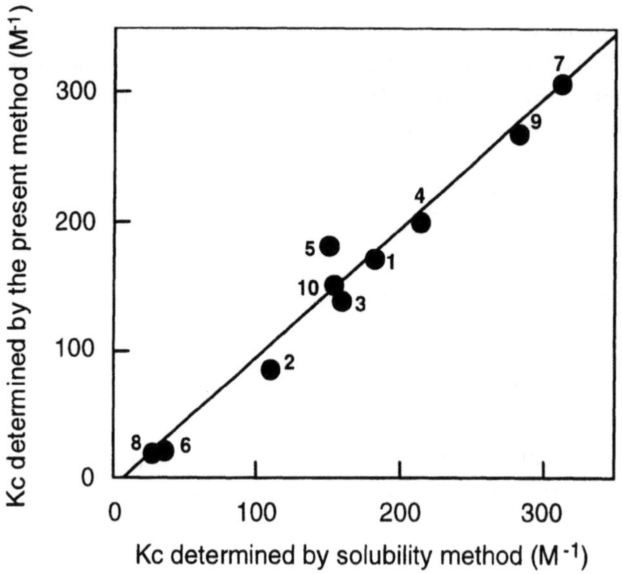

Figure 3. Relationship between Stability Constants (Kc) of Benzoic Acids/β-CyD Complexes Determined by Membrane Permeation Method and Those Determined by Solubility Method in Isotonic Phosphate Buffer (pH 7.4) at 37°C

Therefore, the unknown parameters, k and Kc, can be calculated by applying nonlinear least-squares method to Eq. 8, assuming that $[CyD]_f$ is equal to $[CyD]_{total}$. As shown in Figure 3, the Kc values for benzoic acid derivatives determined by the present method were in good agreement with those determined by the solubility method. For example, the Kc value of phenacetin/β-CyD complex determined by the present method was 172 M^{-1}, while those determined by the solubility and kinetic methods were 182 M^{-1} and 183 M^{-1}, respectively.

4. Conclusion

The main advantage of the membrane permeation method developed here was that the Kc value can be easily obtained from only one permeation experiment, by analyzing quantitatively the drug concentration in the acceptor phase *versus* time. Furthermore, this method was pertinent to analyze competitive inclusion complexation.

Reference

[1] Hirayama F. and Uekama K. (1987) Methods of Investigating and Preparing Inclusion Compounds, in Duchene, D.(ed.), *Cyclodextrins and Their Industrial Uses*, Editions de Sante, Paris, pp.131-172.

[2] Higuchi, T. and Connors, K.A. (1965) Phase-solubility Techniques, in Reilly, C.(ed.), *Adv. Anal. Chem. Instr.*, Wiley Interscience, New York, pp.117-212.

EFFECTS OF HYDROPHILIC CYCLODEXTRINS ON AGGREGATION OF RECOMBINANT HUMAN GROWTH HORMONE

S. TAJIRI, T. TAHARA, K. TOKIHIRO, T. IRIE, and K. UEKAMA

Faculty of Pharmaceutical Sciences, Kumamoto University, 5-1, Oe-honmachi, Kumamoto 862-0973, Japan

1. INTRODUCTION

Our previous studies have shown that hydrophilic cyclodextrins (CyDs) including 6-*O*-maltosyl-β-CyD (G2-β-CyD) and 2-hydroxypropyl-β-CyD (HP-β-CyD) significantly inhibit the adsorption of bovine insulin to hydrophobic surfaces of containers and its aggregation by interacting with accessible hydrophobic side chains of the peptide [1]. Following up these studies, the present contribution deals with the effects of CyDs on the tertiary structure of recombinant human growth hormone (hGH), a globular protein with a molecular mass of ~22 kDa and a pI near 5.3, in aqueous solution. Recent studies on equilibrium denaturation of hGH have demonstrated the presence of self-associated intermediates, leading to precipitation [2]. Thus, we address how hydrophilic CyDs impact upon the aggregation of hGH during refolding from its intermediates produced by a denaturant guanidium hydrochloride (GuHCl).

2. MATERIALS AND METHODS

2.1. Materials

hGH (2.96 IU/mg) was a gift from Novo Nordisk Pharma Co. Ltd. (Tokyo, Japan). G2-α-CyD, G2-β-CyD and 6-*O*-glucosyl-β-CyD (G1-β-CyD) were donated by Ensuiko Sugar Refining Co. Ltd. (Yokohama, Japan). HP-α-CyD, HP-β-CyD, and HP-γ-CyD with average degrees of substitution of 4.1, 4.8, and 4.8, respectively, were donated by Nihon Shokuhin Kako Co. Ltd. (Tokyo, Japan). Maltohexaose were purchased from Sigma Chemical Co. (St. Louis, MO, USA). Glucose, maltose, Tween 20, Tween 40, Tween 80, and GuHCl were purchased from Nacalai Tesque, Inc. (Kyoto, Japan).

2.2. Methods

The electrospray ionization mass spectra of hGH solutions were measured by a M-1200H LC/MS system (Hitachi, Tokyo, Japan). To evaluate the effects of additives on aggregation of hGH during the refolding, the protein with or without additives was first equilibrated for 12 h in phosphate buffer (pH 5.0) containing 4.5 M GuHCl at 25 ℃, at which the protein exists as molten-globule-like intermediates. Each sample was then diluted 20 times by adding the buffer solution with or without additives and was equilibrated at 25 ℃ for 4 h. Insoluble aggregates were monitored by an increase in absorbance of the solution at 350 nm. After filtration through a 0.22 nm membrane filter (Nihon Millipore, Japan), the concentrations of intact hGH and its dimeric form in the filtrate were analyzed by a size exclusion HPLC (TSK-GEL G3000SW, Tokyo, Japan). The circular dichroism (CD) experiments were performed on a Jasco J-40S recording spectropolarimeter (Tokyo, Japan) at 25 ℃.

3. RESULTS AND DISCUSSION

3.1. Interaction of the native hGH with CyDs

ESI mass spectrum of hGH in an acidic medium gave a broad charge distribution. The higher charge state distribution was observed for hGH with CyDs, suggesting less compact conformation of the protein. Furthermore, in the presence of CyDs, new signals were observed, which correspond to the 1:1 and 1:2 adducts of the ionized hGH with CyDs. Fluorescence and CD spectroscopies revealed that CyDs changed the tertiary structure of hGH but had no effect on the secondary structure of the protein, showing a molten globule-like state. ^{1}H-Nuclear magnetic resonance signals of tyrosine residues in the hGH molecule were largely shifted, with a rise in the CyD concentrations, indicating that CyDs reduce the intramolecular hydrophobic interaction of the protein.

3.2. Effects of CyDs and various additives on the aggregation of hGH upon refolding

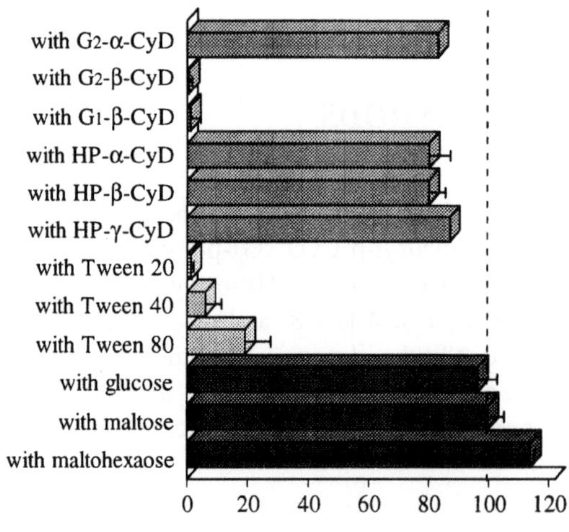

Figure. 1. Effects of Additives (50 mM) on Aggregation of hGH during Refolding from Molten Globule-like Intermediates[a)] in Phosphate Buffer (pH 5.0) Containing 4.5 M GuHCl at 25 ℃

Each point represents the mean ± S.E. of 2-5 experiments.
a) The initial concentration of hGH was 200 μM.

When hGH was refolded from a 4.5 M GuHCl solution, insoluble aggregates were formed, as indiciated by an increase in absorbance of the solution at 350 nm. As shown in Figure 1, G1-β-CyD and G2-β-CyD, as well as Tweens, significantly reduced the aggregation of hGH during refolding from the molten globule-like intermediates, while G2-α-CyD, HP-CyDs, and non-cyclic saccharides showed no noticeable inhibitory effect. This indicates that the β-CyD cavity with a branched sugar moiety is most preferable to prevent the aggregation of hGH. In the presence of G2-β-CyD, the analysis of the hGH retained in solution after refolding also revealed that the percent of monomer was the highest of the hydrophilic CyDs tested (Figure 2). Weak inetractions of CyDs with the molten globule intermediate of hGH may enhance the solubility of the protein by masking the exposed hydrophobic resisudes, therby possibly assisting the refolding of the proetin.

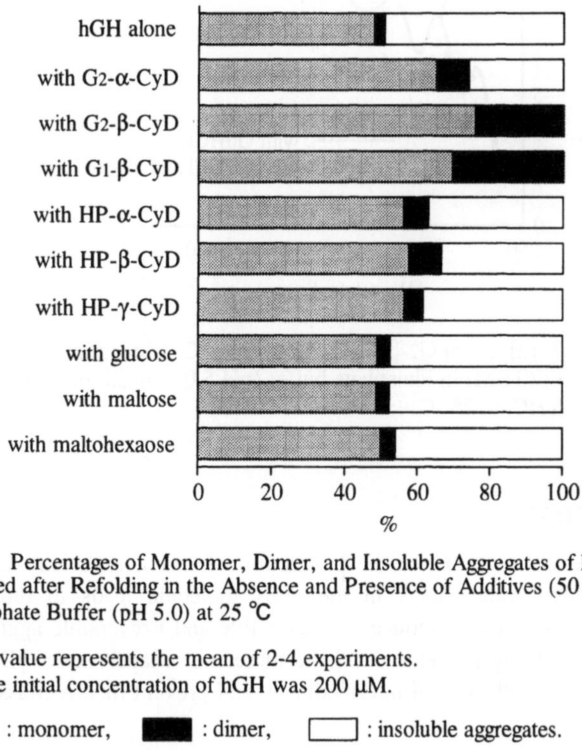

Figure. 2. Percentages of Monomer, Dimer, and Insoluble Aggregates of hGH[a)] Formed after Refolding in the Absence and Presence of Additives (50 mM) in Phosphate Buffer (pH 5.0) at 25 °C

Each value represents the mean of 2-4 experiments.
a) The initial concentration of hGH was 200 μM.

▒ : monomer, ■ : dimer, □ : insoluble aggregates.

When G2-β-CyD was added to the hGH solution containing 4.5 M GuHCl, the far-UV·CD spectrum of the solution did not change, indicating that G2-β-CyD does not affect the secondary structure of the protein (data not shown). However, upon the addition of G2-β-CyD, the near-UV·CD spectrum of the molten globule-like intermediates of hGH changed significantly, with an increase in the intensity of the vibrational bands in the 260-280 nm region reflecting a change in the aromatic moieties, phenylalanine and tyrosine. On the other hand, the addition of HP-CyDs resulted in an insignificant change in the near-UV·CD spectrum of hGH. This indicates that G2-β-CyD interacts with hydrophobic aromatic amino acid residues of the hGH intermediates, and thus prevents the aggregation by eliminating intermolecular hydrophobic contacts. Furthermore, the addition of G2-β-CyD to the hGH solution resulted in substantial decrease in the 300 nm band, which is diagnostic of the associated equilibrium intermediates.

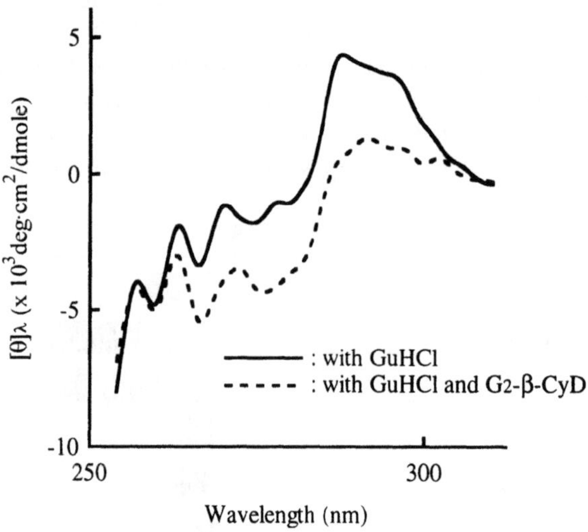

Figure. 3 Effects of G2-β-CyD (50 mM) on CD Spectrum of hGH (20 μM) in Phosphate buffer (pH 5.0) Containing 4.5 M GuHCl at 25 °C

4. CONCLUSION

The present results suggest that the interaction of CyDs, especially G2-β-CyD, with accessible hydrophobic side chains in the hGH molecule leads to the less compact conformation of the protein and reduces its reversible and irreversible aggregation during the unfolding and refolding processes. Like molecular chaperones, CyDs may provide surfaces to stabilize partially folded intermediates, and prevent incorrect association of the protein.

5. REFERENCES

[1] Tokihiro, K., Irie, T., Uekama, K., Pitha, J. (1995) Potential use of maltosyl-β-cyclodextrin for inhibition of insulin self-association in aqueous solution, *Pharm. Sci.*, **1**, 49-53.

[2] Bam, N. B., Cleland, J. L., and Randolph, T. W. (1996) Molten globule intermediate of recombinant human growth hormone: stabilization with surfactants, *Biotechnol. Prog.*, **12**, 801-809.

PREPARATION AND PHARMACEUTICAL EVALUATION OF HEPTAKIS(2,6-DI-*O*-METHYL-3-*O*-ACETYL)-β-CYCLODEXTRIN

S. MIEDA, F. HIRAYAMA, and K. UEKAMA
Faculty of Pharmaceutical Sciences, Kumamoto University,
5-1 Oe-honmachi, Kumamoto 862-0973, Japan

1. Introduction

Many kinds of cyclodextrin (CyD) derivatives have been prepared to extend the inclusion capacity and to improve the pharmaceutical properties of parent CyDs. Among the chemically modified CyDs, methylated CyDs have received considerable attention, because their physicochemical properties are largely changed and the inclusion behavior is magnified. For example, heptakis(2,3-di-*O*-methyl)-β-CyD (DM-β-CyD) is extremely soluble both in water and in organic solvents, less hygroscopic than the parent CyD, and highly surface active, and its aqueous solubility is inversely proportional to temperature increase. However, DM-β-CyD exhibits rather high hemolytic activity and local tissue irritancy due to the high membrane interacting ability. In this study, we report the preparation of heptakis(2,6-di-*O*-methyl-3-*O*-acetyl)-β-CyD (DMA-β-CyD) in which all remaining hydroxyl groups at the 3-position of DM-β-CyD are acetylated. Furthermore, the inclusion and pharmaceutical properties such as solubilization and hemolytic activities were investigated in comparison with DM-β-CyD, heptakis(2,3,6-tri-*O*-methyl)-β-CyD (TM-β-CyD) and parent β-CyD.

2. Experimental

Materials: β-, DM-β- and TM-β-CyDs were donated by Nihon Shokuhin Kako Co. Ltd. (Tokyo, Japan). DMA-β-CyD was prepared by acetylating the remaining hydroxyl groups of DM-β-CyD by acetic anhydride in pyridine, purified by silica gel chromatography (eluent: $CHCl_3$ to $CHCl_3/CH_3OH$ (15:2)), and characterized by nuclear magnetic resonance and fast-atom bombardment (FAB) mass spectrometries, and elemental analysis. Analytical data: m.p. 113-117°C; R_f = 3.0 (TLC silica gel, eluent $CHCl_3$-CH_3OH 15:2); FAB mass [M+*m*-nitrobenzylalcohol-H]$^-$ m/z 1777; ^{1}H-NMR ($CDCl_3$) δ 5.16 (t, H3), 5.00 (d, H1), 3.91-3.87 (m, H5 and H6a), 3.89 (t, H4), 3.54 (d, H6b), 3.37 (s, 6-CH_3), 3.33 (s, 2-CH_3), 3.23-3.20 (dd, H2), 2.04 (s, 3-CH_3). All other materials and solvents were of analytical reagent grade.

Solubility studies: Solubility measurements were carried out according to Higuchi and Connors [1]. Excess amounts of guests were added to aqueous solutions containing various concentrations of CyDs and were shaken at 25± 0.5℃. After equilibration was attained, an aliquot was centrifuged and pipetted through a membrane filter (Advantec, DISMIC 3CP). A portion of sample was assayed by high performance liquid chromatography (HPLC). An apparent stability constant, *Kc*, was calculated from the initial straight line portion of phase solubility diagrams according to the following equation.

$$Kc = \text{slope}/\{\text{intercept} \cdot (1\text{-slope})\}$$

In the case of the Ap-type solubility diagram where the solubility of guests deviated positively from the initial straight line, the ascending curvature was analyzed according to the method of Higuchi and Kristiansen [2] to obtain the stability constants of 1:1 and 1:2 (guest : host) complexes.

Hemolysis Assays: Hemolytic activities of CyDs were assessed as described previously [3]. From freshly drawn citrated rabbit blood, erythrocytes were separated by centrifugation at 1000 x g for 5 min, washed three times with phosphate buffered saline (154 mM sodium chloride, 10 mM phosphate, pH 7.4) and resuspended in the buffer solution to give a hematocrit of 5 %. The cell suspension (0.2 mL) was added to the buffer solution (2 mL) containing CyDs. After 30 min-incubation at 37℃, the release of hemoglobin from the cells was measured spectrophotometrically at 543 nm.

3. Results and Discussion

3-1. CHARACTERIZATION OF β-CyD DERIVATIVES

Table 1. Some Physicochemical Properties of Dimethylacylated β-CyDs

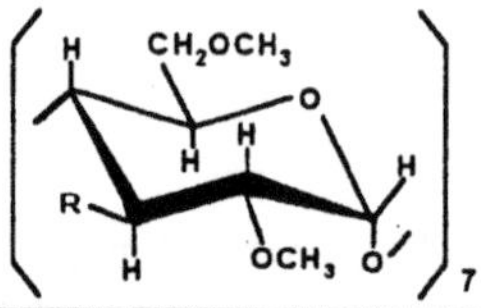

Compound	R	Melting point (℃)	$[\alpha]_D$ [a]	Solubility [b] (mg/dL)
DM-; *O*-Dimethyl-β-CyD	H	280	+159 [c]	57
DMA-; *O*-Dimethyl-*O*-acetyl-β-CyD	$COCH_3$	113-117	+126	>20
DMB-; *O*-Dimethyl-*O*-butyryl-β-CyD	COC_3H_7	108-111	+109	- [e]
DMO-; *O*-Dimethyl-*O*-octanoyl-β-CyD	COC_7H_{15}	- [d]	+96	- [e]

a) Specific rotation in chloroform. b) In water at 25℃. c) In water.
d) Oily substance. e) Could not be determined due to the low solubility.

Table 1 shows some physicochemical properties of dimethylacylated β-CyDs and DM-β-CyD. The melting point, $[\alpha]_D$, and solubility of dimethylacylated β-CyDs decreased with increasing alkyl chain length. The solubility of DMA-β-CyD in water at 25°C was > 20 %, the order being DM-β-CyD (57 %) > DMA-β-CyD (> 20 %) > TM-β-CyD (20 %) > β-CyD (1.8 %), and decreased with increase in temperature, in analogy to those of DM- and TM-β-CyDs.

3-2. SOLUBILIZATION

The inclusion complexation of DMA-β-CyD with various esters of *p*-hydroxybenzoic acid (parabens) having different lengths of the alkyl chain was investigated by the solubility method, and compared with those of DM- and TM-β-CyDs. The guest molecules having short alkyl chains such as methyl, ethyl and propyl groups exhibited typical A_L type phase solubility diagrams in the host concentration range of 0-0.02 M, whereas those having hexyl and octyl groups showed typical A_p type diagrams. Table 2 summarizes the Kc values calculated by the method described in the experimental section. The Kc value of the three host systems increased with increase in the alkyl chain length of guest molecule. The inclusion ability decreased generally in the order of DM-β-CyD > β-CyD > DMA-β-CyD≈ TM-β-CyD. The $K_{1:1}$ value of the octyl ester/DM-β-CyD complex was larger than the $K_{1:2}$ value, suggesting that DM-β-CyD has an ability high enough to include the guest within one host molecule. On the other hand, the $K_{1:2}$ values of the octyl ester/DMA- and TM-β-CyD complexes were much higher than those of the $K_{1:1}$ values, suggesting a cooperative inclusion effect of two host molecules.

Table 2. Stability Constants (M^{-1}) for Inclusion Complexes of various Parabens with β-CyDs in Water at 25℃

Paraben	DM-β-CyD		TM-β-CyD		DMA-β-CyD	
	$K_{1:1}$	$K_{1:2}$	$K_{1:1}$	$K_{1:2}$	$K_{1:1}$	$K_{1:2}$
Methyl Ester	230	-	63	-	35	-
Ethyl Ester	960	-	120	-	140	-
Propyl Ester	5600	-	300	-	470	-
Butyl Ester	slope>1	-	290	-	46	110
Hexyl Ester	3400	90	1000	40	110	120
Octyl Ester	31000	260	94	180000	70	66000
Phenyl Ester	7400	-	1100	-	280	-

3-3. HEMOLYSIS ACTIVITY

Figure 1 shows the hemolytic effect of four types of β-CyD derivatives toward rabbit erythrocytes in phosphate buffered saline (pH 7.4) for 30 min of incubation at 37℃. The hemolytic activity of DMA-β-CyD was significantly lower than those of β-CyD and methylated β-CyDs. No appreciable hemolysis and shape changes of erythrocytes were observed even at 0.1 M DMA-β-CyD solution. Moreover, the amount of cholesterol removed from erythrocytes by DMA-β-CyD was found to be smaller than that of β-CyD.

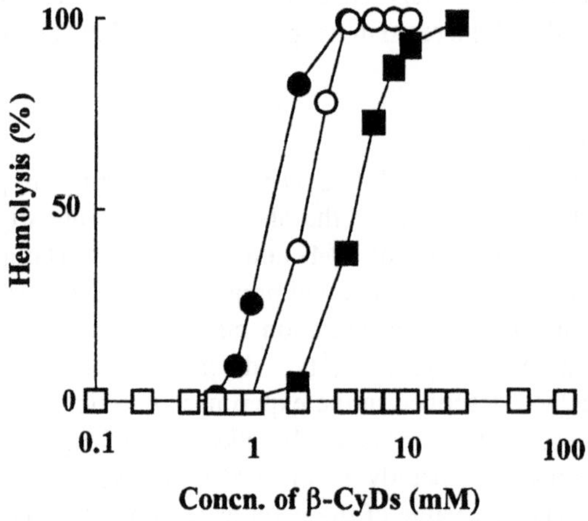

Figure 1. Hemolytic Effects of β-CyDs on Rabbit Erythrocytes in Isotonic Phosphate Buffer (pH 7.4) at 37°C

●: DM-β-CyD, ○: β-CyD, ■: TM-β-CyD, □: DMA-β-CyD.

4. Conclusion

DMA-β-CyD exhibited a low hemolytic activity compared with parent β-CyD and methylated β-CyDs, while it maintained certain inclusion ability comparable to TM-β-CyD. These results suggest that DMA-β-CyD may be useful as a parenteral drug carrier, and safer DM-β-CyD derivatives can be obtained, controlling the degree of substitution of 3-O-acetyl group.

5. References

1. Higuchi, T., and Connors, K. A., (1965) Phase-solubility techniques, *Adv. Anal. Chem. Instr.*, **4**, 117-212.

2. Higuchi, T., and Kristiansen, H., (1970) Binding specificity between small organic solutes in aqueous solution: classification of some solutes into two groups according to binding tendencies, *J. Pharm. Sci.*, **59**, 1601-1608

3. Ohtani, Y., Irie, T., Uekama, K., Fukunaga, K., and Pitha, J. (1989) Differential effects of α-, β- and γ-cyclodextrins on human erythrocytes, *Eur. J. Biochem.*, **186**, 17-22.

CHARACTERIZATION OF ITRACONAZOLE / 2-HYDROXYPROPYL-β-CYCLODEXTRIN INCLUSION COMPLEX IN AQUEOUS SOLUTION

Y. OKAMOTO, M. HIRANO, A. KONDO, K. MIYAKE[1], T. IRIE[1], F. HIRAYAMA[1] and K. UEKAMA[1]

Janssen-Kyowa Co., Ltd., 600-8, Minami-ishiki Nagaizumi-cho Sunto-gun, Shizuoka 411-0932, Japan, [1]Faculty of Pharmaceutical Sciences, Kumamoto University, 5-1, Oe-honmachi, Kumamoto 862-0973 Japan

1. Introduction

Itraconazole is an orally active triazole antifungal agent to inhibit most human fungal pathogens (Figure 1). This drug is practically insoluble in water and soluble only under extremely acidic conditions, leading to a poor oral bioavailability with large individual variation. An oral solution of itraconazole can be prepared using 2-hydroxypropyl-β-cyclodextrin (HP-β-CyD) and propylene glycol as solubilizing agents. The oral solution may have superior bioavailability characteristics and enables a more patient adjusted dose, compared to the oral capsules already on the market. The present contribution deals with the mode of inclusion complexation of itraconazole with HP-β-CyD to gain insight into the mechanism for the solubilization of the drug.

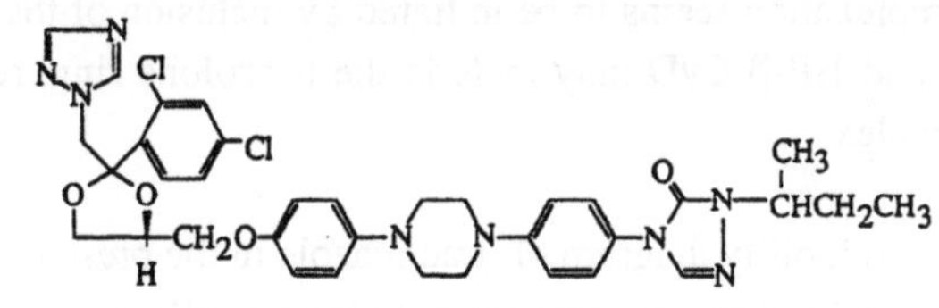

Figure 1. Chemical Structure of Itraconazole

2. Experimental

Nuclear Magnetic Resonance (NMR) Spectroscopy: The ^{1}H-NMR experiments were run using a JNM-α500 (JEOL, Japan) spectrometer operating at 500 MHz. Because of the limited solubility of itraconazole in aqueous solution, the drug and HP-β-CyD were dissolved in DMSO-d_6. The sample solutions were prepared at concentrations of 5 and 10 mM itraconazole and 10, 20 and 50 mM HP-β-CyD, and mixed in various ratios of the host and guest solutions.

Solubility studies: A constant but excess amount of itraconazole was added to acidic solutions (pH 2.0) containing HP-β-CyD at various concentrations, and shaken at 25°C. After equilibrium was attained, the mixtures were filtered and the filtrates were assayed for itraconazole by HPLC. The itraconazole:HP-β-CyD systems containing 10% v/v propylene glycol or all the ingredients of liquid drug preparation were analyzed for the drug in the same manner.

Ultraviolet (UV) Absorption Spectroscopy: The UV absorption spectra of itraconazole in the absence and presence of HP-β-CyD were measured in the acidic solution (pH 2.0, 25°C) containing 10% v/v propylene glycol. Stability constants of itraconazole:HP-β-CyD complex were determined using UV method.

Simulation of dilution process: Using stability constants ($K_{1:1}$ and $K_{1:2}$) determined from the solubility method, a simulation was run in order to estimate the present fractions of the free drug, the 1:1 complex and the 1:2 complex through the process until the liquid itraconazole formulation was diluted 100 times.

3. Results and Discussion

3.1. Stoichiometry in itraconazole:HP-β-CyD complex on NMR spectrum

Upon addition of HP-β-CyD, the proton signals of the triazole ring in the itraconazole molecule were largely shifted to upfield. In a two-dimensional nuclear Overhauser effect spectrum, the cross-peaks connecting the protons between HP-β-CyD and the triazole and triazolone rings in the itraconazole structure were observed. These results indicate that the complexation seems to be initiated by inclusion of the triazole ring into HP-β-CyD, and the second HP-β-CyD may include the triazolone ring, resulting in a 1:2 stoichiometry of the complex.

3.2. Solubility diagram of itraconazole in the presence of HP-β-CyD

The solubility of itraconazole in an acidic solution (pH 2.0) increased with a rise in the HP-β-CyD concentrations, showing a positive deviation from linearity (Figure 2). This solubility curve can be classified as type Ap suggesting the formation of higher-order complexes. The ascending curvature was quantitatively analyzed according to the optimization technique to obtain the stability constants of higher-order complexes ($K_{1:n}$) [1, 2]. As judged from Akaike's information criterion for non-linear regression equation, the 1:1 and 1:2 complexes of itraconazole with HP-β-CyD were assumed to have formed. The $K_{1:1}$ value was remarkably greater than the $K_{1:2}$ value. The addition of propylene glycol to the itraconazole:HP-β-CyD system decreased the $K_{1:1}$ value significantly, while the other ingredients involved in the liquid formulation of the drug did not affect the K values (Table I).

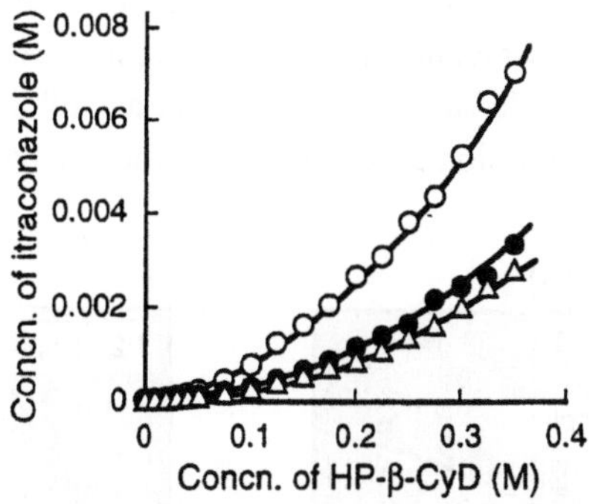

Figure 2. Phase Solubility Diagrams of Itraconazole:HP-β-CyD System in Acidic Solution (pH 2.0) at 25°C
O: without additives, ●: with 10% v/v propylene glycol, Δ: with all the additives.

Table I. Stability Constants of Itraconazole:HP-β-CyD Complex in Acidic Solution (pH 2.0) at 25°C, Determined by Solubility Method

System	S_0 (M)	$K_{1:1}$ (M^{-1})	$K_{1:2}$ (M^{-1})
without additives	1.73×10^{-7}	3350±270	90±35
with 10% v/v propylene glycol	9.94×10^{-7}	170±14	150±13
with all additives	1.81×10^{-6}	180±10	60±16

3.3. Stability constants by UV spectrum method

With increasing concentrations of HP-β-CyD, an absorption maximum of itraconazole at 270 nm was shifted to a longer wavelength with a concomitant increase in the molar absorption coefficient. Similar spectral changes were observed when itraconazole was dissolved in a less polar solvent such as methanol and ethanol, suggesting that the drug chromophore is incorporated into the hydrophobic environment of the HP-β-CyD cavity. The $K_{1:1}$ and $K_{1:2}$ values were calculated by analyzing the biphasic UV changes of itraconazole as a function of HP-β-CyD concentrations. The K values determined using UV method were nearly the same as those determined from the solubility diagram.

3.4. Effect of dilution on dissociation equilibrium of the complex

Using stability constants $K_{1:1}$ and $K_{1:2}$ determined by solubility method, a simulation was run concerning free drug, 1:1 complex and 1:2 complex through the process until the liquid itraconazole formulation was 100 times diluted (Figure 3). Since it appears a non-linear increase in solubility of itraconazole with increasing HP-β-CyD concentrations, precipitation should be considered in such systems at any dilution where the intrinsic drug solubility is lower than the dilution concentration line at a given HP-β-CyD concentration.

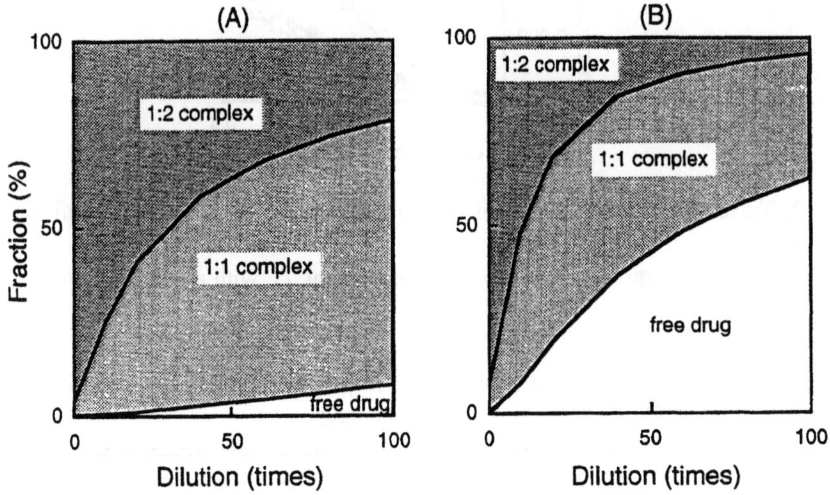

Figure 3. Changes in Fraction of Itraconazole and Its HP-β-CyD Complexes by Dilution of Itraconazole:HP-β-CyD Solution (A) and That Containing All the Additives (B)

4. Conclusion

The present results suggest as follows:

- The stoichiometry of itraconazole:HP-β-CyD complex is 1:2 in aqueous solution at pH 2.0, and the inclusion occurs at triazole and triazolone rings in itraconazole molecule.
- Among the ingredients of itraconazole oral solution, propylene glycol is mainly involved in the competitive inclusion complexation.

5. References

[1] Higuchi, T. and Kristiansen, H. (1970) Binding specificity between small organic solutes in aqueous solution: classification of some solutes into two groups according to binding tendencies. *J. Pharm. Sci.*, 59, 1601-1608.

[2] Uekama, K., Horiuchi, Y., Kikuchi, M., Hirayama, F., Ijitsu, T. and Ueno, M. (1988) Enhanced dissolution and oral bioavailability of α-tocopheriyl esters by dimethyl-β-cyclodextrin complexation. *J. Incl. Phenom.*, 6, 167-174.

SUSTAINED RELEASE AND INTESTINAL ABSORPTION OF DRUG FROM THE HYDROPHOBIC α-, β- AND γ-CYCLODEXTRIN COMPLEXES

KUNIO NAKANISHI, MASATOSHI NISHI, TOHRU MASUKAWA
TANEKAZU NADAI AND KOUICHIRO MIYAJIMA

Faculty of Pharmaceutical Sciences,Setsunan University, 45-1, Nagaotoge-cho Hirakata Osaka, 573-01. Osaka of University Pharmaceutical Sciences., Nasahara, Takatuki Osaka 569-11, Japan*

ABSTRACT

Hydrophobic cyclodextrin derivatives, peracylated α-, β- and γ-cyclodextrin (CyD) with different alkyl chains (acetyl; butanoyl and propanoyl), were used to form a complex with flufenamic acid (FA). The hydrophobic CyD complex formation was demonstrated by differential scanning calorimetry and powder X-ray diffractometry. The release rate of FA from the hydrophobic CyD derivatives were divided into two groups in phosphate buffer pH 6.8, but the release rate in all complexes was significantly retarded compared to that of the FA mixture. An increased mean residence time of FA following the hydrophobic CyD complex administration was observed. These results indicate that the hydrophobic CyD complex may serve as a hydrophobic carrier in sustained-release preparations of FA for oral and transdermal application.

1. INTRODUCTION

Hydrophobic cyclodextrin (CyD) derivatives as slow release carriers for drugs have been employed in attempt to improve the release rate of the drugs.[1,2] The aim of this study was to evaluate the pharmaceutical applications of the hydrophobic CyDs as a sustained-release carrier for drug, compared to the use of glucose as a non-complex excipient. We investigated the drug release behavior from peracylated α-, β- and γ-CyD with different alkyl chains and flufenamic acid (FA) complexes, and the absorption of FA after intestinal administration of the complexes. We also investigated the effect of bile or bile salt on the absorption of FA after intraduodenal administration of the drug-hydrophobic CyD complex.

2. MATERIALS AND METHODS

2.1 Materials

TA-β-CyD (molar substitution 3, purity > 99.9%) was donated by Ensuiko Sugar Refining Co. (Japan). FA was purchased from Wako Pure Chemicals Co. (Japan). Peracetyl α- and γ-CyD, perpropanoyl and perbutanoyl α-, β–and γ-CyD were prepared acylating all hydroxy groups of CyD, using corresponding acid anhydrides in pyridine solution. The identification of these compounds were characterized by NMR and FAB-MS. The inclusion structure of FA and TA-β-CyD complex in various ratios of methanol/H_2O solution was confirmed by ^{19}F-NMR (GE Omega 600).

2.2 Preparation of Solid Complexes

Complexes of FA with the hydrophobic CyD at molar ratio (1:2) were prepared by a kneading method using ethanol as a solvent.

2.3 Differential Scanning Calorimetry (DSC) and Powder X-Ray Diffraction

Complex formation was studied by DSC, using a Shimadzu DSC-50 (Shimadzu Corp.) with DSC crimp cell (sample size 2 mg, heating rate 5 °C/min). The powder X-ray diffraction pattern was determined with a MXP 3VA diffractometer (MAC Science Co.Ltd.).

2.4 In Vitro Release Study

The release rate of drug from the hydrophobic CyD complexes in isotonic phosphate buffers pH 6.8 with or without sodium cholate was measured by the dispersed amount method. The released of FA *in vitro* release experiments was determined spectrophotometrically at 235 nm.

2.5 Absorption Experiment

The absorption experiments were carried out with male Wistar rats (240-260g). The small intestine was exposed, and the duodenal segment was cut, and silicon tubing was inserted. The FA and CyD mixture or the CyD complexes (equivalent to 2.81 mg FA) was administered directly into the intraduodenal lumen with a syringe through the silicone, the blood samples were collected from the femoral artery.

The effect of endogenous bile or bile salt was examined in bile duct-ligated rats. FA in plasma was measured by HPLC according to the method of Dusci and Hacket.[3]

3. RESULTS AND DISCUSSION

3.1 Interaction Behavior of the Hydrophobic CyD Complex in the Solid State

The DSC thermogram of the FA and the physical mixtures showed an endothermic melting peak around 127°C, corresponding to the melting peak of the free FA, and the mixture (1:2) was broadened. The hydrophobic complex showed no endothermic peaks,

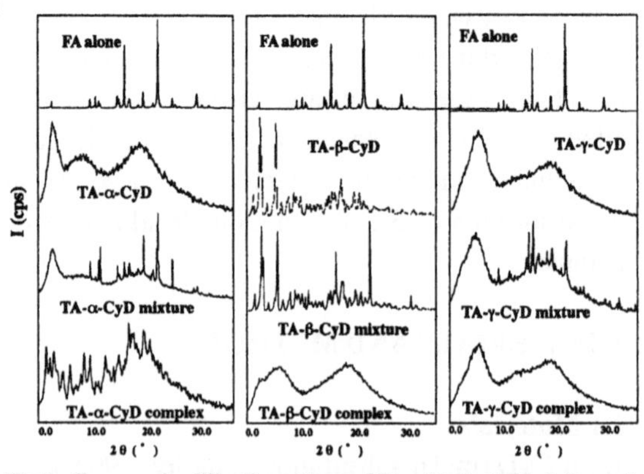

Fig.1 X-ray powder diffraction patterens of FA and hydrophobic CyD complexes

due to the melting of both components. Further, the diffraction peaks of FA also disappeared in all complexes. Figure 1 shows the powder X ray diffraction patterns of FA alone and the hydrophobic CyD complexes. The diffraction patterns of FA and the physical mixture simply reflected the superposition of each of the constituents, that is, sharp peaks due to the drug and CyD were observed. On the other hand, the diffraction peaks of FA, particularly at 2θ=14° and 24°, disappeared on TA-β-CyD and TA-γ-CyD complex formation, but TA-α-CyD complex showed somewhat different peaks compared with TA-β-CyD and TA-γ-CyD complexes.

3.2 Release Rate of FA from the Hydrophobic CyD Complexes

The release rate of FA from the drug-glucose mixture was very fast, due to the high solubility in phosphate buffer pH 6.8, as shown in Fig.2. On the other hand, the release rate of FA from the hydrophobic CyD complexes were divided into two groups, as a relatively faster group (TA-α-CyD, TA-β-CyD, TB-β-CyD, TB-γ-CyD) and a slower group (TP-α-CyD, TB-α-CyD TP-β-CyD, TA-γ-CyD). Sodium cholate enhanced the release rate of the drug in all complexes. The release of drug from TP-α-CyD, TB-α-CyD TP-β-CyD, TA-γ-CyD complexes were less than 10% until 8h.

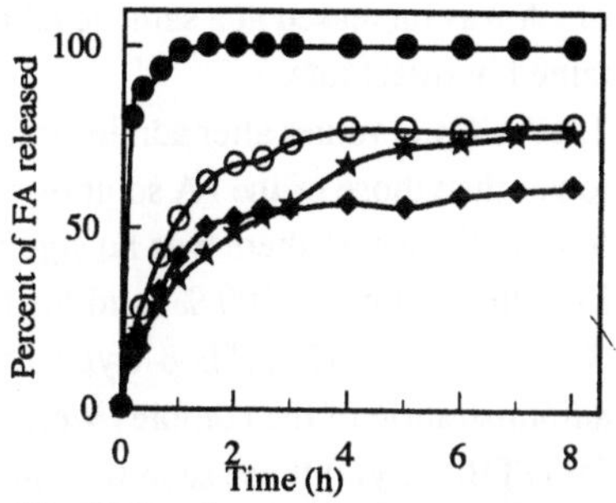

Fig.2 Dissolution profiles of FA from its hydrophobic CyD Complexes
●:FA alone, O:TA-α-CyD, ★ :TA-β-CyD, ◆:TB-β-CyD

3.3 Absorption Experiments

The plasma concentration of FA versus time curves obtained after the intraduodenal administration of powder containing either the FA mixture or the CyD complexes (equivalent to 2.81 mg of FA) to rats is shown in Fig. 3. When the equivalent doses of FA were administered the FA mixture and the CyD complexes, the intestinal absorption of the FA mixture was very fast. On the other hand, the blood level was influenced by administration of the CyD complex used. TA-α-CyD and TA-β-CyD complex did not show a sharp peak plasma concentration compared with the FA mixture, and pro-

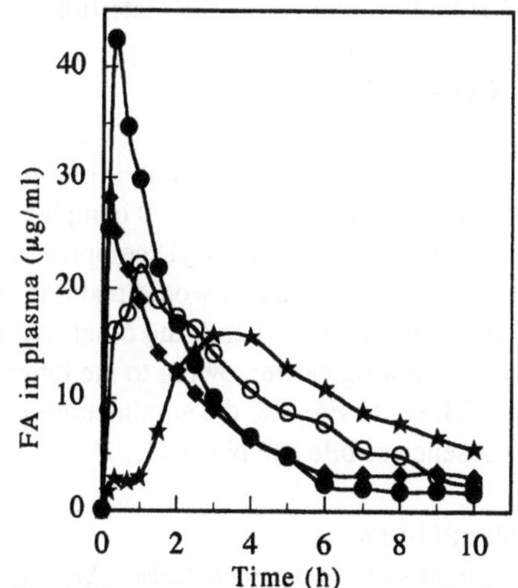

Fig.3 Plasma level time curves of FA after dose of its hydrophobic CyD complexes
●:FA alone, O:TA-α-CyD, ★ :TA-β-CyD, ◆:TB-β-CyD

duced a prolonged plateau plasma level of FA for 6-8 h.

Figure 4 shows the plasma levels of FA after administration of TA-β-CyD complex to bile duct-ligated rats. The plasma level of FA from TA-β-CyD complex was significantly decreased than that of intact rats. When TA-β-CyD complex was administered with sodium cholate in bile duct-ligated rats, the plasma level of FA was increased at a same level obtained in intact rats.

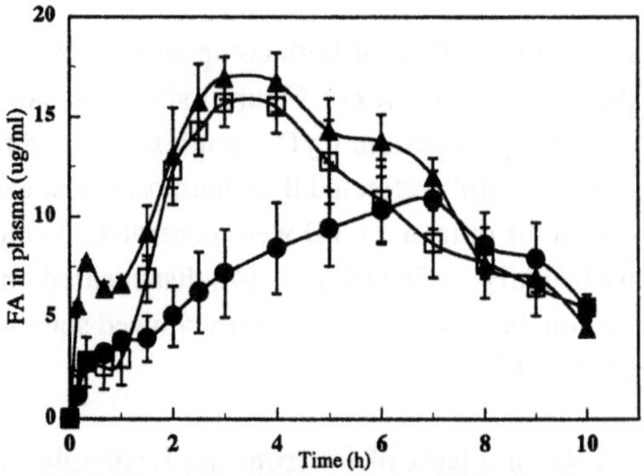

Fig.4 Plasma level time curves of FA after dose of TA-β-CyD complexes
□:with bile, ●:without bile, ▲:with 10mM sodium cholate

The AUC_{0-10} values after administration of the complexes in the powder form were slightly lower than those of the FA solution and the FA mixture following intraduodenal administration. However, there was no significant difference in absolute bioavailability between i.v. administration (100 %) and the FA mixture (88%) or the complexes (TA-α-CyD; 93 %, TA-β-CyD; 92 %,TB-β-CyD; 73 %). The MRT (mean residence time) values after administration of the complex were 1.5-fold (TA-α-CyD), 2.2-fold (TA-β-CyD) and 1.3-fold (TB-β-CyD) those seen with intraduodenal administration of the FA solution.
There was significant relationship between MDT (mean release time) of the complex *in vitro* experiments and MRT obtained *in vivo* experiments. Thus, the increase in MRT value after the intraduodenal administration of the CyD complex was due to the sustained release of the drug from the complex in the intestinal lumen, indicating that the release of FA from the complex was the rate-limiting step.

4. CONCLUSION

The present study show that an initial high plasma peak concentration does not occur after the administration of CyD complexes, suggesting that the side effects may also be reduced. The form of the CyD complex used here appears to be appropriate for practical clinical applications that would result in reduced side effects of the drug and sustained action. On the other hand, the other hydrophobic CyD complexes may be useful as a parenteral drug carrier, owing to the longer sustained release property. Furthermore, the blood level of FA after the administration the complexes was influenced in the presence or absence of bile and bile acid.

REFERENCES

[1] Nakanishi K., Masukawa T., Nadai T., Yoshii K., Okada S., Miyajima K., Sustained release of flufenamic acid from a drug-triacetyl-β-cyclodextrin complex, Biol.Pharm.Bull., **20**, 66-70 (1997)

[2] Uekama K., Horikawa T., Yamanaka M., Hirayama F., Peracylated β-cyclodextrins as novel sustained-release carrier for water-soluble drug, molsidomine, *J.Pharm.Pharmacol.* **46**, 714 -717 (1994).

[3] Dusci L.J. and Hackett L.P., Determination of some antiinflammatory drugs in serum by high-performance liquid chromatography, *J.Chromatogr.*, **172**, 516-519 (1979).

HYDROXYPROPYL GAMMA CYCLODEXTRIN AS A SOLUBILISER AND DISSOLUTION ENHANCING AGENT: THE CASE OF TOLBUTAMIDE—A POORLY WATER-SOLUBLE DRUG

MARÍA-DOLORES VEIGA AND FAKHRUL AHSAN
Departamento de Farmacia y Tecnología Farmacéutica, Facultad de Farmacia, Universidad Complutense de Madrid, 28040-Madrid, Spain

1. Introduction

Hydroxypropyl cyclodextrins are one of the most freely soluble cyclodextrin derivatives. These cyclodextrins have been found very useful in parenteral preparations [1] and in solid dosage forms [2] because of their non-toxic nature, complete and rapid dissolution in water. Hydroxypropyl gamma cyclodextrin can effectively encapsulate a number of bulky drug molecules due to its large cavity diameter. Tolbutamide is a poorly water-soluble hypoglycaemic agent; it forms inclusion compounds with b-cyclodextrin [3,4] and hydroxypropyl b-cyclodextrin [5]. In this paper attempt has been made to know if the solubility and dissolution of this drug become affected due to the presence of hydroxypropyl gamma cyclodextrin both in kneaded systems and in physical mixtures.

2. Materials and Methods

2.1. MATERIALS

Tolbutamide (TBM) was purchased from Sigma Chemical (St. Louis MO, USA) and Hydroxypropyl g-cyclodextrin (HPGCD) was kindly supplied by Wackers-Chemie GmbH (München, Germany).

2.2. METHODS

2.2.1. *Phase solubility study*

A phase solubility study was performed according to the method reported by Higuchi and Connors [6]. An excess of tolbutamide was weighed out into a series of test tubes. A constant volume of demineralized water or aqueous solutions of cyclodextrin containing increasing concentration of 2-HPGCD (0.002-0.02 M) was added to each test tube. Test tubes were then closed and placed in an oscillating water bath, and solutions were brought to solubility equilibrium at room temperature (25°C), with constant shaking for 5 days. After attainment of equilibrium, the contents of the test tubes were filtered through Whatman filter paper (Type 42). The drug concentration in the filtered solutions was determined from the absorbance at 228 nm using a Beckman DU-6 spectrophotometer and three replicates have been made for each assay. To nullify the absorbance due to the presence of cyclodextrin, the apparatus was calibrated with the corresponding blank during each assay.

2.2.2. *Preparation of physical mixtures and binary systems*

Two types of binary systems were prepared: physical mixtures (PM) and kneaded systems (KS). In both binary systems, drug/2-hydroxypropylg-cyclodextrin ratios were 1:1 and 1:2 mol/mol. Physical mixtures were prepared by homogeneous intermingling of previously sieved and weighed drug and cyclodextrin in a suitable container. Kneaded systems were prepared from physical mixtures by adding a small volume of distilled water. After moistening, the resultant systems were kneaded to produce a homogeneous dispersion. Once a homogeneous slurry was obtained, samples were dried at 40C for 24 hours; all of the dried systems were crushed and sieved, and fractions smaller than 100mm was collected for further study.

2.2.3. *Dissolution Assay*

A Sotax AT-7 dissolution apparatus with paddles was employed to carry out all of the tests. The volume of the dissolution medium, experimental temperature and paddle speed were 1000 ml of distilled water, 37 ± 0.C, and 50 rpm, respectively. Previously powdered and sieved (particle size lower than 100 mm) samples (250 mg of tolbutamide or equivalent amount of binary systems) were used for all dissolution studies. The duration of assay was 3 hours and samples were withdrawn at measured time interval and filtered with a Whatman filter paper (Type 42). Dissolved drug was assayed at a wavelength of 228 nm in a Beckman DU-6 spectrophotometer and three replicates of each dissolution assay were carried out.

3. **Results and Discussion**

3.1. PHASE SOLUBILITY STUDY

A tolbutamide-2-hydroxypropyl-g-cyclodextrin phase solubility diagram is shown in figure 1. Phase solubility diagram (Fig. 1) indicates that drug solubility increases linearly in accord with the amount of cyclodextrin added; the diagram is of A_L type [6]. Recently Rajewski and Stella [7] described that when there is a linear increase in drug solubility with increasing cyclodextrin concentration, cyclodextrin complex of drug results from 1:1 mol/mol interaction. Accordingly, we can assume that a 1:1 mol/mol tolbutamide-2 hydroxypropyl-g-cyclodextrin inclusion compound was formed and the apparent stability constant of which can be calculated from the slope and intercept of the linear portion of the phase solubility diagram by using the equation [8]

$$K_{1:1} = \frac{\text{slope}}{\text{intercept } (1 - \text{slope})}.$$

The stability constant obtained was 12 M^{1}, which was a very low stability constant and indicate that the tolbutamide-2-hydroxypropyl-g-cyclodextrin inclusion compound was not sufficiently stable.

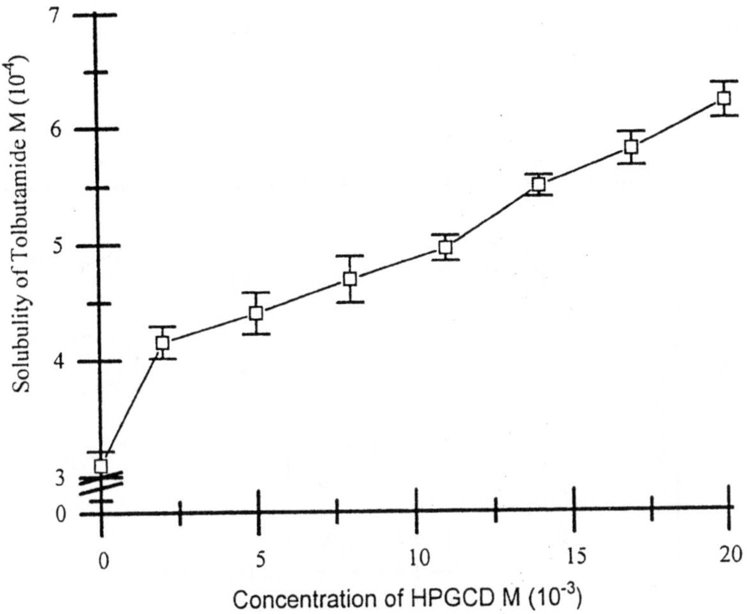

Figure 1: Tolbutamide-2-hydroxypropyl γ-cyclodextrin phase solubility diagram

3.2. DISSOLUTION ASSAY

The results of dissolution assay are presented in table 1 and figure 2. Both 1:1 and 1:2 kneaded systems showed significant improvement in dissolution in comparison to physical mixtures and drug molecule alone. The enhancement in drug dissolution efficiency from kneaded systems was three times as much as those from pure drug. (Table 1). Although within 30 minutes of assay, 125 mg of the drug, out of 250 mg, was dissolved, no further enhancement in the dissolution was observed.

However, the difference in the profiles of 1:1 and 1:2 systems was not significant (Fig. 2); the profiles from physical mixtures were almost similar to those obtained from the pure drug.

Table 1: Dissolution efficiency obtained from pure tolbutamide and tolbutamide/hydroxypropyl-γ-cyclodextrin binary systems.

Systems	Dissolution Efficiency (%)		
	30 minutes	90 minutes	180 minutes
Pure Tolbutamide	14.94	23.43	29.27
TBM/HPGCD Physical mixture 1:1	12.61	22.59	30.25
TBM/HPGCD Physical mixture 1:2	21.36	28.71	34.83
TBM/HPGCD Kneaded systems 1:1	42.05	47.56	48.85
TBM/HPGCD Kneaded systems 1:2	45.72	50.04	51.37

Thus, both the results of solubility study and dissolution test indicate that a tolbutamide-2-hydroxypropy-γ-cyclodextrin inclusion complex was formed. Results indicate that HPGCD can be used effectively to enhance TBM dissolution

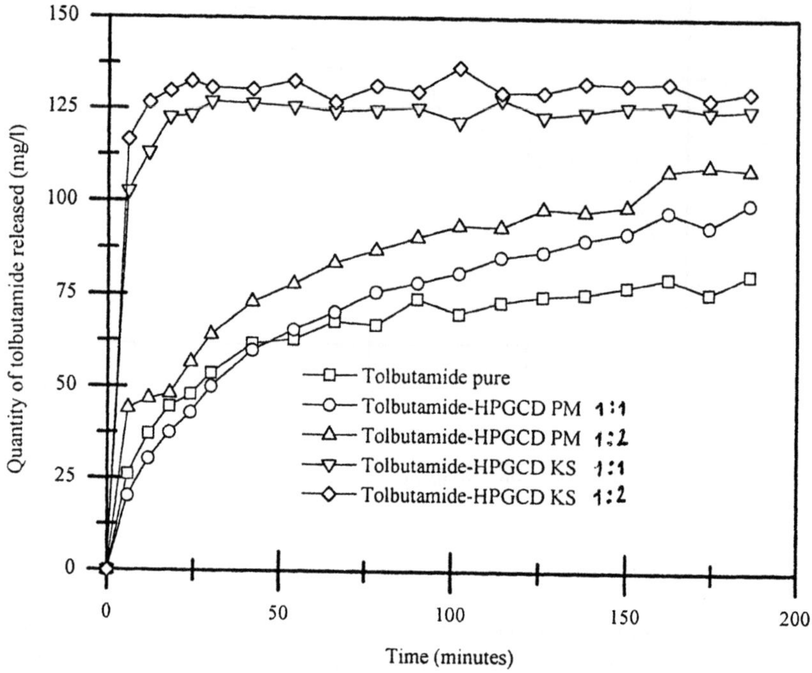

Figure 2: Dissolution profiles of tolbutamide-2-hydroxypropyl γ-cyclodextrin binary systems in water

4. References

1. Brewster, M.E., Simpkins, J.W., Hora, M.S., Stern, W.C. and Bodor, N. J. (1998) The potential uase of cyclodextrin in parenteral formulations. *Parenteral Sci. Technol.* **43**, 231-240.
2. Pitha, J., Harman, S.M. and Michel, M.E. (1986) Hydrophilic cyclodextrin derivatives enable effective oral administration of steroidal hormones. *J.Pharm.Sci.* 75, 165-167.
3. Gandhi, R.B. and Karara, A.H. (1988) Characterisation, dissolution and diffusion properties of tolbutamide-ß-cyclodextrin complex system. *Drug Dev. Ind. Pharm.* 14, 657-682.
4. Veiga, F., Teixeira-Dias, J.J.C., Kedzierewicz, F., Sousa, A. and Maincent, P. (1996) Inclusion complexation of tolbutamide with ß-cyclodextrin and hydroxypropyl ß-cyclodextrin. *Int. J. Pharm.* 129, 63-71.
5. Veiga, M.D. and Ahsan, F. (1998) Solubility study of tolbutamide in monocomponent and dicomponent solution of water. *Int.J. of Pharm.* 160, 143-149.
6. Higuchi, T. and Connors, A. (1965) Phase solubility techniques, in Reilly, C. (ed.) *Advances in Analytical Chemistry and Instrumentation,* Wiley Interscience Publishers, New York, pp. 117-212.
7. Rajewski, R.A. and Stella, V.J (1996) Pharmaceutical applications of cyclodextrin. 2. In vivo drug delivery. *J. Pharm. Sci.* 85, 1142-1169.
8. Thompson, D.O. (1997) Cyclodextrins—enabling excipients: Their present and future use in pharmaceuticals *CRC Crit. Rev. Ther. Drug Carrier Syst.* 14, 1-104.

INTERACTIONS OF SURFACTANTS WITH TOLBUTAMIDE-ß-CYCLODEXTRIN INCLUSION COMPOUND: THE CONSEQUENCE IN DRUG DISSOLUTION.

MARÍA-DOLORES VEIGA AND FAKHRUL AHSAN
Departamento de Farmacia y Tecnología Farmacéutica, Facultad de Farmacia, Universidad Complutense de Madrid, 28040-Madrid, Spain

1. Introduction

Cyclodextrins are now used extensively in all sorts of pharmaceutical formulations such as tablets, capsules, sachets, ointments and suppositories etc [1]. However, studies on the interactions between drug-cyclodextrin inclusion compound and other additives used conventionally to improve dissolution, wettability and solubility of drugs have not yet been undertaken. Use of surfactants in pharmaceutical formulations is a very common practice. The simultaneous presence of a drug and a surfactant in a formulation containing cyclodextrin can modify the release pattern of the drug from the drug-cyclodextrin inclusion compound. Tolbutamide, a poorly water-soluble drug, also forms inclusion complex with ß-cyclodextrin both in the solution [2] and in the solid state [3]. Recently we have shown that the formation of tolbutamide-ß-cyclodextrin inclusion complex was modified due to the presence of different surfactants in the complexing medium: sodium lauryl sulphate (SLS), polysorbate 20, and poloxyl 23-lauryl ether (POE-23) [4]. This paper was designed to study the influence of these three surfactants on the dissolution of a drug from a drug-cyclodextrin inclusion compound.

2. Materials and Methods

2.1. MATERIALS

Tolbutamide (TBM) was purchased from Sigma Chemicals Co. (St. Luis, MO. USA) and ß-cyclodextrin (ßCD) from Janssen (Olen, Belgium). Sodium lauryl sulphate (SLS) and polysorbate 20 and poloxyl 23-lauryl ether (POE-23) were supplied by Panreac (Barcelona, Spain).

2.2. METHODS

2.2.1. *Preparation of Binary Systems*

Two types of binary systems were prepared: physical mixtures (PM) and kneaded systems (KS). In both binary systems, drug/ß-cyclodextrin ratios were 1:1 (17% of tolbutamide and 83% of ßCD) and 1:2 mol/mol (9.30% of tolbutamide and 90.70% of ßCD). Physical mixtures were prepared by homogeneous intermingling of previously sieved and weighed drug and cyclodextrin in a suitable container. Kneaded systems (KS) were prepared from physical mixtures (PM) by adding a small volume of distilled water. After moistening, the resultant systems were kneaded to produce a homogeneous dispersion. Once a homogeneous slurry was obtained, samples were dried at 40°C for 24 hours; all of the dried systems were crushed and sieved, and fractions smaller than 100 µm was collected for further study.

2.2.2. *Dissolution Assay*

A Sotax AT-7 dissolution apparatus with paddles was employed to carry out all of the tests. The volume of the dissolution medium, experimental temperature and paddle speed were 1000 ml of distilled water or aqueous solution of surfactant, 37 ± 0.1°C, and 50 rpm, respectively. All of the systems were assayed both in distilled water and aqueous solution of surfactants. The amount of surfactants incorporated into the media was equimolecular to the amount of tolbutamide used in each study. Previously powdered and sieved (particle size lower than 100 µm) samples (250 mg of tolbutamide or equivalent amount of binary systems) were used for all dissolution studies. The duration of assay was 3 hours and samples were withdrawn at measured time interval and filtered with a Whatman filter paper (Type 42). Dissolved drug was assayed at a wavelength of 228 nm in a Beckman DU-6 spectrophotometer and three replicates of each dissolution assay were carried out.

3. Results and Discussion

Dissolution of the drug from the kneaded systems increased tremendously in comparison with that from the drug alone [5], however, the enhancement in dissolution from the physical mixtures was not significant. (Figures 1-4). In six minutes of assay, 200 mg of tolbutamide, out of 250 mg used in the assay, was dissolved from the 1:2 TBM-ßCD kneaded system (Figure 4).

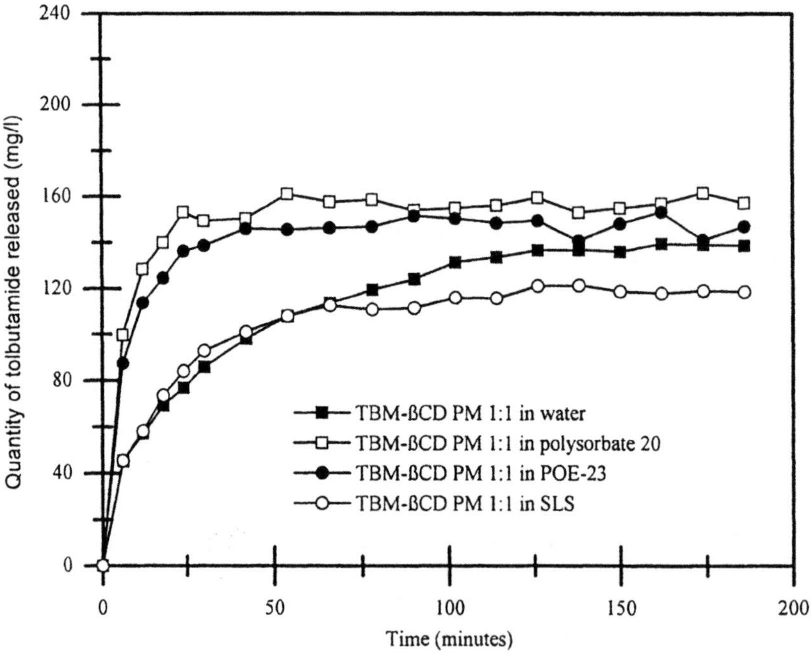

Figure 1: Dissolution profile of 1:1 TBM-ßCD physical mixture in water and in aqueous solution of surfactants

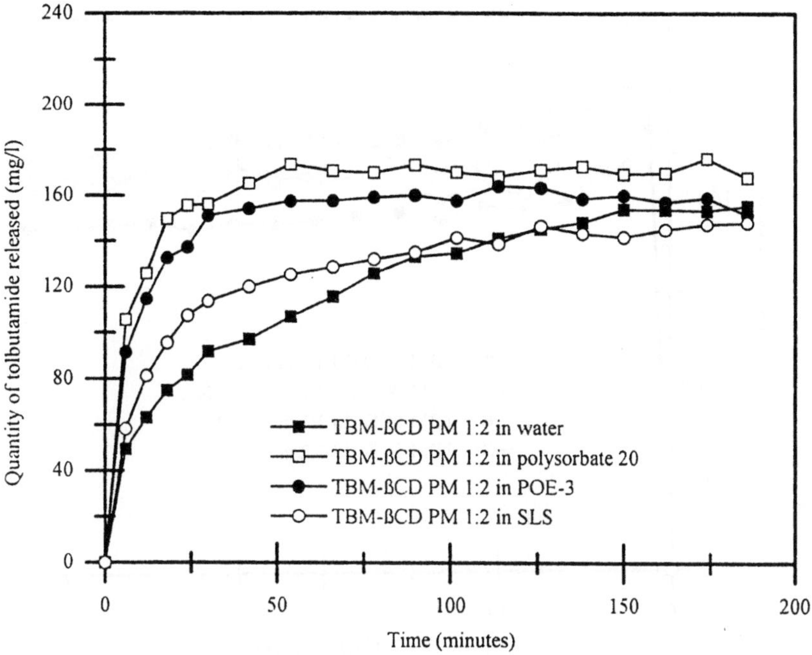

Figure 2: Dissolution profile of 1:2 TBM-ßCD physical mixture in water and in aqueous solution of surfactants

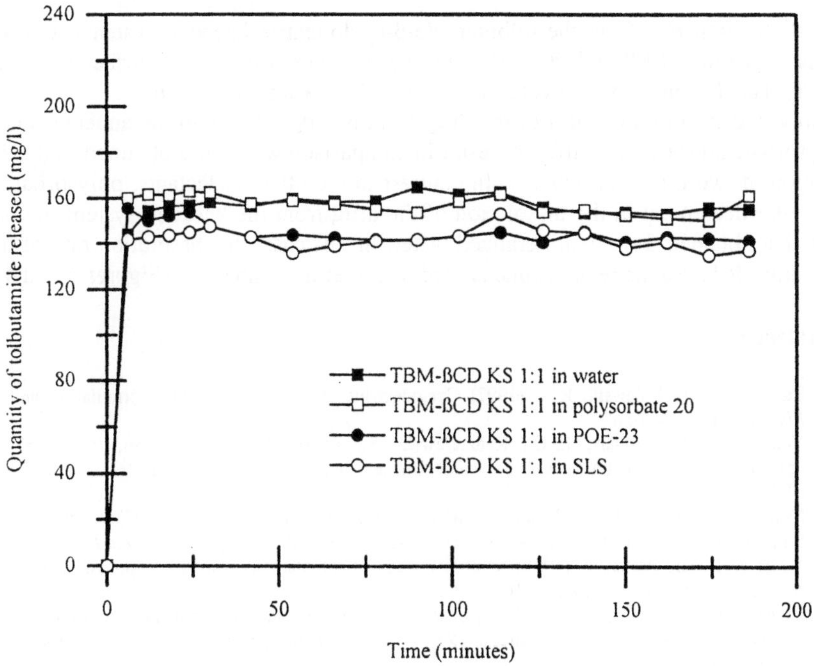

Figure 3: Dissolution profile of 1:1 TBM-ßCD kneaded system in water and in aqueous solution of surfactants

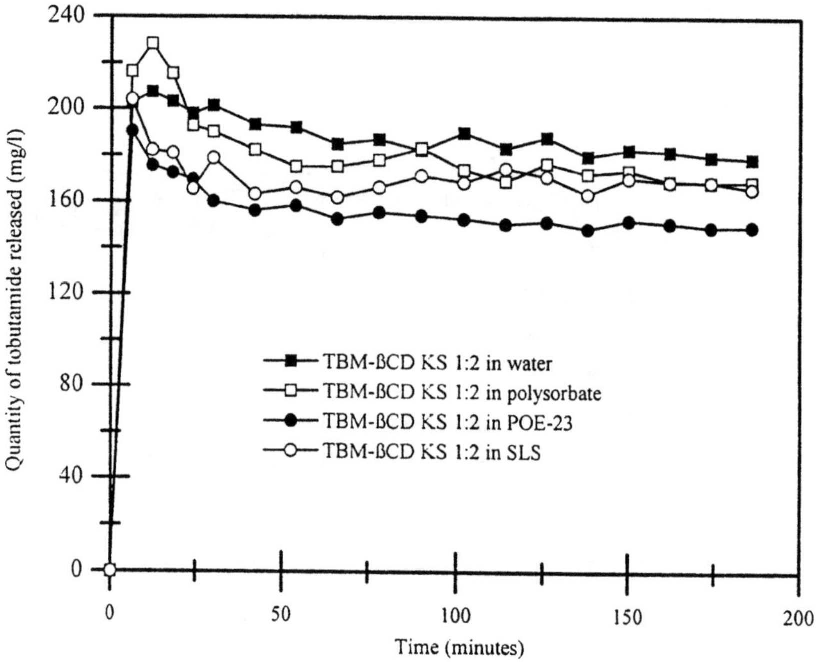

Figure 4: Dissolution profile of 1:2 TBM-ßCD kneaded system in water and in aqueous solution of surfactants

The dissolution rate from the tolbutamide-ß-cyclodextrin kneaded systems was lower in media containing SLS and POE-23 than that in water without any surfactants (Figures 3 and 4). The decrement was greater in the case of 1:2 kneaded system.

However, dissolution profiles of the drug from binary systems in the aqueous solution of polysorbate 20 showed a different result in comparison with those obtained in the aqueous solution of two other surfactants. In contrast to the other surfactants, polysorbate 20 did not cause decrement in the dissolution of the drug from the kneaded system in compared with that obtained from demineralized water. In this case, the dissolution rate of the drug from kneaded systems remains unchanged or sometimes enhanced (Figures 3 and 4).

4. References

1. Stella, V.J. and Rajewski, R.A. (1997) Cyclodextrins: Their future in drug formulation and delivery. *Pharm. Res.* 14, 556-567.
2. Veiga, F., Teixeira-Dias, J.J.C., Kedzierewicz, F., Sousa, A. and Maincent, P. (1996) Inclusion complexation of tolbutamide with ß-cyclodextrin and hydroxypropyl ß-cyclodextrin. *Int. J. Pharm.* 129, 63-71.
3. Nozawa, Y., Nawa, H., Sadzuka, Y., Miyagishima, A. and Hirota, S. (1997) Mechanical complex formation of tolbutamide with ß-cyclodextrin by solid phase roll mixing. *Pharm. Acta Helv.* 72, 37-42.
4. Veiga, M.D. and Ahsan, F. (1998) Solubility study of tolbutamide in monocomponent and dicomponent solution of water. *Int.J. of Pharm.* 160, 143-149.
5. Veiga, M.D. and Ahsan, F. (1997) Study of some bicomponent and tricomponent solid dispersions of ß-CD and two other hydrophilic substances, *1997 Pharmaceutical applications of cyclodextrins conference*, Kansas, USA.

DIFFERENTIAL SCANNING CALORIMETRY AS AN ANALYTICAL TOOL IN DETERMINING THE INTERACTION BETWEEN DRUG AND CYCLODEXTRIN

MARÍA-DOLORES VEIGA, FAKHRUL AHSAN AND MANUELA MERINO
Departamento de Farmacia y Tecnología Farmacéutica, Facultad de Farmacia, Universidad Complutense de Madrid, 28040-Madrid, Spain

1. Introduction

Differential Scanning Calorimetry is a widely used analytical method in the study of multicomponent solid systems to reveal the possible changes during heating [1-3]. Its importance in the characterisation of pharmaceutical solids, polymeric drug delivery systems, usefulness in studying the compatibilities in solid dosage forms has been described in great detail [4,5]. DSC can provide a lot of information on drug/cyclodextrin interactions in the solid state [6]. The aim of this paper is to compare the thermal behaviours of different drug-cyclodextrin binary systems. Drugs selected for this work have been found to form inclusion compound with ß-cyclodextrin: Naproxen [7], Tolbutamide [8-11] and Chlorpropamide [12].

2. Materials and Methods

2.1. MATERIALS

Naproxen, Tolbutamide and Chlorpropamide were purchased from Sigma Chemicals Co. (St. Luis, MO. USA). All the materials were used without any further purification.

2.2. METHODS

2.2.1. *Preparation of binary systems*

Two types of binary systems were prepared: physical mixtures (PM) and kneaded systems (KS). In both binary systems, drug/ß-cyclodextrin ratios were 1:1 and 1:2 mol/mol. Physical mixtures were prepared by homogeneous intermingling of previously sieved and weighed drug and cyclodextrin in a suitable container. Kneaded systems were prepared from physical mixtures by adding a small volume of distilled water. After moistening, the resultant systems were kneaded to produce a homogeneous dispersion. Once a homogeneous slurry was obtained, samples were dried at 40°C for 24 hours; all of the dried systems were crushed and sieved, and fractions smaller than 100 μm was collected for further study.

2.2.2. *Differential Scanning Calorimetry*

Thermograms of pure materials and all binary systems were recorded on a Mettler TA 3000 differential scanning calorimeter (model DSC 20). About 10 mg of sample was placed in a pin-holed aluminum sample pan with lid, and heated at a rate of 10̄C/min in the range of 50-350¯C. The instrument was periodically calibrated with a standard sample of indium

3. RESULTS AND DISCUSSION

The DSC analysis of pure tolbutamide gives rise to a curve of three endotherms. All data are shown in Table 1. The first endotherm is for melting of the drug (Peak A) and the remaining two could indicate the decomposition of the melted drug. DSC trace of ß-CD shows two endothermic peaks corresponding to dehydration (Peak A) and fusion (Peak B). DSC curves of TBM-ßCD physical mixtures show dehydration from ß-CD and fusion of tolbutamide at 121-122°C (Peak A), followed by an endothermic process that does not coincide with ß-CD fusion (Table 1, Peak B). Thermogram of TBM/ß-CD kneaded system 1:1 shows dehydration and fusion at 125.8°C (Peak A). In addition, there is an endothermic peak between 240°C and 300°C corresponding to the melting of the system (Peak B). Thermogram of TMB-ßCD KS 1:2, show also two endothermic peaks at 108.0°C and 264.0°C. In these systems a true inclusion compound was formed, because melting peak of TBM was disappeared and a considerable diminution of ß-CD fusion temperature was noticed.

Table 1: DSC data from Tolbutamide (TBM) and TßM/ß-CD binary systems

Systems	Peak A			Peak B		
	Range (°C)	Peak (°C)	Enthalpy J/g^{-1}	Range (°C)	Peak (°C)	Enthalpy J/g^{-1}
TBM	114.7-137.0	125.7	88.80	--	--	--
ß-CD	60-160	119.3	267	240.340	304.7	402.19
TBM/ß-CD PM 1:1	60-160	122.1	215.93	240-340	250.9	208.25
TBM/ß-CD PM 1:2	60-160	121.5	231.8	240-300	255.4	305.41
TBM/ß-CD KS 1:1	60-160	125.8	130.05	240-300	255.1	253.93
TBM/ß-CD KS 1:2	60-160	108	133.29	240-300	264	246.60

Only one endotherm is observed in the DSC scan of Naproxen. DSC curves of all Naproxen-ßCD (NPX-ßCD) binary systems show all peaks corresponding to pure components: dehydration endotherm, melting of the drug and melting of ßCD (Table 2). This indicates that in the solid state there was no interaction between drug and cyclodextrin. Thermograms corresponding to Naproxen-ß-cyclodextrin binary systems exhibit slight differences between the physical mixtures and kneaded systems.

Table 2: DSC data from Naproxen (NPX) and NPX/ß-CD binary systems

Systems	Peak A			Peak B			Peak C		
	Range (°C)	Peak (°C)	Enthalpy J/g^{-1}	Range (°C)	Peak (°C)	Enthalpy J/g^{-1}	Range (°C)	Peak (°C)	Enthalpy J/g^{-1}
NPX	141.3-165.3	152.6	136.18	--	--	--	--	--	--
ß-CD	60-150.7	119.4	257.02	--	--	--	240-340	304.6	658.19
NPX/ß-CD PM 1:1	40-142.7	111.8	249.00	150-158.5	155.3	22.05	280-330	310.9	294.85
NPXR/ß-CD PM 1:2	42.3-131.7	107.3	247.13	150-160	155.3	19.78	280-330	311.8	375.62
NPX/ß-CD KS 1:1	42.3-145	118.6	231.39	150-165.7	155.8	10.14	260-336	309.4	35.31
NPX/ß-CD KS 1:2	50.3-140	116	233.42	130-165	156.1	13.83	260-336	309.7	435.58

Binary systems prepared with CLP and ß-CD exhibit DSC curves with three endothermic peaks: dehydration (Peak A) and reorganisation (Peak B) of ß-CD , and fusion (Peak C)of the possible inclusion compound upon heating (Table 3). DSC traces of both physical mixtures and kneaded systems were identical.

Table 3: DSC data from Chlorpropamide (CLP) and CLP/ß-CD binary systems

Systems	Peak A			Peak B			Peak C		
	Range (°C)	Peak (°C)	Enthalpy J/g^{-1}	Range (°C)	Peak (°C)	Enthalpy J/g^{-1}	Range (°C)	Peak (°C)	Enthalpy J/g^{-1}
CLP	110-136.3	172.2	91.21	160-280	242.2	296.3	280-340	330.8	131.28
ßCD	44.0-157.3	123.5	281.55	216-232	223.6	0.49	280-334.7	311.1	336.7
CLP/ß-CD PM 1:1	40.2-140	116.8	259.39	194.3-213.3	204.7	3.55	250-290	265.7	244.73
CLP/ß-CD PM 1:2	44.3-150	119.0	248.67	196.3-215.7	205	1.75	250-290	268.6	172.72
CLP/ß-CD KS 1:1	46.3-142.2	117.8	252.06	201-220	213.5	0.75	250-290	268.4	211.65
CLP/ß-CD KS 1:2	54.3-140	117.6	220.5	197.3-216.7	206.7	2.79	250-290	269.8	268.83

4. References:

1. Guillory, J.K., Hwang, S.C. and Lach, J. L. (1969) Interactions between pharmaceutical compounds by thermal methods. *J. Pharm. Sci.* 58, 301-308
2. Ghiron, D. (1986) Application of thermal analysis in the pharmaceutical industry. *J. Pharm. Biomed. Anal.*. 4, 755-770.
3. Ghiron-Forest, D., Goldbronn, C. and Piechon, P. (1989) Thermal analysis methods for pharmacopoeial materials. *J Pharm. Biomed. Anal.* 7, 1421-1433.
4. Ford, J. L. and Timmins, P. (1989) Pharmaceutical thermal analysis: Techniques and applications. Ellis Horword, Chichester.
5. McCauley, J.A. and Brittain, H.G. (1995) Thermal methods of analysis, in Britain H. G. (ed.), Physical characterisation of pharmaceutical solids. Mercel Dekker, New York, pp. 223-251.
6. Veiga, M.D., Díaz, P.J. and Ahsan, F. (1998) Interactions of griseofulvin with cyclodextrins in solid binary systems. *J. Pharm. Sci.* 87, 891-900.
7. Otero-Espinar, F.J., Anguiano-Igea, S., García-Ganzález, N., Vila-Jato, J.L., Blanco-Méndez, J. (1992) Study of the interaction of naproxen with ß-cyclodextrin in solution and in solid state. *Int. J. Pharm.* 79, 149-157.
8. Gandhi, R.B. and Karara, A.H. (1988) Characterisation, dissolution and diffusion properties of tolbutamide-ß-cyclodextrin complex system. *Drug Dev. Ind. Pharm.* 14, 657-682.
9. Veiga, F., Teixeira-Dias, J.J.C., Kedzierewicz, F., Sousa, A. and Maincent, P. (1996) Inclusion complexation of tolbutamide with ß-cyclodextrin and hydroxypropyl ß-cyclodextrin. *Int. J. Pharm.* 129, 63-71.
10. Nozawa, Y., Nawa, H., Sadzuka, Y., Miyagishima, A. and Hirota, S. (1997) Mechanical complex formation of tolbutamide with ß-cyclodextrin by solid phase roll mixing. *Pharm. Acta Helv.* 72, 37-42.
11. Veiga, M.D. and Ahsan, F. (1998) Solubility study of tolbutamide in monocomponent and dicomponent solutions of water. *Int. J. Pharm.* 160, 143-149.
12. Veiga M.D. and Ahsan, F. Unpublished data.

Acknowledgements

F. Ahsan gratefully acknowledges the Spanish Agency for International Co-operation for providing him with a full-bright scholarship to do his Ph.D. at the Complutense University of Madrid.

SOLUBILITY ENHANCER DECREASES THE DISSOLUTION OF COMPLEXED DRUGS: EFFECT OF SODIUM-LAURYL-SULFATE ON DISSOLUTION PROFILE OF COMPLEXED DRUGS

K. KLOKKERS[1], E. FENYVESI[2], L. SZENTE[2], AND J. SZEJTLI[2]

[1]*Hexal A. G. Industrie Str.25., Holzkirchen, Germany*

[2]*Cyclolab R&D Lab., Dombóvári út 5-7, Budapest, H-1117 Hungary*

1. Introduction

The present paper describes the effect of commonly used tabletting excipients on the biopharmaceutical performance of drug/CD complexes. It has long been known that certain pharmaceutical additives (e.g. surfactants) act as competitive complexants, thus can destroy the positive effects of the inclusion complexation [1,2].

Some recent publications have described the effect of surfactants and βCD when added to the dissolution media on the dissolution rate using mequitazine and tolbutamide as model drugs [3,4]. It has been found that the dissolution of the drugs was increased by both the surfactant (sodium lauryl sulfate, SLS) and the CD when added separately to the dissolution medium (binary systems). A decreased dissolution rate was observed, however, when both additives were present (ternary system). These facts suggest that SLS is a competitor for the βCD cavity.

Korean authors reported on the influence of SLS on dissolution rate of entrapped drug from suppository. They found the effect of surfactant positive on Omeprazol release from a HPβCD complex from rectal suppository [5].

This opposite (synergistic) effect can probably be in analogy of the thoroughly studied pyrene/βCD/surfactant systems. It was found that addition of surfactant (e.g. sodium lauryl sulfate) below the critical micelle concentration (c.m.c.), to the pyrene/βCD solution leads to an increase in the hydrophobicity of the environment of pyrene probably in consequence of ternary complex formation. In the presence of a surfactant, no solvent - solute interaction occurs between water and pyrene the latter one being completely wrapped by the CD and tenside together [6].

The effect of surfactants on the dissolution performance of a pre-formed β- or γCD complex has not been studied yet.

In the present work the effect of a frequently employed tabletting additive, sodium-lauryl-sulfate has been studied by registering in vitro dissolution profiles of drug/CD complex alone and in the presence of different amounts of sodium-lauryl-sulfate, applied in a concentration equivalent with that present in tablets. As model drugs Diclofenac-sodium and Doxazosin mesylate were chosen. Both β-cyclodextrin and γ-cyclodextrin complexes of these drugs were prepared and their dissolution behaviour studied.

2. Experimental

2.1. MATERIALS

Sodium lauryl sulfate (SLS) of analytical grade (Fluka) was used. The complexes were prepared according to the usual methods [7]. Their characteristics are listed in Table I.

TABLE I Properties of the complexes

	Doxazosin mesylate/γCD	Diclofenac Na/β-CD	Diclofenac Na/γ-CD
molar ratio	1:1	1:1	1:1
drug content (%)	30.0	21.5	19.3
Loss on drying (3 h at 80 °C) (%)	10	5.1	2.3
Preparation method	kneading	lyophilisation	kneading

2.2. MEASUREMENT OF THE DISSOLUTION RATE

1.675 g Doxazosin mesylate/γCD complex was added to 50 ml 0.1 N HCl solution (pH 1.25) and stirred with 275 rpm at 371.675 g complex was added to 50 ml 0.1 N HCl solution (pH 1.25) and stirred with 275 rpm at 37°C. Samples were withdrawn after 1, 5, 10, 30 and 60 min stirring. The samples were filtered through a membrane of 45 μm pore size and diluted with 1:1 ethanol - water mixture, then measured photometrically.

0.23 g Diclofenac/β- or γCD complex and was added to 20 mL 0.1 N HCl solution under stirring at ambient temperature. Samples were withdrawn after 2, 5, 10 and 30 min stirring. The samples were filtered through a membrane of 45 μm pore size and diluted 3 times with 1:1 ethanol - water mixture, then measured photometrically. The measurements were carried out similarly in the presence of sodium lauryl sulfate (SLS): 0.0125 and 0.0250 g SLS (corresponding to approximately 1:4 and 1:2 SLS to Diclofenac molar ratio) were dissolved in the dissolution medium before adding the complex.

3. Results

It has previously been found that doxazosin-mesylate/γCD complex in a tablet form shows only an insignificant or even negligible enhancement on in vitro dissolution studies, in other words the dissolution performance of tablets made from complex or plain drug are about the same. The binary doxazosin mesylate/γCD complex was found to meet the requirements concerning the dissolution of drug, i.e. an at least 80% extent of dissolution of drug at pH 1.3 from solid complex within five minutes was achieved. Adding SLS to the dissolution medium, however, the dissolution decreases (Fig. 1).

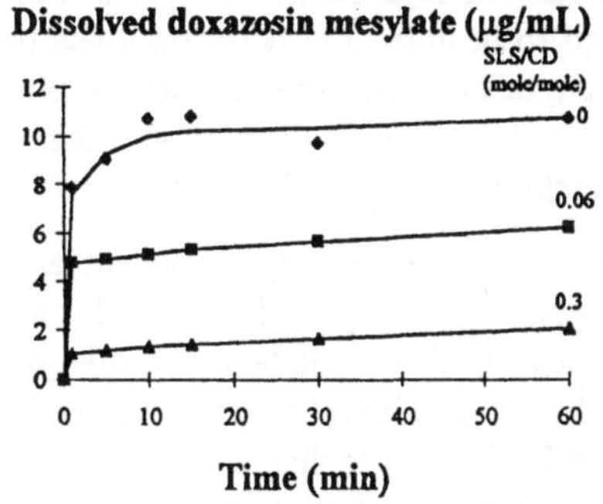

Fig. 1 Dissolution of doxazosin mesylate from its γCD complex in the presence and absence of SLS

The reduction of drug release on the effect of SLS was observed also in case of Diclofenac sodium/β- and γ-CD complexes (Fig. 2 and 3). The extent of this reduction was higher with the γCD complex than with the βCD complex. These results prove that the drug and the surfactant are competitors, and the dissolution of the drug depends on both the drug/CD and surfactant/CD (and probably on the drug/surfactant) interactions.

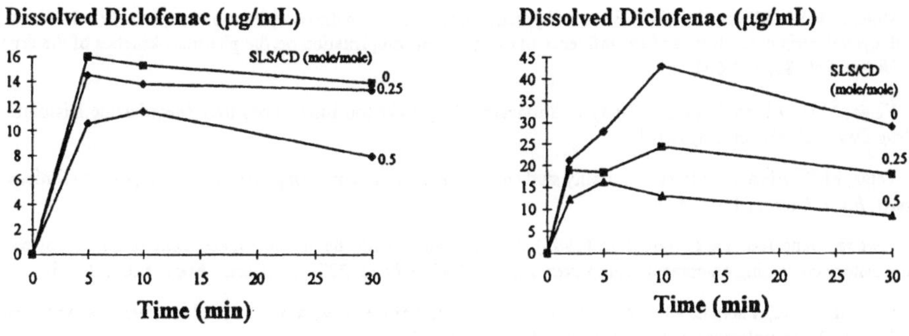

Fig. 2 Dissolution of Diclofenac from its β- *(left)* and γCD *(right)* complex in the presence and absence of SLS

Conclusion:

In case of Doxazosin mesylate/βCD complex even the presence of 1% of sodium-lauryl-sulfate additive (corresponding to as low as 0.06:1 surfactant/βCD molar ratio) in the formulation caused an about 40-50 % reduction of the released drug substance. Diclofenac-sodium/β- and γCD complexes were found to behave similarly, but the extent of decrease of drug release caused by sodium-lauryl-sulfate was less pronounced, than in case of Doxazosin.

The effect of surfactants on the drug release was found to depend both on the type of drug and applied cyclodextrin (although showed the competition and not the synergism). Further studies are required to be able to recommend those type of tabletting additives that do not affect dissolution, or even just contrary, that do affect this process. In certain cases it is desired to reduce the dramatic dissolution enhancement of CD-complexed drugs, to maintain pharmacokinetics of generic formulations.

There arises the possibility for tuning exactly the extent and rate of dissolution of a complexed drug by appropriately dosing sodium lauryl sulfate or other competitive guests in the final pharmaceutical formulation. In this way we could reduce the extremely enhanced dissolution or bioavailability of cyclodextrin complexed generic drug down to a level, which is identical to that of the plain generic drug. (Note that this is of real practical importance from registration standpoint!)

ACKNOWLEDGEMENT

The technical assistance of Zs. Nagy and Zs. Simon is greatly acknowledged.

REFERENCES

1. Kraus, C., Mehnert, W., Frömming, K.H. (1991) Interactions of β-cyclodextrin with Solutol HS 15 and their influence on Diazepam, *PZ Wiss.*, 4, 11-15

2. Mueller, B.W., Albers, E. (1991) Effect of hydrotropic substances on the complexation of sparingly soluble drugs with cyclodextrin derivatives and the influence of cyclodextrin complexation on the pharmacokinetics of the drugs, *J. Pharm. Sci.*, 80, 599-604

3. Veiga, M.D., Ahsan, F. (1997) Study of Surfactants/β-Cyclodextrin Interactions over Mequitazine Dissolution, *Drug Dev. Ind. Pharm.*, 23, 721-725

4. Veiga, M.D., Ahsan, F. (1998) Solubility study of Tolbutamide in monocomponent and dicomponent solutions in water, *Int. J. Pharm.*, 160, 43-49

5. Hwang, Sung-Joo, Park, Sung,Bae (1995) A comparative study on the pharmaceutical properties of rectal suppository containing omeprazole complexes, *Yakche Hakhoechi* 25. 227-237.(Chem. Abstr. 120:226676)

6. Edwards, H. E., Thomas, J.K., A fluorescence-probe study of the interaction of cycloheptaamylose with arenes and amphiphillic molecules, *Carbohydr. Res.*, 1978, 65, 173-82

7. Szente, L. (1996) Preparation of cyclodextrin complexes, in Szejtli, J., Osa, T. (ed.) *Comprehensive Supramolecular Chemistry*, Volume 3, 243-252. Elsevier, Oxford, UK

INTERACTION BETWEEN RANITIDINE HYDROCHLORIDE AND β-CYCLODEXTRIN

László **JICSINSZKY**, Ilona **KOLBE**

CYCLOLAB R&D. Lab. Ltd., Budapest, H-1525 Budapest, P. 0. Box: 435, Hungary

SUMMARY

Stability constants of the ranitidine hydrochloride/β-cyclodextrin (K=134 dm^3/mol) complex was determined in neutral D_20 solution at 30°C. The calculated relatively small stability constant for the complexes suggests a weak interaction between the β-cyclodextrin and the guest molecule. Calculations for the complex stability constants were performed only in cases of 1:1 complexes. Determinations were made in deuterium oxide with and without adding internal standard to the solutions. The loose interaction between the host and guest molecules obtained from molecular modeling studies is confirmed by experimental data.

Introduction

Complex stability constants of cyclodextrin complexes can be determined from the series of chemical shifts recorded in function of the molar ratio of cyclodextrin to the guest. Determination of characteristic chemical shifts and plotting them in a Job's-plot reflects the composition of the complex. In certain cases not only the molar ratio of the complex can be calculated but some assumptions for the binding site of both the cyclodextrin and guest can be made, as well. Appropriately chosen chemical shifts plotted against a transformed concentration value give linear relationship. The slope of the curve gives the stability constant of the complex[1].

In deuterium oxide the stability constants are somewhat different (about 20-30 % higher) from those which can be found in water [2]. In the present case due to the obtained small values it is not necessary to make corrections or confirm the results in water.

The internal standard (DSS) which is usually used in the ^{1}H-NMR experiments can form complexes with cyclodextrins in about the same order of magnitude as the ranitidine hydrochloride therefore the experiments were repeated without adding the standard. The observed chemical shifts in the tables are calculated from the HDO-peak. The HDO is formed from the small amount of absorbed water, the water content of the complex and from the exchangeable hydrogens of the studied molecules.

Experimental

^{1}H-NMR spectra were recorded on a *Varian VXR-400* spectrometer at 400 MHz. Samples of the guest and cyclodextrin were measured into the NMR-tube as dry materials and dissolved in D_20 at 30 °C with ultrasonication. The obtained clear solutions were measured at 30 °C with and without adding 2,2,3,3-tetradeutero-3-trimethylsilyl-propionic acid sodium salt (DSS) as internal reference. On the spectra the chemical shift scale is aligned to DSS but in tables the given values are referred to HDO (δ_{HDO}=4.725 from DSS) due to the lack of internal reference in the samples used in the determination of complex stability constant.

HyperCheM™ 5.1Pro [HyperCube Inc., Gainsville, Fl, USA] and Win-MGM™ 1.01d [Ab Initio Technologies SARL, Obemai France] were used in molecular modeling studies. *Geometry optimization:* MM+ force field, point charge, grad < 0.05, periodic box conditions (~28 Å switched cutoff), 100 Polak-Ribiere cycles then Newton-Raphson optimization (1500 cycles); *Molecular dynamics simulation:* MM+ force field, 650 explicit water molecules, periodic box conditions (~28 A switched cutoff), 20 ps cooling to constant temperature (0 K).

Results

Ranitidine hydrochloride (RAN*HCl) is very well soluble in water but its cyclodextrin complex is less soluble in water. All the solutions were homogeneous in the observed concentration range. On *Fig.* 1. the numbering of the Ranitidine carbons, in **Table 1** the used concentrations are indicated. On the proton spectrum in D_2O the C(10) proton is missing. This is the consequence of the proton-deuterium exchange through tautomeric forms as they are indicated on *Fig. 2.*

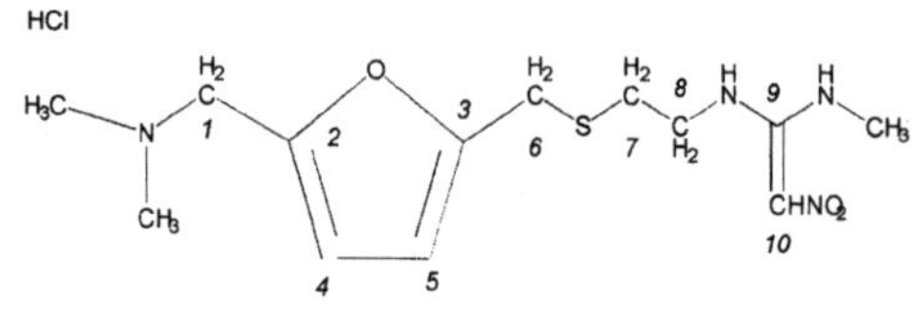

Fig. 1.: Numbering of Ranitidine for the Proton Assigninents

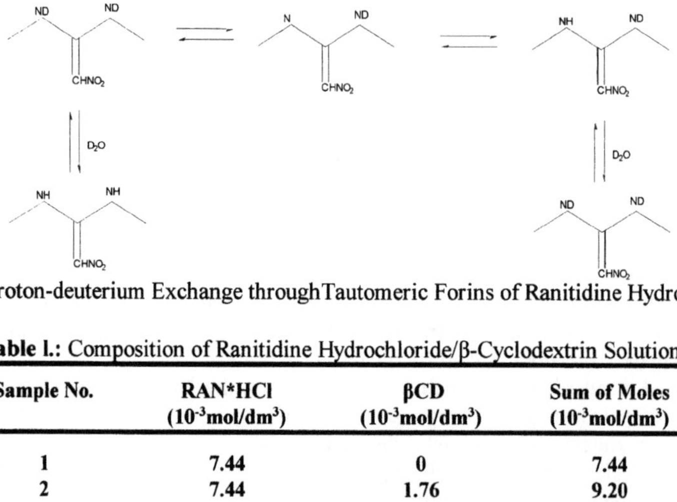

Fig. 2.: Proton-deuterium Exchange throughTautomeric Forins of Ranitidine Hydrochloride

Table l.: Composition of Ranitidine Hydrochloride/β-Cyclodextrin Solutions

Sample No.	RAN*HCl (10^{-3}mol/dm^3)	βCD (10^{-3}mol/dm^3)	Sum of Moles (10^{-3}mol/dm^3)
1	7.44	0	7.44
2	7.44	1.76	9.20
3	7.41	4.54	11.95
4	7.41	7.40	14.81
5	7.41	12.20	19.61
6	7.44	17.46	24.90
7	0	7.40	7.40

In Table 2. the differences of chemical shifts are shown. It is necessary to mention that both for the Job's-plot and the stability constant determinations only a part of data could be used. Interactions between cyclodextrin and guest are restricted only to several parts of the molecules.

Conclusions

From the Job's-plot the 2:1, 1:1, and 2:1 βCD/RAN*HCI complex composition can be also concluded. However, from the measured small differences of the proton chemical shifts the weak interaction between the βCD and RAN*HCI is obvious. The calculated 2:1 and 1:2 complex ratio therefore is the consequence of the "second sphere" complexation, i. e. one RAN*HCI molecule establishes contact with two cyclodextrins and due to the loose interaction two RAN*HCI molecules can be also found in the proximity of one cyclodextrin (at different hydroxyl side).

Table 2.: Differences of the Chemical Shifts for Ranitidine Hydrochloride/βCD Complexes

Sample N°	1	2	3	4	5	6	7
Ranitidine HCl							
3-H	0	0.003	-0.001	0.003	-0.005	-0.012	-
4-H	0	0.003	0.001	0.000	0.000	-0.001	-
$2\text{-}CH_2$	0	0.005	0.004	0.014	0.011	0.007	-
NMe_2	0	0.004	0.008	0.018	0.021	0.023	-
$6\text{-}H_2$	0	0.001	0.001	0.001	-	-	-
$7\text{-}H_2$	0	0.001	-0.003	-0.005	-0.010	-0.013	-
$8\text{-}H_2$	-	-	-	-	-	-	-
NH-ME	0	0.001	-0.003	-0.005	-0.010	-0.014	
β-CD							
H-1	-	0.007	0.001	0.003	-0.001	-0.0020	0
H-2	-	0.012	-0.005	0.000	-0.007	-0.0110	0
H-3	-	0.020	0.020	0.020	0.018	0.0150	0
H-4	-	0.003	-0.001	0.001	-0.004	-0.0050	0

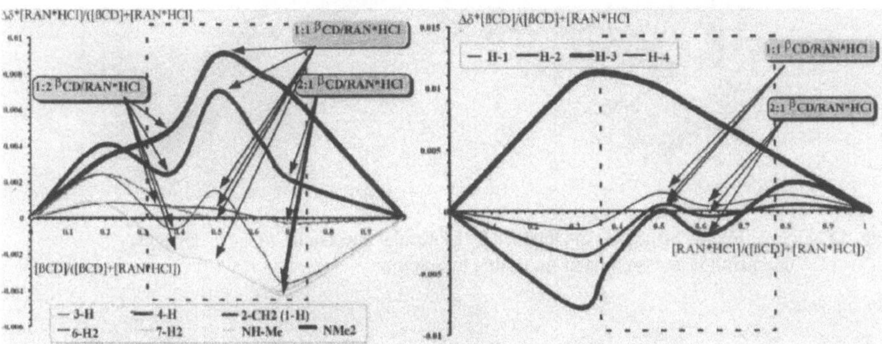

Fig. 3.: Job's-plot for Ranitidine Hydrochloride Protons and β-Cyclodextrin Protons

Another consequence of the weak interactions is that not all the ranitidine protons showed significant and evaluable variations. The number of the cyclodextrin,protons in similar positions of the glucopyranoside (Glcp) units is relatively large (the number of unchanged protons are 3-6 times larger than the changed protons), therefore the variations are rather irrelevant. The interference of the different shifts in the chemical shifts may also result in inutilizable variances what also means the lack of dominant complex ratio(s).

In *Figs. 3.* the significantly evaluable proton shift differences are incorporated into the Job's plot. Using the standard linear regression fit procedure[2] for the assumed 1:1 complex ratio the complex stability constant can be also calculated.

The stability constant for the ranitidine hydrochloride/β-cyclodextrin complexes in D_20 calculated from *Fig. 4.* is:

K= 134 dm^3/mol (D_2O); pK_D = 2.12 at 30°C

Non-linear regression for calculation the complex stability constants for 2:1 and 1:2 complexes is meaningless, due to the calculated low value.

Both molecular mechanics optimization and molecular dynamics simulations suggest that the RAN*HCI is located close to both primary and secondary hydroxyls but real inclusion does not occur. Although the molecular mechanics geometry optimization showed the secondary side second sphere "complexation" the MD simulation resulted arrangement of molecules, gave more complex structure than 1: 1 complex composition.

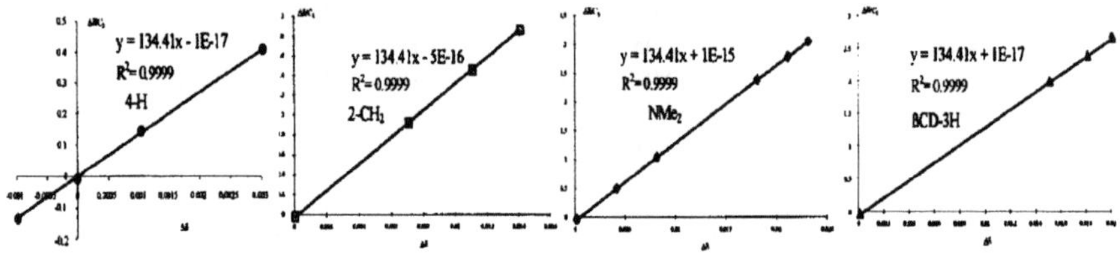

Fig. 4.: Determination of the Stability Constant for Ranitidine Hydrochloride/β-Cyclodextrin Complexes[1]

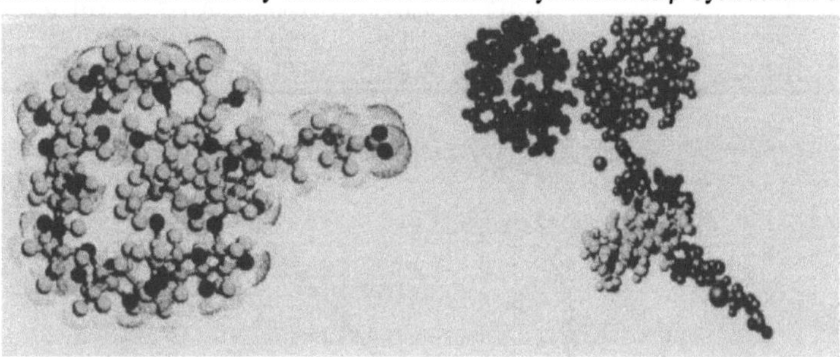

Fig. 5.: Minimal Energy Structures of Obtained by Molecular Mechanics *in vacuo* Geometry optimization and Simulated Annealing in Solution

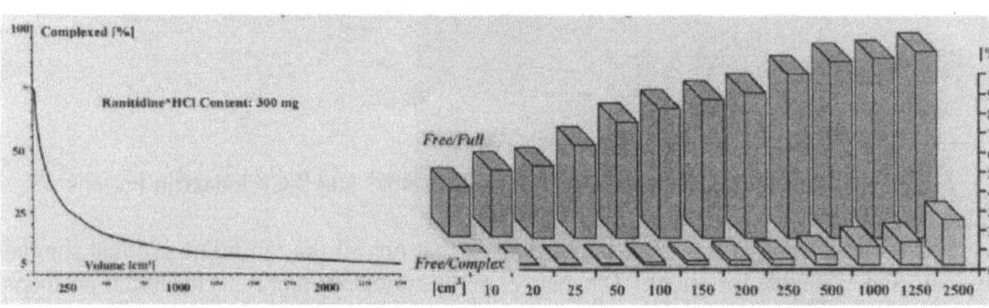

Fig. 6.: Composition of βCD/RAN*HCl/Water systems upon dilution

References

[1] Bekkers, O. *et al* (1991) Inclusion Complex Formation of ..., *J. Incl. Phenom. Mol Recogn.*, **11** 185-193

[2] Wang, S., Matsui, Y. (1994) Solvent isotope effect on .... *Bull. Chem. Soc. Japan,* **67**,2917-2920

PHYSICAL AND CHEMICAL CHANGES IN THE PROPERTIES OF β-CYCLODEXTRIN ON INCLUSION COMPLEX FORMATION

Ágnes Buvári-Barcza and Lajos Barcza
L. Eötvös University, Institute of Inorganic and Analytical Chemistry,
H-1518 Budapest, Hungary

In spite of the fact that cyclodextrins are mentioned often as enzyme mimicing agents (being really very good and relatively simple model molecules) and the changes in physical and chemical properties of guests (solubility, volatility, light-sensitivity, etc.) during the complex formation are of primary interest for practical purposes (and mainly these changes are investigated and utilized), hardly anything is known about the the changes in reactivity of β-cyclodextrin (the most important native representative of the whole family) when any guest molecule is included in its cavity. [A single exception seems to be the strong inhibitory effect found with some guest molecules on acid catalysed ring-opening (*Hirayama et al., 1993*), but the phenomenon has been explained by steric hindrance, i.e. by simple competition.]

A lot of data are published on the increasing solubility of the guests in presence of β-cyclodextrin, but data about the solubility of inclusion complexes can be found (mostly hidden in figures) only when it causes limitation in the solubilizing effect. (These values are significantly lower in all known cases than that of the parent β-cyclodextrin.)

The solubility of β-cyclodextrin in water is the lowest among the common cyclodextrins, and the low solubility is connected to its rather particular hydrogen bonded system stabilizing the solid phase. The modified solubility of complexes means - as the hydrogen bonding abilities are connected to the hydroxy groups - that the properties of the hydrophilic domain of the host must change during the inclusion complex formation.

The aim of this study is to find connection between the properties of inclusion complex and its building units [based on some earlier experiences (*Buvári-Barcza et al., 1996*)]. A lot of precise solubility data have been collected and the correlations calculated among the solubility parameters and different constants [acid-base characteristics of the guest, formation (stability) constant(s) of the inclusion complex, etc.].

The detailed analysis of data in systems containing both undissolved guest and precipitated inclusion complex in the solid phase has shown the existence of the limited (constant) concentration of the host (β-cyclodextrin), characterized as

$$c_{H,lim} = \Sigma\Sigma \; p \; \beta_{pq} \; [H]_{lim}{}^{p} \; [G]_{0}{}^{p}$$

where p and q are stoichiometric factors, β represents the stability constant(s), $[G]_0$ is the solubility of the guest (at the given temperature, while $[H]_{lim}$ is the constant equilibrium concentration of the host itself.

The solubility enhancement of the guests analyzed changed between 1.1-4.4 (always increase!), but the equilibrium concentration of (free) β-cyclodextrin is decreased about one hundredth because of inclusion complex formation.

Rather surprisingly, the solubilities of β-cyclodextrin inclusion complexes can be best correlated with the solubility of the guest itself, as if the guest would enforce its solubility upon the β-cyclodextrin. The phenomenon experienced on this special field of supramolecular chemistry can be named as *guest enforced solubility* (GES).

Some attempt is made to prove the extremely interesting effect made by host-guest interaction on the hydrophilic domain of β-cyclodextrin, including the formation of ternary complexes (*Buvári, Á. et al., 1982*) in β-cyclodextrin - benzoic acid - (benzene) - acetic acid systems, as well as in systems of β-cyclodextrin - p-nitroanailine and malic, tartaric or citric acids (*Buvári-Barcza, Á. et al., 1998*).

Acknowledgements: We thank Hungarian Research Foundation (OTKA 19493) for financial support of this work.

References

Hirayama, F., Kurihara, M., Utsuki, T. and Uekama, K. (1993) *J. Chem. Soc. Chem. Commun.*, 1578-1580
Buvári, Á. and Barcza, L. (1979) *Inorg. Chim. Acta*, L179-181
Buvári-Barcza, Á. and Barcza, L. (1996) *J. Incl. Phenomena*, 26, 303-309
Buvári-Barcza, Á., Pap, V. and Barcza, L. (1998) *to be published*)

A NEW NON-LINEAR METHOD ON DETERMINATION OF THE STABILITY CONSTANT FOR STEROID-CYCLODEXTRIN COMPLEX

S.M. KHOMUTOV, I.A. SIDOROV, D.V. DOVBNYA, M.V. DONOVA
Institute of Biochemistry and Physiology of Microorganisms, Russian Acad. Sci., Pushchino, Moscow reg., 142292, Russian Federation

1. Introduction

The use of cyclodextrins at the process of microbial synthesis of steroids is widespread. CD_S-mediated solubilization of steroid products is one of the major factors determining the enhancement of microbial sterol conversion in the presence of CD_S [1]. This process is characterised by a stability constant (K_A) of inclusion CD complex.

In spite of the variety of the approaches used for K_A determination the phase-solubility technique and linear model for constant estimation is applied preferably in practice [2]. The most sufficient NMR-spectroscopy [3] can hardly be recommended for these purposes due to a relatively low sensibility (10^{-3}-10^{-4} M) and low solubility of steroids in water.

Phase-solubility technique includes a time-consuming procedure to attain a balance between a crystalline and soluble forms of steroid. The equilibrium between the CD complex and free form of steroid in solution can be achieved in a short time and the spectra of complex and free steroid can be analysed immediately after the dilution of steroids in CD-solutions.

A non-linear model for direct estimation of K_A using data of absorption competitive method was developed. This model is free from the limitations and drawbacks of the widely known procedures of competitive method [4].

2. Experimental procedure

Randomly methylated β-cyclodextrin ($RAMEB_W$) (DS 1.8) from Wacker-Chemie (Germany), methylated β-CD (MCD_C) (DS 12.7) from Cerestar (USA), Triton X-100, Brij 35, Tween 80, Pluronic F-68, Pluronic L-64, Methyl Orange (MetOr) from Sigma (USA) were used. 9α-hydroxyandrost-4-ene-3,17-dione (9α-OH-AD), androsta-1,4-diene-3,17-dione (ADD), androst-4-ene-3,17-dione (AD), 20-hydroxymethyl-pregna-1,4-diene-3-one (HMPD) were isolated during cell conversion of sitosterol and identified as described in [1]. TLC was carried out using the following solvent system: chloroform : aceton : formic acid : ethanol = 68:16:8:8 (v/v). The measurement of absorbance was performed by spectrophotometer Specord-M-40-UV VIS using cuvettes with a thermostalled cell holder. The position of absorption maximum was of 505 nm (pH 2.67, PBS 0.1 M, 30°C).

3. Theoretical consideration

The competitive absorption method for determination of K_A is based on the spectral changes of the dye during complex formation with CDs. When MetOr is used as a dye these changes is followed by a shift of prototropic equilibrium and decrease of pK. The shift of absorption maximum was similar to that observed during the decrease of pH. The measurement of absorbance were carried out at fixed wave length and pH providing a maximal difference in absorbance of free form and CD-complex of dye.
Consider a system of three components: cyclodextrin (D), steroid (A), and MetOr (F):

$$D + A \xleftrightarrow{K_A} DA\,,\; D + F \xleftrightarrow{K_F} DF\,,\ \text{where } K_A = \frac{[D][A]}{[DA]},\; K_F = \frac{[D][F]}{[DF]} \tag{1}$$

where [] means the equilibrium concentration of the following substances: DA and DF complexes and unbound species. The last two equations can be rewritten easily by the use of the mass-balance equations:

$$[A]_0 = [A]+[DA],\; [F]_0 = [F]+[DF],\; [D]_0 = [D]+[DA]+[DF], \tag{2}$$

so that

$$K_A = \frac{([A]_0 - [DA])([D]_0 - [DA] - [DF])}{[DA]},\; K_F = \frac{([F]_0 - [DF])([D]_0 - [DA] - [DF])}{[DF]} \tag{3}$$

where $[A]_0$, $[F]_0$, $[D]_0$, are the analytical concentrations of A, F, and D, respectively. So, the problem is to find the $[DF]$ and $[DA]$ using the values of $[A]_0$, $[F]_0$, $[D]_0$, K_A, and K_F. After this, the concentration of unbound species can be calculated by mass-balance equations (2).
The value of $[DF]$ can be found by solving the equation:

$$[DF]^3 C_3 + [DF]^2 C_2 + [DF] C_1 + C_0 = 0 \tag{4}$$

where

$$C_3 = 4\left(S_A + [F]_0 - S_F - [A]_0\right), \tag{5}$$

$$\begin{aligned} C_2 = {} & \left(2S_F - [F]_0 - S_A\right)^2 - 2[F]_0\left(S_A - 2[D]_0\right) - \\ & - S_A^2 + 4[A]_0[D]_0 + 2[F]_0\left(4[A]_0 - 2S_A\right) - [F]_0^2, \end{aligned} \tag{6}$$

$$\begin{aligned} C_1 = {} & 2[F]_0\left(2S_F - S_A - [F]_0\right)\left(S_A - 2[D]_0\right) + \\ & + 2[F]_0\left(S_A^2 - 4[A]_0[D]_0\right) - [F]_0^2\left(4[A]_0 - 2S_A\right), \end{aligned} \tag{7}$$

$$C_0 = [F]_0^2\left(\left(S_A - 2[D]_0\right)^2 - S_A^2 + 4[A]_0[D]_0\right) \tag{8}$$

and

$$S_A = [A]_0 + [D]_0 + K_A,\; S_F = [F]_0 + [D]_0 + K_F, \tag{9}$$

After this one can calculate $[DA]$ and $[A]$, $[F]$, $[D]$ using the expressions defined above. The optical density of solution E can be calculated as:

$$E = \varepsilon_{DF}[DF] + \varepsilon_F[F], \tag{10}$$

where ε_{DF} and ε_F are some coefficients. Function E depends on the following parameters:

$$E = E\left([A]_0, [D]_0, [F]_0, K_A, K_F, \varepsilon_{DF}, \varepsilon_F\right). \tag{11}$$

In the absence of steroid A ($[A]_0$=0) the optical density E^* depends on the reduced set of the parameters:

$$E^* = E^*\left([D]_0, [F]_0, K_F, \varepsilon_{DF}, \varepsilon_F\right). \quad (12)$$

To fit the model to experimental data the least squares method with the following objective function:

$$Q = \sum_{i=1}^{N} n_i \left(\frac{E_i - Y_i}{\sigma_i}\right)^2 + \sum_{i=1}^{N^*} n_i^* \left(\frac{E_i^* - Y_i^*}{\sigma_i^*}\right)^2 \quad (13)$$

was used. Here: E_i, Y_i, theoretical and experimental mean value of optical density for i-th point; n_i, σ_i, sample size and standard deviation of sample for i-th experimental point; N, number of experimental points (i=1,N). Asterisk means that value corresponds to the experiment in the absence of steroid. It should be noted that argument of function E is $[A]_0$ and argument of E^* is $[D]_0$. So, the simultaneous fitting of one function with different arguments to two sets of experimental points is used. The values of following parameters are calculated as a result of fitting: K_A, K_F, ε_{DF}, ε_F.

4. Results and discussion

The products of microbial sterol side chain oxidation with different hydrophoby were examined for the ability of forming inclusion complexes with the methylated CD derivatives by competitive spectrophotometric method.

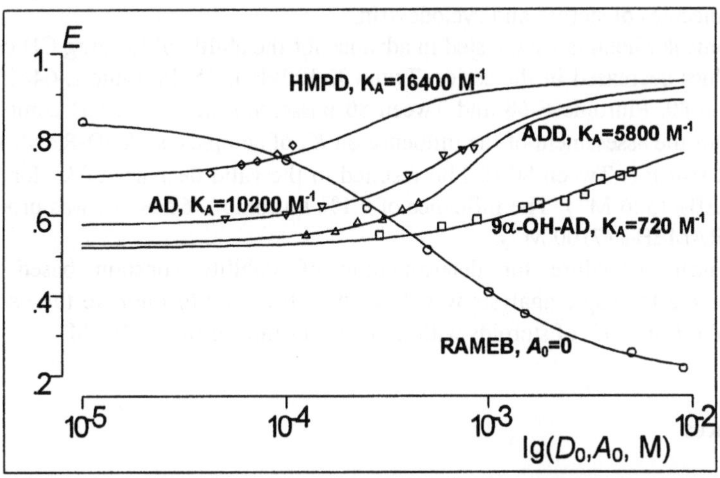

Figure. Dependence of absorbance (E, 505 nm) of MetOr (2.5×10^{-5} M) on: analytical concentration of $RAMEB_w$ (D_0) with no steroid (A_0=0); and different steroid concentrations (A_0) with fixed RAMEB concentration (D_0): 5×10^{-4} M, 9α-OH-AD; 5×10^{-4} M, ADD; 4.5×10^{-4} M, AD; 2.25×10^{-4} M, HMPD. Experimental conditions: pH 2.67, PBS 0.01 M, 30°C.

The model presumes the stoichiometry of CD-steroid complex to be 1:1 (molar ratio) and fits the experimental data with $r^2 \geq 0.99$. The variation of K_A values did not exceed 10% (n=6). The values of K_A increased with the hydrophoby of steroids in the series:

9α-OH-AD < ADD < AD < HMPD (Figure). The influence of fraction composition of RAMEB on K_A was evaluated by the use of commercial CD from different sources. Besides, the method had been applied for a quality control of RAMEB regeneration at microbial cell conversion of β-sitosterol in the CD medium (see Table).

TABLE. The stability constant of steroid-RAMEB inclusion complexs

Steroids	TLC fraction characteristics	K_A, M^{-1} RAMEB$_W$ commercial	RAMEB$_W$ I*	RAMEB$_W$ III*	MCD$_C$ commercial
	R_F 0.47	-**	-	-	+
	R_F 0.26	+	+	+	++
	R_F 0.11	++	++	++	+
	R_F 0.06	+	-	-	-
ADD		5800	5370	5780	5605
AD		10200	-	-	9260
HMPD		16400	-	-	16300

* regenerated RAMEB$_W$ during 1(I) and 3(III) cycles of cell conversion of β-sitosterol
** -/+, absence/presence of fraction for given R_F

The variation of fractions does not lead to valid changes in K_A values for the major sterol biotransformation products. To verify K_A values obtained the model experiment using phase-solubility technique [2] was carried out. The result of experiment is the following: K_A=6700 M^{-1} for [RAMEB$_w$-ADD] complex (it corresponds to one showed in the Table). The effect of pH on K_A (ADD-RAMEB$_w$) was studied within pH 2.6-7.0. Despite of the small difference in extinction coefficients for the bound and free dye forms at neutral pH the method provided a reliable estimation of K_A. The limited decrease of K_A values observed did not exceed the error of the measurements. This pointed to the absence of a considerable influence of pH on the equilibrium of not charged molecules of steroid and cyclodextrin.

The non-ionic surfactants were tested in advance for the ability of forming CD complex. The K_A values decreased in the series: Triton X-100>Brij 35>Pluronic L-64>Pluronic-F-68>Tween 80. Pluronic F-68 and Tween 80 possessing negligible CD complexation were used for the assessment of the influence on K_A of complexes: ADD-RAMEB$_W$, 9α-OH-AD- RAMEB$_W$. Tween 80 (0.1%) resulted in the valid decrease of K_A for 9α-OH-AD- RAMEB$_W$ (526 M^{-1}). The influence of 0.1% Pluronic F-68 on K_A was pronounced for ADD- RAMEB$_W$ (7100 M^{-1}).

So, non-linear procedure for determination of stability constant based on the competitive spectroscopic analysis was developed. It allows to improve the use of this method and estimate K_A of steroids within the values ranged in 10^3-10^4 M^{-1}.

5. References

1. Donova, M.V., Dovbnya, D.V., and Koshcheyenko, K.A., (1996) Modified CDs-mediated enhancement of microbial sterol sidechain degradation. Proceedings of 8-th International Symposium on Cyclodextrins, The Netherlands, Kluwer Academic Publishers, pp.527-530.
2. Higuchi, T. and Connors, K. (1965) Phase solubility techniques, in Reilly, C. (ed.), *Advances in Analytical Chemistry and Instrumentation*, Wiley Interscience, New York, pp.117-212.
3. Djedaini, F., Perly, B. (1993) New Trends in Cyclodextrins and Derivatives, Edition de Sante, Paris.
4. Szeitly, J. (1982) (Ed.) Cyclodextrins and Their Inclusion Complexes, Akademiai Kiado, Budapest.

2,4-DICHLOROPHENOXYACETIC ACID α- AND β-CD INCLUSION COMPLEXES. A ^{1}H-NUCLEAR MAGNETIC RESONANCE STUDY

J.I. PÉREZ-MARTÍNEZ, M.J. ARIAS, J.R. MOYANO, E. MORILLO[1], A.M. RABASCO and J.M. GINÉS
Departamento de Farmacia y Tecnología Farmacéutica, Facultad de Farmacia, Universidad de Sevilla, C/ Profesor García González s/n, 41012-Sevilla, SPAIN.
[1]*Instituto de Recursos Naturales y Agrobiología, Consejo Superior de Investigaciones Científicas (CSIC), Apdo. 1052, 41080 - Sevilla, Spain*

1. Introduction

In the last years pharmaceutical modification of drug molecules as guest by inclusion complexation with cyclodextrins (CD) as host, has been extensively developed to improve their dissolution rate, chemical stability and/or volatile reduction [1]. In a similar way, pesticides can be complexed into the hydrophobic cavity with the subsequent increase of their biological effectiveness [2].

2,4-dichlorophenoxyacetic acid (2,4-D) is a systemic herbicide widely used for weed control in cereals and other crops. It shows a low aqueous solubility, and the 2,4-D molecule is characterized by the presence of two possible complexing groups: the aromatic ring and the aliphatic chain.

Complex formation of 2,4-D with α- and β-CDs in solution has been studied by phase solubility technique [3] and proton nuclear magnetic resonance (^{1}H-NMR) spectroscopy. The ^{1}H-NMR spectroscopy is one of the more precise techniques to study the complexation phenomena. The first advantage of this technique is that allows the identification of the atoms that interact between two molecules [4]. We will to use the ^{1}H-NMR technique to study the complexation between α- and β-CDs and 2,4-D. Since the ^{1}H-NMR allows a clear distinction between true inclusion and any other possible external interaction between CD and guest, the technique has been employed in order to gain insight the complexation mode of the 2,4-D with the CDs under assay. In the presence of CDs, the signals of the host and guest are shifted due to the steric perturbation through inclusion complexation.

2. Experimental

2.1. MATERIALS

2,4-D was supplied by Sigma (St Louis Missouri, USA) and α- and β-CD by Ringdex (Paris, France). D_2O was purchased from SDS (Barcelona, Spain). All other materials were of analytical reagent grade.

2.2. METHODS

2.2.1. Phase solubility studies

The solubility studies were carried out according to the method reported by Higuchi and Connors [3]. An excess of 2,4-D was accurately weighed into each 50 mL Erlenmeyer flasks to that were added 10 mL of water containing various concentrations of α-CD (0.01-0.1 M) and ß-CD (0.002 - 0.014 M). These flasks were sealed and shaken at 25 °C for one week. This time is considered sufficient to reach the equilibrium. After equilibrium, the samples were filtered with syringe through a 0.22 μm Millipore cellulose nitrate membrane filter, properly diluted and analyzed spectrophotometrically at 284 nm. Finally, we proceed to calculate the stoichiometry and the apparent stability constant (K_c) from the plateau portion of the phase solubility diagram.

2.2.2. ^{1}H-NMR spectroscopy

Proton NMR experiments were run at 298 K using a Bruker AC 200 spectrometer operating at 200 MHz. First, we registered the spectra corresponding at 2,4-D and pure CDs, and the binary systems. The concentrations employed were: 0.5 mg/ml of 2,4-D; 4.88 mg/ml of α-CD; 3.0mg/ml de β-CD; 5.38 mg/ml of 2,4-D-α-CD binary system (1:2) and 3.5 mg/ml of 2,4-D-β-CD binary system (1:1). The conditions were as follows: acquisition time 2.818 μs; pulse width, 5 μs; time domain 1.6 K; spectral width 2906.97 Hz.
Continuous variation method or Job's plot [4] has been performed in order to confirm the results that the former studies showed about the stoichiometry of the complexes.

3. Results and Discussion

The phase solubility diagrams obtained for 2,4-D with α- and β-CD are showed in Figures.1 and 2. According to Higuchi and Connors, both diagrams can be classified as Bs type.

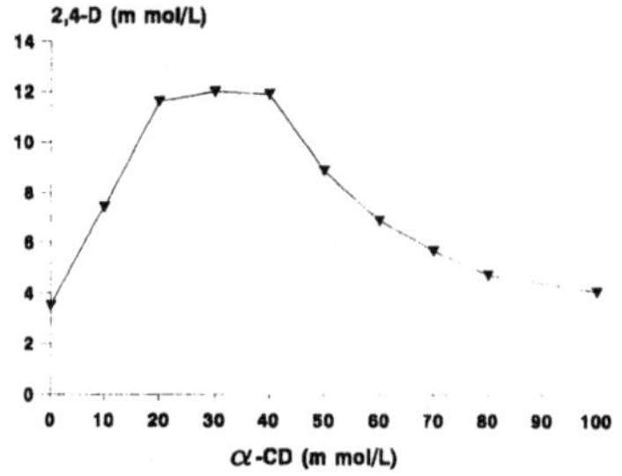

Figure 1. Phase solubility diagram of the 2,4-D–α-CD system in water at 25°C.

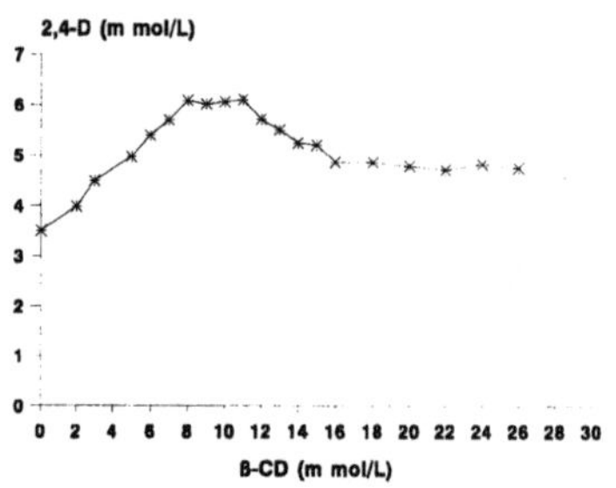

Figure 2. Phase solubility diagram of the 2,4-D-β-CD system in water at 25°C.

From this study, Kc and the molar stoichiometry of the complexes was found to be 1:1 ($K_{1:1} = 336\cdot10^{-3}$ M^{-1}) for the system with β-CD and 1:1 and 1:2 ($K_{1:1} = 94.5\cdot10^{-3}$ M^{-1} and $K_{1:2} = 3.48\cdot10^{-3}$ M^{-2}) for the α-CD one.

Figure 3 summarizes the peak assignments of pure 2,4-D by ^{1}H-NMR spectroscopy. The molecular structure is characterized by the presence of two complexing groups that are potentially able to interact with the CD cavity. In the Table 1 are present the chemical shift values of 2,4-D, α-CD and the binary system.

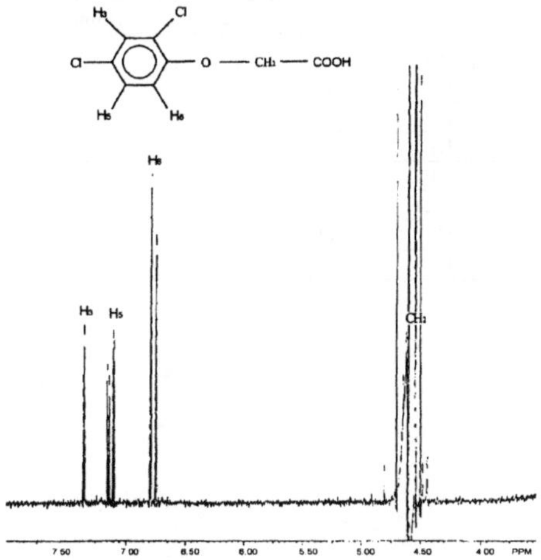

Figure 3 Structure of 2,4-D molecule and assignments of the protons.

In the Table 1 are present the shift values of 2,4-D, α-CD and binary system

TABLE 1. Chemical shifts corresponding to 2,4-D, α-CD and binary system.

Protons	δ_{free}	$\delta_{complex}$	Δδ (ppm)
2,4-D			
H_3	7.361	7.381	-0.020
H_5	7.147	7.186	-0.039
H_6	6.799	6.768	0.031
CH_2	4.544	4.494	0.050
α-CD			
H_3	3.760	3.686	0.074
H_5	3.616	3.666	-0.049

The results show the upfield displacement of the signal corresponding to H3 of the CD. This displacement is due to the anisotropic magnetic effect induced by the presence of the aromatic group of the guest molecule. Quite similar results found Fronza *et al.* [5] for the piroxicam-β-CD system. Besides, it appreciates a downfield shift for the H5 signal of the CD. Similar results were found by Ueda and Nagai [6] for the tolbutamide and the chlorpropamide-β-CD systems. The chemical shifts corresponding to the H3 and H5 of 2,4-D goes downfield. The signals of H6 and aliphatic chain go upfield due to local polarity of the 2,4-D molecule.

Thus, looking the displacement it may request that the aromatic ring of 2,4-D penetrates partially on the cavity of the CD where the diameter is longer (H3). On the contrary, the aliphatic chain may penetrate for the small diameter (H5). Only the signals of the aliphatic chain are displaced in the β-CD system, revealing that one molecule of 2,4-D interacts with only one molecule of β-CD (Table 2).

Using the Job´s plot technique we could confirm the complex stoichiometry (Figure 3). The β-CD complex displayed its maximal displacement at r = 0.5, showing this diagram a highly symmetrical shape. It verifies that the maximal interaction occurs at 1:1 mol:mol ratio. In contrast, for the α-CD complex, the maximum displacement value appears at r = 0.333, establishing the maximal interaction at 1:2 mol:mol ratio.

TABLE 2. Chemical shifts corresponding to 2,4-D, β-CD and binary system.

Protons	δ_{free}	$\delta_{complex}$	**Δδ (ppm)**
2,4-D			
H_3	7.360	7.360	0.000
H_5	7.158	7.160	-0.002
H_5	6.790	6.775	0.015
CH_2	4.521	4.500	0.021
β-CD			
H_3	3.789	3.770	0.019
H_5	3.664	3.708	-0.045

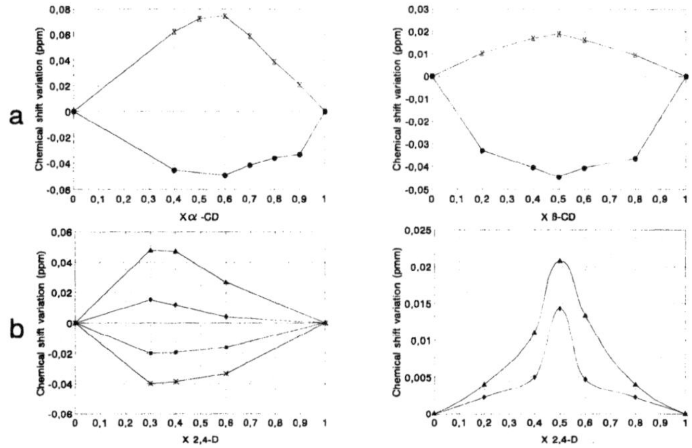

Figure **4** Continuous variation plots for: (a) H3 (✱) and H5 (●) CD protons; (b) H_3 (■), H_5 (✱), H_6 (◊) and CH_2 (▲) 2,4-D protons.

4. Conclusion

The results suggest that 1:2 (mol:mol) 2,4-D-α-CD and 1:1 (mol:mol) 2,4-D-β-CD complexes are formed in aqueous solution, being stronger the interactions recorded for the β-CD system, as the Kc values revealed.

REFERENCES

1. Duchêne, D. and Wouessidjewe, D. (1990) The current state of ß-cyclodextrin in pharmaceutics, Acta Pharm. Technol. 36, 1-6.
2. Szejtli, J. (1982) *Cyclodextrins and their inclusion complexes*, Akadémiai Kiadó, Budapest.
3. Higuchi, T. and Connors, K.A. (1965) Phase-solubility techniques. *Adv. Anal. Chem. Instr.* 4, 117-212.
4. Djedaïni, F. and Perly, B. (1991) Nuclear Magnetic Resonance investigation of the stoichiometries in β-CD: steroid inclusion complexes, *J. Pharm. Sci.* 80, 1157-1161.
5. Fronza, G., Mele, A., Redenti, E. and Ventura, P. (1992) Proton nuclear magnetic resonance spectroscopy studies of the inclusion complex of piroxicam with β-cyclodextrin, *J. Pharm. Sci.* 81, 1162-1165.
6. Ueda, H. and Nagai, T. (1980) Nuclear Magnetic Resonance (NMR) spectroscopy of inclusion compounds of tolbutamide and chlorpropamide with β-cyclodextrin in aqueous solution. *Chem. Pharm. Bull.* 28, 1415-1421.

INVESTIGATION OF THE INCLUSION COMPLEX AND STOICHIOMETRY OF OMEPRAZOLE WITH β-CD BY ^{1}H NMR SPECTROSCOPY

J.R. MOYANO, M.J. ARIAS, M.C. ORTIZ[a], M.A. GARRIDO[b], J.M. GINES, F. GIORDANO[c]
Department of Pharmacy and Pharmaceutical Technology. Faculty of Pharmacy. University of Seville. Seville (Spain). [a]Department of Organic Chemistry. Faculty of Chemistry. University of Seville. Seville (Spain). [b]NMR Service of the Faculty of Pharmacy of Seville. University of Seville. Seville (Spain). [c]Pharmaceutical Department. Faculty of Pharmacy. University of Parma. Parma (Italy)

1. Introduction

NMR techniques are widely employed for the study of the inclusion phenomena of many drugs with CDs [1, 2]. In our case, these ones were used for the elucidation of the nature of interaction between Omeprazole (OME) and β-CD.
OME is a gastric antisecretory widely used in the treatment of gastric and duodenal acid ulcers. OME blocks the gastric acid pump by specific inhibition of the H^+/K^+ ATPase enzyme system at the secretory surface of the parietal cell. However, the OME molecule shows a low solubility in aqueous gastric fluids and hence a slow dissolution rate, accompanied with a low physicochemical stability [3], which cause its low bioavailability (about 50 %) [4]. In order to resolve these drawbacks, the application of CD complexation was expected, being used in many similar cases [5].

2. Materials and Methods

OME was supplied by Andrómaco S.A. (E-Madrid) and β-CD by Ringdex (F-Paris). D_2O was purchased from SDS (E-Barcelona) and NaOD solution from Merck (E-Barcelona). Proton NMR experiments were run at 313K using a Bruker AMX-500 spectrometer operating at 500 MHz. The chemical shifts were referred to an external sodium tetramethylsilane (TMS) at 0 ppm, with calibration using the residual solvent signal (HDO of the D_2O = 4.75 ppm). The solvent employed was a pD 14 solution of NaOD in D_2O, in order to dissolve and stabilize the OME [3]. Thus, an OME sodium salt was formed, being carried out all the NMR experiences at this pD value. Complementary to the study of the interaction between OME and CD, the calculation of the 1:1 and 1:2 association constants was performed by titration of a OME solution (which was keep constant, at $2.23 \cdot 10^{-2}$ M) with β-CD solutions at increasing ratios. The 1:1 and 1:2 association constant was calculated by non-linear curve fitting of the chemical shift variations observed for several giving protons of the ligand.

3. Results and Discussion

The OME structure and peak assignment is reported in Figure 1. Table 1 summarizes the chemical shift values of OME and β-CD protons in the free and bounded states in $NaOD/D_2O$ solution, as well the difference between both signals. In our case, the observed downfield shifts (positive difference) of the guest protons in presence of β-CD indicate that the pyridinic moiety interacts preferentially with the CD cavity, being lower the interaction for the benzoimidazole one, which is reflected in the low variations on their chemical shifts. For the host protons, the study of their chemical shifts variations protons showed that the most affected were the H5 and H6 ones, e.g. those situated on the narrow side of the cavity. The interaction between OME and CD was completed with the calculation of the 1:1 and 1:2 association constants, performed by measurement of the chemical shift variations of selected protons of the guest and the non-linear fittings of the resultant curves (Figure 2). The calculation of both constants was carried out on the basis of a previous phase solubility study, where an OME:β-CD precipitate was isolated, and after analysis revealed a 1:2 stoichiometry [6]. This study is not finished yet, and the calculation of the association constants by this technique will be reported in a future work.
The yielded average values by NMR were of 92.6 M^{-1} for the 1:1 constant and 4.4 M^{-2} for the 1:2 one. This indicates that interaction between the guest and the first CD molecule is stronger than with the second one.

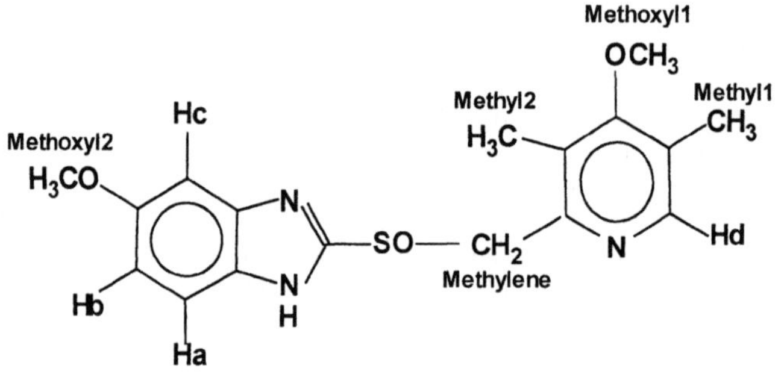

Figure 1. Structure of OME.

The only-slight upfield shifts observed for the inner CD protons may be related to a poor penetration of the aromatic rings. A deep penetration of a aromatic ring would induce a strong anisotropic effect with higher shielding effects by the presence of their π electrons and ring current, and consequently higher chemical shift variations for the CD protons situated in the cavity, e.g. H3, H5 and H6. In our case, the low penetration of the aromatic rings is due to the steric hindrance of methyl and methoxyl substituents, which really interacts with the inside of the CD cavity, being this fact in accordance with the calculated association constant values. As commented before, the most affected β-CD protons are the H5 and H6 ones, which may indicate that the small-diameter side is the most probable point of access of the OME groups in the CD cavity [7, 8]. Also, this side provides more suitable adjustment for the entrance of two ring substituents as methoxyl and methyl or methyl and proton at the same time. Thus, various simultaneous complex geometries can be contemplated.

TABLE 1. [1]H chemical shifts of OME and β-CD, in free and complex states in $NaOD/D_2O$ solution.

Protons	δ_{Free} (ppm)	$\delta_{Complex}$ (ppm)	Δδ (ppm)
OME			
H_a	7.173	7.171	-0.002
H_b	6.827	6.856	0.029
H_c	7.523	7.502	-0.021
H_d	8.121	8.181	0.060
Methyl-1	1.864	1.958	0.094
Methyl-2	2.138	2.253	0.115
Methoxyl-1	3.838	3.930	0.092
Methoxyl-2	3.838	3.930	0.092
Methylene	3.505	3.634	0.129
β-CD			
H1	4.949	4.969	0.020
H2	3.513	3.531	0.018
H3	3.875	3.878	0.003
H4	3.427	3.448	0.021
H5	3.802	3.765	-0.037
H6	3.870	3.803	-0.067

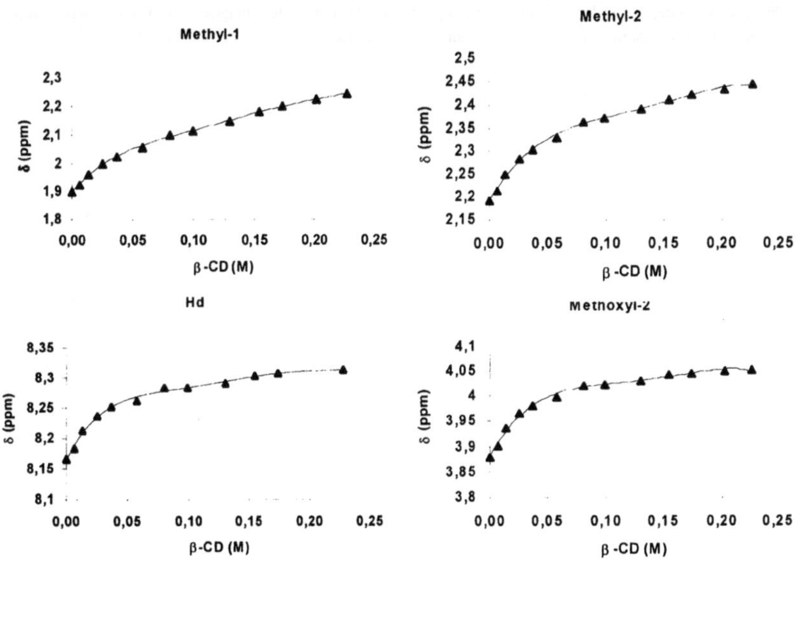

Figure 2. Chemical shift variation of selected OME protons by titration of OME with β-CD solutions at increasing concentrations.

4. Conclusions

The above results suggest that OME does not present a strong interaction with β-CD, reflected in the values corresponding to the calculated association constants. This fact may be related to the experimental conditions, where the high pD value causes the salification and hence the diminution of the hidrophobicity of the guest and, consequently, its affinity for the CD cavity. A second factor seems to be the steric hindrance of the substituents of the side rings of the OME molecule, which block the correct entrance of the aromatic groups and the establishment of non-covalent interactions, e.g. the driving forces of the complexation process.

REFERENCES

1. Matsubara, K., Irie, T. and Uekama, K. (1997) Spectroscopic characterization of the inclusion complex of a luteinizing hormone-releasing hormone agonist, buserelin acetate, with dimethyl-β-cyclodextrin, *Chem. Pharm. Bull.*, 45, 378-383.
2. Moyano, J.R., Arias, M.J., Ginés, J.M., Rabasco, A.M., Pérez-Martínez, J.I., Mor, M. and Giordano, F. (1997) NMR investigations of the inclusion complexation of gliclazide with β-cyclodextrin, *J. Pharm. Sci.* 86, 72-75.
3. Mathew, M., Das Gupta, V. and Bailey, R.E. (1995) Stability of omeprazole solutions at various pH values as determined by high-performance liquid chromatography, *Drug Dev. Ind. Pharm.* 21, 965-971.
4. Martínez-Gorostiaga, J., Alfaro, M.J., Betrán, M.A., Idoipe, A. and Mendaza, M. (1992) *Farm. Hosp.* 16, 33-40.
5. Loftsson, T. and Brewster, M.E. (1996) Pharmaceutical applications of cyclodextrins. 1. Drug solubilization and stabilization, *J. Pharm. Sci.* 85, 1017-1025.
6. Arias, M.J. Muñoz, P., Moyano, J.R. and Ginés J.M., *Unpublished results*
7. Nakajima, T., Sunagawa, M.,Hirohashi, T. and Fujioka, K. (1984) Studies of cyclodextrin inclusion complexes. I. Complex between cyclodextrins and bencyclane in aqueous solution, *Chem. Pharm. Bull.*, 32, 383-400.
8. Fronza, G., Mele, A., Redenti, E. and Ventura, P. (1992) Proton nuclear magnetic resonance spectroscopy studies of the inclusion complex of piroxicam with β-cyclodextrin, *J. Pharm. Sci.* 81, 1162-1165.

INFLUENCE OF CYCLODEXTRINS ON THE CHEMICAL STABILITY OF SALMON CALCITONIN IN AQUEOUS SOLUTION

J. F. SIGURJÓNSDÓTTIR, M. MÁSSON AND T. LOFTSSON.
Department of Pharmacy, University of Iceland, PO Box 7210, IS-127 Reykjavik, Iceland

1. Introduction

Calcitonin (CT) is a peptide hormone, discovered by Copp and colleagues in 1961. It is produced by the parafollicular cells of the thyroid gland in mammals and the ultimobranchial gland of birds and fish. It decreases blood calcium levels and inhibits bone resorption by directly affecting osteoclast activity. CTs of a different origin are used clinically in the treatment of osteoporosis and Paget's disease, and in the management of hypercalcaemia. Salmon calcitonin (sCT) is about 40 times as potential as human CT in lowering blood calcium levels, and is therefore widely used in formulations for treatment of bone diseases [1-2].
Cyclodextrins (CDs) have been shown to increase the stability and solubility of peptides in various formulations [3-5]. Methylated CDs have been shown to enhance absorption of sCT in nasal formulations in rats and rabbits [6]. The mechanism by which CDs increase stability and enhance absorption is not known, but it has been speculated that they may directly stabilize the peptide drug, prevent aggregation or inhibit enzymatic breakdown.
The aim of this work, was to study the effect of CDs on the chemical and physical stability of sCT in aqueous solution.

2. Materials and Methods

2.1 MATERIALS

Synthetic salmon calcitonin from Mallinkrodt Chemicals (Missouri, USA) was a gift from Hexal AG, Holzkirchen, Germany. Cyclodextrins were obtained from various sources: β-cyclodextrin (β-CD) from Nihon Shokuhin Kako Co. (Tokyo, Japan), gamma-cyclodextrin (γ-CD), carboxymethyl-β-cyclodextrin (CM-β-CD), hydroxypropyl-β-cyclodextrin (HP-β-CD), randomly methylated β-cyclodextrin (RM-β-CD) and hydroxytrimethylammoniopropyl-β-cyclodextrin (TMA-β-CD) from Wacker Chemie GmbH (München, Germany) and maltosyl-β-cyclodextrin (G_2-β-CD) from Pharmatec Inc. (Florida, USA). Ninhydrin reagent solution (N-1632) and pruified leucine aminopeptidase (L-5006, E.C. 3.4.11.2) were from Sigma-Aldrich (Dorset, UK). All other materials were of analytical grade.

2.2 KINETIC STUDIES

Stock solutions of the CDs were prepared in citrate-phosphate-borate buffer at the desired pH. Test solutions were prepared by diluting a stock solution of sCT to a final concentration of 0.05 mg/ml sCT in CD-buffer solution. Equal aliquots were transferred to 5 vials, which were sealed tightly and incubated at 55°C. At various time intervals, one vial was removed from the incubator and frozen. Remaining sCT in each vial was determined by reversed-phase HPLC.

2.3 AGGREGATION STUDIES

CD stock solutions were prepared in buffer. Test solutions of sCT were made to a concentration of 10 mg/ml sCT in CD-buffer solution. Equal aliquots of the test solution were transferred to 3 vials, which were then incubated at 55°C. All solutions were clear when put in the incubator. At a 5 day interval, one vial was removed from the incubator and frozen. The samples were filtered through a 0.20 μm cellulose acetate membrane to remove precipitation. Protein concentration in the filtrate was measured by UV-absorption and remaining sCT was determined by HPLC. The amount of soluble aggregates was asessed by native polyacrylamice gel electrophoresis (PAGE).

2.4 ENZYMATIC STUDIES

A stock solution of sCT (2 mg/ml) was prepared in a 10 mM phosphate buffer at pH 3. Stock solutions of CDs were prepared in a 50 mM Tris buffer at pH 7.4. Nasal washings and plasma were obtained from human volunteers. Stock solution of purified leucine aminopeptidase was 100 mU/ml in Tris buffer. The reaction mixture was prepared by mixing 200 μl of CD stock solution (or Tris buffer as a reference) with 200 μl of plasma, nasal washings or LAP solution in a vial. This solution was pre-heated on a water bath at 37°C for 15 min. The reaction was initiated by adding 50 μl of pre-heated sCT stock solution and mixing. Samples of the reaction mixture were then diluted 10 fold in 0.1% TFA (quenching solution) and frozen (t_0 samples). Samples were drawn from the mixture at various time intervals and treated in the same manner as t_0 samples. Remaining sCT in the samples was measured by HPLC.

3. Results and Discussion

3.1 CHEMICAL STABILITY

3.1.1 pH rate profile for sCT.

The degradation of sCT in aqueous solution is pH dependant, with an overall first-order degradation rate constant. The stability is greatest between pH 3 and 4. The pH rate profile for sCT in the citrate-phosphate-borate buffer system is shown in *Figure 1*.

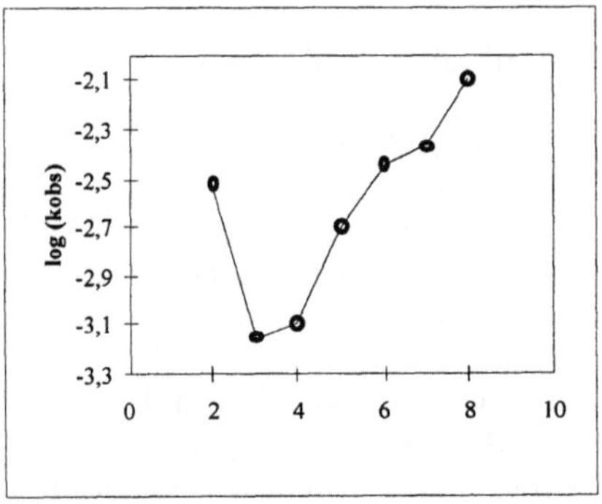

FIGURE 1. The pH rate profile for sCT in citrate-posphate-borate buffer. The initial sCT concentration was 0.05 mg/ml.

3.1.2 Influence of CDs on chemical stability

The influence of various β-CD derivatives at a concentration of 5% (w/v) on the chemical stability of sCT at pH 3 and pH 6 at 55°C was investigated. The results can be seen in *Table 1*. Four CD derivatives were chosen and the influence of different CD concentrations on the stability of sCT at pH 6 and 55°C was investigated. Results are expressed in *Figure 2*. In all cases, initial concentration of sCT was 0.05 mg/ml.

	$t_{1/2}$ (days) / Inhibition (%)	
Cyclodextrin	pH 3	pH 6
None	57.8 /	7.2 /
0.5% β-CD*	54.5 / - 6	6.2 / -14
5% γ-CD	56.6 / - 2	5.6 / -22
5% CM-β-CD	22.8 / -61	9.0 / 25
5% G_2-β-CD	59.0 / 2	6.3 / -13
5% HP-β-CD	60.2 / 4	7.8 / 8
5% RM-β-CD	45.8 / -21	6.3 / -13
5% TMA-β-CD	65.6 / 14	6.8 / - 5

TABLE 1. Chemical stability of sCT in aqueous solution at pH 3 and pH 6 in citrate-phosphate-borate buffer with 5% (w/v) CD at 55°C. The table shows half-lives and inhibition of degradation. (* β-CD has limited solubility.)

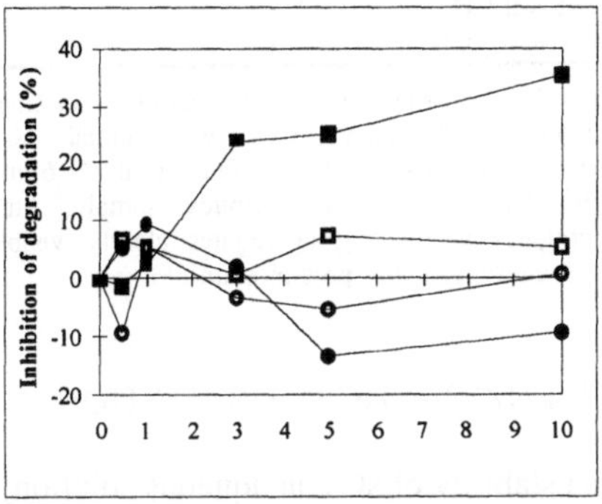

FIGURE 2. Inhibition of degradation of sCT with different CD conc. at pH 6 and 55°C. CM-β-CD (■); HP-β-CD (□); RM-β-CD (●); TMA-β-CD (o).

In general, CDs did not increase the chemical stability of sCT in dilute aqueous solution. At pH 6, the effect of CDs on sCT stability was negligible when the CD concentration was kept below 4% (w/v). The only exception was the negatively charged CM-β-CD derivative, which seemed to greatly increase sCT stability.

3.2 PHYSICAL STABILITY

The influence of four different β-CD derivatives on the physical stability of sCT in concentrated solution at pH 6 and 55°C was investigated. The appearance of sample solutions is described in *Table 2*. Results for measurements of protein and sCT concentrations are in *Figures 3* and *4*. Each sample was also analyzed by native PAGE.

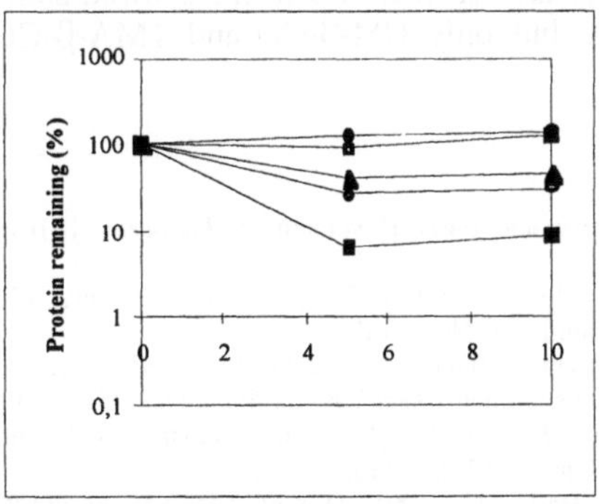

FIGURE 3. Protein concentration measured by absorbance at 280 nm, expressed as percentage of the initial (10 mg/ml). No CD (▲); CM-β-CD (■); HP-β-CD (□); RM-β-CD (●); TMA-β-CD (O).

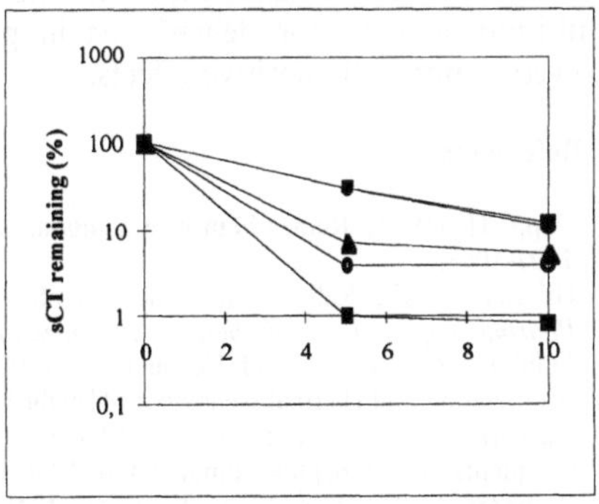

FIGURE 4. Concentration of sCT measured by HPLC, expressed as percentage of the initial (10 mg/ml). No CD (▲); CM-β-CD (■); HP-β-CD (□); RM-β-CD (●); TMA-β-CD (O).

Cyclodextrin	Time at 55°C 5 days	10 days
None	+	++
CM-β-CD	++	+++
HP-β-CD	-	+
RM-β-CD	-	-
TMA-β-CD	-	+

TABLE 2. Apperance of sample solutions containing 5% (w/v) CDs with initial sCT concentration of sCT, after incubation at pH 6 and 55°C. Clear sample (-); cloudy sample, little aggregates visible (+); aggregates clearly visible (++); aggregates and precipitation (+++).

The charged CDs, TMA-β-CD and CM-β-CD, promoted degradation of sCT in concentrated solutions at pH 6. Although CM-β-CD was perivously shown to increase the chemical stability of sCT in dilute solutions at pH 6, it accelerated aggregation and precipitation in concentrated solutions at pH 6. On the other hand, HP-β-CD and RM-β-CD did not only inhibit degradation and aggregation, but they also solubilized the dimers which were formed in the test solutions, and thereby increased the physical stability of the sCT solution.

3.3 ENZYMATIC DEGRADATION

The stability of sCT in aqueous solution, containing 5% (w/v) CD, towards enzymatic degradation was investigated at 37°C and pH 7.4. The results are shown in *Table 3.*

Cyclodextrin	Half-life, $t_{1/2}$ (days) Without enzymes	LAP 40 mU/ml	Nasal washings (50%)	Plasma (50%)
None	25.9	0.6	15.7	6.0
γ-CD	2.4	0.5	1.8	4.3
CM-β-CD	7.4	0.4	5.0	12.3
G_2-β-CD	3.1	0.5	1.2	6.7
HP-β-CD	7.2	0.6	1.1	7.6
RM-β-CD	23.9	0.7	15.0	8.1
TMA-β-CD	41.4	0.8	1.9	20.4

TABLE 3. Half-lives for sCT in enzyme solutions, with and without 5% (w/v) CDs. LAP: leucine aminopeptidase.

In most cases, CDs promoted enzymatic degradation of sCT *in vitro*. However, RM-β-CD did not adversely affect the sCT stability. Nearly all the CDs tested showed some inhibition of enzymatic degradation in plasma, but only CM-β-CD and TMA-β-CD showed significantly positive effects.

4. References

1. Copp, H. (1992) Remembrance: Calcitonin: Discovery and Early Development, *Endocrin.* 131(3), 1007-1008.
2. Gilman, A. G., Rall, T. W., Nies, A. S., and Taylor, P. (1990) *Goodman and Gilman's The Pharmacological Basis of Theraputics, 8th ed.*, Pergamon Press, New York.
3. Tokihiro, K., Irie, T., and Uekama, K. (1997) Varying Effects of Cyclodextrin Derivatives on Aggregation and Thermal Behaviour of Insulin in Aqueous Solution, *Chem. Pharm. Bull.*, 45(3), 525-531.
4. Haeberlin, B., Gengenbacher, T., Meinzer, A., and Fricker, G. (1996) Cyclodextrins - Useful excipients for oral peptide administration? *Int. Jour. Pharm.*, 137, 103-110.
5. Brewster, M. E., Hora, M. S., Simpkins, J. W., and Bodor, N. (1991) Use of 2-Hydroxypropyl-β-cyclodextrin as a Solubilizing and Stabilizing Exipient for Protein Drugs, *Pharm. Res.*, 8(6), 792-795.
6. Schipper, N. G. M., Verhoef, J. C., Romeijn, S. G., and Merkus, F. W. H. M. (1995) Methylated b-Cyclodextrins Are Able to Improve the Nasal Absorption of Salmon Calcitonin, Calcif. Tissue Int., 56, 280-282.

COMPLEXATION PROPERTIES OF β-CYCLODEXTRIN SULFOBUTYLETHER SODIUM SALT

THORSTEINN LOFTSSON AND MÁR MÁSSON
Department of Pharmacy, University of Iceland, PO Box 7210, IS-127 Reykjavik, Iceland

1. Introduction

β-Cyclodextrin sulfobutylether sodium salt (SBEβCD) is an anionic β-cyclodextrin derivative with a sodium sulfonate salt separated from the hydrophobic cavity by a butyl ether space group. The space group should minimise interference of the sulfonate group on the complex formation [1]. SBEβCD has high intrinsic solubility and good complexing abilities in aqueous solutions [1-3]. The purpose of this study was to evaluate the solubilizing and stabilising properties of SBEβCD and to compare them with those of 2-hydroxypropyl-β-cyclodextrin (HPβCD), randomly methylated β-cyclodextrin (RMβCD) and trimethylammoinumpropyl-β-cyclodextrin (TMAPβCD).

2. Materials and Methods

2.1 MATERIALS

Naproxen USP was obtained from Icelandic Pharmaceuticals (Iceland), calcipotriol and ETH-615 from Leo Pharmaceutical Products (Denmark), chlorambucil from Wellcome Foundation (UK), and indomethacin from Sigma Chemical Co. (USA). SBEβCD sodium salt (Captisol™) was kindly donated by CyDex (USA), TMAPβCD hydroxide DS 0.5 and RMβCD DS 0.6 by Wacker-Chemie (Germany), and HPβCD DS 2.7 was obtained from Cyclolab (Hungary). All other chemicals used in this study were commercially available products of special reagent grade.

2.2 SOLUBILITY STUDIES

Solubilities were determined by adding excess amount of the drug to be tested to water or aqueous buffer solutions containing various amounts of different kinds of cyclodextrins with or without a polymer. The suspensions formed were heated in an autoclave in sealed containers to 120-140°C for 20-40 min. After cooling to room temperature (approx. 23°C) small amount of solid drug was added to each container to promote precipitation. Then the suspensions were allowed to equilibrate for at least three days at room temperature. After equilibration was attained, an aliquot of the suspension was filtered through a 0.45 μm membrane filter (Nylon Acrodisc® from Gelman, USA), diluted with 70% (v/v) methanol in water and analysed by HPLC. The stability constants were calculated from the phase-solubility diagrams [4].

2.3 DRUG DEGRADATION STUDIES

Stock solution of the drug to be tested was added to an aqueous buffered cyclodextrin solution which was kept on a temperature-controlled-sample-rack, and the disappearance of the drug was monitored by HPLC. The first order rate constants for the drug degradation in the cyclodextrin solutions (k_{obs}) or the pure buffer solutions (k_o) were obtained from linear regression natural logarithm of the peak heights versus time plots. The stability constant of the drug-cyclodextrin complex (K_C) and the degradation rate constant for drug degradation within the complex (k_c) was obtained by non-linear fit of the data [5].

3 Results and Discussion

3.1 DRUG SOLUBILIZATION

3.1.1 Naproxen

Naproxen is a weak acid with pKa of 4.2 and, thus, the molecule is uncharged at pH below the pKa-value but negatively charged at higher pH. The stability constants of the naproxen-cyclodextrin (1:1) complex was determined at room temperature (Table 1).

TABLE 1. The effect of drug ionisation on the stability of the 1:1 complex of naproxen with three different cyclodextrins at room temperature.

pH	Fraction of naproxen in the ionised from*	Stability constant (M^{-1})		
		TMAPβCD	HPβCD	SBEβCD
2.5	0.02	1,400	2,800	4,900
4.3	0.56	870	860	1,700
5.4	0.94	890	470	530
6.1	0.99	1,600	240	250

*Calculations based on the pKa-value of the free form. Cyclodextrin complexation appears to increase the pKa-value of naproxen.

Of the three cyclodextrins tested the negatively charged SBEβCD formed the most stable complex with the unionised form of naproxen (i.e. was the best solubilizer of the unionised water-insoluble form). SBEβCD was followed by the uncharged HPβCD but the positively charged TMAPβCD formed the least stable complex with unionised naproxen. However, TMAPβCD formed the most stable complex with the negatively charged naproxen (i.e. the cationic TMAPβCD was the best solubilizer of the anionic naproxen) but HPβCD and SBEβCD were equal but less effective.

3.1.2 ETH-615

ETH-615 is an amphoteric drug which forms a zwitterion at pH between approximately 5.5 and 9 and an anion at pH above 10. At pH 7.0 and 22°C the solubility of ETH-615 in 10% (w/v) aqueous SBEβCD, HPβCD, RMβCD and TMAPβCD was 0.30, 1.86, 2.24 and 1.00 mg/ml, respectively [4]. The solubility of the zwitterion in pure aqueous solution under these same conditions was determined to be approximately 1 μg/ml. However, SBEβCD and HPβCD were about as effective complexing agents at pH 5 and 10 but TMAPβCD was less effective at pH 10 (Table 2). Addition of a water-soluble polymer significantly enhanced the complexation.

TABLE 2. The effect of pH on the stability of the 1:1 complex of the zwitter ionic drug ETH-615 with three different cyclodextrins at room temperature.

pH	Stability constant (M^{-1})		
	TMAPβCD	HPβCD	SBEβCD
5.0	0.2	0.2	0.7
10	9.5	30	25

3.1.3 Calcipotriol

Calcipotriol is a very lipophilic and, consequently, its aqueous solubility is low or only 1.3±0.7 μg/ml at 23°C [3]. Cyclodextrin complexes of very lipophilic drugs such as calcipotriol have frequently limited aqueous solubility resulting in Higuchi's B-type phase-solubility diagrams. Thus, the phase solubility of calcipotriol in aqueous HPβCD solutions was a B-type diagram. However, the phase-solubility diagram of calcipotriol in aqueous SBEβCD was linear (i.e. of Higuchi's A_L-type) resulting in better overall solubilization of the drug (Figure 1).

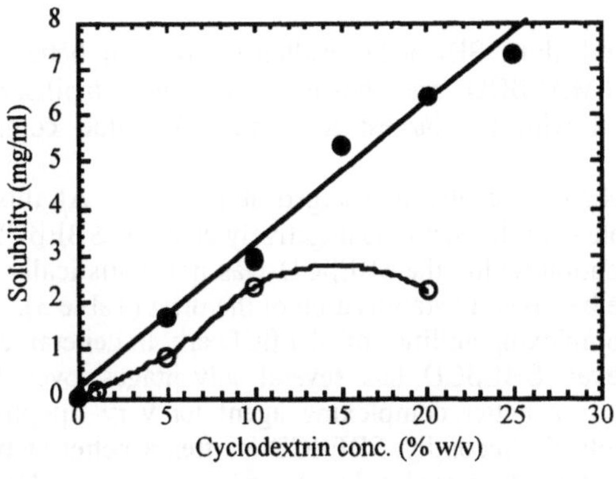

FIGURE 1. The phase-solubility diagram of calcipotriol in aqueous HPβCD (O) and aqueous SBEβCD (●) solutions at approximately 23°C.

3.2 DRUG STABILISATION

Cyclodextrin complexation of chemically (or physically) unstable drugs can frequently result in significant stabilisation, the degradation constant being much smaller within the complex (k_C) than outside it (k_O). The degree of stabilisation does not only depend on the degradation rate constant within the complex but also on the stability constant (K_C) of the drug-cyclodextrin complex formed [5, 6]. Chlorambucil (pKa 5.8) is anionic at pH 7.35. At this pH the stability constant of both the chlorambucil-HPβCD complex and the chlorambucil-RMβCD complex was about 2.5-times larger than that of the chlorambucil-SBEβCD complex (Table 3). However, chlorambucil degraded about 3.5-times slower within the SBEβCD complex than within the HPβCD or

TABLE 3. The stability constants (K_C) and the degradation constant of chlorambucil and indomethacin within the cyclodextrin complex (k_C) at 40°C. The degradation constants for the free chlorambucil (k_O) under these same conditions was determined to be 65.3 min^{-1} and that of indomethacin to be 8.53 min^{-1}.

Cyclodextrin	K_C (M^{-1})	k_C (min^{-1})	k_C/k_O
Chlorambucil (at pH 7.35)			
SBEβCD	1400	1.0	0.015
HPβCD	3400	3.6	0.055
RMβCD	3550	3.3	0.051
TMAPβCD	4250	1.2	0.018
Indomethacin (at pH 9.8)			
SBEβCD	260	≈0 [a]	≈0 [a]
HPβCD	690	3.6	0.055
RMβCD	770	3.3	0.051
TMAPβCD	[b]		

[a]Not statistically different from zero.
[b]No complexation could be detected

RMβCD complex and, thus, SBEβCD resulted in over all better stabilisation. The positively charged TMAPβCD was, however, the best stabiliser of the negatively charged chlorambucil with the largest K_C, and a k_C-value comparable to that of SBEβCD.

Indomethacin (pKa 4.5) is negatively charged at pH 9.8. At this pH indomethacin formed the least stable complex with the negatively charged SBEβCD. However, since indomethacin degradation within the SBEβCD was not statistically different from zero, SBEβCD offered the best overall stabilisation of the drug (Table 3).

In conclusion, the complexing abilities of SBEβCD are, in general, comparable to those of HPβCD. However, SBEβCD has several advantages over HPβCD. Firstly, SBEβCD is frequently a better complexing agent for very lipophilic water-insoluble drugs such as calcipotriol. Secondly, SBEβCD is often a better stabiliser than HPβCD or RMβCD. Frequently drug molecules degrade at a much slower rate within the SBEβCD complex than within the HPβCD complex. Finally, SBEβCD is frequently a better complexing agent than HPβCD if the drug molecule carries a positive charge.

4. References

1. Thompson, D.O. (1997) Cyclodextrins-enabling excipients: their present and future use in pharmaceuticals. *Critical Reviews in Therapeutic Drug Carrier Systems, 14*, 1-104.
2. Jarho, P., Järvinen, K., Urtti, A., Stella, V.J. and Järvinen, T. (1995) Modified β-cyclodextrin (SBE7-β-CyD) with viscous vehicle improves the ocular delivery and tolerability of pilocarpine prodrug in rabbits. *J. Pharm. Pharmacol., 48*, 263-269.
3. Loftsson, T. and Petersen, D.S. (1997) Cyclodextrin solubilization of water-insoluble drugs: calcipotriol and EB-1089. *Pharmazie, 52*, 783-785.
4. Loftsson, T. and Petersen, D.S. (1998) Cyclodextrin solubilization of ETH-615, a zwitterionic drug, *Drug Devel. Ind. Pharm., 24*, 365-370.
5. Másson, M., Loftsson, T., Jónsdóttir, S., Friðriksdóttir, H. and Petersen, D.S. (1998) Stabilisation of ionic drugs through complexation with non-ionic and ionic cyclodextrins. *Int. J. Pharm., 164*, 45-55.
6. Loftsson, T. and Brewster, M.E. (1996) Pharmaceutical applications of cyclodextrins. 1. Drug solubilization and stabilization. *J. Pharm. Sci., 85*, 1017-1025.

HOW DO CYCLODEXTRINS ENHANCE DRUG PERMEABILITY THROUGH BIOLOGICAL MEMBRANES?

MÁR MÁSSON, THORSTEINN LOFTSSON AND EINAR STEFÁNSSON
Department of Pharmacy and Ophthalmology, University of Iceland, PO Box 7210, IS-127 Reykjavik, Iceland

1. Introduction

Biological membranes, such as skin, eye cornea and the oral mucosa, form a lipophilic barrier towards drug permeation. Relatively lipophilic drugs are able to permeate these membranes by passive diffusion. However, passive diffusion is driven by high concentration of dissolved drug in the aqueous membrane exterior (e. g. tear fluid or saliva) or in aqueous drug vehicle (e. g. hydrogels or hydrophilic creams). Thus, greatest permeability should be obtained with drugs which are both lipophilic and highly soluble in the aqueous exterior. Hydrophilic water-soluble drugs are frequently unable to permeate the lipophilic membranes due to their inability to partition into the membrane. Lipophilic water-insoluble drugs are frequently unable to permeate the lipophilic membranes, or do it very slowly, due to very low concentration of dissolved drug in the aqueous exterior. It is possible to increase the permeability of lipophilic water-insoluble drugs through formation of water-soluble complexes [1]. For example, through formation of hydrophilic water-soluble cyclodextrin complexes of lipophilic water-insoluble drugs. This approach has been applied successfully in aqueous dermal formulations [2], mouthwash solutions [3], nasal spray formulations [4] and eye drop solutions [5].

It is generally recognised that cyclodextrins act as true carriers by keeping the drug molecules in solution and delivering them biological membranes. However, it has been shown that the permeability of drugs through skin and eye cornea will decrease when cyclodextrin is added in excess of what is needed to fully dissolve the lipophilic water-insoluble drug [2,5]. Cyclodextrins have also been used to mask the taste of various compounds [6] which indicates decreased availability at the oral mucosa. Mechanistic model which fully explains this dual effect of cyclodextrins has not been available.

In the present work we have developed equations which describe the effect of cyclodextrins on drug permeability. These equations are based on the classical model of membrane and diffusion control drug permeability. The resulting equations were fitted to experimental data.

2. Materials and Methods

2.1 MATERIALS

Hydrocortisone was obtained from Norsk Medisinaldepot (Norway) and triamcinolone acetonide from Akza (the Netherlands). γ-Cyclodextrin (γCD) and randomly methylated-β-cyclodextrin (RMβCD) MS 1.8 was kindly donated by Wacker-Chemie (Germany), and maltosyl-β-cyclodextrin (MAβCD) by Ensuiko Sugar Refining Co., Ltd. (Japan). All other chemicals used were commercially available products of special reagent or analytical grade. Semi-permeable cellophane membrane (Spectrapor® membrane tubing no. 2) was obtained from Spectrum Medical Industries (USA). Female hairless mice (3CH/Tif hr/hr) were obtained from Bommice (Denmark).

2.2 SOLUBILITY AND PERMEABILITY STUDIES

Solubility determinations of hydrocortisone in the aqueous cyclodextrin solutions, and in vitro determinations of the drug flux (and permeability) through semi-permeable membrane or hairless mouse skin (Franz-diffusion cell) has been reported previously [7].

3. Theory

In the Franz-diffusion cell system the drug (D), the cyclodextrin (CD) and the drug-cyclodextrin complex (D•CD) will all present in the aqueous donor phase, and their relative concentrations will be determined by the stability constant K_d of the drug-cyclodextrin complex in the donor phase.

$$K_d = \frac{[D \bullet CD]_d}{[D]_d [CD]_d} \qquad (1)$$

Where $[D \bullet CD]_d$ is the concentration of the drug-cyclodextrin complex in the donor phase, $[D]_d$ is the concentration of free drug in the donor phase and $[DC]_d$ is the concentration of free cyclodextrin in the donor phase. When a drug molecule diffuses from an aqueous donor phase through a lipophilic membrane to a receptor phase it will experience two types of resistance. The resistance to drug diffusion in the aqueous phase, or the aqueous diffusion layer, ($\mathbf{R}_{AQ}$) and the resistance to diffusion through the membrane ($\mathbf{R}_M$). The total resistance to diffusion is the sum of $\mathbf{R}_M$ and $\mathbf{R}_{AQ}$. The permeability coefficients (**P**) are the reciprocals of the resistance. The drug flux ($\mathbf{J}_D$) through the barrier is obtained by multiplying the permeability with the drug concentration, as shown by Equation 2.

$$\boldsymbol{J}_D = \boldsymbol{P}_D[D]_d = \left(\boldsymbol{R}_M + \boldsymbol{R}_{AQ}\right)^{-1}[D]_d = \left(\frac{1}{\boldsymbol{P}_M} + \frac{1}{\boldsymbol{P}_{AQ}}\right)^{-1}[D]_d = \left(\frac{\boldsymbol{P}_M \boldsymbol{P}_{AQ}}{\boldsymbol{P}_{AQ} + \boldsymbol{P}_M}\right)[D]_d \quad (2)$$

where $\mathbf{P}_D$ is the total drug permeability coefficient. The drug permeability coefficient through the lipophilic membrane ($\mathbf{P}_M$) is a function of the partition coefficient between the membrane and the aqueous donor phase ($\mathbf{K}_p$), the membrane diffusion coefficient ($\mathbf{D}_M$) and the membrane thickness (h_M) (see Equation 3).

$$\boldsymbol{P}_M = \frac{\boldsymbol{D}_M \boldsymbol{K}_p}{h_M} \quad (3) \qquad\qquad \boldsymbol{P}_{AQ} = \frac{\boldsymbol{D}_{AQ}}{h_{AQ}} \quad (4)$$

The drug permeability coefficient through the aqueous diffusion layer (**P**$_{AQ}$) is a function of the aqueous diffusion coefficient (**D**$_{AQ}$) and the thickness of the aqueous diffusion layer (h_{AQ}) (see Equation 4). When drug is in suspension then $[D]_d$ is constant and equal to the intrinsic aqueous solubility (S_o) of the drug. Equation 5 is obtained by combining Equations 1 through 4:

$$J_D = P_D [D]_d = \frac{D_M K_p' D_{D \bullet CD} K_{D \bullet CD} [D \bullet CD]_d}{D_M K_p' h_{AQ} + D_{D \bullet CD} K_{D \bullet CD} K_d [CD]_d h_M} \qquad (5)$$

where **K**$_{D\bullet CD}$ is partition coefficients for the partitioning of the drug-CD complex from the donor phase into the aqueous diffusion layer, and **D**$_{D\bullet CD}$ and **D**$_D$ are the diffusion coefficients for the drug-CD complex and drug in the aqueous diffusion layer. Equation 5 should include an additional term describing the drug flux in absence of CD. This final term was omitted as experiments showed that drug flux from aqueous solutions containing no CD was essentially zero.

4. Results and Discussion

Figure 1 shows the diffusion of hydrocortisone through hairless mouse skin from aqueous donor phases containing MAβCD or RMβCD. The data was fitted to Equation 5 (broken line). The $[D \bullet CD]_d$ concentration was determined by phase-solubility studies. Good agreement was between the experimental MAβCD data obtained and values calculated according to Equation 5. The correlation between the RMβCD data and the calculated values was not as good, but it followed the general shape of the curve. The lipophilic RMβCD is known to permeate into lipophilic membranes which can affect the drug permeability through the membrane. This could explain the deviation from experimental values.

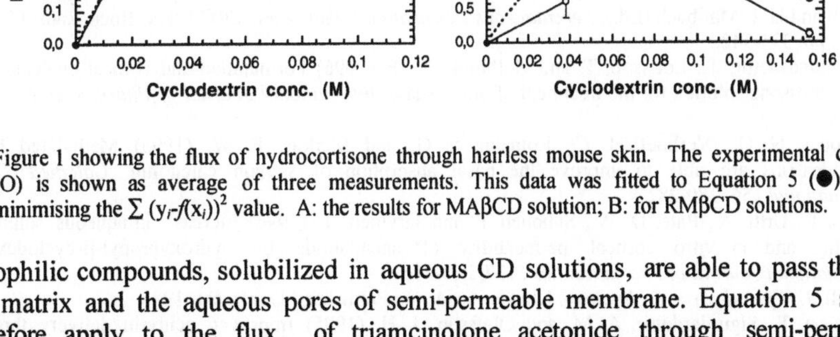

Figure 1 showing the flux of hydrocortisone through hairless mouse skin. The experimental data (O) is shown as average of three measurements. This data was fitted to Equation 5 (●) by minimising the $\sum (y_i - f(x_i))^2$ value. A: the results for MAβCD solution; B: for RMβCD solutions.

Lipophilic compounds, solubilized in aqueous CD solutions, are able to pass through the matrix and the aqueous pores of semi-permeable membrane. Equation 5 should therefore apply to the flux of triamcinolone acetonide through semi-permeable cellophane membrane. Figure 2 shows data obtained for the triamcinolone acetonide flux through semi-permeable cellophane membrane from donor phase containing γCD.

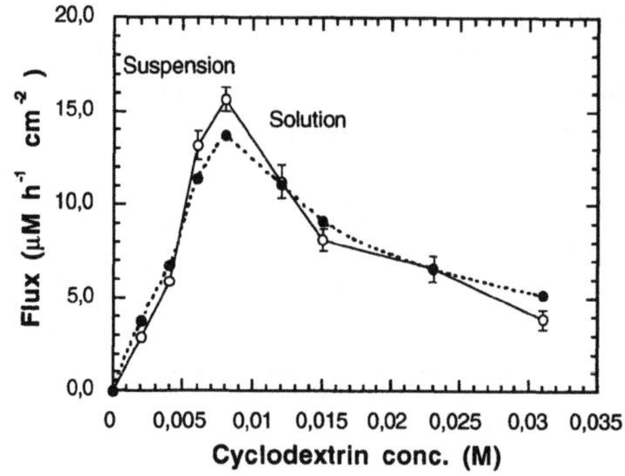

Figure 2. showing the flux of triamcinolone acetonide through semi-permeable membrane. The experimental data (O) and values calculated according to Equation 5 (●) by minimising the $\Sigma\,(y_i - f(x_i))^2$ value.

The theoretical prediction is in good agreement with the experimental data. The structure of semi-permeable membrane is simpler and more uniform than the structure of the mouse skin, which could explain the enhanced correlation between experimental and theoretical data. Similar results were obtained when drug permeability through the eye cornea was investigated.

In conclusion, our results indicate that lipophilic drug molecules encounter two different types of barriers when they permeate from an aqueous CD containing solutions through a lipophilic biological membrane. First the drug molecules must permeate an aqueous barrier (i.e. an aqueous diffusion layer) at the membrane exterior to reach the lipophilic (or membrane) barrier which is located within the biological membrane. CDs act as true carries by delivering the lipophilic water-insoluble drug molecules through the aqueous barrier to the surface of the lipophilic barrier. When the molecules reach the lipophilic barrier they partition from the aqueous barrier into the lipophilic barrier. Finally, the drug molecules permeate through the barrier.

5. References.

1. Loftsson, T. and Brewster, M. E. (1996) Pharmaceutical Applications of Cyclodextrins. 1. Drug Solubilization and Stabilization. *Journal of Pharmaceutical Sciences*, 85, 1017-1025
2. Loftsson, T. and Bodor, N., The effect of cyclodextrins of percutaneous transport of drugs. In: E. W. Smith and H. I. Mailbach (Ed.), *Percutaneous Penetration Enhancesrs*, CRC Press, Boca Rato, Florida, 1995, pp. 335-342.
3. Kristmundsdottir, T., Loftsson, T. and Holbrook, W. P. (1996) Formulation and clinical evaluation of hydrocortisone solution for the treatment of oral disease. *International Journal of Pharmaceutics*, 139, 63-68
4. Schipper, N. G., Verhoef, J. C., Romeijn, S. G. and Merkus, F. W. (1995) Methylated beta-cyclodextrins are able to improve the nasal absorption of salmon calcitonin. *Calcified Tissue international*, 56, 280-282
5. Jarho, P., Urtti, A., Pate, D. W., Suhonen, P. and Jarvinen, T. (1996) Increase in aqueous solubility, stability and in vitro corneal permeability of anandamide by hydroxypropyl-β-cyclodextrin. *International Journal of Pharmaceutics*, 137, 209-216
6. Szejtli, J.: *Cyclodextrin Technology*. Kluwer Academic Publishers, Dordrecht, 1988.
7. Loftsson, T., Sigurdardottir, A. M. and Olafsson, J. H. (1995) Improved acitretin delivery through hairless mouse skin by cyclodextrin complexation. *International Journal of Pharmaceutics*, 115, 255-258

DIFFERENTIAL SCANNING CALORIMETRY ANALYSIS OF CRYSTALLINITY CHANGES OF NAPROXEN IN GROUND MIXTURES WITH MALTOHEXAOSE, THE NON CYCLIC ANALOG OF ALPHA-CYCLODEXTRIN

G. P. BETTINETTI[1], M. SORRENTI[1], A. NEGRI[1], P. MURA[2], M. T. FAUCCI[2], M. SETTI[3]

[1]*Dipartimento di Chimica Farmaceutica, Università di Pavia, Viale Taramelli 12, I-27100 PV, Italy*

[2]*Dipartimento di Scienze Farmaceutiche, Università di Firenze, Via G. Capponi 9, I-50121 FI, Italy*

[3]*Dipartimento di Scienze della Terra, Università di Pavia, Via Ferrata 1, I-27100 PV, Italy*

Abstract

Differential scanning calorimetry supported by X-ray powder diffraction has been applied to the analysis of cogrinding-induced crystallinity changes of naproxen in mixtures with maltohexaose. Relevant factors were the mixture composition and the duration of mechanical treatment which did not affect the chemical integrity of the drug.

1. Introduction

Naproxen ((S)-(+)-6-methoxy-α-methyl-2-naphthaleneacetic acid, NAP) is a non steroidal anti-inflammatory drug substantially insoluble in water (about 27 $mg \cdot L^{-1}$ at 25 °C) which is not easily transformable into the amorphous state by freeze-drying or spray-drying. The drug can be amorphized by cogrinding its mixtures with cyclodextrin derivatives [1,2], maltoheptaose [3] or other linear maltooligomers [4]. In this paper the amorphization capacities of maltohexaose (M6), the non-cyclic analog of alpha-cyclodextrin, toward NAP were investigated in depth by testing with differential scanning calorimetry and X-ray powder diffractometry the NAP-M6 mixtures at 0.67, 0.50, 0.33 and 0.25 mole fraction of drug after grinding times of 0, 10, 20 and 30 min.

2. Materials and Methods

2.1. MATERIALS

Naproxen (NAP) purchased from Sigma Chemical Co (St. Louis, MO, USA) was recrystallized from ethanol. Maltohexaose (M6) was kindly provided by Nihon Shokuhin Kako Co Ltd (Tokyo, Japan).

Physical mixtures of NAP (75-250 μm sieve granulometric fraction) with M6 at 0.67, 0.50, 0.33 and 0.25 mole fraction of NAP (i.e. 0.30, 0.18, 0.10 and 0.067 mass fraction of NAP, respectively) were prepared by turbula mixing for 10 min. Grinding was carried out manually using an agate mortar with a pestle on ≈ 100 mg specimens which were tested by DSC and XRD after grinding times of 0, 10, 20 and 30 min. The same mechanical treatment was performed on crystals of pure NAP for control purposes.

2.2. METHODS

Temperature and enthalpy values were measured with a METTLER STAR® system equipped with a DSC821® Module (3-5 mg samples, 30-300 °C temperature range) and a Mettler TA 4000 apparatus equipped with a DSC 25 cell (6-10 mg samples, 30-180 °C temperature range) in open Al pans at the heating rate of 10 $K{\cdot}min^{-1}$ under static air atmosphere. The fraction of NAP transformed from the crystalline to amorphous state at a prescribed grinding time (t, min) was estimated by Equation (1)

$$\text{Crystallinity \%} = \frac{\Delta H_{GM,t}}{\Delta H_{PM}} \text{x}100 \qquad (1)$$

where $\Delta H_{GM,t}$ and ΔH_{PM} are the heats of fusion of NAP calculated in the ground mixture after t min of mechanical treatment and in the initial physical mixture, respectively. The measurements were taken three times for each sample (coefficient of variation < 4%).
X-ray diffraction patterns were recorded with a computer-controlled Philips PW 1800/10 apparatus equipped with a specific PC-APD software. Wavelengths: $CuK_{\alpha,1}$ = 1.54060 Å, $CuK_{\alpha,2}$ = 1.54439 Å. Scan range: 2-50 °2θ. Scan speed: 0.02 $°2\theta{\cdot}s^{-1}$. Monochromator: graphite crystal. The crystallinity of NAP in the ground mixtures was estimated by Equation (2), where HWHM is the half width at half maximum [5] of the

$$\text{Crystallinity \%} = \frac{HWHM_{PM}}{HWHM_{GM,t}} \text{x}100 \qquad (2)$$

peak in the 6.71-6.75 °2θ range and subscripts PM and GM,t refer to the starting physical mixture (100% crystallinity) and the mixture ground for t min, respectively.
Thin layer chromatography was run on TLC aluminium sheets coated with silica 60 (F_{254} Merck) which were developed in an acetic acid:tetrathydrofuran:toluene (1:3:30) solution to check the presence of degradation products of NAP in the ground samples.

3. Results and Discussion

A sharp endothermal effect (T_{onset} = 153.4±0.3 °C, T_{peak} = 156.7±0.4 °C, fusion enthalpy 140±5 $J{\cdot}g^{-1}$ (4 runs)) was associated with melting of anhydrous crystals of pure NAP (Fig. 1a), whilst M6 displayed a broad endothermal effect due to dehydration (7.7±0.2% mass loss as mass fraction) which was followed by a glass transition [6] (T_{onset} = 184.4±0.3, $T_{midpoint}$ = 187.9±0.2 °C (3 runs)) before thermal decomposition (Fig. 1b). The melting peak of NAP was substantially unaffected in its shape and area by blending with M6 so that the drug maintained its original crystallinity in the physical mixtures (Fig. 1c). Crystallinity was also unaffected by grinding crystals of the pure drug, but was clearly altered by cogrinding. The relative enthalpy change of NAP melting with respect to the starting physical mixture was considered to be a measure of the apparent degree of crystallinity. Quantitative data extracted from the DSC curves in Fig. 1c are plotted in Fig. 1d. The higher extent of drug amorphization (55%) was brought about in the equimolar mixture after 30 min of grinding, as observed for other maltooligomers [3,4], while 27% and 47% amorphous NAP was found in the mixtures at higher carrier-to-drug ratios, i.e. containing 0.33 and 0.25 mole fraction of NAP, respectively. The mixture with excess of NAP (i.e., 0.67 mole fraction of drug) contained only 16% of amorphous NAP after 30 min of grinding. The asymmetry of the melting peak of NAP after 30 min of grinding was evident by the gradual decrease in the onset temperatures in the mixtures at ≤ 0.50 mass fraction of drug (Fig. 1e). Since TLC indicated that chemical degradation of NAP did not occur under the grinding conditions, the effect can be attributed to melting of very fine NAP crystals embedded in a M6 matrix as a result of cogrinding [7]. Actually, in spite of peak asymmetry, in all combinations the peak temperature remained substantially that of pure NAP. Amorphization by cogrinding with hydroxypropyl alpha-cyclodextrin [2] was instead associated with considerable drops of the fusion peak temperature of NAP, suggesting the formation of a true drug-cyclodextrin inclusion complex in the solid state.

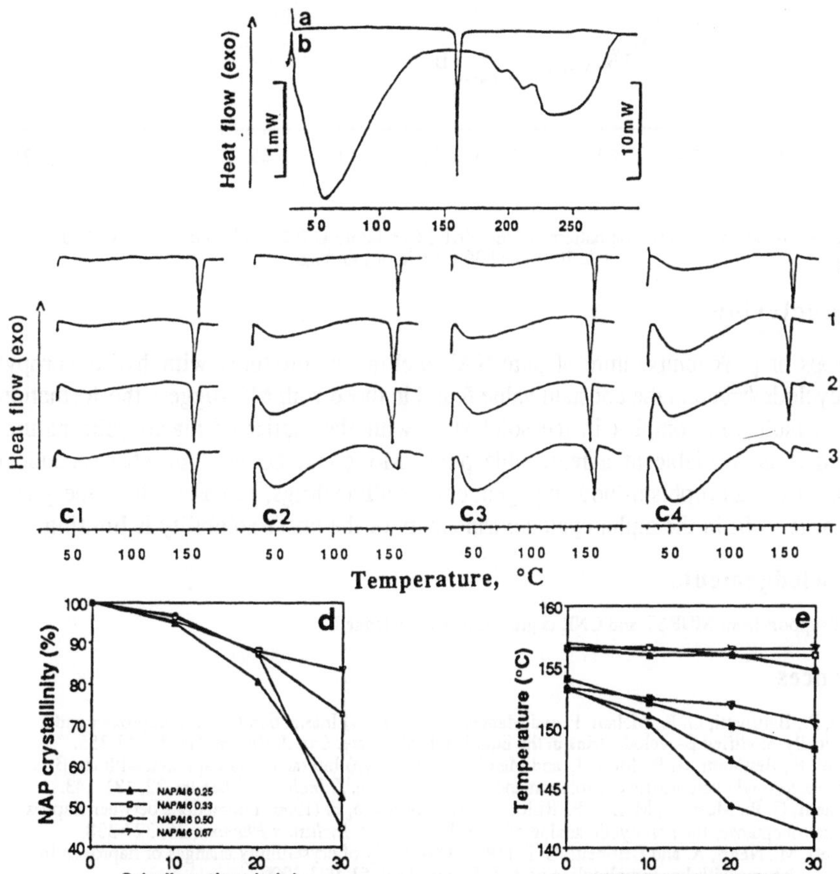

Figure 1. DSC curves of NAP (a), M6 (b), NAP-M6 mixtures at 0.67 (c1), 0.50 (c2), 0.33 (c3) and 0.25 (c4) mole fractions of drug (grinding times (min) on the curves). (d) Decrease in NAP crystallinity with grinding time in the NAP-M6 mixtures at the 0.67 (∇), 0.50 (O), 0.33 (□) and 0.25 (Δ) mole fractions of drug. (e) Effect of grinding time on NAP melting peak (open symbols) and onset (closed symbols) temperature values in the NAP-M6 equimolar mixtures.

The general trend of NAP crystallinity to decrease at increasing grinding times can be observed in the X-ray diffraction patterns (Fig. 2). In the mixture containing 0.25 mole fraction of NAP ground for 30 min, for example, 41% amorphous NAP was calculated from the HWHM increase of the peak in the 6.71-6.75 °2θ range with respect to the starting physical mixture, in reasonable agreement with the DSC results (47%).

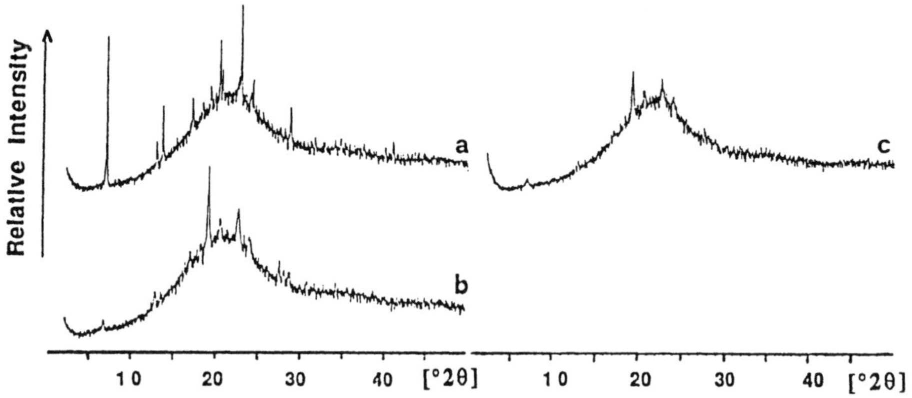

Figure 2. X-ray powder diffraction patterns of the NAP-M6 mixture at 0.25 mole fraction of NAP after 0 (a), 10 (b) and 30 (c) min of grinding.

4. Conclusion

The drops of peak temperature of pure NAP melting in mixtures with hydroxypropyl alpha-cyclodextrin and the constant value found in those with M6 suggest the formation of a true inclusion complex in the solid state with the carrier of macrocyclic nature. Provided it is available at a reasonable price, M6 could be used in pharmaceutical applications as amorphism-inducing agent of crystalline drugs, in cases where the guest is too large to fit in the alpha-cyclodextrin cavity or the host's solubility is limiting.

Acknowledgements

Financial support from MURST and CNR is gratefully acknowledged.

References

1. Mura, P., Bettinetti, G. P., Melani, F. and Manderioli, A. (1995) Interaction between naproxen and chemically-modified β-cyclodextrins in the liquid and solid state. *Eur. J. Pharm. Sci.* 3, 347-355.
2. Melani, F., Bettinetti, G. P., Mura, P. and Manderioli, A. (1995) Interaction of naproxen with α-, β-, and γ-hydroxypropyl cyclodextrins in solution and in the solid state. *J. Inclusion Phenom.* 22, 131-143.
3. Bettinetti, G. P., Mura, P., Melani, F., Rillosi, M. and Giordano, F. (1996) Interactions between naproxen and maltoheptaose, the non-cyclic analog of β-cyclodextrin. *J. Inclusion Phenom.* 25, 327-338.
4. Sorrenti, M., Negri, A. and Bettinetti, G. P. (1998) DSC study of crystallinity changes of naproxen in ground mixtures with linear maltooligomers. *J. Therm. Anal.* 51, 993-1000.
5. Bettinetti, G. P., Rillosi, M., Setti, M. and Mura, P. (1996) The amorphous state of α-cyclodextrin. *Minutes Formul. Poorly-available Drugs Oral Admin.* 311-314. Eur. Symp. (Paris). APGI.
6. Orford, P. O., Parker, R. and Ring, S. G. (1990) Aspects of the glass transition behaviour of mixtures of carbohydrates of low molecular weight. *Carbohydr. Res.* 196, 11-18.
7. Oguchi, T., Terada, K., Yamamoto, K. and Nakai Y. (1989) Molecular state of methyl p-hydroxybenzoate in the solid dispersion prepared by grinding with α-cyclodextrin. *Chem. Pharm. Bull.* 37, 1886-1888.

DISSOLUTION RATE AND THERMAL PROPERTIES OF NAPROXEN IN MIXTURES WITH AMORPHOUS OR CRYSTALLINE DIMETHYL BETA-CYCLODEXTRIN

G. P. BETTINETTI[1], M. SORRENTI[1], A. NEGRI[1], P. MURA[2], M. T. FAUCCI[2]
[1]*Dipartimento di Chimica Farmaceutica, Università di Pavia, Viale Taramelli 12, I-27100 PV, Italy*
[2]*Dipartimento di Scienze Farmaceutiche, Università di Firenze, Via G. Capponi 9, I-50121 FI, Italy*

Abstract

Dissolution rate enhancements of naproxen (NAP) in mixtures with amorphous (DS 1.8 randomly methylated) or crystalline (2,6-di-O-methyl) βCd were directly and linearly related to the carrier-to-drug ratios. Substantially the same increase in dissolution efficiency (4.5- to 19-times that of pure NAP) was carried out by both carriers in powder mixtures, while the crystalline carrier was a better dissolution-rate enhancer from non-disintegrating tablets at constant surface area. Differential scanning calorimetry showed a similar heating-induced modification of the solid state of NAP brought about by carriers.

1. Introduction

Solubility and dissolution rate enhancements of naproxen (NAP), a hydrophobic drug with analgesic and antipyretic properties substantially insoluble in water (≈ 27 mg·L^{-1} at 25 °C), which were carried out by kneading, coevaporation or colyophilization with a number of cyclodextrins (Cds) revealed the particular efficacy of amorphous, DS 1.8 randomly methylated βCd (RAMEB) in the equimolar combination with NAP [1-3]. Since some biopharmaceutical advantages can be obtained using simple drug-Cd binary mixtures [4] in drug formulation, it seemed of interest to investigate the performanceof RAMEB physically mixed with NAP at various carrier-to-drug ratios, as well as that of crystalline heptakis(2,6-di-O-methyl) βCd (DIMEB) in mixtures with NAP of the same compositions. The solubilizing effectof each carrier was evaluated from the respective phase-solubility diagram in water at 25 °C, while the dissolution behaviour in water at 37 °C of each mixture was determined according to the dispersed amount and rotating disc methods. Differential scanning calorimetry (DSC), supported by X-ray powder diffraction, was used to characterize the solid combinations of NAP with the amorphous or crystalline methylated βCd and to shed light on possibile interactions in the solid state between the drug and both carriers

2. Materials and Methods

2.1. MATERIALS

Naproxen (NAP) from Sigma (St. Louis, MO, USA) recrystallized from ethanol and heptakis(2,6-di-O-methyl) βCd (DIMEB) from Cyclolab (Budapest, HU) were used. Randomly methylated βCd at degree of substitution per anhydroglucose unit (DS) 1.8 (RAMEB) was kindly provided by Wacker Chemie GmbH (München 70, FRG).
Physical mixtures of NAP (75-250 μm sieve granulometric fraction) with RAMEB or DIMEB at 0.20, 0.33, 0.50, 0.60, and 0.67 mole fraction (i.e. 0.59, 0.73, 0.85, 0.90, and 0.92 mass fraction) of carrier were prepared by turbula mixing for 15 min.

2.2. METHODS

Solubility measurements of NAP were carried out at 25±0.5 °C by adding 30 mg of drug to 30 mL of water or aqueous solution of DIMEB or RAMEB in the 5 to 25 $mmol{\cdot}L^{-1}$ concentration range and following the procedure described elsewhere [2]. Each experiment was performed in triplicate (coefficient of variation CV 5%). The apparent binding constant of the NAP-Cd complex was calculated from the slope and intercept of the straight line of the phase-solubility diagram, in terms of Equation (1) [2].

$$K_{(1:1)} = \frac{\text{slope}}{\text{intercept}(1-\text{slope})} \qquad (1)$$

Dispersed amount experiments were performed at 37±0.5 °C by adding 60 mg of NAP or NAP equivalent to 75 mL of water under the experimental conditions described elsewhere [2]. Each test was repeated 4 times, CV < 1.5%. In the rotating disc method, samples of 300 mg were compressed (disc area 1.33 cm^2) and tested at 37±0.5 °C in 150 mL of water as described elsewhere [2]. Each test was repeated 4 times, CV < 8%.
Temperature and enthalpy values were measured with a METTLER STAR^e system equipped with a DSC821^e Module on 3-5 mg samples in open Al pans at the heating rate of 10 $K{\cdot}min^{-1}$ in the 30-180 °C temperature range under static air atmosphere.
X-ray diffraction patterns were taken with a computer-controlled Philips PW 1800/10 apparatus equipped with specific PC-APD software. Wavelengths: $CuK_{\alpha,1}$ = 1.54060 Å, $CuK_{\alpha,2}$ = 1.54439 Å. Scan range: 2-50 °2θ. Scan speed: 0.02 $°2\theta{\cdot}s^{-1}$.

3. Results and Discussion

A_L-type phase-solubility diagrams in aqueous solution at 25 °C indicate an increase in solubility of two orders of magnitude that of pure NAP in the presence of 0.025 $mol{\cdot}L^{-1}$ of RAMEB or DIMEB (Fig. 1). The same solubilizing efficiency of the carriers was reflected by the stability constant values of their equimolar complexes with NAP, which were 6778 (±10%) $L{\cdot}mol^{-1}$ and 6200 (±10%) $L{\cdot}mol^{-1}$ for RAMEB and DIMEB, respectively.

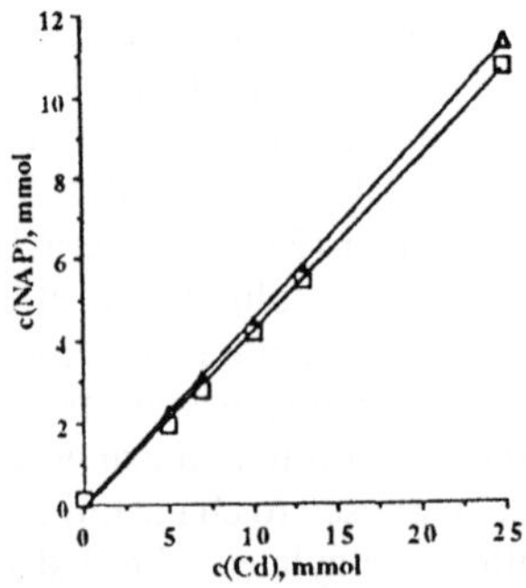

Figure 1. Phase-solubility diagram in aqueous solution (25 °C) of NAP with RAMEB (Δ), DIMEB ( ).

Dispersed amount experiments revealed an increase in the dissolution efficiency (area under the dissolution curve with t=60 min (measured using the trapezoidal rule) expressed as a percentage of the area of the rectangle described by 100% dissolution in the same time [3]) at increasing carrier contents. The improvement was substantially equivalent for the NAP-RAMEB and NAP-DIMEB mixtures of the same composition (no statistically significant differences at P > 0.1) (Figs. 2a and 2b). Figure 2c shows the increase in dissolution efficiencyas a function of the relative amount of DIMEB in the mixture.

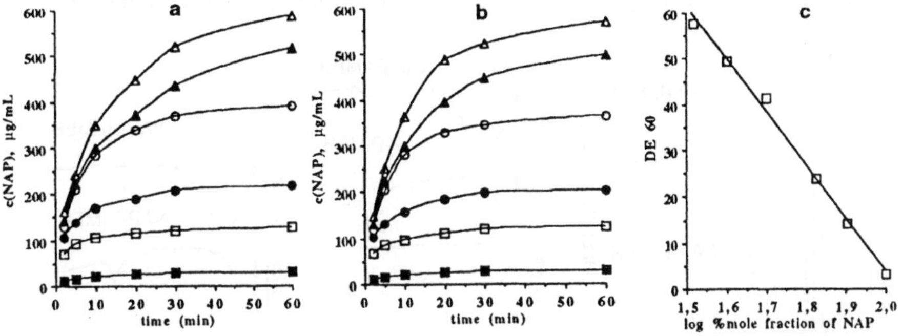

Figure 2. Mean dissolution curves (dispersed amount method) of NAP (■) and (a) NAP-RAMEB, (b) NAP-DIMEB mixtures at 0.20 (□), 0.33 (●), 0.50 (○), 0.60 (▲), and 0.67 (Δ) mole fraction of carrier; (c) dissolution efficiency as a function of DIMEB mole fraction.

Dissolution rates of NAP from non-disintegrating tablets at constant surface area calculated from the linear portion of dissolution profiles showed a parallel increase at increasing carrier contents, but in this respect crystalline DIMEB was slightly more effectivethan amorphous RAMEB (Fig. 3). A statistically significant difference (P = 0.05) in terms of dissolution rate constants was however found only between the NAP-DIMEB and the respective NAP-RAMEB tablets containing excess of drug with respect to the equimolar composition.

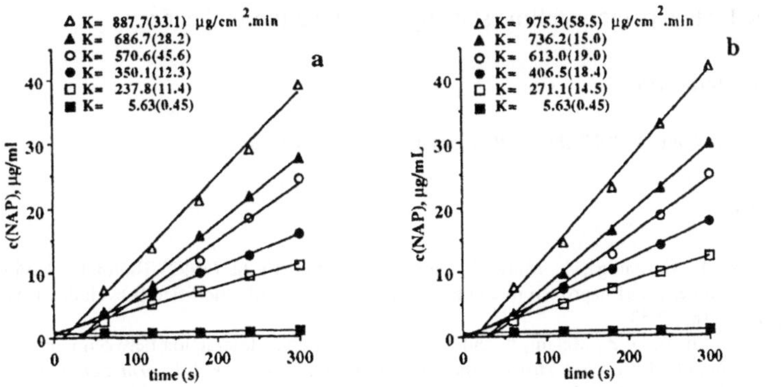

Figure 3. Dissolution rate (rotating disc method) of NAP (■) and (a) NAP-RAMEB, (b) NAP-DIMEB mixtures at 0.20 (□), 0.33 (●), 0.50 (○), 0.60 (▲), and 0.67 (Δ) mole fraction of carrier.

Alterations of the thermal properties of pure NAP (mp 156.7±0.3 °C, fusion enthalpy 140±6 $J \cdot g^{-1}$) are evident in the DSC curves of physical mixtures with RAMEB and DIMEB (Fig. 4). Since the XRD characteristics of the individual components were maintained, the supply of thermal energy during the DSC scan can be considered the driving force of the modification of physical state of NAP responsible for changes in melting endotherm. Such changes were more profound in ground mixtures, probably due to a more intimate physical contact between the components and/or a partial amorphization brought about by mechanical treatment. A total loss of the endotherm associated with NAP melting can be seen in the mixtures at higher carrier contents.

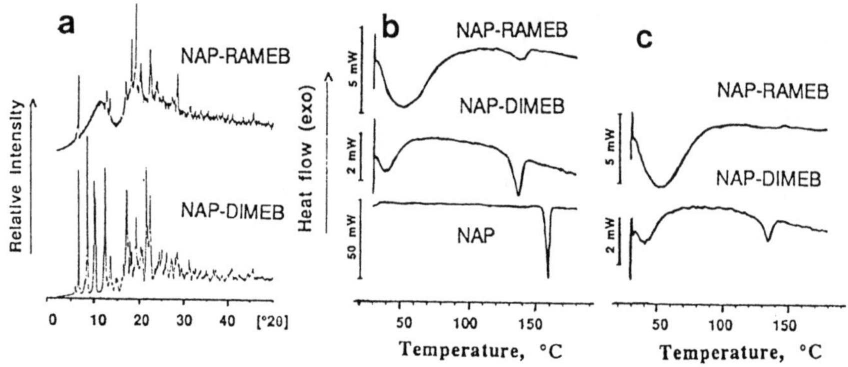

Figure 4. XRD (a) and DSC curves of the physical (b) and ground (c) mixtures of NAP with RAMEB and DIMEB of equimolar composition.

4. Conclusion

Some biopharmaceutical advantages can be obtained using physical mixtures of NAP with RAMEB or DIMEB. The efficacy of both carriers as solubilizer and dissolution rate enhancer for NAP is very similar, though DIMEB seems to be slightly more effectivein enhancing the drug dissolution rate from tablets. The choice of the amorphous carrier in pharmaceutical formulations is thereforesuggested mainly by economic reasons. DSC reveals similar drug-carrier interactions in the solid state which make NAP prone to be transformed into a paracrystalline state by supplying thermal energy.

Acknowledgements

Financial support from MURST and CNR is gratefully acknowledged.

References

1. Bettinetti, G. P., Gazzaniga, A., Mura, P., Giordano, F. and Setti, M. (1992) Thermal behaviour and dissolution properties of naproxen in combinations with chemically modified β-cyclodextrins. *Drug Dev. Ind.. Pharm.* 18, 39-53.
2. Mura, P., Bettinetti, G. P., Melani, F. and Manderioli, A. (1995) Interaction between naproxen and chemically-modified β-cyclodextrins in the liquid and solid state. *Eur. J. Pharm. Sci.* 3, 347-355.
3. Melani, F., Bettinetti, G. P., Mura, P. and Manderioli, A. (1995) Interaction of naproxen with α-, β-, and γ-hydroxypropyl cyclodextrins in solution and in the solid state. *J. Inclusion Phenom.* 22, 131-143.
4. Fromming, K. and Szejtli, J: *Cyclodextrins in Pharmacy*, Kluwer Acad. Publ., Dordrecht, 1994, pp. 143-146.

IMPROVEMENT OF ECONAZOLE SOLUBILITY IN MULTICOMPONENT SYSTEMS WITH CYCLODEXTRINS AND ACIDS

P. MURA, G. FRANCHI, M.T. FAUCCI, A. MANDERIOLI, G. BRAMANTI
Dipartimento di Scienze Farmaceutiche, Università di Firenze, , Via G. Capponi 9, I-50121 Firenze

Abstract

Econazole is an imidazole antifungal agent very poorly water soluble. The combined effects of different acids (nitric, citric, lactic and malic) and cyclodextrins, both natural (α- and γ-) and derivative (statistically-hydroxypropylated β-cyclodextrin), on the enhancement of aqueous solubility of drug was investigated. Multicomponent complex formation was always more effective than salt formation or binary complexation in enhancing the aqueous solubility of econazole. The best result was obtained with the combination of econazole with α-cyclodextrin and lactic acid, in the respective molar ratios 1:1:2.5, which gave an increase of solubility of more than 4200 times in comparison with the pure drug

1. Introduction

Econazole is an imidazole antifungal agent suitable for the treatment of many micotic infections, however its very low water solubility (about 5 μg/mL at 25°C) limits both its therapeutic application and efficacy. Cyclodextrin complexation has been widely used to improve solubility and dissolution rate of a number of hydrophobic drug molecules. Improvement of both water solubility and antifungal activity of econazole was obtained by combining it with α- and ß-cyclodextrins [1]. Nevertheless, the usefulness of natural cyclodextrins, particularly of ß-cyclodextrin, has been limited by their relatively low aqueous solubility. Some papers have recently been published showing that drug-cyclodextrin complexation in the presence of acids often resulted more effective than the corresponding classic two-component complexes in improving the solubility characteristics of base-type active principles [2, 3]. The purpose of the present study was to investigate the combined effects of different acids (nitric, citric, lactic and malic) and cyclodextrins, both natural (α- and γ-) and derivative (statistically-hydroxypropylated β-cyclodextrin), on the enhancement of aqueous solubility of econazole.

2. Experimental

2.1. MATERIALS

Econazole (1-[2-(4-chlorophenyl)methoxy]-2-(2,4-dichlorophenyl)ethyl)-1H-imidazole, ECO) was kindly donated by Italfarmaco (I-Genova), γ-cyclodextrin (γ-Cd) and hydroxypropyl- β-cyclodextrin (average substitution degree 0.9 per anhydroglucose unit, HPβCd) were a gift of Wacker Chemie (D-München). Commercial α-cyclodextrin (α-Cd) and the various acids were purchased from Sigma (USA, St. Louis, MO).

2.2. SOLUBILITY STUDIES

Water drug solubility, alone or in the presence of equimolar concentrations of each studied acid and/or Cd, was determined by adding suitable amounts of ECO, acid, and/or Cd to aqueous solutions which were electromagnetically stirred until equilibrium at 25 °C. The solutions were then filtered (0.45 μm filter pore size) and assayed for drug concentration by second derivative UV spectroscopy [1] using a spectrophotometer Perkin Elmer Mod. 552S (USA-Norwalk). The presence of Cd or acid did not interfere with the spectrophotometric assay. Each test was performed in triplicate (C.V. <5%).

2.3. PHASE SOLUBILITY STUDIES

Solubility equilibrium diagrams in water at 25 °C were obtained according to Higuchi and Connors [4]. In binary systems, excess drug was added to aqueous solutions containing increasing concentrations of Cd (or acid) in sealed glass containers and electromagnetically stirred at constant temperature (25 °C) until equilibrium. The procedure was the same also in ternary systems, with the difference that the excess drug was added together with an equimolar amount of acid (or Cd). When equilibrium was achieved (2 d), the solutions were filtered (0.45 μm filter pore size) and assayed for drug concentration by UV spectroscopy as described above. Experiments were performed in triplicate (C.V. <5%).

3. Results and discussion

Aqueous solubility of ECO in binary and ternary equimolar systems with Cds and acids are shown in Table 1. Salt formation was always more efficacious than Cd complexation, except in the case of nitric salt (even though it is the only ECO salt actually used in pharmaceutical formulations). However the highest solubility obtained by salt formation (0.35 % w/v with citric acid) was not enough for practical utilization and optimal drug efficacy. In fact its pharmaceutical dosage forms (i.e. solutions, sospensions, lotions, ointments, etc.) are all marketed at 1% w/v. On the other hand, as regards the studied cyclodextrins, α-Cd showed the best solubilizing efficacy, followed by HPβCd, whereas γ-Cd was practically inefficacious. The combined use of acid and Cd showed a synergistic effect and was always clearly more effective in enhancing the aqueous solubility of ECO in comparison with either salt formation or binary complexation. Ternary systems containing α-Cd were again the best, all allowing the desired drug solubility level (10 mg/mL) to be reached and even exceeded, except in the combination with nitric acid. However, interestingly, high enhancement of drug solubility was achieved also with ternary systems containing γ-Cd, despite the very poor solubilizing power observed for this cyclodextrin in binary combination with ECO. Probably the presence of a third

component provided a best fit of ECO into the cavity of γ-Cd, too large to include drug alone.
Further enhancement of drug solubility, even though of different degree with the various acids, was obtained passing from 1:1:1 to 1:1:2.5 drug-Cd-acid molar ratio (except for malic acid). The best result was obtained with ECO-α-Cd-lactic acid at 1:1:2.5 molar ratio, where the drug solubility arrived at 22.5 mg/mL.

Table 1. Aqueous solubility of econazole (ECO), its 1:1 nitrate, lactate, citrate and malate salts, its 1:1 complexes with cyclodextrins, and ternary systems at 1:1:1 and 1:1:2.5 ECO-Cd-acid molar ratios.

	ECO solubility, mg/mL						
		1:1:1 ternary systems			1:1:2.5 ternary systems		
Acid	-----	α-Cd	γ-Cd	HPβCd	α-Cd	γ-Cd	HPβCd
-----	0.005	0.620	0.007	0.260	0.620	0.007	0.260
nitric	0.480	5.60	1.90	3.10	---	---	---
lactic	3.50	11.6	6.80	11.4	22.5	17.7	22.0
citric	3.30	12.9	9.10	12.2	19.2	16.9	18.3
malic	1.70	12.8	8.60	9.90	13.3	6.50	9.15

Phase-solubility studies of ECO in binary and ternary systems with α-Cd and acids were then performed to obtain some more information about the multicomponent complex formation. Solubility diagrams obtained by adding increasing amounts of α-Cd to 1:1 mol/mol ECO-acid mixtures, all showed a linear increase of drug aqueous solubility as a function of Cd concentration and can be classified as A_L type according to Higuchi and Connors [4], indicating the formation of soluble complexes (Fig. 1, A).

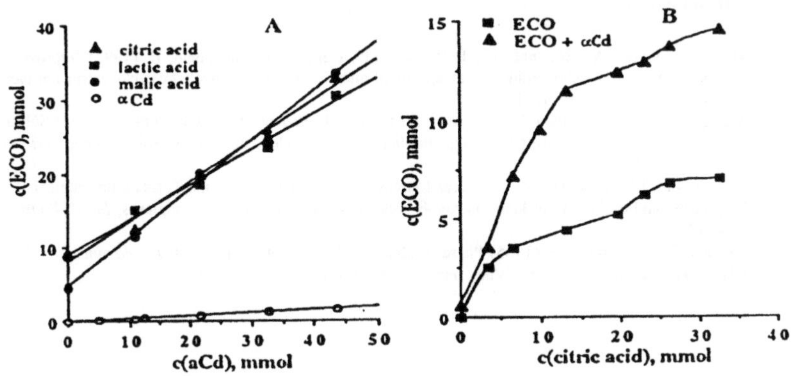

Figure 1. Phase-solubility diagrams of ECO, alone or in the presence of equimolar amounts of acid (A) or α-Cd (B), as a function of α-Cd (A) or acid (B) concentration

The values of 1:1 stability constants, calculated from the slope of the initial straight portion of the phase-solubility diagrams [4], were 2630 M^{-1} for binary system with α-Cd, and 450, 105 e 130 M^{-1} for multicomponent complexes with malic, lactic and citric acids respectively.

The sharp reduction of interaction between drug and α-Cd in the presence of various acids was explained on the basis of increased ionization of ECO (pH 3-4) and the generally less affinity of ionized drugs for the apolar Cd cavity. On the other hand, the differences between K values found with the various acids might be due to the different fit they provide for the drug with α-Cd. However it is evident at a glance that, in spite of the lowering of K values, the solubilizing efficiency of multicomponent complexes was much higher than that of the binary complex.

Analogously, phase solubility diagrams of ECO as a function of acid concentration (Fig. 1, B), showed a further increase of drug solubility when α-Cd (in 1:1 mol/mol ratio with the drug) was present, confirming the synergistic effect of acid and cyclodextrin in promoting ECO solubility.

4. Conclusion

Multicomponent complex formation between a base-type drug, such as ECO, cyclodextrin and hydroxy acid as third component, showed to be an optimal tool for enhancing the drug solubility. However the proper acid and the proper molar ratio for a given host and guest should be experimentally selected. In fact the solubilizing efficacy of a given multicomponent system was not related either to the corresponding salt or binary complex solubility and the optimal drug:Cd:acid molar ratio was different for the various systems. The best combination was the ternary system with αCd and lactic acid which gives, in the molar ratio 1:1:2.5, an increase of solubility of more than 4200 times in comparison with the pure drug.

Acknowledgements

Financial support from MURST and CNR is gratefully acknowledged.

5. References

1) Mura, P., Liguori, A., Bramanti, G., Bettinetti, G.P., Campisi, E. and Faggi, E. (1992) Improvement of dissolution properties and microbiological activity of miconazole and econazole by cyclodextrin complexation, *Eur. J. Pharm. Biopharm.*, *38*, 119-123.

2) Szente, L, Szejtli, J., Vikmon, M., Fenyvesi, E., Pasini, M., Redenti, E. and Ventura, P. (1995) Solution for insolubility problems of base-type drugs: multicomponent cyclodextrin complexation, *Proc. 1st Meeting APGI/APV*, 579-580

3) Esclusa-Diaz, M.T., Gayo-Otero, M., Pérez-Marcos, M.B., Vila-Jato, J.L. and Torres-Labandeira, J.J. (1996) Preparation and evaluation of ketoconazole-ß-cyclodextrin multicomponent complexes, *Int. J. Pharm.*, *142*, 183-187.

4) Higuchi, T. and Connors, K (1965) Phase solubility techniques, in Reilly, C. (ed.), *Advances in Analytical Chemistry and Instrumentation*, Wiley Interscience, New York, pp. 117-212.

PHASE-SOLUBILITY ANALYSIS AND MOLECULAR MODELING IN THE STUDY OF NAPROXEN-CYCLODEXTRIN INTERACTION

F. MELANI, P. MURA, B. TRACUZZI
Dipartimento di Scienze Farmaceutiche, Università di Firenze,
Via G. Capponi 9, I-50121 Firenze

Abstract

This paper describes the use of molecular modeling techniques to get a better understanding of the mechanism of complex formation between cyclodextrins and guest molecules. Naproxen was used as model drug molecule, whereas natural cyclodextrins and β-cyclodextrins statistically alkylated were selected as carriers. a statistically significant correlation was found between the stability constants of the complexes, experimentally determined from solubility phase experiments, and the docking energy of the interacting molecules, calculated at a minimum value of conformational energy by a suitably modified force field. The important role of water molecules in the complexation process was pointed out.

1. Introduction

Cyclodextrins are widely used in the pharmaceutical field due to the their potential to form inclusion complexes with appropriated drug molecules, thus improving some of drug physicochemical properties, particularly water solubility and stability [1]. Moreover the ability to discriminate in complexation towards enantiomers, to catalyse reactions on substrates, or to mime the enzyme behaviour leads cyclodextrins to be considered as very interesting model compounds. A better knowledge of the interactions at molecular level between host and guest molecules should be of great help in enhancing and optimizing the cyclodextrin applications. Molecular modeling techniques have been recently proposed as a powerful tool to obtain the three-dimensional structures of the complexes and study the complexation phenomenon. We undertook the present investigation to study whether it is possible to predict the stability of cyclodextrin complexes using computer-aided molecular modeling. On this basis, we examined the possibility of establishing a correlation between the complex stability and the docking energy of the interacting molecules, calculated through a suitable theoretical model. Naproxen was selected as model drug molecule, whereas natural cyclodextrins and β-cyclodextrins statistically alkylated were selected as carriers. Association constants, assumed as a quantitative measure of the complex stability, were determined by phase-solubility experiments [2].

2. Experimental

2.1. MATERIALS

Naproxen ((*S*)-(+)-6-methoxy-α-methyl-2-naphthaleneaceticacid, NAP) and α- (αCd), β- (βCd) and γ- (γCd) were purchased from Sigma Chemical Co. (St. Louis, MO, USA); methyl- (MeßCd), hydroxyethyl- (HEßCd) and hydroxypropyl- (HPßCd) ß-cyclodextrins, with average substitution degree per anhydroglucose unit of 1.8, 1.6, and 0.9 respectively were kindly donated by Waker-Chemie GmbH (Münich, G).

2.2. PHASE-SOLUBILITY STUDIES

A known excess of NAP was added to unbufferedaqueous solutions (pH≈5) containing various known concentrations of Cd and the resulting mixtures were magnetically shaken at 25 °C until equilibrium was achieved (3 days). Samples were then withdrawn with a syringe-filter (0.45 μm) and spectrophotometrically assayed by a second derivative method [3].

2.3. MOLECULAR MODELING

Analysis and modeling of the structures of the NAP-Cd complexes were carried out using the INSIGHT II 95.0 program [4]. The ßCd derivatives were built-up by adding to ßCd (base molecule) 12 methyl (DS 1.8), 6 hydroxypropyl (MS 0.9) or 11 hydroxyethyl (MS 1.6) groups. Various patterns of substituent distribution were examined for each substituted ßCd; no statistically significant differencesin docking energy values were observed by varying the relative position of substituted glucoses. NAP was fitted into the Cd cavity in an axial orientation, with the carboxyl group directed toward both the widest and the narrowest rim of the cavity [5]. Solvated structures, for calculating the interaction energy with water molecules, were obtained by inserting NAP, Cd, or NAP-Cd complex in a water shell with a 10 Å radius. Each structure (both solvated or not) was subjected to a simulated annealing process from 900 to 0 K (AMBER force field, DISCOVER 2.9.7 program [4]), performing iterations up to a minimum constant value of conformational energy.

3. Results and Discussion

Our theoretical approach was based on the search for a correlation between the experimentally determined stability constant of the complex and the docking energy between the interacting molecules in solution, obtained from the molecular modelling approach.
The NAP solubility (0.12 mmol L^{-1} in unbuffered,pH≈5, aqueous solution, at 25 °C) linearly increased (A_L type phase-solubility diagram [2]) in the presence of all tested Cds (Figure 1).

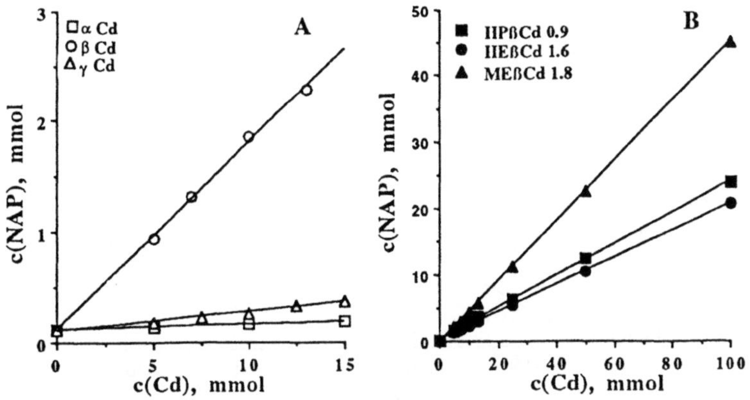

Figure 1. Phase solubility diagrams of NAP with natural (A) and ß-derivative (B) cyclodextrins at 25 °C.

The 1:1 apparent stability constants (Ks) of the complexes were calculated from the initial straight line portions of corresponding phase-solubility diagrams [2] (see Table 1). Docking energies of the various complexes at 0 K were calculated at a stable minimum value of conformational energy by the suitably modified AMBER force-field. This made it possible to obtain the individual contributions to the interaction energy of hydrogen bond and electrostatic and steric terms.. Stochastic methods based on molecular dynamic principle (simulated annealing) were used to find the stable conformations. Entropic contributions to the interaction were not taken into account in this preliminary approach. We calculated the docking energy values of the complex obtained when the guest was fitted into the cavity of each Cd, both in vacuum (CI) and inside a water shell (CHI), as well as the interaction energies between Cd and the water shell (CH) (Table 1).

Table 1

Stability constants Ks (M^{-1}) and docking values (kcal/mol) of NAP-Cd complexes (in vacuum (CI) and inside a water shell (CHI)) and Cd-water interaction energy (CH)

Cyclodextrin	Ks	CI	CHI	CH
αCd	40	-21.7	-6.3	-173.8
ßCd	1700	-27.8	-11.5	-64.8
gCd	146	-27.8	-10.6	-85.2
MeßCd	6890	-33.7	-24.1	-60.8
HPßCd	2580	-23.7	-24.7	-51.5
HEßCd	2145	-23.6	-7.8	-36.8

Correlations between the logarithm of the stability constants (Ks) of the complexes and the docking energies were statistically significant ($r>0.85$). Analysis of linear regression equations indicated that the complex stability was strongly influenced by the interaction between cyclodextrins and water: the higher the Cd-water interaction (CH), the less the stability constant of the corresponding NAP-Cd complex.

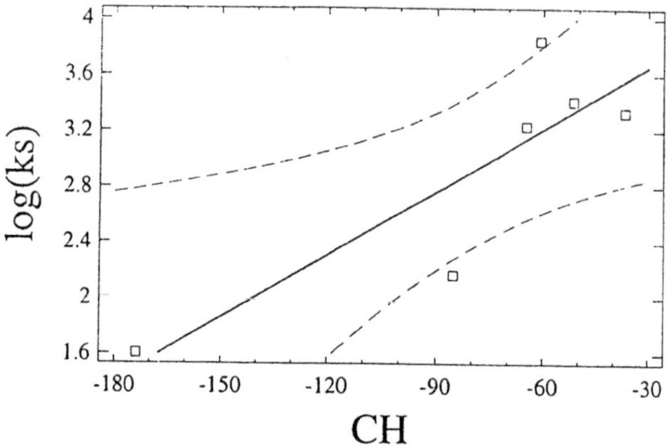

Figure 2. Relation between complex stability constant (Ks) and water-cyclodextrin docking energy (CH)

Correlation by Multiple Regression Analysis of the logarithm of the apparent stability constants with the docking energies CHI and CH (Figure 2) was statistically significant ($r = 0.921$; $s = 0.43$; $F = 8.44$) and confirmed that to obtain a stable drug-Cd complex in solution, not only a strong interaction between host and guest molecules is necessary but also the lowest water-Cd interaction.

4. Conclusions

The predictive power of the method can be considered satisfactory, and even though it is not possible to establish the accuracy of the predicted complex stability constant value but only its statistical significance, this approach would be useful for a rapid screening of a large number of possible complexes to identify the best candidates to subject both to experimental characterization and more accurate simulation. Moreover, we will consider the possibility of finding other variables that could further improve the prediction accuracy of the method.

Acknowledgements

Financial support from MURST and CNR is gratefully acknowledged.

5. References

1. Duchêne, D., *Cyclodextrins and their industrial uses*, Ed. De Santé, Paris, 1987.
2. Higuchi, T. and Connors, K. (1965) Phase solubility techniques, in Reilly, C. (ed.), *Advances in Analytical Chemistry and Instrumentation*, Wiley Interscience, New York, pp. 117-212.
3. Bettinetti, G.P., Mura, P., Liguori, A., Bramanti, G. and Giordano, F. (1989) Solubilization and interaction of naproxen with cyclodextrins in aqueous solution and in the solid state. *Il Farmaco, 44*, 195-213.
4. Biosym/MSI, 9865 Scranton Road, S. Diego, CA 92121-2777
5. Bettinetti, G. P., Melani, F., Mura, P., Monnanni R. and Giordano, F. (1991) Carbon-13 NMR study of naproxen interaction with cyclodextrins in solution *J. Pharm. Sci., 80*, 1162-1170

POTENCY MODIFICATION OF ANTIBACTERIAL ADAMANTANE DERIVATIVES BY COMPEXATION WITH βCD AND HP-βCD. STUDY OF THEIR THERMOTROPIC PROPERTIES IN DIPALMITOYL PHOSPHATIDYLCHOLINE BILAYERS CONTAINING CHOLESTEROL AND ^{13}C NMR STUDY OF THEIR STRUCTURES

Antoniadou-Vyza E[1]., Xitiroglou E.,[2] Papadopoulos A[3], Mayromoustakos T.[3]

[1]*Department of Pharmaceutical Chemistry, School of Pharmacy, University of Athens, Panepistimiopolis Zografou, Athens 15771, Greece*

[2]*Department of Microbiology School of Medicine, University of Athens, 75 M. Asias Goudi, Athens 11527, Greece*

[3]*Institute of Organic and Pharmaceutical Chemistry, The National Hellenic Research Foundation, 48 Vas. Constantinou, Athens 11523, Greece*

1. Introduction

The quarternary ammonium derivatives, octyl- and dodecyl bromide salts of 2-(3-dimethylaminopropyl)-tricyclo [3.3.1.1^{3,7}] decan-2-ol, ADM-8 and ADM-12 correspondingly were synthesized in our laboratory, and found to be active antibacterials with MIC values in the range of 1×10^{-5} M (1). As inclusion in Cyclodextrins, (CDs) is a convenient alternative to solve the problems encountered in the solubilization of hydrophobic drugs, βCD and HP-βCD complexes of the active compounds have been prepared, and as expected, they enhance the solubility of these agents about ten times. The complexes have been prepared according to the precipitation and the freeze drying methods respectively and have been characterized in solid state (2).

In this work we tried to investigate any possible improvement in the biological activity, as a result of complexation with CDs, by a comparative study of their antibacterial effectiveness in both state. This study shows significant differences between the free and the complexed form of the same substances, and also between the two different, compounds under investigation. In an attempt to clarify these differences we were interested in studying the thermal properties of the amphipathic molecules ADM-8 and ADM-12 in free or complexed form in dipalmitoylphosphatidylcholine (DPPC) bilayers containing cholesterol. Finally the determination of the accurate structure of new complexes, under study, with NMR Spectroscopy proved to be very useful for further understanding of the observed unexpected changes. Although in recent literature a plethora of drug/CD complexes have been prepared and studied for their improved solubility, chemical stability and other physicochemical properties, there is very limited information about the interaction of these supramolecules with phospholipid bilayers. Membrane cholesterol is a major determinant of bilayers fluidity Cholesterol molecules intercalate among phospholipids and prevent the fatty acid chains from packing together and crystallizing, a process that drastically reduce the membrane fluidity.

2. Materials and methods

The adamantanol derivatives were synthesized and purified in our laboratory, (1). Their cyclodextrin complexes are prepared according the precipitation method (2). βCD was obtained from Sigma Chemicals (St. Louis, MO). HPβCD was obtained from Janseen (Janseen Biotech), it has a degree of substitution (DS) 0.4 and a relative molecular mass calculated to be 1300. DPPC and cholesterol was purchased from Avanti Polar Lipids, Birmingham, AL and CHCl3 from Aldrich (99% pure, Aldrich Chemical Co., Milwaukee, Wisconsin). *Nuclear Magnetic Resonance Spectroscopy :* The ^{1}H NMR spectra were recorded at 200 MHz on a Bruker AC 200 instrument using D_2O as solvent and Tetramethylsilane (TMS) as external reference. Typical conditions were 16 K data points with zero fitting sweep width 1,4 KHz, giving a digital resolution of 0,34 Hz $poin^{-1}$, pulse width 2 (90 deg. pulse 5,5) acquisition time 2,9 s. Gaussian enhancement was used for the displayed spectra (GB = 0,2, LB =-2). *Microbiological assays :*For the antibacterial assays the compounds were first dissolved in DMF and then diluted with H_2O at the required quantities. In order to ensure that the solvent had no effect on bacterial growth, a control test was also performed. Inoculated petri dishes containing only DMF at the same dilutions as in our experiments were found inactive in culture medium.The compound suspensions were added in the desired concentrations into molten Muller-Hinton agar. After solidification, 1μl of the final suspension of 10^8 bacteria / ml were applied with a multipoint inoculator. Cultures were incubated for 24 h at 37° C. The lowest concentration of compounds that completely inhibited growth was considered to be the minimum inhibitory concentration (MIC) expressed in μg/ml. MIC value was the mean value of 3 measurements. *DifferEntial Thermal Calorimetry assays :*

DPPC alone or with the appropriate amount of adamantanol derivative in a simple or complex form with βCD, with or without cholesterol were dissolved in chloroform. After mixing the solvent was evaporated using an O_2-free N_2 stream and the samples were dried under high vacuum for 6 h. After adding distilled water (50% w/w), a portion of the sample (5 mg) was sealed in a stainless steel capsule. Thermograms were obtained on a Perkin-Elmer DSC-7 instrument. Prior to scanning, the samples were held above their phase transition temperature for 1-2 min to ensure complete equilibration. All samples were scanned at least twice until identical thermograms were obtained using a scanning rate of 2.5 °C/min. The temperature scale of the calorimeter was calibrated using fully hydrated DPPC and indium as standard samples.

3. Results

All compounds and CD complexes were studied for their antibacterial activity, against *Staphylococcous aureus, Streptococcous faecalis, Bacillus subtilis, Escherischia coli*, by maesuring the MIC values of the tested compounds (Table 1). The presence of βCD in DPPC/CHOL. bilayers results in a thermogram similar to that of pure DPPC bilayers. The incorporation of either drug molecules produces lowering of the phase transition temperature and increase in the breadth of the phase transition. The thermograms of the drug containing preparations are similar except that the ADM8 containing preparation contains a phase transition (37.5 °C) with a small distinct peak sitting at its onset. Significant thermotropic differences between the three preparations were observed in DPPC/CHOL. bilayers. While at low concentrations of cholesterol ADM8 caused less significant thermotropic effects than its ADM12 congener, at high cholesterol content caused drastic changes in the bilayer structure. Thus, in the high cholesterol content bilayers the presence of ADM8 squeezes out some of DPPC from the bilayers. This effect is not observed with ADM12.

The study of ^{1}H NMR and ^{13}C NMR spectra of the compounds in free and complexed form show remarkable chemical shift changes. (Table 2-4). These changes ($\Delta\delta = \delta^{free} - \delta^{obs}$) are of the range of - 0.08 ppm until + 1.38 ppm, and are located in three region of the obtained supramolecule: The adamantane ring moiety of the guest molecule, the cyclodextrins interior cavity and the aliphatic chains end point. In the case of both active molecules ADM-8 and ADM-12 the first two significant changes are indicative of the complexation and they arise as a consequence of the incorporation of the molecule from the adamantane ring side in the cyclodextrin cavity. Contrary the shifts attributed to the alkyl chains end point methyl groups are detected only in the case of ADM-12 molecule. These observations lead as to the possible conclusion that the guest molecules, in this case, extent from the CD molecules wide edge, and connects with the next complexes narrowest edge. The subsequent drug cyclodextrin supramolecules have such a position that allows them having a maximum distance, within each supramolecule is in contact with the next one, probably the end point of the guests long akyl tail, is inserted in the narrow edge of the next supramolecules CD cavity, suggesting that they are arranged in such a manner that they are producing nematoid assemblies. However, further work is needed to elucidate in a more detailed manner our suggestions.

4. Conclusion

The prepared complexes of the ADM8 and ADM10 molecules exert a significantly different thermotropic effect in DPPC bilayers and in antibacterial activity as a consequence of being in differently interacting with the other existing supramolecules. The local skin irritating effect, as well as the allergies. which appeared, after repeated treatment are common disadvantages limiting the utility of these cationic antibacterials. Therefore possibly compexation with CDs may be beneficial for their use, preventing the penetration of these agents from the treated surface and may reduce the unwanted effects.

Table 1 : Minimal inhibitory concentration (MIC) of the antibacterial compound ADM-8, ADM-10, ADM-12 and their different CD complexes. MIC (μg . ml^{-1})

	ADM 8	ADM 10	ADM 12	ADM8: βCD	ADM 10: βCD	ADM 12: βCD	ADM12: HPβCD
Staphylococcous aureous	12	4.2	2.4	8	1.5	4.2	4.2
Streptococcous faecalis	28	9	4.2	22	7	2.4	4.2
Bacilus subtilis	14	2,4	3	12	1.8	2,4	2.4
Escherischia coli	69	50	>50	>50	>50	>50	>50

TABLE 2. Chemical Shifts δ(ppm) of ADM12: β-CD in the Free and Complex State.

	Proton*	δ_f(free)	δ_c(comp)	$\Delta\delta$ (δ_c-δ_f)
ADM-12	Ad(d,2H,4eq+9eq)	1,545	1,639	+0,094
	Ad(m,10H)	1,615	1,692	+0,077
	Ad(d,2H,4ax+9ax)	2,086	2,269	+0,183
		2,146	2,333	+0,187
	$N(CH_3)_2$(s,6H)	3,081	3,085	+0,004
β-CD	H3(x7H)	3,878	3,829	-0,049
		3,949	3,874	-0,075
		4,043	3,920	-0,123
	H5(x7H)	3,540	3.530	-0,010
		3,586	3,583	-0,003
	Anomeric(x7H)	5,080	5,056	-0,024
		5,098	5,074	-0,024

TABLE 3. Chemical Shifts δ(ppm) of ADM12 : β-CD in the Free and Complex State.

Carbons	δ_f(free)	δ_c (complex)	$\Delta\delta$ (δ_c-δ_f)
CH3	14.15	13.98	-0.17
CH2	15.53	15.45	-0.08

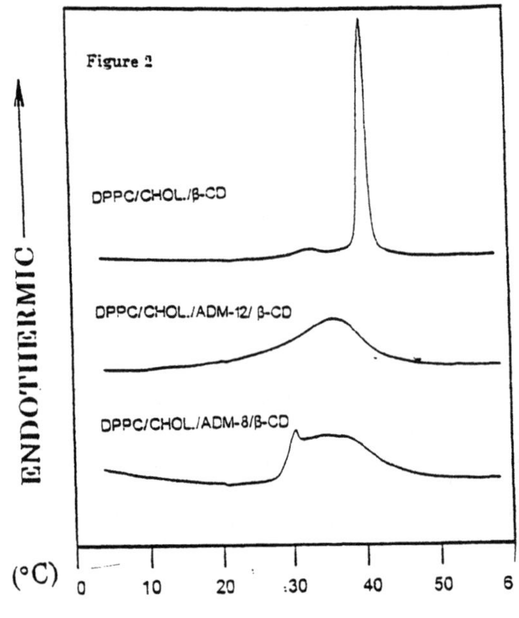

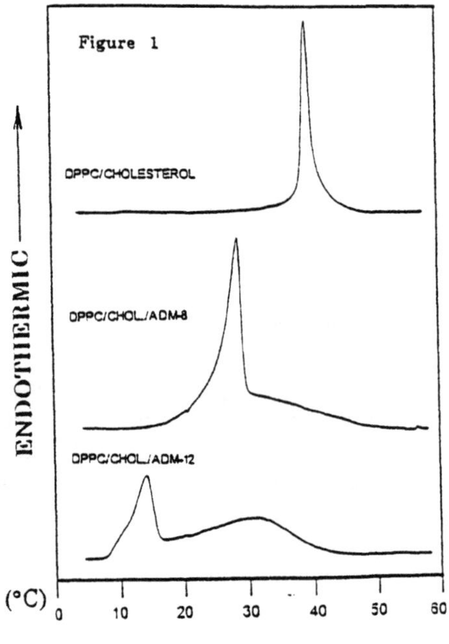

References

1. Antoniadou-Vyza E., Tsitsa P., Hytiroglou E., Tsantili-Kakoulidou A. *Eur J Med Chem (1996) 31, 105-110*
2. Mayromoustakos T., Papadopoulos A., Theodoropoulou E., Dimitriou C., Antoniadou-Vyza E. *Life Sciences (1998)*
3. Perrakis A., Antoniadou-Vyza E., Hamodrakas S., *Carbohydrate Res (in press*
4. Hammond S A, Morgan G R, Russel A D, (1987) *Journal of Hospital Infections* 9,255-264
5. M.J. Janiak, D.M., Small and G.G. Shipley, *Biochemistry* 15 4575- 4580 (1976).
6. S. Mabren, P.L. Mateo, J.M. StUrtevant, *Biochemstry* 17, 2464-2468 (1978)
7. T. Mavromoustakos, A. Papadopoulos, E. Theodoropoulou, C. Dimitriou, E. Antoniadou-Vyza, *Life Sci.* in press.

PHARMACOLOGICAL INVESTIGATIONS OF NEW PEPTIDO-CYCLODEXTRINS

C. PÉAN [a], A. WIJKHUISEN [b, c], F. DJEDAÏNI-PILARD [a], C. CRÉMINON [b], J. GRASSI [b] and B. PERLY [a]
a: DRECAM/SCM, CEA-Saclay, F-91191 Gif sur Yvette, (France)
b: DRM/SPI, CEA-Saclay, F-91191 Gif sur Yvette, (France)
c: UFR de Biologie, Université PARIS 7, F-75251, Paris, (France)

1. INTRODUCTION

Cyclodextrins (CD's) could be used as molecular carrier dedicated to drug targeting[(1)]. In a previous work, we synthesized and characterized height new different peptido-cyclodextrins. They are composed with a β- or γ-CD part, and a peptidic part constituted of the neuropeptide Substance P (noted SP, H-Arg-Pro-Lys-Pro-Gln-Gln-Phe-Phe-Gly-Leu-Met-NH_2)[(2)] or one of its derivatives, the SP 4-11. Products obtained were β- or γ-CD(Lys3)-SP, β- or γ-CD(Arg1)-SP, di-β- or di-γ-CD(Arg1, Lys3)-SP (noted CD-SP) and β- or γ-CD-SP 4-11. More, we demonstrated the preservation of the inclusion properties of CD's part of these compounds. In this communication, we reporte *in vitro* pharmacological investigations of these CD-SP. We demonstrate the recognition properties of the diverse conjugates by SP receptor mimics (anti SP polyclonal antibodies), by recombinant human NK1 receptor (using binding experiments on CHO transfected cells) and the production of the second messengers inositolphosphates induced by fixation of the CD-SP on the NK1 receptor.

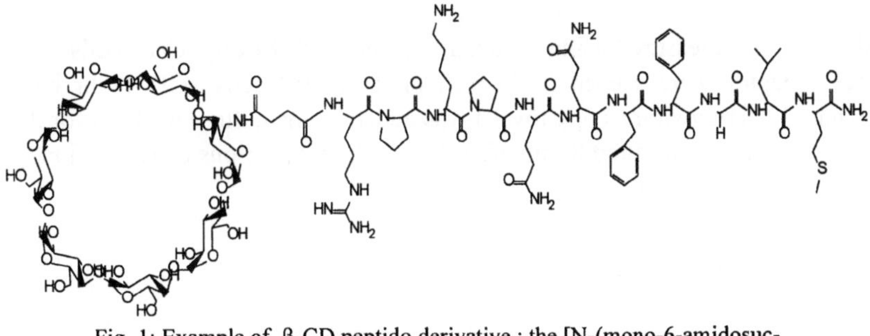

Fig. 1: Example of β-CD peptido derivative : the [N-(mono-6-amidosuccinylamido-6-deoxy-β-cyclodextrin)Arg[1]]-Substance P (noted β-CD(Arg1)SP)

2. MATERIALS AND METHODS

2.1 MATERIALS

Immunoassays were performed into 96 wells microtiter plates (Maxisorb Nunc, Denmark), using enzyme labelling[(3)]. CHO cells expressing the NK1 receptor were cultured in Ham-F12 medium supplemented with 10% fetal calf serum, in a 37°C, 5% CO_2 and humidified atmosphere. Bolton-Hunter substance P ([^{125}I]-BHSP, 75 Ci/mmol) and *myo*-[2-^{3}H]inositol (17 Ci/mmol) were purchased from Amersham Corp. (Les Ulis, France). ^{125}I-SP was used at 15 pmol.l^{-1} concentration in each binding experiment.

2.2 IMMUNOASSAYS

Using either anti-β- and anti-γ-CD polyclonal antibodies or anti-SP monoclonal and polyclonal antibodies, we wanted to check for possible modification generated by the grafting of CD's on the SP backbone (for memory, Nα-Arg1 or Nε-Lys3 or Nα-Arg1, Nε–Lys3 were the grafting positions). Antibodies thus act as a molecular probes to estimate these conformational modifications. They can be considered as an imperfect NK1 receptor mimic,since the interacting site of SP with either the receptor or the antibodies concerns the same part of the peptide (the 6 C-terminal residues). Immunoassays were performed in a competitive format by comparing the displacement of enzyme-labelled SP by SP and CD-SP adducts, or the displacement of CD tracer by CD and CD-SP adducts, using standart immuno-enzymatic assays protocol [(3)].

2.3 BINDING EXPERIMENTS

Binding experiments were performed on intact CHO cells expressing the human NK1 receptor. [^{125}I]-BHSP was used at 15 pmol.l^{-1} with various concentrations of unlabelled CD-SP derivatives, as described in standart protocol[(4)]. All determinations were performed in triplicate, and non specific binding was determined in the presence of 1μmol.l^{-1} native SP.

2.4 MEASUREMENT OF PHOSPHATIDYL INOSITOL (PI) HYDROLYSIS

PI hydrolysis was measured using a standart protocol[(4, 5)]. Briefly, CHO cells expressing the NK1 receptor were seeded in 24-well plates (5.10^4 cells/well) 48 hr before the assays. *myo*-[2-^{3}H]inositol (0.5μCi/well) was added to the culture medium for 24 hr. PI hydrolysed level was mesured 8 min after the addition of various concentrations of CD-SP conjugates on the cells.

3. RESULTS

3.1 IMMUNOASSAYS

For both SP and CD assays, standard curves (fig. 2) reveal comparable recognition by the antibodies, thus demonstrating the absence of significant modifications of the molecular structure. Moreover, the use of immunometric assay (where CD-SP adducts were "sandwiched" between an anti-CD antibody ensuring the capture of the molecule and an anti-SP enzyme labeled antibody as tracer) confirms these results furthermore, allowing us to check for the ratio of CD grafted on SP. These experiments comfort us to evaluate the binding and the pharmacological characteristics of these new compounds towards the NK1 receptor on cellular model.

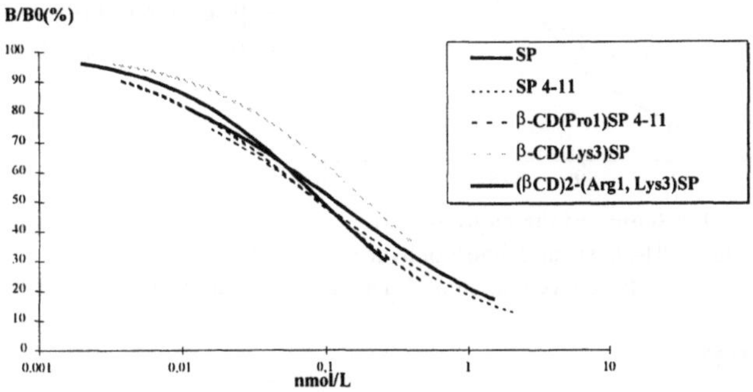

fig 2: Competitive immunoassays curves of the different β-CD-SP adducts using anti SP polyclonal antibodies

3.2 BINDING EXPERIMENTS

All CD-SP derivatives have a lower affinity towards the human NK1 receptor compared to the natural SP, with differences between CD-SP compounds.

Table 1: IC_{50} values toward the human NK1 receptor determined for each CD-SP products from binding experiments

Compounds	IC50 (nmol/L)	Compounds	IC50 (nmol/L)
Substance P	7	(β-CD)2(Arg1, Lys3)SP	115
Substance P 4-11	85	γ-CD(Pro1)SP4-11	299
β-CD(Pro1)SP4-11	137	γ-CD(Arg1)SP	180
β-CD(Arg1)SP	31	γ-CD(Lys3)SP	180
β-CD(Lys3)SP	30	(γ-CD)2(Arg1, Lys3)SP	178

It can be noted that affinities (~ IC_{50} values) quite similar to that of SP were obtained with the monosubstituted β-CD-SPs, and that a decreasing affinity was observed with the disubstituted β-CD derivatives as for all γ-CD derivatives. These results could be closely related to the greater steric hindrance of these last compounds. Concerning the SP4-11 derivatives, IC_{50} values were decreased compared to the native SP4-11, which is also decreased compared to the entire SP.

3.3 MEASUREMENT OF PHOSPHATIDYL INOSITOL (PI) HYDROLYSIS

Effects of β-CD-SP moities in term of their pharmacological potenties were also investigated. As shown in Fig.3, all β-CD-SP conjugates possess potency effects for stimulating PI hydrolysis, with sligh differences compared to SP. Thus, SP stimulated PI production with an EC_{50} value in the nanomolar range, whereas EC_{50} value for the β-CD-SP conjugates is generally slightly decreased.

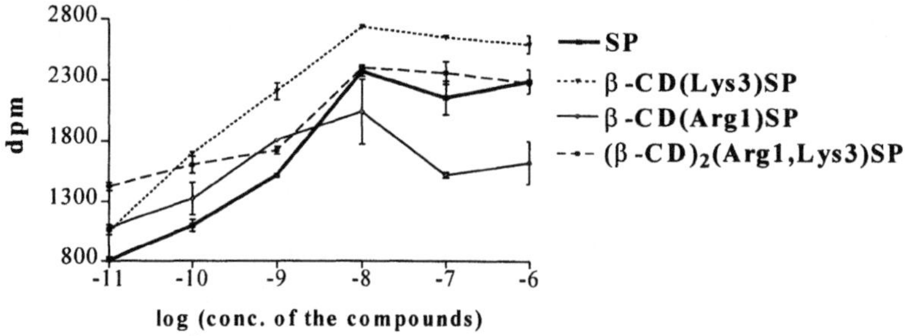

fig. 3: ^{3}H-phosphatidylinositol hydrolysis production as function of the concentration of different CD-SP conjugates

4. CONCLUSION

With the aim to use cyclodextrins in the drug-targeting domain, we investigated 8 new peptido-cyclodextrins such as $(CD's)_n$-Sub. P (or SP 4-11), to evaluate their respective pharmacological properties. We demonstrated *in vitro*, on cells cultures, the good recognition properties (compared to the native SP) of each compounds towards the recombinant human NK1 receptor. More, we demonstrated the agonist properties of these products by evidenced second messengers production. These *in vitro* evaluations permit us to take up the next step of this work, the *in vivo* studies of the CD-SP conjugates, and the evaluation of their potential uses as molecular carrier dedicated to drug targeting.

5. REFERENCES

(1) : F. Djedaini-Pilard, J. Désalos and B. Perly, *Tetrahedron. Lett.*, **34**, 2457-2460 (1993)

(2) : Y. Q. Cao, P. W. Mantyh, E. J. Carlson, AM. Gillespie, C. J. Epstein and A. I. Basbaum, *Nature*, **392**, 390-394 (1998)

(3) : C. Créminon, F. Pilard, J; Grassi, B. Perly and Ph. Pradelles, *Carb. Res.*, **258**, 179-186 (1994)

(4) : S. Sagan, G. Chassaing, L. Pradier and S. Lavielle, *J. Pharm. Exp. Ther.*, **276**, 1039-1048 (1996)

(5) : Y. Torrens, JC. Beaujouan, M. Saffroy and J. Glowinski, *Peptides,* **16(4)**, 587-594 (1995)

This work was supported by the European Commission (DGXII) under the FAIR Programme CT95-0300.

CHARACTERIZATION OF CYCLODEXTRIN COMPLEXES OF (S)-NAPROXEN BY X-RAY AND THERMAL METHODS OF ANALYSIS

M.R. CAIRA AND V.J. GRIFFITH
Department of Chemistry, University of Cape Town, Rondebosch 7701, South Africa

Abstract

Preparation and characterization of the following crystalline cyclodextrin (CD)/naproxen complexes are reported: β-CD · (S)-naproxen · $10.9H_2O$ (**1**), β-CD · (S)-naproxen$^-$ Na^+ · $13.4H_2O$ (**2**) and γ-CD · (S)-naproxen · $15.1H_2O$ (**3**). Thermal analysis showed essentially single-step dehydration for **1** and **3** which contain the neutral naproxen molecule as guest while multi-step dehydration was revealed for **2**. Single crystal X-ray methods were used to confirm an unusual solution-phase transformation of **1** into **2** and to deduce probable crystal packing arrangements in the complexes.

1. Introduction

Because of the potency of naproxen as an anti-inflammatory drug and the improvements in its performance resulting from CD-inclusion, naproxen/CD interactions in both solution and solid states have been studied extensively. Recent reports described such interactions with various CDs including β-CD [1, 2], trimethyl β-CD [3], and α-, β- and γ-hydroxypropyl-CDs [4]. Here we report the preparation and characterization by thermal and X-ray analyses of two distinct solid complexes between β-CD and (S)-naproxen species, namely β-CD · (S)-naproxen · $10.9H_2O$ (**1**) and β-CD · (S)-naproxen$^-$ Na^+ · $13.4H_2O$ (**2**), and of the complex γ-CD · (S)-naproxen · $15.1H_2O$ (**3**). Our interest in comparing the behaviours of **1** and **2**, which contain neutral and ionised forms of the drug respectively, was prompted by our earlier studies of related complexes [5, 6]. Crystallographic data for **1-3** were also recorded to establish structural information.

2. Materials and Methods

β-CD (Chinoin-Reanal, Hungary), γ-CD (Cyclolab, Hungary) and (S)-naproxen (Syntex, USA) were used as received. Complex **1** was prepared by dissolving 0.30 mmol drug and 0.30 mmol β-CD in 2 ml distilled water at 70°C and the pH was adjusted with 0.1M NaOH until almost all of the drug had dissolved. The solution was filtered, diluted 1 in

1.5 and cooled to room temperature over several days. Complex **2** was obtained by dissolving 0.15 mmol drug and 0.15 mmol β-CD in 2 ml distilled water at 50°C and adjusting the pH with 0.1M NaOH to dissolve the drug. Crystals of **2** appeared after one year. Crystals of **3** were obtained by dissolving 0.10 mmol drug and 0.15 mmol γ-CD in 2 ml distilled water at 70°C, filtering the solution and cooling to room temperature over several days. TG and DSC traces were recorded on a Perkin-Elmer PC7 Series Thermal Analysis System at a scanning rate of 10° min^{-1} under N_2 gas-purge. UV spectrophotometry (Philips PU8700 instrument, 230 nm) was used to determine host:guest ratios. Elemental analysis was used to confirm the complex compositions. Unit cell and space group data were obtained by single crystal X-ray precession photography using CuKα-radiation.

3. Results and Discussion

3.1 COMPLEX COMPOSITIONS AND THERMAL STABILITIES

UV data yielded host:guest ratios of 1:1 for **1-3**. Combining these data with TG and elemental analysis data led to the formulations given in the Introduction. The TG and DSC traces for **1** and **3** (containing the neutral drug molecule) were unremarkable, in each case showing essentially one-step mass loss and a single-step dehydration endotherm. In contrast, as shown in Figure 1, the DSC trace for **2** (containing naproxen$^-$ and Na^+ ions), yielded three resolved dehydration endotherms (A, B, C) whose DTG trace reflects successive losses of approximately 4, 4 and 5 water molecules respectively per complex unit.

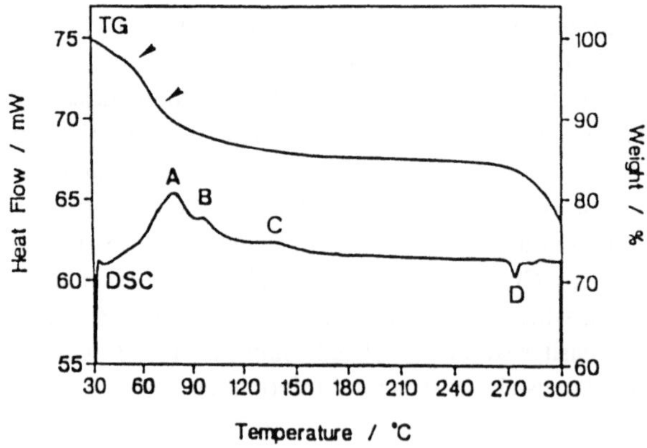

Figure 1. Combined TG/DSC Traces for Complex **2**

Notably, the last water molecules lost are released at a significantly higher temperature than those in **1** and **3** which contain the neutral drug molecule. DSC data for **1-3** are listed in Table 1. Over the complex decomposition temperature range, the DSC trace for **1** shows no prominent features while the decomposition of **2** is preceded by an exotherm (D) and the decomposition of **3** is accompanied by a prominent, broad endotherm (B).

TABLE 1. DSC Data for the Complexes

Complex	Event	Range (°C)	Onset (°C)	Peak (°C)
1	endotherm A	30 - 147	32	72
2	endotherm A	30 - 90	59	76
	endotherm B	90 - 116	90	94
	endotherm C	115 - 159	124	133
	exotherm D	261 - 273	263	266
3	endotherm A	40 - 148	58	94
	endotherm B	297 - 381	311	342

3.2 CRYSTAL TRANSFORMATION AND X-RAY STUDIES

Crystals of **1** were unstable in mother liquor and converted to the thermally more stable species **2** after several weeks. Since this type of transformation is very unusual for CD complexes, its occurrence was confirmed not only by DSC and microscopic observations but also by single crystal X-ray diffraction. This yielded the space group and unit cell data listed in Table 2, which includes data for complex **3**.

TABLE 2. Crystal Data for the Complexes

Complex	System	Space group	Z	a/Å	b/Å	c /Å	β/°
1	Monoclinic	C2	4	15.7(1)	24.5(1)	19.5(1)	110.0(3)
2	Orthorhombic	$P2_12_12$	8	30.2(1)	32.0(1)	15.5(1)	
3	Tetragonal	$P42_12$	6	23.8(1)		23.3(1)	

By analogy with known crystal structures of CD complexes [7], we infer from the above data that complex **1** contains dimeric β-CD units arranged in channels which accommodate the guest naproxen molecules. No complex of the unsubstituted β-CD has yet been reported in the space group $P2_12_12$ and therefore details of the drug inclusion mode and crystal packing in **2** await complete X-ray analysis, for which crystal quality has thus far been inadequate. In complex **3**, the host γ-CD molecules pack in channels with fourfold rotational symmetry, requiring the included naproxen molecules to be disordered around the tetrad.

3.3 STRUCTURE-STABILITY CORRELATION

Multi-step dehydration spanning a wide temperature range, as observed for complex **2**, can be attributed to the presence of Na^+ ions in the crystals. The higher thermal stability of CD complexes containing drug anions and Na^+, K^+ or Cs^+ counterions relative to those containing neutral guest molecules has been noted recently [6]. The behaviour of **2** on heating resembles that of the β-CD · diclofenac sodium · $11H_2O$ complex whose X-ray structure and thermal analyses [5, 6] revealed that some of the water molecules are strongly retained in the complex by coordination to Na^+ ions and are therefore the most likely ones to be released in the final dehydration step.

The strong ionic interactions existing in such complexes are not necessarily unfavourable from the viewpoint of drug delivery; it was recently reported [8] that an oral formulation of the β-CD ·diclofenac sodium complex [5] had a significantly higher *in vivo* absorption rate compared with that of a commercially available rapid release oral preparation of the drug which contained no CD.

4. Conclusions

Significant differences in thermal stabilities, dehydration behaviour and crystal packing modes of CD complexes of naproxen species have been identified by a combination of thermal analysis and X-ray diffraction. The higher stability of the β-CD complex containing naproxen$^-$ Na$^+$ provides a further example of CD-complex stabilisation imparted by the presence of cations in the crystal.

5. References

1. Loftsson, T., Olafsdottir, B., Frioriksdottir, J. and Jonsdottir, S. (1993) Cyclodextrin complexation of NSAIDs: physicochemical characteristics, *Eur. J. Pharm. Sci 1*, 95-101.

2. Ganza-Gonzalez, A., Vila-Jato, J.L., Anguiano-Iges, S., Otero-Espinar, F.J. and Blanco-Mendez, J. (1994) A proton nuclear magnetic resonance study of the inclusion complex of naproxen with β-cyclodextrin, *International Journal of Pharmaceutics, 106,* 179-185.

3. Caira, M.R., Griffith, V.J., Nassimbeni, L.R. and Van Oudtshoorn, B. (1995) X-ray structure and thermal analysis of a 1:1 complex between (S)-naproxen and heptakis(2,3,6-tri-O-methyl)-β-cyclodextrin, *J. Inclusion Phenomena and Mol. Recognition in Chemistry, 20,* 277-290.

4. Melani, F., Bettinetti, G.P., Mura, P. and Manderioli, A. (1995) Interaction of naproxen with α-, β-, and γ-hydroxypropyl cyclodextrins in solution and in the solid state, *J. Inclusion Phenomena and Mol. Recognition in Chemistry, 22,* 131-143.

5. Caira, M.R., Griffith, V.J., Nassimbeni, L.R. and Van Oudtshoorn, B. (1994) Synthesis and X-ray crystal structure of β-cyclodextrin diclofenac sodium undecahydrate, a β-CD complex with a unique crystal packing arrangement, *J. Chem. Soc., Chem. Comm.* 1061-1062.

6. Caira, M.R., Griffith, V.J. and Nassimbeni, L.R. (1998) Desorption of water from CD/drug inclusion complexes: thermal behaviour-crystal structure correlation, *J. Thermal Analysis and Calorimetry, 51,* 981-991.

7. Cambridge Crystallographic Database and Cambridge Structural Database System, Version 5.14, October 1997, Cambridge Crystallographic Data Centre, University Chemical Laboratory, Cambridge, England.

8. Penkler, L.J., Whittaker, D.V., Glintenkamp, L.A. and Van Oudtshoorn, M.C.B. (1996) Enhanced pharmacokinetic properties of oral and parenteral diclofenac-cyclodextrin delivery systems, in Szejtli and Szente, L. (eds.), *Proceedings of the Eighth International Symposium on Cyclodextrins,* Kluwer Academic Publishers, Dordrecht, pp.481-486.

CD-MEDIUM CONTROL OF MICROBIAL STEROL SIDECHAIN CLEAVAGE

D.V. DOVBNYA, S.M. KHOMUTOV, V.M. NIKOLAYEVA, and M.V. DONOVA,
Institute of Biochemistry & Physiology of Microorganisms, Rus.Acad.Sci.
142292 Pushchino, Moscow Region. Russia

1. Introduction

Microbial sidechain cleavage of natural sterols is a best way to produce important precursors for syntheses of a number of steroidal pharmaceuticals [1]. Recently, a significant enhancement of the process by mycobacteria in the presence of modified β-cyclodextrin derivatives was reported [2]. CD-mediated increase of poor soluble sterol conversions and product molar yields was accompanied by a substantial shift in a ratio of bioconversion products.

A multifunctional mechanism of CDs action on sterol bioconversions is not totally clear and proposed to include different aspects of sterol, steroids and biocatalyst interactions with CDs.

The present study was undertaken to evaluate a role of CD-mediated solubilization of steroid products at microbial sterol sidechain cleavage. Bioconversion of β-sitosterol by *Mycobacterium sp.* VKM Ac-1816D in the presence of randomly methylated β-CD (MCD) was used as a model process for the investigations.

2. Materials and methods

Sitosterol (Ultra grade, 91.4% of β-sitosterol) was obtained from Kaukas (Finland); randomly methylated β-CD with D.S. 1.69 (MCD) was purchased from Wacker-Chemie GmbH (Germany); C_{19}-steroids (analytical grade) were purchased from Serva (Germany).

Mycobacterium sp. VKM Ac-1816D was obtained from All-Russian Collection of Microorganisms (IBPhM RAS). Culture maintenance and precultivation was carried out as described earlier [2].

Bioconversions of β-sitosterol (12 mM) were carried out on a rotary shaker in flasks at 220 rpm, 30°C, in the presence of MCD at the range of concentrations 0-24 mM as described earlier [2].

Steroids were separated by HPLC using C_{18}-column, acetonitrile and water (70/30, v/v) as a mobile phase and detected at 240 nm.

Values of $S_0 \cdot K_A$ for the bioconversion products were obtained using phase solubility technique in MCD solutions (0-190 mM) at 30°C according to [3].

Sitosterol micronization was performed by liquid phase extrusion; particles size was followed using light microscopy and photon correlation spectrophotometer Coulter N4. For the freeze-fracture electron microscopy the specimen frozen with Freon F22 was prepared using Balzers device BA 360 M equipped with electron beam evaporator. The replicas were analysed by a JEM-100 B electron microscope.

3. **Results and discussion**

3.1. SITOSTEROL BIOCONVERSION IN THE PRESENCE OF MCD

Androsta-1,4-diene-3,17-dione (ADD) as the major product, 20-hydroxymethylpregna-1,4-diene-3-one (HMPD) and androst-4-ene-3,17-dione (AD) as the major by-products were accumulated in the medium during bioconversions as a result of β-sitosterol side chain oxidation by living cells of *Mycobacterium sp*. The accumulation rates and yields of the products were found to increase with the MCD concentrations (*Fig. 1*). The most of AD formed at the first phase of the process then easy converted to ADD at the conditions used. No accumulation of total product was observed after 72-96 h of transformation at the range of MCD concentrations of 0 - 12 mM in spite β-sitosterol remained non-converted in the medium. Increase of MCD concentrations higher than equimolar to β-sitosterol resulted in almost quantitative substrate conversion.

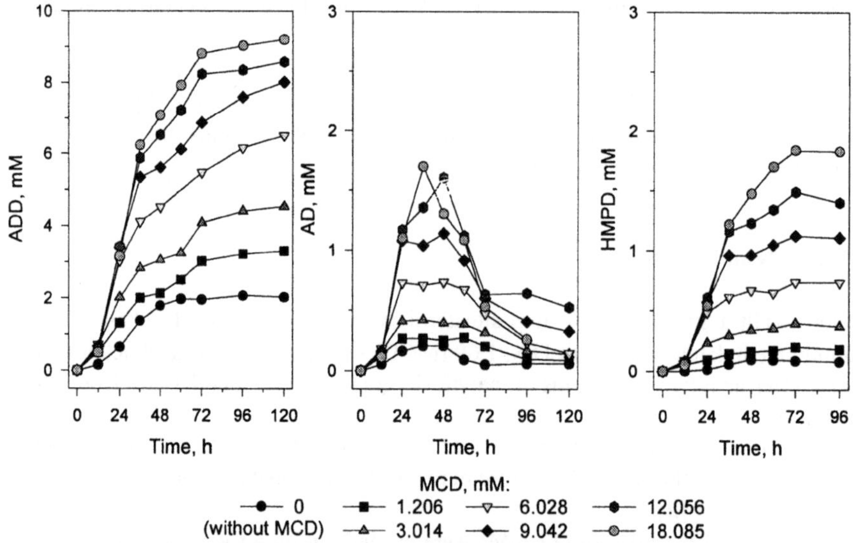

Figure 1. Accumulation of products of β-sitosterol (12 mM) bioconversion by *Mycobacterium sp*. VKM Ac1816D in MCD media

3.2. CELL-SITOSTEROL INTERACTIONS IN MCD-MEDIUM

An uptake of sterol substrate during microbial transformations (in a non-CD media) was shown to take place via direct contact between mycobacterial cells and solid substrate particles [4]. The special investigations were undertaken to estimate character of cell-sitosterol interactions during sterol bioconversion in the presence of MCD. Optical and electron-microscope observations showed cells adsorbed on the surface of substrate particles and cells growing into substrate microcrystallite similar to those observed at the transformation in the medium without MCD.

Expansion of available surface of sitosterol microcristallite by micronization resulted in the increase of conversion rates but did not take essential effect on the product yields both in the presence or absence of MCD.

The results indicate that possible MCD-mediated substrate solubilization is not a superior promoting factor for β-sitosterol bioconversion process.

3.3. ESTIMATION OF BIOCONVERSION PRODUCTS SOLUBILITIES IN MCD MEDIUM

The value of steroid solubility (S_S) in water solution of MCD with concentration C_{MCD} can be expressed as:

$$S_S = S_0 + [S\text{-}MCD] \qquad [1]$$

where S_0 is steroid solubility in water and [S-MCD] is the concentration of soluble steroid-MCD inclusion complex. The equilibrum state of MCD:steroid (1:1) system is expressed by the association constant K_A:

$$K_A = \frac{[S\text{-}MCD]}{S_0 \cdot [MCD]} \qquad [2]$$

where [MCD] is the concentration of a free MCD in a water solution, and

$$[MCD] = C_{MCD} - [S\text{-}MCD] \qquad [3]$$

where C_{MCD} is the total concentration of MCD. Substituting [3] into [2] it is obtained:

$$[S\text{-}MCD] = \frac{K_A \cdot S_0 \cdot C_{MCD}}{1 + K_A \cdot S_0} \qquad [4]$$

Consequently the solubility of single steroid in MCD solution can be expressed as:

$$S_S = S_0 + \frac{K_A \cdot S_0 \cdot C_{MCD}}{1 + K_A \cdot S_0} \qquad [5]$$

The major steroid products of β-sitosterol bioconversion accumulating in the medium are competitive for MCD. For a few steroids simultaneously present in the solution the solubility of anyone individual steroid (S_{Si}) can be expressed as:

$$S_{Si} = S_{0i} + \frac{K_{Ai} \cdot S_{0i} \cdot C_{MCD}}{1 + K_{A1} \cdot S_{01} + K_{A2} \cdot S_{02} + \ldots + K_{Ai} \cdot S_{0i}} \qquad [6]$$

Value of S_S for individual steroid product in the β-sitosterol bioconversion medium was proposed to limit the product formation by the microbial culture. Therefore value of S_{Si} can be understood as the capacity of MCD medium for one of simultaneously present steroid products (ω_i):

$$\omega_i = S_{0i} + \frac{K_{Ai} \cdot S_{0i} \cdot C_{MCD}}{1 + \sum_{i=1}^{N} (K_{Ai} \cdot S_{0i})} \qquad [7]$$

where N is number of simultaneously present steroids.

The phase solubility curves obtained in MCD solutions for ADD, AD and HMPD were of A_L type and corresponding values of $K_A \cdot S_0$ were calculated as 11.2, 1.44 and 1.59 correspondingly. The phase solubility curve obtained for β-sitosterol was of S-shape that propose complex stohiometry differed from 1:1. Nevertheless the rough $K_A \cdot S_0$ value for β-sitosterol calculated as 0.0319 let suggest the comparatively negligible contribution of β-sitosterol to a diminution of ω_i values for the bioconversion products.
The correlation of the individual product yields with the correspoding ω_i values was examined. Maximum product yields at sterol bioconversions were found to correlate directly with ω_i values calculated for the corresponding steroids in the range of MCD concentrations (*Fig.2*). The deviation of yield curves from the linear form obtained at high concentrations of MCD are explained by bioconversion limitation on the substrate. The role of ω_i as one of the major factors responsible for the yields and accumulation kinetics of hydrophobic steroidal products at the processes of sterol bioconversion in cyclodextrin media is discussed.

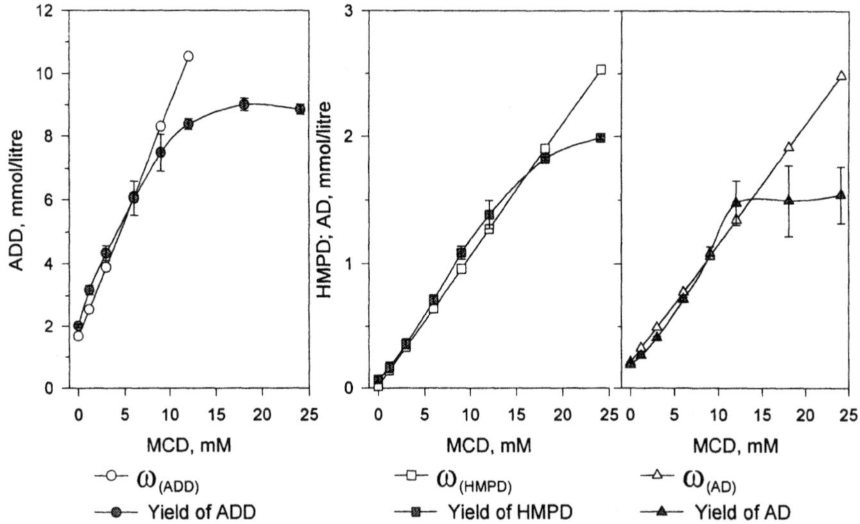

Figure 2. Correlation of the product yields at β-sitosterol (12 mM) bioconversions by *Mycobacterium sp.* VKMAc1816D with MCD medium capacity values for the corresponding steroids (ω_i)

4. References

[1] Kieslich, K. Microbial side-chain degradation of sterols. (1985) *J. Basic Microbiol.*, **25**, 7, 461-474
[2] Donova, M.V., Dovbnya, D.V., Koshcheyenko, K.A. (1996) Modified CDs-mediated enhancement of microbial sterol side chain degradation. Proc. 8-th Int. Symp. on Cyclodextrins, Kluwer Acad. Publishers, Netherlands, 527-530
[3] Higuchi, T. and Connors, K. (1965) Adv. Anal. Chem. Instrum., Wiley Interscience, New York, 117-212.
[4] Atrat, P., Hosel, P., Richter, W., Meyer, H.W., Horhold, C. (1991) Interactions of *Mycobacterium fortuitum* with solid sterol substrate particles. *J.Basic Microbiol.*, **31**, 6, 413-422

COMPRESSIONAL PROPERTIES OF CYCLODEXTRINS

A. Muñoz-Ruiz[1] and P. Paronen[2]. [1] Department of Pharmaceutical Technology. Faculty of Pharmacy. University of Seville. E-41012 Sevilla. SPAIN. [2]Department of Pharmaceutics. University of Kuopio. SF-70211 Kuopio. FINLAND

1. Introduction

There is little information about powder and binding properties of cyclodextrins (CDs). The earliest paper in relation with binding properties of a β-cyclodextrin polymer pilot product was published by Fenyvesi et al. (1984). The β-cyclodextrin polymer was found to be well suitable for direct compression, owing to its relatively advantageous binding and disintegration properties. Giordano et al. (1990) and more recently Pande and Shangraw (1995) studied the compaction capacity of βCD and the influence of water content.

The aim of this study was to evaluate compressional properties of plain cyclodextrins and the effect of applied pressure and compression speed on CD-tablet properties.

2. Materials and Methods

2.1 MATERIALS

The cyclodextrins studied were: αCD, βCD, γCD, all manufactured by Cyclolab, Budapest, Hungary, and hydroxypropyl-β-cyclodextrin (HPβCD, Encapsin® HPB), manufactured by Janssen, Stockholm, Sweden.

2.2 METHODS

2.2.1 *Compression*

The compression was carried out using a compaction simulator (Puuman Ltd., Kuopio, Finland). Quantities of powder were manually filled into the die (10.00 mm diameter) to produce tablets having a theoretical thickness of 1.4 mm at zero porosity. Single sided saw-tooth profiles, i.e. constant velocity punch movement, were selected. Punch velocities were 3 and 300 mm/second. At both velocities, flat tablets of CDs were compressed at 25, 75, 100, 150, and 200 MPa of applied pressure. The tablets in the compaction experiments were made with die wall lubrication. This was accomplished by manually treating the punch and die wall with a cotton swab using a solution of 5% w/v magnesium stearate. In every case, four parallel tablets were compressed.

2.2.2 *Physical testing*

The physical testing of the tablets was performed 24 hours after ejection. Weight loss of tablets was measured in friability test for 4 min at 25 rpm with Erweka TA3 apparatus

(Erweka, Heusenstamm, Germany). Breaking strength of the tablets was measured using a CT5 Tonne Testing Machine (Engineering Systems, Notthingham, England) equipped with a 50 Kg load cell and linear voltage displacement transducer. Loading speed in measurements was 1.0 mm/min. Tensile strength was calculated from breaking strength and the dimensions of the tablets. Work of failure was also calculated by numerically integrating the area under the diametral compression testing load-displacement curves. The disintegration of six tablets was measured individually using distilled water in the USP 23 apparatus without disks.

3. Results and discussion

3.1 FRICTION PROPERTIES

The ratio of the lower to the upper punch force (R) during maximum compression has often been used as measure of the frictional effect at the die wall. In order to compare friction behavior of the materials, however, it appears necessary to calculase punch force ratio at the same loads and tablets dimensions (Hölzer and Sjögren, 1978). In this paper we used to compare friction properties the Järvinen and Juslin (1974) equation for precise friction work calculations, taking into account the actual movement of theoretical action point of friction.

Table 1 shows the mean fricton work of tablets done at slow and fast compression of the four cyclodextrins under study. Analysis of variance (ANOVA) was used in the evaluaton of the statistical significance of the results. The compression speed and the type of cyclodextrin were significant sources of variation ($p < 0.01$) in the value of friction work. As expected, friction work increased as well as compression speed was increased. The order of the material according to friction during tableting was the following: HPβCD > αCD > γCD and βCD being this later the material with the lower friction. The lubricating propertes of the HPβCD was previously studied by Giordano et al. (1990), but only in terms of the role played by the water content.

Table 1. Frictional properties of cyclodextrins

Material	Compression speed	Parameters	
		Work of friction(J)	Work of ejection (J)
αCD	Slow	0.361 (0.026)	2.098 (0.199)
	Fast	0.455 (0.016)	1.919 (0.200)
βCD	Slow	0.129 (0.027)	0.376 (0.198)
	Fast	0.314 (0.026)	0.819 (0.199)
γCD	Slow	0.210 (0.017)	0.316 (0.020)
	Fast	0.394 (0.026)	0.522 (0.167)
HPβCD	Slow	0.557 (0.121)	0.467 (0.067)
	Fast	0.735 (0.017)	0.799 (0.198)

Table 1 shows the mean ejection work measured for slow and fast compression. Results showed a clear difference between αCD and the other cyclodextrins (LSD, $p < 0.01$). The differences in the ejection and friction during compression may by due to the stickness of αCD, this support previous results (MuñozRuiz et al., 1993) found for directly compressible maltodextrins, which are oligosacchañdes produced also by enzymatic degradaron of starch, but connected by linear α-(1,4)bonds. The maximum ejection force was only higher than the 750 Newtons, stated as limit for acceptable ejecton properties, for αCD in slow compression.

3.2 BINDING PROPERTIES

The mean tensile strengths with confidence intervals (p = 0.05) of tablets at the several applied pressures and the two compression speed are shown in Figure 1. Tablet formed at high compaction rates were weaker than those formed at low compaction rate according to previous results from Garr and Rubinstein (1991). The linear relationship between breaking force and mean applied force was reported by Newton et al. (1971). These authors found the same relation when the tensile strength was computed instead the breaking load. The formaton of permanent bonds was rather weak for βCD, remaining the breaking strength below 20 Newtons even at very high compressional pressures. HPβCD and αCD, showed profiles with a maximum tensile strength at an applied pressure between 150 and 200 MPa, thus the bonding between particles of these cyclodextrins was not possible to be computed due to the low correlation coefficient of the linear regression between tensile strength and applied pressure. This behaviour was previously observed for modified starch (Monedero et al., 1994) with a maximum tensile strength between 200 and 300 MPa. The maximum tensile strength attained at certain applied pressure depends on the compression speed for βCD and HPβCD. These results, support the finding of Marshall et al (1993) for ibuprofen formuladons with a limited tensile strength of tablets which decreased as well as punch velocity was increased.

The mean values of work of failure showed in general higher deviations than those obtained for tensile strength. For αCD tablets there is not a limit in the value of work of failure as occurs in the tensile strength profile. The profile of βCD was almost horizontal with a small decrease in the work of failure above 200 MPa, this profile verified that formation of permanent bonds was rather weak for βCD, remaining the value below 2.5 mJ, even at very high compressional pressures. The profile of γCD showed a high increase in work of failure as well as applied pressure was increased. HPβCD showed the same profile than observed for tensile strength with a limited maximum value of work of failure at 150 MPa. Thus, the applied pressure will be a critical factor in the tableting of formulations containing HPβCD. The differenoes between the work of failure of HPβCD and the other CDs is considerably higher than the differences in tensile strength, in the pressure range up to 150 MPa. This fact can be explained because for tablets of equal tensile strength, the tablets which deform more

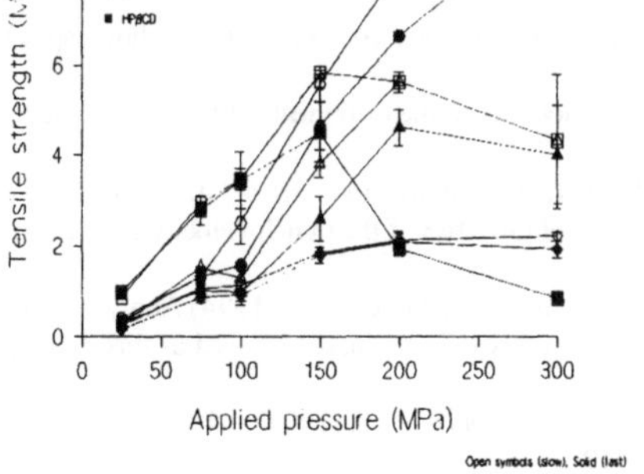

Figura 1. Tensile strength of CDs

extensively under load showed higher work of failure. Since, HPβCD tablets deform more than other CD tablets.

3.3 TABLET PROPERTIES

Friabiity values decreased from the lowest pressure to an applied pressure between 100 and 150 MPa depending on the CD under study. Above certain pressure, a further increase in applied pressure had no effect on friability for αCD, βCD, γCD and values remaining practically constant. At the highest pressure, however, for βCD and γCD at fast compression speed, a very pronounced capping tendency was manifestad itself by a friability clearly enlarged of tablets at fast compression speed in comparison with those compressed at slow speed. Material that cap can often made into tablets below some critical pressure (Garr and Rubinstein, 1991). This critical pressure is related with compression speed. Here, the critical pressure at which capping is initiate for these materials is between 200 and 300 MPa at the fast compression speed. The behaviour of HPβCD was clearly different after the 100 MPa of applied pressure a pronounced capping tendency was observed, which it is possible consider as the critical pressure for HPβCD. Thus, the application of suitable pressure was a decisive factor to obtain firm tablet, but also compression speed was a significant factor to consider, even more at high applied pressures.

βCD compressed into tablets showed practically instantaneous disintegration. Here, due to the mentioned negligible formaton of permanent bonds for βCD, on the basis of the very low values of tensile strength. The disintegration behavior was independent of compressional pressure. HPβCD showed also a few dependence on applied pressure, however, longer disintegradon times were obtained for tablets compressed at slow speed. Disintegration times of αCD and γCD increased with increasing applied pressure, and a further increase in applied pressure (above 150 MPa) had no effect on tablet disintegration time. Again, capping of tablet was manifested, for αCD at fast speed..

4. References

Fenyvesi, E., Shirakura O., Szejtli, J. Nagai, T. (1984)Properties of cyclodextrin polymer as a tableting aid. *Chem. Pharm. Bull.*, **32**, 665-669.

Garr, J.S.M., Rubinstein, M.H., (1991) An investgabon into capping tendency of paracetamol at increasing speed of compression. *Int. J. Pharm.* 72, 117-122

Giordano, F., Gazzaniga, A. Bettinetti, G.P.(1990) The influence of water content on the binding capacity of β-cyclodextrin. *Int. J. Pharm.*, 62, 153-156,

Hölzer, A.W., Sjögren J. (1978) The influence of the tablet thickness on measurements of friction during tableting. *Acta Pharm. Suec.*, 15, 59-66.

Järvinen, M.J., Juslin, M.J. (1974) On fricctonal work during tablet compression. *Farm. Aikak*, 74 1-8.

Marshall, P.V., York, P., MacLaine, J.Q. (1993) An investgation of the effect of the punch velocity on the compaction properties of ibuprofen. *Powder Techn.* 74, 71-77.

Monedero M.C., Muñoz-Ruiz, A., Velasco M.V., Muñoz, N., Jiménez-Castellanos, M.R. (1994) Analysis comparadve of methods to evaluate consolidaron mechanisms in plasfic and viscoelasfic matedais used as direct compression excipient. *Drug Dev. Ind. Pharm.* 20, 327-342.

Muñoz-Ruiz, A., Monedero M.C., Velasco M.V., Jiménez-Castellanos, M.R.(1993) Physical and rheological properdes of raw matedais. *S. T.P. Pharma Sci.*, **3**, 307-312

Newton, J.M., Rowley, G., Fell, J.T., Peacock, D.G., Rigdway, K. (1971) Computer analysis of the relation between tablet strength and compaction pressure. *J. Pharm. Pharmacol* 23, 195S-201S.

Pande, G.S., Shangraw, R.F. (1995).Characterization of β-cyclodextrin for direct compression tableting II The role of moisture in the compactability of β-cyclodextrin. *Int. J. Pharm.*, 124, 231-239

IN VITRO AND *IN VIVO* INVESTIGATIONS OF THE SPECIFIC BINDING OF SUBSTANCE P-γ-CYCLODEXTRIN ADDUCTS TO RAT BRAIN NK1 RECEPTORS.

C. PEAN[a] J. FISCHER[b], S. DOLY[b], A. WIJKHUISENE[c,d], F. DJEDAINI-PILARD[a], R. SHIGEMOTO[e], C. CREMINON[c], B. PERLY[a] and M. CONRATH[b].

[a]DRECAM/SCM, CEA-Saclay, 91191 Gif sur Yveite (France), [b]Laboratoire de Cytologie, Université Paris 6, CNRS UMR 7624, 7 Quai St Bernard, 75005 Paris (France), [d]DRM/SPI, CEA/Saclay, 91191 Gif sur Yvette (France) (Université Paris 7, UFR de Biologie, 75252 Paris (France), [e]Department of Morphological Brain Science, Faculty of Medicine, Kyoto University, Kyoto 606-01 (Japan)

1. Introduction

The grafting of peptides as signal molecule on γ-cyclodextrin adducts is expected to provide useful tools to target included drugs towards a specific biologically-relevant site (Djedaïni-Pilard et al., 1993, Pean et al. this issue). We show here, using autoradiography, that the neuropeptide substance P (SP) coupled to γ-cyclodextrin (γCD-SP) specifically binds rat brain NK1 substance P receptor (NK1R) *in vitro.* To test the binding of γ-CD-SP *in vivo,* intracerebral injections were performed in the striatum, a region rich in NK1 receptors. The subsequent γ-CD-SP-induced internalization of the NK1R was studied with immunocytochemistry using a specific anti-NK1R antibody.

2. *In vitro* investigations

Autoradiography with ^{125}I-SP was performed on brain and spinal cord coronal sections from three male Wistar rats. Briefly, sections were incubated for 2 h in 0.125 nM ^{125}I-SP in Tris pH 7.4 containing a cocktail of peptidase inhibitors (0.05

μg/ml bacitracin, 0.3 μg/ml benzamidine, 20 μg/ml leupeptin, 20 μg/ml chymostatin, 0.03 μg/ml phenylmethylsulfonylfluoride), 3 mM $MnCl_2$ and 0.2 g/l bovine serum albumin. To test the relative potencies of SP and γ-CD-SP to displace ^{125}I-SP binding, 10^{-5} to 10^{-9} M cold SP or γ-CD-SP were added in competition with ^{125}I-SP in the incubation medium. After washing and drying, sections were exposed on Amersham ^{3}H Ultrofilm for 3-6 days.

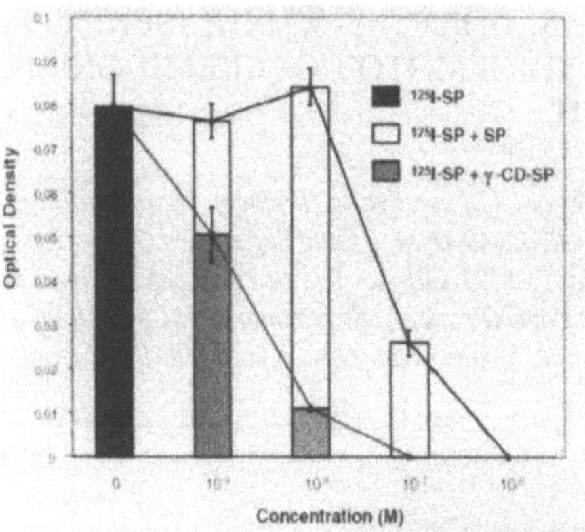

Figure 1: Inhibition ^{125}I-SP binding by γ-CD-SP and cold SP. Four concentrations of SP or γ-CD-SP (10^{-6} to 10^{-9} M) were used to displace 0.125 nM ^{125}I-SP. Binding is indicated in optical density measured in lamina I of the cervical spinal dorsal horn. Values are the mean of four or more determinations for each experimental condition (S.E.M. are indicated).

γ-CD-SP displaced dose-dependently the labeling observed in all NK1-rich brain nuclei: olfactory bulb, amygdala, striatum, parabrachial nucleus, inferior colliculus, locus coeruleus, parabrachial and solitary tract nuclei, and dorsal horn of the spinal cord. Quantitative analysis have shown that 10^{-6} M γ-CD-SP totally abolished ^{125}I-SP labeling and that the potency of γ-CD-SP to displace ^{125}I-SP labeling was about ten fold lower than that of cold SP (Figure l). These results showed that γ-CD-SP recognized the brain NK1 receptor *in vitro* with a high affinity.

*3. **In vivo*** investigations

Under anesthesia, 12 male Wistar rats received a stereotaxic injection of 1 μl of 10^{-4} to 10^{-8} M γ-CD-SP in synthetic cerebrospinal fluid in the striatum. Ninety second to 12 min later, intracardiac perfusion of 4% paraformaldehyde were made. Immunocytochemistry was performed on 50 μm vibratome sections, using the avidin-biotin peroxidase method. At the injection site, a massive translocation of

NK1R immunoreactivity was observed in cell bodies from the plasma membrane to endosomes (figure 2, B) compared to a normal striatum (figure 2, A). Moreover, dendrites underwent the marked morphological reorganization previously describes (Mantyh et al., 1995) showing large swollen varicosities filled up by densely packed immunoreactive endosomes (figure 2, b compared to 2, a).

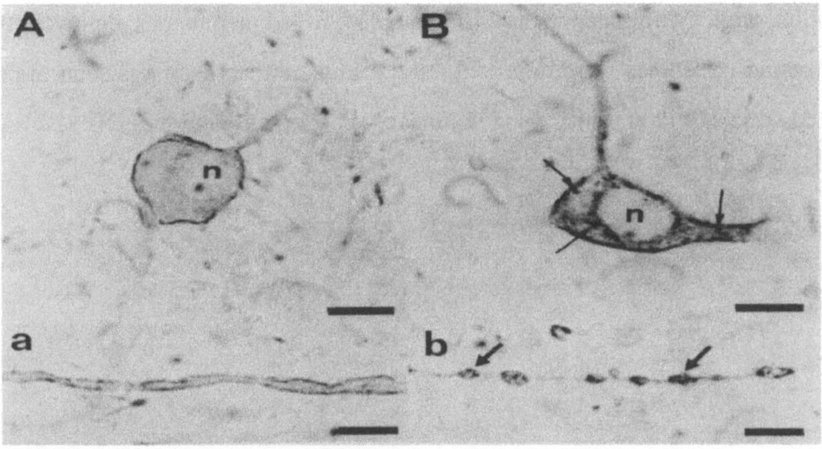

Figure 2: γ-CD-SP-induced internalization of NK1R immunoreactivity.
In a normal striatum (Aa), NKIR immunoreactivity is mostly restricted to the plasma membrane. Distal dendrites (a) are linear and regular in diameter. Six minutes after injection of 10^{-6} M γ-CD-SP, numerous immunoreactive endosomes (thin arrows) are observed in cells bodies (B) and dendrites (b). Distal dendrites (b) show a drastic morphological rearrangement and appear as large swollen varicosities (arrows) connected by thin segments. Varicosities are filled up by numerous immunoreactive endosomes (n:nucleus). Scale hars: A,B = 50 µm; a,b = 150 µm.

The level of internalization, was dependent on the concentration of injected γ-CD-SP. Six minutes after injection of 10^{-4} M γ-CD-SP, almost all dendrites and cell bodies showed endosome labeling and all distal dendrites displayed morphological changes in the ipsilateral striatum. Six minutes after a 10^{-8} M γ-CD-SP injection, only distal dendrites showed marked internalization with endosome labeling and morphological changes (swelling).

For control, NK1R immunoreactivity was observed in the non-injected striatum. In addition, injection of 10^{-4} M spantide II, a specific NK1 antagonist, with 10^{-7} M γ-CD-SP greatly decreased the internalization ratio observed with 10^{-7} M γ-CD-SP alone. Injection of cerebrospinal fluid alone and placing the needle without injection did not produce any internalization of the NK1R immunoreactivity.

Under the electron microscope, numerous dendrites exhibited morphological features typical of endocytosis: coated pits, coated vesicles and endosomes. At early stages after injection (90 seconds), numerous coated vesicles and primary endosomes, were observed (figure 3, A). Primary endosomes were mainly localized near the plasma membrane. After 10-12 minutes, many perinuclear endosomes were observed. In addition, large cytoplasmic inclusions containing rolled membranes, multivesicular bodies and lysosomes were observed. NK1R immunoreactivity was seen in both coated vesicles and in primary and secondary endosomes (Figure 3, B).

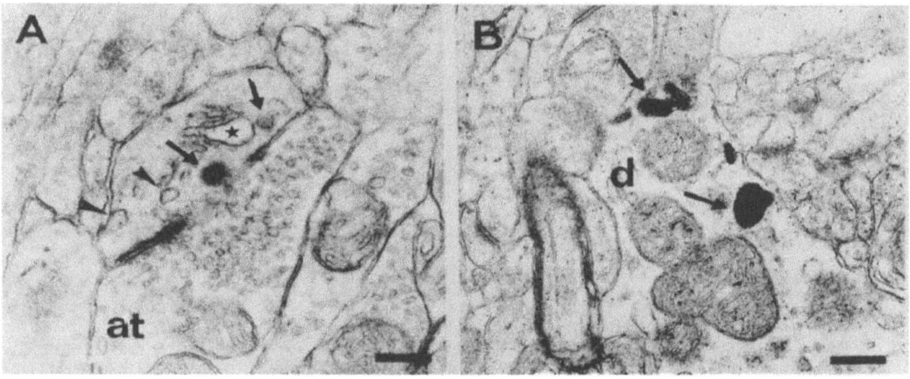

Figure 3: γ-CD-SP-induced internalization at the ultrastructural level.
Ninety second after 10^{-4} M γ-CD-SP injection (A), numerous dendrites exhibit intense internalization as seen by the frequency of coated vesicles (arrows), primary (arrow heads) and secondary endosomes (star) (at: axon terminal). Inununolabeling of NK1R (B) shows an intense staining of endosomes (arrows) in a large swollen dendrite (d) six minutes after injection of 10^{-6} M γ-CD-SP. Scales bars: 150 nm.

In conclusion our results show that grafted SP specifically targets γ-CD to cells bearing NK1 receptors *in vitro and in vivo.* They strongly suggest that γ-CD-SP, like free SP, may enter the cell through receptor-mediated internalization, although the presence of γ-CD into the cells remains to be determined. The targeting and potential entrance of a guest drug included into the γ-CD is now under investigation.

4. References

Djedaini-Pilard, F., Desalos, J. and Perly, B. (1993) Synthesis of a new molecular carrier: N-(Leu-enkephalin)yl 6-amido-deoxy-cyclomaltoheptaose. *Tetrahedron Lett.*, 34, 2457-2460.

Garland, A.M., Grady, E.F., Payan, D.G., Vigna, SR., and Bunnett, N.W. (1994) Agonist-induced intemalization of the substance P (NK1) receptor expressed in epithelial cells. *Biochem.* J 303, 177-186.

Mantyh, P.W., Allen, C.J., Ghilardhi, J.R., Rogers, S.D., Mantyh, C.R., Liu, H., Basbaum, A.I., Vigna,SR. and Maggio, J.E. (1995) Rapid endocytosis of a G-protein-coupled receptor: substance Pevoked intemalization of its receptor in the rat striatum in *vivo. P.N.A.S.*, 92, 2622-2626.

Péan, C., Djedaini-Pilard, F., Créminon, C., Wijkhuisen, A., Grassi, J., Guénot, P., Jéhan, P and Perly, B.Synthesis and characterization of peptido-cyclodextrins dedicated to drug targeting. *(this issue).*

STABILIZATION OF RETINOL WITH γ-CYCLODEXTRIN

T. WIMMER, M. REGIERT, J.-P. MOLDENHAUER
Wacker-Chemie GmbH, Johannes-Heß-Str. 24,
D-84489 Burghausen, Germany

Abstract

β- and γ-cyclodextrin inclusion compounds of retinol were prepared under nitrogen atmosphere by known methods. Generally a molar ratio of 2 : 1 (CD:retinol) was found. Comparative storage studies of different complexes and physical mixtures with lactose were performed using day light and UV radiation. The best stabilization was obtained using γ-cyclodextrin which leads to new potential uses for γ-CD also in health care applications.

1. Introduction

Retinol (vitamin A) and its esters are not only important in the human diet, they also reveal some valuable properties in skin care products. In topical cosmetic antiaging formulations retinol reduces wrinkles and helps to restore UV damaged tissue.

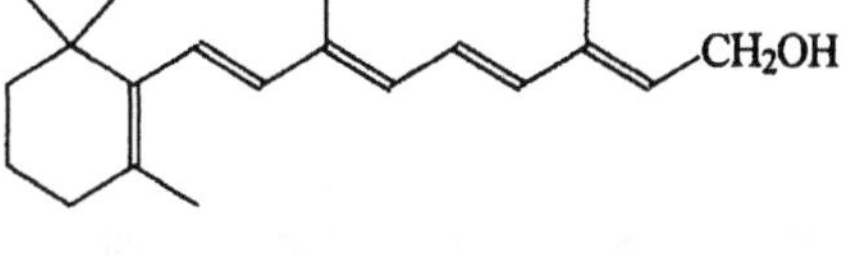

Nevertheless the use of retinol is limited due to its high instability. Especially under the influence of UV light and in the presence of oxygen the fat-soluble vitamin tends to rapid oxidation and polymerization. During the oxidation some peroxidic toxic intermediates are formed.
Complexes of retinol acetate and retinol palmitate with β-cyclodextrin are well known [1, 2]. Soluble complexes of retinol with branched CDs are describes in [3].
Stabilization of vitamin A with methyl-β-CD is covered by a Japanese patent **[4]**. The binding constants with β-CD and DIMEB are reported in [5]. In this study complexation with γ-CD and the stability of its complexes were investigated.

2. Materials and Methods

Materials: Retinol was purchased from Aldrich. β- and γ-Cyclodextrin are products of Wacker-Chemie GmbH, Munich, Germany.

Methods: The inclusion compounds were prepared from concentrated cyclodextrin solutions or by the kneading method with the exclusion of light and under nitrogen atmosphere. The completes were isolated by filtration, followed by a washing step with nitrogen saturated deionized water and finally dried at 30-40 °C in vacuo.
The ratio of retinol / γ-CD was measured by proton NMR.
Stability tests were performed in open petri dishes. For comparison physical mixtures of lactose and retinol with a retinol content of 10 % by weight were prepared. The samples were exposed to day light or irradiated with UV light (366 nm). The Retinol content in the stored samples was measured by HPLC.

3. Results and Discussion

3. 1. COMPLEXATION

Retinol was found to form inclusion compounds with β-CD and γ-CD. In both cases solid completes were easily prepared by precipitation in a 1:2 molar ratio (guest/CD). Due to the low water solubility of the complex unreacted γ-CD can be easily removed by a washing step. Using the solution method the complexation was found to be complete within 24-48 hours, whereas the complexation time could be reduced to 4-6 hours when kneading was applied.
The γ-CD complex (1:2) was characterized by DSC thermograms shown in Figure 1. In the physical mixture an endothermic peak at 63 °C indicates the melting point of free retinol. The peak disappeared in the case of the complex.

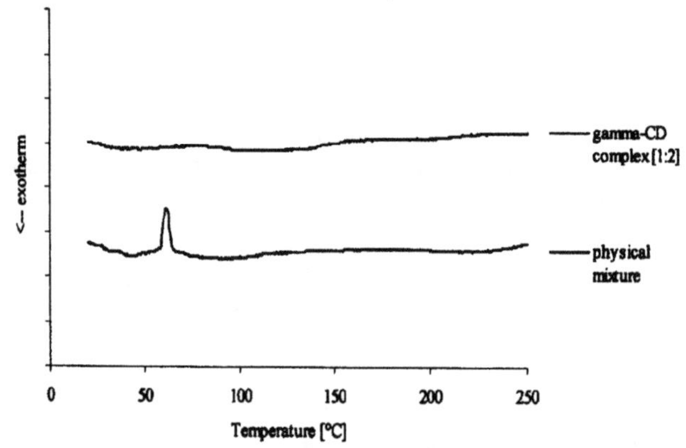

Figure 1. DSC thermograms

3.2. STABILITY TESTS

Under the influence of UV light (366 nm) and oxygen a very poor stability of retinol was found in case of its β-CD complex. Similar results were found for the inclusion compound of β-CD and retinol acetate [6]. Figure 2 demonstrates a much better stabilizing effect of γ-CD against the oxidation and polymerization of retinol. No difference could be found on the methods used for complexation.

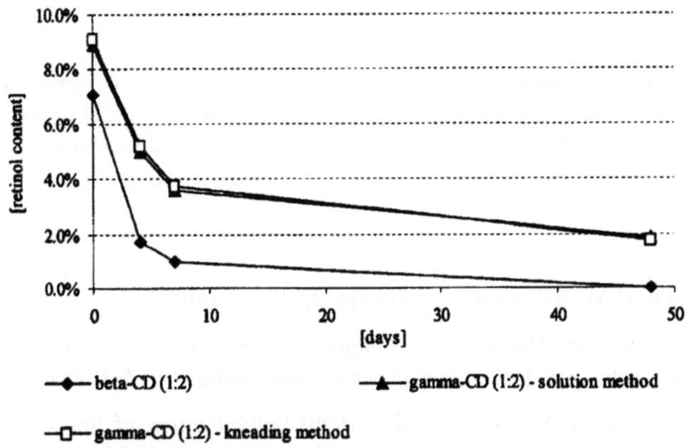

Figure 2. Stability of retinol with UV irradiation

The stability of retinol as its γ-CD complex was also compared with free retinol mixed with lactose. The initial retinol concentration in both cases was about 10 % by weight.
The samples were exposed to daylight over a period of 7 weeks. Samples were analyzed after 4, 7, 21 and 48 days for retinol content by HPLC. A good stabilization for the vitamin was observed as shown in figure 3. In the lactose mixture a rapid degradation of the uncomplexed retinol is evident.

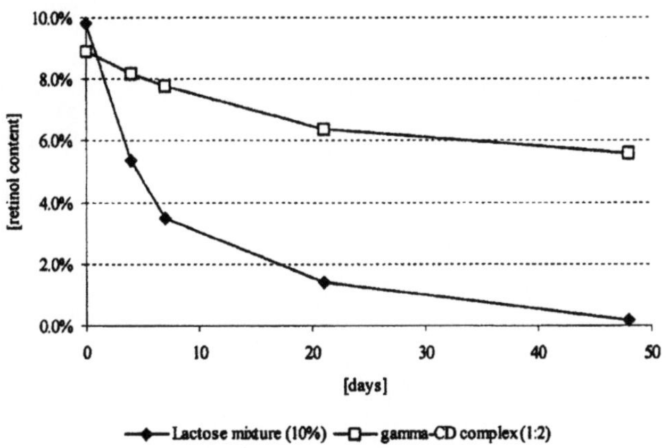

Figure 3. Stability of retinol under day light conditions

4. Conclusion

β-CD and γ-CD form poorly soluble complexes with retinol in a 1:2 molar ratio. Kneading is the preferred method for the preparation of γ-CD inclusion compounds. γ-CD proved to be the most effective cyclodextrin in stabilizing the vitamin against light induced degradation.

References

[1] Schlenk H., Sand D.M., Tillotson J. A., US Pat. 2,827,452 (1958)

[2] Palmieri G.F.; Wehrle P.; Duportali G.; Statmm A..;(1992), Inclusion complexation of vitamin A palmitate with beta-cyclodextrin in aqueous solution, *Drug Dev, Ind. Pharm* 18(19), 2117-21

[3] Okada, Y.; Tachibanna, M.; Koizumi,K.; (1990) Solubilization of lipid-soluble vitamins by complexation with glucosyl-beta-cyclodextrin, *Chem. Pharm. Bull.* 38 (7), 2047-9

[4] Koide, M.; Hozumi, S.; (1994) Stable opthalmic solutions containing vitamin A, *Jpn Kokai Tokkyo Koho,* JP 06293638

[5] Guo, Q.-X.; Ren, T.; Fang, Y.-P.; Liu, Y.C.; (1995) Binding of vitamin A by beta-cyclodextrin and heptakis (2,6-O-dimethyl)-beta-cyclodextrin, *J. Inclusion Phenom. Mol. Recognit Chem.* 22 (4) 251-6

[6] Froemming, K.H.; Gelder, T.; Mehnert, W.; (1988) Inclusion compound of beta-cyclodextrin and vitamin A acetate, *Acta Pharm. Technol.* 34, 152.

INCLUSION STUDY IN ß-CYCLODEXTRIN AND DIMETYL-ß-CYCLODEXTRIN OF ANTIPARASITARIES IN SOLUTION AND THE SOLID STATE

L. NIETO-REYES*, M.E. VILLAR-LÓPEZ*, J.A. CASTRO HERMIDA**, E. ARES-MAZÁS**, F. OTERO-ESPINAR*, J. BLANCO-MÉNDEZ*

*Dpto. Farmacia y Tecnología Farmacéutica, Facultad de Farmacia, Universidad de Santiago de Compostela, Santiago de Compostela, (Spain)

**Lab. Parasitología, Dpto. Microbilogía y Parasitología, Facultad de Farmacia, Universidad de Santiago de Compostela, Santiago de Compostela, (Spain)

1. Introduction

The protozoan *Cryptosporidium parvum* is an agent causative of diarrhoea disease. Unfortunately, there is not effective therapy available for this parasite due, in part to variability in the response of antiparasitaries used because of the problems of solubility and stability (Woods K. M. et al., 1996). Diloxanide furoate is very slightly soluble in water, and is hydrolysed in intestinal fluids before been absorbed. Nalidixic acid and Norfloxacin are also insoluble drugs (Martindale, 31st Edn.). The aim of this work was to improve the physic-chemical properties, e.g. solubility and stability. Thus, cyclodextrins are appropriated for this goal (Loftsson T. Et al, 1996, Palmieri G. F. et al 1997).

2. Materials

Diloxanide furoate (DF) was a generous gift from Laboratorios Knoll, S.A. (Spain), Nalidixic acid (NA) and Norfloxacin (N) was purchase from Sigma (Spain). Dimetyl- ß-cyclodextrins was purchase from Cyclolab (Hungary) and ß-cyclodextrin was a generous gift from Roquette (Spain). All other chemicals were of analytical grade. Distilled water was used throughout the study.

3. Methods

Solubility studies were carried out according to the method of Higuchi and Connors (Higuchi et al., 1965). The apparent 1:1 stability constant (K_c) was calculated from the linear part of the phase solubility diagram. Different molar ratios of DF, NA and N with ß-CD and DIMEB, respectively, were prepared by kneading. Besides, DF with both cyclodextrins was prepared by coevaporation, spray-drying and freeze-drying using ethanol as cosolvent The complexes were analysed by DSC (Shimazdu DSC-50), and X-Ray (Siemens D-500) and dissolution studies (Prolabo Dissolutest) were carried out. The hydrolytic degradation of DF was followed without and with ß-CD at three concentration.

4. Results and discussion

The interactions between DF, NA, N and ß-CD or DM-ß-CD in solutions were investigated by constructing phase solubility diagrams.

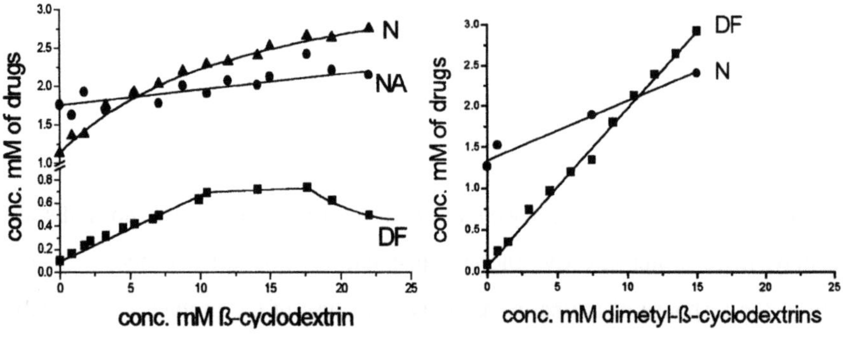

Figure 1: Diagrams of solubility of drugs and cyclodextrins.

As show in Fig. 1, the solubility of DF increased as a function of concentration of ß-CD, showing a B_s type solubility curve (k_c 550 M^{-1}) the complex stechiometric is 1:1; with DIMEB gives an A_L type diagram (k_c 2684 M^{-1}). Norfloxacin shows an A_N shape with ß-

CD; with DIMEB gives an A_L type diagram with a low stability (k_c 45.87). The formation of the complex with nalidixic acid can not be confirmed.

In Figure 2 is shown the DSC curves of complexes prepared. DF exhibits its characteristic endothermic peak at 115 °C, due to the melting of the drug. As would be expected, the interaction of DF and CD is accompanied by disappearance of this peak as a function of concentration of CD. Same results are presented for N.

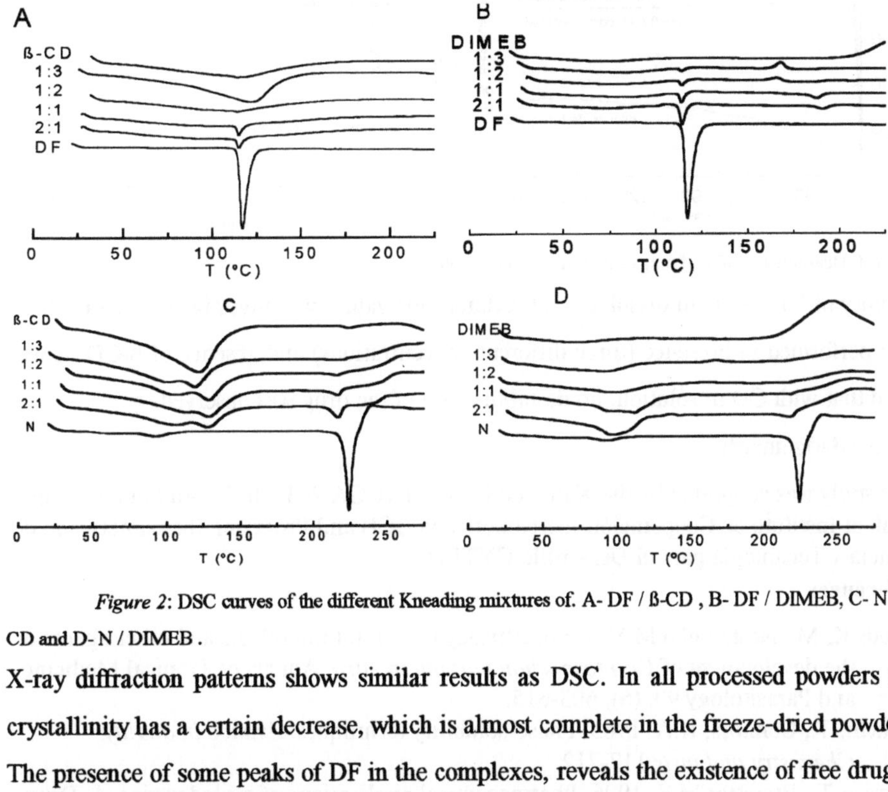

Figure 2: DSC curves of the different Kneading mixtures of. A- DF / ß-CD , B- DF / DIMEB, C- N / ß-CD and D- N / DIMEB .

X-ray diffraction patterns shows similar results as DSC. In all processed powders the crystallinity has a certain decrease, which is almost complete in the freeze-dried powders. The presence of some peaks of DF in the complexes, reveals the existence of free drug in the mixture. Free drug disappears at higher proportions of CD.

The IR spectrum displays that for CD there are no absorption bands in the region of carbonyl group of the ester (DF). Nevertheless, due to the presence of this band, the spectrums show a blocking of DF in CD cavities for all the molar ratios and methods, except for freeze-drying.

Figure 3 displays that in simulated intestinal fluid there is a great increase in the solubility and the rate of dissolution for all the complexes of DF. The complex N-ßCD increase the rate dissolution but not solubility of drug.

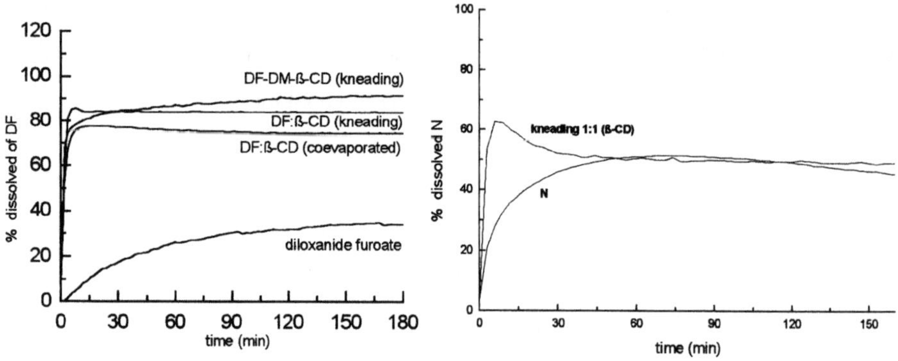

Figure 3. Dissolution profile of drugs and its inclusion complexes.

Stability of DF (ester) in dissolution at different pH values was investigated. Assays have been performed in presence (three different concentrations) and absence of ß-CD. It was noted that with CD in solution, no destabilisation of the drug was observed.

Acknowledgements

This study was supported by the Xunta de Galicia (XUGA 20310b97). Authors gratefully thank at Instituto de Cooperación Iberoamericana (ICI) and Programa Iberoamericano de Ciencia y Tecnología para el Desarrollo CYTED

References

Woods K. M., Nesterenko M.V. 1996. Efficacy of 101 antimicrobials and other agents on the development of *Cryptosporium parvum in vitro*. Annals of Tropical Medicine and Parasitology 90, (6), 603-615.

Higuchi, T., Connors, K.A. 1965. Phase-solubility techniques. *Avance in anality Chemistry and instr* 117-212.

Loftsson T., Brewster M.E. 1996. Pharmaceutical applications of cyclodextrins. 1. Drug solubilization and stabilization. Review article. *Journal of Pharmaceutical Sciences* 85, (10), 1017-1025.

Palmieri G.F., Giovannucci G., Antonini I., Martelli S. 1997. Inclusion complexation of fenifibrate with ß-cyclodextrin and hydroxypropyl-ß-cyclodextrin. Evaluation of interactions in solution and solid complex characterization. *S. T. P Pharma Sciences* 7, (2), 174-181.

STUDY OF INCLUSION COMPOUND OF TRIAMCINOLONE ACETONIDE

M.E. VILLAR-LÓPEZ, L. NIETO-REYES, F. OTERO-ESPINAR, J. BLANCO-MÉNDEZ

Dpto. Farmacia y Tecnología Farmacéutica, Facultad de Farmacia, Universidad de Santiago de Compostela, Santiago de Compostela, (Spain)

1. Introduction

Triamcinolone Acetonide (TA), used in the treatment of ulcerative colitis and Crohn´s disease (Martindale, 31st Edn.), is an almost water-insoluble molecule.
ß-cyclodextrin (ß-CD) is a natural cyclic oligosaccharide with hydrophilic characteristic (Palmieri G.F. et al., 1997), and it is known for its ability to form inclusion complexes with many lipophylic drugs. As a result of this complexation major physico-chemical properties (eg. solubility, stability) of the included drug might change (Loftsson T. et al., 1996; Frijlink H.W. et al., 1991).
We carried out the present investigation to find out whether it is possible to obtain an inclusion complex between TA and ß-CD.

2. Materials

Triamcinolone Acetonide was purchased from Roig-Farma (Spain) and ß-cyclodextrin (Roquete) was a generous gift from Roquette-Laisa (Spain). All other chemicals were of analytical grade. Distilled water was used throughout the study.

3. Methods

Solubility studies were carried out according to the method of Higuchi and Connors (Higuchi et al., 1965). The apparent 1:1 stability constant (K_c) was calculated from the linear part of the phase solubility diagram. Inclusion complexes of TA and ß-CD in different molar ratios were prepared by kneading, coevaporation and freeze-drying. They were analyzed by DSC (Shimazdu DSC-50) and X-Ray (Siemens D-500) and dissolution studies were carried out. Samples have been seen under the microscope and photos have been taken in an Olympus SZ60 stereomicroscope.

4. Results and Discussion

1. Inclusion complex in aqueous solution

The interactions between TA and ß-CD in solution was investigated by constructing phase solubility diagrams. As shown in Fig. 1, the solubility of TA increased linearly as a function of concentration of TA-ß-CD showing an A_L type solubility curve (Higuchi et al., 1965).

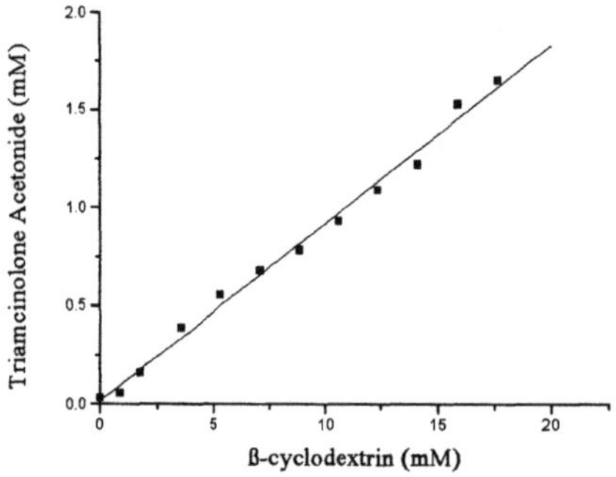

Figure 1. Higuchi and Connors solubility curve for TA.

The apparent 1:1 stability constant (K_c) for TA with ß-CD was calculated as 2800 M^{-1}.

2. Characterization of solid complex

Different instrumental techniques were used to examine and characterize complexation of TA and ß-CD. Fig. 2 shows the differential scanning calorimetry curves of powders prepared with ß-CD, compared with those of the pure TA and ß-CD.

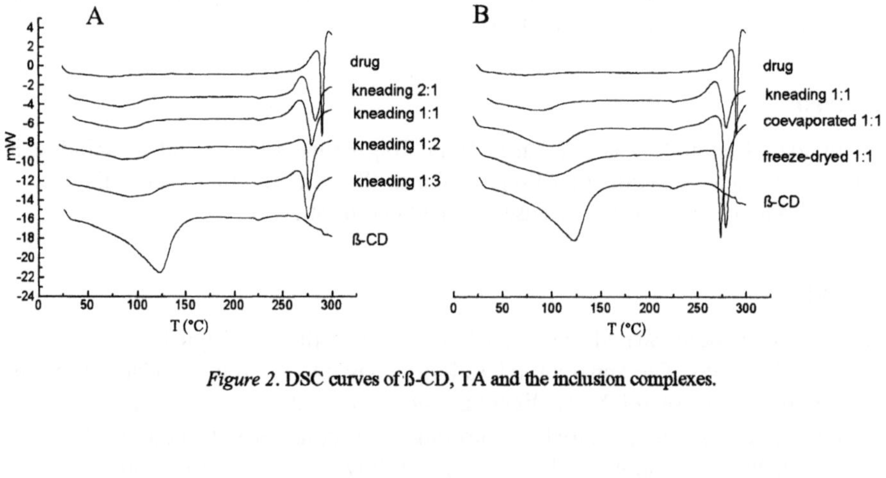

Figure 2. DSC curves of ß-CD, TA and the inclusion complexes.

DSC curves demostrate an excess of free drug in the different inclusion complexation methods.

The DSC curve of TA cristals shows an endothermic peak at 289.62°C due to the melting of TA. ß-CD displays no peak at this temperature. When the complex is prepared by kneading, the drug melting peak do not desappear in none of the ratios prepared with ß-CD (Fig. 2A); this suggest no total complexation of drug in these powders. Hence, kneading is not an appropiate preparation method to obtain inclusion complexes between TA and ß-CD. Same happens with coevaporated and freeze-dryied powders (Fig. 2B). Nevertheless, a decrease in the fusion temperature and the broading of the endothermic peak of the TA, suggest the formation of a solid solution. It is a result of the dispersion of the drug in the hydrophilic net of the ß-CD when kneading and freeze-drying are used.

Fig. 3 shows the X-ray diffraction pattern of pure TA and ß-CD, of solid complexes obtained by kneading in the molar ratios of 1/1, 1/2, 1/3 and 2/1 and the ones obtained by coevaporation and freeze-drying in the equimolar ratios.

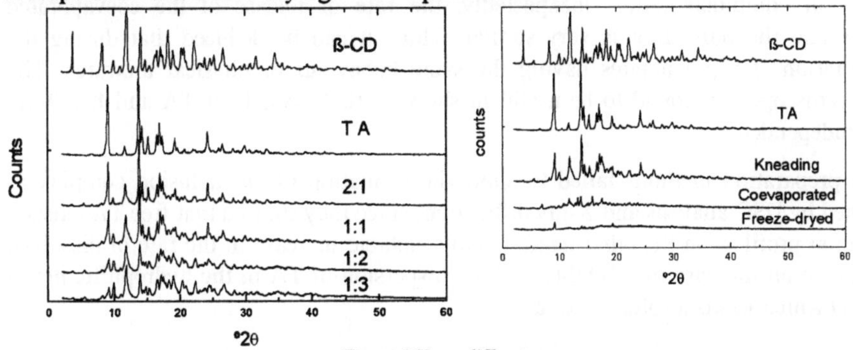

Figure 3 X-ray diffractograms

Diffractograms of kneading powders show drug peak even with a TA/ß-CD molar ratio of 1/3. Obviously, these peaks are greater in the 2/1 molar ratio powder, but the fact is that they are still present.

When comparing diffractograms of solid complex in equimolar ratios with those of pure drug and ß-CD, a certain decrease in crystallinity in all processed powders can be observed. The reduction is almost complete in the freeze-dried powders.

Fig. 4A presents the dissolution profile of TA and the inclusion complex prepared by kneading in the molar ratios 1/1, 1/2, 1/3 and 2/1 in phosphate buffer at 37°C. A molar ratio of 2/1 shows a low release because of the insufficient amount of ß-CD in their composition. In fact, with a double quantity of ß-CD the release of TA is total.

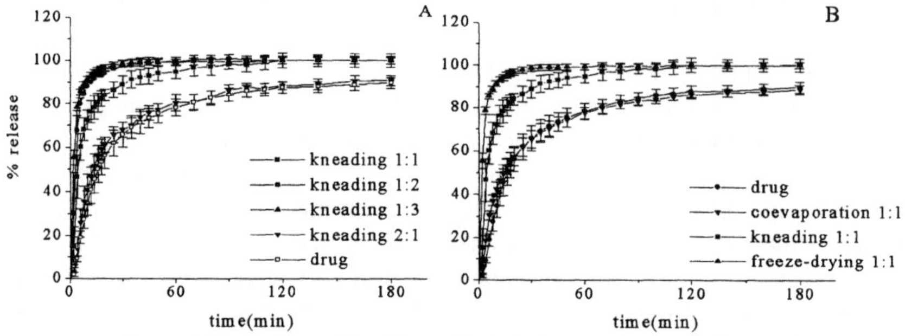

Figure 4. Dissolution profile of TA and its inclusion complexes at different molar ratios.

Fig. 4B shows that the freeze-dried powders in equimolar ratio exhibits, at the same time, total release of TA, although the dissolution occurs more rapidly than the kneading complex in equimolar ratio. Unexpectedly, the rate of release of the coevaporated complex and the pure drug is very similar. Thus, it can be deduced that during the coevaporation, TA precipitates having the same behaviour of physical mixtures. The stereomicroscope was found to be useful in showing that crystals of TA and ß-CD are totally independent.

All the preparation methods failed to show the formation of an inclusion complex as confirmed by DSC analysis and X-ray diffraction, since they showed that free TA exists. Dissolution profiles, on the other hand, led to a notable increase in the rate of dissolved drug, due to an increase in solubility and the dispersion of TA in the hydrophilic net of the ß-CD which forms a solid solution.

5. Acknowledgments

This work was supported by a grant from the Xunta de Galicia (XUGA 20301A95). We thank the Xunta de Galicia (DOG 2-XII-97) for a fellowship for MEVL.

6. References

Frijlink H.W., Eissens A.C., Schoonen A.J.M., Lerk C.F. 1991. The effect of cyclodextrins on drug release from fatty suppository bases. II. In vivo observations. *Eur. J. Pharm. Biopharm* 37, (3), 183-187.

Higuchi, T., Connors, K.A. 1965. Phase-solubility techniques. *Avance in anality Chemistry and instr* 117-212.

Loftsson T., Brewster M.E. 1996. Pharmaceutical applications of cyclodextrins. 1. Drug solubilization and stabilization. Review article. *Journal of Pharmaceutical Sciences* 85, (10), 1017-1025.

Palmieri G.F., Giovannucci G., Antonini I., Martelli S. 1997. Inclusion complexation of fenifibrate with ß-cyclodextrin and hydroxypropyl-ß-cyclodextrin. Evaluation of interactions in solution and solid complex characterization. *S. T. P Pharma Sciences* 7, (2), 174-181.

COMPLEXATION OF BILE SALTS BY β-CYCLODEXTRIN AND DERIVATIVES.

P. RAMOS, E. ALVAREZ PARRILLA, F. MEIJIDE, J. A. SEIJAS, E. RODRÍGUEZ NÚÑEZ AND J. VÁZQUEZ TATO. *Universidad de Santiago de Compostela. Facultad de Ciencias, Lugo (Spain).*

1. INTRODUCTION

Cyclodextrins are known to form inclusion complexes with a variety of molecules, a property used to increase the bioavailability of poorly soluble drugs[1]. Bile salts are biosurfactants involved in the cholesterol metabolism, and used as drugs in gallstone diseases treatments[2].

Different methods have been published for the determination of the equilibrium constants for the formation of inclusion complexes of cyclodextrins. Among them, Nuclear Magnetic Resonance Spectroscopy (NMR) is one of the most useful techniques because it also provides microscopic information on the complex structure[1].

The purpose of this study was to evaluate the stoichiometry, binding constants and structure of the complexes formed between two bile salts (Sodium Cholate, NaC, and Sodium Deoxycholate, NaDC) and *β*-cyclodextrin and two derivatives, using NMR techniques.

2. EXPERIMENTAL SECTION

2.1 GENERAL PROCEDURE.

Commercial bile salts (Sigma-Aldrich) and β-cyclodextrin (Roquette) and synthetized derivatives were dried in a vacuum oven. Other chemicals were of high quality and used without further purification. Thin layer chromatography (TLC) was performed on aluminium-backed silica gel plates eluting ethyl acetate: isopropyl alcohol: water: concentrated NH_4OH (2:3:4:0.3) and visualized with ultraviolet light, 5% H_2SO_4 in MeOH or 0.2% Ninhydrin in EtOH sprays followed by charring. Melting points were obtained on a Gallenkamp melting point apparatus. ^{1}H, ^{13}C and DEPT 135 NMR spectra were recorded on a Brucker spectrometer at 300 and 75 MHz at 273,1 (± 0,1) K. The Rotating-Frame Overhauser Effect Spectroscopy (ROESY) were recorded on a Brucker spectrometer at 500 MHz. All NMR experiments were carried out in D_2O.

Using different volumes of two equimolar solutions (Host and Guest) in D_2O, samples of different molar fractions were prepared.

2.2 SYNTHESIS.

2.2.1 *6-deoxy-6-amino-β-cyclodextrin(β-CDNH$_2$).*

This compound was synthesized by a modified method of Fragoso et al[3]. 6-O-tosyl-β-cyclodextrin[4] (1 g, 0,77mmol) was dissolved in a 25% ammonium solution (25 mL) and stirred at 50° C over night. After concentration under reduced pressure, the resulting white solid was redissolved in water and purified with a Sephadex C-25 cationic column using water and 0.1 M NH_4HCO_3 as eluents, to give product in 45-50% yield. R_f 0.2; mp 202-203°C; ^{1}H NMR: δ 4.97(s, 7, H-1), 3.44-3.89 (m, H-2,H-3, H-4, H-5, H-6), 3.36 (t,1,H-4'), 3.24 (dd, 1, CH_2-NH_2), 2.94 (dd, 1, CH_2-NH_2); ^{13}C and DEPT 135: δ 104.45 (C-1), 85.82 (C-4'), 83.89 (C-4), 75.71 (C-3), 74.43 (C-2), 74.37 (C-5), 62.93 (C-6, negative signal in DEPT), 42.82 (C-6', negative signal in DEPT).

2.2.2 *Dimer I*

Dimer I was synthesized by the reaction of 6-deoxy-6-amino-β-cyclodextrin (β-CDNH$_2$) (0.5 g, 0.44 mmol) with 1,2,4,5-benzene tetracarboxylic dianhydride (0.038 g, 0.176 mmol) in DMF (50 mL) (Scheme 1). The reaction was stirred for 48 h at 50° C. The solvent was removed under reduced pressure at low temperature, and the resulting solid was redissolved in water. Water was removed (3-4X) until no DMF was observed, and finally purified through a Sephadex C-25 column using water and 0.1 M NH_4HCO_3 as eluents, obtaining a white solid (0.31 g, 62%). R_f 0.6; mp 185-190°C (Dec.); ^{1}H NMR δ7.52 (m, 1, aromatic), 4.95(s, 7, H-1), 3.47-3.89 (m, 42, H-2,H-3, H-4, H-5, H-6); ^{13}C and DEPT 135 δ 176.09 (COOH, signal disappear), 173.79 (CONH, signal disappear in DEPT), 141.38, 140.51, 139.4, 138.51, 137.44 substituted aromatic carbons, signals disappear in DEPT), 130.2, 129.79 (CH aromatic ring), 104.54 (C-1), 85.38 (C-4'), 83.75 (C-4), 75.69 (C-5), 74.73 (C-3), 74.38 (C-2), 62.83 (C-6, negative signal in DEPT), 43.14 (C-6', negative signal in DEPT).

Scheme 1

2.3 STOICHIOMETRIES AND STABILITY CONSTANTS DETERMINATION

Equation [1], represents the equilibrium formation of a n:m complex between a host (Cyclodextrin) and a guest (Bile Salt) :

$$m\ BS + n\ CD \rightleftarrows \text{Complex} \quad ; \quad K = \frac{[\text{Comp}]}{[CD]^n[BS]^m} \qquad [1]$$

The observed chemical shift of a given nucleus of the host or guest in the equilibrium, depends on the binding rate. Under fast exchange conditions (transition between free and complexation state of the observed molecule is faster than the absorption of energy from the radiofrecuency field) the observed chemical shift, δ_{obs}, is the average of the chemical shifts of the nucleus in free and complexed states, weighted by the fractional occupancy of these states[5]:

$$\delta_{obs} = f_{CD}\,\delta_{CD} + n\, f_{Comp}\,\delta_{Comp} \qquad [2]$$

The stoichiometry of the inclusion complexes formed were provided using the continuous variation technique (Job's Plot)[6] based on the difference observed in the chemical shifts of different carbons of the Cyclodextrin (or derivatives) in the presence of increasing amounts of the bile salt. The plot of $\Delta\delta_{obs} \cdot X_{HOST}$ against the mole fraction of HOST or GUEST shows a maximum at $X_{CD} = n / (n + m)$ or $X_{BS} = m / (n + m)$ respectively.

From equations [1] and [2] a relationship between the equilibrium concentration of the species and the chemical shift displacement is deduced:

$$\Delta\delta_{obs} = \frac{[Complex]}{[CD_0]} \cdot \Delta\delta_m \qquad [3]$$

were [Complex] is obtained solving the equation (for a 1:1 stoichiometry):

$$[Complex]^2 - \left([CD_0] + [BS_0] + \frac{1}{K}\right) \cdot [Complex] + ([CD_0] + [BS_0]) = 0 \qquad [4]$$

Fitting data to this equation, the stability constants were calculated.

3. RESULTS AND DISCUSSION

Chemical shifts of carbon number 1 (for which the highest displacement is observed) of β-CD and derivatives were used to obtain the Job's plot (see Figure 1). Similar results were obtained from carbons 3 and 4. Complexes formed between β–CD and β–$CDNH_2$ with NaC and NaDC, showed 1:1 and 2:1 stoichiometries respectively, in agreement with Tan and Lindebaum[7]. Dimer I showed a 1:2 stoichiometry with NaC and n:n stoichiomety with NaDC.

Table 1 Stability constants and δ_m values obtained for NaC/CD complexes.

Host	Carbon 1		Carbon 4	
β-CD	K = 8333 M	$\delta_m = 0.4139$	K = 7669 M	$\delta_m = 0.5744$
β-CD-NH_2	K = 11547 M	$\delta_m = 0.3898$	K = 9951 M	$\delta_m = 0.5553$

Stability constants (obtained for chemical shifts of C1 and C4 carbons of cyclodextrins) and δ_m values for 1:1 complexes are reported in Table 1 (see fittings in Figure 2).

Correlation between K and δ_m is observed and consequently, their variances are abnormally high.

As Guo *et al*[8]. reported previously it is possible to calculate the association constants of the 2:1 complexes by the use of the chemical shift data, but the fitting of four different parameters (K_{11}, K_{21},δ_{m1} and δ_{m2}) is only precise enough if we have a large number of experimental data to be fitted.

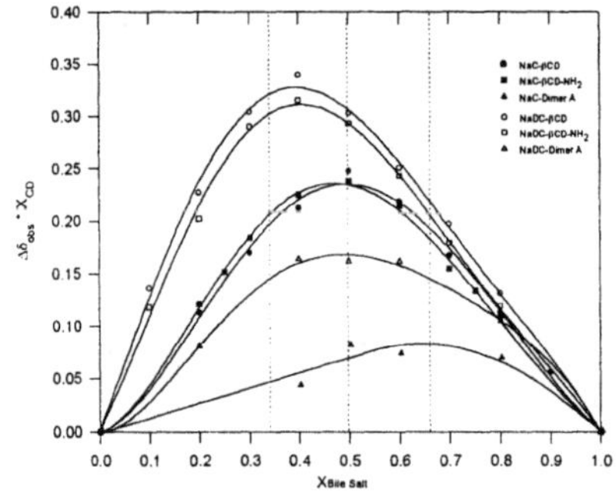

Figure 1

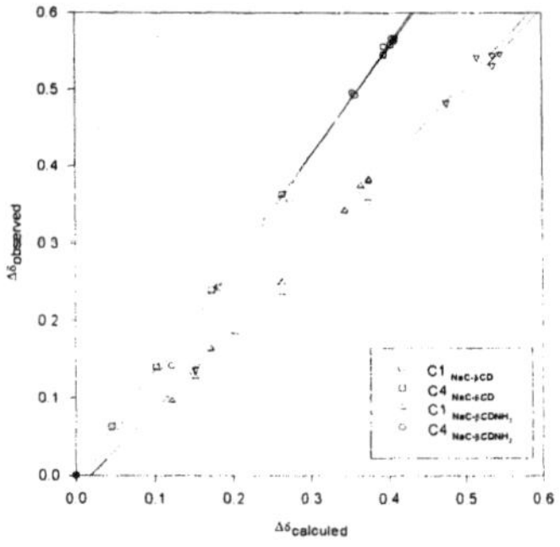

Figure 2

ROESY spectra show that bile salts enter with the 5-C ring of its steroid body forward into the inner cavity of cyclodextrins by the side of the secondary hydroxy groups. Variations on the insertion degree and side chain position of bile salts were observed for the different complexes due to the presence of amine groups.

Acknowledgement. E.A.P. thanks CONACYT (México) for a research scholarship. We thank Xunta de Galicia (Proyect XUGA 26201b96) and CYTED (Project VIII.3) for financial support.

References

(1) Szejtli, J., 1996, "Comprehensive Supramolecular Chemistry"; Vol. 3: Cyclodextrins. Szejtli & Osa Ed. Pergamon Press, U.K.

(1) Mucci, A. Schenetti, L. Vandelli, M.A., Forni, F, Ventura, P. and Salvioli, G. (1996). "One- and two-dimensional NMR study of complexation of ursodeoxycholic acid with β-cyclodextrin". *J. Chem. Soc. Perkin. Trans.* 2, 2347

(3) Fragoso, A. Cao, R. D'Souza V.T. (1997). "Influence of positively-charged guests on the superoxide dismutase mimetic activity of copper (II) β-cyclodextrin dithiocarbamates[1]". *J. Carbohydrate Chemistry.* 16(2), 171.

(4) Matsui, Y. Okimoto, A. (1978). "The binding and catalytic properties of a positively charged Cyclodextrin". *Bull. Chem. Soc. Jpn.*, 51,3030.

(5) Connors K.A.,(1987). "Binding Constants: the Measurement of Molecular Complex Stability" chap. 5. Wiley, New York.

(6) Gil, V.M.S. and Oliveira N. C. (1990). "On the use of the method of continuous variations". *J. Chem.Educ.*, 67, 473.

(7) Tan, X. and Lindenbaum, S. (1991)."Studies on complexation between β-Cyclodextrin and Bile Salts". *Int. J. Pharm.*, 74, 127.

(8) Guo, W., Fung, B. M. and Christian, S. D. (1992). "NMR study of Cyclodextrins inclusion of Fluorocarbon surfactants in solution". *Langmuir*, 8, 446.

^{1}HNMR CONTRIBUTION TO PROVE THE FORMATION OF INCLUSION COMPLEX BETWEEN NIMESULIDE AND β-CYCLODEXTRIN AND ITS HYDROPHILIC DERIVATIVES

WOUESSIDJEWE D. [1] AND ROSELLI C. [2]
1) Université Joseph Fourrier de Grenoble, EP 811 du CNRS, Faculté de Pharmacie, 5 avenue de Verdun, F-38243 Meylan cédex, France
2) DRECAM, Service de Chimie Moléculaire, CEA Saclay, F-91191 Gif sur Yvette, France

Many pharmaceutical companies now include, in an early stage of galenic development, the use of cyclodextrins as formulation aid excipient. On the other hand, there are many ways to associate the active ingredient with cyclodextrin, i.e. solution, semi-solid and solid (Wouessidjewe et al. 1993). The main role of cyclodextrins is to enhance the poor water solubility of a molecule by forming inclusion complexes (Duchêne 1987). However, tests performed to measure the increase of the solubility does not give any information on the nature of interaction (inclusion or not) between the drug and cyclodextrin.

Nimesulide is a poorly water soluble non steroidal anti-inflammatory drug (NSAID). Its solubility and dissolution were shown to be increased in the presence of cyclodextrin. This suggests an interaction between the drug and the cyclodextrin probably through the formation of inclusion compounds in liquid medium (Woussidjewe et al. 1997, Wouessidjewe 1998). These results prompted us to perform ^{1}HNMR experiments to evidence and structurally characterize an inclusion complex between nimesulde and the cyclodextrin.

In a first step, we have tested the solubility of nimesulide in the presence and absence of several cyclodextrins. Then, we have characterized the formation of a soluble complex with β-cyclodextrin by NMR.

Phase solubility diagram study was first carried out in distilled water at 37 °C between nimesulid (4' - nitro - 2'- phenoxymethanesulphoanilide) and several hydrophilic cyclodextrins: β-cyclodextrin (β-CD), 2-hydroxypropyl-β-cyclodextrin (HPβCD), partially methylated-β-cyclodextrin (PMβCD), and sulphobutylether-β-cyclodextrin (.Captisol®)The results showed that in all cases, the aqueous solubility of the drug increases as a function of cyclodextrin concentration until the solubility limits of the CD are reached. The solubilizing effect increased as the following: PMβCD > Captisol® > Beta W7 HP 0.9 > βCD [2].

Mini compacts were prepared from ground mixtures in a ball mill of nimesulide/cyclodextrins (molecular ratio NIM/CD was 2/1). The cyclodextrins used were βCD, Captisol® and PMβCD. The complete formulation and procedure are described elsewhere [3]. The dissolution tests on the solid devices were carried out according to the USP XXIII paddle method in the USP artificial intestinal juice without enzyme (pH 7.5). The results showed that the dissolution rates of nimesulide were largely enhanced when ground with CD before compacting. The compacts with PMβCD lead to a much higher and faster release of the drug than the other solid devices: about 50 % of nimesulide was dissolved within 5 min and almost 100 % within 25 min (Wouessidjewe 1998).

To better understand the increase of solubility of nimesulide in the presence of cyclodextrins, we have performed NMR experiments to characterize the formation of a complex. For that, we have focused on non substituted β-cyclodextrin. Figure 1 shows the partial ^{1}HNMR spectra of βCD alone and βCD in the presence of nimesulide.

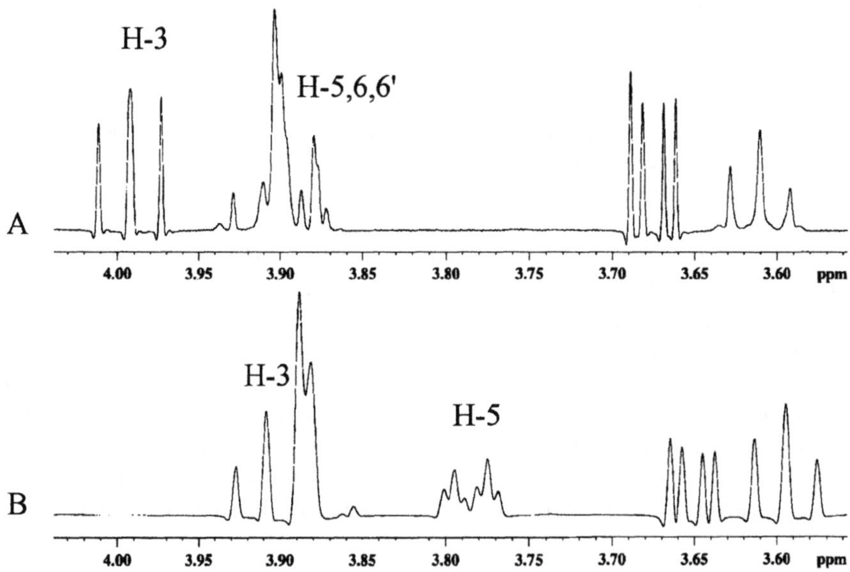

Figure 1: Partial ^{1}H NMR spectra recorded at 298K in a pH= 9, 50mM PO_4 buffer in deuterium oxide, of: 5 mM β-cyclodextrin, **A**; 5 mM β-cyclodextrin and 5 mM nimesulide, **B**.

H-3 and H-5 peaks of the βCD alone are found at 4.00 ppm and 3.90 ppm, respectively. In the presence of nimesulide, they are shifted to 3.91 ppm and 3.78 ppm respectively, suggesting the formation of an inclusion complex. To characterize structurally this inclusion complex, we have performed 2D ROESY experiments (not shown). This should allows us to see the dipolar interactions of the protons of the inner sphere of the βCD, with those of the nimesulide which are included in the βCD. From this 2D experiment we could deduce the structure of the inclusion complex which is shown in Figure 2. Only one of the aromatic group is included in the βCD, the nitro-group being on the H-5 side of the cyclodextrin.

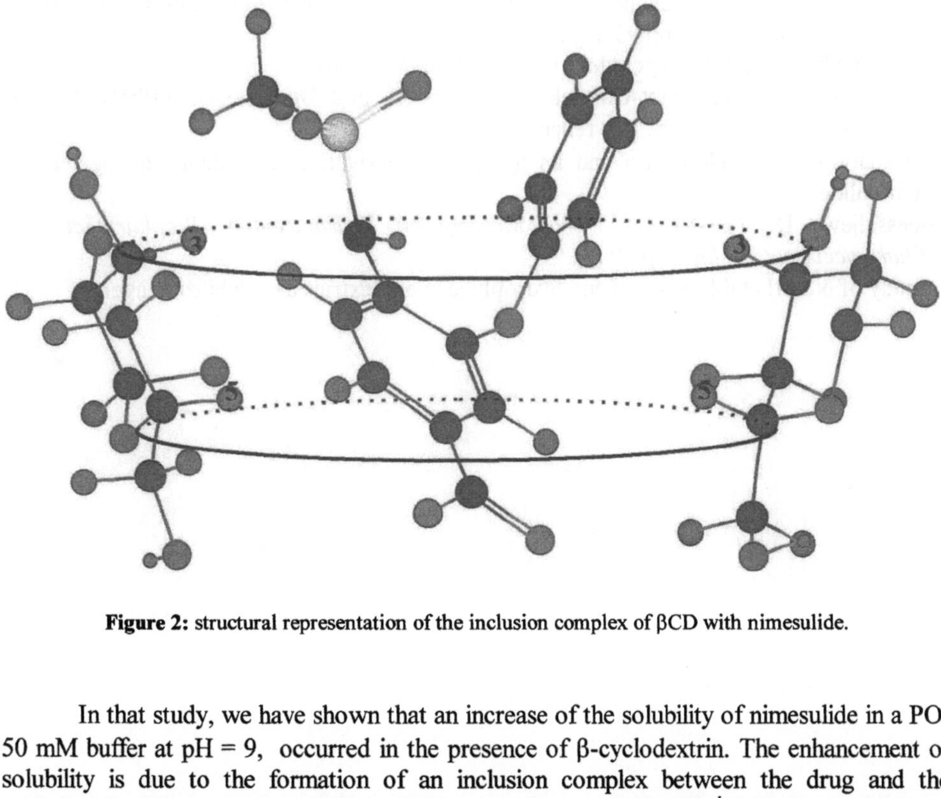

Figure 2: structural representation of the inclusion complex of βCD with nimesulide.

In that study, we have shown that an increase of the solubility of nimesulide in a PO_4 50 mM buffer at pH = 9, occurred in the presence of β-cyclodextrin. The enhancement of solubility is due to the formation of an inclusion complex between the drug and the cyclodextrin. The structure of this complex was characterized with ^1H NMR experiments. The disponibility of nimesulide as a complex especially with hydrophilic derivatives of βCD, may lead to new formulations of this drug for parenteral administration.

β-cyclodextrin was obtained from Kleptose®, Roquette, Lestrem, France; 2-hydroxypropyl-β-cyclodextrin DS 0.9 (Beta W7 HP 0.9) from Wacker, Lyon, France; partially methylated-β-cyclodextrin DS 2.07 was from Orsan, Les Ulis, France; and sulphobutylether-β-cyclodextrin (Captisol®) from CyDex, Overland Park, USA).

References

Duchêne D. in *Cyclodextrins and their industrial uses*, D. Duchêne Ed., Editions de Santé, Paris 1987, pp213-257

Wouessidjewe D. and Duchene D. (1993) , Proc. The first European Pharm. Tech. Conference, Düsseldorf, pp 93-100.
Why and how to prepare cyclodextrin inclusion compounds

Wouessidjewe D., Eggelkraut-Gottanka S., Skipa M. and Duchene D., (1997), *Pharm. Applic. of Cyclodex. Conf.*, Lawrence
Evaluation of β-cyclodextrin and its hydrophilic derivatives as solubilizing agents for nimesulide,.

Wouessidjewe D., (1998), *2nd World Meeting on Pharmaceutics, Biopharmaceutics, Pharmaceutical Technology*, Paris.
Study of NSAID tablets containing hydrophilic cyclodextrins as solubilizing agents,

COMPLEXATION OF AMINO ACIDS BY 6^A-(HYDROXYETHYLAMINO)-6^A-DEOXY-β-CYCLODEXTRIN (β-CDEA) AND THE METALLO-DERIVATIVES IN AQUEOUS SOLUTION

Rusell, N.R., Van Hoof, N. and McNamara, M.

Dublin Institute of Technology, Kevin Street, Dublin 8, Ireland.

Introduction

The presence of a chiral hydrophobic cavity in cyclodextrin molecules renders them efficient hosts for a variety of guest. The most stable complexes are usually formed with hydrocarbon type (hydrophobic/lipophilic) species.
To improve their chiral recognition, it is desirable to have a metal ion center present as well as the chiral cavity of the molecule. Native cyclodextrins are not good coordinating ligands because of intramolecular hydrogen bonding. Their complexing ability may be enhanced by adding a chelating moiety at C(6), e.g. $NH(CH_2)_2$-OH to form 6^A-(hydroxyehtylamino)-6^A-deoxy-β-cyclodextrin (β-CDEA). The multifunctional assets of these complexes (chiral cavity and metal ion centre) can be exploited with a view to the enantiwelective separation of enantiomers of amino acid species.
Hence, the aim of this work was to investigate whether this bifunctional system is capable of enantioselectivity. Several amino acids were chosen for study by potentiometric titration over the pH-range 2.0 to 11.5.
In the 2,0-11.5 pH-range, several complexes formed in the aqueous solutions of βCdea, $M^{2+}$and amino acids. Their stabilities were calculated form the diferences between the pH-profiles arising form titration of the acidified solutions against NaOH. The SUPERQUAD[1] software program was used for these calculations.
Several factors affects the stability and enantioselectivity of these cyclodextrin-amino acid complexes and warrant further investigation. According, we now report a study in which the complexation of tryptophan, phenylalanine and histidine by β-CDea is explored. The roles of the cyclodextrins, divalent metal ions and amino acids affecting complexation is discussed.

Metallo-complexes with β-CDea

The ability of CDs to include hydrophobic species, together with the presence of a multitude of active sites on the cavity rims with potential for hydrogen bonding and indeed coordination justifies their consideration as intriguing ligands for metal ions. The presence of a hydrophobic chiral cavity and an adjacent metal ion centre within the une molecule renders the metallo-cyclodextrin species an ideal candidate for metallo-enzyme modelling. This bifunctional molecule could have molecular recognition properties and therefore be capable of distinguishing between enantiomers. With this in mind, β-CDea was reacted with transition metal ions to form a bifunctional system.
There is wealth of literature evicence for metal ion complexation with ethanolamine[2]. On this basis, and

also on the basis os Atomic Absorption Spectroscopy experiments carried out, two possible structures were proposes for Cu-β-CDea (fig 1). Complex with other metals (Co^{2+}, Ni^{2+} and Zn^{2+}) are assumed to be of similar nature.

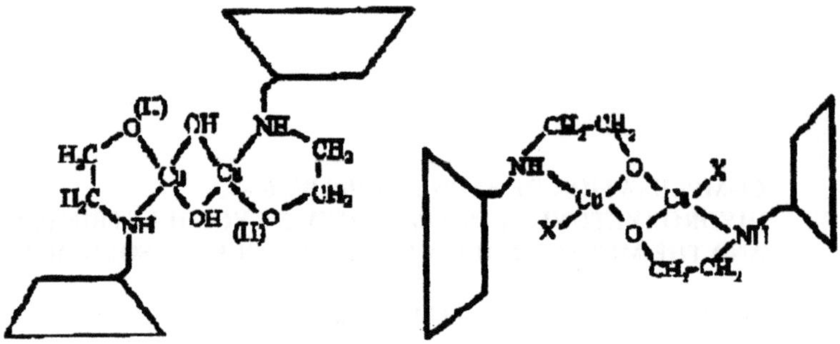

Complex (1) M2A2H2H-2 = M2A2 OR M2A2H-2

Complex (2) [M2A2]

Fig. 1

Ternary system

The aim of this work was to investigate whether this bifunctional system is capable of enantioselectivity. The formation of a binary metallo-cyclodextrin through the coordination of a metal ion by a functionalised CD, and the formation of a ternary metallo-cyclodextrin through the further binding of a substrate, offers an opportunity of study the effects of metal centre and CD cavity interactions on the stability and the enantioselectivity of the ternary complex.
In order to investigate the stability and enantioselectivity of these systems, a ternary system was set up, consisting of amino acids as substrate, metal ion and β-CDea .
The binary metallo-cyclodextrin can partly encapsulate a substrate via the hydrophobic cavity, which

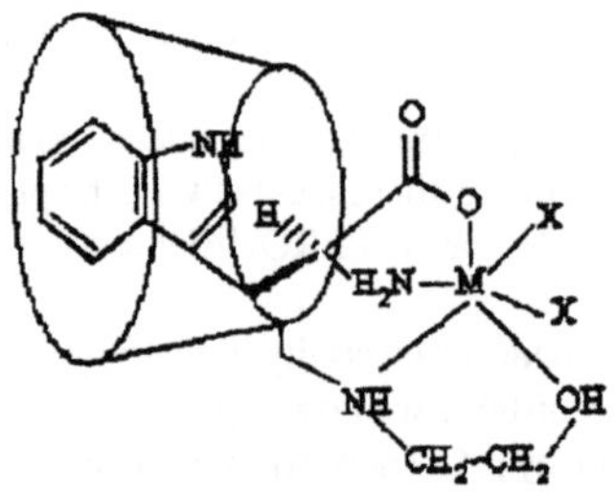

Fig. 2

also interacts with the adjacent metal centre (fig. 2). Natural and modified CDs exist in single enantiomeric forms and, when acting as host molecules, may preferentially complex one enantiomer of a chiral agent. The degree of enantioselectivity varies substantially with the nature of the guest and Cds.

In this case, the enantioslectivity depends on two factors, firstly, the inclusion of the hydrocarbon ring moiety of the amino acid into the CD cavity and secondly, the ease of coordination of the polar substituents of both host and guest to the metal ion. R- and S-guest experience different geometric and electrostatic interaction with the CD which may generate different stabilities.

Experimental

The complexation of amino acids by metallo-β-CDea was studied by means of potentiometric titration and stability constants were determined using the computer program SUPERQUAD.

Several amino acids, all having a hydrocarbon ring (tryptophan, phenylalanine and histidine) were chosen for study, using Mettler DL-25 automatic titrator equipped with a Mettler DG-111-SC-pH electrode. During all titrations, a fine stream of nitrogen gas was passed through the solution to prevent CO_2-adsorption, which can caused an extensive drift in the EMF in the pH-region 6-8. The solution was mechanically stirred and maintained at 25°C in a titration vessel that was closed to the atmosphere except for the nitrogen release. The titrations were performed over the pH-range 2-11.5, by titrating acidified solutions containing different combinations of the complexing species against NaOH.

The secuence of titrations was: (i) pK_a determinations of the amino acids followed by determination of the stability constants of complexes in solution of (ii) β-CDea and either (*R*)- of (*S*)-amino acid, (iii) M^{2+} and the amino acid and (iv) M^{2+}, β-CDea and either (*R*)- or (*S*)-amino acid. The pK_a's determined in (i) together with the pK_a of β-CDea and the stability constants for complexes between M^{2+}, β-CDea determined under the same conditions were used as constants in the determination of stability constants in (ii)-(iv). The stability constants determined in (ii) and (iii) were employed as constants in the determination of stability constants in (iv).

The respective pK_a's and pK's are calculate using SUPERQUAD. SUPERQUAD simulates a titration curve using a sugested model for the equilibrium system and then refines this model to fit the experimental data, therefore refining the suggested stability constants to values consistent with both the model and the experimental data. The criterion for selection of a model is the 955 confidence level; it is therefore necessary to have a consistent model for the equilibrium system.

SUPERQUAD was also able to distinguish between the two proposed for the metallo-β-CDea (fig. 1). There are three possible forms in which β-CDea can exist, the protonated β-CDea, referred to as AH2 because it can be lose two protons, the neutral AH and the deprotonated form A. In the model, and OH-group is refered to as an H-1 group. The possible structures for complex (1) are M2A2H-2 and M2A2 (=M2A2H2H-2). The possible structure for complex (2) is M2A2. All these structures were tried as possible models in the SUPERQUAD program and only M2A2H-2 gave an acceptable result for the stability constant of the complex. Therefore, complex (1) in which the β-CDea is deprotonated is put forward as the most likely structure to exit in solution.

Results

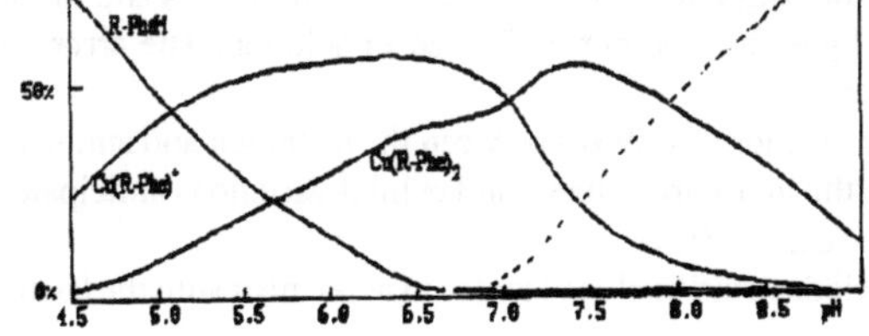

Fig.3

	tryptophan		phenylalanine		histidine	
	R	S	R	S	R	S
pK_a	9.39	9.39	9.19	9.19	9.21	9.21
	2.37	2.37	2.70	2.70	6.11	6.11
					1.99	1.99
Cu	7.65	7.65	7.80	7.80	10.45	10.45
	7.67	7.67	6.92	6.92	8.71	8.71
Ni	5.78	5.78	5.19	5.19	8.54	8.54
	4.97	4.97	4.39	4.39	6.90	6.90
Co	4.37	4.37	4.42	4.42	6.96	6.96
	3.57	3.57	3.44	3.44	5.42	5.42
Zn	5.26	5.26	4.39	4.39		
	4.77	4.77	4.02	4.02		
CDea	4.82	4.81	3.91	4.17	3.72	3.54
Cu-CDea	8.43	8.68	7.87	7.75	8.91	8.98
Ni-CDea	7.02	7.75	5.95	6.58	7.49	8.01
Co-CDea	6.14	6.39	5.85	5.53	6.44	6.72
Zn-CDea	7.20	7.20	6.10	6.10		

(a)

* 45°C

	Cdea
Cu	5.84
Ni	3.72
Co	2.69
Zn	3.41

(b)

Table 1

	tryptophan		phenylalanine		hystidine	
	R	S	R	S	R	S
β-CD*	2.33	2.33	2.91	2.83	–	–
β-CDea**	4.82	4.81	3.91	4.17	3.72	3.54
Cu-β-CDea**	8.43	8.68	7.87	7.75	8.91	8.98
Co-β-CDea**	6.14	6.39	5.85	5.53	6.44	6.72
Ni-β-CDea**	1.02	7.75	5.95	6.58	7.44	8.01

* literature, ** this work

Table 2

Discussion

Fig. 3 shows the distribution of the species existing in solution for the ternary system formed between Cu^{2+}, R-phenyalanine and β-CDea. Similar distribution diagrams are obtained for all the systems that were studied. Table 1 shows the pK_a-values of the respective amino acids and the stability constants, expresed as pK-values for all possible complexes formed in solution. The error on the pK_a and pK values range from 0.05 to 0.15.

The stability constants determined for complexes between the metal ion and amino acids (table 1a) are in reasonable agreement with the literature values[3], and exhibit variations anticipated from the Irving-Williams series ($Co^{2+} < Ni^{2+} < Cu^{2+} > Zn^{2+}$).

The stability of the metallo-cyclodextrins (table 1b) is also in agreement with the Irving-Williams series.

The relative stabilities of the β-CDea complexes with the amino acids decreased in the sequence β-CDea.Trp⁻ > β-CDea.Phe⁻ > β-CDea.His⁻ . The most probable structures of β-CDea.Trp⁻ and β-CDea.Phe⁻ place the phenyl group inside the CD annulus where hydrophobic interactions occur, and the amino acid moieties in the vicinity of the ethanolamine substituent of β-CDea, where hydrogen bonding interactions occur. The greater stability of β-CDea.Trp⁻ and β-CDea.Phe⁻, relative to that of β-CD.Trp⁻ and β-CD.Phe⁻, is consistent with those two interactions being additive in stabilising the β-CDea.Trp⁻ and β-CDea.Phe⁻ complex (table 2). The greater stability of β-CDea.Trp⁻ compared to β-CDea.Phe⁻ may be attributed to the greater length of Trp⁻ allowing an optimization of the two interactions in β-CDea.Trp⁻. However, in the case of the native β-CD, the β-CD.Phe⁻ complex is more stable than the β-CD.Trp⁻ complex. The Trp⁻is now hydrated by bulk solvent in contrast to the situation in β-CDea.Trp⁻ where it interacts preferentially with the ethanolamine moiety. In β-Cdea.Phe⁻, the Phe⁻ is almost entirely included minimizing solvent hydration effects. The complexation of His⁻ by β-CDea is the least stable. It appears that although the His⁻ rign is flat and possesses aromatic character, the ability of both the ring and the amino acid function of His⁻ to hydrogen bond with water, and possibly the smaller size of the ring, engender lesser stability in β-CDea.His⁻ .

The stability constants for the ternary complexes show that the complexes formed with Cu^{2+} are the most stable, whereas the complexes formed with Ni^{2+} are the most enationselective. The higher stabilities of the ternary complexes by comparison with those of the binary system formed between β-Cdea and amino acid, demonstrate that coordination to M^{2+} strengthens the complexation of the amino acid. Nevertheless, the similar or slightly higher stabilities of the ternary complexes by comparison with those of the binary metal amino acid complexes, indicate that the factors stabilizing complexation of the amino acis do not reinforce each other (table 1).

The variation of stability with the nature of M^{2+} coincides with the variation of the ionic radii of six coordinate Co^{2+} , Ni^{2+} , Cu^{2+} and Zn^{2+}. However, stability and enantioslectivity do not go hand in hand 8table 2). Enantioselectivity varies with the geometric constraints arising from ligand field effects in Co^{2+}, Ni^{2+} and Cu^{2+}, and the lack of such constraints in d^{10} Zn^{2+}. Tetragonally distorted d^8 Ni^{2+} has crystal field restrictions that can lead to diamagnetic complexes with increased CFSE. It is expected, therefore, that a Ni^{2+} centre should show some discrimitation between S and R enantiomers. By the same argument, Zn^{2+} , with d^{10} configuration, has a CFSE equal to zero and is therefore not expected to discriminate between the enantiomers. The smaller enantioselectivity observed in the more table complexes demonstrate that increasing complex stability does not necessarily induce a corresponding increase in enantioselectivity.

Finally, the presence of a metal ion can either reinforce or reverse the enantioslectivity 8table 2). In the case of Cu^{2+} and Co^{2+}, the enantioselectivity is reversed. If the β-CDes is enantioselecrive toward the R-isomer, the metallo-β-CDea is enantioselective toward the S-isomer and vice versa. However, it is not as simple as it seems, because with Ni^{2+} and its stereochemistry are particularly appropriate in engendering enantioselectivity for the S-isomer over the R-isomer.

References

1. Gans, P.; sabatino, A. and Vacca, A. *J. Chem. Soc, Dalton Trans.*, 1985, 1195-1200
2. a) C.W. Davies, B.N. Patel. *J. Chem. Soc. (A)*, 1968, 1824-1828

   b) R. Tauler, E. Casasas, b.M. Rode. *Inorganica Chimica Acta*, **114**, 1986, 203-209

   c) D.G. Brannon, R.H. Morrison, J.L. Hall, G.L. Humprey, D.N. Zimmerman. *J. Inorg. Nucl. Chem.*, **33,** 981-990
3. *Critical Stability Constants*, ed. R.M. Smith and A.E. Martell. Plenum Press, New York, 1975, Vol 1.
4. S.E. Brown, c.A. Haskard, C.J. easton, S.F. Lincoln, *J. Chem., Soc., Faraday Trans.*, 1995, **91**, 1013-1018

EFFECT OF THE COMPLEXATION OF CIPROFLOXACIN AND NORFLOXACIN WITH CYCLODEXTRIN DERIVATIVES ON ITS DISSOLUTION CHARACTERISTICS

Piñeiro Martínez, M.C.; Carro Martínez, A.; De Labra Piñón, P., Pérez Marcos, M.B., Vila Jato, J.L. and Torres Labandeira, J.J.

Department of Pharmacy and Pharmaceutical Technology. School of Pharmacy. University of Santiago de Compostela. Campus Universitario Sur. E-15706 Santiago de Compostela. Spain.

INTRODUCTION

Norfloxacin and ciprofloxacin are antibacterial agents that belong to the fluoroquinolones family. Their spectrum of antimicrobial activity covers both gram-negative and gram-positive organisms. But this widely antimicrobial spectrum is seen decreased by their low oral bioavailability, due to their low solubility at physiological pH. In fact, norfloxacin solubility at the pH of the intestine (≈7) is rather to 0.40 at 25 °C and 0.75 at 37°C. On the other hand, ciprofloxacin solubility is rather to 0.09 at 25 °C and 0.15 at 37 °C.

In order to increase their solubility, both drugs were formulated as inclusion complexes with β- and hydroxypropyl-β-cyclodextrins. The aim of this study was to prepare inclusion complexes between norfloxacin and ciprofloxacin with β-cyclodextrin (β-CD) and hydroxypropyl β-cyclodextrin (HPβCD) and evaluate the effect of complexation on the solubility and dissolution rate of norfloxacin and ciprofloxacin in artificial enteric juice

MATERIALS AND METHODS

Materials.

Norfloxacin (1-ethyl-6-fluoro-1,4-dihydro-4-oxo-7-[piperazin-1-yl] quinolone-3-carboxylic acid], ciprofloxacin HCl (1-cyclopropyl-6-fluoro-1,4,-dihydro-4-oxo-7-[piperazin-1-yl] quinolone-3-carboxylic acid], β-cyclodextrin from Roquette (Lestrem, France) and hydroxypropyl-β-cyclodextrin was a generous gift from Janssen Pharmaceutiche (Belgium). All other reagents were of analytical reagent grades.

Preparation of physical mixtures.

The physical mixtures of an appropriate amount of drugs and cyclodextrins in the 1:1 and 1:2 molar ratios are obtained by pulverizing and subsequent mixing in turbula T2C mixer (5 minutes at 30 rpm).

Preparation of inclusion complexes.

The inclusion complexes are prepared using the freeze-drying method, in the same 1:1 and 1:2 molar ratios. Drug and cyclodextrin are dissolved in water and frozen by immersion in liquid nitrogen. Freeze-drying will be completed in 24 hours in a Lyph-lock 6 equipment.

Characterization of the solid state inclusion complexes.

Thermal analysis. Differential Scanning Calorimetry (DSC) was performed on a Shimadzu DSC-50 system with a DSC equipped with a computerized data station TA-5 WS/PC. General conditions: scanning rate 10°C/min^{-1}, scanning temperature range 50-250 °C.

X-ray. X-ray powder diffraction patterns were recorded on a Philips X-ray diffractometer (PW 1710 BASED) using Cu-K_α radiation

Dissolution studies.

In vitro dissolution studies of pure drug, physical mixtures and the inclusion complexes were carried out placing the corresponding amount of the product in a hard shell colorless gelatin capsule in simulated gastric fluid (USP23). The capsule was placed in a stainless steel cylinder to avoid its flotation. Powdered samples containing 50 mg of drug or its equivalent in complexed or physically mixed form in the gelatin capsule were placed in 900 ml of the dissolution medium in a beaker at 37 °C for 180 min and shaken at 500 rpm. At predetermined time intervals, samples were taken for spectrophotometric determination of drug concentration (Ciprofloxacin HCl λ=273 nm, $E_{1\%, 1cm}$ = 940.70; Norfloxacin: λ=273 nm, $E_{1\%, 1cm}$ = 994.90) following filtration. All samples were analyzed in triplicate. Dissolution efficiencies after 180 min.(DE_{180}) were calculated according to Khan[1]. The effects of drug formulation on dissolution efficiency at each pH were investigated by one-way analysis of variance with the Student-Newman-Keuls test for multiple comparisons.

RESULTS AND DISCUSSION

Characterization of the solid complexes.

The diffraction patterns (figures no shown) of the physical mixtures correspond to the superimposed diffractograms of the drugs and the cyclodextrins. This is more evident in the β-CD systems, because of the amorphous characteristics of te hydroxypropyl derivative. Those plots corresponding to the complexes show fewer and less intense peaks. In fact, both inclusion complexes show an amorphous path.

The DSC thermograms (figures not shown) of both drugs show a significant peak that decrease with the cyclodextrins, and more with the inclusion complexes.

These results indicate that inclusion of the drugs within the β-CD and HPBCD cavities can be achieved by freeze-drying process

Dissolution behavior

★ ***Ciprofloxacin***

Figure 1 illustrates the dissolution profiles of ciprofloxacin containing β-CD systems. One analysis of variance (ANOVA) of the dissolution efficiency calculated indicates that the factor formulation has a significant effect on 0-180 minutes dissolution efficiency ($F_{4,10}$ = 43.08, α<0.01).

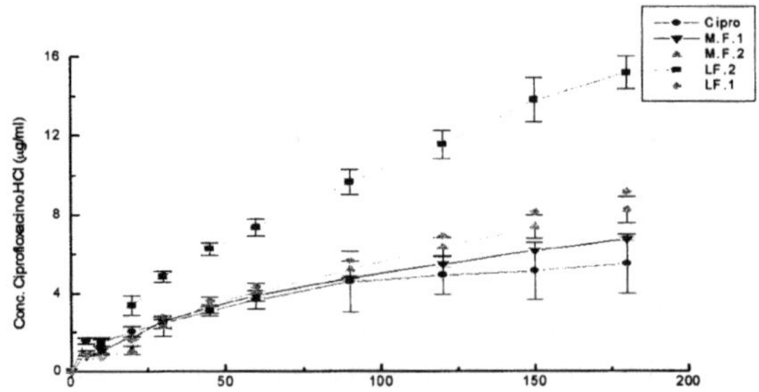

Fig 1. Dissolution profiles of Ciprofloxacin / β-CD systems

[1] KHAN, K.A.- The concept of dissolution efficiency.- J. Pharm. Pharmacol., **27**, 48-49, 1975.

The Student-Newman-Keuls method for pairwise multiple comparison grouped the preparations in the following order (from lower to higher DE_{180})

CIP PM 1:1 (mol:mol) PM 1:2 (mol:mol) FD 1:1 FD 1:2

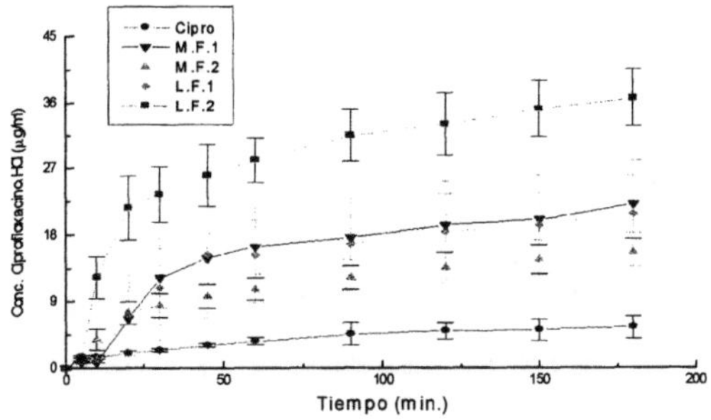

Fig. 2. Dissolution profiles of Ciprofloxacin / HP β-CD systems

Figure 2 shows the dissolution behavior of cirpofloxacin HPBCD systems. The ANOVAindicates that the factor formulation has a significant effect on 0-180 minutes dissolution efficiency ($F_{4,10}$ = 18.87, $\alpha<0.01$). The Student-Newman-Keuls method for pairwise multiple comparison grouped the preparations in the following order (from lower to higher DE_{180}).

CIP PM 1:1 (mol:mol) PM 1:2 (mol:mol) FD 1:1 FD 1:2

★ ***Norfloxacin***

Figure 3 shows the dissolution profiles of norfloxacin, physical mixtures and inclusion complexes. ANOVA indicates that the factor formulation has a significant effect on 0-180 minutes dissolution efficiency ($F_{4,10}$ = 18.21, $\alpha<0.01$).

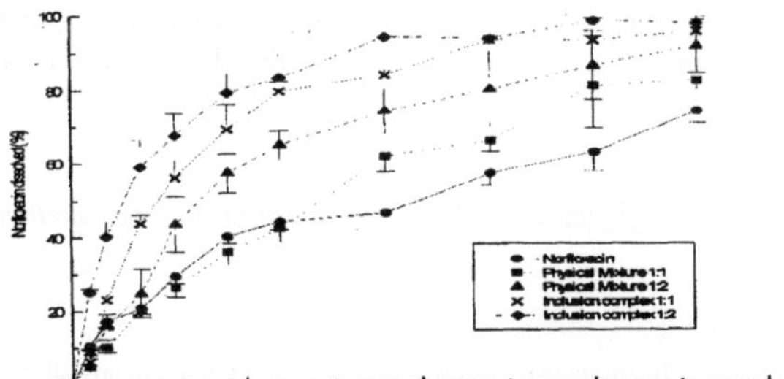

Fig.3. Dissolution profiles of Norfloxacin / β-CD systems

The Student-Newman-Keuls method for pairwise multiple comparison grouped the preparations in the following order (from lower to higher DE_{180})

CIP PM 1:1 (mol:mol) PM 1:2 (mol:mol) FD 1:1 FD 1:2

Thus, no increase in the dissolution efficiency of norfloxacin was found after simple mixing in the 1:1 molar ration with β-CD. However, if the amount of cyclodextrin rose to 1:2 mol:mol increased dissolution efficiency, founding no significant difference with both inclusion complexes.

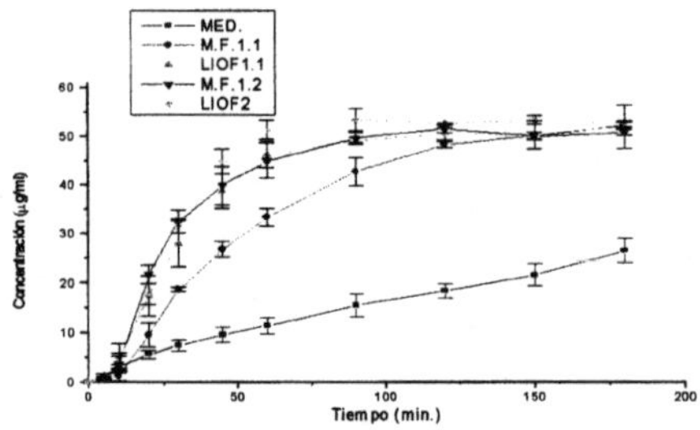

Fig. 4. Dissolution profiles of Norfloxacin / HPβ-CD systems

Figure 4 shows the dissolution profiles of norfloxacin, physical mixtures and inclusion complexes. ANOVA indicates that the factor formulation has a significant effect on 0-180 minutes dissolution efficiency ($F_{4,10} = 83.86$, $\alpha<0.01$). The Student-Newman-Keuls method for pairwise multiple comparison grouped the preparations in the following order (from lower to higher DE_{180}).

CIP   PM 1:1 (mol:mol)   P M 1:2 (mol:mol)   FD 1:1   FD 1:2

Thus, HPBCD increases the dissolution behavior of norfloxacin, but there is neither effect on the characteristic of the system (physical mixture or inclusion complex) nor the amount of the cyclodextrin in the system.

CONCLUSIONS

Freeze drying is a suitable method for the preparation of solid inclusion complexes of ciprofloxacin and norfloxacin with β- and HPβ-CD. The formulation of both drugs with the cyclodextrin derivatives used in this study increases their dissolution characteristics. The effect of hydroxypropyl-β-cyclodextrin is higher in both cases than the corresponding of β-CD

*****

To Dr. Tsunegi Nagai and Dr. Josef Szjetli in their 65th Birthday

*****

ACKNOWLEDGMENTS

Special thanks to Janssen Pharmaceutiche (Belgium) for the hydroxypropyl-β-cyclodextrin used in this study. This work was supported by a research grant from Xunta de Galicia XUGAA20320B96.

EFFECT OF HYDROXYPROPYL-β- CYCLODEXTRIN ON THE STABILITY OF PILOCARPINE IN SOLUTION. INFLUENCE OF THE COMPLEX PREPARATION

Echezarreta- López, M.; Esclusa-Díaz, T.; Vila-Jato, J.L. and Torres-Labandeira, J.J.

Department of Pharmaceutical Technology. Faculty of Pharmacy. University of Santiago de Compostela. E-15706 Santiago de Compostela. Spain

INTRODUCTION

Pilocarpine hydrochloride is a widely used drug for the treatment of primary (open-angle) glaucoma. It is also used as a model drug in studies aimed at developing controlled-release ocular drug delivery systems. In such applications, pilocarpine is typically formulated in an aqueous system (Desai y col., 1992). In aqueous medium, pilocarpine can hydrolyze to pilocarpic acid and epimerize to isopilocarpine which, in turn, can hydrolyze to isopilocarpic acid. The extend of the degradation products depends on pH, temperature and time. Since this degradation results in deterioration of the pharmacological effects, ophthalmic aqueous pilocarpine formulations have limited stability (Noordam y col., 1981).

Recently, hydroxypropyl-ß derivatives (HPBCD) have been suggested to increase ocular bioavailability of pilocarpine in rabbits. The maximum increases in peak miotic intensity of 0.1% pilocarpine solution (4.8 mM, pH 4.5) was achieved when the molar ratio of HPBCD to pilocarpine was 1:1 (Freedman y col., 1993). However, Järvinen y col., (1994) using the same derivative do not find any effect on the miotic intensity of pilocarpine.

The aim of this work was to analyze the effect of the method of preparation of the inclusion complexes P:HPBCD d.s. 2.7 (1:1) on the stability of pilocarpine hydrochloride. Two methods were employed: freeze-drying and *in solution in situ*. Solid inclusion complexes were characterized by differential scanning calorimetry (DSC) and Fourier Transformation Infrared (FT-IR). The structure of the complex in solution was determinated by nuclear magnetic resonance studies (^{1}H-NMR).

MATERIALS AND METHODS

Materials:

Pilocarpine hydrochloride (P) was a generous gift from Laboratorios Cusí, SA. (Spain) and hydroxypropyl-ß-cyclodextrin (HPBCD) with molar degree of substitution 2.7 was purchased from Cyclolab Ltd., (Hungary). All chemicals used were of analytical reagent grades.

Methods:

❑ Preparation of inclusion complexes P:HPBCD

- Freeze-drying method: P (50 mg) and HPBCD, in a molar ratio 1:1, were dissolved in double distilled water (25 mg) and frozen by immersion in liquid nitrogen. Freeze-drying was completed in 24 h in a Lyph-lock 6 apparatus (Labconco, USA-Kansas, MO).

- *In solution in situ* method: Inclusion complexes of P:HPBCD were prepared in a molar ratio 1:1 by dissolving the required amount of pilocarpine and cyclodextrin in the same medium that was realized the study (phosphate buffer solution and deuterated water), without mechanically agitation.

❑ Characterization of the inclusion complexes P:HPBCD

- *Thermal analysis*. DSC was performed with a DSC-4 apparatus (Perkin Elmer, USA, Norwack, CO) at a scanning rate of 10 °C min^{-1}.

- *Fourier Transformation Infrared*. The IR spectra (in KBr disk) were performed in a Perkin-Elmer 1330 IR spectrophotometer.

- *Nuclear Magnetic Resonance studies* .^{1}H-RMN were obtained with a Bruker WN 300 spectrometer (Bruker, Anal. Messtechnik GmbH, D-Rheinstetten) at 25°C. Samples were dissolved in deuterated water. The samples from inclusion complexes in state solid were prepared by dissolving the required amount of pilocarpine (8 mg). The internal reference was the peak due to small amounts of DHO and D_2O present as impurity assigned a value δ=4.8 ppm.

❑ Stability studies

Stability studies were carried out in a phosphate buffer solution pH 6.6 (0.05 M, ionic strength, 0.5) at 45°C. Pilocarpine solutions with or without HPBCD were prepared by dissolving the corresponding amount of a drug (500 mg) and HPBCD, in a molar ratio 1:1, in 50 ml of phosphate buffer solution pH 6.6, either *in solution in situ* or as freeze-dried complex for 90 days. The solutions were stored in ampules at constant temperature. The remaining P and its degradation products were determined by HPLC (Sternike y col., 1992). The pseudo-first order rate constant (K_{obs}) for degradation of pilocarpine were determined from linear plots of the logarithm of remaining pilocarpine against time.

RESULTS AND DISCUSSION

➤ **Complexation in the solid state**. Figure 1 shows the DSC curves corresponding to the solid system. Thermograms of the freeze-dried and no freeze-dried pilocarpine have been found similar. Both show an endothermic peak at 202 °C, which is not present in the corresponding DSC of the physical mixture and solid complex.

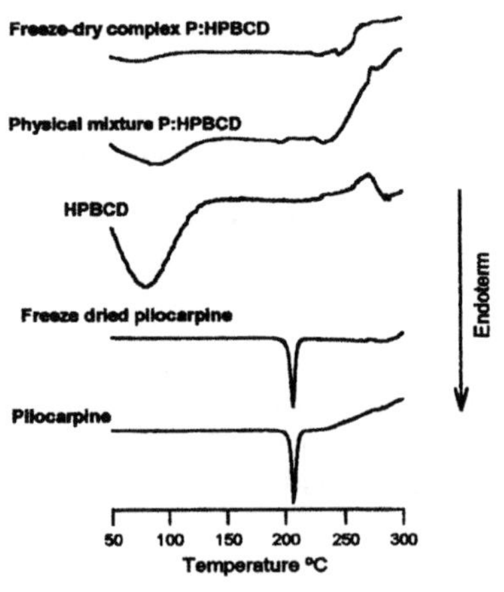

Figure 1. DSC thermograms of the different P and HPBCD systems.

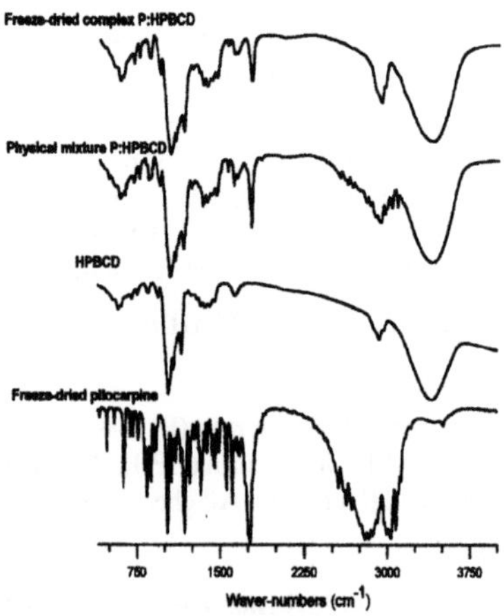

Figure 2. FT-IR spectra of the different P and HPBCD systems

These systems presents the same aspect but their interactions are different as it can be deducted from the FT-IR results (Figure 2).

The freeze-dried complex does not show the bands of P in the in the interval of wavenumbers 3100-2550 cm^{-1}. On the other hand, the absorption band of the lactone ring at 1740 verifies the presence of the P in the complex, whereas the physical mixtures produce spectra which are simple overlaying of single spectra (P and HPBCD). These results indicate the inclusion of the P and it is the imidazole part which is inside the cavity.

➤ **Complexation in the dissolution state:** The effects of HPBCD at different concentrations on the ^{1}H-NMR spectrum of P are shown in figure 3. Changes in the chemical shifts of the protons of the guest molecule indicate the complex formation. No new peaks appeared which could be assigned to the pure complex.

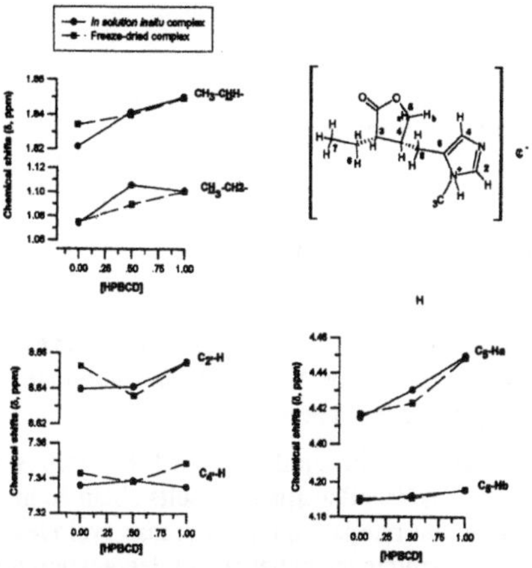

Figure 3. Structure and chemical shifts of th pilocarpine in presence of HPBCD

In solution, freeze-dried and *in solution in situ* complexes show the same path. It is evidence that all the protons of the P are affected by the presence of HPBCD. Nevertheless, slight differences in the chemical shifts of the protons of the ethyl group (CH_3-CH_2-) to downfield suggest that this group is interacting with the hydroxypropyl radicals of this cyclodextrin. Besides, the imidazole ring (C_2-H, N-CH_3, C_4-H) undergoes in the non-polar environment into the cavity or edge of cyclodextrin structure. This is confirmed by the chemical shifts shows by proton C_5-H_a.

➤ **Effect of the complex formation on the stability of pilocarpine.**

Table 1 shows the pseudo first-order rate constants (k_{nobs}) and shelf-lives ($t_{90\%}$) for the overall degradation of pilocarpine (0,046 M) in phosphate buffer solutions pH 6.6 from the inclusion complex P:HPBCD (1:1).

Stability studies of Pilocarpine Hydrochloride			
Formulation	**k_{obs} 10^3 (day^{-1})**	**$t_{90\%}$ (day)**	**k_{obs}/k_o**
Pilocarpine	11.5 ± 4.02	10	1
Freeze-dried Pilocarpine	10.3 ± 3.14	11	0.8956
***In solution in situ* complex**	9.72 ± 0.52	11	0.8452
Freeze-dried complex	10.43 ± 0.06	10	0.9069

k_{obs} (with HPBCD) / k_o (without HPBCD)

Tabla 1. Stability constant observed (k_{obs}) and shelf-lives ($t_{90\%}$) for the degradation of pilocarpine.

In Figure 4, it can be observed that the degradation of pilocarpine is not affected by freeze-drying process. On the other hand, k_{obs} calculated confirm that the presence of HPBCD show a slight effect n the stability of pilocarpine in aqueous solution

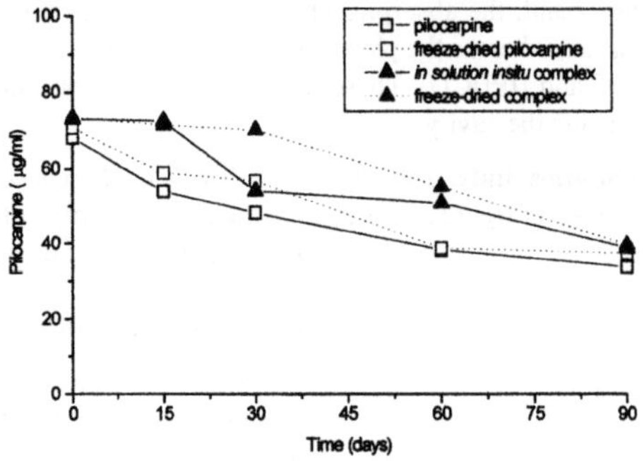

Figure 4. Stability profiles of pilocarpine with or without HPBCD.

According to the ^{1}H-RMN results, the structure of both P:HPBCD inclusion complexes - *in solution in situ* and freeze-dried - is the same in solution. Therefore, because the *in solution in situ* is easy to obtain a stable complex, the freeze-drying process is not necessary.

As the hydrolysis of pilocarpine is a reversible reaction, the equilibrium will be established between P and its degradation products, pilocarpic acid and isopilocarpine. Results obtained in the degradation process of pilocarpine suggest that the initial rate is modified because cyclodextrin influenced the equilibrium to favor the pilocarpine, and therefore the initial rate of degradation does not increase with the presence of HPBCD (Masson y col., 1998).

CONCLUSIONS

★ Pilocarpine forms inclusion complexes in solution and in solid state with HPBCD

★ In the inclusion complex, the imidazole group of the structure is included in the cavity.

★ The presence of HPBCD - either as freeze-dried or *in solution in situ* complexes - improves the stability of pilocarpine in solution, but no differences have been found between both systems.

*****
To Dr. Tsunegi Nagai and Dr. Josef Szjetli in their 65th Birthday
*****

ACKNOWLEDGMENTS

This work was supported by Laboratorios Cusí (Spain) and a research grant from Xunta de Galicia XUGAA20320B96.

REFERENCES

1. **Desai, S.D.; Blanchrdt, J.** J. Chrom. Sci., **30,** 149, (1992).
2. **Noordam, A.; Maat, L.; Beyerman, H.C.** J. Pharm. Sci., **70,** 1, (1981)
3. **Freedman, K. A.; Klein, J.W.; Crosson, C. E.** Curr. eye Res., **12,** 641, (1993)
4. **Järvinen K, Järvinen, T.; Thompson, D.O.; Stella, V.** Curr. Eye Res., **13,** 897, (1994)
5. **Masson, M.; Loftsson, T.; Jónsdottir, S.; Fridriksdóttir,H.; Petersen, D.S.** Int. J. Pharm., **164,** 45 (1998).

DISSOLUTION BEHAVIOR OF DICLOFENAC SODIUM- β- AND HYDROXYPROPYL-β-CYCLODEXTRIN INCLUSION COMPLEXES

Pose Vilarnovo, B., Santana Penín, L., Pérez-Marcos, M.B., Vila-Jato, J.L. and Torres-Labandeira, J.J.

Department of Pharmaceutical Technology. Faculty of Pharmacy. University of Santiago de Compostela. E-15706 Santiago de Compostela. Spain

INTRODUCTION

Diclofenac sodium is a widely used nonsteroidal anti-inflammatory and analgesic drug. Its limited water solubility is low, specially in gastric juice (about 15 μg/ml) and, it is unstable in aqueous solution. This limited solubility in acidic medium gives problems in its oral bioavailability and it is a drawback in its formulation in controlled release devices.

The aim of this study was to improve the solubility of diclofenac sodium in artificial gastric juice pH 1.2 by its complexation with β-cyclodextrin (βCD) and hydroxypropyl-β-cyclodextrin (HPβCD). Phase solubility diagrams were obtained to characterize the interaction between drug and CD in this dissolution medium. Solid inclusion complexes of diclofenac sodium/cyclodextrin were prepared by freeze-drying. X-ray diffractometry, differential scanning calorimetry were used to characterize the systems prepared. The influence of the complexation on drug dissolution behavior was also analyzed.

MATERIALS AND METHODS

Materials.

Diclofenac sodium (*2-[(2,6-dichlophenyl) amino] benzeneacetic acid monosodium salt*) was purchased from Sigma Chemical Co. (St. Louis, MO, USA), β-cyclodextrin from Roquette (Lestrem, France) and hydroxypropyl-β-cyclodextrin was a generous gift from Janssen Pharmaceutiche (Belgium). All other reagents were of analytical reagent grades.

Phase solubility diagrams

Solubility diagrams were obtained according to Higuchi and Connors (1) in gastric juice of pH=1.2. The apparent stability constant of the Diclofenac-β-CD and Diclofenac-HPβCD complexes, assuming 1:1 stoichiometry, were calculated from the slope of the initial straight portion of the solubility diagram.

Preparation of the physical mixtures

The physical mixtures of an appropriate amount of diclofenac/β-CD and diclofenac/HPBCD in the 1:1 molar ratios were obtained by pulverizing and subsequent mixing in a Turbula T2C mixer (5 min at 30 rpm).

Preparation of the inclusion complexes

The solid inclusion complexes of diclofenac with β-CD and HPBCD (1:1 mol:mol) were prepared using the freeze-drying method. Both components were dissolved in 0.2 N aqueous ammonium hydroxide. The solution was filtered (0.45 μm) and frozen by immersion in liquid nitrogen. Freeze-drying was completed in 48 h in a Lyph-lock 6 equipment (Labconco).

Characterization of the solid state inclusion complexes.

Thermal analysis. Differential Scanning Calorimetry (DSC) was performed on a Shimadzu DSC-50 system with a DSC equipped with a computerized data station TA-5 WS/PC. General conditions: scanning rate 10°C/min^{-1}, scanning temperature range 50-250 °C.

X-ray. X-ray powder diffraction patterns were recorded on a Philips X-ray diffractometer (PW 1710 BASED) using Cu-K_α radiation

Dissolution studies.

In vitro dissolution studies of pure drug, physical mixtures and the inclusion complexes were carried out placing the corresponding amount of the product in a hard shell colorless gelatin capsule in simulated gastric fluid (USP23). The capsule was placed in a stainless steel cylinder to avoid its flotation. Powdered samples containing 50 mg of Diclofenac or its equivalent in complexed or physically mixed form in the gelatin capsule were placed in 900 ml of the dissolution medium in a beaker at 37 °C for 180 min and shaken at 500 rpm. At predetermined time intervals, samples were taken for spectrophotometric determination of Diclofenac concentration (λ=276 nm, $E_{1\%,\ 1cm}$ = 283.85) following filtration. All samples were analyzed in triplicate. Dissolution efficiencies after 180 min.(DE_{180}) were calculated according to Khan (2). The effects of drug formulation on dissolution efficiency at each pH were investigated by one-way analysis of variance with the Student-Newman-Keuls test for multiple comparisons.

RESULTS AND DISCUSSION

Phase solubility diagrams

Phase solubility profiles of diclofenac sodium with both cyclodextrins are shown in figure 1. Both diagrams can be classified as A_L type according to Higuchi and Connors (1).

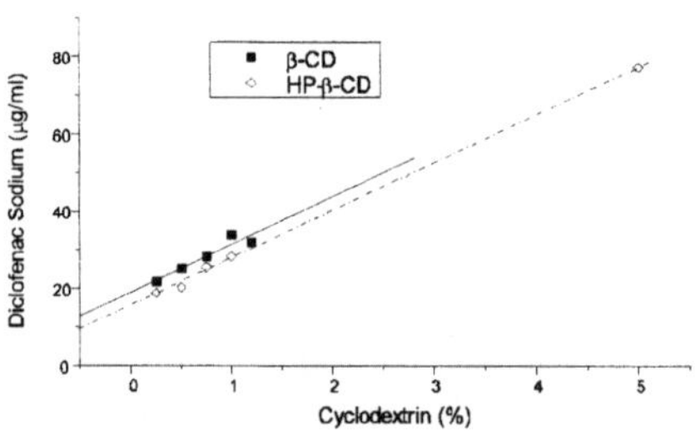

Figure 1. Solubility diagrams. Plot of Diclofenac concentration *vs* cyclodextrin concentration

This indicates that, within the cyclodextrin concentration range tested, a soluble complex is formed. On the other hand, because both straight lines have a slope less than unity, it was assumed that the increase in solubility was due to the formation of a 1:1 mol:mol complex. The values of the stability constant, were 100.6 M^{-1} for β-CD and 115.8 M^{-1} for HPBCD which indicates a similar interaction between the drug and both cyclodextrin derivatives in the conditions used in the study.

Characterization of the solid complexes

The X-ray diffractograms of formulations are shown in figure 2. The diffraction patterns of the physical mixtures correspond to the superimposed diffractograms of the drug and the cyclodextrins. This is more clear in the β-CD system, because of the amorphous characteristics of the hydroxypropyl derivative. Those plots corresponding to the complexes show fewer and less intense peaks. In fact, both inclusion complexes show an amorphous path.

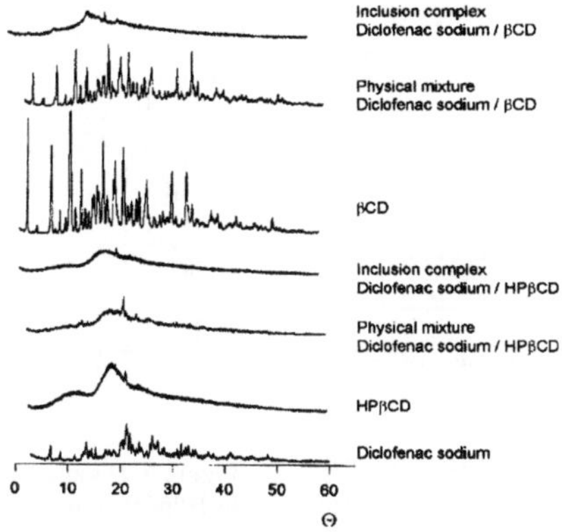

Figure 2. X-Ray diffractograms corresponding to the indicated products

Figure 3 illustrates the DSC thermograms of the preparations. The drug does not have any significant peak in the studied temperature range because the melting and decomposition point is 283-285°C.

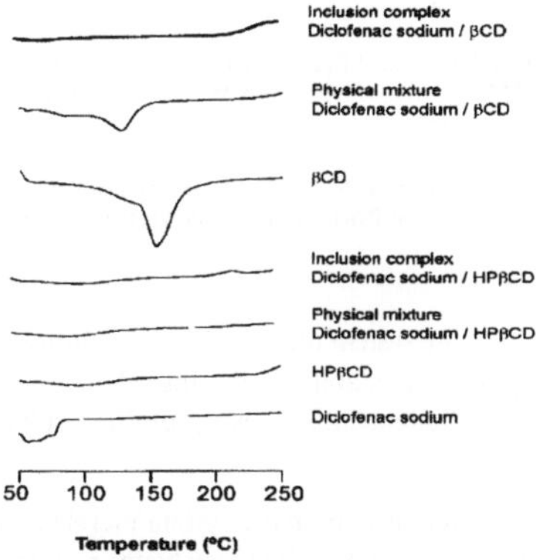

Figure 3. DSC curves corresponding to the indicated products

These results indicate that inclusion of the diclofenac within the β-CD and HPBCD cavities can be achieved by freeze-drying process

Effects of complexation on the dissolution behavior of the drug

Figures 4 and 5 show the dissolution profiles of Diclofenac, physical mixture and inclusion complexes in artificial gastric juice.

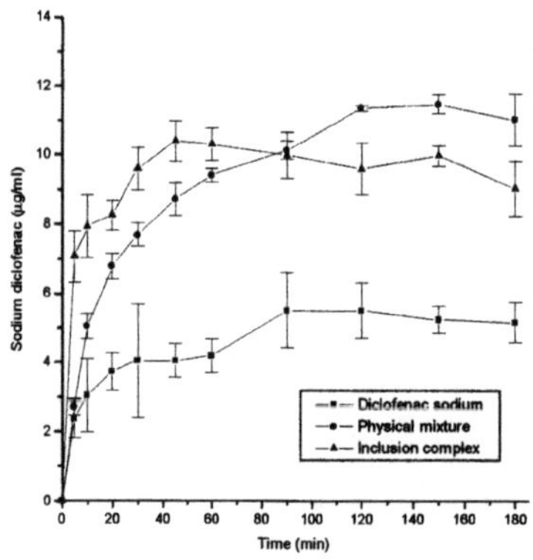

Figure 4. Dissolution profiles of Diclofenac and its βCD systems in artificial gastric juice without enzymes

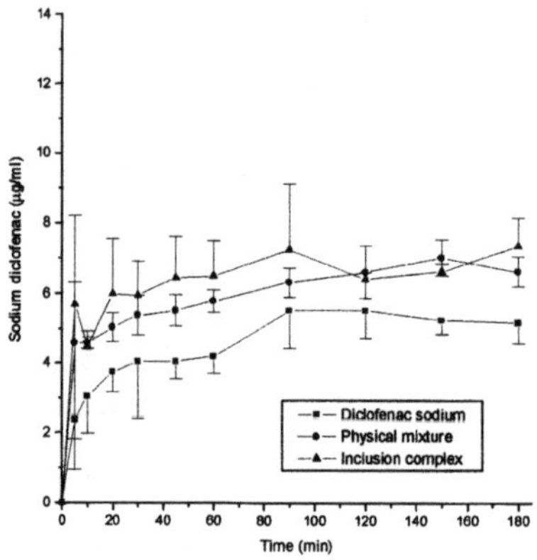

Figure 5. Dissolution profiles of Diclofenac and its HPβCD systems in artificial gastric juice without enzymes

One way analysis of variance indicates that the factor formulation has a significant effect on 0-180 min dissolution efficiency ($F_{4,10}$ = 81.9, α<0.01). The Student-Newman-Keuls test for pairwise multiple comparison grouped the preparations in the following order (from lower to higher DE_{180}):

Diclofenac   PM- HPβCD   FD-HPβCD   PMβCD   FDβCD

Key: **PM-HPβCD :** Physical Mixture Diclofenac-HPβCD; **PM-BCD**: Physical Mixture Diclofenac-BCD; **FD-HPβCD**: Freeze dried Diclofenac-HPβCD complex; **FD-BCD**: Freeze dried Diclofenac-BCD complex

The presence of cyclodextrin in the system increases the dissolution properties of the drug. This effect is higher with β-CD in spite of the stability constant calculated from the solubility diagram is slightly smaller for the natural cyclodextrin. However, similar results were found with a physical mixture and the inclusion complex.

In conclusion, the presence of cyclodextrin in the system increases the dissolution properties of the drug. However, the systems that contain β-CD show better results.

*****
To Dr. Tsunegi Nagai and Dr. Josef Szjetli in their 65th Birthday
*****

ACKNOWLEDGMENTS

Special thanks to Janssen Pharmaceutiche (Belgium) for the hydroxypropyl-β-cyclodextrin used in this study. This work was supported by a research grant from Xunta de Galicia XUGAA20320B96.

REFERENCES

1. HIGUCHI, T. and CONNORS, K.A. - Phase solubility techniques.- Adv. Anal. Chem. Instr. **4**, 117-212, 1965
2. KHAN, K.A.- The concept of dissolution efficiency.- J. Pharm. Pharmacol., **27**, 48-49, 1975.

EFFECT OF $(SBE)_{7M}$-β-CD ON METHYLPREDNISOLONE TRANSPORT ACROSS ETHYLCELLULOSE MICROPOROUS MEMBRANES

E. A. ZANNOU[1], S. SHIRAISHI[2], V. M. RAO[1] and V. J. STELLA[1]
[1]*The center for Drug Delivery Research and the Department of Pharmaceutical Chemistry, The University of Kansas, Lawrence, Kansas, 66047, USA;* [2]*Wakunaga Pharmaceutical Co., Ltd, Osaka, Japan.*

1. Introduction

$(SBE)_{7M}$-β-CD have been used as a solubilizing and osmotic agent for the controlled release formulation of poor water soluble drugs [1-3]. One of the key of the success of any controlled porosity osmotic pump is the formulation of the rate controlling microporous membrane. Experiments aimed at the characterization of an ethylcellulose microporous membrane which has been used in the Methylprednisolone - $(SBE)_{7M}$-β-CD controlled porosity osmotic pump tablet are reported here.

2. Materials and Methods

2.1. MATERIALS

Methylprednisolone (MP) and sodium Chloride were obtained from Sigma Chemical Co. (St. Louis, MO). Ethylcellulose (EC, Ethocel[R] Standard 10 Premium) and Polyethylene Glycol (PEG 1450, Carbowax[R]) were donated by Dow Chemical Co. (Midland, MI) and Union Carbide (Danbury, CT) respectively.

2.2. METHODS

2.2.1. *Membrane Preparation*
The coating solution containing 4.6% of EC (semipermeable polymer: permeable to water, not to the solute), 3.4% PEG 1450 (pore former and plasticizer, water soluble), 1.5% of water and 90.5% of ethanol, was manually air sprayed on Teflon™ inserts under constant air flow (40°C). The membranes were dried overnight (50°C), checked for cracks and flaws, and thickness measured using a micrometer. The membranes were then placed between the Side-Bi-Side™ diffusion cells.

2.2.2. *Diffusion Studies*
Two types of studies were conducted using Side-Bi-Side™ diffusion cells at 37°C under constant stirring (600 rpm). In both cases, the driving force to the transport of MP and its $(SBE)_{7M}$-β-CD complex was diffusion only, there was no difference in osmotic pressure across the membrane. In each experiment, the donor cell contained 90 μg/ml MP in an aqueous solution of $(SBE)_{7M}$-β-CD, and sink conditions were respected throughout. In the first type of experiments, same concentration $(SBE)_{7M}$-β-CD solutions (ranging from 0 to 50 mM) were placed on both sides of the cell. In the second case, the donor cell contained MP in a 50 mM $(SBE)_{7M}$-β-CD solution and the receiver cell contained a sodium chloride solution of same osmotic pressure. At specific time intervals, 200 μl

aliquots were sampled from the receiver cell and replenished with fresh medium. Samples were analysed for MP by HPLC with UV detection at 254 nm.

2.2.3. *Influence of Osmotic Pressure*
These experiments were also conducted in Side-Bi-Side™ diffusion cells at 37°C under constant stirring (600 rpm); the donor cell contained 90 μg/ml MP in a 50 mM $(SBE)_{7M}$-β-CD solution and the receiver cell contained double distilled water. The donor cell was sealed to minimize the net water flux due to the difference in osmotic pressure across the membrane and no volume changes in the cells were observed. Sink conditions were respected throughout. Sampling and MP detection were conducted as described in 2.2.2.

3. **Results and Discussion**

3.1. DIFFUSION STUDIES

During diffusion across Side-Bi-Side™ diffusion cells, unstirred or boundary layers have been reported to provide significant resistances to the solute transport [4]. Since experiments were conducted in aqueous media, some resistance to the solute transport from the aqueous boundary layer (ABL) and the membrane (MB) itself is expected. Assuming the ABL resistance to be equivalent on both sides of the membrane, the solute permeability (P) across the membrane can be described by equation (1):

$$\frac{1}{P} = R = \frac{2}{P_{ABL}} + \frac{h}{DK} = \frac{2}{P_{ABL}} + \frac{h}{P_{eff}} \qquad (1)$$

with P_{ABL} and P_{eff} the solute permeability through the ABL and the membrane respectively, h the membrane thickness, D the diffusion coefficient and K the partition coefficient. The permeability values obtained from the diffusion experiments in the presence of $(SBE)_{7M}$-β-CD solutions of various concentrations were plotted as a function of the membrane thickness as indicated in (1) (see Figure 1). P_{ABL} and P_{eff} were derived from this plot using the intercept and slope values respectively. As shown in Figure 1, the ABL resistance to the solute transport was negligible. The expression describing the solute transport rate was then simplified to equation (2) and P_{eff} was derived at the various $(SBE)_{7M}$-β-CD concentrations (see Figure 2.a).

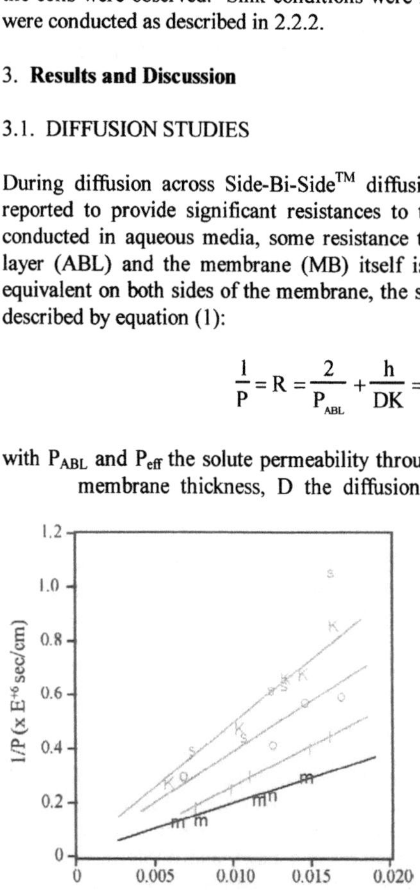

Figure 1: Methylprednisolone effective membrane permeability as a function of the membrane thickness (○ no $(SBE)_{7M}$-β-CD, ● 1 mM $(SBE)_{7M}$-β-CD, ❑ 5 mM $(SBE)_{7M}$-β-CD, ▲ 10 mM $(SBE)_{7M}$-β-CD, ✧ 50 mM $(SBE)_{7M}$-β-CD)

$$\frac{dM}{dt} = \frac{P_{eff} A C_d}{h} \qquad (2)$$

As shown in Figure 2.a, P_{eff} decreased with increasing $(SBE)_{7M}$-β-CD concentration and appeared to plateau at higher concentrations. The binding constant of MP with $(SBE)_{7M}$-β-CD has been determined to be 700 M^{-1} (phase solubility study at 25°C). Thus in the range of 0 to 50 mM $(SBE)_{7M}$-β-CD, MP changed from a free species to a almost completely complexed one (the 90 μg/ml of MP are 97% in the complexed form in the 50 mM $(SBE)_{7M}$-β-CD solution.) The second type of diffusion experiment conducted with a sodium chloride solution in the receiver side also indicated that, in the presence of 50 mM of $(SBE)_{7M}$-β-CD, MP was mainly transported in the complexed form. We then assumed that the solute was present in two forms: the free drug and the complex. P_{eff} could then be deconvoluted into these two components as shown in equation (3):

$$P_{eff} = (P_{free} - P_{complex})f_{free} + P_{complex} \tag{3}$$

The results are plotted in Figure 2.b. The effective membrane permeability for the free drug and the complex were calculated from the slope and the intercept respectively; they were determined to be $5.30.10^{-8}$ cm^2/sec and $1.86.10^{-8}$ cm^2/sec. The lowering of MP permeability upon complexation was expected due to the increase in molecular size and consequent decrease in the effective diffusion coefficient.

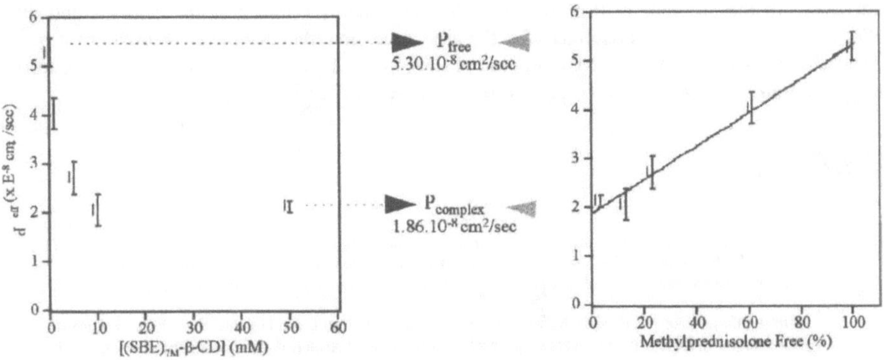

Figure 2: Effective membrane permeability: a. from equation (2), b. from equation (3)

3.2. INFLUENCE OF THE OSMOTIC PRESSURE

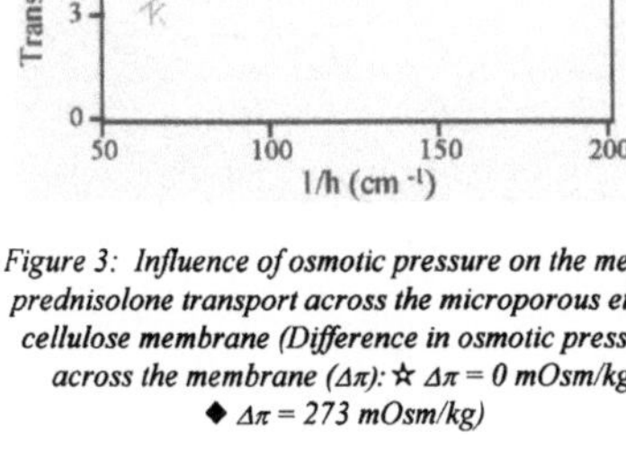

Figure 3: Influence of osmotic pressure on the methyl-prednisolone transport across the microporous ethyl-cellulose membrane (Difference in osmotic pressure across the membrane (Δπ): ☆ Δπ = 0 mOsm/kg, ◆ Δπ = 273 mOsm/kg)

The driving forces to the drug transport from a typical controlled porosity osmotic pump have been identified as being primarily osmosis with some diffusion contribution [5]. In an attempt to characterize the MP - $(SBE)_{7M}$-β-CD osmotic pump tablet, the Side-Bi-Side™ diffusion cells were used to estimate the relative contribution of these two forces (even though transport conditions in the tablet were not possible to reproduce at this state). The results are shown in Figure 3 and Table 1 and indicated that osmosis seemed to be the main driving force to MP transport across the microporous ethylcellulose membrane.

Table 1: Relative contribution of osmosis versus diffusion for the methylprednisolone transport across the membrane

h (cm)	Transport Rate (x E^{-13} mol/sec)			Osmosis / Diffusion
	Total	Diffusion	Osmosis	
0.0124	13.3	3.8	9.5	2.5
0.0135	11.1	3.5	7.6	2.2
0.0147	10.9	3.1	7.8	2.5
0.0152	8.8	3.4	5.5	1.6

4. Conclusion

The microporous ethylcellulose membrane used in the formulation of the MP - $(SBE)_{7M}$-β-CD controlled porosity osmotic pump tablet has been shown to be the main resistance to the drug transport. In the presence of $(SBE)_{7M}$-β-CD, MP was mainly released as the complex and the effective permeability was quantified for the free drug and the complex. Moreover, it was demonstrated that even at a low concentration of $(SBE)_{7M}$-β-CD, osmosis is the main driving force to the MP transport across the microporous ethylcellulose membrane.

5. References

[1] Stella, V.J., Uekama, K., Irie, T., Rao, V.M., Zannou, E.A., Rajewski, R.A., Shiraishi, S., and Okimoto, K. (1998) The use of $(SBE)_{7M}$-β-CD (CAPTISOL™) as a solubilizing and osmotic agent for controlled and complete oral delivery of poorly water soluble drugs, *Proceedings of the Ninth International Symposium on Cyclodextrins.*

[2] Okimoto, K., Miyake, M., Ohnishi, N., Rajewski, R.A., Stella, V.J., Irie, T. and Uekama, K. (1998) Design and Evaluation of an osmotic pump tablet for prednisolone, a poorly water soluble drug, using $(SBE)_{7M}$-β-CD, *Pharm. Res.*, Accepted for publication.

[3] Okimoto, K., Rajewski, R.A. and Stella, V.J. (1998) Release of testosterone from an osmotic pump tablet utilizing $(SBE)_{7M}$-β-CD as both a solubilizing and an osmotic pump agent, *J. Control. Release*, Submitted for publication.

[4] Friedman, M.H. (1986) Free diffusion, in Springer-Verlag Berlin Heidelberg (ed.), *in Principles and models of biological transport*, pp. 22-43.

[5] Zentner, G.M., Rork G.S. and Himmelstein K.J. (1985) The controlled porosity osmotic pump, *J. Controlled Release*, 1, 269-282.

OPTIMIZATION OF ENTRAPMENT OF METRONIDAZOLE IN AMPHIPHILIC β-CYCLODEXTRIN NANOSPHERES

M. Skiba[1], S. Shawky-Tous[2], D. Wouessidjewe[3] and D. Duchêne[3]

[1]Rouen University, I.U.T. d'Evreux, 43, rue St-Germain 27000 Evreux France
[2]Assiut University, Faculty of Pharmacy, Department of Pharmaceutics, Assiut, Egypt
[3]Paris XI University, Faculty of Pharmacy, URA CNRS 1218, Chatenay-Malabry 92296

1. Introduction

Metronidazole is a nitro-imidazole which has been used as an antiprotozoal and antimicrobial agent for many years. It is the first-line drug used in the treatment of extra-intestinal involvement of amoeba (hepatic abscess) (1).

The potential use of nanospheres as drug carriers has been exploited with success to reduce the toxic side effects of several drugs, thus improving their therapeutic indexes.

Furthermore, the preferential uptake of nanospheres by liver macrophages opens up important therapeutic perspectives in the particular case of hepatic abscess.

Entrapment of metronidazole in nanospheres may be a convenient approach to improve its therapeutic index, permiting an enhancement of drug delivery to infected sites and avoiding the tissues in which the drug produces toxic effects.

Recently, a new colloidal carrier system prepared from modified cyclodextrins was described (2).

These nanospheres have been characterized and visualized by freeze-fracture electron microscopy (3). The self-assembling structural properties of several amphiphilic cyclodextrins and the internal organization of the amphiphilic cyclodextrin nanospheres have been described (4).

These modified cyclodextrins will be here after be called β-CD-C6 .

2. Materials

Hexyl-β-cyclodextrin ester (β-CD-C6) was obtained by a synthetic route (5)

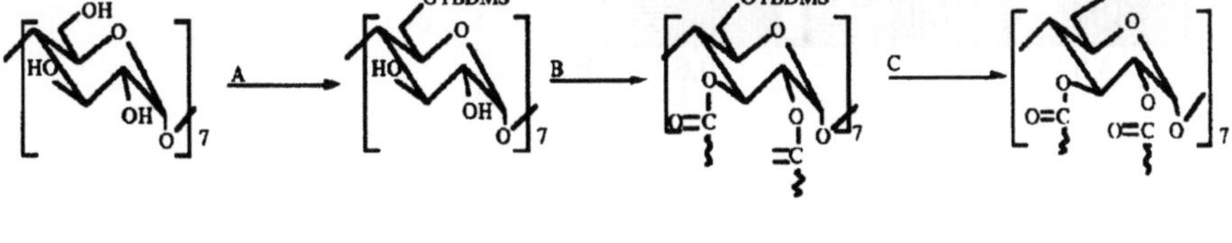

A: Protection B: Grafting C: Deprotection

Drug: Metronidazole (Sigma, St Quentin, France)
Surfactants: Pluronic F68® (ICI, Clamart, France)

3. Preparation of βCD-C6 nanospheres

The nanocrystallization method consisted of injecting an acetonic solution of βCD-C6 into an aqueous phase containing Pluronic F68 surfactant at several concentrations (or the reverse).
Metronidazole was associated with the nanospheres by addition of the drug to the βCD-C6 solution before nanocrystallization .
In both cases, the water-miscible solvent was totally removed under reduced pressure and the suspensions were concentrated to the desired final volume in the same way.

4. Particle size determination

The particle mean diameter and the size distribution of the nanospheres were determined by the quasi-elastic light scattering method (QELS) with a nanosizer N4MD apparatus (Coultronics, Margency, France).

5. Determination of drug loading

Metronidazole content was determined by the reversed-phase HPLC method with a spectro-photometric detector set at 320 nm. The chromatography analysis was performed under the following conditions: column, mBondex C18 (300 x 4.6 mm, SFCC, France); mobile phase, buffer acetate
0.05 M / methanol (70/30, v/v); flow rate, 0.8 ml/min. Metronidazole concentration was determined in all the suspensions (total drug) after dissolution of the nanospheres in acetonitrile and in the supernatants (free drug) after ultracentrifugation at 120,000 g for 1 h at 20 °C.
The association of the drug (%) in the nanospheres was calculated from the difference between the total and free drug. The drug loading, expressed as micrograms of fixed metronidazole per milligram of βCD-C6 , was calculated.
The metronidazole entrapment efficiency (%) was then estimated from the drug content found in the nanospheres and the initial drug content added in the formulations.

6. In vitro metronidazole release

Metronidazole release kinetics from nanospheres was carried out at 37 °C under mechanical stirring after dilution of the colloidal suspensions. These dilutions were performed in an isotonic phosphate buffer solution pH 7.4. In order to separate the particles from the medium, centrifugal ultrafiltration technique as employed. 400 ml of the diluted suspensions was deposited in the Ultrafree MC unit (100,000 NMWL, Polysulfone membrane type, Millipore, France) and subjected to centrifugation at 5,000 g for 5 min. The associated metronidazole (%) was determined by the HPLC method described above.

7. Results and discussion

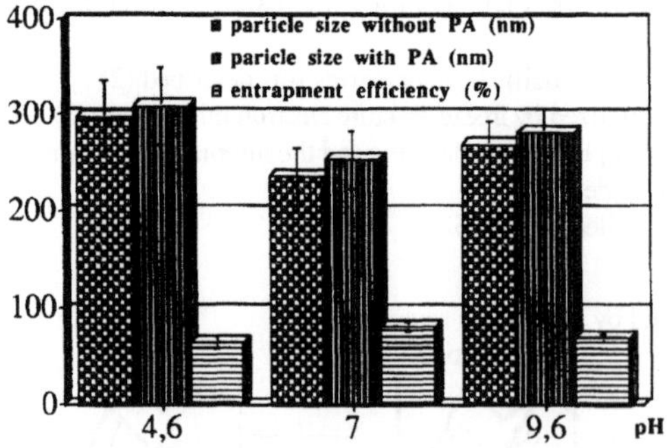

Figure 1 shows that the pH of the buffer used for the preparation did not influence either the percentage of metronidazole associated or the particle size of nanospheres.

Figure 2

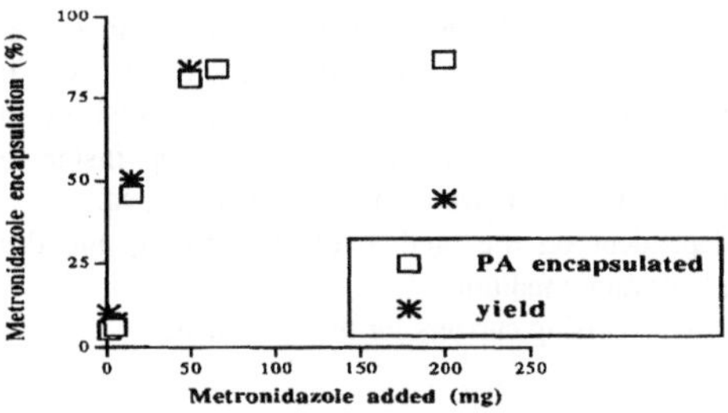

Figure 2 shows the results of metronidazole loading in the amphiphilic βCD-C6 cyclodextrin nanospheres. In this figure, the influence of the initial content of the drug added in the aqueous phase is demonstrated. The maximum association of metronidazole with the cyclodextrin nanospheres was reached when 66 mg of the drug was added in the formulations. In this case, the entrapment efficiency was almost 84 % The excess metronidazole precipitated with the evaporation of the organic solvent and was eliminated by filtration.

Figure 3 allows us to compare the nanocrystallization process already described (injecting an acetonic solution of βCD-C6 into an aqueous phase containing Pluronic F68 surfactant at several concentrations) with another process (injecting an aqueous phase containing Pluronic F68 surfactant into an acetone solution of βCD-C6 several concentrations) . Metronidazole was dissolved in the aqueous phase. The second procedure led to lower drug encapsulation probably because the solubility limit of βCD-C6 was reached before the whole volume of the aqueous phase had been injected and the nanospheres formed prematurely.

FIGURE 3

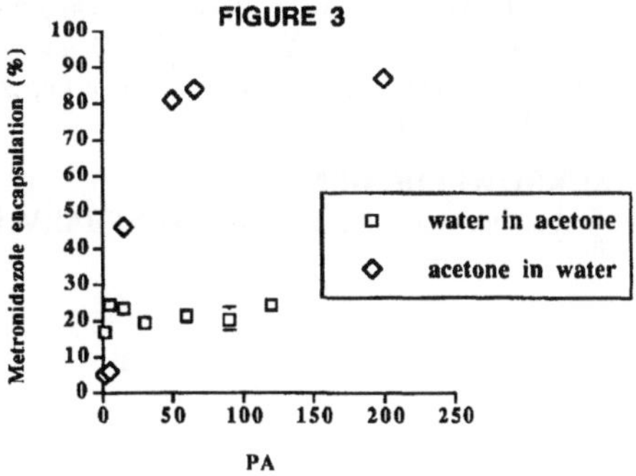

Figure 4

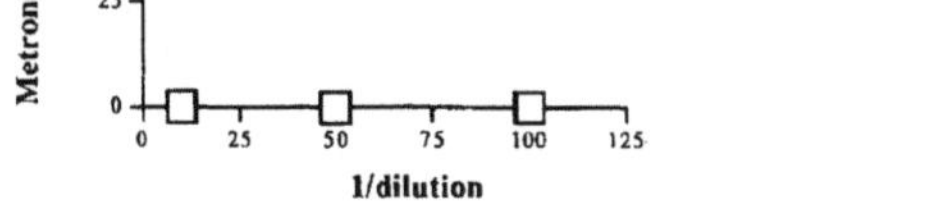

Figure 4 shows the release of metronidazole from nanospheres prepared by the two different procedures. When the acetonic phase was added to the aqueous phase, metronidazole was released only progressively with dilution, showing a strong association with the particles. In contrast, wher nanosperes were prepared in the reverse fashion, the drug was completely dissociated by a ten-fold dilution, suggeshing that it was simply adsorbed on the surface.

8. CONCLUSION

The most suitable parameters for the entrapment of metronidazole in nanospheres were determined as a preliminary step for their use as pharmaceutical carriers in the treatment of hepatic abscess. Nanospheres made of amphiphilic β-cyclodextrin containing metronidazole were prepared by adding an acetonic solution of amphiphilic cyclodextrin to an aqueous solution of metronidazole with or without Pluronic PE68© as the surfactant. An optimized formulation with high encapsulation efficiency, with the drug inside the nanosphere matrix and a particle size appropriate for intravenous administration, was developed.

The entrapment of metronidazole was strongly dependent on the method of preparation, and drug concentration, but was independent of the pH of the hydration medium.

These nanospheres prepared by nanocrystallization are promising carriers for metronidazole.

9. REFERENCES

1. Gordeeva, L.M., Trop. Dis. Bull. 62, 1115-11122, 1965
2. Skiba M. Wouessidjewe D., Coleman A. W., Fessi H., Devissaguet J-Ph., Duchêne D., and Puisieux F. PCT Applications FR 93/00594 (1993)
3. Skiba M., Wouessidjewe D., Puisieux F., Duchêne D. and Gulik A., Int. J. Pharm. 142, 121-124, 1996.
4. Gulik A. , Delacroix H., Wouessidjewe D. and Skiba M. Langmuir, 14, 1050-1057 (1998)
5. Zhang P., Ling C.C., Coleman A ., Parrot-Lopez H. and Galons H. Tetrahedron Letters, 32, 2769-2770 (1991).

10. ACKNOWLEDGMENT

The first author would like to thank ETHYPHARM for partly supporting this work.

CYCLODEXTRIN POLYSULFATES IN CELL BIOLOGY AND THERAPEUTIC PHARMACOLOGY

P.B. WEISZ[a1], M.M. JOULLIE[a2], P. PORTONOVO[a2] E.I.MARX[b], J.M. TARBELL[b], H. KAJI[c], R.L. WILENSKY[a3], E. MACARAK [a4]

[a] *University of Pennsylvania, Philadelphia, PA.*
[1] *Department of Bioengineering, 19104-6392.*
[2] *Department of Chemistry, 19104-6323*
[3] *Presbyterian Medical Center, 19104-2689*
[4] *School of Dental Medicine, Dept. Anat. & Histolology. 19104-6002*
[b] *Pennsylvania State University, Department of Chemical Engineering, University Park, PA 168O2;*
[c] *Jefferson Medical College, Department of Pharmacology, Philadelphia, PA 19104*

ABSTRACT

Cyclodextrin polysulfates possess a variety of cell modulating properties possessed analogous to those of structurally highly complex and heterogeneous heparin, The same structural element, a critical intramolecular sulfate density appears to be basic to a multitude of cell biological effects. This has implications to the fundamental role of polyanions (e.g. glycosaminoglycans) in biology and presents new potentials to pharmacology, medicine and biotechnology.

BACKGROUND

Folkman *et al* [1] discovered that simultaneous application of heparin and hydrocortisone would inhibit blood vessel generation towards tumors (angiogenesis), This observation led us to the hypothesis that the hydrophobic interior of the helix formed by heparin in aqueous media was transporting the cortisone by inclusion. It suggested cyclodextrin (CD) as an alternative to heparin. This proved totally ineffective but led to an inquiry as to "what other properties" of heparin might we add to CD to obtain that activity. Indeed, we discovered that the next step undertaken, namely the addition of sulfates - and no more - was sufficient to obtain anti-angiogenic action with the steroid, and with other angiostatic agents [2]. equal or better than with heparin. This led to investigations of other cell-biological properties, of the mechanisms involved and the implications to pharmacology and medicine.

SPECTRUM OF CELL MODULATING PROPERTIES AND CRITICAL STUCTURAL, FEATURES

We have found that the same polysulfated cyclodextrins (CDSS) possess a variety of cell biological activities [3]: inhibition of smooth muscle cell proliferation, promotion of endothelial cell proliferation, and cell protection from virus invasion [3,4]. They also protect erythrocytes against hemolysis [5]. Furthermore, in this wide spectrum of seemingly different properties, the only structural requirement is one and the same critical degree of sulfation, of at least about 9-10 sulfate groups on β-CD [6]. Electrostatic complexing of the anion cluster of the cyclodextrin sulfates with multiple cationic amino acid sites of proteins is clearly the common denominator [3]. Such complexing is also involved in the common interaction with dyes, such as in Azure A metachromasia [7], which parallels the biological activities. Fig. 1 demonstrates the commonality of these phenomena, i.e. the activity vs. number of sulfates on β-CD [6]. We have placed the relative degree of activity for the various types of behavior onto the same plot.

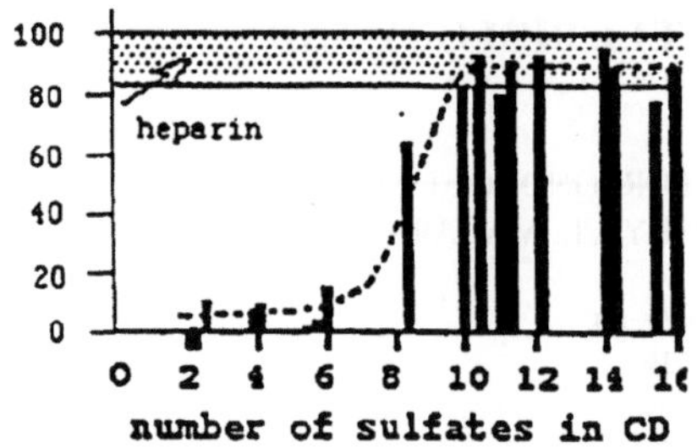

Fig. 1. Relative cell-biological activities of CDS-sulfates vs sulfate number: Inhibition of smooth-muscle cells, promotion of endothelial cells, anti-(HIV)virus activity, and Azure A metachromasia (See ref. 6).

Although the idea of inclusion within the heparin helix stimulated the beginnings of this chain of work, we soon found that neither hydrocortisone nor other molecules were readily included in these CDS compounds. The very need for a minimum number of sulfate groups automatically leads to steric obstruction of the entrance ports of the CD.

All indications point to the critical parameter to be the sulfate *density* rather than the *number* of sulfate groups on the active molecule [6]. On heparin, we have sequences of sugar units with 3, 1, and 2 neighboring sulfate groups on adjacent units of the foldable, flexible chain. On β-cyclodextrin we have 7 positions in close proximity on one "entrance" side of the toroid ring, and 14 of the other. Random sulfation of a 10 sulfates bearing CD will distribute these between the two sides, i.e. involve no more than 6 on one side. However, whatever fraction is found on either side will have multiple close neighboring "ring" positions. By comparison, cyclohexane hexasulfate (myo-inositol), while having six sulfates attached to a single ring, has these anion positions separated in a three dimensional manner, rather than in planar vicinity. Fig. 2 shows that, indeed, both the cell modulation activity (smooth muscle cell inhibition) and the polyionic complexing measured by Azure A metachromasia is weak.

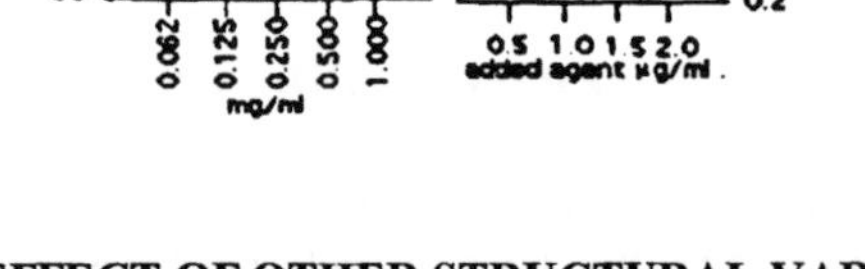

Fig. 2. Comparison of activity of CDS and cyclohexane hexasulfate; for smc inhibition and Azure A metachromasia (left)

EFFECT OF OTHER STRUCTURAL VARIANTS

As to the basic requirement for cell biological activity, the detailed structure of the polyanion molecule can vary widely, as seen by the very large compositional and structural difference between a 15 to 20,000 molecular weight heparin and a CDS of just seven sugar units with no other substituents whatsoever. On the other hand, we can obtain additional tuning of by additional structure. E.g. we have observed a longer time of resistance to enzymatic degradation of a CD-sulfate as compared to a dextran sulfate, as seen in anti-viral essays [6]. Others have found antiviral enhancement by aryl group additions to CDS [8]. We again find parallelisms in the effect of such modifications for different biological targets. Fig. 3 shows relative efficacy of a normal CDS and one with added he pta-thiooctyl substituents (CDS-HO), for anti(HIV)-viral and for smooth muscle cell inhibitory activity. The effect of the substituent is similar for these two different phenomena.

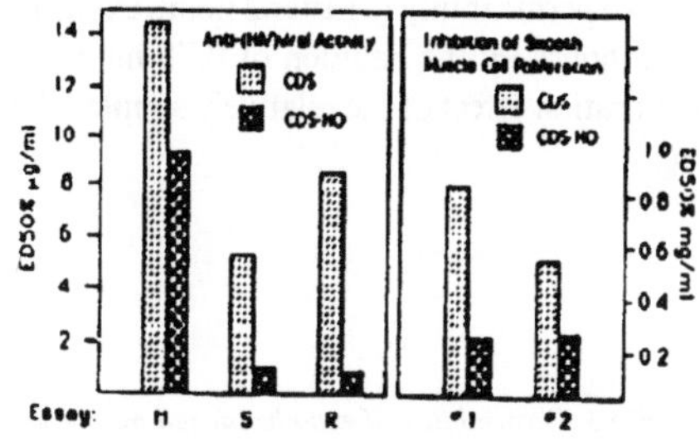

Fig. 3, Comparison of CDS and hepta-thiooctyl CDS, for anti-(MV)viral activity(left) and for inhibition of smc growth (right). Synthicium Inhibtion, R = Reverse transcrioptase; (right, see ref. 14): 2 cultures of human umbilical smooth muscle cells.

SOLID CDS POLYMERS

Solid polymers containing CDS monomers become ready complexing media for equilibrium absorption (storage) and desorption (donation) for polycationic peptides or proteins [6,9]. Fig. 4 shows adsorption behavior of 2-FGF from solution, and redesorption kinetics from that same polymer sample into solution free of the FCF.

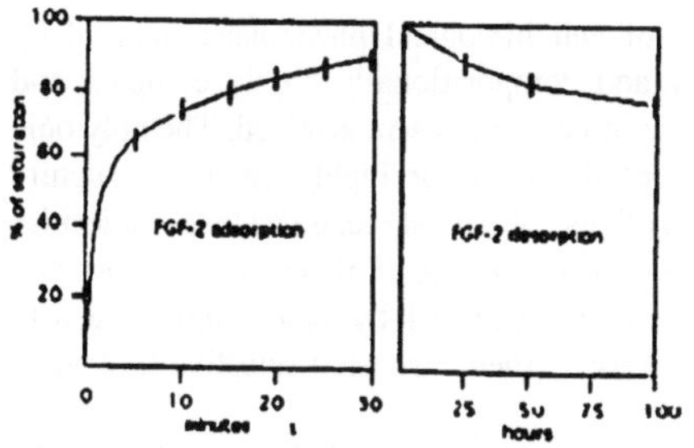

Fig. 4. Complexing (sorption) and de-complexing dynamics (desorption) of FGF-2 and cyclodextrin sulfate polymer.

IN VIVO STUDIES OF CDS FOR INHIBITION OF VASCULAR HYPERPLASIA

The capability of CDS to inhibit smooth muscle cell proliferation *in vitro* has been followed by demonstrating of potential for inhibiting hyperplasia leading to restenosis after balloon angioplasty in rabbit and rat [9,10]. An extension of this work to study this potential therapeutic application has progressed to a study of the use of CDS in a porcine angioplasty model involving 78 pigs (25-30 kg). This study has shown a successful and highly statistically significant inhibition of intimal area of 45% for a 14 day intravenous infusion of 100 mg of CDS/kg/day. Also an indication of some 20% inhibition was achieved for oral intake of 300 mg/kg/day (detailed results of the full study to be published elsewhere).

EFFECTS ON CELL AND TISSUE MEMBRANES

Our demonstration that the same concentration of CDS applied to erythrocytes provides the same protection to attack by a soluble organic hemolytic agent a-s by a solid mineral agent suggests a common mechanism related to a "stiffening" of cell membrane, i.e. of cohesion against penetration. Similarly, the anti-viral phenomenon of CDS (and other polyanions such as dextran sulfate), becoming recognized as involving resistance to virus invasion into the cell, would appear to fall into the same category, Furthermore, a role of heparinic agents in maintaining tissue membrane sufficiency has been generally observed [e.g.:11,12]. We have examined the action of CDS on the control of permeability (hydraulic conductivity) of a confluent layer of retinal microvascular endothelial cells. Fig, 5 shows the relative permeability (a) after 2 hours exposure to 0. 15 mg/ml protamine, and its restoration after 5 hours by the addition of 0.75 mg/ml of heparin (b) or of CDS (c). We note the membrane restoration effect of the relatively simple CDS molecule to be at least equivalent to that of heparin.

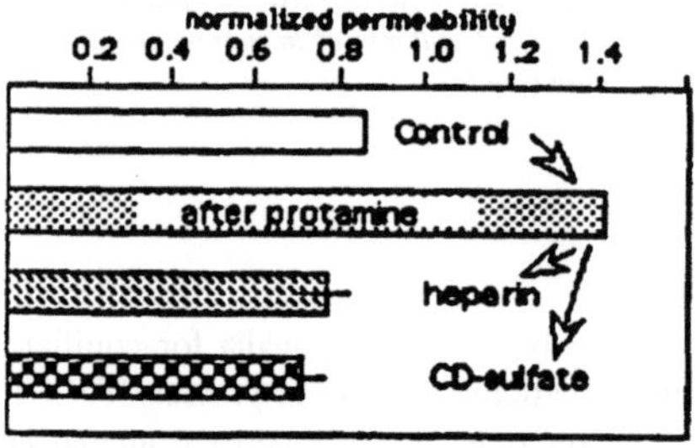

Fig.5. Permeability of endothelial cell membrane (in vitro) on exposure to protamine, and recovery by heparin or CDS.

DISCUSSION AND CONCLUSION

The qualitative equivalence of the great variety of cell biological phenomena induced by cyclodextrin polysulfates and by the structurally and compositionally highly complex and heterogeneous heparin points to a structurally simpler common element involved: The polyionic clustering of sulfate groups. We commonly accept the need for highly detailed structure dependence in biological agents such as proteins and others. We must recognize these molecular agents of critical anionic density (MACADs) [6] as an important class of bioactive molecules that are not primarily detailed structure dependent, but exercise their task by ionic complexing with (polycationic portions of) proteinic partners that exercise their structure-selective biological functions.

It follows that the critically anionic agents play a current and potential role in biomedicine. The potentials include two broad categories of application: 1. Cellular processes that involve ionicically complexable proteins such as growth factors to modify, promote or inhibit cellular or tissue generation or inhibition; 2. Modification of cell or tissue membrane permeability.

The solid polymeric form of cyclodextrin sulfate, as describes above, illustrates the ability, e.g. in colloidal dispersion, to provide the means for targeted absorption or delivery of proteinic agents. Table 1 illustrates areas of therapy where removal or supply of such proteinic factor is involved.

Table 1. Examples of Potentials in Targeted Therapeutic Uses

Anti-proliferative Therapies Use of: CDS dispersión CDS-antiangiogenic-protein complex	**Pro-proliferative Therapies** Use of: CDS-growth-factor-protein complex
Oncology (neopiasms, tumors) Dermatology (psoriasis, mastocytosis -topical use)	Wound healing Ischaemia-revascularization Implant, Transplant acceptance

In contrast to the broad area of "inclusion technology" of cyclodextrins these phenomena and their applications do not depend on inclusion of a guest molecule, but involve association of a guest protein by external ionic complexation. They point to a basic family and role in cell biology and to potential applications in biomedicine and pharmacology in relation to cell growth modulation as well as to issues of tissue membrane sufficiency.

REFERENCES

[1] Folkman J., Langer R., Linhardt R..J., Haudenschild, C. and Taylor, S. (1983), Angiogenesis inhibition and tumor regression caused by heparin or heparin fragment in the presence of cortisone, *Science* **221**, 719-725

[2] Folkman J., Weisz P.B., Joullie M.M., Li W.W. and Ewing W.R. (1989), Control of angiogenesis with synthetic heparin substitutes, *Science* **243**, 1490-149

[3] Weisz P.B., Hernnann H.C, Joullié M.M., Kumor K., Levine E.M., Macarak E.J., Weiner D., (1 992) Angiogenesis and GAG-Mimics, in *Angiogenesis - Key/Principles-Science-Technology-Medicine,* R.Steiner, P.B.Weisz, R. Langer, Eds., Birkhaeuser/Springer, Basel and New York, pp. 107-117; and Weisz P.B., Angiogenesis - The Interdisciplinary Challenge; *ibid, pp. 14-19.*

[4] Weiner D.B., Williams W.V., Weisz P.B., Greene M.I. (1992)., Synthetic Cyclodextrin Derivatives Inhibit HIV Infection in vitro, *Pathobiology,* **60** (4), 206-211

[5] Macarak E., Kumor K., Weisz P.B (1991), Sulfation and Hemolytic Activity of Cyclodextrin, *Biochem. Pharmac.,* **42**, 150

[6] Weisz P.B., Joullié M.M., Hunter C.M., Kumor K.M., Zhang Z., Levine E., Macarak E., Wiener D., Barnathan E. S. (1997), A basic compositional requirement of agents having heparin-like cell-modulating activities, *Biochem. Pharmacol.,* **54**, 149-157

[7] Grant A.C., Linhardt R-J., Fitzgerald G.L., Park J.J. and Langer R. (1984), Metachromatic activity of heparin and heparin fragments, *Anal. Biochem.* **137**, 25-32

[8] Moriya, T., Kurita H., Matsumoto T.O., Mori H., Morimoto M., Ueba N. and Kurata N. (1991), *J. Med. Chem., 2301-2304.*

[9] BachinskyW.B., Bamathan E.S., LiuH.,Okada S.S., Raghunath P.N., Muttreja M., Caron R., Tomaszewski J.E., Weisz P.B., Golden M.A (1995), Sustained inhibition of intimal thickening: in vivo and in vitro effects of polymeric β-cyclodextrin sulfate; *J. Clin. Invest.* **96**, *2583-2592*

[10] Herrmann H.C., Okada S. S., Hozakowska E., LeVeen R., Golden M.,. Tomaszewski J. E., Weisz P.B., Barnathan E. S. (1993), lnhibition of experimental angioplasty restenosis by oral administration of the heparin mimic β-cyclodextrin tetradecasulfate, *Arteriosclerosis and Thrombosis,* **13**, *924-931*

[11] Parsons C.L., Bouchul D., Hurst R. and Callahan, H. (1990), Bladder surface glycosaminoglycans: An epithelial permeability barrier, *J. Urol.,* **143**, *139-142*

[12] Gambaro G, and Baggio B, (1992), Role of glycosaminoglycans in diabetic retinopathy, *Acta Diabetologica, 149-155*

[13] Assay used was as in: Okada S. S., Kuo A., Muttreja R., Hozakowska E., Wcisz, P.B. and Bamathan E.S. (1995), Inhibition of human vascular smooth muscle cell migration and proliferation by β-cyclodextrin tetradecasulfate, *J. Pharmac. And Exper. Therap.* **273**, *948-954*

[14] Assay used was as in:: Fujihashi T., Sakata T., Kaji A. and Kaji H. (1995), Antiviral action of oligodeoxyguanylic acides against HIV virus type 1, *AIDS Res. & Human Retroviruses* **11**, *461-471*

THE USE OF $(SBE)_{7M}$-β-CD (CAPTISOL™) AS A SOLUBILIZING AND OSMOTIC AGENT FOR CONTROLLED AND COMPLETE ORAL DELIVERY OF POORLY WATER SOLUBLE DRUGS

V. J. STELLA[1], K. UEKAMA[2], T. IRIE[2], V. M. RAO[1], E. A. ZANNOU[1], R. A. RAJEWSKI[1], S. SHIRAISHI[3] and K. OKIMOTO[3]

[1]*The center for Drug Delivery Research and the Department of Pharmaceutical Chemistry, The University of Kansas, Lawrence, Kansas, 66047, USA;* [2]*Department of Pharmaceutics, Faculty of Pharmaceutical Sciences, Kumamoto University, Kumamoto, Japan;* [3]*Wakunaga Pharmaceutical Co., Ltd, Hiroshima, Japan*

1. **Introduction**

The objective of this article is to demonstrate *via* examples, the use and advantages of $(SBE)_{7M}$-β-CD for the controlled and complete oral delivery of poorly water soluble drugs.

2. **Materials and Methods**

2.1. MATERIALS

2.1.1. *Sustained Release Tablets*

Methylprednisolone (MP) was obtained from Sigma Chemical Co. (St. Louis, MO). Ethylcellulose (EC, Ethocel[R] Standard 10 Premium) and Polyethylene Glycol (PEG 3350 and PEG 400, Carbowax[R]) were donated by Dow Chemical Co. (Midland, MI) and Union Carbide (Danbury, CT) respectively.

2.1.2. *Delayed Release Tablets*

Dipyridamole (DP) and Citric Acid were both obtained from Sigma Chemical Co. (St. Louis, MO). Cellulose Acetate (CA, CA-320S) and Hydroxypropylmethyl Cellulose Phthalate (HPMCP) were both donated by Eastman (Kingsport, TN). Triethyl Citrate (TEC) was obtained from Aldrich Chemical Co., Inc. (Milwaukee, WI).

2.2. METHODS

2.2.1. *Sustained Release Tablets*

Methylprednisolone (MP), a neutral steroid (intrinsic aqueous solubility: 0.08 mg/ml) was chosen as the model compound. The tablet core was prepared with a 1:7 molar ratio of MP to $(SBE)_{7M}$-β-CD. This ratio was calculated from the phase solubility determined binding constant (700 M^{-1}) so as to have sufficient $(SBE)_{7M}$-β-CD to solubilize all the MP present. Both physical (co-grinding of MP and $(SBE)_{7M}$-β-CD powders under low humidity conditions) and freeze-dried (freeze-drying of a MP and $(SBE)_{7M}$-β-CD aqueous solution) mixtures were studied. The powders were sieved through a 75 μm screen and tablets were compressed in a dissolution die (see Figure 1) using a Carver Laboratory Press. The coating solution containing 4.5% of EC (semipermeable polymer: permeable to water, not to the solute) and PEG 3350 (pore former, water soluble), 0.9% of PEG 400 and 90.1% of ethanol, was manually air sprayed on Teflon™ inserts under constant air flow (40°C). The membranes were dried overnight (50°C), checked for cracks and flaws, measured using a micrometer and secured on the top of the tablet contained in the dissolution die.

2.2.2. *Delayed Release Tablets*

Dipyridamole (DP), a basic drug (intrinsic aqueous solubility: 3.5 μg/ml) was chosen as the model compound. The tablet core contained a 1:9:3 molar ratio of DP, $(SBE)_{7M}$-β-CD and citric acid. Citric acid was used as an acidifying agent to further enhance the drug dissolution. A physical mixture of DP, $(SBE)_{7M}$-β-CD and citric acid was prepared, sieved and tableted under the same conditions as in 2.2.1. The coating solution containing 2.5% of CA (semipermeable polymer) and HPMCP (enteric polymer dissolving at pH > 5.5), 1% of TEC and 94% of a 50:50 mixture of methylene chloride and methanol, was directly coated on the tablet surface under constant air flow (70°C). The coated tablets were dried an additional 15 min. at 70°C and overnight at room temperature. The thickness of the membrane was assumed to be the difference in thickness of the tablet after and before coating, and was measured using a screwgauge micrometer.

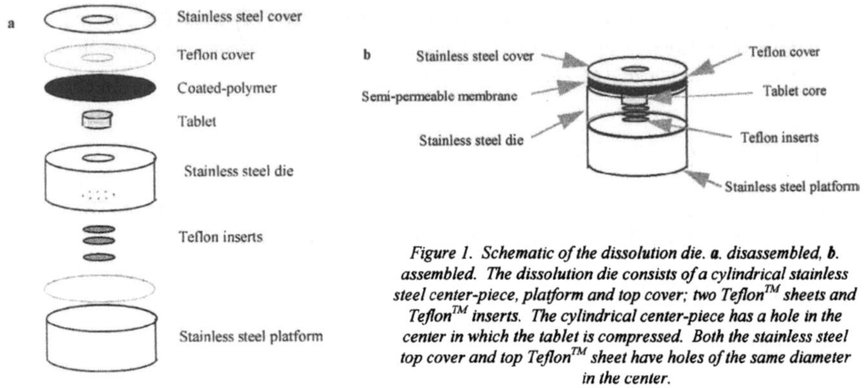

*Figure 1. Schematic of the dissolution die. **a.** disassembled, **b.** assembled. The dissolution die consists of a cylindrical stainless steel center-piece, platform and top cover; two Teflon™ sheets and Teflon™ inserts. The cylindrical center-piece has a hole in the center in which the tablet is compressed. Both the stainless steel top cover and top Teflon™ sheet have holes of the same diameter in the center.*

2.2.3. *In Vitro Release Studies*

In vitro release studies for both sustained and delayed release formulations were conducted by placing the dissolution die in a USP dissolution apparatus II (37°C, 100 rpm). The sustained release studies were conducted in 350 ml double distilled water. The delayed release studies were conducted at pH 1.5 (HCl) for the first two hours and at pH 6.8 (0.15 M phosphate buffer) for the rest of the study (450 ml); these changes in the dissolution medium pH were intended to model the pH changes in the gastrointestinal tract (GIT). Samples were collected periodically and replaced by the same volume of dissolution medium. The 100 % release was determined by removing the membrane from the die and allowing the drug dissolution to be complete. Both MP and DP were detected by HPLC. $(SBE)_{7M}$-β-CD was indirectly detected using TNS and the enhancement of its fluorescence upon inclusion in the cyclodextrin cavity (since it is itself fluorescent, DP was first removed from the samples by solid phase extraction).

3. **Results and Discussion**

3.1. SUSTAINED RELEASE

The controlled porosity osmotic pump developed by Zentner et al. [1] and shown in Figure 2 is mainly intended for the controlled release of water soluble drugs. This approach has been shown to be quite successful; it is however not applicable to pharmaceutical compounds which exhibit poor water solubility. Cyclodextrins and their derivatives have been used widely and very successfully to improve drug solubility [2]. One of these derivatives, $(SBE)_{7M}$-β-CD, has the added advantage of being charged and to contain associated sodium ions. Our hypothesis was to use this material both as a solubilizing and osmotic agent for our controlled porosity osmotic pump. A typical release profile, shown in Figure 3, exhibited an apparent zero order release during 10-15 hours for both MP and $(SBE)_{7M}$-β-CD. Figure 3 also shows that no MP could be detected in the dissolution medium if $(SBE)_{7M}$-β-CD was replaced by another osmotic agent (50:50 mixture of fructose and lactose). The effect of the relative amount of MP and $(SBE)_{7M}$-β-CD was also studied: the amount of MP released corresponded to the amount solubilized by the $(SBE)_{7M}$-β-CD contained in the tablet core. These results indicated that both the solubilizing and osmotic effects of $(SBE)_{7M}$-β-CD were required for MP sustained release from the device.

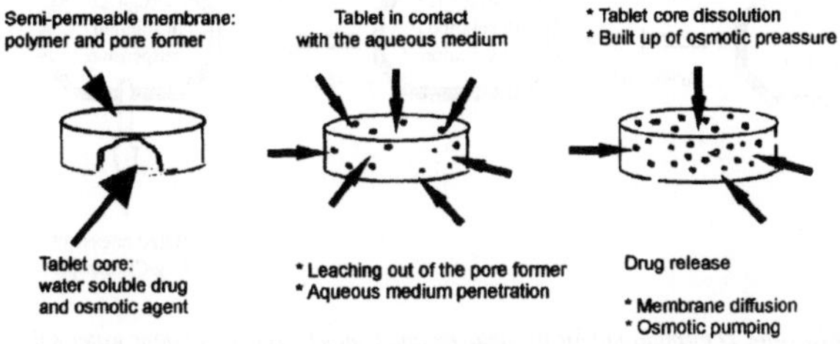

Figure 2: The controlled porosity osmotic pump: composition and release mechanisms

Further investigation of the formulation indicated that the MP release is dependent on the membrane thickness and that the release rate was the same for MP and $(SBE)_{7M}$-β-CD independently of the mixture type (physical or freeze-dried). The microporous membrane was thus rate-controlling as expected, and MP seemed to be released as the $(SBE)_{7M}$-β-CD complex.

In controlled porosity osmotic pumps, the usual driving forces to the release are diffusion and osmosis (Figure 2). Two main methods to probe the relative contribution of these driving forces were used. The first probe consisted in modifying the osmotic pressure of the dissolution medium so as to decrease the difference in osmotic pressure across the membrane. The MP release rate was significantly decreased as a consequence. The second method was using Side-Bi-Side™ diffusion cells in an attempt to decouple the relative contribution of the two driving forces. The results of this study are detailed elsewhere [3]. Both studies indicated that osmosis was the main driving force to the MP release with diffusion being a minimum contributor.

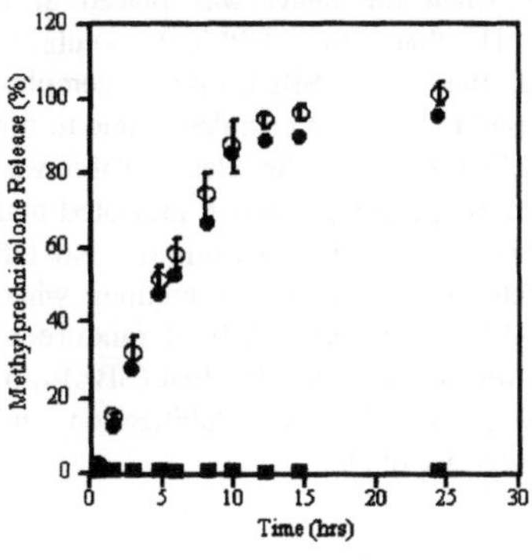

Figure 3: Typical methylprednisolone release rate profile (n = 5 tablets, membrane thickness: 140 µm, ● physical mixture, ○ freeze-dried mixture, ■ fructose-lactose tablet)

The use of $(SBE)_{7M}$-β-CD both as solubilizing and osmotic agent in a controlled porosity osmotic pump has been shown to be successful *in vitro* as well as *in vivo* with different drugs and membrane compositions [4-5]. The main advantage of this technique, other than the fact that it allows poor water soluble drugs to be formulated for sustained release, is that it can be tailored to the drug and the desired release rate by modifying the $(SBE)_{7M}$-β-CD content and / or the membrane characteristics.

3.2. DELAYED RELEASE

Delayed release formulations are very appropriate for drugs which do not require sustained release but are subject to saturable processes (first pass effect, efflux system, etc.), degradation in acidic medium or are in need of GIT site-specific delivery. Similarly to sustained release formulations, delayed release tablets are common for water soluble drugs and problematic for poor water soluble drugs. $(SBE)_{7M}$-β-CD was used to palliate to the solubility issue. In the type of formulations developed (see Figure 4) , the delayed release was provided by the characteristics of the tablet coating. More specifically for the formulation we are reporting in this paper, water penetration through CA in acidic conditions provided the lag time and allowed DP to be solubilized by $(SBE)_{7M}$-β-CD and citric acid. Once exposed to the basic medium, the DP - $(SBE)_{7M}$-β-CD complex was released rapidly and completely.

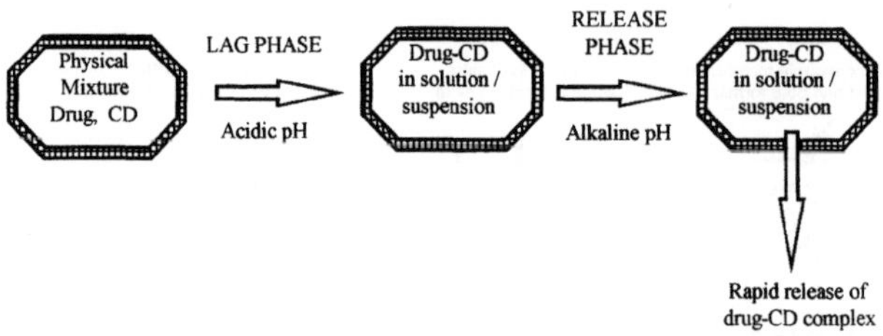

Figure 4: The $(SBE)_{7M}$-β-CD formulation for the delayed and complete release of poor water soluble drugs

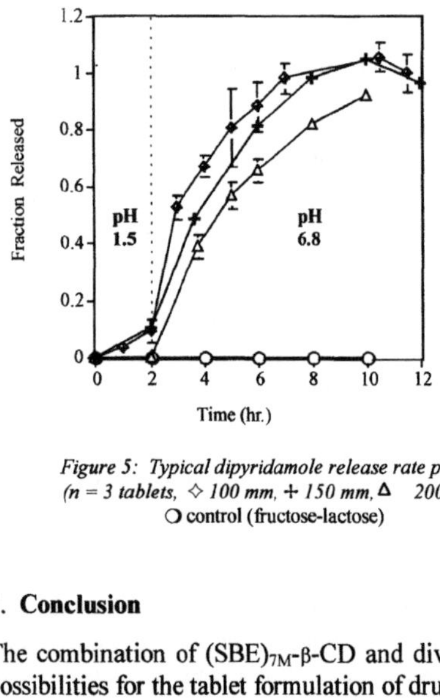

Figure 5: Typical dipyridamole release rate profile (n = 3 tablets, ◇ 100 mm, + 150 mm, Δ 200 mm, ○ control (fructose-lactose)

A typical release profile is shown in Figure 5. The lag phase lasted two hours which corresponds to the length of time the tablet was maintained under acidic conditions. As expected, when the tablet was placed at a higher pH than the HPMCP solubility threshold, the DP - $(SBE)_{7M}$-β-CD complex was released rapidly and completely due to the creation of large pores. The release profile was also thickness dependent, which indicated that the membrane was rate controlling. No DP was detected in the dissolution medium when $(SBE)_{7M}$-β-CD was replaced by a mixture of fructose and lactose, indicating that $(SBE)_{7M}$-β-CD was a must for the solubilization and complete release of DP.

4. Conclusion

The combination of $(SBE)_{7M}$-β-CD and diverse pharmaceutical technologies opens a wide array of possibilities for the tablet formulation of drugs with solubility related limitations.

5. References

[1] Zentner, G.M., Rork G.S. and Himmelstein K.J. (1985) The controlled porosity osmotic pump, *J. Controlled Release*, 1, 269-282.

[2] Loftsson, T. and Brewster, M.E. (1996) Pharmaceutical applications of cyclodextrins. 1. Drug solubilization and stabilization, *J. Pharm. Sci.*, 85, 10, 1017-1025.

[3] Zannou, E.A., Shiraishi, S., Rao, V.M. and Stella, V.J. (1998) Effect of $(SBE)_{7M}$-β-CD on methylprednisolone transport across ethylcellulose microporous membranes, *Proceedings of the Ninth International Symposium on Cyclodextrins.*

[4] Okimoto, K., Miyake, M., Ohnishi, N., Rajewski, R.A., Stella, V.J., Irie, T. and Uekama, K. (1998) Design and Evaluation of an osmotic pump tablet for prednisolone, a poorly water soluble drug, using $(SBE)_{7M}$-β-CD, *Pharm. Res.*, Accepted for publication.

[5] Okimoto, K., Rajewski, R.A. and Stella, V.J. (1998) Release of testosterone from an osmotic pump tablet utilizing $(SBE)_{7M}$-β-CD as both a solubilizing and an osmotic pump agent, *J. Control. Release*, Submitted for publication.

PREPARATION AND CHARACTERIZATION OF INCLUSION COMPLEXES OF CYCLODEXTRINS AND TUBERCULOSIS PRIMARY TREATMENT DRUGS

Heron 0. S. Lima, Angela M. Moraes and Maria Helena A. Santana
School of Chemical Engineering. State University of Campinas. P. 0. Box 6066,- CEP 13081-970. Campinas, SP - Brazil; e-mail: lena@feq.unicamp. br
Flávio F. de Moraes and Gisella M. Zanin
Chemical Engineering Department, State University of Maringá,- Av. Colombo, 5 790, E 46-09, Maringá, PR - Brazil, CEP 87020-900.

1. Introduction

Tuberculosis is an infectuous disease of slow recovery that is the second cause of death according to the World Health Organization. Tuberculosis has affected about one-third of the world's population, killing about three million patients each year and beeing the single most important infectious cause of death on earth (Bloom, 1992). It is common to observe in the conventional treatment side effects, systemic toxicity, resistance of the strains to various antibiotics, and short time in between medicine intake. Hard-to-follow long-term treatments have shown poor adhesion by patients and this limits effectiveness and cure. On the other hand, when followed correctly, conventional therapies are successful but not free from toxicity. Nowadays, some different approaches have been proposed for the therapy of tuberculosis, using encapsulated drugs or delivery systems. Cyclodextrins (CDs) are products of high added value that have found applications as carriers for therapeutic drugs. Inclusion complexes of drugs with cyclodextrins may modify drug properties such as physical and chemical stability and bioavailability (Szejtli, 1988). Otherwise inclusion complexes in aqueous solutions can be associated to other drug delivery systems such as liposomes, increasing the efficacy of healing and decreasing unwanted side effects of drugs (Loukas et al., 1995). These properties make cyclodextrins attractive candidates for improving the treatment regimes for serious diseases characterized by slow healing such as tuberculosis.

The aim of this study was to evaluate the application of O-cyclodextrin (O-C) as a complexing agent of drugs used in the primary treatment of tuberculosis such as pyrazinamide and isoniazid.

2. Materials and Methods

β-cyclodextrins (β-CD) and HEPES (N-[2-Hydroxyethyl]piperine-N'-[2-ethanesulfonic acid], buffer, were purchase from Sigma Chemical Co. (St. Louis, MO, USA). The drugs pyrazinamide and isoniazid were obtained from Aldrich Chemical Co. (Gilligam, Dorset, UK). Deuterium oxide (99.9%) was from Fluka (Poole, Dorset, UK). Milli-Q water was used throughout. All other reagents were of analytical grade,

Preparation and characterization of the Inclusion Complexes

The inclusion complexes of pyrazinamide or isoniazid drugs with β-cyclodextrin were prepared in aqueous medium (7). Briefly, Pz or Iz (1 mM) were dissolved in 10 mL of s.aline buffer (IHEPES 10 mM/NaCl 120 mM). The clear solution was added to β-CD aqueous solutions (10 mL with 1 and

4 mM). The solution was allowed to stir in the dark at 25 'C for 2 days. The kinetic behavior of pyrazinamide and isoniazid drugs inclusion in β-CD was monitored by absorbance measurements of the solutions. The final solution was lyophilized and stored in a desiccator. The inclusion complexes formed were characterized in aqueous phase by nuclear magnetic resonance. This method has been used in literature to confirm the presence of inclusion complex in β-CDs using other drugs as describes by Wood, 1977, Djgdaine and Perly, 1990 and Nishiro et al.,1995, 1997. ^{1}H-NMR spectra in D_20 were recorded in an INOVA-500 spectrometer connected to an Aspect 3000 computer. The chemical shifts were related to residual solvent signal (hydrogen-deuterium oxide = 4.62 ppm at ambient). Typical conditions were pulse 32.0 degrees, act. time 4,000 sec, width 8000,0 Hz and 345 repetitions.

RESULTS AND DISCUSSION

Kinetic behaviour of Pyrazinamide and Isonizid inclusions in β-CD

The kinetic behaviour of Pyrazinamide and Isoniazid inclusions in β-CD, PZ:β-CD and Iz:β-CD, is shown in Figures 1 and 2. The absorbance of solutions increases when the drugs are included in β-CD. Curves reach a plateau indicating that increased β-CD concentration in the complexing solutions increases the saturation level of samples, suggesting that β-CD is the limiting compound in the inclusion process.

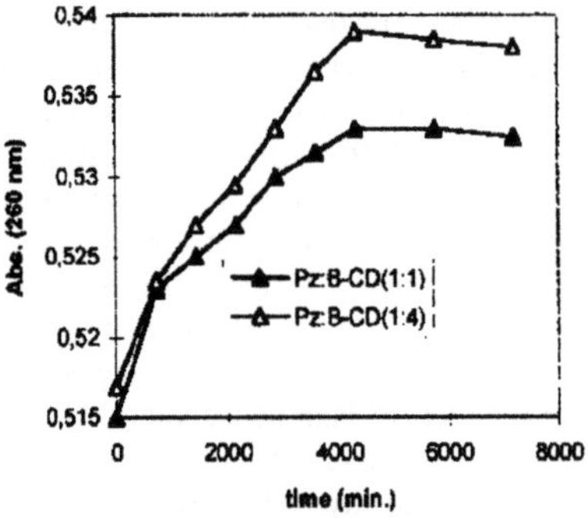

Figure 1 - Kinetic behaviour of Pyrazinamide inclusion in β-Cyclodextrins at Pz:β-CD proportions 1:1 and 1:4.

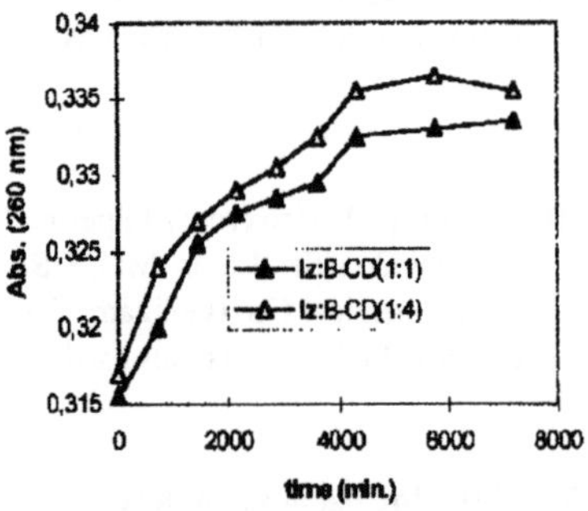

Figure 2 - Kinetic behaviour of Isoniazid inclusion in β-Cyclodextrins at Pz: β-CD proportions 1:1 and 1:4.

Characterization of the Inclusion Complexes by Nuclear Magnetic Resonance of Protons ('H-NMR)

Due to the magnetic field generated by aromatic structures, the PZ:β-CD and Iz:β-CD inclusion complexes are particularly susceptible to detection by Nuclear Magnetic Resonance of Protons. This feature was utilized as an auxiliary technique for the determination of the presence of true inclusion complexes in aqueous solution.

β-CD has primary and secondary hydroxyl groups in the terminal regions of the toroidal structure. The H-3 and H-5 protons are lead to the interior of the cavity, while H-1, H-2, and H-4 are localized in its exterior (Fig. 3). So, ¡t is expected that, if the inclusion really occurs, the protons localized in the cavity (H-3 and H-5) or next to ¡t (H-6) must be covered. Alternatively, if the association occurs in the exterior of the toroidal structure, H-1, H-2 or H-4 must be affected.

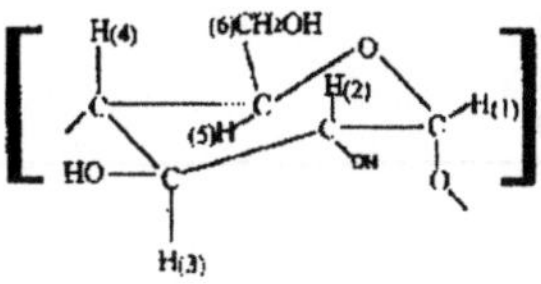

Figure 3 - Molecular structure of β-CD

The presence of the inclusion complexes in solution can be evidenced by NMR chemical shifts of the protons of β-CD free and after the addition of drugs, as shown in Table 1. For PZ:β-CD complex, the sign of H-5 is dislocated from its initial position to higher flelds. The remaining signs do not have meaningful changes. In the situation of the lz:β-CD complex the dislocations of signs H-3, H-5 and H-6 lo higher fields are clearly observable, and they are not observed in the remaining protons of the toroidal structure.

Table 1 - ^{1}H-NMR Chemical shifts of the protons of β-CD free and inclusion complexes states

	δ		
Proton	**β-CD**	**PZ:β-CD**	**Iz:β-CD**
H1	4.873	4.876	4,881
H2	3.463	3.464	3,463
H3	3.756	3.764	3.738*
H4	3.384	3.390	3,390
H5	3.653	3.611	3,591
H6	3.675	3.679	3.611

* δ altered protons.

The comparison of the changes of signs in free drugs with their respective complexes also shows evidence of formation of inclusion complexes (Table 2). It is also noticeable that the sign variation is related only to the protons that are a part of the aromatic structure, confirming the evidences observed before of a possible formation of true complexes.

The chemical structures of the Pz and Iz molecules, and the behavior of chemical dislocation describes above suggest that, in both situations, the aromatic ring is the part of the drug that is placed in the interior of the β-cyclodextrin and the π aromatic ring is responsable for covering the protons.

Table 2 - NMR Chemical shifts of the protons of drugs free and inclusion complexes states

Proton	δ	
	Pz	Pz:β-CD
H1'	8.950	8.985*
H2'	8.529	8.552*
H3'	8.586	8.605*
H4'	4.626	4.605
	δ	
	Iz	Iz:β-CD
H1''	8.392	8.495*
H2''	7.401	7.514*
H3''	4.580	4.550

CONCLUSIONS

From these results we conclude that β-CD can be used as a complexing agent of pyrazinamide and isoniazid drugs. NMR chemical shifts of 'the protons and absorbance measurements characterized the presence of Pz and lz:β-CD inclusion complexes in solution. The kinetic profiles of the inclusion of drugs in β-CD indicated that β-CD is the limiting compound in the complexation process. The successful formation of inclusion complexes containing Pyrazinamide and Isoniazid in β-cyclodextrins represents an unconventional approach to the development of new therapeutics for tuberculosis treatment.

ACKNOWLEDGEMENTS

The authors would like to acknowledge the financial support received from FAPESP agency.

REFERENCES

Bloom, B.R. and Murray, C.J. L. *Science* pp. 257-1055, 1992.

Szejtli, J. *Cyclodextrin technology,* Kluwer Academic Publishers, Dordrecht, pp. 79 - 185, 1988.

Loukas, Y.; Gregoriadis, G. and Jayasekera, P. *Journal of Physical Chemistry,* **99,** 1103 5 - 11040, 1995.

Saenger, W. *Angew. Chem.,* Int. Ed. Engl. **19,** 344 - 362, 1980.

Wood, D. J. J *Am. Soc.,* **99,** 1735 - 1742, 1977.

Djedaine, F. and Perly, B. *Magn. Reson.* Chem, **28,** 372 - 3 80, 1990.

Nishijo, J.; Nagay, M.; Yasuda, M.; Ohno, E. and Ushiroda, Y. (1995). *Journal of Pharmaceutical Sciences,* 84,1420 -1426.

Nishijo, J.; Ushiroda, Y.; Ohbori, H.; Sugiura, M. and Fujii, N. *Chem. Pharm. Bull.,* **45,** 899 - 903, 1997.

INFLUENCE OF DIFFERENT CYCLODEXTRINS ON PHYSICO-CHEMICAL AND PHARMACOLOGICAL PROPERTIES OF ETODOLAC

B. CAPPELLO, M. IERVOLINO, M. I. LA ROTONDA AND A. MIRO

Universitá degli Studi di Napoli "Federico II". Facoltà di Farmacia. Dipartimento di Chimica Farmaceutica e Tossicologica. 80131 Naples.Italy

1. Introduction

Etodolac (Et), an indolacetic acid derivative, is an useful non steroidal antiinflammatory drug (NSAID) marketed for the treatment of pain and inflammation associated with various forms of arthritis; like other NSAIDS, it is scarcely water soluble and highly lipophilic. The dissolution rate is often the limiting step for oral absorption of poorly soluble drugs and can also strongly affect the bioavailability. Therefore, the optimization of the pharmaceutical formulation of Et should be considered to improve drug dissolution rate in order to achieve an improvement of both oral availability and onset rate of analgesic effect.

The inclusion of cyclodextrins (CDs) in pharmaceutical dosage forms has been demonstrated to be an effective tool for increasing dissolution rate of drugs. Moreover, different preparation methods of drug/CD mixtures also play an important role because they can cause different physico-chemical characteristics at solid state which in turn will affect the behaviour in solution [1].

The aim of this work was to study the influence of both naturally occurring and chemical modified cyclodextrins on the physico-chemical and pharmacological properties of Et formulations.

2. Experimental Methods

2.1 MATERIALS

Etodolac was a gift from C.F.M. S.p.A. (Milan, Italy), β-cyclodextrin (β-CD) and hydroxypropyl-β-cyclodextrin (Hp-β-CD) were kindly supplied by Roquette Frères (Lestrem, France) and γ-cyclodextrin (γ-CD) was a commercial sample from Fluka (Buchs, Switzerland). All substances were used without further purification. All chemicals were of analytical reagent grade. Double distilled water was used throughout the study.

2.2 PREPARATION OF SAMPLES

Physical mixtures and colyophilized samples were prepared according to the usual experimental procedures to carry out inclusion of drugs in CDs [1]; physical mixtures of separately lyophilized components were also prepared.

2.3. INVESTIGATION METHODS

The X-ray diffraction patterns were collected on a Philips PW 3710 diffractometer; Kα radiation of Cu was generated at 40 kV and 30 mA.

DSC measurements were carried out using a Mettler DSC 30 apparatus equipped with TC II at scanning rate of 10°C/min. Dry nitrogen was used as a purge gas.

FT-IR spectra (KBr disc) were obtained on a Bruker Mod IFS-48 apparatus using Fourier transformation of 8 scans.

UV spectra were obtained by a Philips PU 8700 spectrophotometer.

Complex formation between Et and CDs in aqueous solution was investigated according to the "Spectral shift method" [2] in presence of Et and increasing concentrations of CDs (from $2*10^{-3}$ to $1.2 * 10^{-2}$ of β-CD and γ-CD; from $2* 10^{-3}$ to $5*10^{-2}$ of Hp-β-CD). All the aqueous solutions were prepared in phosphate buffer at pH 7.0 UV spectra of Et and Et/CD mixtures were collected at 276 nm.

Solubility studies were performed according to Higuchi and Connors method [3]. An excess amount of Et (5 0 mg) was added to 50 ml of water or CD solution (from $1*10^{-3}$ to $1.5*10^{-2}$ of β-CD; from $1*10^{-3}$ to $2*10^{-2}$ of HP-β-CD and γ-CD) and shaken in screw-capped glass vials at 25°C until equilibrium. An aliquot was filtered and analyzed for Et content by spectrophotometry.

^{13}C-NMR spectra were measured by a Bruker AMX-500 spectrometer at 20.0±0.1°C. Tetramethylsilane was used as an external reference and no correction was made for susceptibility of the capillary. Solutions containing Et at a constant concentration ($2*10^{-2}$M) and variable concentrations of each CD (from 0 to $8*10^{-2}$ M) were made up in D_20 by adding NaOD, 0.1 M.

The lipophilicity value (expressed as log D) of Et and its complex with β-CD was determined according to the "shake flask" procedure [4]. n-octanol was used as lipophilic phase whereas the aqueous phases were a phosphate buffer 0.05 M at both pH.7.4 and 4.5 and a KCl/HCl buffer at pH 2.

The dissolution rate of both Et and different Et/CD mixtures was evaluated according to the USP23 method. 100 mg of Et, or equivalent amounts of Et/CD mixtures, were added to 1 L of water at 37.0±0.1 °C in an Erweka DT Apparatus. Suitable aliquots were removed at different time intervals, filtered and spectrophotometrically analyzed for Et content.

2.4. PHARMACOLOGICAL STUDIES

The acetic acid writhing test was used to evaluate the analgesic activity. Et and Et/β-CD colyophilized sample were orally administered in water suspension to groups of 6 male mice weighing 25-30 grams (Harlan Nossan, Italy). Acetic acid solution 0.06% in saline (0.5 ml/mouse) was injected intraperitoneally 15, 30, 60 and 90 min after oral administration of the samples. The writhing movements of each animal were recorded 20 min after the irritant injection. The analgesic effect of the compounds was expressed as number of contractions and protection percentage compared to the control group.

3. Results and Discussion

The IR spectra of Et/CD mixtures show appreciable infrared absorption in the region (1800-1300 cm^{-1}) which can be attributed to Et, since CDs have not functional groups active in the same region. The spectra of the physical mixtures are the superimposition of CD and Et ones. In these spectra the strong absorption band centered at 1744 cm^{-1} (characteristic C=O stretching band of Et) is of diagnostic value and it disappears in the spectra of the colyophilized samples. Hence the absence of this band in the colyophilized samples can be related to the occurrence of an inclusion compound. Therefore, we can suppose that the colyophilization represents an effective method to produce an inclusion complex of Et with β-CD, HP-β-CD and γ-CD and existing in the solid state.

Consistent results were obtained with DSC and the powder X-ray diffraction analysis. The UV absorption curves of Et in the presence of CDs show spectral bathocromic shifts and increases in absorption value as a function of CDs concentration that are indicative of the formation of Et/CD complexes in solution.

Solubility studies have shown that all CDs have a significant influence on drug solubility, particularly Hp-β-CD.

NMR analysis according to the mole ratio method [5] shows a concentration dependency of CD-induced chemical shift changes in Et. The chemical shift behaviour of Et carbons is qualitatively consistent with the introduction of its benzenic portion into the β-CD cavity. In addition the pattern of chemical shift changes indicates that also the alkylic chain of Et interacts with Hp-β-CD, giving a more strong host-guest interaction (the highest shifts are recorded in this case) and that γ-CD could fit the whole Et molecule.

The lipophilicity studies have shown that the partition coefficient of Et decreases in presence of β-CD (freeze-dried sample).

The dissolution profiles of Et alone and its different mixtures with CDs are reported in figure 1.

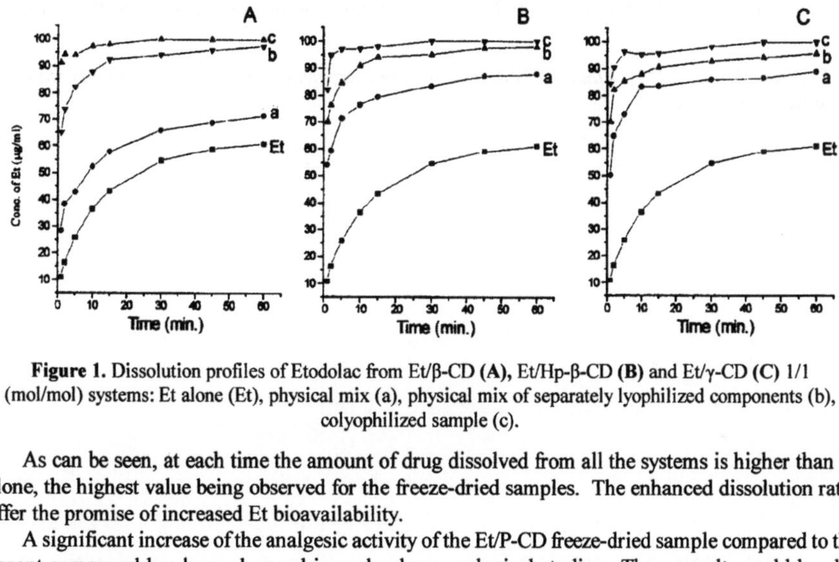

Figure 1. Dissolution profiles of Etodolac from Et/β-CD **(A)**, Et/Hp-β-CD **(B)** and Et/γ-CD **(C)** 1/1 (mol/mol) systems: Et alone (Et), physical mix (a), physical mix of separately lyophilized components (b), colyophilized sample (c).

As can be seen, at each time the amount of drug dissolved from all the systems is higher than Et alone, the highest value being observed for the freeze-dried samples. The enhanced dissolution rates offer the promise of increased Et bioavailability.

A significant increase of the analgesic activity of the Et/P-CD freeze-dried sample compared to the parent compound has been observed in early pharmacological studies. These results could be also ascribed to the fact that the higher complex dissolution rate than the free drug determines a faster onset of the analgesic effect.

4. Conclusion

In this work we have demonstrated that Et can form inclusion compounds with β-CD, HP-β-CD and γ-CD. The preparation procedure influences both the physico-chemical properties of the products and the dissolution behaviour in aqueous solution. Freeze-drying results the method of choice to obtain inclusion compounds at the solid state.

The significant increase of the analgesic activity of the Et/β-CD freeze-dried observed in preliminary pharmacological studies motivates further tests, also regarding to the other Et/CD prepared systems, presented in this paper. These studies, including ulcerogenic activity and *in vivo* bioavailability, should be performed in order to evaluate the best suitable candidate for a solid dosage form.

5. References

1. Duchéne, D., (1987): Cyclodextrins and their industrial uses, Editions de Santé, Paris.
2. Connors, K.A., Mollica, J.A., (1966) Theoretical analysis of comparative studies of complex formation, J *Pharm. Sci.,* **55**, 772-780.
3. Higuchi, T., Connors, K.A., (1 965) Phase solubility techniques, *Adv. Anal. Chem. Instr.,* **4**, 117-212
4. Leo, A.J., Hansch, C., Elkins, D., (1972) Partition coefficients and their uses, *Chem* Rev.,71, 525-616.
5. Connors, K.A., (1987): *Binding Constants: the measurement of molecular complexes stability,* Wiley, New York.

SOLID-STATE NIFEDIPINE-CYCLODEXTRIN PHOTOSTABILITY STUDIES.

M.S. WORTHINGTON, B.D. GLASS, L.J. PENKLER*

School of Pharmaceutical Sciences, Rhodes University, P.O. Box 94, Grahamstown, 6140, R.S.A.

* *Druggists Group Research, South African Druggists Ltd, P.O Box 4002, Korsten, Port Elizabeth,6014,R.S.A..*

1. Introduction

The past decade has seen a number of researchers attempt to alter the solid-state photostability of nifedipine via cyclodextrin (CD) inclusion complexation [1, 2, 3]. While Mielcarek *et al.* [2] observed a 5-fold improvement in solid-state nifedipine photostability for a kneaded β-CD complex, others noted only marginal improvements for α-CD, β-CD, γ-CD, heptakis(2,3,6-tri-O-methyl)-β-CD (TM-β-CD), heptakis(2,6-di-O-methyl)-β-CD (DM-β-CD) and β-CD polymer inclusion complexes [1, 3]. This preformulation study continues to investigate the solid-state photostability of nifedipine in β-CD, γ-CD, 2HP-β-CD, randomly methylated-β-CD (RM-β-CD) and DM-β-CD kneaded products in order to establish the feasibility of employing these cyclodextrins as photoprotecting agents for nifedipine.

2. Materials and methods

2.1 MATERIALS

Nifedipine, β-CD, γ-CD and 2HP-β-CD (Average Degree of Substitution, D.S. 4.8) were kindly donated by South African Druggists, Ltd (Port Elizabeth, South Africa). RM-β-CD (D.S. 12.4) and DM-β-CD were purchased from Cyclolab (Budapest, Hungary). All other materials were of analytical reagent grade.

2.2 METHODS

2.2.1 *Preparation of nifedipine-CD binary systems*
Nifedipine-CD binary systems were prepared with β-CD, γ-CD, 2HP-β-CD, RM-β-CD and DM-β-CD in 1:1 molar ratios (2:1 and 1:2 molar ratios additionally for γ-CD and RM-β-CD) using a kneading method. An ethanol : water (50:50 % v/v) mixture was used to knead nifedipine with β-CD and γ-CD, while ethanol alone was used as solvent for the remaining cyclodextrin derivatives. The slurry was kneaded for 90 minutes and the paste obtained dried under reduced pressure at 33 ± 1 °C for 24 hours and screened through a 315 μm sieve. Physical mixtures were prepared by simple blending of the individual components in a mortar. Verification of solid-state inclusion complexation was undertaken using differential scanning calorimetry (DSC) and infrared (IR) spectroscopy.

2.2.2 *Solid-state Nifedipine-CD Photostability*

The powdered nifedipine-CD binary systems and physical mixtures were evenly spread (1-2 mm layer thickness) on a flat surface and simultaneously exposed to window-filtered daylight on an east-facing window sill for between 80 and 100 minutes. Dark controls were wrapped in foil and placed alongside the authentic samples during photodegradation studies. No thermal degradation was not observed in these dark controls. Only a single comparative photodegradation study was performed each day, in order to minimize the impact of variable light intensity between the morning and afternoon. All studies were performed in triplicate and were initiated at 09h00 on clear, cloudless days. Luxmeter measurements indicated that minimal fluctuations in visible light intensity occurred over the course of a mornings exposure.

3. Results and discussion

The industrially applicable kneading method yielded binary systems with spectral and thermal characteristics similar to their respective physical mixtures (data not shown), implying weak solid-state inclusion complexation. These findings were not entirely unexpected, since the complex stability constants ($K_{1:1}$), previously determined by phase solubility analysis, were low (< 300 M^{-1}) for the nifedipine-CD binary systems [4]. An amorphous nifedipine product was achieved by heating a physical mixture of nifedipine and RM-β-CD up to 200°C, followed by rapid cooling of the melt in an ice-bath. The amorphization occurred at a low stoichiometric ratio (1 : 0.25 nifedipine : RM-β-CD), suggesting that monomolecular dispersion of nifedipine in the cyclodextrin matrix was the dominant mechanism, rather than inclusion complexation.

Nifedipine photodegradation obeyed first-order kinetics in all the binary systems studied. Photodegradation half-lives were generally less than 60 minutes, irrespective of the nifedipine - CD molar ratio, the cyclodextrin used or the time of year during which the experiment was performed. A semilogarithmic plot of the percentage nifedipine remaining versus irradiation time for the 1:2 molar ratio nifedipine : γ-CD physical mixture and kneaded product is shown in figure 1.

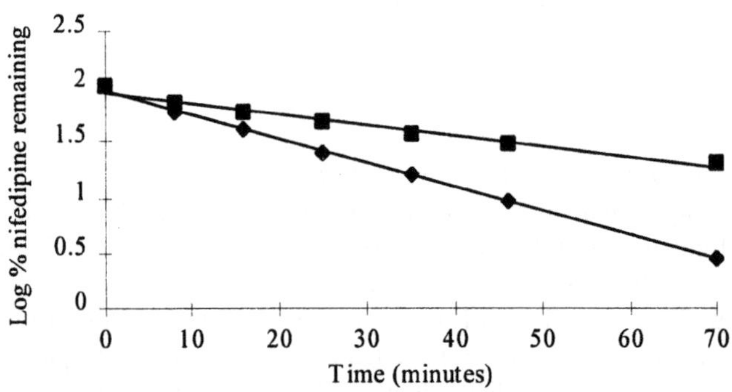

Figure 1. Semilogarithmic plot for the solid-state photodegradation of a 1:2 molar ratio nifedipine : γ-CyD kneaded product (■) and physical mixture (◆) upon exposure to window filtered morning daylight on a sunny window sill.

The degree of photostabilization (or destabilization) was evaluated relative to the physical mixtures and was expressed as a 'stability indicating ratio' (R_{si}), calculated by dividing the first-order rate constant of the physical mixture (k_{pm}) by that of the kneaded product (k_{kp}). The 'stability indicating ratios' (R_{si}) for the nifedipine-CD binary systems studied are listed in table 1 in descending order of magnitude.

TABLE 1. Stability indicating ratios(R_{si}) for the nifedipine - cyclodextrin kneaded products.

Description	Molar ratio	R_{si}*
Nifedipine : γ-CD	1 : 2	2.28
Nifedipine : γ-CD	1 : 1	1.93
Nifedipine : γ-CD	2 : 1	1.81
Nifedipine : β-CD	1 : 1	1.69
Nifedipine : RM-β-CD	2 : 1	1.12
Nifedipine : DM-β-CD	1 : 1	1.09
Nifedipine : RM-β-CD	1 : 2	1.05
Nifedipine : RM-β-CD	1 : 1	0.96
Nifedipine : 2HP-β-CD	1 : 1	0.80

* R_{si}: stability indicating ratio = k_{pm} / k_{kp}.
where k: rate constant
pm: physical mixture
kp: kneaded product

A slight improvement in nifedipine photostability was observed for the β-CD and γ-CD kneaded products prepared in 1:1 molar ratios. Nifedipine photostability in these systems increased by a factor (R_{si}) of 1.69 and 1.93, respectively, relative to their physical mixtures. Kneading with 2HP-β-CyD appeared to produce a slight destabilising effect (R_{si} = 0.80), while neither a stabilising nor destabilising effect was observed for the 1:1 nifedipine : RM-β-CD (R_{si} = 0.96) and nifedipine : DM-β-CD (R_{si} = 1.09) kneaded products. The impact of changing nifedipine-CD stoichiometries on nifedipine stability was investigated for the 2:1 and 1:2 nifedipine : β-CD and γ-CD kneaded products. Stability indicating ratios for the nifedipine : γ-CD binary systems decreased in the order 1:2 (R_{si} = 2.28) > 1:1 (R_{si} = 1.93) > 2:1 (R_{si} = 1.81), and hence the observed stabilising effect could be attributed to the increasing γ-CD content.

Altering the molar ratio of the nifedipine : RM-β-CD kneaded product from 2:1 to 1:2 produced little change in nifedipine photostability. Colour changes observed in the β-CD and γ-CD kneaded products may serve to explain the photostabilising effect of these cyclodextrins. Whereas the surface colour of the 2HP-β-CyD, RM-β-CD and DM-β-CD kneaded products and their respective physical mixtures changed from bright to dark yellow, the β-CD and γ-CD kneaded products had assumed a creamy-white colour by the end of the exposure period. Dark controls did not show similar changes. Since the extent of light penetration into a powder bed has been shown to influence the rate photodegradation [5], it was hypothesized that the photochemically induced opacity changes observed in the β-CD and γ-CD kneaded products decreased light transmission into the powder bed and thus reduced the extent of nifedipine photodegradation.

4. Conclusion

The most potent photostabilizer of nifedipine in the solid-state was found to be γ-CD, with the cyclodextrin derivatives, 2HP-β-CyD, RM-β-CD and DM-β-CD, offering little photoprotection when employed as single-component photostabilizers. Kneading has been recognized as one of the more suitable techniques for preparing solid-state drug-CD inclusion complexes on an industrial scale, since it is generally cost effective and permits ease of scaling up using freely available conventional mixers or blenders. The yield can however vary considerably. This was particularly evident for the nifedipine-CD kneaded products prepared in this study, which exhibited thermal and spectroscopic characteristics similar to their respective physical mixtures, suggestive of a low inclusion complex yield.

Although stabilization was observed with the β-CD and γ-CD kneaded products, the photoprotecting effect was not an independant and inherent function of the cyclodextrins, but was rather an effect created by a combination of light exposure and formulation composition. If the opacity changes were capable of providing complete photoprotection to a tablet, it is conceivable that considerable batch to batch variability in tablet colour could arise depending on the duration of light exposure during the manufacturing process or during storage of the dosage form by the patient, thus producing inconsistencies in tablet colour and an unacceptable aesthetic appaerance. The option of using a colourant or opacifier in a coating formulation may therefore be a more efficacious, convenient and conventional approach for dealing with nifedipine photoinstability in a tablet dosage form.

5. Acknowledgements

Financial support from South African Druggists, Ltd and the Foundation for Research Development is greatfully acknowledgement.

6. References

1. Tomono, K.; Gotoh, H; Okamura, M.; Ueda, H.; Saitoh, T.; Nagai, T. (1988) Effect of β-cyclodextrin and its derivatives on the photostability of photosensitive drugs, *Yakuzaigaku*, *48*, 322-325.
2. Mielcarek, J.; Sadaj, A. (1994) Inclusion compounds of nifedipine and other 1,4-dihydropyridine derivatives with β-cyclodextrins II. Photochemical stability of the inclusion compounds formed by nifedipine with β-cyclodextrins, *Acta. Pol. Pharm.*, *51*, 21-24.
3. Thoma, K. (1996) Photodecomposition and stabilization of compounds in dosage forms, in Tønnesen, H.H. (ed.), *The Photostability of Drugs and Drug Formulations*, Taylor and Francis, London, pp 111-140.
4. Worthington, M.S.; Glass, B.D.; Penkler, L.J. (1996) Phase solubility analysis in studying the interaction of nifedipine with selected cyclodextrins in aqueous solution, *J. Inclusion Phenom. Mol. Recognit. Chem.*, *25*, 153-156.
5. Sande, S.A. Mathematical models for studies of photochemical reactions, in Tønnesen, H.H. (ed.), *The Photostability of Drugs and Drug Formulations*, Taylor and Francis, London, pp 323-339.

INCLUSION COMPLEXATION OF NIMESULIDE WITH β-CYCLODEXTRINS AND ITS FORMULATION INTO TABLETS

P. R. VAVIA AND NISHARANI A. ADHAGE,
*Pharmaceutical Division, University Department of Chemical Technology (Autonomous), University of Mumbai,
Mumbai -400 019, INDIA.*

1. Introduction

Nimesulide (NM), a NSAID, is chemically a sulphonanilide, has poor aqueous solubility. It is used in the treatment of osteoarthritis, oncology and rheumatoid arthritis[1]. The present study describes the complexation of NM with β-cyclodextrin (β-CD) and Hydroxy propyl β-cyclodextrin (HPβ-CD). The complexes with β-CD were further formulated into tablet dosage forms.

2. Materials

NM was supplied by Wave Pharma Ltd., Hyderabad, India, β-CD and HPβ-CD were generously donated by Nihon Shokuhinkako Co. Ltd., Japan and AMAIZO, USA, respectively.

3. Methods

3.1 PHASE SOLUBILITY STUDIES

Solubility measurements were conducted according to Higuchi and Connors[2]. Filtrate was analyzed using UV Spectrophotometer (CECIL CE 2020 2000 series) at 394 nm for NM content.

3.2 PREPARATION OF INCLUSION COMPLEXES

The freeze drying technique[3] (LABCONCO, Freeze Dry System, Freezone 4.5) was used to prepare inclusion complexes of NM and cyclodextrins in molar ratio of 1:1. Physical mixtures of NM and respective cyclodextrins were also prepared in the same molar ratio by mixing the powders in geometric proportions.

3.3 CHARACTERIZATION OF INCLUSION COMPLEXES

3.3.1 *Fourier Transformed Infra Red (FTIR) Spectral Studies*

The spectra of NM, physical mixtures and inclusion complexes were recorded in a KBr pellet using JASCO FT/IR-5300 Spectrophotometer.

3.3.2 *Differential Scanning Calorimetry (DSC) studies*

The samples were subjected to DSC studies using Perkin Elmer DSC 7 model. Alumina was used as a reference materials and samples were scanned at the rate of 10°C/min.

3.3.3 *Powder X-ray diffraction (XRD) Studies*

The XRD patterns were recorded using Philips X-ray generator (PW 1729) and automatic X-ray diffractometer model PW 1710 unit.

3.3.4 *In-vitro Dissolution Studies*

The dissolution studies of samples were performed according to USP XXIII type II apparatus in water containing 0.02%w/v Tween 80[4]. The temperature was maintained at 37±0.5°C and the rotation speed was 100 rpm. The samples were withdrawn at various time intervals and analyzed Spectrophotometrically.

3.3.5 *Tablet Preparation*

Tablets were made using NM, NM β-CD physical mixture and NM β-CD complex. The tablets were directly compressed using CADMACH single tablet press machine equipped with 14mm punches. The tablets contained 100mg of TFA. Final weight of the tablets was adjusted using Vivacel® when necessary. Magnesium stearate 1% was used as a lubricant.

4. Results And Discussion

Figure.1

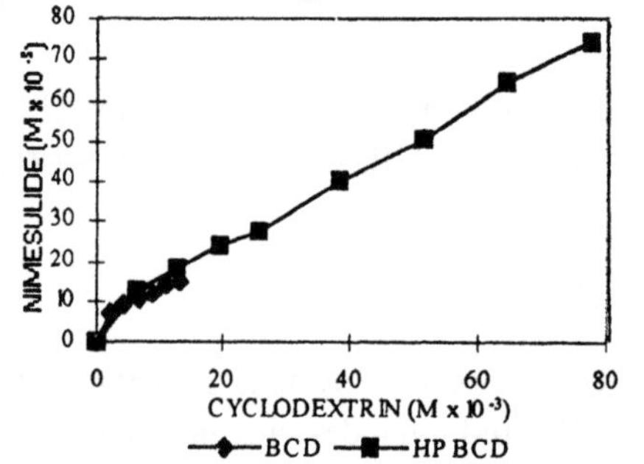

NM solubility in water was found to be 0.01mg/ml. Phase solubility studies of NM (Fig.1) confirms the solubility enhancement capabilities of the CDs. The solubility profiles for NM with β-CD(K=93.78M^{-1}) and HPβ-CD(K=123.15M^{-1}) in aqueous medium are of the A_L type. The FTIR spectra of the complexes of NM with β-CD and HPβ-CD (Fig.2) shows the appearance of a intense broad peak at 3364.16 and

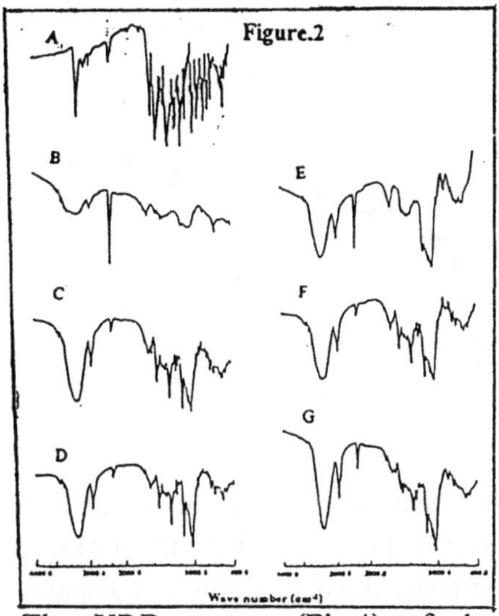

3366.09cm^{-1} respectively. This broad peak indicates a possible hydrogen bonding between NM and cyclodextrins. The DSC thermograms (Fig.3) of NM showed a sharp endothermic peak at 159.03°C for NM. This endothermic peak was also observed at 158.36°C and 157.61°C in the physical mixtures with β-CD and HPβ-CD respectively. The slight shift is probably due to a weak interaction between the host and guest molecules. The complexes showed absence of NM endotherm indicating the formation of cyclodextrin complexes.

The XRD pattern (Fig.4) of the physical mixture showed several peaks corresponding to the crystalline form of NM. The inclusion complexes showed broad diffuse pattern confirming formation of new solid phase.

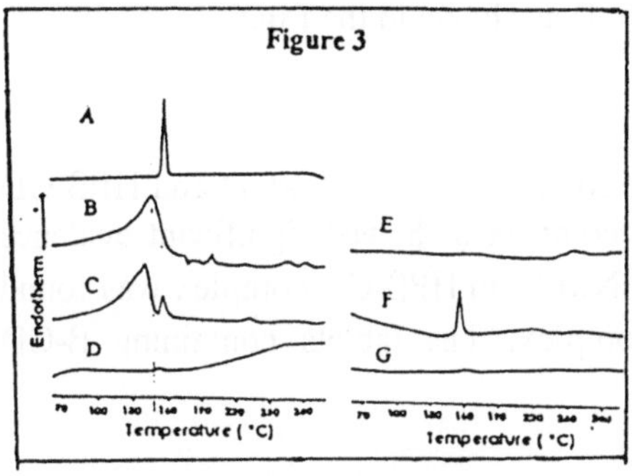

A-NM; B-β-CD; C-NM β-CD physical mixture; D-NM β-CD complex; E-HPβ-CD; F-NM HPβ-CD physical mixture; G-NM HPβ-CD complex. (Applies throughout the paper)

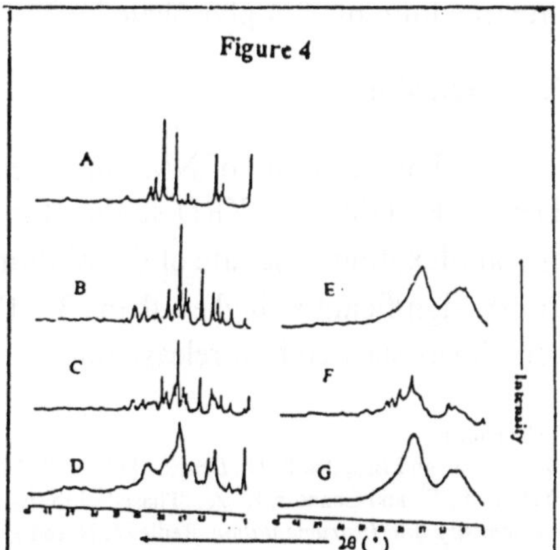

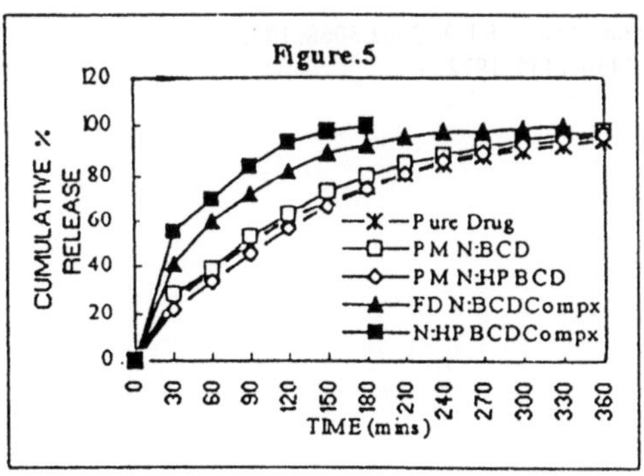

In-vitro dissolution of NM, physical mixtures and inclusion complexes are illustrated in Fig.5. A comparison of the $T_{25\%}$, $T_{50\%}$, $T_{75\%}$ and $T_{90\%}$ are shown in Table 1. The dissolution of NM in pure form, from the physical mixture and inclusion complexes obeyed Hixson Crowell cube root

Table.1

Powders	$T_{25\%}$ MINS	$T_{50\%}$ MINS	$T_{75\%}$ MINS	$T_{90\%}$ MINS	K $\%^{1/3}$/MIN
NM	30.6	91.2	177	297	0.0098
P M	32.8	91.4	187	284	0.0109
NM :					
βCD					
P M	33.0	104	180	273	0.0096
NM :					
HPβCD					
Complex	18.0	44.5	102	165	0.0174
NM :					
βCD					
Complex	13.5	27.0	60.0	109	0.0266
NM :					
HPβCD					

K = Hixson Crowell Cube Root Dissolution Rate Constant.

Figure.6

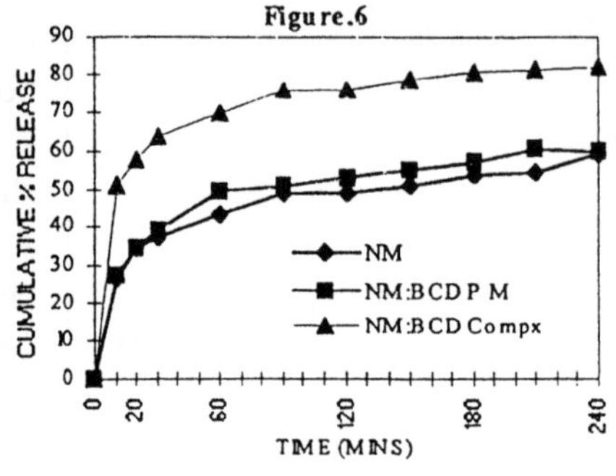

dissolution rate equation for powders. This equation was used to obtain the corresponding dissolution rate constants given in Table 1. The dissolution rate of drug from HPβ-CD complex appears to have a faster dissolution rate as compared with the rate profile from β-CD complexes. This may be due to the increased solubility and wettability along with the decrease in crystallinity caused by complex formation. The tablets containing β-CD complexes showed best release rate (medium pH 7.4 phosphate buffer 900ml) as shown in the Fig.6.

5. Conclusion

The solubility of NM improved by complexation with β-CD and HPβ-CD. The FTIR, DSC and XRD studies for the complexes showed significant evidence of complexation. The rate of dissolution of NM from HPβ-CD complex was found to be significantly higher than β-CD complex. The tablets containing β-CD complexes showed best release rate.

References

1) Davis, R. and Brogden R. N., *Drugs*, 48(3), 732, 1994.
2) Higuchi, T. and Connors, K. A., "Phase-Solubility Techniques" in *Advances in Analytical Chemistry and Instrumentation*, Reilley C.N. (ed.), John Wiley and Sons, New York, Vol. 4, pp. 117-212, 1965.
3) Kurozumi, M., Nambu, N. and Nagai, T., *Chem. Pharm. Bull.*, 23(12), 3062-3068, 1975.
4) Monkhouse, D.C. and Lach, J. L., *J. Pharm. Sci.*, 61,1430-1435, 1972.

INCLUSION COMPLEXATION OF TOLFENAMIC ACID WITH β-CYCLODEXTRINS AND ITS FORMULATION INTO TABLETS

P. R. VAVIA AND NISHARANI A. ADHAGE,
Pharmaceutical Division, University Department of Chemical Technology (Autonomous), University of Mumbai,
Mumbai -400 019, INDIA.

1. Introduction

Tolfenamic acid (TFA) is a NSAID used in the treatment of rheumatic diseases, dysmenorrhea and migraine[1]. In order to achieve fast relief from pain. e.g. headaches, arthritis and migraine attacks it requires a rapid plasma appearance. However, it has a poor aqueous solubility and dissolution rate. In an effort to achieve this we have complexed the drug with cyclodextrins and formulated these complexes into tablet dosage forms.

2. Materials

TFA was supplied by Secifarma, Italy, β-cyclodextrin(β-CD) and Hydroxy propyl β-cyclodextrin(HPβ-CD) were generously donated by Nihon Shokuhinkako Co. Ltd., Japan and AMAIZO, USA respectively.

3. Methods

3.1 PHASE SOLUBILITY STUDIES

Solubility measurements were conducted according to Higuchi and Connors[2]. Filtrate was analyzed using UV Spectrophotometer (CECIL CE 2020 2000 series) at 289 nm for TFA content.

3.2 PREPARATION OF INCLUSION COMPLEXES

The freeze drying technique[3] (LABCONCO, Freeze Dry System, Freezone 4.5) was used to prepare inclusion complexes of TFA and cyclodextrins in molar ratio of 1:1. Physical mixtures of TFA and respective cyclodextrins were also prepared in the same molar ratio by mixing the powders in geometric proportions.

3.3 CHARACTERIZATION OF INCLUSION COMPLEXES

3.3.1 *Fourier Transformed Infra Red (FTIR) Spectral Studies*

The spectra of TFA, physical mixtures and inclusion complexes were recorded in a KBr pellet using JASCO FT/IR-5300 Spectrophotometer.

3.3.2 *Differential Scanning Calorimetry (DSC) studies*

The samples were subjected to DSC studies using Perkin Elmer DSC 7 model. Alumina was used as reference and samples were scanned at the rate of 10°C/min

3.3.3 *Powder X-ray diffraction (XRD) Studies*

The XRD patterns were recorded using Philips X-ray generator (PW 1729) and automatic X-ray diffractometer model PW 1710 unit.

3.3.4 *Nuclear Magnetic Resonance Spectrometry (NMR)*

The ^1H NMR measurements were performed with a Bruker FTNMR spectrometer (recordings in DMSO-d_6) operating at 200 MHz. Trimethyl silane was used as an external reference.

3.3.5 *In-vitro Dissolution Studies*

The dissolution studies of samples were performed according to USP XXIII type II apparatus. Phosphate buffer pH 7.4 was employed as dissolution medium at temp 37 ± 0.5°C. The rotation speed was 100 rpm. The samples were withdrawn at various time intervals and analyzed Spectrophotometrically.

3.3.6 *Tablet Preparation*

Tablets were made using TFA, TFA β-CD physical mixture and TFA β-CD complex. The tablets were directly compressed using CADMACH single tablet press machine equipped with 14mm punches. Each tablet contained 100mg of TFA. Final weight of the tablets was adjusted using Vivacel®. Magnesium stearate 1% was used as a lubricant.

4. Results And Discussion

TFA solubility in water was found to be 0.0065mg/ml. Phase solubility studies of TFA (Fig.1) confirms the solubility enhancement capabilities of the cyclodextrins. The solubility profiles for TFA with β-CD and HPβ-CD were B_S(K=11.398 M^{-1}) and A_L(K=99.49 M^{-1}) type respectively.

The FTIR spectra of TFA, physical mixtures and inclusion complexes are shown in Fig.2. The inclusion complexes showed a displacement of carbonyl stretching of TFA from 1660.86 cm^{-1} to 1612.64 cm^{-1}. This indicates a possible hydrogen bonding between the C=O group and hydroxyl function of cyclodextrins.

Figure 1.

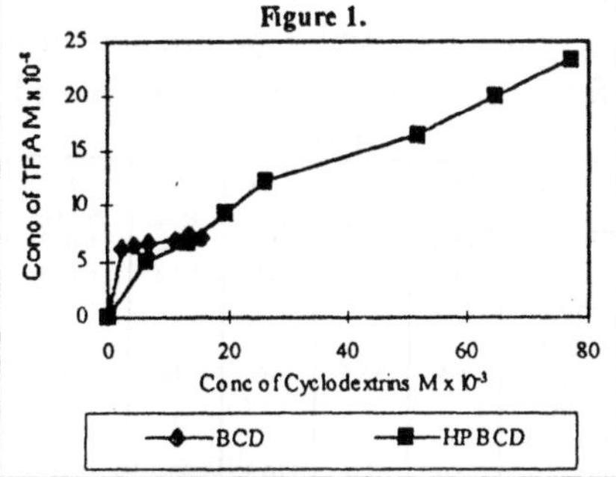

The DSC thermograms (Fig.3) revealed the endothermic peak of TFA at 221.81°C. This peak was absent in the complexes indicating formation of cyclodextrin complexes The pure TFA is of crystalline form as demonstrated by the sharp and intense diffraction peaks (Fig.4). The XRD pattern of the physical mixture showed several peaks corresponding to the crystalline form of TFA. The inclusion complexes showed broad diffuse pattern confirming formation of new solid phase.

Figure 2

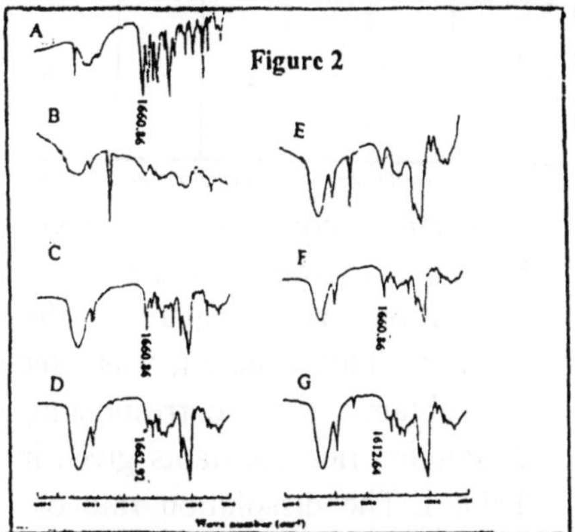

Figure 3

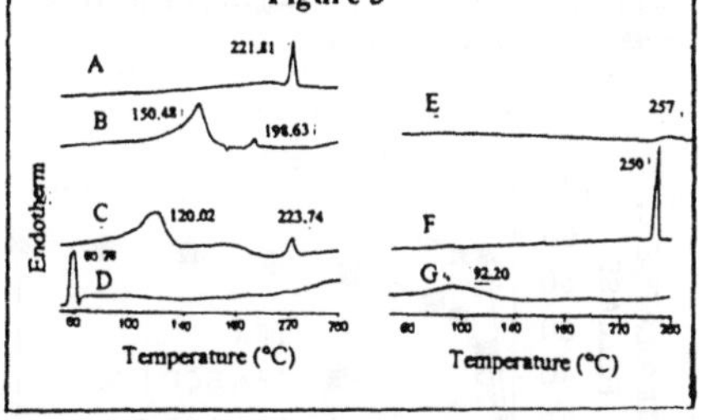

A-TFA; B-β-CD; C-TFA β-CD physical mixture; D-TFA β-CD complex; E-HPβ-CD; F-TFA HPβ-CD physical mixture; G-TFA HPβ-CD complex. (Applies throughout the paper)

^{1}H NMR spectra also proved the inclusion of TFA in the cyclodextrin cavity (Fig.5) The inclusion of TFA is shown by the change in the chemical shift of some of the guest and host protons in comparison with the chemical shifts of the same protons in the free components.

Figure 4

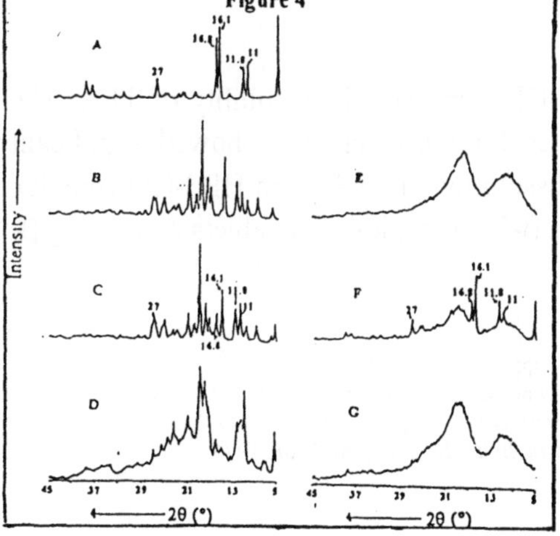

Figure 5

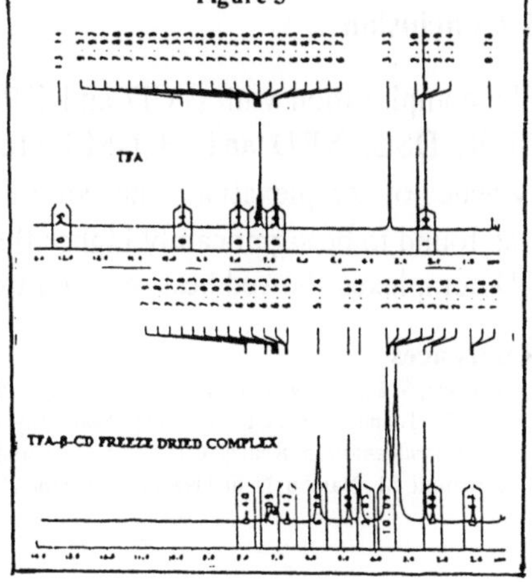

In-vitro dissolution of TFA, physical mixtures and inclusion complexes are illustrated in Fig.6. A comparison of the $T_{25\%}$, $T_{50\%}$, $T_{75\%}$ and $T_{90\%}$ are shown in Table 1. The dissolution of TFA in pure form, from the physical mixture and inclusion complexes obeyed Hixson Crowell cube root dissolution rate equation for powders. This equation was used to obtain the corresponding dissolution rate constants given in Table 1. The dissolution rate of drug from HPβ-CD complex appears to have a faster dissolution rate as compared with the rate profile from β-CD complexes. This may be due to the increased solubility and wettability along with the decrease in crystallinity caused by complex formation. The tablets containing β-CD complexes showed best release rate (medium pH 7.4 phosphate buffer 900ml) as shown in the Fig.7.

Figure.6

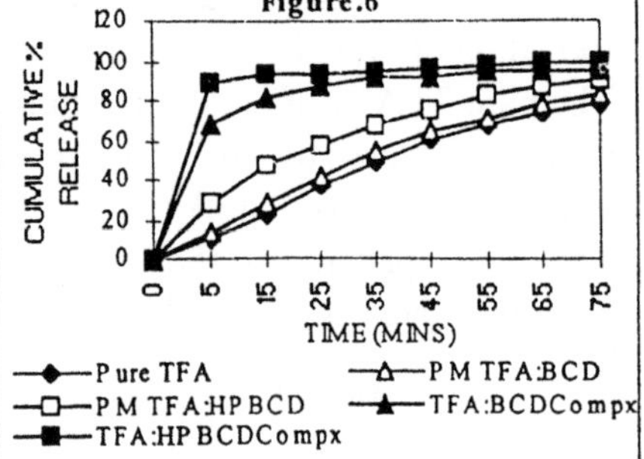

Table:1

Powders	$T_{25\%}$ MINS	$T_{50\%}$ MINS	$T_{75\%}$ MINS	$T_{90\%}$ MINS	K $\%^{1/3}$/MIN
TFA	17.5	35.5	68.5	122	0.026
PM TFA : βCD	14.5	30.5	59	97.5	0.030
PM TFA : HPβCD	4.5	15.5	44	75	0.044
Complex TFA : βCD	2.0	3.7	10.5	30.0	0.119
Complex TFA : HPβCD	1.5	3.0	4.2	7.5	0.166

K = Hixson Crowell Cube Root Dissolution Rate Constant.

Figure.7

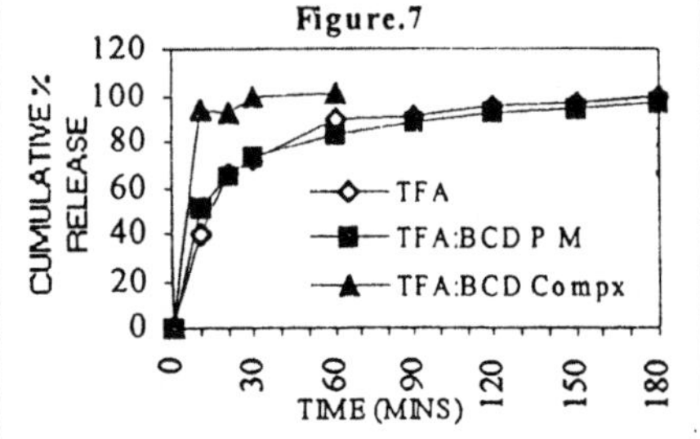

5. Conclusion

The complexation with β-CD and HPβ-CD improved the solubility of TFA. The FTIR, DSC, XRD and ^{1}H NMR studies for the complexes showed significant evidence of complexation. The rate of dissolution of TFA from HPβ-CD complex was found to be significantly higher than β-CD complex. The tablets containing β-CD complexes showed best release rate.

References

1) Pedersen, S. B., *Pharmacology and Toxicology*, **75**, Suppl. 11, 22-32, 1994.

2) Higuchi, T. and Connors, K. A., "Phase-Solubility Techniques" in *Advances in Analytical Chemistry and Instrumentation*, Reilley, C.N. (ed.), John and Sons, New York , Vol. 4, pp. 117-212, 1965.

3) Kurozumi, M., Nambu, N. and Nagai, T., *Chem. Pharm. Bull.*, **23**(12), 3062-3068, 1975.

PREPARATION BY EXTRUSION/SPHERONIZATION OF TRIAMCINOLONE / ß-CYCLODEXTRIN PELLETS AS A FAST RELEASE DOSAGE FORM

M.E. VILLAR-LÓPEZ, F. OTERO-ESPINAR, J. BLANCO-MÉNDEZ

Dpto. Farmacia y Tecnología Farmacéutica, Facultad de Farmacia, Universidad de Santiago de Compostela, Santiago de Compostela, (Spain)

1. Introduction

Oral controlled release products consisting of spherical pellets are being used with greater frequency, and the extrusion/spheronization technology has been shown to give high quality pellets for pharmaceutical applications (Hellen et al., 1992). Nearly sphericity and a smooth surface are required if a film coat is to be applied to provide targeting of the drug (Podczeck et al., 1995).
The aim of this investigation was to study the preparation of pellets in a 5% of Triamcinolone acetonide and ß-Cyclodextrin as a fast release dosage form.

2. Materials

Triamcinolone Acetonide (TA) was purchased from Roig-Farma (Spain), microcrystalline cellulose (MCC) (Avicel® PH101) from FMC International (UK) and ß-cyclodextrin (ß-CD) (Roquete®) was a generous gift from Roquette-Laisa (Spain). All other chemicals were of analytical grade. Distilled water was used throughout the study.

3. Methods

The dry excipients and drug were mixed in a Turbula T2C mixer. Addition of wetting agent took place in an orbital mixer. The paste was extruded in a Caleva Model 10 basket extruder (Caleva Ltd., UK) and the extrudates were spheronized in a Caleva Model 120 spheronizer (Caleva Ltd., UK). Moisture sorption in pellets were analysed by TGA (Shimadzu TGA 50) and dissolution studies were carried out in simulated intestinal fluid. Samples have been seen under the microscope and photos have been taken in an Olympus SZ60 stereomicroscope.

4. Results and Discussion

ß-CD has been shown to be a suitable excipient in the production of pellets (Gazzaniga et al., 1995). It can be confirmed in Table 1 where datas show a good circularity of particles.

TABLE 1. Circularities and Feret diameters of TA pellets and the proportion of drug release after 2h in a standard dissolution test.

MCC(%)	ß-CD(%)	Wetting agent ml	Circularity	Feret diameter (μm)	% release after 2 h
95	—	42	0.92±0.032	769.9±72.52	8.37±0.65
85	10	48	0.91±0.035	1242±9735	7.52±0.561
45	50	30	0.91±0.037	949±71.51	35.25±2.04
15	80	20	0.93±0.04	638.3±184.45	48.6±0.521
5	90	30	—	—	—

Figure 1 shows the TA release profile of pellets produced with Avicel® alone and with three different ratios of Avicel/ß-CD (85/10, 45/50, 15/80).

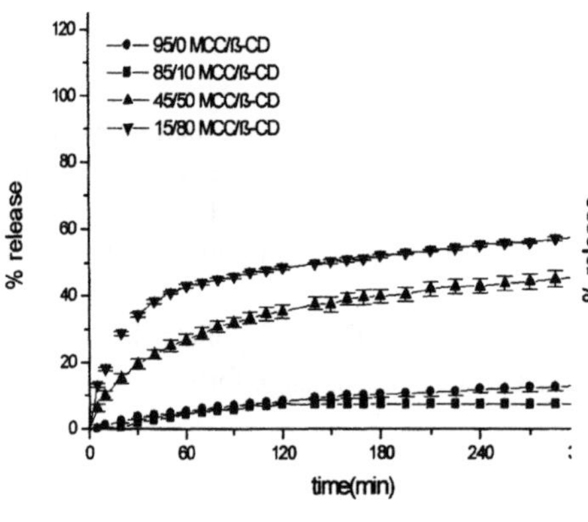

Figure 1 Dissolution profile for TA pellets with different ratios Avicel®/ß-CD

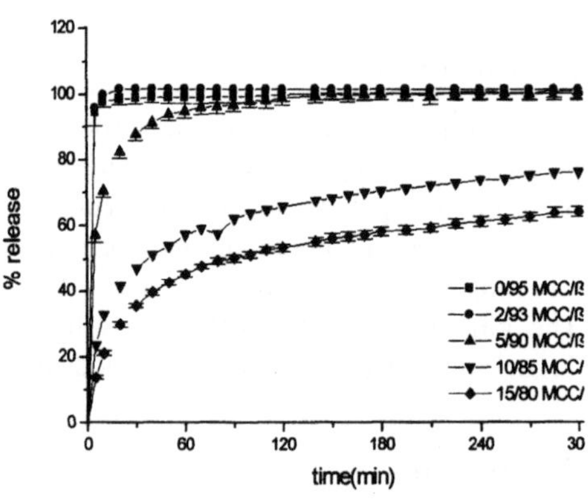

Figure 2 Dissolution profile for TA pellets with different ratios Avicel®/ß-CD (kneading)

Inclusion of ß-CD in pellets modified the properties of the TA-Avicel® mass, particles obtained with 85/10 presented a significant increase in granule size, but this ratio failed to improve TA release. Nevertheless, the release was higher than MCC alone. It was observed that when higher proportions of ß-CD were employed, granule size fell and drug release improved, reaching almost 50% within 2 h in the ratio 15/80 Avicel®/ß-CD. Ratios 5/90 Avicel®/ß-CD failed to produce pellets. Villar-López et al.(Villar-López et al., 1998) have shown that TA is able to form soluble complexes with ß-CD in solution. As the aim of this work is to produce particles as a fast release dosage form, we have to favour the formation of this complex of inclusion in solid state. Thus, kneading of TA/ß-CD was included in the elaboration of pellets. As shown in Table 2, this step led to the inclusion of higher percentages of hydrophilic excipient, even ß-CD alone was successful in the formulation of pellets of TA. Incorporation of ß-CD notably improved drug release in the simulated intestinal fluid. Nevertheless, cores with Avicel® contents less than about 5% disintegrated in the dissolution medium. The drug release profile is displayed in Figure 2.

TABLE 2 Circularities and Feret diameters of TA pellets formulated by kneading, and the proportion of drug release after 2h in a standard dissolution test.

MCC(%)	ß-CD(%)	Wetting agent (ml)	Circularity	Feret diameter (µm)	% release 2 h
15	80	42.5	0.91±0.037	1041.6±51.75	53.4±1.08
10	85	25+5	0.92±0.036	1129.1±63.13	65.65± 0.72
5	90	25	0.94±0.035	1091.7±165.5	98.83±2.10
2	93	25	0.90±0.04	480.2±168 65	101.47±1.17
—	95	28.5	0.92±0.036	850.4±183.89	99.87±1.94

Particles containing 90/5 Avicel®/ß-CD stored for 6 days at room temperature, displayed a decrease in profile of release. This variations are reported in Figure 3, and it can be related to the moisture sorbed due to the amorphous structure of Avicel® (Stubberud et al., 1995). Storing pellets at 60°C allows water expulsion, then, the original profile of release was observed. It can be demostrated in Figure 4 which, reports the sorption of water vapour onto pellets (95/5) for 6 days, and the curves of TGA.

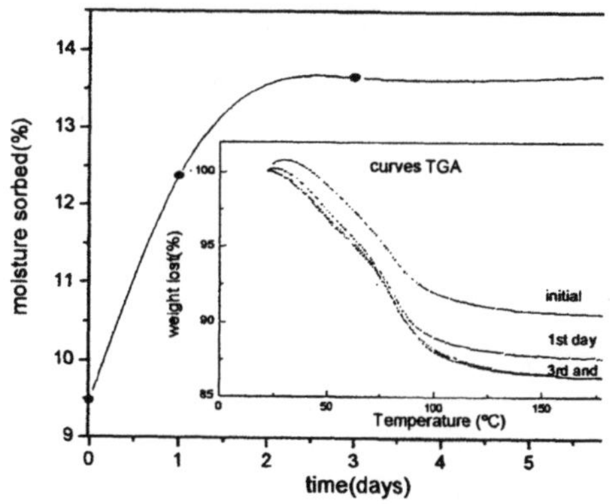

Figure 4 Moisture sorption profile and curves of TGA

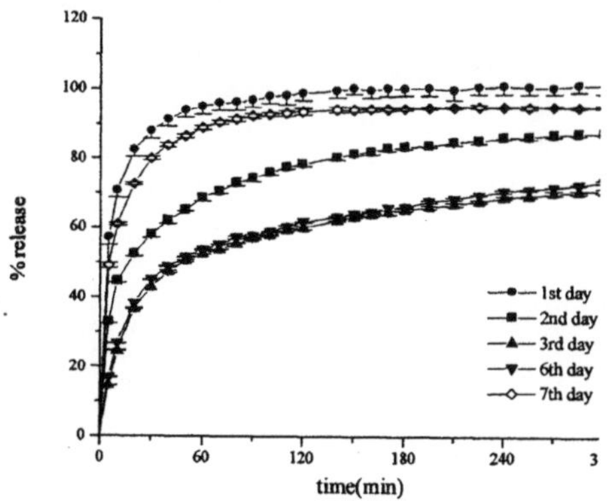

Figure 3 Dissolution profiles for TA pellets stored at room temperature for 6 days (solid symbol) and dried at 60°C (open symbol).

5. Acknowledgements

This work was supported by a grant from the Xunta de Galicia (XUGA 20301A95). We thank the Xunta de Galicia (DOG 2-XII-97) for a fellowship for MEVL.

6. References

Gazzaniga, A., Sangalli, M.E., Rillosi, M., Bruni, G., Vecchio, C., Giordano, F. 1995. Evaluation of betacyclodextrin as pelletization agent. *Proc. 1st World Meeting APGI / APV Budapest,* 371-372.

Hellen, L., Ritala, M., Ylirunsi, J., Palmroos, P., Kristoffersson, E. 1992. Process variables of the radial screen extruder: Part I-Production capacity of the extruder and properties of the extrudate. *Pharm.l Techn. Intern.,* 50-55.

Podczeck, F., Chopra, R., Newton, J.M. 1995. The use of a two-and a three-dimensional shape factor to characterize the sphericity of pellets. *Proc. 1st World Meeting APGI / APV, Budapest,* 351-352.

Stubberud, L., Arwidsson, H.G., Graffner, C. 1995. Effect of moisture on tensile strength of tablets based on microcrystalline cellulose and polyvinyl pyrrolidone. *Proc. 1st World Meeting APGI/APV, Budapest,* 161-162.

Villar-López, M.E., Nieto-Reyes, L., Otero-Espinar, F., Blanco-Méndez, J. 1998. Study of inclusion compound of Triamcinolone acetonide. *9th International Symposium on Cyclodextrins, Santiago de Compostela.*

Chapter 4

CYCLODEXTRINS IN CHEMISTRY

RENATURATION OF SDS- AND THERMALLY DENATURED CELLULASE SYSTEM BY α-CYCLODEXTRIN

B. ANDI AND R.YAZDANPARAST*, *Institute of Biochemistry and Biophysics, University of Tehran, P.O.Box 13145-1384, Tehran, Iran, Fax: 0098-021-6404680*

1. Summary

The influence of α-cyclodextrin, as a protein folding assistant, in the refolding of the cellulase system of *Trichoderma reesei*, a multicomponent protein, under aggregating conditions was studied. The enzyme was first denatured by either SDS or heat. The denatured enzyme was then diluted by the refolding buffer containing different concentrations of α-cyclodextrin. The activity of the SDS-denatured enzyme was recovered by almost 100% using α-cyclodextrin at 150 mM concentration.However, the activity of the heat-denatured cellulase system was not recovered by the same approach of artificial chaperone-assisted refolding.

2. Introduction

Many of recombinant protein pharmaceuticals are produced today by microorganisms in the form of cytoplasmic aggregates or inclusion bodies (non-native or misfolded proteins) and therefore functionally in inactive forms [1, 2]. To improve the recovery of the biologically active molecules, the protein content of the inclusion bodies is first solubilized with a denaturant. The unfolded protein is then refolded by removal of the denaturant by dilution or dialysis. However, during refolding, many proteins tend to aggregate, possibly due to the exposure of hydrophobic surfaces [2]. In that respect, a number of folding aids, such as polyethylene glycol, sugars and surfactants have been reported to prevent aggregation. Recently, a new process:"artificial chaperone-assisted refolding" has been evaluated as an effective means of preventing protein aggregation [3, 4]. This process, inspired by the mechanism of GroEL/S chaperone protein [5], consists of application of a detergent to prevent aggregation and then of a stripping agent to remove the detergent. This new technique has been applied successfully in refolding of a monomeric and a dimeric enzyme [6, 7]. The aim of this investigation was to evaluate the extent of refolding of a multicomponent enzyme system such as cellulase by the new technique.

**To whom all correspondance should be addressed.*

3. Materials and Methods

Reagents. All chemical reagents were of high quality grade and were used without further purification. Cellulase (*Trichoderma reesei*) was purchased from Merck and used without further purification. Protein concentration was determined according to the method of Lowry and the enzyme activity was measured according to filter paper method [8]. The activity was expressed in millifilter paper unit (mFPU) defined as μmol reducing sugar (glucose equivalent) produced per min.
Turbidimetric measurements. Protein aggregation was measured at 450 nm at room temperature on a Beckman MVI spectrophotometer.

4. Results and Discussion

As it is shown in Figure 1, Dilution of the native cellulase system with the refolding buffer containing 30 mM α-cyclodextrin, resulted in no changes in the absorption pattern of the protein for 20 min.However, significant changes were observed for the denatured enzyme under the same experimental conditions. Based on Figure 2 and 3, it is evident that the refolding of the SDS-denatured enzyme by the α-cyclodextrin-containing refolding buffer is totally dose dependent. Low concentration of α-cyclodextrin (30 to 90 mM) in the refolding media caused turbidity in the enzyme solution followed by precipitation of the aggregated enzyme. However, at higher α-cyclodextrin concentrations (more than 100 mM), the extent of enzyme aggregation was suppressed and no turbidity was observed. In addition, the activity recovery, with respect to the native enzyme, in the presence of α-cyclodextrin was almost 100% (Figure 3).

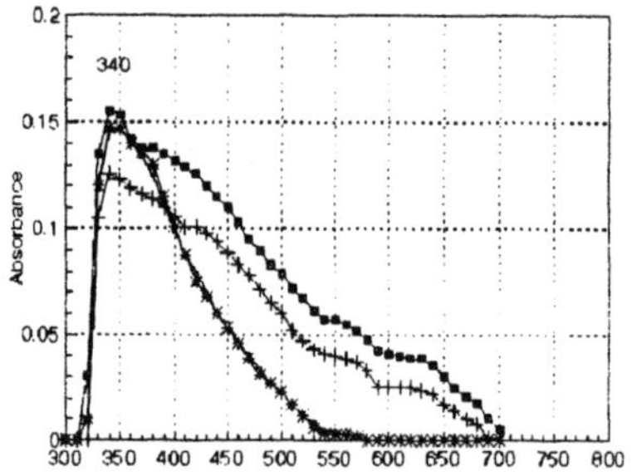

Figure 1. Absorption pattern of the native and denatured cellulase upon dilution with refolding buffer containing α-cyclodextrin (30 mM) with respect to time. The enzyme (20 mg/ml) denatured with SDS (7 mM) was diluted ten fold in citrate buffer (pH 4.8) containing α-cyclodextrin. Zero min after dilution (▫, native enzyme; +, denatured enzyme); twenty min after dilution (*, native enzyme; ■, denatured enzyme).

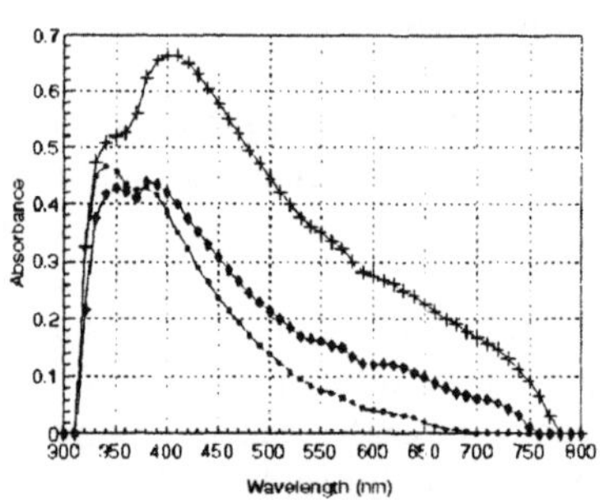

Figure 2. Renaturation of SDS-denatured enzyme in the presence of different concentrations of α-cyclodextrin, 30 min after dilution with the refolding buffer. The experimental details as in Figure 1 (enzyme 60 mg/ml). α-cyclodextrin concentration (mM): zero, ▫; 30, +; and 150, ♦.

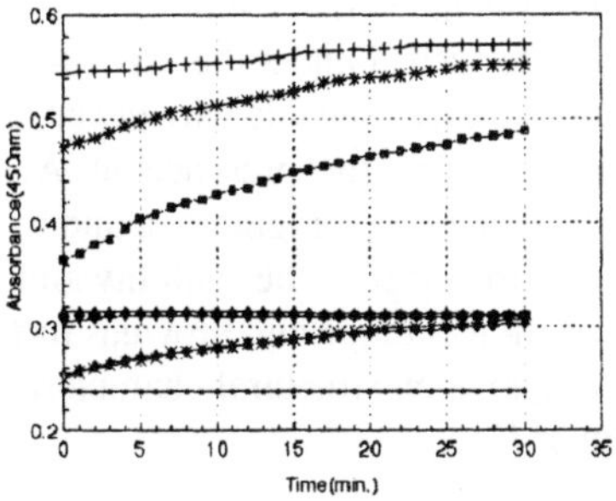

Figure 3. Renaturation of SDS-denatured cellulase system with different concentrations of α-cyclodextrin. The experimental details as in Figure 1 (enzyme 60mg/ml). α-cyclodextrin concentration (mM): zero, ▫ (denatured) and Δ (native); 30, +; 60, *; 90, ■; 120, ×; 150, ♦ (denatured) and ∇ (native).

Despite these observations, Figure 4 (b) indicate that full activity recovery by the renaturing buffer containing α-cyclodextrin (150 mM) is not achieved when the cellulase system is denatured by heat alone (Figure 4b). These observations strongly supports the interaction between α-cyclodextrin and SDS or the stripping of SDS by α-cyclodextrin during the process of refolding of this multifunctional enzyme system.

On the other hand, inspection of data presented in Figure 4 and 5 indicate that α-cyclodextrin is also capable of inhibiting the enzyme activity by almost 33%. To find out the type of inhibition caused by α-cyclodextrin, we investigated the effect of various α-cyclodextrin

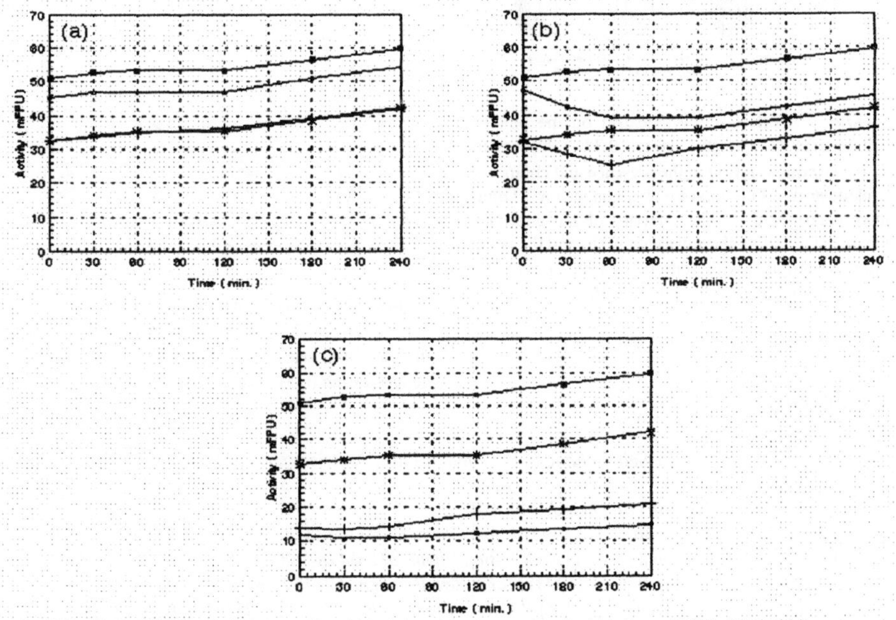

Figure 4. Effect of α-cyclodextrin on the activity regeneration of (a) SDS-, (b) heat-, (c) SDS and heat-denatured cellulase system. Experimental details as in Figure 1 (enzyme 60 mg/ml). α-cyclodextrin concentration (mM): zero, ▫ (denatured); 150, + (denature); zero, ■ (native); 150, * (native).

concentrations on the enzyme activity at various substrate concentrations. According to data presented in Figures 5 and 6, it may be concluded that α-cyclodextrin is a mixed-type (noncompetitive) inhibitor for the cellulase system under the experimental conditions used. Based on this conclusion, it becomes understandable why the enzyme inhibition by α-cyclodextrin is less (15%, Figure 5c) in the presence of Avicel compared to the 33% enzyme inhibition in the absence of Avicel (Figure 5, a and b).
In conclusion, our work provides further evidence to support the validity and the generality of the artificial chaperone-assisted refolding technique and provides further evidence concerning the independency of this technique upon structural characteristics of proteins as stated by Couthon, F. and his colleagues [7].

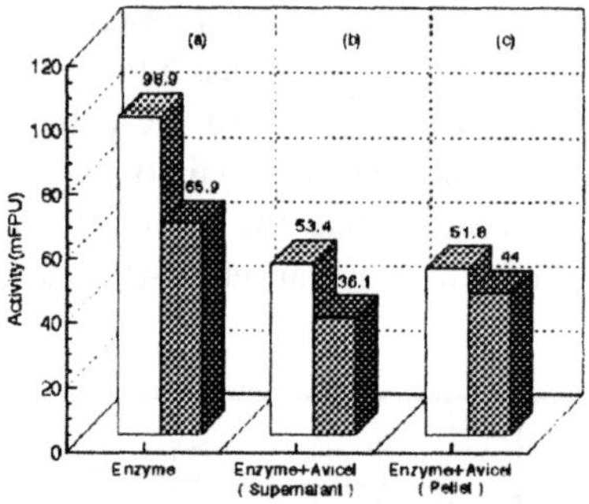

Figure 5. Effect of cyclodextrin on the binding of cellulose to its substrate. Avicel (0.3%, w/v) was added to the enzyme solution in the presence and absence of α-cyclodextrin. The mixture kept on ice for 30 min and then centrifuged. The activity of the enzyme was measured in both the supernatant (b) and the pellet (c) after desorption with citrate buffer. α-cyclodextrin concentration (mM): zero, □; 150, ■.

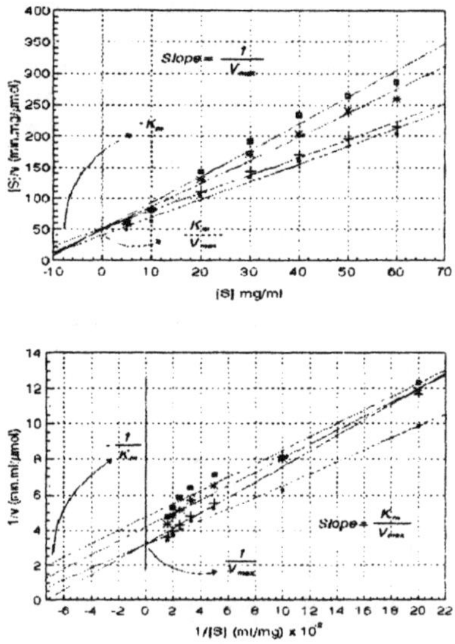

Figure 6. The inhibitory effect of different concentrations of α-cyclodextrin on the cellulase activity in the first ten min of catalysis. α-cyclodextrin concentration (mM): zero, ▫; 30, +; 60, *; 90, ■.

5. References

1. Cleland, J.L. (1993) *Protein Folding in Vivo and In Vitro*, American chemical society, Washington, D.C., pp. 1-21.
2. Marston, F. A. O. (1986) *Biochem J.* 240, 1-12.
3. Rozema, D., and Gellman, S. H. (1995) *J. Am. Chem. Soc.* 117, 2373-2374.
4. Rozema, D., and Gellman, S. H. (1996) *J. Biol. Chem.* 271, 3478-3487.
5. Hendrick, J. P., and Hartl, F. U. (1993) *Annu. Rev. Biochem.* 62, 349-384.
6. Karuppiah, N. and Sharma, A. (1995) *Biochem. Biophys. Res. Commun.* 211, 60-66.
7. Couthon, F., Clottes, E. and Vial, C. (1996) *Biochem. Biophys. Res. Commun.* 227, 854-860.
8. Wood, W. A. and Kellogg, S.T. (1988) *Methods in Enzymology*, Academic press, Sandiego, CA, pp. 92-112.

ABSORPTION AND RECOVERY OF VOLATILE CHLORINATED HYDROCARBONS BY USE OF AQUEOUS SOLUTIONS OF BRANCHED CYCLODEXTRINS

I. UEMASU, S. KUSHIYAMA, and R. AIZAWA
National Institute for Resources and Environment
16-3 Onogawa, Tsukuba, Ibaraki 305-8569, Japan

1. Introduction

We have been studying the use of aqueous solutions of branched cyclodextrins (CDs) as the absorbents of volatile chlorinated hydrocarbons such as trichloroethylene, chlorobenzene, etc.[1,2]. Since they are toxic and cause environmental pollution, it is necessary to control their emission from industrial sources.

CDs generally have a strong affinity for chlorinated hydrocarbons and unsubstituted CDs readily form solid inclusion complexes with them. The solid complexes are usually so fine and sticky that it is very difficult to handle them in a continuous process. On the other hand, branched CDs have high water solubilities and their inclusion complexes with chlorinated hydrocarbons hardly precipitate. Therefore, aqueous solutions of branched CDs are expected to be effective absorbents of chlorinated hydrocarbons. We report here the results of absorption experiments conducted using apparatus in which chlorinated hydrocarbon gas diluted with N_2 contacted aqueous solutions of branched CDs in a bottle. In addition, two ways of recovering the chlorinated hydrocarbons absorbed in the CD solution were examined.

2. Materials and Methods

2.1 Materials

Glucosyl-α-CD mixture (G-α-CD mixture) and maltosyl-β-CD mixture (M-β-CD mixture) were obtained from Ensuiko Sugar Refining Co., Ltd. (Yokohama, Japan. Their compositions are listed in Ref. 2.). Trichloroethylene, tetrachloroethylene, monochlorobenzene, diethyl ether, hexane, and heptane were purchased from Wako Pure Chemical Industries, Ltd. (Tokyo, Japan). They were all guaranteed reagents and used without further purification. Distilled and deionized water was used.

2.2. Methods

2.2.1. ABSORPTION OF CHLORINATED HYDROCARBON GAS IN A CONTINUOUS FLOW SYSTEM

The absorbing bottle used here is illustrated in Figure 1(See Ref. 2 as for the generator of chlorinated hydrocarbon gas.). The aqueous solution of a branched CD mixture (flow rate: 10ml/min) was continuously introduced into the bottom of the absorbing bottle (solution volume: 250ml), contacted chlorinated hydrocarbon gas (flow rate: 100ml/min) at 25°C, and flowed out of the upper exit of the bottle. The concentration change of the gas after contact with the aqueous solution of branched CD mixture was monitored with a GC every 3 min. For comparison, the experiments using pure water instead of CD aqueous solutions were performed, too. The removal rate(R %) was defined as follows:

$$R = 100\frac{W - C}{W}$$

where C is the concentration of chlorinated hydrocarbon after contact with CD solution and W is that after contact with pure water. The values of concentrations at quasi-stationary state were adopted.

2.2.2. RECOVERY OF CHLORINATED HYDROCARBON ABSORBED IN CD SOLUTIONS

Two ways were examined. One is heating the CD solution saturated with chlorinated hydrocarbon at 70°C, placing the bottle of the CD solution in a bath thermostated at the temperature. N_2(flow rate: 100ml/min) was continuously introduced into the solution to drive out the dissociated chlorinated hydrocarbon, which was lead into a condenser and recovered. The other way is extraction by organic solvents at room temperature. Diethyl ether, hexane and heptane were examined.

3. Results and Discussion

3.1. Absorption

Figure 2 shows the result of experiments absorbing 3,000ppm tetrachloroethylene gas. The removal rate was 44% in the case of 0.10kg/L G-α-CD mixture solution and 66% in

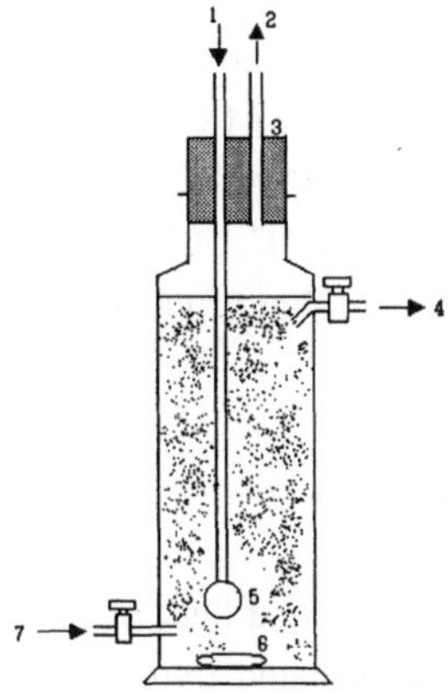

Figure 1. Absorbing Bottle
1) Chlorinated hydrocarbon gas(in)
2) to GC, 3) Silicone stopper, 4) CD solution(out), 5) Ball filter, 6) Magnetic stirrer, 7) CD solution(in)

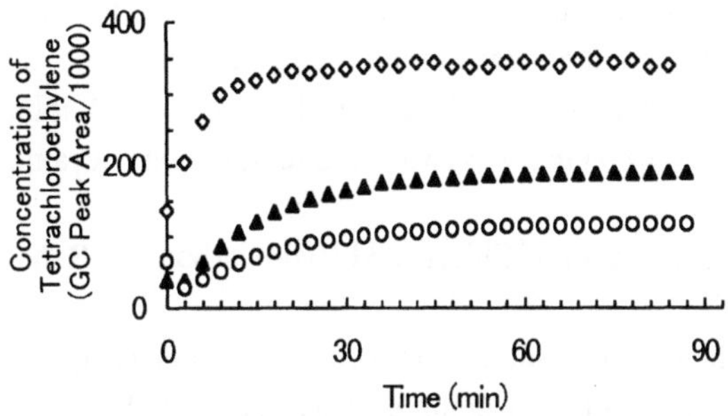

Figure 2. Concentration Change of Tetrachloroethylene Gas after Contact with G-α-CD Mixture Solution (▲), M-β-CD Mixture Solution(○) and Pure Water (◊)

the case of 0.10kg/L M-β-CD mixture solution. In the experiment where 3,000ppm gas (flow rate: 100ml/min) and 0.10kg/L CD mixture solution (flow rate: 10ml/min) were employed, the efficiency of absorption was in the order: monochlorobenzene > tetrachloroethylene > trichloroethylene, as listed in Table 1. When 500ppm trichloroethylene was introduced, the removal rate was 68% in the case of 0.10kg/L G-α-CD mixture solution and 58% in the case of 0.10kg/L M-β-CD mixture solution. The concentration of trichloroethylene after contact with CD solutions was 90ppm in the case of G-α-CD mixture solution and 140ppm in the case of M-β-CD mixture solution, respectively. The concentration increase of CD solutions improved removal or trapping of chlorinated hydrocarbons. For example, in the system of G-α-CD mixture solution and 3,000ppm trichloroethylene, the removal rate was 29% in the case of 0.05kg/L CD solution, 42% in the case of 0.10kg/L solution and 59% in the case of 0.15kg/L solution. With increasing the flow rate of CD solutions, the removal rate became higher. In the

TABLE 1. Removal Rate in Absorption of Trichloroethylene, Tetrachloroethylene and Monochlorobenzene with Aqueous Solutions of Branched CD Mixtures

	Trichloroethylene	**Tetrachloroethylene**	**Monochlorobenzene**
G-α-CD Mixture	42 %	44 %	80 %
M-β-CD Mixture	22 %	66 %	76 %

See text as for Removal Rate.

system of 0.10kg/L G-α-CD mixture solution and 3,000ppm trichloroethylene gas(100 ml/min), the removal rate was 25% at the flow rate of 5ml/min, 42% at the flow rate of 10ml/min and 51% at the flow rate of 20ml/min. When the temperature of CD solution was raised, the removal rate became low. In the system of 0.10kg/L M-β-CD mixture solution and 500ppm trichloroethylene, the removal rate was 59% at 15°C, 50% at 25°C and 43% at 35°C. When the flow rate of CD solution was kept and that of chlorinated hydrocarbon gas was twice, the removal rate became less than a half.

3.2. Recovery of Chlorinated Hydrocarbon and Renewal of CD Solution

When the CD solution absorbing chlorinated hydrocarbon was heated and kept at 70°C, the CD inclusion complex was dissociated. The dissociated chlorinated hydrocarbon can be carried out on the stream of nitrogen gas and then easily recovered through a condenser. Although water vapor also condenses, the organic layer clearly separates from water. In this way, even tetrachloroethylene and monochlorobenzene, whose vapor pressures are higher than that of water, were recovered. On the other hand, the CD aqueous solution is renewed, cooled to room temperature and then can be provided for absorption. This method, however, has a disadvantage that heating and cooling of the CD aqueous solution require energy and time.

Extraction by organic solvents can dissociate the CD inclusion complex, extract chlorinated hydrocarbons into the organic layer and renew the CD aqueous solution. All of these can be done at room temperature except for the separation of chlorinated hydrocarbons from the organic layer. Among three organic solvents examined, diethyl ether was found the most efficient and the two other solvents were inferior a little. For practical use, hexane or heptane seems to be better from the viewpoint of safety. The problem is to find a method for contacting the CD aqueous solution with an organic solvent so effectively that the CD solution can be renewed in a continuous process.

4. References

1. Uemasu, I. and Kushiyama, S. (1994) Inclusion complexation of volatile chlorinated hydrocarbons in aqueous solutions of branched cyclodextrins, *J. Incl. Phenom.,* 17, 177-185.
2. Uemasu, I., Kushiyama, S., and Aizawa, R. (1996) Capture of volatile chlorinated hydrocarbons by aqueous solutions of branched cyclodextrins, *J. Incl. Phenom.,* 25, 221-224.

OXIDATION OF HYDROPHOBIC *O*-DIPHENOLS BY LIPOXYGENASE IN THE PRESENCE OF CYCLODEXTRINS

E. NUÑEZ-DELICADO, M. SOJO, A. SANCHEZ-FERRER AND F. GARCIA-CARMONA
Department of Biochemistry and Molecular Biology-A, Faculty of Biology, University of Murcia, Campus de Espinardo, E-30071 Murcia, Spain.

INTRODUCTION

Cyclodextrins (CDs) are cyclic oligosaccharides composed of six to eight glucopyranose residues linked by α (1→4) glicosidic bonds in a cylinder-shaped structure. The most important feature of cyclodextrins is their ability to complex a variety of guest molecules into their hydrophobic cavities in aqueous solutions [2]. The physicochemical properties of these inclusion complexes have been extensively studied and several analytical techniques have been used to determine their formation constants [3]: spectrophotometry , fluorimetry, calorimetry, circular dichroism, phase solubility, etc.

In this paper, the determination of the complex formation constant using the enzymatic activity over the guest in described. This method yields highly precise constants because of the great sensitivity of the enzymatic reactions.

MATERIALS AND METHODS

Electrophoretically pure soybean lipoxygenase-L1 (EC 1.13.11.12) type V was purchased from Sigma. 4-*tert*-octylcatechol (TOC) and H_2O_2 were obtained from Aldrich and 4-*tert*-butylcatechol (TBC) was a product from Fluka. 2-Hydroxypropyl-β-cyclodextrins were kindly supplied by Amaizo (Hammond, IN). All other chemicals used were of analytical grade.

The oxidation of TBC and TOC to their corresponding quinones was followed spectrophotometrically at the absorption maximum of the quinones (400 nm) at 25 °C. The absorption coefficient of the TBC quinone used was (ε_{400}=1150 $M^{-1}cm^{-1}$) [4] and its of TOC quinone was calculated using periodate as a chemical oxidant (ε_{400}=1155 $M^{-1}cm^{-1}$). The hydrogen peroxide and lipoxygenase solutions were freshly prepared everyday and their concentrations were calculated using ε_{240}=39.4 $M^{-1}cm^{-1}$ and ε_{280}=160,000 $M^{-1}cm^{-1}$ [5], respectively.

RESULTS AND DISCUSSION

The enzymatic activity of lipoxygenase (LOX) over TBC has been used to describe an enzymatic method to determine the complex formation constant K_c between the substrate and cyclodextrins. TBC was an ideal substrate for this purpose because its aqueous solubility permitted its concentration in the reaction medium to reach enzyme saturation. Moreover, its hydrophobic side chain allowed it to form inclusion complexes with cyclodextrins.

When TBC was incubated with lipoxygenase in the presence of H_2O_2, a maximum at 400 nm was developed, which corresponds with the TBC quinone product previously described in the enzymatic oxidation of this compound by polyphenol oxidase [6]. The high aqueous solubility of TBC permitted the kinetic characterisation of the system TBC/ H_2O_2/LOX. The k_{cat} and K_M values for TBC were found to be 1522.5 min^{-1} and 1.4 mM, respectively.

In the presence of cyclodextrins in the reaction medium, a decrease in enzymatic activity was observed as cyclodextrin concentration was increased, at fixed TBC concentration (Fig. 1). Depending on the degree of enzyme saturation, the inhibition curve had a more or less sigmoidal form.

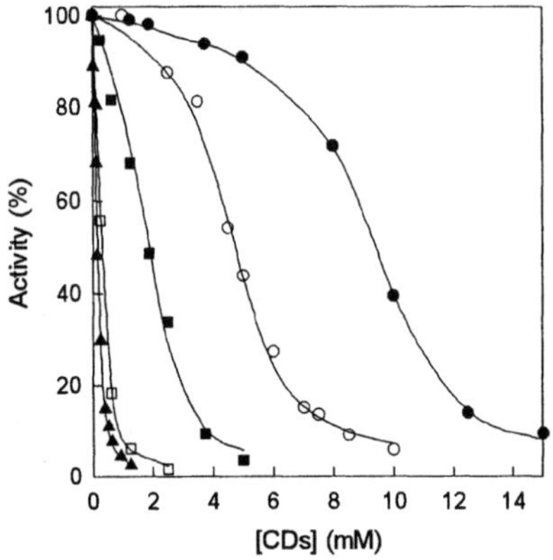

Figure 1: Effect of cyclodextrin concentration on oxidation of different diphenols by LOX in the presence of H_2O_2. The reaction medium at 25°C contained 100 mM sodium phosphate buffer, pH 7.4, 1.3 mM H_2O_2, 0.821 nM lipoxygenase and increasing concentrations of 2-hydroxypropyl-β-cyclodextrins (0-15 mM). (|) 10 mM dopamine, (=) 10 mM TBC, (O) 5 mM TBC, (■) 2.5 mM TBC, (□) 0.5 mM TBC, (▲) 0.25 mM TOC.

The inhibition shown in Fig. 1 could be explained by the formation of inclusion complexes between TBC and CDs (TBC-CD) which cannot be oxidised by LOX, unless they dissociate. Then, Michaelis-Menten velocity equation had to be expressed as a function of free TBC concentration:

$$v = \frac{V_m[TBC]_f}{K_M + [TBC]_f} \tag{1}$$

This, $[TBC]_f$ can be expressed as a function of the only two known parameters $[TBC]_t$ and $[CD]_t$, where the subscript t stands for overall compound concentration. Assuming that only one molecule of TBC may enter into a cyclodextrin molecule (stoichiometry 1:1), the equilibrium can be expressed as:

$$[TBC]_f + [CD]_f \xleftrightarrow{K_c} [TBC - CD] \tag{2}$$

where the complexation constant, K_c, is usually defined as:

$$K_c = \frac{[TBC-CD]}{[TBC]_f[CD]_f} \tag{3}$$

Taking into account the mass balance and after rearrangement of the equations, the following quadratic relationship was obtained:

$$K_c[TBC]_f^2 - ([CD]_t K_c - [TBC]_t K_c + 1)[TBC]_f - [TBC]_t = 0 \tag{4}$$

From this, $[TBC]_f$ can be obtained:

$$[TBC]_f = \left(-([CD]_t K_c - [TBC]_t K_c + 1) + \sqrt{([CD]_t K_c - [TBC]_t K_c + 1)^2 + 4K_c[TBC]_t}\right)/2K_c \tag{5}$$

and substituted in eq. 1 to give eq. 6:

$$v = \frac{V_m\left[\left(-([CD]_t K_c - [TBC]_t K_c + 1) + \sqrt{([CD]_t K_c - [TBC]_t K_c + 1)^2 + 4K_c[TBC]_t}\right)/2K_c\right]}{K_M + \left[\left(-([CD]_t K_c - [TBC]_t K_c + 1) + \sqrt{([CD]_t K_c - [TBC]_t K_c + 1)^2 + 4K_c[TBC]_t}\right)/2K_c\right]}$$

Equation 6 shows a non linear relationship between v and $[CD]_t$ as in all curves of Fig. 1. Fitting the data by non linear regression, a value of 18,000 M^{-1} was obtained for K_c. The kinetic parameters k_{cat} and K_M obtained in all cases agreed with those obtained in the saturation curve, demonstrating for the first time the validity of this method for the whole saturation curve, including the linear portion, in which only k_{cat}/K_M ratio could be calculated, using the equation 7:

$$v = {}^{V_m}\!/_{K_M}\left[\left(-([CD]_t K_c - [TBC]_t K_c + 1) + \sqrt{([CD]_t K_c - [TBC]_t K_c + 1)^2 + 4K_c[TBC]_t}\right)/2K_c\right]$$

In order to clarify whether LOX was only working with free TBC, the data of figure 1 were replotted in Fig. 2, as a function of free TBC, using eq. 5.

It is important to note that the points of the different curves in Fig. 1 which represent the same activities also have the same $[TBC]_f$. Thus, Figure 2 clearly shows that the enzyme was sensitive to free TBC, given a Michaelis-Menten representation of the data with the same values of k_{cat} and K_M as those found in absence of CDs.

Once the validity of the enzymatic method had been demonstrated along the saturation curve using TBC, another diphenolic compound was used: 4-*tert*-octylcatechol (TOC), a diphenolic compound in which the hydrophobic side chain is longer than in TBC. This compound was also oxidised by LOX in the presence of H_2O_2 and generated a stable quinone but, due to its low aqueous solubility, the enzyme could not be saturated and only the k_{cat}/K_M ratio could be calculated (638.2 $min^{-1}mM^{-1}$). When CDs were added to the reaction medium, a drastic decrease in enzymatic activity was observed as cyclodextrin concentration was increased (Fig. 1, filled triangles). Using this enzymatic method, the Kc value between TOC and CDs could be calculated in the linear portion of the saturation curve, as previously described for other low soluble substrates of LOX [7, 8]. The value obtained, 35,000 M^{-1}, indicated that TOC formed complexes with cyclodextrins easily than TBC does because of its high hydrophobic character. In this case the k_{cat}/K_M ratio obtained also agreed with that obtained in the absence of cyclodextrins, indicating the validity of Kc value obtained.

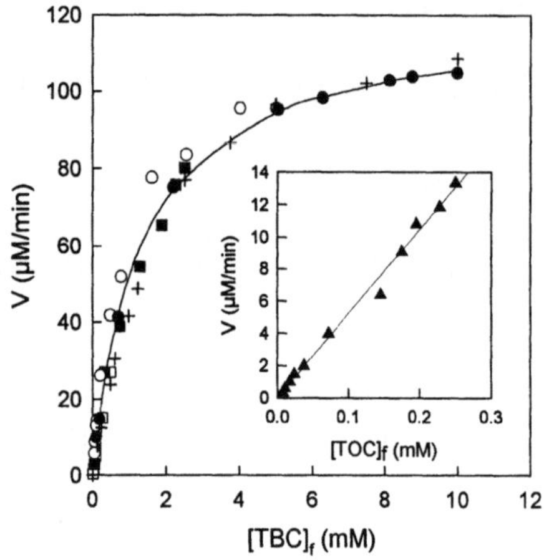

Figure 2: Effect of free TBC concentration on lipoxygenase activity. The free PMC concentrations were calculated from data shown in Fig. 1, using eq. 5 (see text for details). (+) Increasing concentration of TBC in the absence of cyclodextrins. Inset: Effect of freeTOC on lipoxygenase activity.

In addition, when the data of Fig.1 (filled triangles) where replotted into Fig. 2, inset, as a function of free TOC, a straight line was obtained. The value of the k_{cat}/K_M ratio was the same as that calculated in the absence of CDs, demonstrating that LOX was only working with free TOC, and not with the complexed (TOC-CD).

In conclusion, this work demonstrates for the first time the validity of using the enzymatic method for the whole saturation curve, including the linear portion.

ACKNOWLEDGEMENTS

This work was partially supported by DGES (MEC) (PB95-1024, PB97-1032) and Caja Murcia (Murcia, Spain). E.N.D. is a holder of research grant from Caja Murcia. M.S. is a holder of PFPI grant (MEC).

REFERENCES

1. Szejtli, J.: *Cyclodextrins and Their Inclusion Complexes*; Akademiai Kiado; Budapest, Hungary, 1982.
2. Connors, K.A.: *Binding Constants-The measurement of Molecular Complex Stability*; John Wiley & Sons; New York, 1987.
3. Núñez-Delicado, E., Bru, R., Sánchez-Ferrer, A. and García-Carmona, F. (1996) Triton X-114-aided purification of latent tyrosinase. *J. Chrom. B* 680, 105-112.
4. Nelson, D.P. and Kiesow, L.A. (1972) Enthalpy of decomposition of hydrogen peroxide by catalase at 25 °C with molar extintion coefficients of H_2O_2 solutions in the UV. *Anal. Biochem.* 49, 474-478.
5. Núñez-Delicado, E., Sánchez-Ferrer, A. and García-Carmona, F. (1997) Cyclodextrins as secondary antioxidants: synergism with ascorbic acid. *J. Agric. Food Chem.* 45, 2830-2835.
6. López-Nicolás, J.M., Bru, R. and García-Carmona, F. (1997) Enzymatic oxidation os linoleic acid by lipoxygenase forming inclusion complexes with cyclodextrins as starch model molecules. *J. Agric. Food Chem.* 45, 1144-1148.

INCLUSION PROPERTIES, METAL ION COORDINATION ABILITY AND ANALYTICAL APPLICATIONS OF TREHALOSE CAPPED CYCLODEXTRINS

V.CUCINOTTA[a], G.GRASSO[b], G.MACCARRONE[a], A.MAZZAGLIA[a], G.VECCHIO[a]

a) *Dept Scienze Chimiche, Università di Catania, viale A. Doria, 6 - 95125 Catania, Italy*

b) *ISSN, CNR, viale A. Doria, 6 - 95125 Catania, Italy*

Abstract

A new capped derivative of β-cyclodextrin was synthesized. The capping unit comprises a trehalose molecule, bridged to A and D primary groups of β-cyclodextrin by nitrogen atoms. This compound, together to a similar compound previously synthesized, constitutes a new class of compounds, which we call *hemispherodextrins*. The synthesis, the characterization and the interaction with anthraquinone-2-sulfonic acid (ACS), as well as preliminary results on the interaction with proton and with copper(II) of this compound are reported and discussed.

1. Introduction

This work regards a new class of A,D capped derivatives of β-cyclodestrins. Recent developments in this field are directed towards the use of capping units that should confer additional characteristics to these compounds with respect to parent cyclodextrins, following a criterion of complementarity. In other words, the goal is to obtain something like a symbiotic molecule coupling the properties of the cyclodextrin cavity to the capping unit properties. Following this criterion, aromatric or heteroaromatic, cyclopeptides, etc. were used by different researchers [1-5].

The idea underlying this work is different: the goal is to amplify the peculiar properties of cyclodextrins. In this respect, the capping unit can have an important role. This capping unit in A,D derivatives, like a handle of a basket, mainly develops on a plane roughly perpendicular to the cavity axis. Therefore, if we insert a saccharidic unit in the capping moiety, the resulting molecule shows a more extended saccharidic system. If the spherical cap, placed on the frustum of cone of cyclodextrin, comprises a saccharidic system, an almost hemispherical carbohydrate is obtained, which may advantageously be called *hemispherodextrin*. In this kind of host, the included molecule should be less exposed to the solvent, thus increasing the hydrophobicity of the cavity.

In the derivatives synthesized by us, the saccharidic system comprised in the cap is a α, α'- trehalose molecule, bonded to the primary groups of A and D glucopyranosinic rings by bridge groups. The heteroatoms of these bridges both increase the selectivity of these receptor, and give them the capability to coordinate metal ions.

The first example of this class of receptors was that in which the bridge group is cysteamine (CDTHCM) [6]. The study of the interaction of CDTHCM with different guests has shown its receptor characteristics. Its enantiomeric separation ability was shown in CE experiments carried out on dansyl-derivatives of aminoacids.

Here, we report about a new member of this class of compounds, having the trehalose unit directly bridged via the amino nitrogens (CDTHNH), in chart I. In principle, from a structural point of view, the much shorter bridging unit should result in a better match between the capping unit and the cyclodextrin cavity.

2. Experimental

2.1 MATERIALS AND METHODS

Commercially available reagents were used unless otherwise noted. TLC was carried out on silica gel plates 60F-254 (Merck). CD drerivatives were detected with UV light and anisaldehyde reagent. Merck Lichroprep RP-8 (40-63 μm) was used for reverse phase column chromatography . ^{1}H NMR spectra were recorded in D_2O solutions on a Varian Inova 500 spectrometer without a reference compound.

Circular dichroism spectra were recorded on a JASCO J-600 spectropolarimeter at 25°C, on freshly prepared aqueous solutions. Quartz cuvettes of 0.1 cm pathlength were used. Experiments at pH 6 were carried out in phosphate buffer solution (0.015 mol dm^{-3}, pH=6).

2.2 SYNTHESIS

Synthesis of 6A,6D-di-deoxy-6A,6D-[6,6'-diamino-α-α'-trehalose]-β-cyclodextrin (CDTHNH). 6, 6'-dideoxy-6,6'diamino-α-α'-trehalose (THDNH) [7] (0.350 g, 0.260 mmol) dissolved in anydrous DMF (20 ml) was added under stirring in 1 h to a solution of 6A, 6D-dideoxy -6A, 6D- diiodo-β-cyclodextrin ($ADCDI_2$) [8] in anydrous DMF (60 ml) under N_2 at 80 ^{0}C. After 48 h the solvent was evaporated and the resulting solid was purified by CM-Sephadex C-25 column (NH_4^+ form), using a linear gradient 0-0.2 M of NH_4HCO_3 solution as the eluent. The final product CDTHNH (0.070 g, Yield 19 %, Rf=0.12, PrOH/AcOEt/H_2O/NH_3; 5:1:3:2) was obtained ES/MS [M=1439].

3. Results

The new compound CDTHNH was synthesized by the reaction of $ADCDI_2$ with the trehalose derivative. The compound was characterized by ^1H NMR spectroscopy. The spectrum, assigned by COSY and TOCSY spectra show signals due to the two rings of trehalose (TH) show a loss of equivalence in this symmetric molecule due to the bonding to the cavity: A large diastereotopicity of the 6-H of TH together with the presence of two signals both for the two H-1 and for two H-4 of TH is evident. Also the corresponding protons of the two functionalized rings of CD show different chemical shift as a consequence of the capping reaction.

The CDTHNH has two amino groups and its protonation was investigated by potentiometry (log K_1 = 8.1, log K_2 = 7.3). By the comparison of these values with typical values for amines, it clearly appears the decrease of basicity of the amino groups, due to the interaction with the cavity. Furthermore, the proximicity of the values regarding the first and the second step of protonation precludes the possibility to have a pH range in which the monoprotonated species is prevailing.

The ^1H NMR spectrum of the diprotonated species (pH 6) appears as being more complex than that of the unprotonated species. The spectrum shows the increased asymmetry of the cavity, probably due to the formation of hydrogen bonds between the protonated nitrogen atoms and upper rim hydroxyl groups.

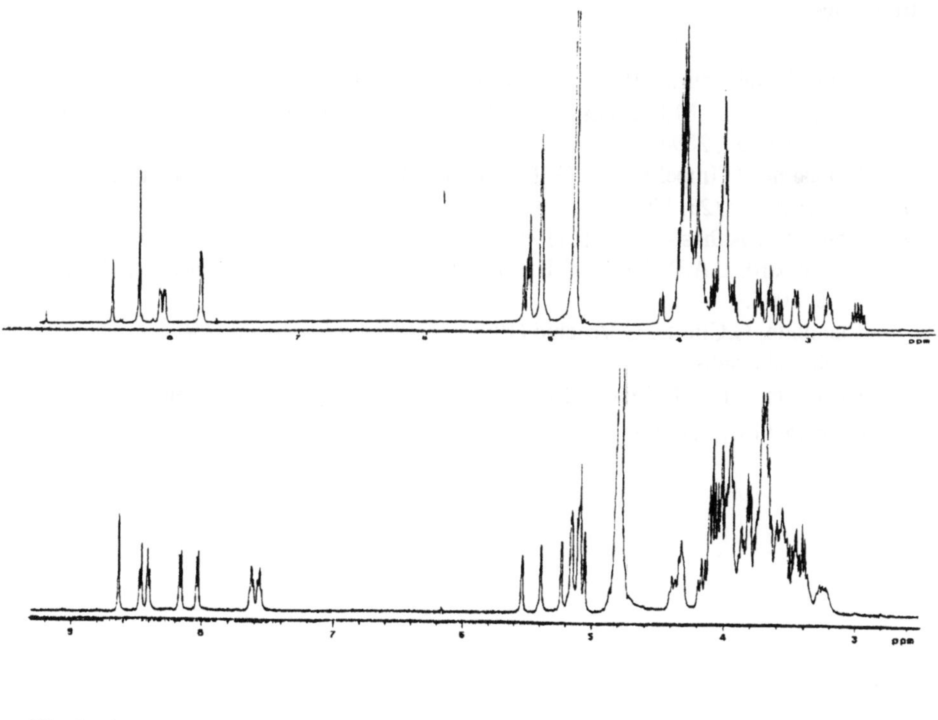

Fig.1 – ^1H NMR spectra of CDTHNH – ACS system, in D_2O, at 500 MHz; a) pH = 9; b) pH = 6

The ^{1}H NMR spectra of the CDTHNH-ACS (anthraquinone-2-sulfonic acid) in D_2O (pH=9 and pH=6) have been carried out (see figures 1a, 1b). The effect of inclusion is evident particularly at pH 6 where a low field shift can be seen. Interestingly, TH proton chemical shifts are also influenced, suggesting the deep inclusion of ACS within the cavity. The ROESY spectrum at pH 6 shows correlations between the 1-H and 3-H of the ACS and the 3-H and 5-H region of the CDTHNH, suggesting an orientation of the ACS in the cavity probably due to the electrostatic interaction with the protonated amino groups. The ROESY spectrum at pH 9 shows correlations between the 3-,5-H region of the cavity with all the protons of the ACS, suggesting that this ring is included in the CD cavity but an equilibrium between different orientations can be suggested.

The c.d. spectra of CDTHNH-ACS system at pH =6 and pH=9 show an induced cotton effect (positive at about 260 nm) in both the cases which increases when the ACS concentration increases. These spectra suggest axial inclusion of ACS in the cavity.

Preliminary results regarding the system formed by CDTHNH with copper(II) ion suggest the formation of two different species, i.e. the simple complex with the ligand (1:1:0 species) and an hydroxylated one (1:1:-1 species) which is prevailing in a large pH range. In this species, an additional bond of the ligand by a deprotonated primary hydroxylic group might be present.

4. References

1. T.Kato, Y. Nakamura, (1988) *Heterocycles, 27*, 973
2. M.F.Acquavella, M.E.Evans, S.W.Farraher, C.J.Nevoret, C.J.Abelt (1994) *J.Org.Chem.*, 59, 2894.
3. R.P.Bonomo, G.Impellizzeri, G. Pappalardo, E. Rizzarelli, G. Vecchio, (1993) *Gazz. Chim. It., 123*, 593
4. A.Ueno, T. Takahashi, T. Osa, (1981) *J.Chem.Soc., Chem.Commun.*, 94.
5. B. K.Hubbard, L.A.Beilstein, C.E.Heath, C.J.Abelt., (1996) *J.Chem.Soc., Perkin Trans 2*, 1005.
6. V.Cucinotta, G.Grasso, G.Vecchio (1998) *J.Inclusion Phenom., 31 43.*
7. unpublisced results
8. V.Cucinotta, F. D'Alessandro, G. Impellizzeri, G. Vecchio, (1992) *Carbohydr.Res.*, 224, 95.

COVALENT AND NON-COVALENT ADDUCTS BASED ON CYCLODEXTRINS AND POLY(ETHYLENEOXIDE)

I. N. Topchieva

Department of Chemistry, Lomonosov State University,
119899 Lenin Hills, V-234, GSP-3 Moscow. Russia.

1. Introduction

Supramolecular chemistry as a subject has developed rapidly in the last two decades and is now an interdisciplinary branch of chemistry which has reached a high degree of diversity and complexity. Synthesis of supramolecular structures is based on the principle of molecular recognition and molecular self-assembly. Las time the attention is drawn to a new topological type of supramolecular structures comprising a sequence of cyclodextrin (CD) molecules threaded onto polymer chains, "molecular necklaces" [1]. The stability of the complexes depends on the complementarity between the internal cavity of CD and the thickness of the "guest"macromolecule. Thus, poly(ethylene oxide) (PEO) forms crystalline complex with six-membered α-CD, whereas poly(propylene oxide) (PPO) - with seven-membered β-CD. A step forward in this direction was made by using diverse PEO-containing compounds namely, PEO-PPO block copolymers (Proxanols) [2], linear and branched non-ionic detergents [3] and star-shaped supermolecules (conjugates based on proteins and PEO)

2. Results and Discussion

2.1. COMPLEXES OF PEO-PPO BLOCK COPOLYMERS WITH CYCLODEXTRINS

We used diblock copolymers (PEO-PPO, with molecular masses from 2 lo 6 kDa. Fig.1 gives a schematic representation the complexes obtained by selective interaction of α- or β-CD with PEO or PPO blocks of copolymers. CD content in complexes consistent with the stoichiometric ratio of one CD molecule per two monomer units. The

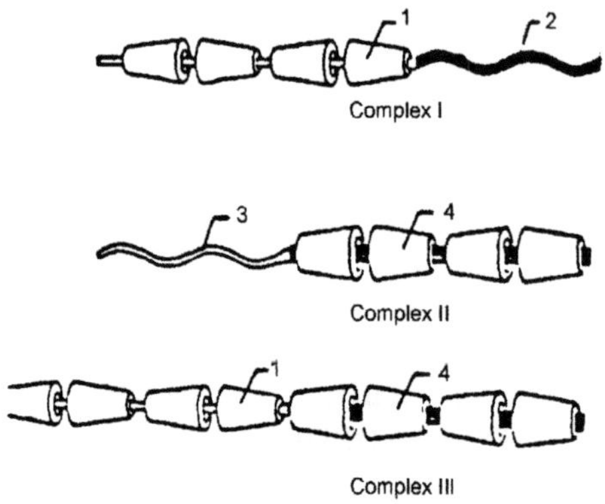

Figure 1. Complexes based on PEO-POP copolymer
1. α-CD; 2. PPO; 3. β-CD; 4. PEO

composition of the complexes is independent of the CD/copolymer ratio, which indicates the stoichiometric nature of the complexes. These complexes represent novel block copolymers, with a rigid block formed by a molecular necklace, and a flexible one, by a free polymer. Note, that PEO blocks can crystallize, whereas PPO blocks have an amorphous structure. X-ray power diffraction studies of complexes 1 and 2 showed that both are crystalline compounds differing from each other and the initial components by crystal lattice type. The X-ray powder diffraction pattern of complex 2 does not exhibit reflections typical of PEO, and the pattern of complex 1 does not contain an amorphous halo typical of PPO.

Therefore we have shown that the formation of (α- or β-CD complexes of PEO-PPO copolymers changes morphological properties not only of the block interacting with CD, but of the free PAO block as well. It seems interesting to study the interaction between diblock copolymers. With the mixture of α-CD and β-CD as host molecules. The X-ray pattern of the resulting precipitate shows only the well-pronounced reflections characteristics of molecular necklaces based on β-CD. One may anticipate that either β-CD preferentially interacts with PPO block of the copolymer or both CDs form inclusion complexes in which only one type of molecular necklaces is able to crystallize. To solve this question, the complex was destroyed, and its composition was analyzes using HPLC technique. It was show that the reaction mixture contains both types of CDs, that testifies the formation of the complex 3 (Fig. 1). Thus, the similarity between the crystalline structure of this copolymer and the structure of molecular necklaces based on β-CD suggest that only one type of molecular necklaces with a higher molecular volume is able to crystallize, whereas molecular necklaces based on α-CD are unable to produce crystalline phase because of the steric reasons.

2.2 COMPLEX BETWEEN NONIONIC SURFACTANTS AND CYCLODEXTRINS

Using of PEO-surfactans insted of PEO and PEO-PPO block copolymers shows that PEO blocks of non-ionic surfactants may play a role of the filaments for the construction of molecular necklaces based on α-CD. For this propose we used a number of PEO-

PEO-surfactants including Tritons. PEG-1000-monostearate, Brif 35 and Tween 60. Mixing of PEO-surfactants with α-CD results in the formation of a white crystalline precipitate. The crystal structure parameters and composition of the complexes formed are similar to those of the complexes obtained using PEO. Important information on the character of interaction of surfactant with α-CD and its localization is provide by studies of the effect of CD on the crystal micellization concentration (CMC). The bend on the plot of the CMC versus the CD/surfactant molar ratio allows to determine the stoichiometric composition of the complex (Fig. 2). Composition of the crystalline complex and the complex in solution coincide.

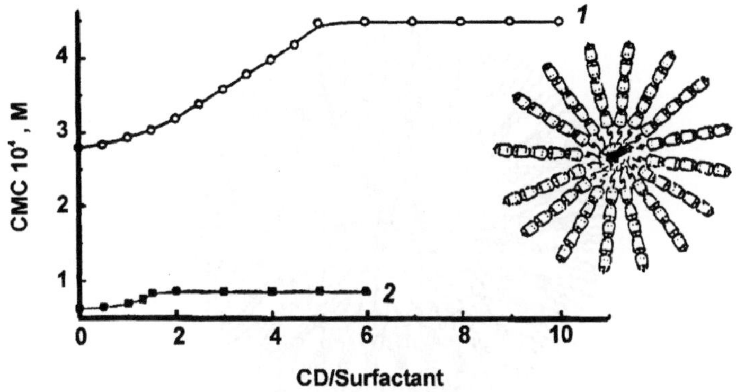

Figure 2. The plot of CMC against CD/surfactant molar for aqueous solutions of (1) Triton X-100 and (2) Triton X-45.

It is know that molecules of γ-CD are able to be threaded onto two chains of PEG or PEO-PPO block copolymers [1]. Mixing aqueous solutions of two components leads to the precipitation of a crystalline compound. Crystalline structure of complexes is similar to those of the complexes obtained from PEO and γ-CD. The composition of complexes corresponds to one γ-CD molecule per four ethylene oxide units. The stoichometric ratio of complex in solution was determined from the plot of CMC value against the molar ratio CD/surfactant. Stoichiometric complexes present novel amphiphilic compounds containing double stranded complex as hydrophylic part and double tailed hydrophobic part.

2.3. DENDRITIC STRUCTURES BASED ON PEO-DERIVATIVES OF CDs AND PEO-STAR-MOLECULES

The interaction between PEO-surfactant and 2,3,6-derivatives of CDs (PEO-CD) [4] was studied by measuring of the surface tension of surfactant in the presence of PEO-CD. The structure of micelles based on modified surfactants resembles dendrimer (Fig. 3).

Similar type of dendritic structures were obtained from star-shaped PEO-protein conjugates and PEO-CDs.

Figure 3. Schematic structure of the complex between PEO-surfactant and PEO-CD.

3. References

1. Harada A., Li J. and Kamachi M. (1993) Preparation and properties of inclusion complexes of poly(ethylene glycol) with α-cyclodextrin, *Macromolecules,* **26**, 5698-5703.

2. Topchieva, I.N., Gerasimov V.I., Panova I.G., Karezin, K.I. and Efremova, N.V. (1997) Nanostructures based on poly(ethylene oxide)-cyclodextrin complexes, Polymer Science, **A 40**, pp.310-318

3. Topchieva I.N., Karezin K.I., Panova I.G. and Gerasimov V.I. (1997) New type of micellar species prepared by molecular self-assembling of cyclodextrins and non-ionic surfactants, Doklady Phys.Chein., **355,** pp.233-236.

4. Topchieva I.N., Polyakov V.A., Elezkaya S.V., Bystrizky G.I. and Karezin K.I. (1997) One-pot synthesis of cyclodextrins, modified with poly(ethylene oxide), Polymer Bull., **38**, pp.359-364.

THE EFFECT OF CYCLODEXTRINS ON THE HYDROLYSIS OF BENZYL HALIDES

Francesco Trotta and Davide Cantamessa
Dipartimento di Chimica Inorganica, Chimica Fisica e Chimica dei Materiali dell'Università.
Via Pietro Giuria 7, 10125 Torino - Italy.

Abstract

Carrying out the hydrolysis of benzyl halides in the presence of suitable amount of cyclodextrins or cyclodextrin derivatives, higher reaction rates were reached in the production of the corresponding benzyl alcohol. No detectable amount of symmetric ethers were observed.

1. Introduction

Reactions involving reagents soluble in non miscible phases proceed only with low reaction rate. As the reactions between organic compounds and inorganic anions are very common, the problem becomes very important since these latter are soluble preferentially in water.
Phase Transfer Catalysis (PTC)[1], as it is well known, represents the better way to carry out under mild reaction conditions many reactions between molecules soluble in water and compounds soluble in a immiscible organic phase. Up to now thousands papers and patents have expanded the use of PTC to a wide range of reactions and processes. In any case the PTC catalyst (usually a onium salt or a crown ether) allows to transfer the activated anion in the organic layer.
On the contrary, by working under Inverse Phase Transfer Catalysis (IPTC)[2]conditions, the organic molecule is transferred in the aqueous phase where the reaction takes place.
Recently cyclodextrins (non reducing, stable and non toxic cyclic oligosaccharides,) were successfully used as IPTC catalysts.
Particularly interesting results were obtained in the oxidation of terminal alkenes[3] (Wacker process) and in the carboxylic acid esters hydrolysis reaction.[4]
The hydrolysis of the alkyl halides (Reaction 1) to form the corresponding alcohol is a

1) 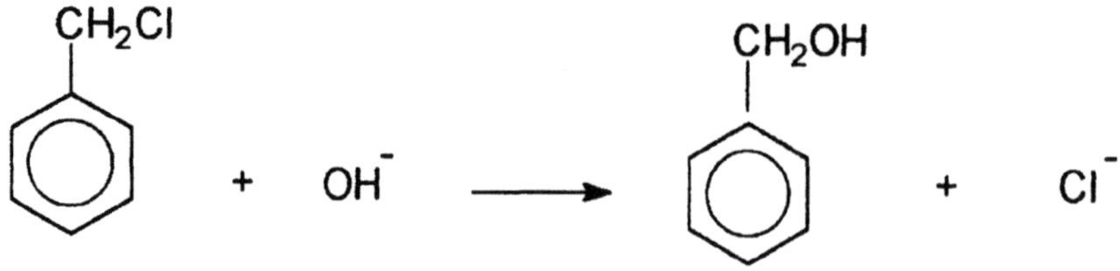

well known reaction in organic chemistry, but industrially it is of negligible importance since, generally, alkyl halides are obtained starting from the parent alcohol.[5] The main exception is represented by the synthesis of benzyl alcohol carried out industrially hydrolysing benzyl chloride being this latter easily obtained from chlorination of toluene.[6]
Because benzyl chloride is insoluble in water, the reaction rate is very low. On the other hand under classical PTC conditions the hydrolysis of benzyl halides is fast, but due to the strong anionic activation, the reaction leads to the formation of the corresponding symmetric ethers.[7]
We have found that carrying out the hydrolysis of benzyl chloride or benzyl bromide in the presence of cyclodextrins or cyclodextrin derivatives, the reaction proceeds faster under milder reaction conditions, providing exclusively benzyl alcohol as reaction product. No organic solvent is required.

2. Materials and Methods

β-cyclodextrin was kindly supplied from Roquette-Italia (Cassano Spinola - Italy). Other cyclodextrins were gifted by Wacker-Chemie (Germany). All solvents and reagents employed (ACS grade) were bought from Aldrich (U.S.A) and used without further purification. GC-MS analysis were performed on a HP 5890A spectrometer. Magnetic stirring rate was measured with a Cole-Palmer 08199 phototachometer. Reaction environment was thermostated with a Lauda RC6 thermostat.
In a typical experiment to 30 ml of a K_2CO_3 solution (20 % p/v) the required amount of catalyst was dissolved. Then 0.1 ml of benzyl chloride was added and the mixture allowed to react at 50°C and 500 rpm. Once the reaction was over, the reaction mixture was extracted with diethyl ether and the organic phase analysed to GC-MS. Definitive confirmations were obtained by comparison with authentic samples.

3. Results and Discussion

Since alkyl halides are insoluble in water, their hydrolysis is a reaction extremely low even under drastic stirring of the biphasic system.
The use of PTC catalysts increases the reaction rate, but leads to the formation of the corresponding ethers.

Benzyl halides hydrolysis reaction could be profitably carried out in the presence of cyclodextrins as IPTC catalysts.
An accepted possible mechanism is sketched in Figure 1. The cyclodextrins are able to form stable inclusion compounds with the benzyl halide. This latter could be transferred in aqueous phase where the reaction takes place. Actually the cyclodextrin acts as co-catalyst being the hydroxyl anion the true catalyst in the hydrolysis reaction.
Table 1 reports some data obtained in the hydrolysis of benzyl chloride in the presence and without the cyclodextrins as IPTC catalyst. It can be note that by adding 10 wt% of cyclodextrin to the reaction mixture, higher reaction rate were detected. In particular the methyl-β–cyclodextrin acts as better catalyst in comparison with the parent β-cyclodextrin.
This fact could be ascribed to its higher solubility in both phases. As a consequence it is a better phase transfer catalyst of the organic molecule in aqueous layer.
The formation of the inclusion compound between the CD and the benzyl halide seems to be a prerequisite for the catalysis. In fact, by using linear oligosaccharides such as maltose, it was observed only a moderate enhancement in the reaction rate.
Because the hydrolysis of benzyl halide leads to quite lypophilic benzyl alcohol having a low water solubility, greater amount of the IPTC catalyst, in comparison with other IPTC reactions, is required to show significant enhancement in the reaction rate.

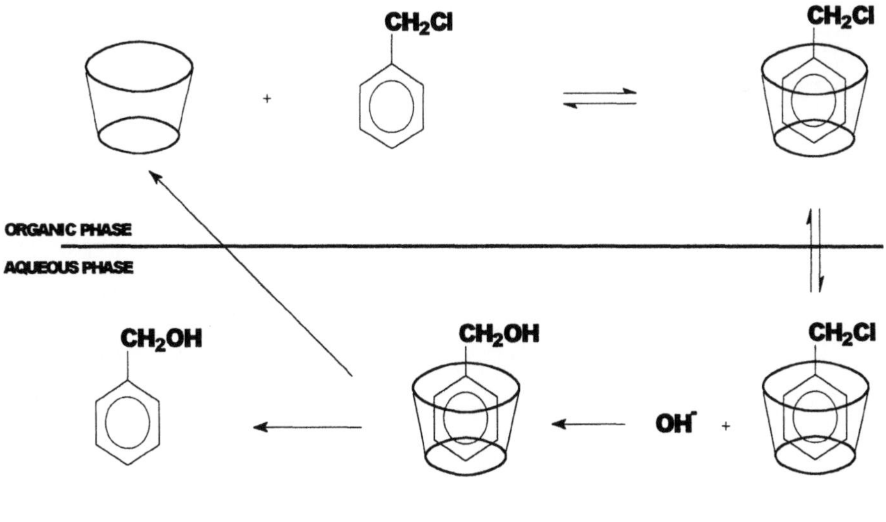

Figure 1 : Possible mechanism for the hydrolysis of benzyl chloride in the presence of cyclodextrin.

Catalyst	Time (h)	Conversion % Catalyzed (a)	Uncatalyzed (b)	a/b ratio
Metyl-ß-cyclodextrin 10%	4	85,9	26,1	3,3
ß-cyclodextrin 10%	4	42,4	26,1	1,6
maltose 10%	4	36,9	26,1	1,4

Table 1 – Hydrolysis of benzyl cloride under IPTC conditions

References

1. F. Montanari, D. Landini and F. Rolla, Top Curr. Chem., 101, 14-200 (1982)
2. L. J. Mathias and R.A. Viadya, J. Am. Chem. Soc., 108, 1093 (1986)
3. E. Monflier et al., Tetrah. Lett., 36, 387 (1985)
4. F. Trotta, G. Moraglio and A. Rapposelli, J. Inclusion Phenom., 20, 353 (1995)
5. Kirk-Othmer "Concise Encyclopedia of Chemical Technology" p.275 Wiley (1985)
6. Kirk-Othmer "Concise Encyclopedia of Chemical technology" p. 163 Wiley (1985)
7. E. V. Demhlow "Phase Transfer Catalysis" p. 182 Verlag (1983)

REGIOSELECTIVE SUBSTITUTION OF CYCLODEXTRINS WITHOUT USING PROTECTING GROUPS

László **JICSINSZKY**

CYCLOLAB R&D. Lab. Ltd., H-1525 BUDAPEST, P.O.B: 435, XI. Dombóvári út 5-7 HUNGARY

Summary

The different reactivity of the primary C(6)-OH and the secondary C(2)-OH and C(3)-OH groups offer the possibility of regioselective substitution without using protecting group(s). The general principles for substitution rules are summarized together with the general substitution rules of the most common cyclodextrin derivatives. Due to the competitive,, reactions on the various hydroxyls, these rules are only informative and their validity is restricted to low degrees of substitution (DS). Quantum chemical tools were used for the explanation of regioselective substitution of cyclodextrins on a simplified model, Michael addition of acrylamide to methyl glucopyranoside. The results of theoretical calculations are in accordance with the ones observed at the preparation of carboxyethylated cyclodextrins.

Introduction

Regioselective substitution is always a big challenge for a synthetic chemist particularly when ¡t involves such a large number of active groups as in case of cyclodextrins. For cyclodextrin hydroxyls the number of really applicable protecting groups is limited. In most cases the otherwise good reagents result in mixtures of regioisomeric products and their purification requires usually at least one chromatographic step, which limits the batch size, and usually this virtual regioselectivity is reported in the literature. On the other end of publications the several tenth percent fragments of the substitution pattern are reported.

It is known that in moderately basic solutions the hydroxypropylation of cyclodextrins prefers the secondary hydroxyls [1]. Based on this observation some publications (e. g. [2]) declared the absolute and unique (and false) strategy for functionalization of the 2-position of CDs involving deprotonation of CD by sodium hydride followed by a nucleophilic attack of the resultant CD oxyanion, the hardly believable simplicity of this method forced the researchers to study the substitution pattern of (2-hydroxy)propylated cyclodextrins. Carboxymethylation reaction runs as usual: the substitution pattern of the product show similarities to that of the corresponding celluloses using the standard alkylation procedures. There are two principal ways for the preparation of the carboxyethylated cyclodextrin derivatives: a) the classical reaction of 3-halogeno propionic acid (salt); b) Michael addition of cyclodextrin to an activated double bond (acrylonitrile or acrylamide). Extensive study of substitution pattern revealed that even at very low concentration of base, some substitutions occur at the primary hydroxyls [3,41 as in case of alkylation of cellulose. Low concentration of base and low temperature favors the secondary side substitution using an epoxide in the alkylation reaction. Sometimes a high regioselectivity can be achieved by technical tricks, e. g. slow addition of base, or reversed addition of reagents and reactants, while in other cases these knacks have no effects on regioselectivity. In some cases "abnormally" high regioselectivities were observed despite the applied out-of-rule conditions. It was found that these 'irregularities" has also some common features: these reactions are more under electronic controls than the usual nucleophile substitutions. A typical electronic factors' controlled reaction is the Michael addition that can be successfully utilized in the preparation of some widely used cyclodextrin derivatives, like cyanoethyl-, aminopropyl-, carbamoylethyl-, carboxyethyl-cyclodextrins with low DS.

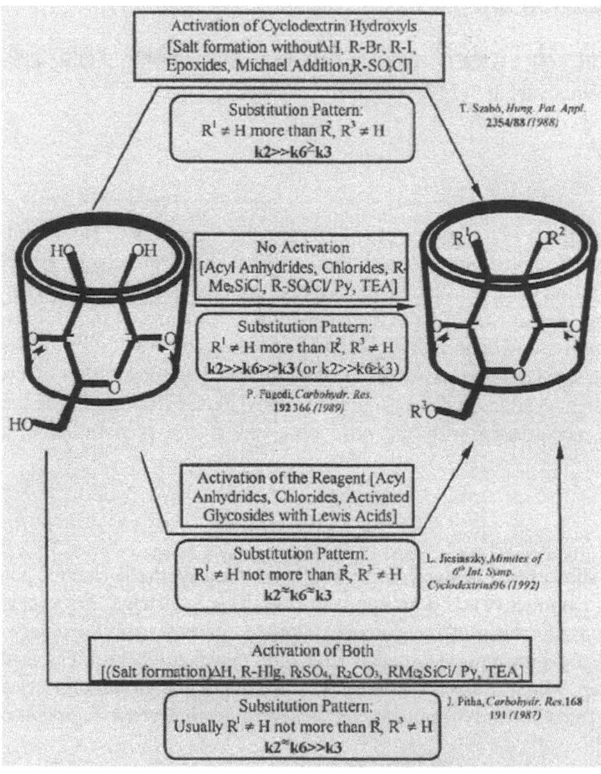

***Fig.* 1.:** The general substitution roles of cyclodextrin derivatives having low DS

The possible mechanism and the energetic relations of the Michael addition were studied on a simplified model. The assumed reaction paths resulted in characteristic substitution feature of methyl (α-D-Glc*p* with acrylamide: O(2) > (O6) > O(3). The obtained results for methyl α-D-Glc*p* substitution explains the observed high regioselectivity in the preparation of cyanoethyl-, carbamoylethyl-, carboxyethyl-, and partly the (2-hydroxy)propyl-cyclodextrins (which is, although, not Michael addition, but electronically governed nucleophil addition).

Table 1. *In Vacuo* Heat of Formation (Kcal/mole)[1] of the Different Cyclodextrins

	O(2)	O(3)	O(6)
Non-ionized α-cyclodextrin		-1410.0	
α-cyclodextrin nonanion (ΔH)	-1398.6 (11.4)	-1391.8 (18.2)	-1391.8 (18.8)
Non-ionized β-cyclodextrin		-1654.4	
β-cyclodextrin nonanion (ΔH)	-1642.4 (12.0)	-1639.6 (14.8)	-1638.4 (16.0)
Non-ionized γ-cyclodextrin		-1898.7	
γ-cyclodextrin nonanion (ΔH)	-1886.1 (10.6)	-1870.1 (26.6)	-1872.7 (24.0)
pK_a (α,- β-, and γ-cyclodextrins) [7]	12.1-12.7 **α-CD**	12.2-12.8 **β-CD**	14-15 **γ-CD**
ΔH exp [7]	8.4	10.0	11.2

Table 2. Heat of Formation (in *Vacuo*) of the Reaction Fragments [Kcal/mole]

	O(2)	**O(3)**	**O(6)**
Non-ionized		-304.1	
Anion	-302.4	-301.9	-296.3
Non-ionized (hydrated [external], non ionized methyl-α-D-Glpx*ps*		-1034.6	
Anion (hydrated [external], non ionized methyl-α-D-Glpx*ps*	-792.8	-792.4	-791.9
TS (min) [1]	-160.6	-140.3	-146.6
TS (max1) [2]	-114.0	-92.8	-98.1
TS (max2) [3]	-49.3	-25.5	-36.2
Product	-336.8	-337.0	-339.2

[1] TS (min): minimal energy in the transition state
[2] TS (maxl): maximal energy in the transition state before the reaction takes place
[3] TS (max2): maximal energy in the transition state before the molecule relaxes.

In theoretical calculations the following presumption were used [61:

- Methyl (α-D-Glcp is suitable for modelling of cyclodextrin glucopyranoside units [5]
- HOMOs of the reagent must have similar (same) symmetry as the LUMOs of the reactant with minimal movements of atoms to reach the TS
- The molecular orbitals (MOs) must preserve their symmetry in the transition state (TS) and MOs of the product have similar symmetry as in the TS
- Minimal energy differences between the HOMO of reagent (CD/MeGlcp anion) and LUMO of reactant (acrylamide)
- Usually the product must be in lower energy state than the TS

Experimental

Preparative:1 mole of the cyclodextrin is dissolved in 2000 cm 3 of water containing 6 mole sodium hydroxide. At room temperature 5 mole of acrylamide is added. The solution is stirred for 24 hrs, then refluxed for 2 hrs. Ethanol addition to the concentrated reaction mixture results in the precipitation of the crude product. The precipitate is dissolved in water, treated with 1.5 kg strong ion-exchanger (H^+-form) (pH=> 2.5-4). Filtration, clarification with 5 w% charcoal and freeze drying result in the product containing 3-4 carboxyethyl groups on one cyclodextrin ring (average DS:~3-4), and >90 % of substituents are located on the secondary hydroxyl side. Yields: αCD:75 %; βCD: 75%, γCD: 80 %.

Computational. - HyperChem® 5.1Pro [8], P-100 (HyperCube Inc., Gainsville, Florida, USA).

Results

The general substitution rules of cyclodextrins in various substitution/addition reactions are summarized on *Fig.* 1. **Table 1** summarizes the heat of ion formation of cyclodextrins. **Table 2** and *Fig. 2.* summarize the energetic relations of the assumed reaction path of Michael addition of acrylamide to methyl a-D-Glc*p*.

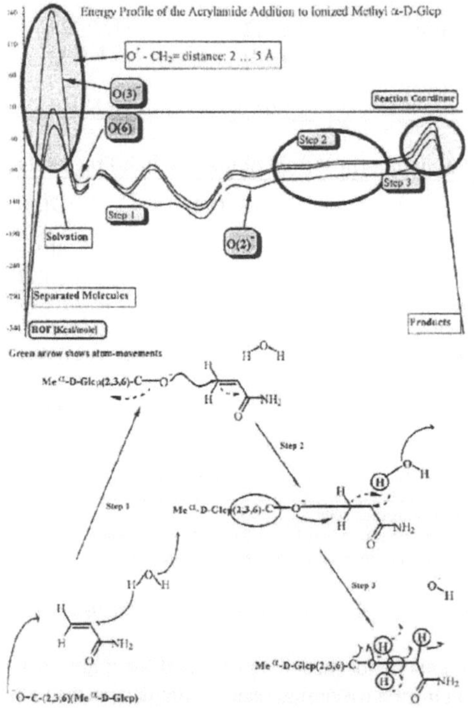

Fig. 2.: Reaction path of the acrylamide addition to methyl a-D-Glc*p*

References

[1] Tóth, A., Horváth, G., Komár, P., Peterdi, V., Szabó, T., Szejtli, J., *HU 202'889 (1988)*

[2] Rong, D., D'Souza, V. T., (1990), A convenient method .... *Tetrahedron Lett.*, **31** (30), 4275-8

[3] Lindberg, B., Pitha, J., *WO 9'012'035 (1990)*

[4] Pitha, J., Rao, C. T., Lindberg, B., Seffers, P., (1990), Distribution of substituents in 2-hydroxypropyl ethers of cyclomaltoheptaose, *Carbohydr. Res.*, **200**, 429-35

[5] Bakó, I., Jicsinszky, L., (1994), Semiempirical calculations .... *J Incl. Phenom. Mol. Recogn.*, 18(3),275-289

[6] Epiotis, N., **Theory of Organic Reactions,** *Springer Verlag,* 1978

[7] Van Hooidonk, C., Groos, C. C., Model studies for..., *Rec. Trav. Chim. Pays-Bas,* **89** 845 (1970), Gelb, R. I.; Schwartz, L. M.; Laufer, D. A., Acid dissociation of.... *Bioorg. Chem.* 9 299, 9 450 *(1980)*, **11**, 274 *(1982)*

MOLECULAR IMPRINTING SYSTEM USING CYCLODEXTRIN

TETSUYA TANABE and AKIHIKO UENO*
Department of Bioengineering, Faculty of Bioscience and Biotechnology,
Tokyo Institute of Technology, 4259 Nagatsuta-cho, Midori-ku, Yokohama
226-8501, Japan

Abstract

Cyclodextrins (CDs) form inclusion complexes with various organic molecules and the selectivity for the guests have been determined. We report here the change of the selectivity using a newly developed molecular imprinting method. The template molecule is included in the CD cavity and the template molecule is imprinted by the reaction of dicarbonyl chlorides with the template-CD complexes.

1. Introduction

In nature many molecular recognition systems have high selectivities, as enzyme, antibody etc. Particularly the antibody's system is interesting because it has controlled selectivity to various antigens. This work was performed to test the molecular imprinting CD to approach the artificial antibody system. Several investigations have been conducted in making high selectivity polymer by use of molecular imprinting method [1-2]. But the molecular imprinting system using host molecule such as CD has not been reported with some exceptions [3].

On this basis, we have attempted to produce a simple system in this study. We adopted naphthalene, 1-methoxynaphthalene, 2-methoxynaphthalene and 2-ethoxynaphthalene as the guest molecules (Fig. 1), and then 1-naphthaleneacetic acid and 2-naphthaleneacetic acid which have similar shape and high solubility as the template molecules.

2. Method

The method of molecular imprinting is schematically drawn in Fig 2. β-CD and the template molecules were dissolved in phosphate buffer (pH 9.0) at room temperature, and then added malonyl chloride, succinyl chloride and glutal chloride to the stirred buffer solution. The reaction mixture was poured into acetone. The crude product was dissolved in pure water, and then the solution was desalted with a Asahikougyo Co. Ltd. S1 system equipped with an AC-220-20 until conductivity of the solution become constant. The desalted solution was poured into acetone, and then the product was collected on a membrane filter and dried.

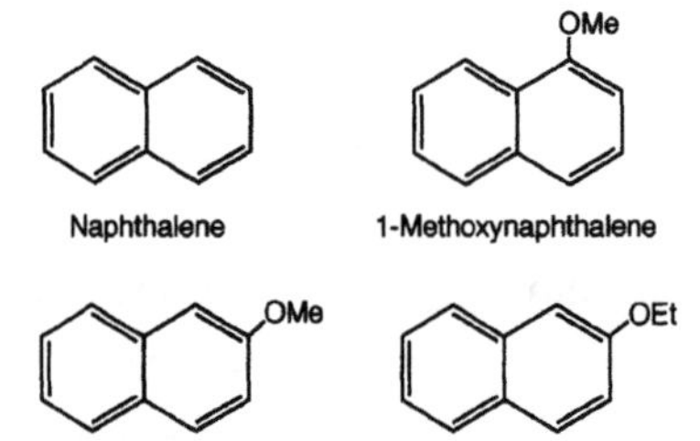

Fig. 1: The structure of guest molecules.

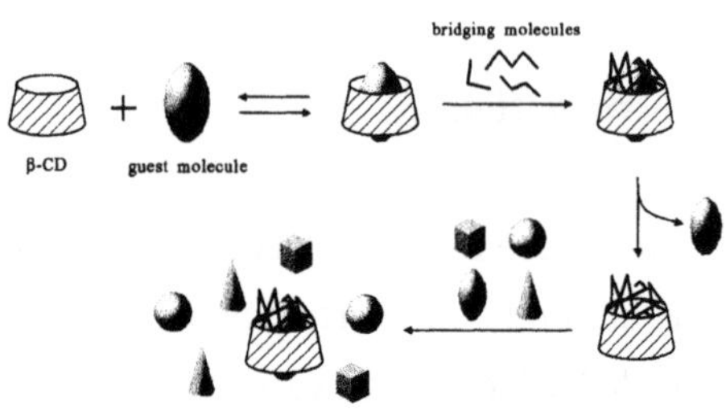

Fig. 2: The method for producing imprinted β-CD derivatives.

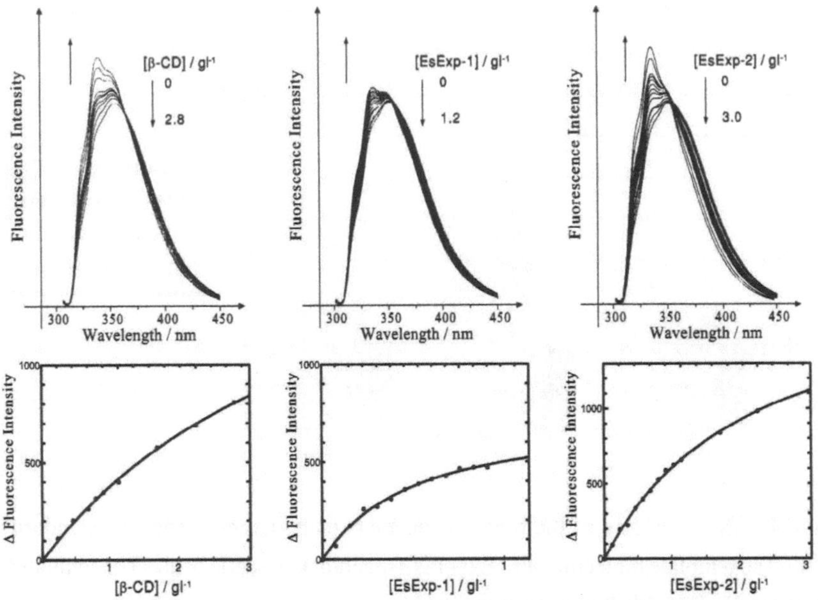

emission wavelength 278nm, phosphate buffer pH. 7, KIO^3 20 mM

3. Results and discussion

The CDs imprinted by 1-naphthaleneacetic acid and 2-naphthaleneacetic acid were defined as EsExp-1 and EsExp-2, respectively. The estimation of the selectivity in imprinting has been done by the fluorescence intensities enhanced by the host molecules. The change of the intensity increases with increasing host molecules. We analyzed the variation based on the 1:1 complex model. Fig. 3 is the spectra of 1-methoxynaphthalene and CDs. Each spectrum has an isoemissive point at 365 nm and varies in the same manner. In the case of the other guests, the spectral intensity also increases with increasing concentration of the host molecules.

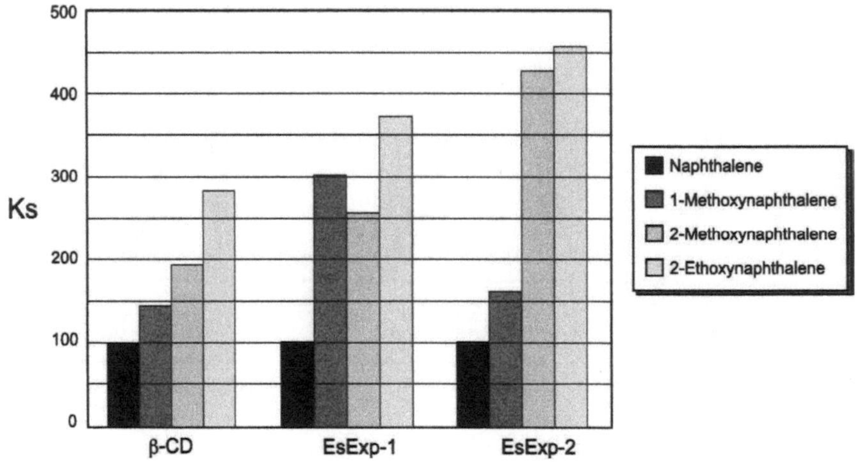

Figure 4: Ks is the ratio of the binding constants using naphthalene as a standard (100). The template molecule of EsExp-1 is 1-naphthaleneacetic acid. The template molecule of EsExp-2 is 2-naphthaleneacetic acid.

Then, we defined Ks, which is the ratio of the binding constant using naphthalene as a standard (100), as the selectivity parameter. Fig. 4 shows the values of Ks for β-CD and imprinting molecules (EsExp-1, EsExp-2), indicating that the imprinted β-CD derivatives have guest selectivities different from those of β-CD, this result reflecting the shape of the imprinting molecule in each case.

4. References

[1] T.Takeuchi and J.Matsui (1996) Molecular imprinting: an approach to "tailor-made" synthetic polymers with biomimetic functions, Acta Polymer., 47, 471-480.

[2] Borje Sellergren (1997) Noncovalent molecular imprinting:antibody-like molecular recognition in polymeric network materials, trends in analytical chem., 16, 6, 310-320.

[3] H.Asanuma, M.Kahazu, M.Shibata, T.Hishiya and M.Komiyama (1997) Molecularly imprinted polymer of β-cyclodextrin for the efficient recognition of cholesterol, Chem. Commun.,1971-1972.

ENERGY TRANSFER IN NAPHTHALENE-CONTAINING ROTAXANE AND POLYROTAXANE

MAKIO TAMURA, DE GAO and AKIHIKO UENO
Department of Bioengineering, Faculty of Bioscinece and Biotechnology, Tokyo Institute of Technology, 4259 Nagatsuta-cho, Midori-ku, Yokohama 226, Japan

1 Introduction

A rotaxane is a member of supermolecues in which macrocyclic component is threaded by a linear molecular chain bearing bulky stoppers at its ends. Because of its singular structure, there are a lot of reports about rotaxane and polyrotaxane [1,2]. In particular, cyclodextrins have been widely used in the synthesis of rotaxane and polyrotaxane since many organic molecules can be included in their cavities[3,4,5]. Although the structure of rotaxanes and polyrotaxanes have a potential to be scaffolds to make functional supermolecues, this type of research has begun only recently[6,7,8]. In nature, excitation energy transfer plays an important role in photosynthesis in which the antenna units containing a large number of chromophoric groups absorb the incident light and transfer the excitation energy to reaction centers in photosystems I and II[9]. In recent years, many artificial molecular devices have been constructed to harvest solar energy to mimic the natural antenna systems[10,11,12]. There have been attempts to use polymeric systems with suitable chromophores to harvest solar energy radiation. The rod-like architecture may also be used as a suitable framework for the support of donor-acceptor functionality[13]. Therefore, we designed the rotaxane **1** and attempted to construct an energy transfer system as the basic research for the study of artificial energy harvesting systems using rotaxane and polyrotaxane. The theoretical analysis was then applied to calculate the energy transfer ratio and the distance between naphthalene and dansyl units.

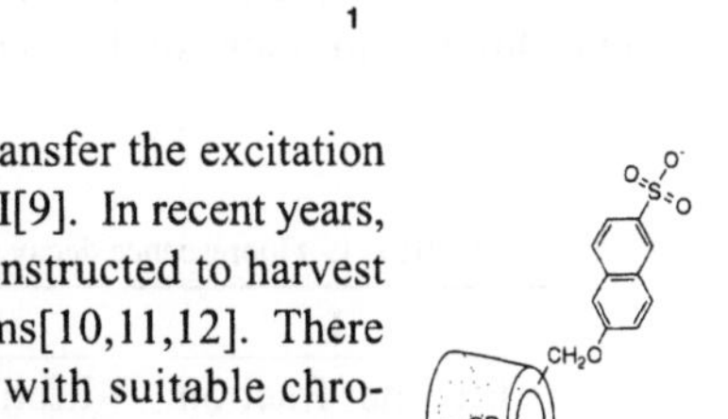

1

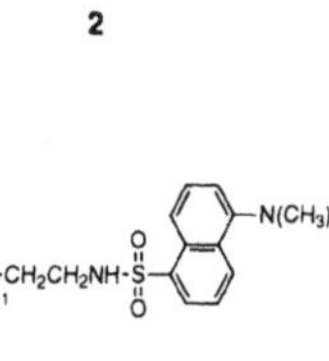

2

3

2 Materials

Synthesis of **2** was performed by a reaction of sodium 6-hydroxy-2-naphthalenesulfonate and 6-O-(2-naphthalenesulfonyl)-α-CD in dimethylsulfoxide with sodium hydride. The mixture of **2** (100 mg, 84.8 μmol) and ω,ω'-diamino-poly(ethylene glycol) (DP=12, 3.84 mg, 7.07 mmol), which had been prepared according to the procedure reported by Harada et al., was allowed to react with dansyl chloride in aqueous buffer solution at pH 9.5, and formed rotaxane **1** was separated and purified by column chromatography with Sephadex G-15 (yield 6.6%). Dansyl-terminal poly(ethylene glycol) (**3**) was also prepared by a reaction of ω,ω'-diamino-poly(ethylene glycol) and dansyl chloride. The identification of **1** was performed by ^{1}H-NMR (500-MHz) and UV-vis absorption spectra.

3 Results and Discussion

3-1 FLORESCENCE LIFETIME MEASUREMENT

Table 1 shows the lifetimes of **1**, **2** and **3**. The two lifetimes 5.7 ns and 10.4 ns of **2** suggest that the naphthyl moiety is located in two different environments. In the longer lifetime species, the naphthyl moiety may interact with the mouth of the hydrophobic cavity while in the shorter lifetime species the naphthyl moiety may be located in the bulk water environment apart from the CD unit. It is interesting that **3** has two lifetimes 3.8 ns and 13.0 ns in spite of the fact that the two dansyl moieties are located at both ends of one linear chain. It might reflect that poly(ethylene glycol) takes a folded conformation in which the dansyl is involved in the hydrophobic region of the folded chain or exposed to bulk water outside of the folded structure. On the other hand, rotaxane **1** exhibits four lifetimes 0.9, 3.6, 7.1, and 13.1 ns. Among them, 3.6 ns and 13.1 ns may be attributed to the lifetimes of the dansyl moiety. On this basis, we may attribute other lifetime species to those of the naphthyl unit quenched by energy transfer to the dansyl units. These two cases might be related to the different locations of the CD ring in **1**.

TABLE 1. Fluorescence decay parameters of **1**, **2** and **3** in H_2O at 25 ℃[a].

1		**2**		**3**	
lifetime (ns)	fraction (%)	lifetime (ns)	fraction (%)	lifetime (ns)	fraction (%)
0.9	48.1	5.7	53.8	3.8	44.4
3.6	21.5	10.4	46.2	13.0	55.6
7.1	20.1				
13.1	10.3				
χ^2=1.04		χ^2=1.09		χ^2=1.10	

[a]Concentrations of **1**, **2** and **3** are 12 μmol dm^{-3}. λex=276 nm; λem≧320 nm.

3-2 THEORETICAL ANALYSIS

Fluorescence lifetime without quenchers (τ) and with quencher (τ') are measured as follow

$$\tau=1/(k_r+k_{rf}) \quad (1)$$

$$\tau'=1/(k_r+k_{nr}+k_T) \quad (2)$$

k_r is radiation transition kinetics from S_1 state to S_0 state, k_{nr} radiationless transition kinetics and k_T is the energy transfer kinetics. Eenergy transfer efficiency(E_T) is can be obtain by equaion

$$E_T=k_T/(k_r+k_{nr}+k_T)=1-(\tau'/\tau) \quad (3)$$

In the rotaxane **1**, it is satisfactory to consider that 0.9 ns and 7.1 ns are lifetimes shortened by the energy transfer from 5.7 ns and 10.4 ns of **2**, respectively, so the corresponding energy transfer efficiencies are ca. 80 % and ca. 30 %. Förster has developed a quantitative expression for the rate of energy transfer due to dipole-dipole interactions. The efficiency of Förster-type energy transfer is usually expressed in terms of a critical radius R_0, the distance of separation of the donor and acceptor at which the rate of intermolecular energy transfer is equally the sum of the rates for all other donor deexcitation processes.

$$R_0 = 8.798\times10^{-25}\kappa^2\phi_D n^{-4} J \quad (4)$$

n is the refractive index of the solvent, κ^2 is an orientation factor which equals 2/3 for random distribution of donor and acceptor molecules. A value of 1.33 was used for refractive index. ϕ_D is the quantum yield of donor emission in the absence of the acceptor. Quantum yield of sodium 6-methoxy-2naphthalenesulfonate was 0.20 and this value was used for ϕ_D[14]. J is a spectral overlap integral normalized for the excitation coefficient of the acceptor.

$$J=\int F_D(\nu)e_A(\nu)\nu^{-4}d\nu \quad (5)$$

ν is the wave number, $F_D(\nu)$ is the spectral distribution of the donor emission in quanta normalized to unity, $e_A(\nu)$ is the molar extinction coefficient for acceptor absorption. Using the information above, we obtained a critical radius of 17.5 Å for the interaction of naphthalene unit and dansyl units. The efficiency for energy transfer by Förster mechanism can be related to actual distance R of donor and acceptor by the expression

$$k_T=\left(\frac{R_0}{R}\right)^6\left(\frac{1}{\tau}\right) \quad (6)$$

With these theory, actual distance of naphthalene unit and dansyl units can be estimated by following equation

$$\frac{1}{1-E_T}=1+\left(\frac{R_0}{R_1}\right)+\left(\frac{R_0}{R_2}\right) \quad (7)$$

R_1 is the distance of one side and R_2 is the distance of the other side. If the efficiency for energy transfer is 80 %, the smaller value of R_1 and R_2 can be evaluated about 13.5 Å and if the efficiency is 30 %, the smaller value of R_l and R can be evaluated about 22.0 Å. It is assumed that rotaxane have two dominant conformation, one may be globular conformation and the other may be longitudinal.

4 Conclusion

We construct the energy transfer system using rotaxane **1**. In this study, relationship between the efficiency for energy transfer and the conformation in the rotaxane **1** has been explored as to investigate the potentiality of rotaxane as the scaffold for construction of energy harvesting systems. Further study for using polyrotaxnae is necessary. This promising field of research has just begun.

References

[1] Cardenas, D. J., Gavina, P., Sauvage, J.-P., (1997) Construction of interlocking and threaded rings using two different transition metals as templating and connecting centers: catenanes and rotaxanes incorporating Ru(terpy)2-units in their framework, *Journal of the American Chemical Society*, 119, 2656-2664.

[2] Cardenas, D. J., Gavina, P., Sauvage, J.-Pierre., (1996) A rotaxane with two Ru(terpy)2 derivatives as stoppers, *Chemical Communications*, 16, 1915-1916.

[3] Ogino, H., (1981) Relatively high-yield syntheses of rotaxanes. syntheses and propreties of compounds consisting of cyclodextrins threaded by α,ω-diaminoalkanes coordinated to Cobalt(III) complexs,*Journal of the American Chemical Society*, 103, 1303-1304.

[4] Wenz, G., Keller, B., (1992) Threading cyclodextrin rings on polymer chains, *Angew. Chem. Int. Ed. Engl*, 31, 197-199

[5] Harada, A., Li, J., Kamachi, M., (1992) The molecular necklace: A rotaxane containing many threaded α-cyclodextrins, *Nature*, 356, 325-327.

[6] Linke, M., Chambron, J.-C., Heitz, V., Sauvage, J.-P., (1997) Electron transfer between mechanically linked porphyrins in a [2]Rotaxane, *Journal of the American Chemical Society*; 119; 11329-11330. (Communication)

[7] Murakami, H., Kawabuchi, A., Kotoo, K., Kunitake, M., Nakashima, N., (1997) A light-driven molecular shuttle based on a rotaxane, *Journal of the American Chemical Society*; 119; 7605-7606. (Communication)

[8] Ooya, T., Yui, N., (1996) Synthesis and characterization of biodegradable polyrotaxane as a novel supramolecular-structured drug carrier, *Journal of Biomaterials Science - Polymer Edition*, 8, 437-438.

[9] McDermott, G., Prince, S. M., Freer, A. A., Hawthornthwaite-Lawless, A. M., Papiz, M.Z., Cogdell, R.J., Lsaacs, N.W., (1995) Crystal structure of an integral membrane light-harvesting complex from photosynthetic bacteria, *Nature*, 374, 517-521

[10] Richard, W. W., Jonathan, S. L., (1994) A molecular photonic wire, *Journal of the American Chemical Society*; 116, 9759-9760. (Communication)

[11] Gian, M. Stewart., Marye, A. F., (1996) Chromophore-labeled dendrons as light harvesting antennae, *Journal of the American Chemical Society*; 118, 4354-4360. (Communication)

[12] Krasnovsky Jr, A A., Bashtanov, M. E., Drozdova, N. N., Liddell, P. A., Moore, A. L., Moore, T. A., Gust, D., (1997) Porphyrin and pyropheophorbide phosphorescence in synthetic molecules that mimic photosynthetic triplet energy transfer, *Journal of Photochemistry and Photobiology - Chemistry Section*, 102, 157-162.

[13] Aric, B.-H., Joseph, K., Raoul, K., (1997) Dendrimers as controlled artificial energy antennae, *Journal of the American Chemical Society*; 119, 6197-6198. (Communication)

[14] David, M. G., James, E. G., (1993) Synthesis and photophysics of a water-soluble, naphthalene-containing β-cyclodextrin, *Journal of the American Chemical Society*; 115, 5970-5974.

KINETIC STUDY OF THE OXIDATION OF LINOLEIC ACID BY LIPOXYGENASE IN PRESENCE OF β-CYCLODEXTRIN

J.M. LÓPEZ-NICOLÀS[(1)], R. BRU[(2)], J.M. LÓPEZ-ROCA[(1)] AND F. GARCÍA-CARMONA[(3)]

[(1)] E.T.S. Ingenieros Agrónomos. Department of Food Technology. Campus of Cartagena. University of Murcia. Spain. [(2)] Department of Agrochemistry and Biochemistry. Faculty of Science. University of Alicante. Spain. [(3)] Department of Biochemistry and Molecular Biology. Faculty of Biology. University of Murcia. E-30080. Spain

1. INTRODUCTION

Lipoxygenases (LOX; linoleate: oxygen oxidoreductase, EC 1.13.11.12) are a group of enzymes that catalyze the dioxygenation of polyunsaturated fatty acids (PUFAs) containing one or more 1,4-*cis,cis*-pentadiene systems to conjugated hydroperoxy fatty acids (1). In solution, PUFAs exhibit an aggregation beheviour that, depending on the pH, may lead to the formation of true micelles or merely to a dispersed oil phase. At high pH (above 9), the aggregates form true transparent micellar solutions whilst the preparations of PUFA at neutral pH are turbid suspensions. Most LOXs display optimum activity at pH values around neutrality (1) but in these conditions PUFAs are quite insoluble. For this reason the kinetic properties of LOX are well known at pH 9-10 (2) and practically unknown at neutral or slightly acid pH values. The ability of cyclodextrins (CDs) to enhance the aqueous solubility of many different compounds (3) makes them an ideal alternative system for performing enzyme-catalyzed conversions of poorly water-soluble substrates compared with organic media or two-phase system. In a previous study (4) the complexes between β-CD and PUFAs were shown to have a stoichiometry of 2:1 (β-CD: PUFAs). In this study the oxidation of linoleic acid entrapped in β-CD by lipoxygenase has been characterized and a model for enzyme catalysis in a CD medium is proposed.

2. MATERIALS AND METHODS

Linoleic acid (LA) was purchased from Cayman chemical Co. (Paris, France). β-CD was obtained from Sigma (Madrid, Spain). Lipoxygenase (type V) from soybean were obtained from Sigma (Madrid, Spain). 5-Lipoxygenase was purified from potato tubers (var. Desireé) (27μmole/min/mg protein). The complexes LA-β-CD were prepared by dissolving β-CD in 0.1 M potassium phosphate buffer pH 6.3 containing 1%v/v EtOH, followed by addition of fatty acid prepared in the same buffer. The samples were flushed with N_2 to prevent oxidation of LA during preparation. LOX activity was assayed by monitoring the increase in absorbance at 234nm of the forming hydroperoxides. The reaction was started by adding 5 μL LOX to 1 ml of LA-β-CD complex.

3. RESULTS AND DISCUSSION

3.1. KINETIC MODEL FOR THE ENZYMATIC OXIDATION OF LA BY LOX IN THE PRESENCE OF β-CD

The oxidation of PUFAs by LOX in presence of β-CD is based in the scheme of enzyme catalysis in the presence of cyclodextrins previously proposed by us (5,6) (Scheme 1).

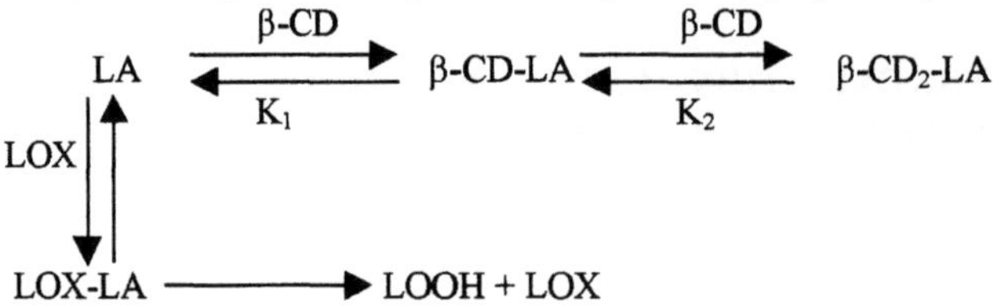

SCHEME 1. Scheme of oxidation reaction of LA by LOX in the presence of β-CD forming 1:2 complexes with LA

Thus, the free LA is the only form of substrate that LOX is able to bind and thus to react. Accordingly, the reaction rate equation should be:

$$V = \frac{V_{max} * LA_f}{K_m + LA_f} \qquad (1)$$

Considering the mass balance and that the concentration of the complex is negligible with respect to the total concentration of β-CD, the non-complexed LA can be expressed as:

$$LA_f = \frac{LA_t}{1 + K_1 * \beta CD_t + K_1 * K_2 * \beta CD_t^2} \qquad (2)$$

where subscript f and t stand for free and total respectively. Carrying out the substitution of equation (2) into equation (1), we have:

$$V = \frac{V_{max} * LA_t}{K_m^{app} = K_m * (1 + K_1 * \beta CD_t + K_1 * K_2 * \beta CD_t^2) + LA_t} \qquad (3)$$

This kinetic model predicts a quadratic dependence of the reciprocal of the enzyme activity with respect to β-CD concentration.

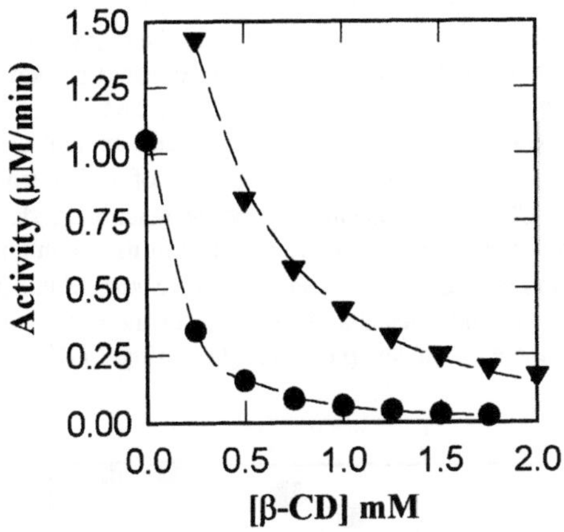

Figure 1. *Dixon Plot of soybean-LOX catalyzed LA oxidation at pH 6.3 in the presence of β-CD at different LA concentrations: (●) 7μM; (▼) 55.2 μM.*

As shows Figure 1 these expectations were experimentally fulfilled and as a consequence, the equilibrium constants can be determined by n.l.r. of these data to the double reciprocal of equation 3 using the previously determined kinetic parameters as constant coefficients. The parameters which best fit the data in fig.1 are K_1 =11.2±1.2 and K_2=1.7 ± 0.4 mM^{-1}

3.2. " THE CYCLODEXTRIN ASSAY"

On the basis of Scheme 1, it is expected that when the concentration of effective free substrate is constant ,the enzyme activity is constant too, independently of the total substrate concentration. To ascertain this, an experiment called the "cyclodextrin assay" was designed in which, by using the equilibrium constants, LA_t is calculated so that LA_{free} remains constant, independently of the β-CD concentration used. Consequently, the enzymatic reaction rate should be constant regardless of total β-CD and LA concentrations. The equlibrium constants determined as above using soybean LOX at pH 6.3 were applied to perform the β-CD assay with potato LOX since them must be the same regardless of which enzyme is used in their determination. As shown in Table 1, activity is independent of the total LA concentration, thus proving our expectations concerning the mechanism of inhibition: the apparent inhibition of LOX activity by β-CD is due to depetion of effective substrate by complex formation. In absence of β-CD the few soluble substrate is quickly exhausted so making the study of the enzymatic reaction difficult. The total concentration of soluble substrate can be increased in the presence of cyclodextrin at the time the free substrate concentration can be set through the equilibrium constants. The reaction rate holds constant for longer periods of time since the substrate converted is replenished from the CD-LA pool. This leads to a longer duration of linear product accumulation (DLPA) (Table 1) and, thus, the analysis of the enzymatic reaction is improved.

TABLE 1. Potato-LOX catalyzed oxidation of LA reaction at pH 6.3. The β-CD and LA concentrations were calculated to yield 7 μM free LA using the set of constants determined. Data of activity, DLPA and lag were obtained from the curves of reaction progress.

$[LA]_t$ (μM)	$[\beta\text{-CD}]_t$ (mM)	Activity (μM/min)	DLAP (seg)	Lag (seg)
7	0	3.30	248	17
79.05	0.5	3.24	460	50
139.81	0.75	3.32	553	62
217.07	1	3.35	777	95

3.3. COMPLEXES BETWEEN HYDROPEROXIDES AND β-CYCLODEXTRIN

Table 1 also shows the length of the characteristic lag phase of the dioxygenation of PUFA catalyzed by LOX. Such lag phase is shortened and even eliminated by addition of trace amounts of the hydroperoxide product in the reaction media and the achievement of maximum reaction rate depends on the accumulation of a certain amount of product (2). The fact that the lag phase increases with the concentration of β-CD may be due to competition between the free β-CD and the enzyme to bind the reaction product. As shown in Figure 2, for a given concentration of product there is a shift in its maximum absorbance wavelength of a few nanometers as β-CD concentration is increased. Molar absorption is practically unchanged. This result indicates that LOOH-β-CD interaction occurs and it might be responsible for the increase in lag phase length during LOX reaction in the presence of β-CD.

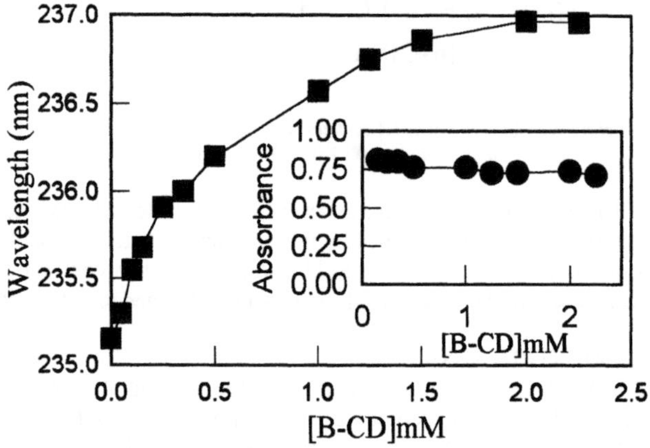

Figure 2. Effect of β-CD concentration on the maximum wavelength of complexes LOOH-β-CD. Inset: Effect of β-CD concentration on complexes LOOH-β-CD absorbance at 236nm.

4. REFERENCES

1. Vliegenthart, J.F.G. and Veldkin, G.A. (1982) Lipoxygenases. In *Free Radicals in Biology*, Pryor, W.A., De.; Academic Press: New York; Vol. 5, pp 29-64.

2. Schilstra, M.J., Veldkin, G.A., Verhagen, J. and Vliegenthart, J.F.G. (1992) Effect of hydroperoxide on lipoxygenase kinetics. *Biochemistry* 31, 7692-7699.

3. Saenger, W. (1980) Cyclodextrin inclusion compounds in research and industry. *Angew. Chem. Int. De. Engl.* 19 344-362.

4. López-Nicolás, J.M., R. Bru, Sánchez-Ferrer, A. and F. García Carmona (1995). Use of soluble lipids for biochemical processes: linoleic acid-cyclodextrin inclusion complexes in aqueous solutions. *Biochem J.* 308, 151-154.

5. Lòpez-Nicolàs, J.M., Bru, R. and Garcia-Carmona, F. (1997) Kinetic characteristics of the enzymatic conversion in presence of cyclodextrins: study of the oxidation of polyunsaturated fatty acids by lipoxygenase. Biochim. Biophys. Acta 1347, 140-150.

6. Lòpez-Nicolàs, J.M., Bru, R. and Garcia-Carmona, F. (1997) Enzymatic oxidation of linoleic acid by lipoxygenase forming inclusion complexes with cyclodextrins as starch model molecules. *J. Agric. Food Chem.* 45, 1144-1148.

5. ACKNOWLEDGEMENTS

This work has been supported in part by research grants from CICYT (Project 1FA97-0197) and ALFA (Tecmusa 6.0259.8). J.M.L.N. is a holder of a grant from CajaMurcia.

ESTEROLYSIS REACTION CATALYSED BY BETA-CYCLODEXTRIN : INFLUENCE OF THE TACTICITY OF WATER-SOLUBLE POLYMERS.

C. MARIE, L. LECLERCQ, M. MORCELLET and B. MARTEL
Laboratoire de Chimie Macromoléculaire, UA CNRS 351, Université des Sciences et Technologies de Lille, 59655 Villeneuve d'Ascq cedex, France

1. Introduction

It is well known that the kinetics of esterolysis of nitrophenol esters is strongly increased in the presence of β –cyclodextrin (βCD) and this effect has been attributed mainly to the formation of an inclusion complex and to the cleavage of the ester function by the hydroxyl groups surrounding the cavity of βCD [**1**]. Some studies about the kinetics of p-nitrophenol acetate (PNPA) hydrolysis in the presence βCD and some polymeric chains in aqueous solution have been reported [2-4]. Seo et al.[**2**] have shown that poly(styrene sodium sulfonate) (NaPSS) has a pronounced acceleration effect attributed to the concentration of both βCD and PNPA near to the chain of the macromolecule through inclusion effects and hydrophobic interactions. In a recent work[**5**], we have shown that NaPSS is actually able to include its aromatic substituent into the cavity of βCD. Poly(methacrylic acid) (PMA), another amphiphilic polymer, on the contrary, is reported to exhibit an inhibition effect because of selective interactions between βCD and the polymer[**2**]. The later study has been extended to the study of the influence of the tacticity of PMA and its benzylated derivatives (PMB) on the kinetics of the above mentioned reaction, a parameter that has not been investigated up to date to our knowledge.

2. Materials and methods

The synthesis and the characterization of the poly(methacrylic) esters of different tacticities and their benzylated derivatives has been previously reported [**6**]. For example, IPMA, IPMB 8 and IPMB 18 are the terms used to name isotactic derivatives with respectively 0% 8% and 18% of benzylated monomer units. The kinetic measurements followed the procedure described in reference [**4**]. The viscosimetric measurements were performed in tris buffer (pH= 8.74) using a capillar viscosimeter (Schott Gerate AVS400) at T°= 25°C and [polymer] = 10^{-2}M.

3. Results and discussion

The first part of the study consisted to measure the pseudo first order constants of esterolysis upon progressive addition of the polymers. Two main phenomena are observed : 1) the addition of syndiotactic PMA provokes the inhibition of the reaction whilst that of isotactic PMA has an acceleration effect (figure 1). Considering both tacticities, the benzylated derivatives have an inhibition effect, depending on the degree of substitution. The overall effect of benzylation is the inhibition of the reaction.

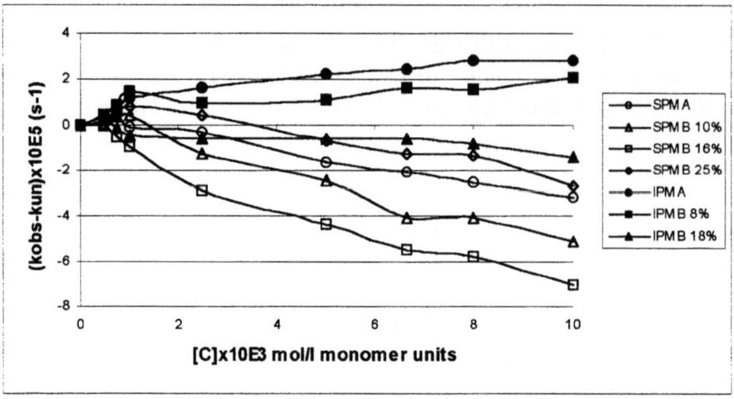

Figure 1 : Variation of the first order constant rate upon addition of PMA or PMB

The viscosimetric curves in fig.2 show that the benzylation of isotactic and syndiotactic polymers has two different effects. The viscosity of isotactic polymers and copolymers remains more unchanged than that of syndiotactic ones upon increase of the benzylation degree. It means that the isotactic macromolecules have the tendency to keep a general extended conformation on the contrary of the syndiotactic polymers. This results in the difference of the properties of hydrophobic microdomains of the polymers : isotactic polymers have an opened microdomain that allows the approach of the hydroxide ions near to the ester so that the reaction rate is increased. On the other hand, the syndiotactic polymers have a closed microdomain that insulates the ester and prevents its esterolysis. The increase of the benzylation degree tends to the former characteristics.

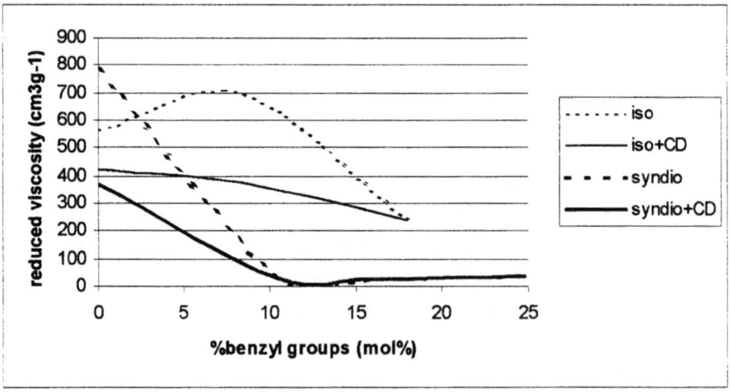

Figure.2 : Evolution of the reduced viscosity of the polymers upon their degrees of benzylation.

The PNPA hydrolysis was catalysed by βCD kept at constant concentration (10^{-2}M) while the polymers were added gradually (up to 10^{-2}M in monomer units). In the first case, the inhibition effect clearly increases with the syndiotactic PMB concentration (fig.3). It is worth to note that syndiotactic PMA has a negligible effect compared to its benzylated derivatives (see also fig. 4).

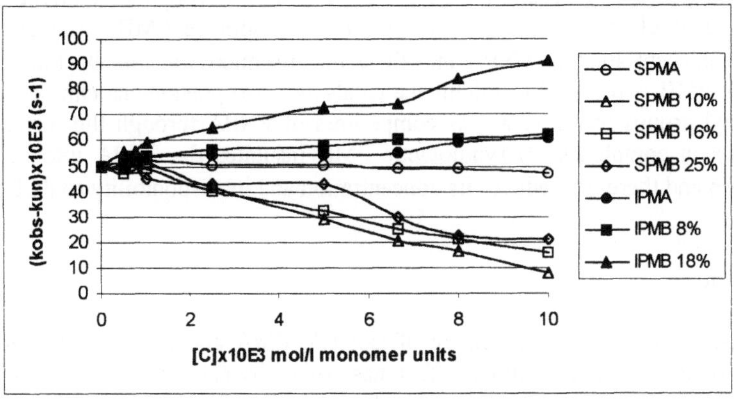

Figure 3 : variation of the kinetic constant upon addition of PMA and PMB; [βCD] kept constant at 10^{-2}M

On the other hand, the addition of the isotactic polymers have a detectable positive effect on the esterolysis rate; especially for PMB18 (fig.3). From these data, we can emphasise that isotactic and syndiotactic PMA have a reduced influence on the catalytic role of cyclodextrin compared to their benzylated derivatives. That means that the interaction between βCD and the polymers increase with their degree of hydrophobicity, what is expectable when considering the ability of cyclodextrins to interact with aromatic species.

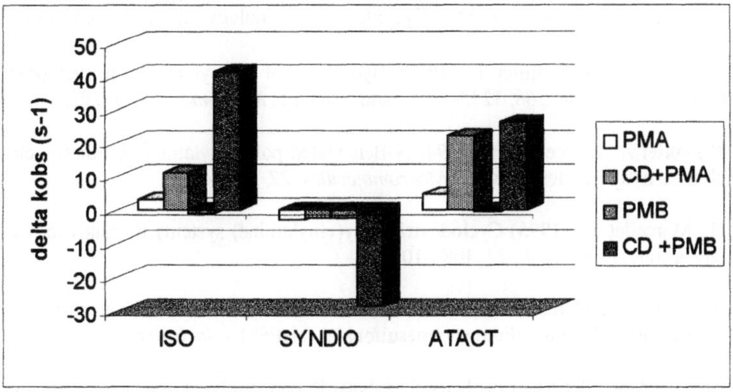

Figure 4 : Variation of the first order kinetic constant upon addition of PMA or PMB ([C]=10^{-2}M in monomer units). IPMB 18% ; SPMB 25% ; APMB 45 % in the presence (=10^{-2}M) or in absence of βCD.

Fig. 4 represents the value of the variation of the first order constant rate when PMA or PMB are added to a concentration equal to 10^{-2}M in the reaction vessel. The main phenomenon depicted is the opposite effect of isotactic and syndiotactic PMB on the esterolysis rate in the presence of βCD. In both cases, the measured effects are considerably more important than in the absence of cyclodextrin. This remark clearly proves the existence of interactions between both species. A cooperative effect is at the origin of the acceleration effect in the presence of isotactic PMB : the copolymer favours the concentration of PNPA and βCD into its microdomain. On the contrary, syndiotactic PMB is reported to inhibit the catalysis : two phenomena can be involved 1) the benzyl groups probably act as competitors of PNPA through the formation of stable inclusion complexes. 2) syndiotactic PMB insulates the PNPA into its closed microdomain and therefore reduces its concentration and its disponibility to βCD.

4. Conclusion

We have observed a general tendency of isotactic PMA and PMB to accelerate the esterolysis of PNPA. The contrary effect has been observed in the presence of the corresponding syndiotactic polymers. This difference has been interpreted in terms of the influence of the tacticity on the conformation and solubilizing properties of the macromolecules. An open microdomain able to gather the catalytic hydroxydes of the bulk solution or βCD is attributable to the isotactic systems whilst a closed microdomain that selectively encloses the PNPA can be deduced from our results in the case of syndiotactic systems. At last, the atactic polymers which correspond to a mixture of iso and syndiotactic triads, behave rather like isotactic polymers (fig. 4).

References

1. Bender M. L. and Komiyama M. (1977). Cyclodextrin Technology; Springer, NewYork
2. Seo, T., Kajihara T. and Iijima T. (1987) Hydrolysis of phenyl esters in cyclodextrin-polymers systems, *Makromol. Chem. 188*, 1295-1304, and also ibid., *Makromol. Chem. 191*, 1665-1676
3. Martel B., Pollet A., Morcellet M. (1994) N-Benzylated poly(vinylamine):synthesis, characterisation and catalytic activity in ester cleavage, *Macromolecules, 27*, 5258-5262
4. Martel B., Morcellet M. (1995) Cyclodextrin-poly(vinylamine) systems II: catalytic hydrolysis of p-nitrophenylacetate, *Eur. Pol. J., 31*, 1089-1093
5. Leclercq L., Bria M., Morcellet M., Martel B (1998) Beta cyclodextrin-polyelectrolyte interaction in aqueous solutions : I Poly(4-sodium styrenesulfonate) (NaPSS) *J. Inclusion Phenom., 30*, 215-226
6. Bottiglione V., Morcellet M., Loucheux C. (1980) Polymer-solvent interactions in mixed solvents, *Makromol.Chem, 181*, Part 1 : 469-484, part 2 : 485-495

CYCLODEXTRIN ASSISTANCE IN THE ENZYMATIC DEGRADATION OF THE *MORINGA* GLUCOSINOLATE.*

C. ROSELLI [a], B. PERLY [a], S. CASSEL [b], P. ROLLIN [b],
R. IORI [c], L. MANICI [c], S. PALMIERI [c]
a) DRECAM/SCM, CEA-Saclay, F-91191 Gif sur Yvette (France)
b) ICOA, UPRES-A 6005, Université d'Orléans, BP 6759, F-45067 Orléans (France)
c) Istituto Sperimentale per le Colture Industriali - MiPA, via di Corticella 133, I-40129 Bologna (Italia)

Moringa oleifera (Moringaceae) is a "multipurpose" tree which has now become indigenous in many tropical or equatorial regions (Ramachandran *et al.* 1980). Crushing of *Moringa* seeds provides an oil of good quality whereas the presscake - which is currently used in Africa for water treatment - displays interesting biological properties. The seeds indeed contain a high proportion (8-10%) of the unusual glucosinolate **1** (GSL).

Controlled enzymatic degradation of **1** with myrosinase (EC 3.2.3.1) results in the formation of the corresponding isothiocyanate (ITC) **2** according to the following scheme :

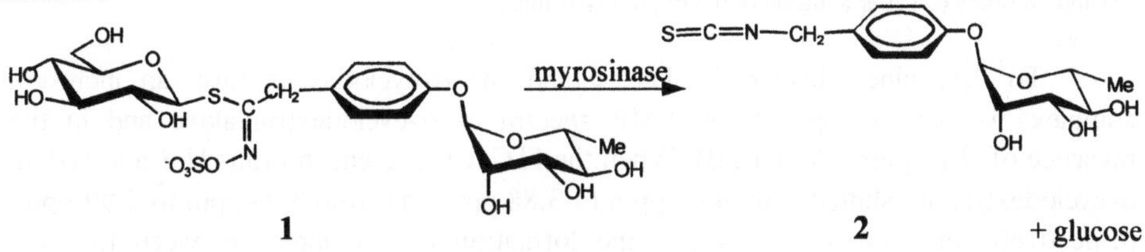

2 is known to exhibit fungistatic and bactericidal properties (Eilert *et al.* 1981, Manici *et al.* 1997). However, owing to surfactive properties, its solubility in water is poor. Soluble formulations are therefore expected to improve the bio-availability of **2** and other structurally-related isothiocyanates.

Since cyclodextrins can strongly improve the solubility of hydrophobic compounds in water, the formation of inclusion complexes between α-cyclodextrin and **2** was investigated using NMR. The isothiocyanate being produced by enzymatic reaction of myrosinase on compound **1**, we have first recorded the NMR spectrum of a mixture of α-cyclodextrin and GSL **1** dissolved in a pH = 6.5 PO_4 buffer (Figure 1A).

For structural reasons, the glucosinolate **1** it is not expected to form an inclusion complex with α-cyclodextrin. Indeed, the NMR spectrum of α-cyclodextrin is not perturbated (Figure 1A) by the presence of the GSL, therefore assessing the absence of inclusion. Upon addition of myrosinase, several NMR signals decrease in intensity (4.75 ppm and 5.58 ppm) and new features appear at 4.70 ppm and 5.70 ppm (Figure 1B). After 18 hours (Figure 1C), the conversion is completed since signals arising from the GSL have totally disappeared and the new signals assigned to H-1 protons of the ITC and of α- and β-D-glucopyranoses are found at 5.70 ppm, 5.25 ppm and 4.70 ppm, respectively.

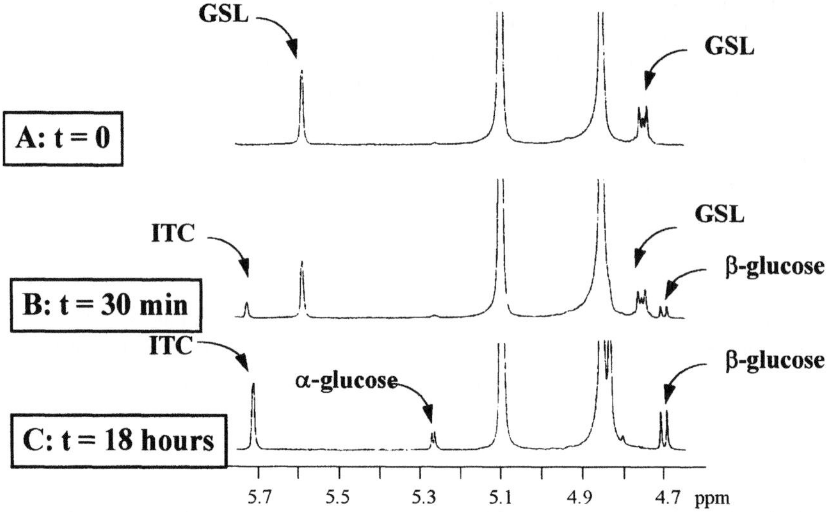

Figure 1: Partial ^{1}HNMR spectra of a solution of 5 mM α-CD and 5 mM GSL in 50 mM PO_4 buffer in deuterium oxide, pH= 6.5 at 298K as a function of time: without myrosinase (A); recorded 30 minutes (B) and 18 hours (C) after addition of 0.5 mg of myrosinase.

To determine whether the ITC **2** and the α-cyclodextrin form an inclusion complex, we have compared the NMR spectra of α-cyclodextrin alone and in the presence of **2** (Figure 2A and 2B). When the ITC **2** is present, protons H-3 and H-5 in α-cyclodextrin are shifted from 4.02 ppm to 3.88 ppm and from 3.93 ppm to 3.99 ppm, respectively, therefore demonstrating the formation of a complex between the two partners. 2D ROESY experiments (not shown) confirmed the inclusion of the ITC in the cyclodextrin and allowed to derive the structure of the latter complex as depicted in Figure 3. The NCS function of the ITC is found at the vicinity of the wider rim of the cyclodextrin.

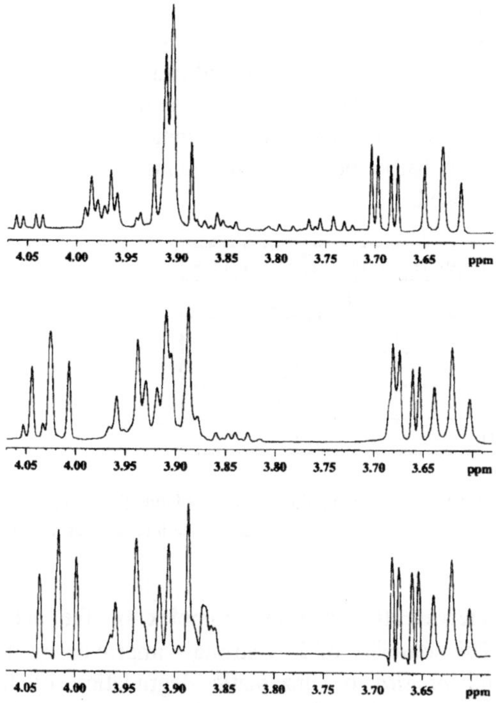

Figure 2: Partial ^{1}H NMR spectra recorded at 298K in a pH= 6.5, 50mM PO_4 buffer in deuterium oxide, of: 5 mM α-cyclodextrin, A; 5 mM α-cyclodextrin and 5 mM GSL, B; 5 mM α-cyclodextrin and 5 mM ITC obtained by enzymatic degradation of GSL, C.

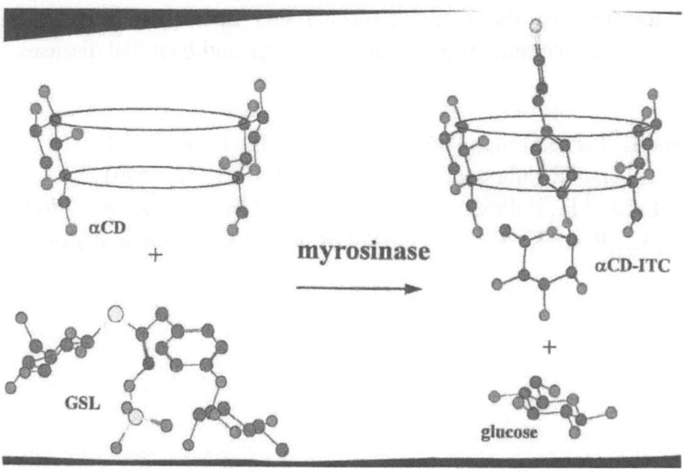

Figure 3: Schematic representation of the enzymatic degradation of the parent glucosinolate and of the inclusion of the ITC metabolite in the cyclodextrin as deduced from NMR data.

Improving the bio-availability of ITC **2** by forming a complex with cyclodextrin is of very high importance for further formulations but it requires that the fungistatic and bactericidal properties of the guest are retained. This was quantified using *in vitro* spore germination inhibition assays comparing the free ITC and the inclusion complex. The fungus tested is *Botrytis cinerea.* Spore germination was rated according to the length of germ tube after germination. Table 1 shows that in the case of the inclusion complex, at least 80% of the activity of the ITC is retained.

Dose	**% of spore germination effectiveness**	
(μmol/ml)	ITC	ITC + α-CD
0.5	61	36
1.6	80	63
3.3	100	76
8.2	100	98

Table 1: Spore germination effectiveness of *Botrytis cinerea* in the presence of the ITC obtained by enzymatic hydrolysis of GSL **1** with and without α-cyclodextrin. (Assay procedure as described in Manici *et al.* 1997).

In this study, we have characterized the inclusion complex of the active isothiocyanate **2** with α-cyclodextrin. The solubility of **2** is strongly improved by this process. The ITC could be produced *in situ* by enzymatic degradation of the corresponding glucosinolate with myrosinase. The liberation of the active ITC is therefore controlled by two processes, the enzymatic conversion and the inclusion of the final product in the host. Moreover, the fungistatic properties of the isothiocyanate are retained when it is included in the cyclodextrin. This approach allows a very easy control of the final liberation of the active molecule and opens the way to new soluble formulations for the processing of plants against fungal and bacterial diseases.

* C. Roselli and B. Perly, French Patent 98 03961, March 31st 1998.
Eilert U., Wolters B. and Nahrstedt A., *Planta Med.* 1981, **42**, 55-61.
Manici L. M., Lazzeri L., Palmieri S., *J. Agric. Food Chem.*, 1997, **45**, 2768-2773.
Ramachandran C., Peter K. V. and Gopalakrishnan P.K., *Economic Botany*, 1980, **34**, 276-283.

KINETIC EFFECTS ON THE DISMUTATION OF SUPEROXIDE RADICAL BY COPPER(II) COMPLEXES OF CYCLODEXTRIN-BASED SOD MODELS

ALEX FRAGOSO[1], ROBERTO CAO[1], YAN RODRÍGUEZ[2]
[1]*Laboratory of Bioinorganic Chemistry, Faculty of Chemistry, Havana University, Havana 10400;* [2]*Faculty of Chemistry, Central University of Las Villas, CUBA.*

1. Introduction

Superoxide dismutases (SOD) [1] are a family of metalloenzymes that catalyze *in vivo* the dismutation of superoxide radical to hydrogen peroxide and oxygen according to the equation:

$$2\,O_2^{\cdot-} + 2\,H^+ \rightarrow O_2 + H_2O_2$$

Several Cu(II) complexes are known to catalyze this reaction but only a few of them involve cyclodextrins (CD) as ligands [2]. In this paper we report our studies on the different kinetic factors affecting the SOD-like activity of a Cu(II) complex using CD derivatives as models of SOD.

2. Experimental

All chemicals used were of high quality and used without further purification. L-Arginine derivatives were prepared by tosyl displacement (primary regioisomer) or 2,3-*manno*-epoxide opening (secondary regioisomer) with L-arginine free base in THF/water (2:1) in the presence of NEt_3 and purified by CM-Sephadex C-25 chromatography.

Mono-6-N$_\alpha$-L-arginyl-6-deoxy-β-CD: ^{13}C-NMR (62.5 MHz): 184.2 (COO), 159.4 (guanidine carbon), 104.6 (C-1), 85.9 (C-4'), 83.8 (C-4), 75.8-74.5 (C-2,C-3,C-5), 73.7 (C-5'), 66.1 (C_α), 63.0 (C-6), 49.9 (C-6'), 43.7, 32.1, 27.2 (arginine CH_2). HRMS: Calc. for $C_{48}H_{83}N_4O_{36}$: m/z 1291.4787 $[M+H]^+$. Found: m/z 1291.4796.

Mono-3-N$_\alpha$-L-arginyl-3-deoxy-α-CD: ^{13}C-NMR (62.5 MHz): 184.0 (COO), 159.3 (guanidine carbon), 106.2-102.8 (C-1), 84.0-82.5 (C-4), 79.4-72.7 (C-2,C-3,C-5,C_α), 63.4-62.3 (C-6), 59.5 (C-3'), 43.43, 31.93, 27.20 (arginine CH_2).

The SOD-like activity was measured according to Beauchamp and Fridovich [3] with the same specifications as reported elsewhere [4,5]. Binding constants were calculated as reported in ref. 5.

3. Results and Discussion

1. Comparison of the SOD-like Activity of Primary and Secondary Cyclodextrins

The copper(II) complexes of β-cyclodextrin derivatives depicted below were studied as SOD-models. The results of the SOD-like activity are reported in Table 1.

In the cases of *DTC*, *en* and *hm* complexes, the CD-containing ligands were more active than non-CD-containing analogs, except in the case of Cu-C6hm. This fact suggests that the CD moiety plays an important role in the catalytic activity of these complexes by fixing the substrate via H-bonding with the hydroxyl groups of CD and providing a hydrophilic surrounding around the active site of the complexes which promotes proton transfer [4].

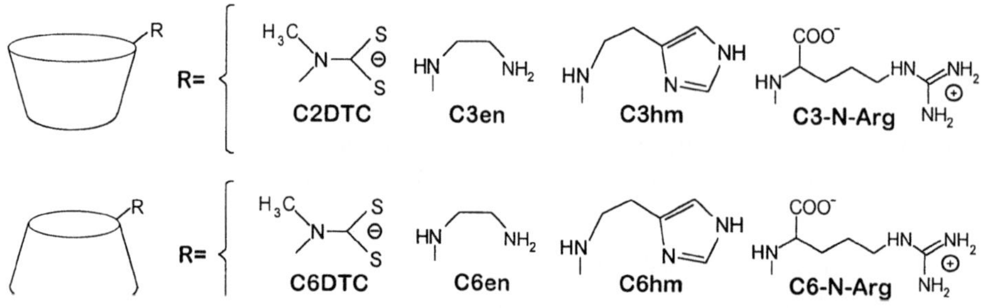

Table 1. IC_{50} Values of the Studied Complexes.

COMPLEX	IC_{50} (μM/L)
$Cu(C2DTC)_2$	4.6
$Cu(C6DTC)_2$	60
$Cu(DDTC)_2$	>50*
$Cu(C3en)_2^{2+}$	0.14
$Cu(C6en)_2^{2+}$	0.29
$Cu(en)_2^{2+}$	0.74
$Cu(C3hm)_2^{2+}$	0.10
$Cu(C6hm)_2^{2+}$	0.27
$Cu(hm)_2^{2+}$	0.26
$Cu(C3\text{-}N\text{-}Arg)_2^{2+}$	0.43
$Cu(C6\text{-}N\text{-}Arg)_2^{2+}$	0.46
$Cu(L\text{-}Arg)_2^{2+}$	0.49
$CuHPO_4$	2.2

*The complex precipitates at higher concentrations

The effect of the substitution pattern of the CD derivatives on the SOD-like activity was analyzed by measuring the activity of primary and secondary regioisomers. Secondary regioisomers were found to be 2 to 13-fold more active than primary regioisomers. It is known that the SOD-like activity depends, among other factors, on the degree of distortion of the coordination geometry around the copper center. [2] The more tetrahedrally distorted the complex is, the higher is its activity. EPR spectroscopy has been found to be a valuable tool to study the nature of the coordination sphere in Cu(II) complexes. Therefore, the X-band EPR spectra of frozen aqueous solutions of our complexes were recorded in order to detect possible differences in the geometry around the copper center provoked by the different attachments to the CD-residue in the regioisomers. However, according to the spin-Hamiltonian parameters obtained (Table 2), both primary and secondary regioisomers of *DTC*, *en* and *hm* derivatives seem to have the same distortion patterns, which does not explain the observed differences in the SOD-like activity from the thermodynamic point of view. An explanation of this fact arises from the analysis of the different nature of the hydroxyl groups located at the two rims of CD. It is known that the hydroxyl groups located at the 2-position are the most acidic (pK_a=12.2) and those at the 6-position are most basic (pK_a=15-16). Thus, assuming the formation of a H-bond between the hydroxyl groups and superoxide radical, proton transfer should be favored in the case of secondary regioisomers due to their higher acidity. Alternatively, a decrease in the activity of primary regioisomers can be attributed to the presence of the more bulky hydroxymethyl groups close to the catalytic center which present a steric hindrance to the substrate approach.

These conclusions prompted us to prepare new CD ligands derived from L-arginine since the guanidine moiety is known to have an acidity similar to that of secondary hydroxyl groups (pK_a=12.5). Investigation of the SOD-like activity of their Cu(II) complexes showed that there is no difference in the catalytic behavior of primary and secondary regioisomers. Moreover, it is noteworthy that the CD residue seems to have little or no influence on the activity as revealed by the value of the activity of $Cu(L\text{-}Arg)_2^{2+}$. The case of Arg complexes suggests that superoxide radical 'prefers' a stabilizing electrostatic interaction with the positively charged guanidine moiety that favors H-bonding and proton transfer. This interaction is absent in the other complexes, therefore, hydroxyl groups should play the cooperative function in these cases increasing the catalytic activity in secondary regioisomers with respect to the primary ones.

Table 2. EPR Parameters (X-Band) of Cu(II) Complexes in 0.01 M Aqueous Solutions at 130 K.

COMPLEX	$g_{\parallel}$	$A_{\parallel}$ (cm^{-1})	$g_{\perp}$	f (cm)
$Cu(C2DTC)_2$	2.062	0.0187	2.022	110
$Cu(C6DTC)_2$	2.061	0.0187	2.022	110
$Cu(C3en)_2^{2+}$	2.215	0.0178	2.041	124
$Cu(C6en)_2^{2+}$	2.213	0.0178	2.041	124
$Cu(C3hm)_2^{2+}$	2.234	0.0168	2.055	133
$Cu(C6hm)_2^{2+}$	2.235	0.0168	2.055	133

These results illustrate the importance of cooperative groups in the effectiveness of substituted CDs as enzyme mimics.

2. Modeling of Arg-141

In native SOD, the access of the anionic radical is controlled by the electrostatic field produced by positively-charged amino-acid residues located at the entrance of the cavity. The Arg-141 residue seems to be especially important for this function because of its proximity to the copper center. The strategy employed to mimic its function in native SOD consisted in measuring the SOD-like activity of the copper complexes of CD containing ligands in the presence of positively-charged guests having hydrophobic residues capable of being included in the CD cavity (Fig. 1) [5].

For this purpose, dithiocarbamate derivatives of α- and β-CD were selected since they form neutral complexes with Cu(II), avoiding an electrostatic repulsion with the positive residue of the guest that might result in a low stability inclusion complex. The results, expressed in terms of the percent of acceleration of superoxide radical dismutation relative to non-guest containing assays, are presented in Figure 2. In all cases the dismutation reaction is accelerated by 35-70% in the presence of the guests with respect to the complex alone. These results demonstrate that electrostatic interactions have a significant contribution in controlling the active site accessibility of charged substrates in the activity of the enzyme.

The saturation behavior observed (Fig. 2) as the guest:host molar ratio is increased suggests that the relative acceleration is a function of the binding of the guest which fixes the anionic substrate in the vicinity of the active site. Therefore, a lower guest:host molar ratio is required to attain the maximum activity for substrates with a higher binding constant. Apparent binding constants were estimated for the three hosts from the relative acceleration variations assuming 1:1 complexation stoichiometry and these results are listed in Table 3. As shown, TTMA and TTEA appear to interact more strongly with Cu-βC6DTC and Cu-αC2DTC than CTMA and CTEA, while an opposite behavior was found for Cu-βC2DTC. Particularly, Cu-αC2DTC shows a great selectivity for the formers while β-CD hosts are less selective. These differences are explained by the different size of the hydrophobic moiety and the CD cavity. No direct influence of the N-trialkyl residue on the activity is observed, suggesting that these groups produce no steric hindrance to the substrate approach. In general, higher K values were obtained for both tri-ethyl containing guests than for their tri-methyl analogs which are attributed to an enhanced non-polar interaction between the alkyl chain and the essentially hydrophobic active site.

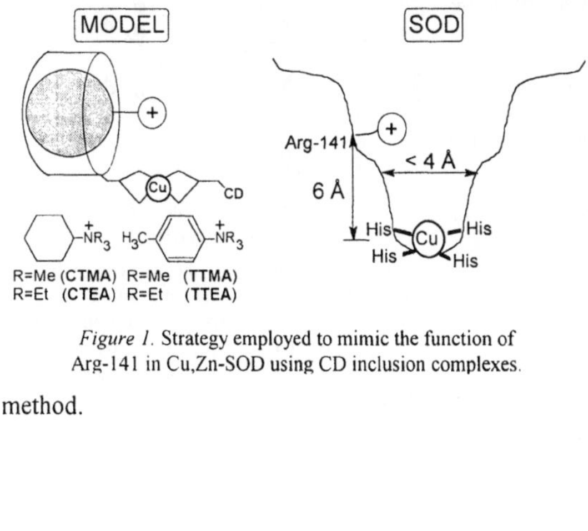

Figure 1. Strategy employed to mimic the function of Arg-141 in Cu,Zn-SOD using CD inclusion complexes.

In order to understand the influence of the magnitude of the positive charge of the guests on the relative acceleration, the net charges of the nitrogen atom were calculated by means of the AM1 semiempirical method.

Table 3. Binding Constants (K) of Cu-βC6DTC, Cu-βC2DTC and Cu-αC2DTC with the Studied Guest.

	K (M^{-1})		
GUEST	**Cu-βC6DTC**	**Cu-βC2DTC**	**Cu-αC2DTC**
CTMA	910	2000	410
CTEA	1100	1800	260
TTMA	1900	750	3600
TTEA	2100	840	3900

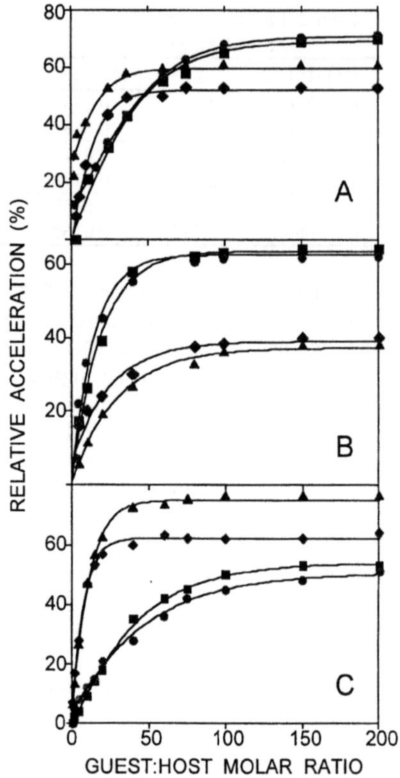

Figure 2: Relative acceleration of superoxide radical dismutation by Cu-βC6DTC (A), Cu-βC2DTC (B) and Cu-αC2DTC (C) in the presence of guests CTMA (■), CTEA (●), TTMA (•) and TTEA (◆).

The values obtained were CTMA: 0.0854, CTEA: 0.0743, TTMA: 0.0288, TTEA: 0.0252. Since the positive charge is fixed by the cavity, its magnitude should affect the extent of the catalytic activity under saturation conditions, that is the maximum relative acceleration observed (A_{max}) when the host exists totally as the inclusion complex. This rule is observed for both β-CD hosts in which the activity increases when the charge becomes more positive which could be explained in terms of stronger electrostatic substrate-catalyst interaction. However, the opposite effect is observed for Cu-αC2DTC. This host interacts weakly with cyclohexyl guests as suggest the K values, therefore a decrease in A_{max} is to be expected (Fig. 3).

As a consequence of the inclusion of the positively-charged guest into the CD cavity, the total charge of the complex becomes positive. Thus, the substrate is drawn towards the active site facilitating their reciprocal interaction. Moreover, when the Cu(II)/Cu(I) reduction step takes place, the active site acquires a negative charge that is compensated by the included guest. This effect, absent without guest, could aid the reoxidation step since the coordination of a second substrate is favored. The final result of this combination of kinetic effects is an enhancement in the catalytic activity.

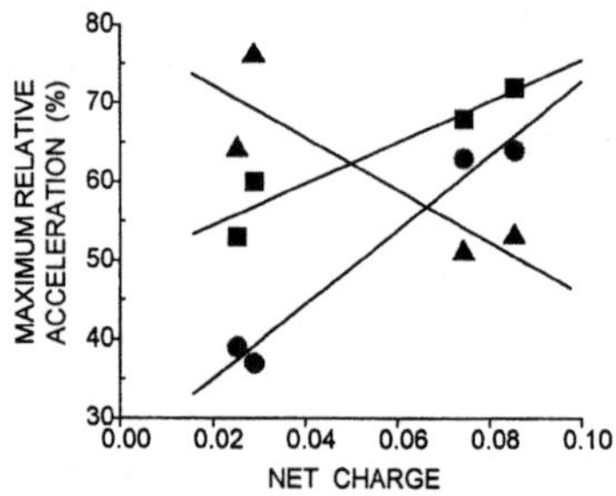

Figure 3: Influence of the net charge of the nitrogen atom on the maximum relative acceleration of superoxide radical dismutation by Cu-βC6DTC (■), Cu-βC2DTC (●) and Cu-αC2DTC (•) in the presence of the studied guests.

Acknowledgments

We thank Dr. Alicia Díaz for recording EPR spectra and helpful discussions. Financial support from Havana University (grant *Alma Mater 1997*) and from *CYTED* Project (Sub-program VIII. 3) is also gratefully acknowledged.

References

1- For the most recent review on Cu-Zn-SOD see: Bertini, I., Mangani, S., Viezzoli, M. S., (1998) *Adv. Inorg. Chem.*, ***45***, 127-250.

2- Bonomo, R. P., Conte, E., De Guidi, G., Maccarrone, G., Rizzarelli, E., Vecchio, G., (1996) *J. Chem. Soc. Dalton Trans.*, 4351-4355.

3- Beauchamp, C., Fridovich, I., (1971) *Anal. Biochem.*, ***44***, 276-280.

4- Fragoso, A., Cao, R., Villalonga, R., (1995) *J. Carbohydr. Chem.*, ***15***, 1379-1386.

5- Fragoso, A., Cao, R., D'Souza, V. T., (1997) *J. Carbohydr. Chem.*, ***17***, 171-180.

CYCLODEXTRINS AS MOLECULAR TEMPLATES

CHRISTOPHER J. EASTON,*,[a] JASON B. HARPER[a] AND STEPHEN F. LINCOLN[b]
[a]*Research School of Chemistry, Australian National University, Canberra ACT 0200, Australia*
[b]*Department of Chemistry, University of Adelaide, Adelaide SA 5005, Australia*

Abstract

The potential of cyclodextrins as molecular templates is demonstrated by the use of *N,N'*-bis(6^A-deoxy-β-cyclodextrin-6^A-yl)urea to bias competing reactions of indoxyl anion to give indigo and indirubin.

Breslow and Chung [1] demonstrated that the extent of cooperative guest binding by linked cyclodextrins [1-4] is dependent on the match of the shape of the guest to the relative orientations of the host cavities. Our studies [4] of the complexation of dyes by cyclodextrin dimers showed a preferred non-linear orientation of the cyclodextrin annuli in the urea **1**. This observation has now been exploited, using the urea **1** as a molecular template, in the formation of a non-linear product from reagents complexed in the cyclodextrin annuli.

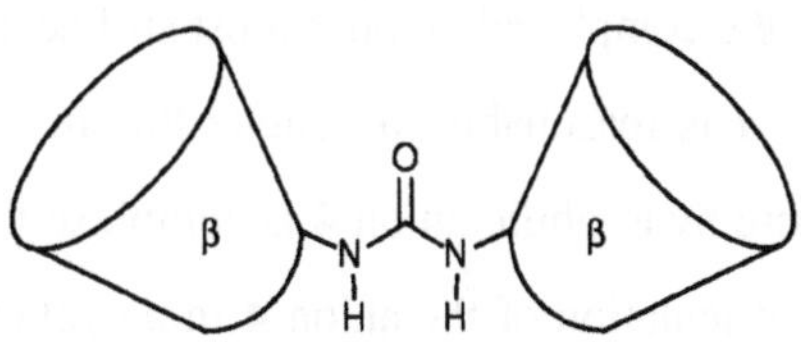

1

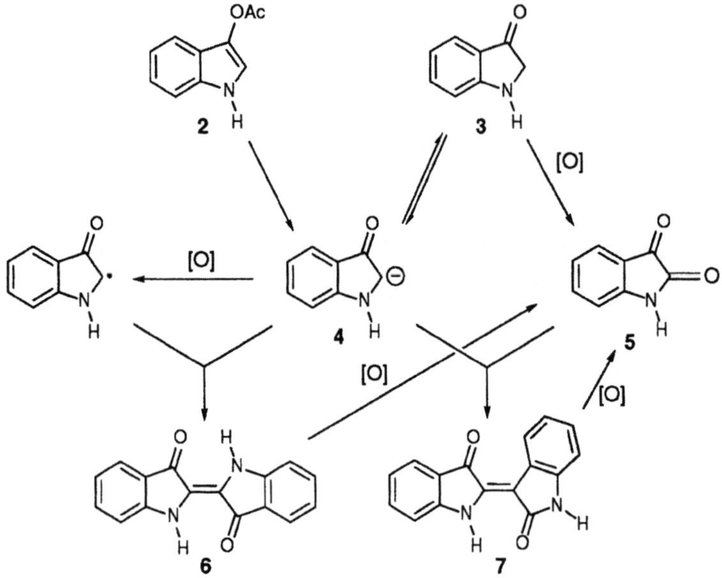

SCHEME 1.

Indigo **6** and indirubin **7** form competitively from the oxidative dimerisation of the indoxyl anion **4** and its condensation with isatin **5**, respectively (Scheme 1) [5]. Isatin **5** is formed in the reaction as an oxidation product of indoxyl **3**, indigo **6** and indirubin **7**. The effect of the cyclodextrin **1** (6.6×10^{-3} mol dm^{-3}) on these reactions was established when indoxyl anion **4** was generated *in situ,* through hydrolysis of the corresponding acetate **2** (9.7×10^{-5} mol dm^{-3}), at pH 10.0 and 298 K, in buffered aqueous solution containing isatin **5** (5.9×10^{-4} mol dm^{-3}). The cyclodextrin **1** sharply reduced the ratio of formation of indigo **6** and indirubin **7**, from 1:1 in the absence of a cyclodextrin, to 1:30 when the cyclodextrin **1** was used. The fact that very little indigo **6** forms shows that most of the indoxyl anion **4** must be complexed by the cyclodextrin **1**, in an orientation that does not allow oxidative dimerisation. However, the complexed anion **4** must still be able to react with isatin **5**, to form indirubin **7**. It is reasonable to assume this involves isatin **5** which is complexed, since the more hydrophilic anion **4** is complexed under these conditions [2]. The most probable orientation of the anion **4** in a cyclodextrin annulus is with the enolate portion protruding from the narrow end (Figure 1). In this orientation, the enolate is most shielded when complexed by the dimer **1**. Oxidative dimerisation of the complexed anion **5**, to give indigo **6**, is therefore disfavoured by the dimer **1**, as a result of the unsuitable geometry of alignment of the cyclodextrin annuli. At the same time, the geometry of the cyclodextrin dimer **1** allows reaction of the anion **4** with complexed isatin **5**, to give indirubin **7** (Figure 1). Therefore, the cyclodextrin **1** serves to preassemble the reagents and act as a molecular template for the formation of indirubin **7**.

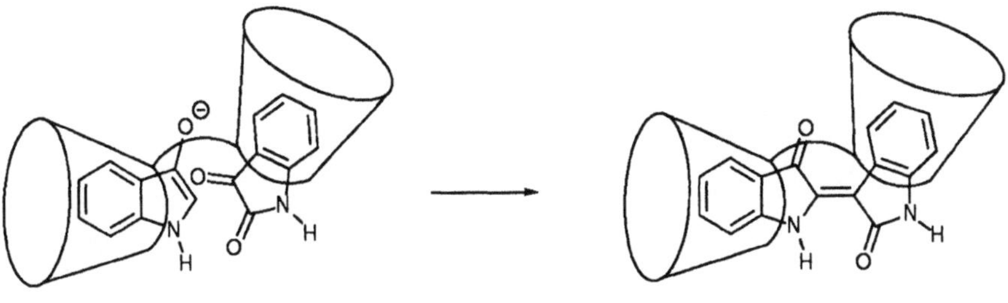

FIGURE 1. Effect of the cyclodextrin **1** as a molecular template in the formation of indirubin **7**.

References

1 For a review see: S. F. Lincoln and C. J. Easton, *Structural Diversity and Functional Versatility of Polysaccharides*, ed. S. Dumitriu, Marcel Dekker, Inc., New York, 1998, pp. 473-511.

2 For examples see: K. Fujita, S. Ejima and T. Imoto, *J. Chem. Soc., Chem. Commun.*, 1984, 1277; R. Breslow, N. Greenspoon, T. Guo and R. Zaryzycki, *J. Am. Chem. Soc.*, 1989, **111**, 8296; R. C. Petter, C. T. Sikorski and D. Waldeck, *J. Am. Chem. Soc.*, 1991, **113**, 2325; F. Venema, C. M. Baselier, E. van Dienst, B. H. M. Ruël, M. C. Feiters, J. F. J. Engenbersen, D. H. Reinhoudt and R. J. M. Nolte, *Tetrahedron Lett.*, 1994, **35**, 1773.

3 R. Breslow and S. Chung, *J. Am. Chem. Soc.*, 1990, **112**, 9659.

4 J. H. Coates, C. J. Easton, S. J. van Eyk, S. F. Lincoln, B. L. May, C. B. Whalland and M. L. Williams, *J. Chem. Soc., Perkin Trans. 1*, 1990, 2619; C. A. Haskard, C. J. Easton, B. L. May and S. F. Lincoln, *J. Phys. Chem.*, 1996, **100**, 14457; C. A. Haskard, B. L. May, T. Kurucsev, S. F. Lincoln and C. J. Easton, *J. Chem. Soc., Faraday Trans.*, 1997, **93**, 279; C. J. Easton, S. J. van Eyk, S. F. Lincoln, B. L. May, J. Papageorgiou and M. L. Williams, *Aust. J. Chem.*, 1997, **50**, 9.

5 G. A. Russel and G. Kaupp, *J. Am. Chem. Soc.*, 1969, **91**, 3851; A. Wahl and P. Bagard, *Bull. Soc. Chim. Fr.*, 1910, **7**, 1090.

CYCLODEXTRIN RADICALS PRODUCED BY PHOTOCHEMICAL HYDROGEN ABSTRACTION BY KETONES AND NITROGEN HETEROCYCLES

Martin G. Bakker, M. N. Lehmann and C. Chin
Department of Chemistry, The University of Alabama,
Tuscaloosa, AL 354 8 7-0336, USA.

1. Introduction

During the last two and half decades there has been considerable interest in the use of cyclodextrins (CDs) as photo- and thermal-stabilizers [1],'micro-reactors' [2] and as a means of stabilizing reactive specics[3-6]. However, it is becoming increasingly clear that CDs cannot be treated as inert hosts which take no part in radical chemistry [7-10]. In previous work we have suggested [9] that some of the stabilization effect of inclusion within CDs may result from trapping of reactive free radicals by CD, thereby reducing the rates of chain reactions. The result of these reactions is the formation of relatively stable CD radicals. We have applied laser flash photolysis of ketones to produce the excited triples states which abstract hydrogen from CD's [11]. Detection of the CD radicals is by Time Resolved Electron Paramagnetic Resonance (TREPR) Spectroscopy. We observe some selectivity in the position of hydrogen abstraction and that the radicals formed depend upon both CD and ketone.

2. Results and Discussion

Figure 1 shows the TREPR spectrum from β- CD resulting from absorption by acetone of 308 nm fight from an excimer laser. The peaks marked (*) are from the 2-hydroxy-2-propyl radical produced when the triplet excited state of acetone abstracts a hydrogen from CD. The other smaller peaks are from the various CD radicals produced. The form of a TREPR spectrum is considerably different from that of normal EPR spectra because of the phenomenon of Chemically Induced Dynamic Electron Polarization (CIDEP) [12]. Radical-radical recombination results in spin-polarization, which is a non-Boltzmann distribution of population density in the electronic and nuclear energy levels. The net result is that the EPR lines can be cither in absorption or emission. The presence of CIDEP increases the intensity of the TREPR signal and also gives some information about the radical processes that generate spin-polarization. In interpreting spectra such as that in Figure 1 it is necessary to simulate the spectra in order to identify the radicals present. Such simulation also helps in estimating the relative amounts of each radical formed. The efficiency of spin-polarization generation depends upon the EPR coupling constants for the differerit radicals and tlús can then be corrected for. Table 1 gives the relative TREPR intensities for the various radicals produced by laser flash photolysis of acetone.

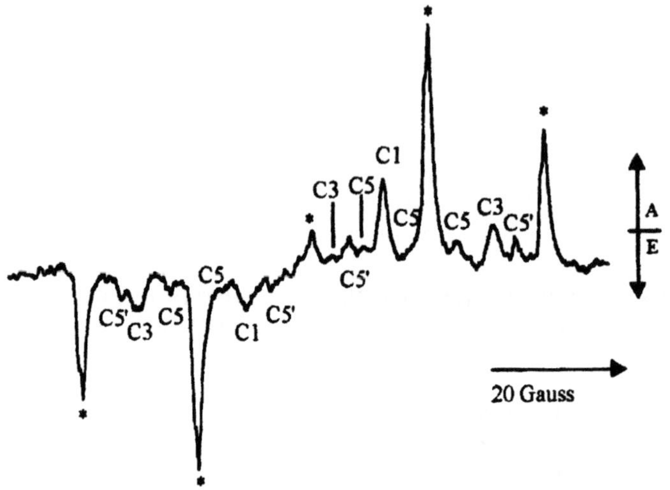

Figure 1. TREPR Spectrum from Photolysis of β-CD/Acetone

The observed spin polarization is produced by two mechanisms. One is the Triplet Mechanism (TM) which is produced in the acetone and passed onto the CD radicals, and is likely to he similar for all the CD radicals. The other mechanism is the radical pair mechanism (RPM) which in this case will he mainly the reaction of CD radicals with 2-hydroxy-2-propyl radicals. From preliminary kinetics data the rates of these reactions appear to fairly similar. The data in Table 1 is therefore expected to reflect the relative amounts of the different radicals formed (within the *ca.* 500 ns time window of the experiment). For α-CD the radicals observed are those from abstraction from within the CD cavity. As the size of the cavity increases less abstraction from within the cavity is observed and the primary site of abstraction shifts to the C1 position which is on the exterior of the CD. This is in agreement with our results for glucose, and maltose for which the major product is the C1 radical. Hence it would appear that as the CD becomes larger the behavior seems to approach that of a linear sugar. It is not clear if the abstraction pattern mirrors the binding of the acetone to the CD or the binding of the triplet excited state, since the lifetime of the triplet state is likely lo be of the order of 100 ns, which would be sufficient for acetone in the triplet excited state lo diffuse from the exterior lo the interior of the cyclodextrin.

TABLE 1. Relative Polarization Intensities for CD/Acetone Derived Radicals

Cyclodextrin	Ketil-Radical	C1	C2	C3	C4	C5	C6	C7
α	59	*	*	14	*	27	*	*
β	50	15	*	12	*	11	*	12
γ	27	16	*	*	*	*	*	57

* no radical detected

The **C5'** radical reported in Table 1, is believed lo be an acyclic C5 centered radical produced from a Cl CD radical by opening of a glucose ring. This type of rearrangement is relatively common for glucose sugars[13]. The yield of the **C5'** radical increases with size, as might be expected because the yield of the parent **Cl** radical also grows with ring size. However that this is the major product for γ-CD is somewhat surprising.

Table 2 summarizes the reactivity patterns for various ketones with α-CD. In all cases the **C3** radical is a major product, but there are clearly substantial differences in radical yield as the parent ketone is varied. The differences in triplet state energy of the ketones are unlikely to be the cause of the yield differences. Instead the structure of the parent ketone must affect the binding of either the ketone or the triplet excited ketone. The growth of the C5 radical relative to the C1 radical suggests that differences in the strength of binding might he involved.

TABLE 2. Relative Polarization Intensities for α-CD Derived Radicals

Ketone	**Ketyl-Radical**	**C1**	**C2**	**C3**	**C4**	**C5**	**C6**
Acetone	59	*	*	14	*	27	*
2-Butanone	96	*	*	4	*	*	*
Pyruvib Acid	41	*	*	20	*	39	*
Levulinic Acid	34	13	*	53	*	*	*

* no radical detected

Decreasing pH has little effect on the yields and lifetimes of the various CD radicals. Raising the pH has considerable effect, totally changing the TREPR spectra. Preliminary results indicate that considerable radical rearrangement occurs giving TREPR spectra more consistent with small radicals than with CD radicals. The exception is β-CD for which a C3 carbonyl with radical the radical center at C2 has been identified. This radical is produced by dehydration of a C3 CD radical.

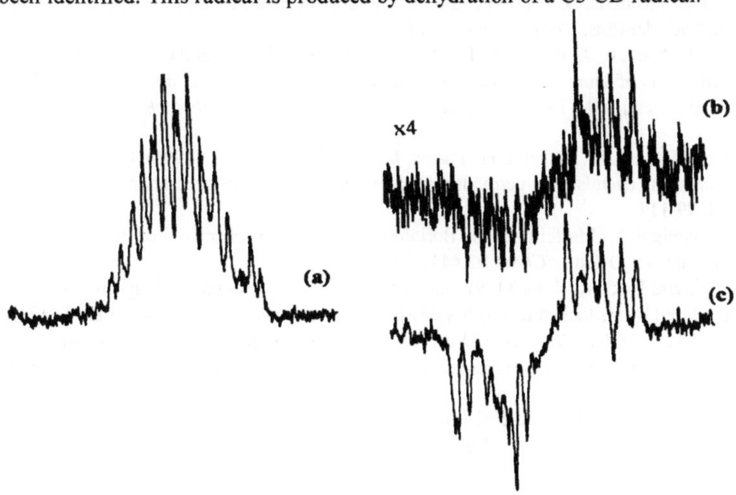

Figure 2. TREPR Spectra of Pyrazine in (a) glucose (b) β-CD (c) α-CD

The triplet excites states of nitrogen containing aromatics such as pyrazine, pyrimidine, quinoline and quinoxaline also abstract hydrogen. Figure 2 shows the TREPR spectra from solutions of pyazine with glucose, α-CD and β-CD. Most striking is the difference in the form of the spectra with glucose and with the CDs. With glucose the spin polarization is produced primarily by the TM which is believed to reflect the high molecular symmetry of pyrazine [14]. Inclusion within a CD must reduce the local symmetry sufficiently that the TM is no longer as efficient in producing spin-polarization. The difference in signal intensity between α-CD and β-CD suggests also that there must he substantial differences in the efficiency of hydrogen abstraction and radical recombination lo produce RPM type spin polarization. In the case of pyrimidine also the TREPR spectrum from α-CD are much more intense than those observes from β-CD. Weak TREPR spectra are also observes from quinoline and quinoxaline, however the spectra are very broad and do not show any clear fine structure due to CD radicals.

Acknowledgments

The donors of the Petroleum Research Fund are gratefully acknowledged for partial support of thi research. The laser used was funded by NSF under grant CHE 8922310. C.C is grateful for support from the NSF-REU program.

References

1. Matsui, Y., Naruse, H, Mochida, K. and Date, Y. (1970) Formation of Inclusion Compounds of Cyclodextrin with Hydroperoxides, *Bull Chem. Soc. Jpn.* 43, 1909.
2. Ramamurthy, V. and Eaton, D. F. (1988) Photochemistry and Photophysics within Cyclodextrin Cavities, *Acc. Chem. Res.* 21, 300.
3. Bardsley, J., Baugh P. J., Goodall, J.I. and Phillips,G.O. (1974) Hydrogen Adduct Radical Formation in g-irradiated α-, β- and γ-Cycloamylose-Benzene Complexes, *J. Chem. Soc. Chem. Comm.* 890-891.
4. Rao, V. P., Zimmit, M.B. and Turro, N. J. (1991) Photoproduction of Remarkably Stable Benzylic Radicals in Cyclodextrin Inclusion Complexes, *J. Photochem. Photobiol. A* 60, 335-360.
5. Kubozono, Y., Ata, M., Aoyagi, M. and Gondo, Y. (1987) The ESR Spectra of p-benzosemiquinone Radical Anion included in Cyclodextrins, *Chem. Phys. Lett.* 137, 467-470.
6. Lucarini, M. and Roberts, B. P. (1996) EPR Spectroscopic Characterization of Transient Organic Radicals Included In Cyclodextrins, *Chem. Comm.* 1577-1578.
7. Aquino, A. M., Abelt, C. J., Berger, K. L., Darragh, C.M., Kelley, S. E. and Cossette, M.V. (1990) Synthesis and Photochemistry of Some Anthraquinone-Substituied β-Cyclodextrins, *J. Am. Chem. Soc.* 112, 5819-5824.
8. Beeby, A. and Sodeau, J.R. (1990) Photochemistry in Cyclodextrins, *J. Photochem. Photobiol. A: Chem.* 53, 335-342.
9. Lehmann, M., Bakker, M. G. Patel, H., Parton, M. L. and Dormady, S. (1995) The effect of inclusion in β-cyclodextrin on the chemistry of peroxides: reactions of radicals with cyclodextrin, *J. Inclu. Phenom. Mol. Recogn.* 23, 99-117.
10. Monti, S., Flamigni, L., Martelli, A. and Boriolus, P. (1988) Photochemistry of BenzophenoneCyclodextrin Inclusion Complexes, J *Phys. Chem.* 92,4447-4451.
11. Lehmann, M. and Bakker, M. G. (1997) Identification by time-resolved EPR spectroscopy of cyclodextrin radicals produced by photochemical hydrogen abstraction, *J. Chem. Soc., Perkin Trans 2* 2131-2133.
12. McLauchian, K. A. (1990) Continuous-Wave Transient Electron Spin Resonance, in Kevan, L. and Bowman, M. K. (ed.), *Modern Pulse and Continuous-Wave Electron Spin Resonance,* John Wiley Sons Place, pp. 285-363.
13. Von Sonntag, C. (1 980) Free-Radical Reactions of Carbohydrates as studied by Radiation Chemistry, in Tipson, R. S. and Horton, D. (ed.), *Advances in Carbohydrate Chemistry and Biochemistry, Academic* Press Place, pp. 7-78.
14. Buckley, C. D. and McLauchlan, K. A. (1984) Flash Photolysis Electron Spin Resonance and CIDEP Studies of Radicals Derived from Nitrogen Heterocycles. II. The Photolysis and Photochemistry of the Methypyrazines, *Chem. Phys.* 86, 323-329.

OXIDATIVE STABILITY OF DOCOSAHEXAENOIC ACID OIL (TRIGLYCERIDE FORM) INCLUDED IN CYCLODEXTRINS

K. Mikuni, Koji Hara, W. Qiong, Kozo Hara, and H. Hashimoto
Bio Research Corporation of Yokohama, 13-46 Daikoku-cho, Tsurumi-ku, Yokohama, 230-0053 Japan

ABSTRACT

In order to improve the oxidative stability of docosahexaenoic acid oil (triglyceride form), storage trials of the inclusion complexes with α-, β- or γ-CD were performed at 25 °C for 25 days in aqueous solutions. The peroxide values of these oils were then analyzed. Our results show that α-CD and γ-CD enhance docosahexaenoic acid oil's stability to autoxidation (where the stability of α-CD > γ-CD) while β-CD has no effect.

1. INTRODUCTION

4,7,10,13,16,19-Docosahexaenoic acid (DHA) is one of the major long-chain polyunsaturated fatty acids and known to have physiological functions such as antithrombotic and cholesterol depressant properties. It has also been reported that a deficiency in DHA was associated with a loss of discriminate learning ability and visual acuity. DHA is chemically quite reactive and have a low stability to heat, light and atmospheric oxygen exposure. Yoshii et al. [1] reported that α-CD enhanced the stability of DHA oil (triglyceride form) more than β-CD. γ-CD was found to be the most favorable CD to complex and stabilize triglycerides of polyunsaturated fatty acid [2]. In the present investigation, we attempted to improve the storage stability of DHA oil (triglyceride form) by performing storage trials of inclusion complexes with α-, β- and γ-CD in aqueous solutions. In addition, we evaluated the difference of CD-DHA complexes formations between triglyceride form and free fatty acid.

2. MATERIALS AND METHODS

2.1 Materials

Reagent grade α-, β- and γ-CD were donated from Ensuiko Sugar Refining Co., LTD. and Wacker Chemicals. 4,7,10,13,16,19-Docosahexaenoic acid (99 % purity) was purchased from Sigma Chemicals Co. and docosahexaenoic acid oils: triglyceride form (DHA: purity 27 % and 45%) were donated from Maruha Co. DHA oils were purified from fish oil and Table 1 outlines the fatty acid compositions (w/w %) of DHA oils.

Table 1. Fatty acid compositions of DHA oils

Fatty acid	double bonds	DHA 27 (%)	DHA 45 (%)
myristic acid	0	2.5	-
palmitic acid	0	14.5	12.7
palmitoleic acid	1	5.7	-
stearic acid	0	3.0	-
oleic acid	1	21.3	15.8
arachidonic acid	4	2.2	2.2
eicosapentacnoic acid	5	6.5	4.3
docosapentaenoic acid	5	1.7	1.9
docosahexaenoic acid (DHA)	6	27.8	47.5
other fatty acid	-	14.8	15.6

2.2 Preparation of CD-DHA complexes and analytical methods

The preparation of solid complexes of DHA with α-, β- or γ-CD was performed by coprecipitation. DHA was added to each solution of CD and shaken at room temperature for 2 h. The mixtures were then centrifuged at 10,000 rpm for 10 min. After washing with distilled cold water and centrifuging again at 10,000 rpm for 10 min, precipitates obtained were freeze-dried. The quantities of CD were determines by a phenol-sulfuric acid method. DHA was analyzed by gas chromatography.

2.3 Stability tests

An equivalent quantity of DHA 27 to CD was added to 10 % (w/w) α-CD, 1.5 % (w/w) β-CD or 10 % (w/w) γ-CD solutions. The mixtures were then homogenized at 10,000 rpm for 10 min. The solutions were stored at 25 °C in a dark enviroment. After an interval of 5 days, the peroxide value (POV) of the DHA oil was quantified by iodometric titration method.

3. RESULTS AND DISCUSSION

3.1 Effect of CD concentration on the formation of CD-DHA triglyceride complexes

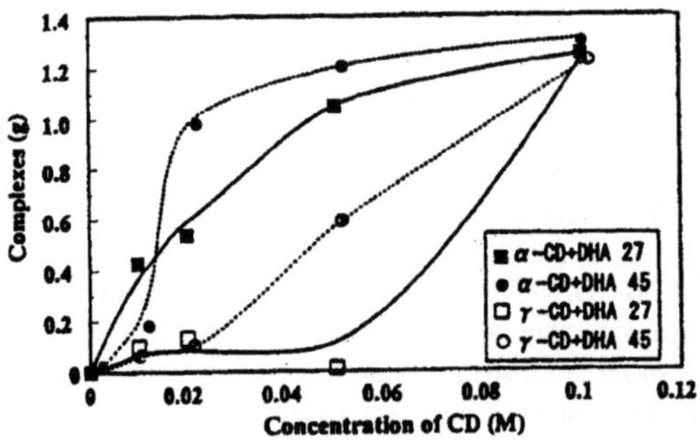

Fig. 1Effect of CD concentration on the formation of CD-DHA triglyceride complexes.

As shown in Fig. 1, α-CD formed complexes well with both DHA 27 and DHA 45. On the other hand, γ-CD formed very small amounts of complexes with DHA 27 within 0.01-0.05 M γ-CD. However, α-CD and γ-CD formed almost same amounts of complexes at 0.1 M CD. DHA 45 was more suitable than DHA 27 in forming complexes up to concentration of 0.05 M with both α-CD and γ-CD.

3.2 Mole ratio of CD-DHA complexes

All three types of CD increased mole ratio of CD/DHA in the precipitates with an increasing mole ratio of CD/DHA in solutions as shown in Fig. 2. Maximum mole ratios of α-CD/DHA, β-CD/DHA and γ-CD/DHA were 6.4, 2.5 and 1.7, respectively. Matsui et al. [3] reported that mole ratios of α-CD/Iinoleic acid, β-CD/linoleic acid and γ-CD/ linoleic acid were 3, 2 and 1.5, respectively, and the cavity of γ-CD could include 2 molecules of linoleic acid. The molecular length of DHA is longer than that of linoleic acid and so that mole ratios of CD/DHA were higher than thoes of CD/linoleic acid. On comparing DHA fatty acid with DHA triglyceride however, it was obvious that the mole ratios of CD/DHA triglyceride were less than those of CD/DHA fatty acid as outlined in Table 2. Although the conformation of triglyceride has not been reported, it has been suggested [4] that acyl chains of 1-stearoyl-2-docosahexaenoyl glycerol were compactly packed. DHA in fish oil is mainly at the second position in the triglycerides. We constructed MM2-minimized model of 1-palmitoyl-2-docosahexaenoyl-3 -oleoyl glycerol using CS Chem3D™ software (CambridgeSoft Co.). From this model, it was easy to hypothesize that the acyl chains couldn't be included in many CD molecules such as the DHA fatty acid.

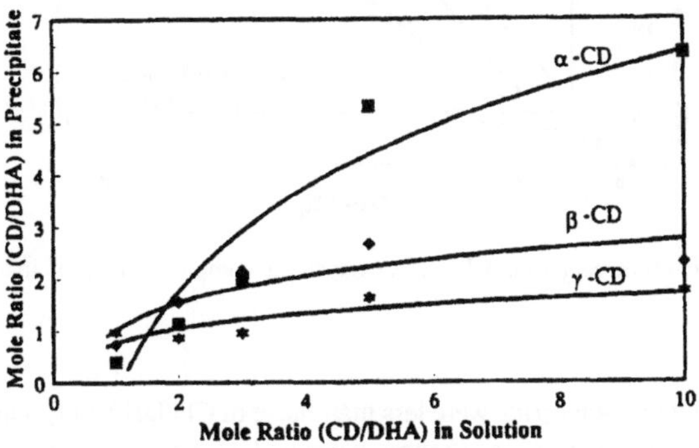

Fig.2 Plots of mole ratio of CD/DHA in the precipitates of CD-DHA inclusion complexes *vs.* the mole ratio of CD-DHA mixtures in aqueous solutions.

Table2. Mole ratio of CD/DHA of triglycerides in the precipitates with excess CD

Cyclodextrin	DHA 27	DHA 45
α-CD	1.2 (0.33*)	1.5 (0.71*)
β-CD	0.53 (0.15*)	1.2 (0.57*)
γ-CD	2.5 (0.70*)	1.8 (0.86*)

* Estimation of the mole ratio of CD/fatty acid of triglycerides calculated from fatty acid compositions (w/w %) outlined in Table 1.

3.3 Stability tests

The autoxidation time courses of CD-DHA triglyceride complexes are illustrated in Fig. 3.α- and β-CD emulsified DHA oil and the emulsion was stable during the storage period. However, γ-CD was found to precipitate soon after homogenization. α-CD was most effective compound against autoxidation whereas β-CD had no antioxidation effect on DHA. Many workers have reported that both α- and β-CD stabilized unsaturated fatty acid. In the present study, storage trials were peformed in aqueous solutions, while in a previously reported study, similar trials were performed using solid complexes. Further the mole ratio of β-CD/DHA 27 was found to be especially low (see Table 2).

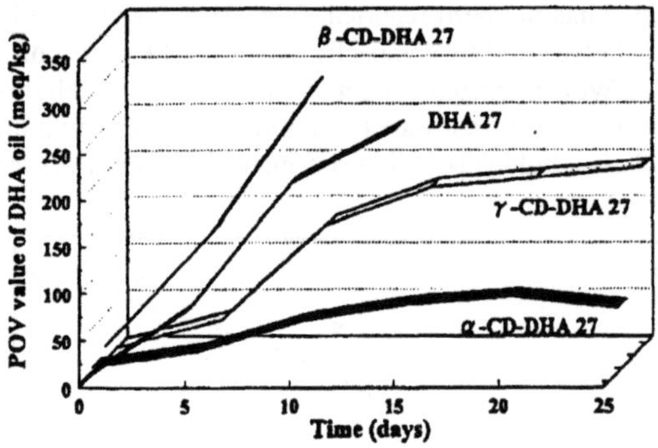

Fig. 3 The autoxidation time courses of CD-DHA triglyceride complexes at 25 °C in aqueous solutions.

4. CONCLUSION

The mole ratios of CD/DHA triglyceride were less than those of CD/DHA fatty acid. The mole ratio of γ-CD/DHA triglyceride was highest among three types of CD. Our results indicate that in aqueous solution, both α-CD and γ-CD enhance the stability of DHA triglyceride against autoxidation (where the stability of α-CD > γ-CD), while β-CD has no stabilizing effect.

5. REFERENCES

[1] Yoshi, H., Furuta, T., Yasunishi, A., Linko, Y.-Y. and Linko, P. (1996) Oxidation stability of cicosapentacnoic and docosahexaenoic acid inciuded in cyclodextrins, *Proceedings of the Eighíh Internalional Symposium on Cyclodextrins,* 579-582.

[2] Regiert, M., Wimmer, T. and Moldenhauer, J.-P. (1996) Application of γ-cyclodextrin for the stabilization and/or dispersion of vegetable oils containing triglycerides of polyunsaturated acids, *Proceedings of the Eighth International Symposium on Cyclodextrins,* 575-578.

[3] Matsui, Y. and Yoneyama, T. (1996) NMR spectroscopy on inclusion complexes of cyclodextrins with unsaturated fatty acids, *Abstracts of the 14th Cyclodextrin Symposium, Japan, 89-90.*

[4] Applegate, K. R. and Glomset, J. A. (1986) Computer-based modeling of the conformation and packing properties of docosahexaenoic acid, *J Lipid Res.,* **27**, 658-680.

MOLECULAR RECOGNITION OF A SELF-ASSEMBLED MONOLAYER OF A POLYDITHIOCARBAMATE DERIVATIVE OF ß-CYCLODEXTRIN ON SILVER

EDUARDO ALMIRALL[1], ALEX FRAGOSO[2], ROBERTO CAO[2]
[1]Department of Chemistry, ISP "Pinar del Rio", PR 20200, [2]Laboratory of Bioinorganic Chemistry, Faculty of Chemistry, Havana University, Havana 10400; CUBA.

1. Introduction

Molecular recognition systems based on cyclodextrin derivatives have been extensively studied and applied in the fields of enzyme modeling, artificial catalyzers, molecular recognition sensors, etc. In particular, self-assembled monolayers (SAM) of thiolated CD derivatives on gold surfaces have attracted great interest due to their potential application as selective electrodes [1-3]. In the present work we describe the synthesis and molecular recognition properties of a new poly-substituted β-CD dithiocarbamate chemisorbed on a Ag surface. This electrode was prepared to discriminate between positional isomers of aromatic compounds containing nitro and carboxylate or hydroxyl groups using cyclic voltammetry.

2. Experimental Part.

All chemicals were of high quality and used without further purification. Hepta-6-amino-6-deoxy-βCD was prepared via the hepta-chloro derivative [4] as reported elsewhere [5] according to the following scheme:

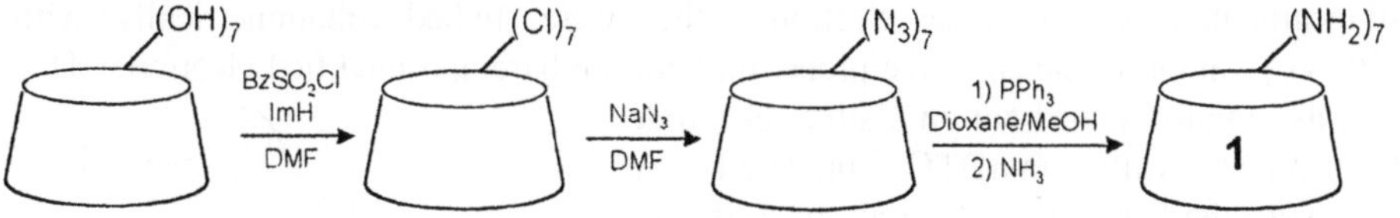

Ammonium poly-6-deoxy-βCD dithiocarbamate was synthesized by treating 1.75 g of **1** with 0.65 mL of CS_2 in a 1:1:1 solution of concentrated ammonia, water and ethanol. This system was mixed during six hours and then

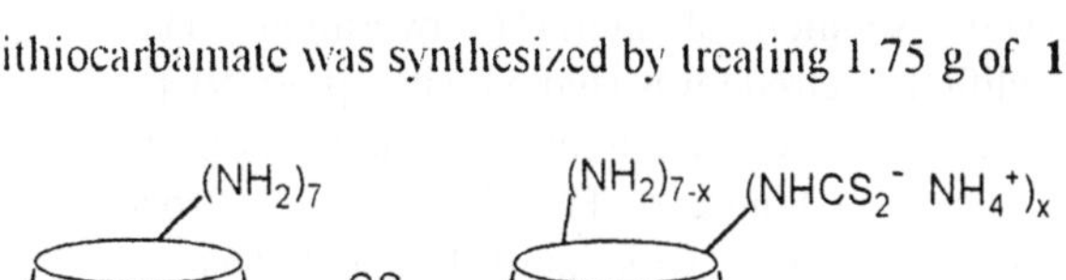

precipitated with acetone, washed with water and acetone and air dried to give **2** (1.3 g). ^{13}C-NMR (62.5 MHz): 214.5 (CSS), 102.1 (C-1), 83.4 (C-4), 70-72 (C-2, C-3, C-5), 56.5, 52.5 (C-6', C-6); UV: λ_{max} (ε), 248 nm (4.33×10^4) ($\pi\pi^*$ CSS), 287 nm (4.74×10^4) ($\pi\pi^*$ NCS).

NMR spectra were recorded on a Bruker AC250 spectrometer in DMSO-d_6. UV spectra were recorded on an Ultrospec III (Pharmacia-LKB) spectrophotometer.

The degree of substitution of the dithiocarbamate groups on the primary rim in compound **2** was determined by two different ways. One method was using the integrated NOE-suppressed ^{13}C-NMR spectra of **2** and calculating the ratio between the integration of the substituted and the (non-substituted + substituted) C-6 signals of **2**. This gave a 64.1 % of substitution. The second method was by titration of 5 mL of a 0.05 M aqueous solution of **2** with a 0.05 M standard solution of iodine until the mixture turns to a purple color. This method gave a 65.3 % of substitution. Therefore, x = 4-5 in the formula of **2**.

Cyclic voltammograms were carried out on a Yanaco P8 polarographic analyzer, using a three electrode cell: working electrode: silver electrode, bare and modified; reference electrode: SCE; counter electrode: platinum wire. The supporting electrolyte was 0.2 M Na_2SO_4. The scan rate was of 0.050 V/s. The diameter of the silver electrode was 2 mm, with 14 mm of total length. It was previously polished and cleaned with diluted nitric acid, water and ethanol. Monolayer preparation was performed in several steps. The bare electrode was immersed in a 10^{-2} M solution of **2** in DMSO during 12 h and washed with water, and afterward it was immersed in a 10^{-2} M solution of sodium morpholyl-dithiocarbamate (morDTC) during 6 h in order to cover the free spaces, as the "sealing" agent. Finally, this double modified electrode was carefully washed with water and ethanol and air dried. This electrode, and the bare silver one, gave a flat background response within the working range (-1.2 to +1.0 V) in 0.2 M Na_2SO_4.

3. Results and Discussion

Up to now only gold electrodes have been used as supports for monolayers of thiolated derivatives of βCDs [1-3]. Therefore, in order to confirm the chemisorption of dithiocarbamates on the silver electrodes, they were studied voltammetrically, with $[Fe(CN)_6]^{3-}$ anion as electroactive probe, on both the bare and modified electrode. This test was initially carried out on a silver electrode chemisorbed with morDTC prepared by immersing the bare silver electrode in a 10^{-2} M DMSO solution of morDTC overnight. This modified electrode was studied with $[Fe(CN)_6]^{3-}$. The resulting voltammogram gave no redox peak, as shown in Figure 1. Therefore, this electrode was chemisorbed with morDTC, but has no molecular recognition properties and hence could successfully be used as the sealing agent.

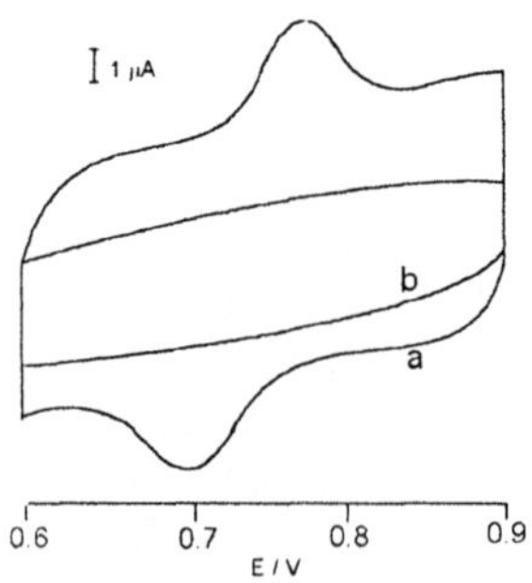

Figure 1: Cyclic voltammograms of $[Fe(CN)_6]^{3-}$ on bare silver electrode (a) and modified electrode (b).

The double modified electrode, chemisorbed with

2 and morDTC, gave a current peak decrease and a peak-to-peak potential splitting increase in $[Fe(CN)_6]^{3-}$ anion redox processes when compared with the bare electrode. The peak-to-peak potential splitting with the bare electrode corresponds to a value of 0.06 V, while when the electrode chemisorbed with **2** and morDTC was used a splitting of 0.24 V was observed. The current peak decrease is due to a reduction in the area available for redox processes in the double modified electrode since morDTC, the sealing agent, covers part of the electrode. The observed increase in the peak-to-peak splitting has been already reported for gold electrodes chemisorbed with thiolate derivatives of β-CDs and is attributed to kinetic factors provoked by the inclusion of the electroactive probe in the CD cavity, but deserves further study [2,3]. When this latter electrode was immersed in a solution of $[Fe(CN)_6]^{3-}$ containing cyclohexanol the peaks disappeared, due to the competing inclusion process of this latter cyclic compound with the iron(III) complex anion. This experiment demonstrates the molecular recognition (MR) properties of the monolayer of **2** chemisorbed on the electrode. From now on this electrode chemisorbed with **2** and morDTC will be called the MR electrode. This electrode was used for several weeks with good reproducibility.

The molecular recognition properties of the MR electrode were studied and used to discriminate between the positional isomers of nitrobenzoate anion and nitrophenol

Table 1. Potentials (V) of reduction peaks of the positional isomers of nitrobenzoate and nitrophenol with both working electrodes.

		E/V vs. SCE	
Working Electrode	**Isomer**	**Nitrobenzoate**	**Nitrophenol**
	ortho	-0.52	-0.94
Bare Silver Electrode	*meta*	-0.40	-0.46
	para	-0.45	-0.15
	ortho	no peak	no peak
MR electrode	*meta*	-0.60	-0.52
	para	-0.45	-0.23

(Table 1). For both compounds the orto isomer does not present any reduction peak attributed to the nitro group [6], while the meta and para isomers do. In the former isomer both functional groups are next to each other provoking a steric effect great enough to avoid the orientation of the nitro group towards the silver surface of the MR electrode. This is schematically shown in the following figure for the ortho (**I**) and meta (**II**) isomers of nitrobenzoate. Both meta (**II**) and para isomers can include into the β-CD cavity in such a way that the nitro group is susceptible to interact with the silver surface. For the three positional isomers of nitro-benzoate anion the

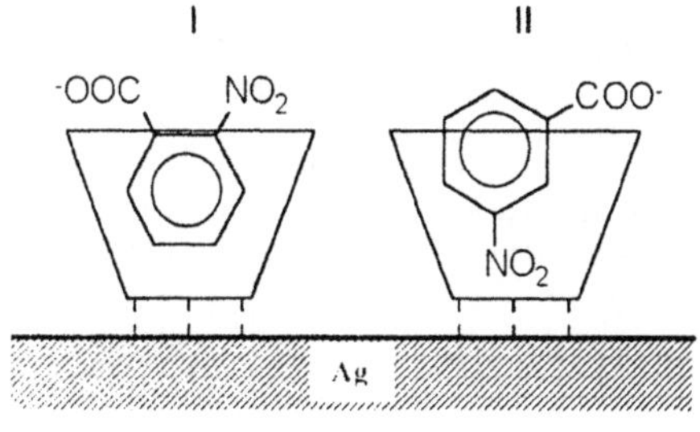

carboxylate group should form H-bonds with the secondary hydroxyl groups of β-CD. It can also be seen from Table 1 that the reduction potentials are shifted to more negative values when the MR electrode is used, similarly to $[Fe(CN)_6]^{3-}$.

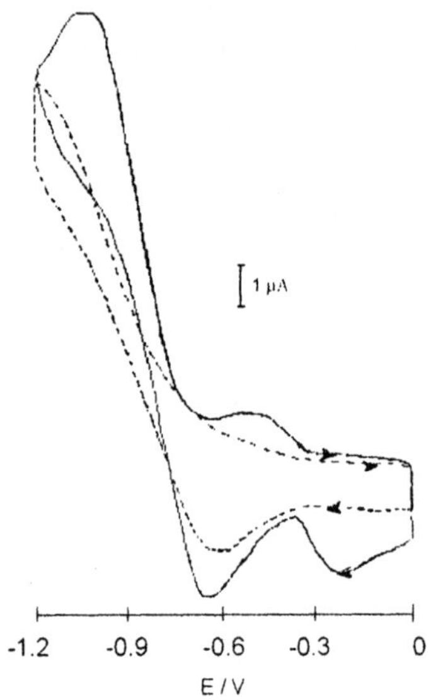

Figure 2. Cyclic voltammograms of meta-nitrobenzoate before (—) and after addition of cyclohexanol (---) using MR electrode.

Experiments with the meta and para isomers of both nitro compounds, in the presence of cyclohexanol, showed a decrease in signal intensity. An example of this is given in Figure 2. This suggests a competitive complexation of cyclohexanol as a guest and, therefore, that the electroactive probe is complexed to the β-CD cavity. These experiments demonstrate that the obtained silver electrode chemisorbed with poly-6-deoxy-βCD-dithiocarbamate (2) behaves as a molecular recognition electrode that allows it to discriminate between the positional isomers of nitrobenzoate anion and nitro-phenol. It should be expected that this type of MR electrode should also recognize the positional isomers of other electroactive aromatic and cyclic compounds.

Acknowledgments

Financial support from Havana University (grant *Alma Mater 1997*) and from *CYTED* Project (*Sub-program VIII. 3*) are gratefully acknowledged.

References

1. Rojas, M. T.; Koniger, R.; Stoddart, J. F. and Kaifer, A. E. (1995) Supported Monolayers Containing Preformed Binding Sites. Synthesis and Interfacial Binding Properties of a Thiolated β-Cyclodextrin Derivative, *J. Am. Chem. Soc.*, **117**, 336-343.
2. Weisser, M.; Nelles, G.; Wohlfart, P.; Wenz, G. and Mittler-Neher, S. (1996) Immobilization Kinetics of Cyclodextrins at GoldSurfaces, *J. Phys. Chem.*, **100**, 17893-17900.
3. Nelles, G.; Weisser, M.; Back, R.; Wohlfart, P.; Wenz, G. and Mittler-Neher, S. (1996) Controlled Orientation of Cyclodextrin Derivatives Immobilized on Gold Surfaces, *J. Am. Chem. Soc.*, **118**, 5039-5046.
4. Kahn, A. R. and D'Souza, V. T. (1994) Synthesis of 6-Deoxychlorocyclodextrin via Vilsmeir-Haack Type Complexes, *J. Org. Chem.*, **59**, 7492-7495.
5. Garcia-Fernandez, J. M.; Ortiz-Mellet, C.; Jimenez-Blanco, J. L.; Fuentes-Mota, J.; Gadelle, A.; Coste-Sarguet, A.; Defaye, J. (1995) Isothiocyanates and Cyclic Thiocarbamates of α,α'-Trehalose, Sucrose and Cyclomaltooligosaccharides, *Carbohydr. Res.*, **268**, 57-71.
6. Organic Electrochemistry: An Introduction and a Guide, Lund, H., Baizer, M., Editors, Marcel Dekker, New York (1990).

INHIBITION BY CYCLODEXTRINS OF NITROSATION REACTIONS

M. SUEIRO, R. VAZQUEZ PICOS, E. ALVAREZ PARRILLA, F. MEIJIDE, P. RAMOS, E. RODRIGUEZ NUÑEZ, J. VAZQUEZ TATO
Departamentos de Química Física y Física Aplicada, Facultad de Ciencias, Universidad de Santiago, Campus de Lugo (Spain)

1. INTRODUCTION

Nitrite esters (RONO) are effective nitrosating reagents in aqueous basic media.[1-5] Previous studies[5] of their reaction with different alifatic and heterocyclic secondary amines, to yield N-nitrosoamines, have shown that the following rate equation

$$v = k_2\,[\text{amine}]\,[\text{RONO}]\,/\,(1 + [H^+]/K_1) \qquad (1)$$

is obeyed. It suggests that the rate limiting step is the attack of the nitrite ester on the free amine. The observed influence of the ionic strength, the kinetic results obtained in water/tetrahydrofuran mixtures and in D_2O and the values of the activation entropy determined, are indicative that the transition state involved in the slow step is tetracentric, as in the figure:

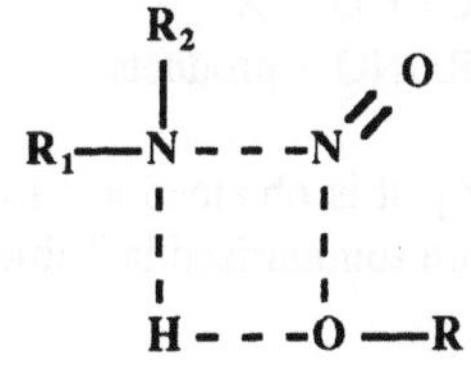

Since the ratio $k_2(H_2O)/k_2(D_2O)$ diminishes with the electronegativity of the group attached in β position of the nitrite esters it was suggested a transition from a concertated to a fast loss of the proton from the transition state.

Cyclodextrins are cyclic molecules with hydrophobic cavities[6] (with sizes varying with the number of linked glucopyranose molecules) that are able to host different kind of molecules giving inclusion complexes.[7] Consequently, the formation of these complexes affect the kinetics and/or mechanisms of different reactions.

In this communication we present the results for the nitrosation of piperidine (by 1-propyl and 2-buthyl-nitrite) and pyrrolidine (by 2-buthyl-nitrite) in the presence of hydroxypropyl-β-cyclodextrin (CD) which has the advantage of a greater solubility with respect to the non-substituted one.

2. EXPERIMENTAL

The reactions were studied spectrophotometrically at 245 nm for pyrrolidine and 250 nm for piperidine and in basic media at acidities in which all amine may be considered as free ($K_1 >> [H^+]$) and, therefore, the rate equation (1) is clearly simplified. Amines were always in excess over nitrite esters.

Other experimental conditions were T = 25°C and 0.25 M of ionic strength (adjusted with NaCl). Nitrite esters were synthesized *in situ* from sodium nitrite in perchloric acid media and the corresponding alcohols. Then, they were transfered to the reaction mixture in the spectrophotometric cell.

3. RESULTS AND DISCUSSION

3.1 KINETICS OF THE REACTIONS

All the reactions were found to be of first order in RONO. For all studied systems the hydroxypropyl-β-cyclodextrin inhibits the reaction (see figures 1 and 2 as examples), according to the rate equation:

$$v = a\,[\text{amine}]_o\,[\text{RONO}] / (1 + b\,[\text{CD}]_o) \qquad (2)$$

where the subscript refers to total concentrations. This equation implies the linearity also shown in figure 2. In order to account for this rate equation, we propose the following mechanism in which part of the amine is not reactive because it forms an 1:1 inclusion complex (X) in a fast equilibrium:

$$\text{amine} + \text{CD} \rightleftharpoons \text{X} \qquad \text{fast equilibrium, K}$$
$$\text{amine} + \text{RONO} \rightarrow \text{products} \qquad \text{slow, } k_2$$

Since $[\text{amine}]_o = [\text{amine}] + [\text{X}]$, it is obtained $a = k_2$ and $b = K$. The values experimentally determined for these constants are summarised in Table 1.

Table 1. Kinetic and equilibrium constants obtained for the systems studied (see text)

Amine	RONO	$k_2/M^{-1}s^{-1}$	K_{kin}/M^{-1}	K_{NMR}/M^{-1}
Piperidine (PIP)	1-propyl-nitrite	0.0193	54.9	40.4
piperidine	2-buthyl-nitrite	0.0060	42.9	40.4
Pyrrolidine (PYRR)	2-buthyl-nitrite	0.0038	32.9	21.0

The lower reactivity of 2-buthyl-nitrite with respect to 1-propyl-nitrite may be explained in terms of the higher steric effects in the attack to the amine by the former. On the other hand, both amines react with 2-buthyl-nitrite at similar rates because they are practically identical in basicity (pKa = 11.30 for PIP and 11.49 for PIRR).

The previous discussion and conclusions are based on a 1:1 stoichiometry for the inclusion complex between amines and hydroxypropyl-β-cyclodextrin and it would be very important to confirm this assumption. Therefore we have carried out NMR measurements, commented on in the following paragraph.

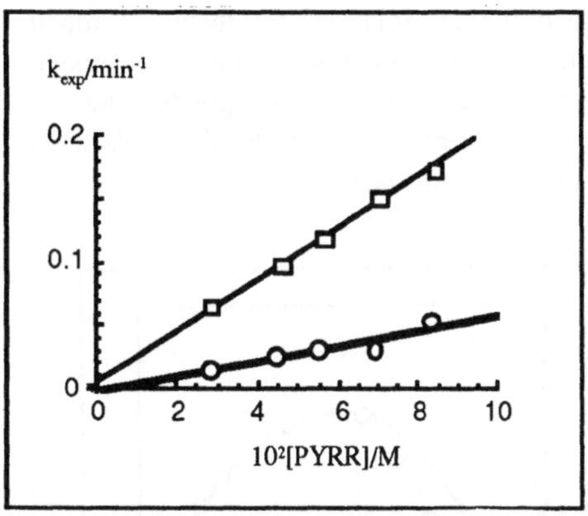

Figure 1. Influence of the amine concentration on the experimental rate constant with (circles) and without (squares) CD for the nitrosation of pyrrolidine by 2-buthyl-nitrite.

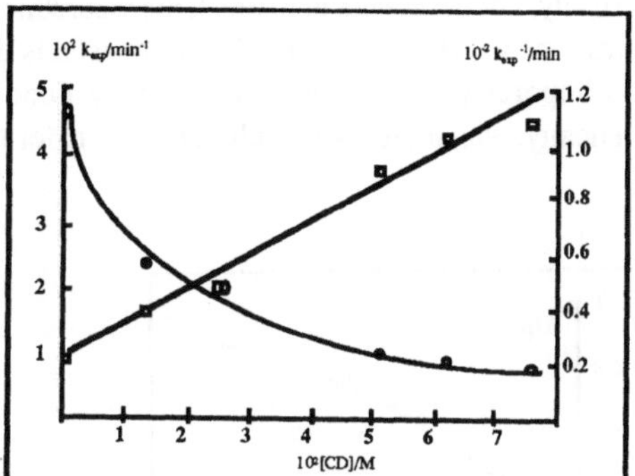

Figure 2. Influence of CD concentration on the experimental rate constant for the nitrosation of piperidine by 1-propyl-nitrite.

3.2 STOICHIOMETRY AND EQUILIBRIUM CONSTANTS BY NMR

It is well-known that the ^1H and ^{13}C spectral signals of host and guest are shifted when these compounds associate in the inclusion complex. Figure 3 show this behaviour for the piperidine-hidroxypropyl-β-ciclodextrin system.

Under conditions of rapid exchange between amine and CD, the observed chemical shift of a specific nucleus of the amine (of constant concentration within the series), δ_{obs}, is given by[8]:

$$\delta_{obs} = f_{amine}\,\delta_{amine} + f_X\,\delta_X \tag{3}$$

where δ_{amine} is the shift of free amine, δ_X the shift of the pure complex and f_i are mole fractions. This equation may be rearranged to:

$$\Delta = \Delta_o\,[X]/[amine]_o \tag{4}$$

where $\Delta = \delta_{obs} - \delta_{amine}$ is experimentally determined. The determination of the stoichiometry of the complex (m and n unities of amine and CD, respectively) is based on the fact that the complex concentration is maximum for a molar ratio amine/CD = m/n. Therefore, the plot of $f_{amine}\Delta$ vs f_{CD} has a maximum at n/(m+n). Figures 4 shows typical results, confirming the supposed stoichiometry.

On the other hand, in equation 4, $\Delta_o = \delta_X - \delta_{amine}$ is a optimizable parameter, and the complex concentration, [X], may be writen in terms of initial concentrations (of both CD and amine) and the equilibrium constant, the second parameter to optimize. Solutions for them are found iteratively. For the system in figure 3, the theorethical curve reproduces very well the experimental points. The values for the equilibrium constants determined in this way are also given in table 1. It is important to note the fact of a very good agreement between them and those obtained kinetically, which means an indirect support for the proposed mechanism.

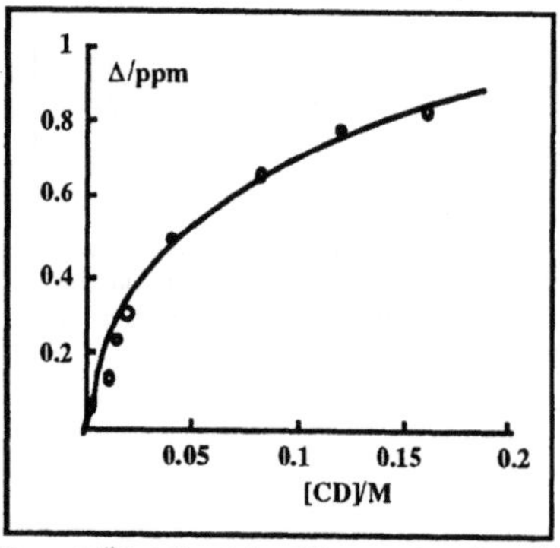

Figure 3. ^{13}C shifts of piperidine as a function of the CD concentration. The line has been drawn with the optimized parameters (K and Δ_o, see text).

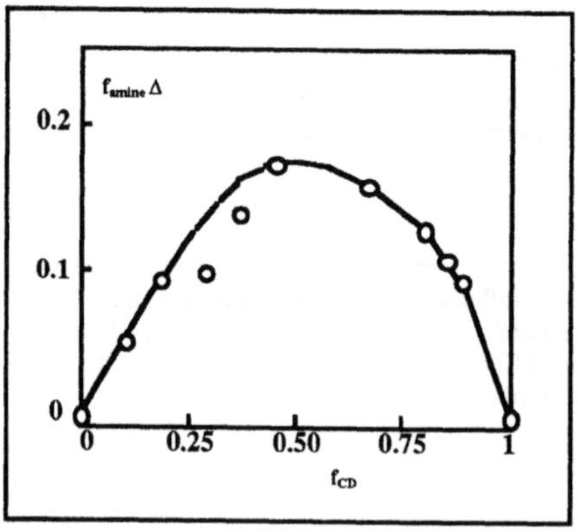

Figure 4. Job's plot showing the 1:1 stoichiometry for the inclusion complex of piperidine and hydroxypropyl-β-ciclodextrin.

4. REFERENCES

1. Oae, S., Asai, N. and Fujimori, K. (1978) *J. Chem. Soc. Perkin Trans. II*, 1124.
2. Yamamoto, M., Yamada, T. and Tanimura, A. (1979) *J. Food Hyg. Soc. Japan*, 20, 15.
3. Challis, B.C. and Shucker, D.E.G. (1979) *J. Chem. Soc. Chem. Comm.*, 315.
4. Challis, B.C. and Shucker, D.E.G. (1980) *Food Cosmet. Toxicol.*, 18, 283.
5. Casado, J., Castro, A., Lorenzo, F.M. and Meijide, F. (1986) Monats. Chem., 117, 335.
6. Szejtli, J.: *Cyclodextrin Technology*, Kluwer Academic Publishers, London.
7. *Comprehensive Supramolecular Chemistry (Cyclodextrins, Vol. 3)*, Szejtli, J. and Osa, T. (eds.), Pergamon, UK.
8. Tsukube, H., Furuta,H., Odani, A., Takeda, Y., Kudo, Y., Inoue, Y., Liu, Y., Sakamoto, H. and Kimura, K. (1996) Determination of Stability Constants in Szejtli, J. And Osa, T. (eds), *Comprehensive Supramolecular Chemistry (Cyclodextrins, Vol. 8)*, Pergamon, UK.

Acknowledgement
We thanck to Xunta de Galicia (Project XUGA26201B96) and CYTED (Project VIII.3) for financial support.

MOLECULAR NECKLACES CONTAINING REPORTER MOLECULES

E. I. POPOVA[1], I. N. KARPOV[2], I. N. TOPCHIEVA[1] and O. I. MIKHALEV[2]

[1]*Department of Chemistry. Lomonosov State University. 119899 Lenin Hills, V-234, GSP-3 Moscow. Russia;* [2]*Photochemistry Center. Russian Academy of Sciences. 117421 St. Novatorov, 7a. Moscow. Russia*

1. Introduction

Linear poly(alkylene oxide)s and cyclodextrins (CDs) can be used in the creation of supramolecular rotaxane-type structures. Complexes of these compounds represent molecular necklaces—structures where many cyclodextrin molecules are threaded on a polymer chain. Their stability is governed by the correspondence between the cyclodextrin cavity size and the diameter of a polymer guest. Thus, poly(ethylene oxide) (PEO) gives complexes with six-membered α-CD, and poly(propylene oxide) (PPO) gives complexes with seven-membered β-CD [1]. γ-CD can form molecular necklaces with 2 chains of PEO or 1 chain of PPO [1, 2]. Following the concept of spatial correspondence, one can predict the occurrence of ternary complexes in which molecular necklaces include low-molecular guest molecules. For example, these complexes can be obtained in the system PEO-β-CD-aromatic compound.

In this work, we pioneered in obtaining and characterizing such ternary complexes in system β-CD – PEO – aromatic compounds (benzene, phenol, benzoic acid, *para*-nitrophenol, *ortho*-nitrophenol, 2,4-dinitrophenol) [3] and investigated complex formation in ternary systems on the base of γ-CD: γ-CD – PEO – spin probes, γ-CD – triblock copolymer PPO – PEO – PPO (PEP) – spin probes.

2. Results and Discussion

2.1. TERNARY SYSTEMS BASED ON β-CD – PEO – AROMATIC COMPOUNDS

Since β-CD does not give complexes with PEO, the presence of PEO in the precipitate can be taken as a criterion for the formation of the ternary complex. PEO was quantitatively identified in the precipitate by using of tritium-containing PEO (^{3}H-PEO). The β-CD content was determined by polarimetry, and aromatic compounds were determined by UV spectroscopy (*p*-nitrophenol) or calculated from the quantities of the other components (complexes with benzene). Data on the composition of the ternary complexes are listed in the table 1.

TABLE 1. Composition of the β-CD – PEO – aromatic compound ternary complexes

Aromatic compound	Composition of the complex					
	Calculated *			Experimental		
	β-CD, %	PEO, %	aromatic compound, %	β-CD, %	PEO, %	Aromatic compound, %
Benzene	87.2	6.8	6.0	84 ± 4	6.5 ± 1.5	10 ± 1
p-nitrophenol	83.3	6.5	10.2	89 ± 4	6 ± 1.5	9 ± 1

*Calculation was performed on the assumption of stoichiometric composition of the complexes β-CD : PEO : aromatic compound = 1 : 2 : 1.

The use of this criterion showed that ternary complexes form in the presence of benzene, benzoic acid, or *p*-nitrophenol with stoichiometric ratio β-CD : PEO : aromatic compound = 1 : 2 : 1, and they do not form in the presence of *o*- or trisubstituted benzenes (dinitrophenol).

Another important criterion for the formation of the ternary complexes is the structures of the precipitates. X-ray diffraction patterns show that the structures of the ternary complexes are identical to the structure of molecular necklaces, which represents channel – like structure. These structures differ substantially from those of binary complexes, which exhibit cage – like structure [4]. It was found that the structures of the precipitates of the ternary complexes with benzoic acid or p-nitrophenol are identical with the structures of molecular necklaces. At the same time, the precipitate obtained in the β-CD-PEO-benzene ternary system is a mixture of the binary and ternary complexes, as judged from the X-ray diffraction pattern. The mixture can be enriched in the ternary complex by multiple reprecipitation from hot water.

To check the possibility for formation of molecular necklaces with various guests, molecular models were calculated with HyperChem release 4 for Windows for the following aromatic compounds: benzene, *o*-nitrophenol, *p*-nitrophenol. and 2,4-dinitrophenol. The modeling was performed for the chain segment consisting of four β-CD molecules aligned in the head-to-head manner and 13 ethylene oxide monomeric units. A guest molecules were placed in the cavity of each β-cyclodextrin. It was found that only benzene and *p*-nitrophenol gave the regular structures, whereas *o*-nitrophenol and 2.4-dinitrophenol tended to go out of the β-CD cavity.

Thus, a combination of analytical and structural methods with computer modeling made it possible to identify ternary complexes in the system β-CD – PEO – aromatic compounds. Last substances play a role of "space – regulator" and permit to form molecular necklaces between CD and polymer, which are not complementary to each other. Besides that small guests of ternary complexes may provide some information on the nature of the interaction between components of the complex. Therefore they may be regarded as reporter – molecules. We have studied the changes in UV and IR spectra of complexes as compared with spectra of initial substances. In the spectra of water – alcoholic solutions of the complexes containing *p*-nitrophenol the shoulder at λ = 400 nm transforms into a new absorption band; for the binary complex, this band is stronger than for the ternary one. It is presumably connected with the formation of charge-transfer complexes between *p*-nitrophenol and the ether oxygen of the β-CD cavity. The difference in the intensities of the charge-transfer bands apparently results from the degree of charge transfer: the charge-transfer complex in *p*-nitrophenol-β-CD system is stronger than in the ternary system. The formation of the tentative charge-transfer complex between the components of the complexes was demonstrated more clearly by IR spectra: the ν_{sym}(N–O) absorption band of Ar-NO_2 shifted from 1354 cm^{-1}, which is characteristic of *p*-nitrophenol, to 1338 cm^{-1} in the ternary complex and 1336 cm^{-1} in the binary complex. IR data also support a higher degree of charge transfer in the binary complex with *p*-nitrophenol as compared to the ternary complex.

2.2. TERNARY SYSTEMS γ-CD – PEO – R (ROH)

More information about interaction of components in ternary complex one can receive by using of spin or fluorescence probes as reporter – molecules. It would allowed us to investigate molecular necklaces by the methods of ESR and fluorescent spectroscopy. But these compounds represent bulky structures and can not be used for obtaining of ternary complexes on the base of β-CD. Therefore we began to work with ternary system on the base of γ-CD: γ-CD – PEO – spin probes: 2,2,6,6-tetramethylpiperidine-1-oxy (R) and 4-hydroxy-2,2,6,6-tetramethylpiperidine-1-oxy (ROH). The investigation of complexation between PPO and γ-CD was performed using water-soluble symmetric block copolymer PEP (share of PPO is 40%, $\overline{M}_n$ =3000). The probability of existence of ternary complexes γ-CD – PEO – ROH and γ-CD – PPO – ROH was confirmed using computer models. It was determined that the precipitate emitted from the system γ-CD – PEO – R doesn't contain PEO. It means that binary complex γ-CD – R is formed. The composition of precipitate obtained in the γ-CD – PEO – ROH coincided with theoretical calculation (1 γ-CD : 2 monomeric units of PEO : 1 ROH). But analysis of ESR-spectra of precipitate testified that it presents a mixture of two binary complexes: γ-CD – two chains of PEO and γ-CD – ROH. Thus, it was proved that ternary complexes don't form in these systems.

2.3. TERNARY SYSTEMS γ-CD – PEP – R (ROH)

The precipitates emitted from both systems contained all three components. Analysis of ESR-spectra showed that spin probes aren't included in the cavities of CDs threaded on polymer chain. It was also shown that a part of spin probes has rotation correlation time lesser then probes in binary complex of γ-CD – R (ROH). These probes are located outside CD-cavities. On the basis of these data we proposed the model of product structure. According to this model γ-CD form molecular necklaces with external PPO-blocks of the copolymer and spin probes are concentrated on free PEO-blocks of complex γ-CD – PEP forming a kind of intercalate complex with PEO. We investigated the system β-CD – PEP – ROH in order to confirm our assumption. The admixture of binary complex β-CD – ROH was removed during recrystallization. X-ray analysis showed that structure of precipitate containing spin probes was identical with the structure of molecular necklace. Thus, using systems β-CD – PEO – aromatic guests and γ-CD – block-copolymer PEP – spin probes two new diverse types of ternary complexes based on molecular necklaces and reporter – molecules were synthesized.

References

1. Harada, A., Li, J. and Kamachi, M. (1990) Complex formation between poly(propylene glycol) and β-cyclodextrin, *J. Chem. Soc. Chem. Commun.*, 19, 1322-1323.
2. Harada, A., Li, J., Kamachi M. (1994) Double-stranded inclusion complexes of cyclodextrin threaded on poly (ethylene glycol) *Nature*, 370, 126-128.
3. Topchieva, I.N., Popova, E.I., Kalashnikov, F.A., Panova, I.G., Avakjan, V.G., Ksenofontov, A.L., Gerasimov, V.I. (1997) β-Cyclodextrin – poly(ethylene glycol) molecular necklaces impregnated with aromatic compounds, *Doklady Chemistry*, 357, c.648 – 651.
4. Szejtli, J. (1982) Cyclodextrins and their inclusion complexes, Acad. Kiado, Budapest.

MOLECULAR DYNAMICS SIMULATIONS OF POLYROTAXANES FORMED BY POLY(OXYTRIMETHYLENE) AND α-CYCLODEXTRINS

J. POZUELO, F. MENDICUTI and E. SAIZ.
Departamento de Química Física, Universidad de Alcalá, 28871 Alcalá de Henares, Madrid, Spain.

1. Introduction

The polyrotaxanes are supermolecules formed by macrocycles threaded by linear polymer chains, with no covalent bonding between them. Harada et al.[1] have prepared and characterized *"channel type"* polyrotaxanes that form spontaneously from polymers and CDs in aqueous solutions. Complexation properties are usually attributed to the size and polarity of the hydrophobic inner CD cavity relative to the cross sectional areas of the polymer chains and their hydrophobic characteristics.

We are employing Molecular Mechanics (MM) and Molecular Dynamics (MD) simulations to study the complexation of CDs with small molecules [2,3] and polymers [4,5]. Here, we report simulations of molecular dynamics of the αCD-POT end-capped complexes of *Figure 1*. One of our purposes is to infer the stabilities and configurations of the complexes from the conformations during the trajectory. Several parameters that characterize the complex are compared with those obtained for isolated αCDs [6] and the isolated chain of POT.

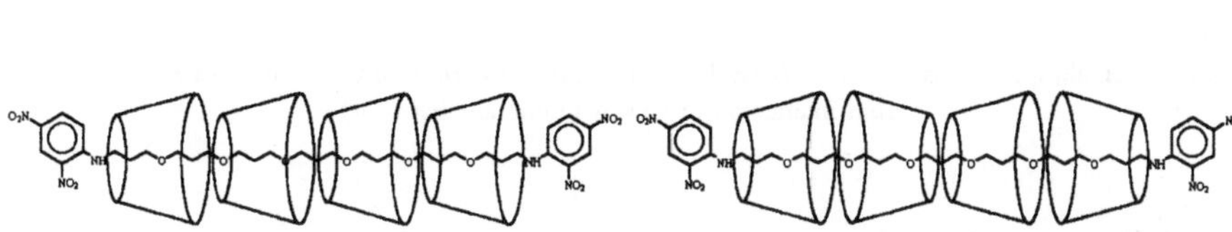

Figure 1. Polyrotaxanes formed by POT and αCDs in position Head-to-Tail (left) RPOT7CD4HT and Head-to-Head (right) RPOT7CD4HH.

2. Methodology for the simulations

MD trajectories of 0.5 ns were computed using Sybyl 6.3 (Tripos Force Field 5.2) at 500K. The molecules studied were isolated end-capped POT7 chains and polyrotaxanes abbreviated as RPOT7CD*m*HT or HH, where 7 denotes the number of oxytrimethylene units, *m* the number of CDs and -HT (-HH) means head-to-tail CD orientation (head-to-head). Characteristics of the simulations were described previously [4-6].

3. Study of the POT:αCD ratio in the rotaxane

Calulations were performed on RPOT7CD*m*HT with *m* in the range 2-5. *Figure 2* depicts the negative values of $E_{binding}$ as a function of *m*. $E_{binding}$ shows that a stabilization of the complex occurs as *m* increases from 2 to 4 and then decreases for *m*=5. We interpreted the negative slope in *Figure 2* to signify that the interaction of one bound CD with its bound neighbor contributes to the stabilization of the complex. For this reason, continued incorporation of additional CDs is favored until the capacity of the chain has been saturated. Similar results were obtained with other PEG:αCD [4] and PPG:βCD [5] polyrotaxanes. The increase of the interaction energy between POT and CDs when adding a fifth CD comes mainly from its repulsive interactions with the bulky end groups. The $E_{binding}$ minimum for POT-CD*m* at *m*=4 implies a preferred stoichiometric composition near the 1.75 oxytrymetilene units per CD, which is similar to Harada's experimental results of 1.5-2 oxytrymetilene units per CD [1].

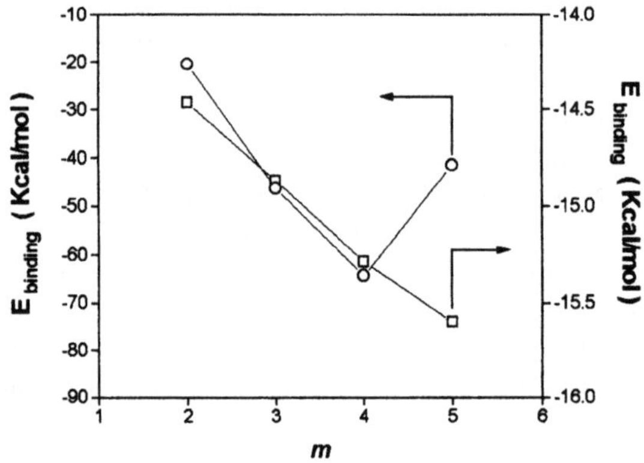

Figure 2. Binding Energies for RPOT7CDmHT (-O-), and that portion of the Binding Energy that arises from Interaction of CDs with one another (-□-)

4. Stabilization of the polyrotaxane

The total potential energy, $E_{binding}$ and its components were obtained for two polyrotaxanes named RPOT7CD4HH and RPOT7CD4HT. Both polyrotaxanes have negative $E_{binding}$ values. The van der Waals between POT and CDs are the most important contributions, with -68.2 Kcal/mol for RPOT7CD4HH and -69.2 Kcal/mol for RPOT7CD4HT. Hardly 2% of the $E_{binding}$ energy is due to electrostatics, the remaining 98% is due to van der Waals. $E_{binding}$ differences between -HH or -HT forms are not very large. However, non-bonded interaction energies between pairs of neighbour CDs are more negative for the head-to-head sequences than for the head-to-tail or tail-to-tail ones.

5. Intra and intermolecular hydrogen bonding interactions of CDs in the polyrotaxanes

The evaluation of the number of hydrogen bonds (HB) was carried out assuming that a HB is formed when the O···H distance is 0.8-2.8 Å and the angle O···H−O is in the range 120-180°. Approximately two intramolecular HBs per CD unit were obtained during the simulation. The most important results are the total number of intermolecular HBs between CDs and the contribution of each pair of CDs. Intermolecular hydrogen bonds between CDs are more numerous with -HH sequences (5.7) than with -HT ones (3.7), thus supporting the conclusions obtained by Harada's group [1]. The largest contribution to the total intermolecular HB for HH sequences comes from the head-to-head interactions between pairs 1-2 and 3-4 (2.3-and 2.8 HB respectively). Tail-to-tail interactions give a small contribution (0.6) to the number of intermolecular HBs between CDs. The intermolecular HBs from the interactions CD-POT and CD-end group are less important.

6. Analysis of the CDs in the polyrotaxanes

The average value for the bond angles τ at the bridging oxygen atom is 117.9°, close to the result of 117.7° obtained from the analysis of isolated αCDs [6]. The distribution functions for the ϕ and ψ torsion angles at the bridging oxygen atom show a single region around the *trans* state. There is no indication of any population of *cis* state as was observed for isolated CDs. This *cis* state was primarily responsible for the distortion of isolated CDs [6]. Any of the χ torsion angles at C(5)-C(6) can visit all three g^+,g^- and *trans* states during the 0.5 ns MD simulation.
Several parameters related to the size and shape of the CD cavities, as well as the distortion and flexibility of CDs were obtained [4,5] for CDs in the complex and they were compared with those for the isolated CDs [6]. Analysis suggests that αCDs in polyrotaxanes adopt a more cylindrical and symmetrical macroring conformation, as compared to the isolated one. Values of standard deviation of the root-mean-square radius of the giration of the six bridging oxygen atoms, which is smaller for CDs in the complex (0.06) than for the isolated one (0.16), shows that its flexibility in the complex is smaller.

7. POT chain in the polyrotaxanes

The end-to-end distance and the radius of gyration of POT chains containing seven monomer units were computed both when the chain was isolated and when it was forming a polyrotaxane. *Figure 3* shows the history of **s** and **r** during the 0.5 ns trajectory MD simulations. The considerably larger values of these parameters when the POT chains are forming polyrotaxanes, as well as the smaller fluctuations, indicate the larger population of *trans* states and the lower mobility of internal bonds for POT chains in the complex as compared to the isolated one. This feature in the complexes is the origin of some crystallographic characteristics of polyrotaxanes.

Quantitative comparison of the distribution function for the four internal dihedral angles at each oxytrimetylene unit permits us to reach similar conclusions. Table 2 collects the average population of *trans* states over all internal bonds for isolated POT chains and the POT in polyrotaxanes.

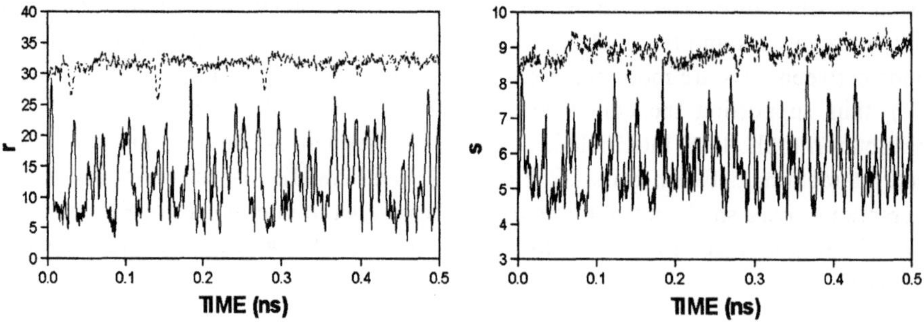

Figure 3. Instantaneous values of the End-to-end Distance (left) and the Radius of Giration (right), for the Isolated POT7 (botton) and RPOT7CD4HH (top) during the 0.5 ns of MD trajectories

TABLE 2. Average Population (%) of the *trans* State for all Internal Torsional Angles of the Isolated POT Chain and POT in the complex. ε_1 (O-CH_2-CH_2-CH_2), ε_2 (CH_2-CH_2-CH_2-O), ε_3 (CH_2-O-CH_2-CH_2) and ε_4(CH_2-CH_2-O-CH_2)

Compound	ε_1	ε_2	ε_3	ε_4
POT7	45	46	64	23
RPOT7CD4HH	82	87	77	72
RPOT7CD4HT	83	91	74	86

8. References

1. Harada, A.,Okada, M. and Kamachi, M. (1995) Complex Formation between Poly(oxytrimethylene) and Cyclodextrins, *Acta Polym.* 46, 453-457.
2. Madrid, J.M., Pozuelo, J., Mendicuti, F. and Mattice, W.L. (1997) Molecular Mechanics Study of the Inclusion Complexes of 2-Methyl Naphthoate with α- and β- Cyclodextrins. *J. Colloid Interface Sci.*, 193, 112-120.
3. Madrid, J.M., Mendicuti, F. and Mattice, W.L. (1998) Inclusion Complexes of 2-Methylnaphthoate and γ-Cyclodextrin: Experimental Thermodynamics and Molecular Mechanics Calculations, *J. Phys. Chem.* B. 102, 2037-2044.
4. Pozuelo, J., Mendicuti, F. and Mattice, W.L. (1997) Inclusion Complexes of Chain Molecules with Cycloamiloses: 2. Molecular Dynamics Simulations of Polyrotaxanes formed by Poly(ethylene glycol) and α-Cyclodextrins, *Macromolecules,* 30(12), 3685-3690.
5. Pozuelo, J., Mendicuti, F. and Mattice, W.L. (1998) Inclusion Complexes of Chain Molecules with Cycloamiloses: 3. Molecular Dynamics Simulations of Polyrotaxanes formed by Poly(propylene glycol) and β-Cyclodextrins, *Polymer J.* 000-000.
6. Pozuelo, J., Madrid, J.M., Mendicuti, F. and Mattice, W.L. (1996) Inclusion Complexes of Chain Molecules with Cycloamiloses: 1. Conformational Analysis of the Isolated Cycloamyloses Using Molecular Dynamics Simulations, *Comput. Theort. Polym. Sci.*, 6, 125-134.

KINETIC BEHAVIOR OF β-CYCLODEXTRINS IMMOBILIZED IN PEEK-WC MEMBRANES

A. GORDANO[1], F. TROTTA[2], C. MANFERTI[2], E. TOCCI[1], E. DRIOLI[1]

[1] Research Institute on Membranes and Modelling of Chemical Reactors IRMERC - CNR, c/o Department of Chemical and Materials Engineering, University of Calabria, I-87030 Rende, Italy

[2] Department of Inorganic, Physical and Material Chemistry, University of Torino. V. San Pietro Giuria, I-10125 Torino, Italy

1. Introduction

Flat sheet membranes made of modified polyetheretherketone known as PEEK-WC, charged with O-octyloxycarbonyl β-cyclodextrins (β-CD) were prepared by the phase inversion method.

Our contribution regards the analysis of the catalytic behaviour of β-CD acyclic carbonate derivative, dispersed in PEEK-WC polymeric membrane, on the rate of the hydrolysis reaction of p-nitrophenylacetate (PNPA) to p-nitrophenol (PNP).

In previous studies [1] the influence of pH, substitution degree (DS) of CD and CD concentration on the trend of hydrolysis reaction in the membrane catalytic reactor has been examined.

In this study we worked in the optimal conditions determined previously [1] examining the effect of temperature and substrate concentration and estimating the reaction rate.

By using immobilized membrane β-CD a significant improvement of reaction rate in comparison with the PEEK-WC membrane has been observed.

To analyze also the catalytic membrane theoretically, a modelling of the amorphous polymer filled with a β-cyclodextrins has been performed.

2. Membrane

The membranes were prepared following the traditional phase inversion process [2]. The membrane forming the system is composed of PEEK-WC (15%), O-octyloxycarbonyl β-CD derivative (7,5%), N,N-Dymethilformammide (DMF) (77,5%) as solvent and water as non solvent.

Both membranes show a linear dependence of water flux on the applied pressure gradient, at constant temperature.

3. Hydrolysis of PNPA

PNPA hydrolysis reaction is an ideal model reaction for its simple mechanism, widely investigated, to obtain information on the catalytic and selectivity properties of CDs [3,4,5].

The hydrolysis of esters occurs spontaneously in alkaline solutions; also in the absence of CD a production of PNP is observed, but the reaction rate is significantly higher when the reaction is carried out in the β-CD carbonate membrane reactor than in the same without β-CD derivative.

A flat cell was used. The solution of PNPA in phosphate buffer, pH 8,4, permeated through the membrane with constant flow rate having a pressure difference as driving force, DP = 0,01 ± 0,002 bar. The membrane area was 133 cm^2.

4. Discussion and conclusion

This study shows the performance of a novel design of catalytic membrane reactor, in which the specific properties of a non conventional catalyst immobilised in a polymeric membrane can promote and extend new applications of these systems.

In Figure is shown the trends of the hydrolysis at optimal concentration of PNPA of 1.2×10^{-4} M in function of different temperatures.

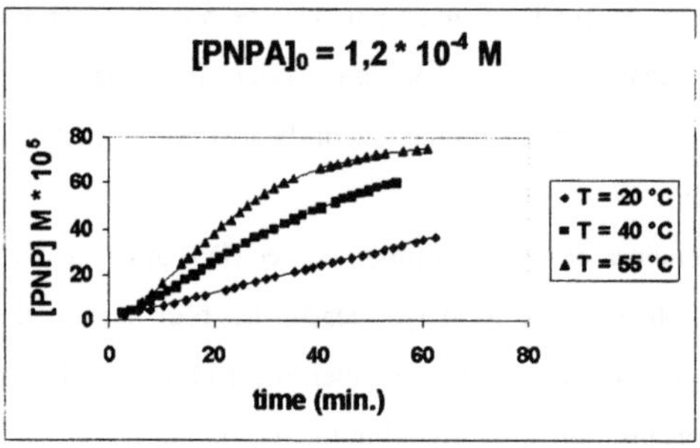

The calculation of the reaction rate produces a pseudo-first order kinetic [5].

In conclusion, the use of a polymeric membrane functionalised with O-octyloxycarbonyl β-CD derivative enhances the hydrolysis reaction rate, optimizes the interaction between CDs and the substrate, increases the chemical stability of the catalyst and allows the reuse of the catalytic membrane.

At the moment we are studying also the hydrolysis reaction of another kind of substrate, the p-nitrophenyldiphenylphosphate (PNPDPP) to PNP. In this case, the reaction doesn't occur without cyclodextrins.

A detailed kinetic study of the system is in progress.

5. References

1. Drioli, E., Natoli, M., Koter I. and Trotta, F. (1995) An Experimental Study on a β-Cyclodextrin Carbonate Membrane Reactor in PNPA Hydrolysis. Biotechnology & Bioengineering, **46**, 415-420.
2. Kesting, R.: (1985) Synthetic Polymeric Membranes, Wiley Interscience, New York.
3. Fujita, K., Akihiro, S., Taiji, I. (1980) Hydrolysis of phenyl acetates with capped β-cyclodextrins, Bioorg. Chem., **4**, 237-249.
4. Kitaura, Y., Bender, M. L. (1975) Ester hydrolysis catalysed by modified cyclodextrins, Bioorg. Chem., **4**, 237-249.
5. Van Etten, R. L., Sebastian, J. F., Clowes, G. A., Bender, M. L., (1967) Acceleration of phenyl ester cleavage by cycloamylose. A model of enzymatic specificity, J. Am. Chem. Soc., **89**, 3242-3252.

Chapter 5

CYCLODEXTRINS IN ENVIRONMENT SCIENCES

APPLICATION OF CYCLODEXTRINS IN NUCLEAR WASTE MANAGEMENT

L. SZENTE, É. FENYVESI, J. SZEJTLI

Cyclolab R&D Lab., Budapest, Dombóvári út 5-7, H-1117 Hungary

1. Introduction

Since the Chernobyl nuclear power plant disaster, authorities have been regularly inspecting the security protection systems -that are mainly air filters- installed in the plants with the special aim to adsorb effectively radioactive iodine vapors emitted upon normal-or minor malfunctions of the nuclear power plants. The malfunction of nuclear power plants in most cases results in the generation of considerable amounts of radioactive elemental and small amounts of organic iodine in form of vapor.

The iodine waste formation in nuclear power plants is related to the fission process of Uranium 235 isotope as described below:

$$U^{235} \rightarrow Te^{135} \rightarrow \mathbf{I^{135}} \rightarrow \mathbf{Xe^{135}} \rightarrow Cs^{135} \rightarrow Ba^{135}$$

As two of fission intermediates are of "gaseous" phase they are the primary targets for effective entrapment and immobilization, to prevent spreading of nuclear contaminants.

It is also of environmental importance, that the half life time values of these gaseous fission products are still in <u>several hour range</u>, while that of the Cesium^{135} isotope is already in the <u>million year range</u>. Consequently if iodine can be effectively immobilized then the radioactivity remains localized, and such wide spreading of radioactivity what happened at Chernobyl nuclear disaster can be prevented.

At present the nuclear security systems installed in most of the running nuclear power plants are simple or coated activated carbon filled air-filters.

Why to employ cyclodextrins in iodine traps? It has long been known that starch and starch derivatives react with elemental iodine and this specific and very sensitive reaction has been routinely used in the analytical chemistry. The reaction of cyclodextrins with iodine has also been known since late forties.(1) Cyclodextrins are known to form very stable inclusion complexes with iodine and tri-iodide ions. (2, 3)

The purpose of this study was to investigate whether aqueous solutions of adequate types of cyclodextrins or water insoluble cyclodextrin bead polymers are useful to effectively entrap emitted iodine from air. This practical use is based on the fact that elemental iodine forms stable inclusion complexes with cyclodextrins in presence of water.

2. Experimental

2.1. MATERIALS

The α-, β- γ-cyclodextrins, and the randomly methylated α- β- and γ-cyclodextrins (RAMEA, RAMEB and RAMEG) were produced by Wacker Chemie (Munich).

The 2-hydroxypropylated-β-cyclodextrin (HPBCD DS=4.7) and heptakis 2,6-di-O-methylated α- and β-cyclodextrins (DIMEA, DIMEB) were prepared by Cyclolab.

Branched-β-cyclodextrin was the product of Ensuiko Sugar Refining Co. (Yokohama)

Elemental iodine of analytical grade was purchased from Merck, Co. (Darmstadt)

Epichlorohydrin cross-linked α-and β-cyclodextrin polymers were produced by Cyclolab. These water insoluble polymers appear as yellowish beads, have an average grain size of 0.05-0.3 mm, and their cyclodextrin content varies between 50-55%. All other reagents and chemicals were of analytical grade.

2.2. METHODS

Solubility studies: The interaction between elemental iodine and CDs in water was studied by using Higuchi-Connors-type phase solubility method. The dissolved iodine concentration in the equilibrated aqueous solutions was determined by sodium-thiosulfate titration.

Assessment of iodine binding capacity of CDs: Iodine vapor was generated by heating solid iodine of analytical grade in chamber connected by a gas-inlet type adapter directly to the absorbing glass tube filled with aqueous cyclodextrin solutions. This glass tube was mounted into another gas washing tube filled with aqueous starch solution to detect the first appearance of escaped elemental iodine from the cyclodextrin solution, in other words to detect the saturation value of the aqueous cyclodextrin solutions by elemental iodine. The generated iodine vapor was bubbled through the aqueous solutions using nitrogen gas, dry- and humid air-stream with a constant speed, and the temperature of absorbing solution was maintained at 25 °C throughout the experimental runs. The iodine concentration of air or N_2 gas was about 0.02 mg/cm^3 . The absorbed amount of elemental iodine vapor in the aqueous cyclodextrin solutions was determined by $Na_2S_2O_3$- titrimetry.

Immobilization of iodine on cyclodextrin polymer-filled columns: The iodine binding potency of α-and β-cyclodextrin bead polymers was tested on an adsorber column filled with 3 g cyclodextrin bead polymers. Iodine vapor was passed through this column of a gel volume of approximately 10 cm^3 with an emission rate of 100 cm^3 per minute at

25°C. The binding potency of both the dry and swollen (wetted with water or with 0.1 N KI solution) polymers was determined. The breakthrough of iodine on the polymer bed was detected as iodine (blue color) appeared in the post column starch solution. The quantitative determination of amount of bound iodine was made by titration.

3. Results

3.1. SOLUBILIZATION POWER OF CYCLODEXTRINS TO ELEMENTAL IODINE

The solubilizing effect of various cyclodextrins on elemental iodine can be regarded as a measure of their "iodine-binding" capacity under laboratory test conditions. Table 1. lists aqueous iodine solubility data in different CD solutions which undoubtedly point to the superiority of methylated CD derivatives.

TABLE 1. Solubility of elemental iodine in aqueous cyclodextrin solutions at 25 °C in mg/mL (each value is a mean of three parallel determinations)

CDs	concentration of applied cyclodextrins (%)						
	0	1.5	5	8	10	20	40
αCD	0.43	0.2*	0.5	-	-	-	-
βCD	0.43	0.2**	-	-	-	-	-
γCD	0.45	0.6	0.2	-	-	-	-
DIMEA	0.45	3.0	12.2	20.2	24.3	40.1	88.0
RAMEA	0.45	2.4	11.9	19.0	21.5	38.8	81.2
DIMEB	0.45	0.9	5.0	7.8	11.1	23.0	48.6
RAMEB	0.45	1.0	4.8	8.0	10.3	21.6	45.8
RAMEG	0.45	0.5	1.0	2.2	3.6	4.8	6.5
$G_2\beta CD$	0.45	0.5	1.3	1.6	1.8	2.9	4.0
HPBCD	0.45	0.6	0.8	1.2	1.4	3.0	3.6

**reaction of iodine results in blue color formation, **reaction of iodine results in greenish gray color formation*

Among the studied CDs the most suitable one for binding of elemental iodine in aqueous system is the methylated α-cyclodextrin, followed by the methylated β-cyclodextrin. It was found that the methylated γ-cyclodextrin does not provide iodine binding of practical relevance.

The application of aqueous solutions of parent cyclodextrins for absorption of elemental iodine vapor has been shown unfeasible, since the solubilization potency of these type of cyclodextrins was significantly inferior to that of the methylated analogues.

3. 2. IODINE VAPOR BINDING CAPACITY OF CYCLODEXTRIN SOLUTIONS

The experimental set up illustrated in Figure 1. was used to measure the iodine vapor binding capacity of aqueous cyclodextrin solutions during simulated iodine emission. The results of the iodine vapor absorption studies are listed in Table 2.

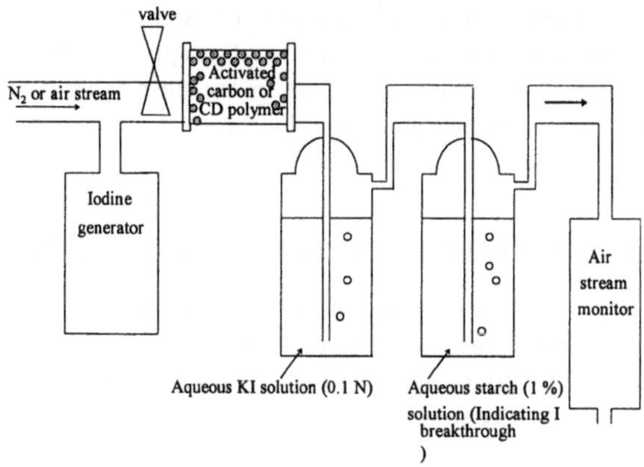

Figure 1. Set up for iodine trap assessment

TABLE 2. Binding of elemental iodine vapor by aqueous cyclodextrin solutions at 25 °C

type of absorber solution	elemental iodine binding (mg iodine per mL of solution)		
	in N_2 gas	in dry air	in humid air (R.H. 95%)
5% RAMEA	7.4	9.2	8.4
40% RAMEA	57.2	61.0	56.8
5% RAMEB	2.8	3.5	3.0
40 % RAMEB	21.3	23.8	19.0

3.3. IMMOBILIZATION OF IODINE VAPOR ON CYCLODEXTRIN POLYMER-FILLED COLUMNS

Both the α- and β-type of epichlorohydrin-cross-linked cyclodextrins were found to show effective iodine sorption capacity under laboratory test conditions. The characterization of the iodine immobilizing capacity of the polymer-filled columns was expressed by breakthrough time which was defined by the time when first iodine escaped from column during continuous iodine emission in circulating air. The results of this test are listed in Table 3.

TABLE 3.Iodine binding performance of cyclodextrin polymer filled columns as air filters

Packing	breakthrough time (hours)	iodine went through (mg)	iodine measured in the solution (mg)	iodine sorbed by the packing (mg)
αCDP dry	1	12.6	6.7	5.9
αCDP wetted by water	9.5	119.7	0.6	119.1
αCDP wetted by 0.1 N KI solution	**18**	226.8	4.9	221.9
βCDP dry	1	12.6	8.8	3.8
βCDP wetted by water	3	37.8	0.9	36.8
βCDP wetted by 0.1 N KI solution	**16**	201.6	8.8	192.8

As can be seen from the above data both the αCD and βCD based bead can be used as effective iodine immobilizers for air filtration, particularly those pre-wetted with KI solution. The solid, iodine saturated CD-polymers were then tested for iodine retention upon storage at elevated temperature. The heat resistance of entrapped iodine in polymer matrix and in the traditional charcoal adsorber is compared in Table 4.

The improved heat resistance of iodine in the cyclodextrin-polymer matrix provides a further advantage of these novel type of iodine traps, as they not only entrap iodine effectively, but also preserve it even at elevated temperature for 12-24 hours.

TABLE 4. Loss of complexed iodine from CD-polymer packing upon storage at 60°C in open dishes. All data are presented in % of the total iodine load in adsorbents.

Packing	**loss of iodine (%) during storage at 60°C**				
	time zero	**4 hours**	**8 hours**	**12 hours**	**24 hours**
dry αCD-polymer	0	6	9	10	10
swollen αCD-polymer	0	9	14	15	15
dry βCD-polymer	0	12	12	21	29
swollen βCD-polymer	0	17	17	27	32
activated carbon	0	28	34	62	72

4. Conclusions

From the results of the above preliminary investigations the following conclusions have been drawn:

•among studied monomer cyclodextrins methylated β- and α-cyclodextrins are the most effective iodine binding agents, thus a solution-type iodine trap can be constructed by employing methylated α- or β-CDs

•the water insoluble CD bead polymers can be used for column packing for effective and selective iodine immobilization in air filters

•CD polymers were found to provide more effective, much more selective iodine binding than the activated carbon

Further studies are in progress to optimize the construction of iodine traps and to assess their iodine binding performance under standardized conditions according to official guidelines using radioactive iodine source.

Acknowledgment

Authors thanks are due to the Hungarian Research Fund (OTKA T 022002) for supporting the present project.

REFERENCES:

1. Thoma, J.A. and French, D. (1958) The Interaction of Cyclohexaamylose with Iodine and Iodide, *J. Am. Chem. Soc.* 80, 6142

2. Sanemasa, I. (1986) Measurement of Association Constant of Iodine with Cyclodextrins, *Bull. Chem Soc . Japan.* 59, 2269-2272
3. Diard, J. (1985) Potentiometric Association Constant Measurements of α- β- and γ-Cyclodextrin Complexes Involving Iodine, Tri-iodide or Iodide Species, *J. Electroanal. Chem. Interfacial Electrochem.* 189, 113-120.

OPTIMIZATION OF FUEL OIL DESULFURATION BY β-CYCLODEXTRIN

M.MARZONA and R. CARPIGNANO
Dipartimento di Chimica Generale e Organica Applicata – Università di Torino – Corso M.D'Azeglio 48 – 10124 Torino – Italy

1. Introduction.

Fuel oils contain 0.1-3% of a number of sulphur aromatic compounds including benzo- and dibenzotiophene which oxidize to SO_2, the main source of the acidic rains. Removal of sulphur aromatic compounds is a topic of great industrial interest.
β–CD is a natural product, cheap and easily avalaible and its inclusion complexes find increasing industrial applications in the extraction of undesiderable compounds, like caffein and cholesterol from foods and beverages.
This work aims to find out if the cyclodextrin complexation could be a effective method to solve the problem of fuel oil desulphuration. As chemical composition of fuel oil and sulphur impurities is rather complex, we chosen a model system constituted by a solution of dibenzothiophene (DBT) in a hydrocarbon, n-esane or n-dodecane, to explore the applicability of β-CD in desulphuration.
The DBT extraction process has been investigated and optimized by using the Experimental Design techniques.

2. Optimization procedure

Optimizing a process means to determine the experimental conditions that give an optimal performance. In the present study the problem can be defined as maximizing the dibenzothiophene extraction by cyclodextrin complexation. Multivariate Experimental Design techniques (1) are appropriate to evaluate the effects on the response of the variables studied as well as of their interactions.
The approach we used is that of carrying out a preliminary study by the strategy of Fractional Factorial Design with the aim of selecting the most important variables that were studied in a following step by a Central Composite Design.

3. Results and Discussion

Factorial Design
Seven variables were investigated in a 2^{7-3} Fractional Factorial Design (1). The variables and their levels are reported in Table 1 The response was the percentage of DBT extracted from the organic solution by an aqueous solution of β-cyclodextrin.
In Table 1 the matrix of the 2^{7-3} design and the response values are reported.

Table 1. 2^{7-3} Fractional factorial design matrix and responses.

run	X_1	X_2	X_3	X_4	X_5	X_6	X_7	Y_{sper} % Extr
1	-	-	-	-	-	-	-	**13.7**
2	+	-	-	-	+	-	+	**30.2**
3	-	+	-	-	+	+	-	**35.1**
4	+	+	-	-	-	+	+	**25.3**
5	-	-	+	-	+	+	+	**49.0**
6	+	-	+	-	-	+	-	**7.7**
7	-	+	+	-	-	-	+	**15.1**
8	+	+	+	-	+	-	-	**25.2**
9	-	-	-	+	-	+	+	**19.6**
10	+	-	-	+	+	+	-	**28.8**
11	-	+	-	+	+	-	+	**46.0**
12	+	+	-	+	-	-	-	**14.7**
13	-	-	+	+	+	-	-	**13.7**
14	+	-	+	+	-	-	+	**13.0**
15	-	+	+	+	-	+	-	**21.4**
16	+	+	+	+	+	+	+	**41.0**

The experiments were performed in random order at each temperature. The total volume (100 ml of organic + aqueous phase) was shaken by a mechanical stirrer.

Table 2. Variables investigated in the fractional factorial design and their levels.

Variables	Levels (-)	Levels (+)
X_1 temp.(C°)	25	35
X_2 time (h)	24	90
X_3 solvent	n-C_6	n-C_{12}
X_4 pH	4	7
X_5 β-CD (g/l)	9.25	18.5
X_6 DBT (g/l)	0.5	3
X_7 β-CD/DBT (mol/mol)	1	10

Table 3. Main effects of the variables

Variables	Main effects
X1	-1.71
X_2	6.64
X_3	-2.70
X_4	0.21
X_5	16.90
X_6	7.70
X_7	9.40

Data were analyzed calculating the main effects and the second order interactions by the Yates algorithm (1). Table 3 lists the calculated effects of the variables.
The results show that the extraction is mainly affected , positively, by β-CD solution concentration(X_5), β-CD/DBT ratio (X_7), DBT concentration (X_6) and time of extraction (X_2), while temperature, pH, and solvent do not seem to be influent. Also interactions, not reported, do not show important values.

Response surface method

Information from the fractional factorial design was used to run a Central Composite Design (1).

Variables studied were DBT concentration and β-CD/DBT ratio. Temperature, pH, time were kept constant at 25°C , 7 and 90 hours respectively. n-dodecane was chosen as solvent and the maximum concentration of aqueous solution of β-CD(18.5 g/l) was used. The Central Composite design with two variables and four replicates at the central point is reported in Tables 4 and 5. Again the response was the percentagey of extraction of DBT.

Table 4. Experimental domain and coding of the variables in the Central Composite Design

Levels	-1.41	-1	0	1	1.41
$R_{CD/DBT}$ (mol/mol)	0.5	1.9	5.25	8.6	10
DBT conc. (g/l)	0.5	1.23	3	4.77	5.5

Table 5. Matrix of the Central Composite Design in two variables and responses

	Variables		Response		Variables		Response
Run	X_1	X_2		Run	X_1	X_2	
1	-1	-1	32.6	7	0	-1.41	26.0
2	-1	+1	31.8	8	0	+1.41	54.0
3	+1	-1	33.5	9	0	0	47.2
4	+1	+1	56.8	10	0	0	41.0
5	-1.41	0	21.5	11	0	0	43.1
6	+1.41	0	45.7	12	0	0	49.0

All the experiments were performed in random order.
With a Central Composite Design it is possible to determine a response surface, the projection of which is called isoresponse diagram. The mathematical form of a response surface is a second order polynomial equation correlating the causal variables x and the response y
The polynomial equation obtained, using the MODE Program package (UMETRI, Sweden) is the following (I) :

$$y = 45.1 + 7.5\, x_1 + 7.8\, x_2 - 5.3\, x_1^2 - 2.1\, x_2^2 + 6.0\, x_1 x_2 \qquad (I)$$

($R^2 = 0.93$; $s = 3.9$, $p < 0.004$)

The response surface shows a maximum and the isoresponse diagram is represented in figure 1. The maximum of extraction is 70.4% and corresponds to the following values: x_1 [β-CD/DBT]=1.8 (real value11.6); x_2 [DBT conc]=2 (real value 6.5 g/l).

Validation of the model

The reliability of the response surface predictions was evaluated by experiments run near the optimum conditions and at the central point of the design.
Differently from the former experiments, the total volume (organic + aqueous phases) was increased to 300 ml and a magnetic stirrer was used for the extractions.
The experimental conditions explored and the corresponding responses are reported in Table 6.

Fig 1. Isoresponse diagram

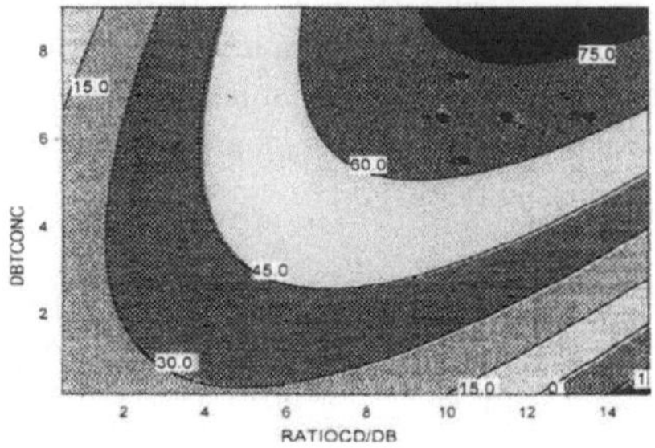

Table 6. Conditions and responses of model validation runs.

Run	X_1 (ratio β-CD/DBT)		X_2 (DBT conc.)		Responses
	Coded values	Real values	Coded values	Real values	E%
1	0	5.25	0	3	48.4
2	1.6	10.6	2	6.5	85.0
3	1.8	11.5	2	6.5	85.7

The results show a conspicuous increase of the yield of extraction, presumably due to a better mixing of the organic and aqueous phases.

To check the validity of the conditions predicted for the maximum the surrounding dominion was explored by Simplex method (2).

The experiments run and the corresponding responses are summarized in Table 7 and depicted in figure 1.

Table 7. Experimental conditions and responses of Simplex design

run	X_1(β-CD/DBT ratio)		X_2(DBT conc.)		Responses
	Coded values	Real values	Coded values	Real values	E%
0	1.8	11.5	2	6.5	85.7
1	2	12.0	3	8.3	74.3
2	1.5	10.3	2.5	7.4	75.2
3	1.25	9.4	2	6.5	80.3
4	1.5	10.3	1.5	5.6	74.9
5	2.4	13.5	2	6.5	81.3

The results confirm the predicted maximum conditions and indicate the importance of a good contact of the two phases for increasing the yield of extraction.

4. Conclusions

The experimental data so far collected by Experimental Design techniques show that a satisfactory desulfuration can be achieved (ca.86%), taking into account the simple procedure required.

5. References

1. Box G.E.P. Box, Hunter W.G., Hunter J.S., Statistics for experimenters, J. Wiley, New York (1978)
2. Carlson R., Design and optimization in organic synthesis, Elsevier, Amsterdam (1992)

INCLUSION OF ENVIRONMENTAL CONTAMINANTS WITH β-CYCLODEXTRIN, AND ADSORPTION AND REMOVAL BY β-CYCLODEXTRIN POLYMER

Kenjiro HATTORI[1] and Shoji MURAI[2]
[1]*Tokyo Institute of Polytechnics, Atsugi, Kanagawa 243-0297, Japan*
[2]*Kanagawa Industrial Technology Research Institute, Ebina, Kanagawa, 243-0435, Japan*

1. Introduction

Removal of some organic compounds by inclusion and adsorption on β-cyclodextrin was investigated for the purpose of environmental protection. The guest compounds, which are considered to be environmental contaminants, were nonionic[1] and ionic surfactants[2], phthalates[3] and chlorinated hydrocarbons. In a previous study, the removal of environmental contaminants from aqueous solution was investigated using activated carbon or mineral clay. With the view of making effective use of natural sources, effluents have to be widely recycled for industrial use and so a new system of recovering organic compounds from water and recycling these compounds should now be established.

2. Materials and Methods

2.1 MATERIALS

The nonionic surfactants were commercially available octylphenol ethoxylates (OPE) and nonylphenol ethoxylates (NPE). Both of them have a hydrophilic ethylene oxide group ranging from 3 to 22 units (mainly 10 units). The structures of OPE and NPE are shown in Fig. 1. Sodium dodecylbenzenesulfonate (DBS) and benzalkonium chloride (BKC) are frequently used anionic and cationic surfactants. Analytical grade dimethyl (DMP), diethyl (DEP), dipropyl (DPP), dibutyl (DBP), diheptyl (DHpP), and di-(2-ethylhexyl)phthalate (DEHP) were commercially available. Chlorinated hydrocarbons, trichloroethylene (TrCE) and tetrachloroethylene (TeCE) were also commercially available. β-Cyclodextrin (β-CD) and the β-cyclodextrin polymer (β-CDP) were supplied from Japan Maize Company. The presumed mole equivalent of β-CDP was 0.37 m mol/g and the spherical diameter was about 0.1-2 mm.

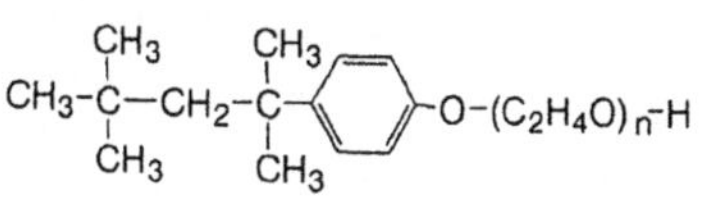

OPE

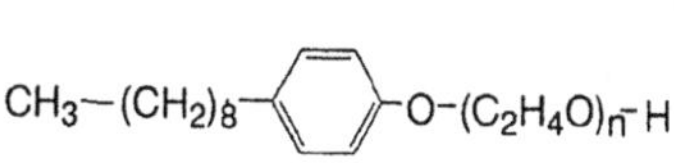

NPE

Fig. 1 Structures of OPE and NPE

2.2 METHODS

Inclusion was certified by the proton NMR spectra by the resonance shifts(δ) with a JEOL JMN-GSE400NMR spectrometer. Stability constants of the guest with β-CD, K, were calculated by plotting of the Benesi-Hildebrand equation with a competitive inhibition method using the fluorescence intensities of TNS. Stoichiometry with β-CD was determined by the Job plots of the UV spectra and conductivity measurements. HLB values were calculated by the equation; HLB = 7+11xlog (M_w/M_0). M_w and M_o are the weights of the hydrophilic and lipophilic parts in each OPE and NPE. The octanol-water partition coefficients, K_{ow},, in Table 2 were cited from the literatures; 1)Ellington J.J. *et al.,EPA Environ.Res.Breif,S.96/006.Sep 1996* and 2)Howard P.H. *et al.,Environ.Toxicol.Chem.*, **4**,653(1985). Adsorption isotherm tests for a guest on β-CDP were conducted with one gram of β-CDP and 100 ml of 5 mM phosphate buffer (*p*H 7.0) by varying the concentration (0-12 mM) of the guest. They were shaken for 24 h at 25 °C to ensure equilibrium. The equilibrium concentration, C, was measured by UV spectrometry or GC. The relationship between adsorption amount q and the logarithm of C could be described by the Freundlich equation; $q = K_f C^{1/n}$. By plotting C/q as a function of C, the graphic presentation showed the Langmuir adsorption isotherm. Breakthrough curves were observed by the column adsorption test, which was done using a Pyrex tube that was 13 mm in diameter and 300 mm long with a water jacket. Five grams of β-CDP was loaded on the column. A solution of 0.2 mM DMP or DPP was fed at a 1.3 ml/min rate into the column at 25.0 °C. Recovery efficiency of the compounds on β-CDP was performed with various methanol-water solutions. One gram of β-CDP with a previously adsorbed amount of the guest was shaken for 2 h with 100 ml of aqueous methanol. The recovery ratio for each solvent was calculated based on the ratio of the recovered guest amount relative to the guest amount previously adsorbed on β-CDP.

3. Results and Discussion

3.1 INCLUSION AND STABILITY CONSTANTS

All of the inspected guest compounds were found to form 1:1 inclusion complexes with β-CD based on Job plots or the linear plots of the Benesi-Hildebrand equation. The stability constants, K, between the guests and β-CD were obtained. The values of K were dependent on the hydrophobic parameters of the guests such as HLB of the nonionic surfactants (Table 1) or the octanol/water partition coefficients, K_{OW}, of the phthalates (Table 2). These results suggests that the simple approximation of K is possible from the structure parameters or partition coefficients of the homologous surfactants or phthalates.

Table 1 Stability constants (K) for β-CD with OPE or NPE, and HLB value

	Nos. of EO unit (n)	K $\times 10^3$ M^{-1}	HLB
OPE	7.5	5.0	10.1
OPE	10	3.1	11.5
OPE	30	1.5	16.9
OPE	40	0.5	18.4
NPE	5	6.2	7.8
NPE	10	2.5	11.1
NPE	18	1.7	14.0

Table 2 Stability constants (K) for β-CD with phtalate and octanol-water partition coefficient (K_{ow})

		logK_{ow}	
phthalate	K(M^{-1})	lit.1)	lit.2)
DMP	82	1.60	1.56
DEP	107	2.42	2.47
DPP	300	3.31	
DBP	1160	4.50	
DHpP	2142	6.95	
DEHP	928	7.54	3.98

3.2 ADSORPTION ISOTHERM; FREUNDLICH TYPE AND LANGMUIR TYPE

The adsorption isotherms of β-CDP were dependent on the substrates. The relation between concentration C and the adsorption amount q of the phthalates (DPP and DMP) and chlorinated hydrocarbon (TeCE) were fitted to the Freundlich isotherm (Fig. 2) and the surfactants (OPE and BKC in Fig. 2) were not fitted in the linear plotting, but were fitted in the Langmuir isotherm in the relation between C *vs.* C/q (Fig. 3). It was confirmed that the adsorption phenomenon of surfactants having a hydrophobic group and a hydrophilic group in its own molecule was mainly caused by inclusion into the CD cavity. Otherwise, phthalates and chlorinated hydrocarbons having a nonpolar character adsorbed on β-CDP presumably in various fashion such as inclusions, interaction on crosslinked bridges or on the outer part of the CD cavity.

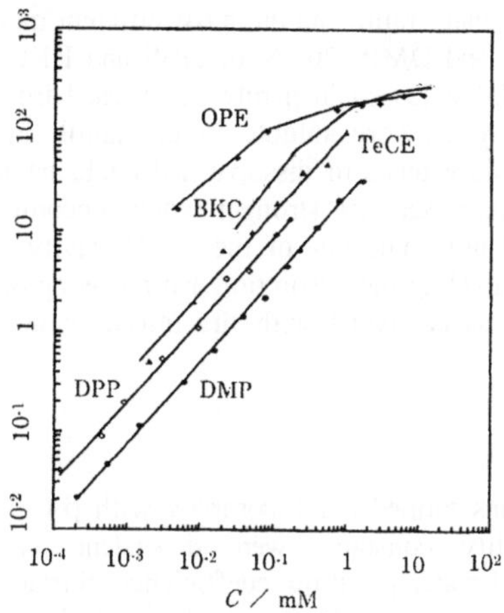

Fig. 2 Freundlich adsorption isotherms

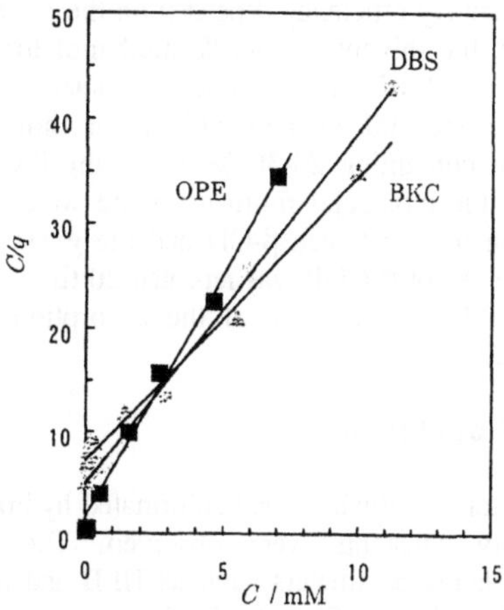

Fig. 3 Langmuir adsorption isotherms

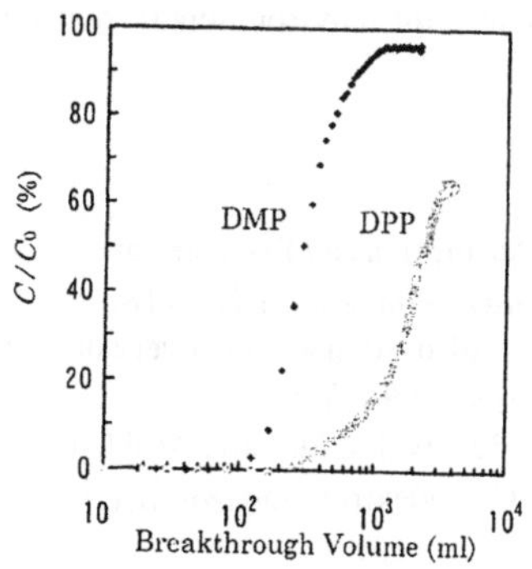

Fig. 4 Breakthrough curves of DMP and DPP

Table 3 Breakthrough volume and capacity of β-CDP with DMP and DPP

	DMP			DPP		
Effluent/Influent C/C_0 (%)	5	10	20	5	10	20
Breakthrough Volume (ml)	137	168	202	370	680	1220
Breakthrough Capacity (μmol g⁻¹)	5.45	6.62	7.70	15.2	26.6	45.2

3.3 BREAKTHROUGH VOLUME

Breakthrough curves of DMP and DPP for β-CDP were observed (Fig. 4). The breakthrough capacity was then calculated (Table 3). The data show that a larger amount of DPP was adsorbed on β-CDP as compared to DMP and the breakthrough capacity ratio of DMP/DPP was in the range of 0.17-0.37 in which C/C_0 was 0.05-0.2. The difference in the breakthrough capacity could be interpreted to be due to the fact that the stability constant of DMP is smaller than that of DPP. This suggests that a compound having a small stability constant such as DMP should be treated over a long contact time with β-CDP by a large column volume or the proper flow rate.

3.4 RECOVERY TEST AND RECYCLING OF β-CDP

The recovery efficiency depended on the methanol/water ratio, and the most efficient mixing ratio of the solvent was 80 % methanol for OPE and DMP, 70 % for DBS and BKC and 100 % for TrCE. The included surfactants and phthalates could hardly be released from β-CDP in the 100 % methanol eluent though they are very soluble in methanol. In the solvent containing 20-30 % water, the β-CD moiety tends to dissolve and could be freed due to the molecular motion; as the water was increased, the stronger would become the binding force between β-CD and the guest could not come out of the β-CD cavity. The recycle use of β-CDP was repeated 20 times by recycling the adsorption and releasing of the guest. More than 90 % of the adsorption ratio as compared with the initial value was assessed.

4. Conclusion

Surfactants, phthalates and chlorinated hydrocarbons formed a 1:1 complex with β-CD and stability constants were observed. The stability constants were dependent on the hydrophobic parameters such as HLB and octanol-water partition coefficients. Surfactants with β-CDP were fitted to the Langmuir type adsorption isotherm equation and phthalates and chlorinated hydrocarbons with β-CDP were fitted to the Freundlich type isotherm equation, reflecting their structural factors. The difference in the breakthrough volume at the column test was due to the stability constants of the guests. The adsorbed compounds were easily recovered by aqueous methanol and the recovery ratio depended on the mixture ratio of methanol. β-CDP was also able to be reused. Therefore, it appeared that the β-CDP was useful for protecting the effusion of contaminants and for environmental remedial technology.

5. References

1. S.Murai,S.Imajo,Y.Maki,K.Takahashi,K.Hattori(1996) Adsorption and Recovery of Nonionic Surfactants by β-Cyclodextrin Polymer: *J.Colloid Interface Sci.*, **183**, 118.
2. S.Murai,S.Imajo,H.Inumaru,K.Takahashi,K.Hattori(1997) Adsorption and Recovery of Ionic Surfactants by β-Cyclodextrin Polymer: *J.Colloid Interface Sci.*, **190**,488.
3. S.Murai,S.Imajo,Y.Takasu,K.Takahashi,K.Hattori(1998) Removal of Phthalic Acid Esters from Aqueous Solution by Inclusion and Adsorption on β-Cyclodextrin: *Environ. Sci. Technol.*, **32**,782.

LUMINESCENT BEHAVIOR OF $Eu(nta)_3dipy$ COMPLEX IN THE PRESENCE OF β-CYCLODEXTRINS

VILA NOVA, S. P; DA SILVA, J. E. C; ALVES JR, S; BATISTA, H. C. N; DE SOUSA, J. D. F; DE SÁ, G. F; MALTA, O. M. L.
Departamento de Química Fundamental, BSTR-CCEN-UFPE, Av. Luiz Freire, s/n Cidade Universitária, 50.740-540, Recife, PE, Brazil

Keywords: luminescence, lanthanide complex, fluorinated β-diketones.

1. INTRODUCTION

The chemical literature attests to the growing importance of rare-earth ions in the development of luminescent materials.[1] Recently, highly luminescent europium (III) adducts with ligands such as fluorinated β-diketones and 2,2'-dipyridyl have been synthesized and suggested as promissing hight-conversion molecular devices (LCMD), as described by Lehn.[2]

Efficient LCMD constitute a new class of materials with potential applications, such as luminescent materials,[3] fluoroimmunoassays[4] and thin film electroluminescent devices.[5]

α-, β-, and γ-cyclodextrins (α-, β-, and γ-CD) are cyclic oligossacarides consisting of six, seven or eight units of glucose. They present cavities of 5-8 Å diameters, with a hydrophobic character. CDs present interesting properties. They can form supramolecular structures such as inclusion complexes (host-guest), nanotubes, etc. It was observed in countless cases that the longevity of chemical properties of molecules associated with CD increased due to stabilization in the CD complex.[6]

With the objective of studying physical properties of lanthanide complexes in the presence of CD's, we present our initial results of luminescent properties of the $Eu(nta)_3dipy$ (nta=4,4,4-trifluoro-1-(2-naphthyl)-1,3-butanedione, dipy = 2,2'-dipyridyl) complex in the presence of β-CD, with tentative explanations for the obtained results.

2. EXPERIMENTAL DETAILS

2.1. Synthesis

Nta and dipyridyl were obtained from Eastman Chemical and used without further purification. Lanthanide oxides Eu_2O_3 (99,99%, Aldrich) and Gd_2O_3 (99,99%, Molycorp) were used to prepare europium (III) and gadolinium (III) chlorides.

Complexes were prepared essentially by the method of ref [7]. Syntheses of $Eu(nta)_3dipy$ and Gd $(nta)_3dipy$ were accomplished by adding a stoichiometric amount of **$LnCl_3.6H_2O$($Ln=Eu^{3+}$ or Gd^{3+})** dissolved in EtOH (1mmol) dropwise with stirring to an ethanolic solution containning dipy (1mmol) and the nta anion (3mmol). pH was controled with ethanolic NaOH to ~6 and the solution refluxed for 12h.

Complexes precipitated under solvent evaporation. $Eu(nta)_3dipy$ or $Gd(nta)_3dipy$ were isolated as powders melting at ~193°C and ~148°C respectively and characterized by IR, UV-VIS and ^{1}H-NMR techniques.
To an aqueous solution of β-CD, we added an ethanolic solution of the lanthanide complex (1/3 mol equivalent). Stirring overnight, evaporation and drying in vacuo at 70°C provided a solute solid.

2.2.Measurements

The absorption spectra of $Eu(nta)_3dipy$ and $Eu(nta)_3dipy.\beta$-CD dissolved in ethanol/water were measured in a UV-visible spectrophotometer LAMBDA 6 model 2688-002. Fig.1 shows the absorption spectra of the $Eu(nta)_3dipy$ in the 200-500 nm spectral range.
For luminescence spectra the Eu^{3+} and Gd^{3+} samples were excited by a 150W Xe-Hg lamp. The wavelengths were selected by a 0.25 monochromator (Jobin Yvon model H-10). Emission spectra were analysed by a Jobin Yvon double monochromator (model U-1000) and the signal, detected by a water-cooled RCA C31034-02 photomultiplier, was processed by a Jobin Yvon Spectralink system.

3.**DISCUSSION**

In absorption spectrum of $Eu(nta)_3dipy$, we observed four bands at 212, 242, 283 and 334 nm. The absorption spectrum of $Eu(nta)_3dipy.\beta$-CD is similar, without change in position or relative intensitys of bands. These correspond to singlet-to-singlet transitions in the ligands, and we can conclude that they are localized in the nta ligands.

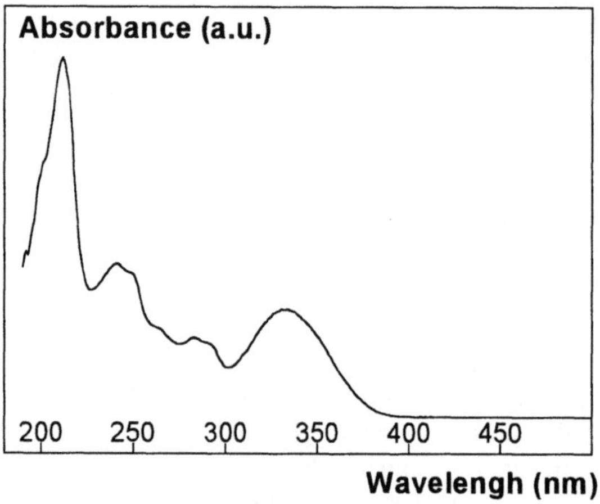

Figure 1: Absorption spectrum of $Eu(nta)_3dipy$

Figure 2 shows the room temperature emission spectra of the $Eu(nta)_3dipy$ (*A*) and $Eu(nta)_3dipy.\beta$-CD (*B*) compounds respectively, in the 560-720nm spectral range. Both of them show the Eu^{3+} ion occupying a low symetry site. The main feature is the high intensity of $^5D_0 \rightarrow ^7F_2$ transitions. These spectra are very similar, except for the unfolding and better definition of $^5D_0 \rightarrow ^7F_2$ transitions in *B*, suggesting no change in symmetry of Eu^{3+} ions since we didn't observe shifts of transitions.

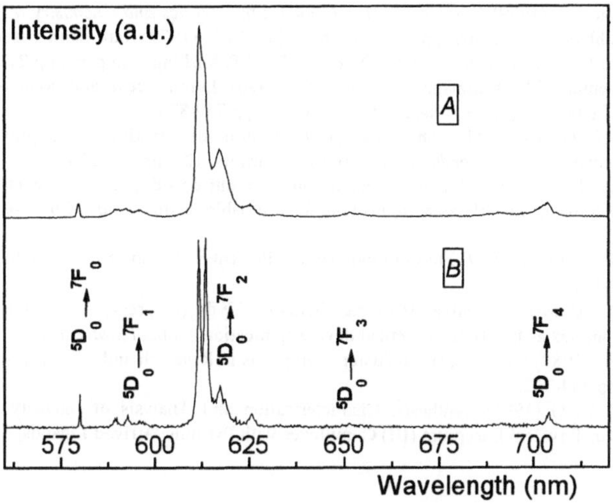

Figure 2: Emission spectra of the Eu(nta)$_3$dipy (*A*) and Eu(nta)$_3$dipy.β-CD (*B*) at room temperature

Some transitions are hypersensitive even to little changes in symmetry around the lanthanide ion, especially the $^5D_0 \rightarrow ^7F_2$ transitions. Therefore, we calculated the hypersensibility (η_{21}) of Eu(nta)$_3$dipy and Eu(nta)$_3$dipy.β-CD. We obtained η_{21}= 12,37 and η_{21}= 8,66 respectively. The lower sensibility previously observed, suggests a partial molecular protection due to CD cavities, without large distortion of molecular symmetry but promotion of unfolding of $^5D_0 \rightarrow ^7F_2$ transitions.

Both compounds were analised at 77K, and their emission spectra were shown to be similar to *B* spectrum. This suggests, that at room temperature, there is suppresion of vibrational transitions due the presence of CD.

Gd(nta)$_3$dipy and Gd(nta)$_3$dipy.β-CD emission spectra at 77K, observed in the 400-700 nm spectral range, are identical. They are centered around 550 nm and correspond to the emission from the lowest triplet states, localized in an nta ligand molecule.

ACKNOWLEDGMENTS
The authors acknowledge CNPq and CAPES (Brazilian agencies) for financial support.

REFERENCES

1. a)Thompson, L.C.(1979) Complexes, *Handbook on the Physics and Chemistry of Rare Earths*, North-Holland Publishing Company, pp. 208-297; b)Weber, M.J., (1979) Rare Earth Lasers, *Handbook on the Physics and Chemistry of Rare Earths*, North-Holland Publishing Company, pp 275-339; c) Balzani,V.; Juris, A;Venturi, M., Campagna, Serroni, S. (1996) Luminescent and Redox-Active Polynuclear Transition Metal Complexes, *Chemical. Reviews*, 96, pp 759-833.
2. Sabbatini, N.; Guardigli, M.; Lehn, J.-M. (1993) Luminescent lanthanide complexes as photochemical supramolecular devices, *Coordination Chemistry Rewiews*, 123, pp 201-228.
3. Silva, J.E.C.da (1997) *Full-color* simulationin a multi-doped glass and controlled quenching of luminescence using Er (III) as a *suppressor* for a tunable device, *Journal of Luminescence* 72-74, pp 270-271.
4. Soini, E.; Hemmilä, I.(1979) Fluoroimmunoassay: Present Status and Key Problems, *Clinical Chemistry* 25, pp 353-361.
5. Sá, G.F.de; Alves-Jr, S.; Silva, B.J.P.da; Silva-Jr, E.F.da (in press) A novel fluorinated Eu(III) β-diketone complex as thin film for optical device application, *Optical Materials.*
6. Saenger, W. (1980) Cyclodextrin Inclusion Compouns in Research and Industry, *Angew. Chem. Int. Ed. Engl.* 19, pp 344-362.
7. SILVA, F. R. G (1995) Synthesis, Characterization and Analysis of Intensity Parameters of 4f-4f transitions in Trivalent Europium (III) Complexes with Pyridine Derived Ligands, Master thesis, UFPE.

STUDY OF COMPLEXATION BETWEEN THE PESTICIDE CHLORPYRIFOS AND β AND HP-β CYCLODEXTRINS

E. MORILLO[1], R. FERNÁNDEZ-URRUSUNO[2], M.J. ARIAS[3], J.I. PÉREZ-MARTÍNEZ[3] and J.M. GINÉS[3]

1. Instituto de Recursos Naturales y Agrobiología, Consejo Superior de Investigaciones Científicas (CSIC), Apdo. 1052, 41080 - Sevilla, Spain
2. Dpto. de Farmacia y Tecnología Farmacéutica. Universidad de Santiago de Compostela. E-15706 Santiago de Compostela, Spain
3. Dpto. de Farmacia y Tecnología Farmacéutica. Univ. de Sevilla. 41012 - Spain.

1. Introduction

Inclusion complexes of pesticides with cyclodextrins (CDs) frequently results in advantageous modification of chemical and physical properties of these compounds. Some papers have been published about these aspects: Kamiya *et al.* (1) have demonstrated that the inclusion effects of β-CD inhibits the photodegradation rate of paration but promotes that of paraoxon; the complexation of the herbicide 2,4-D with β-CD produces a new inclusion compound in solution and in solid state, improving the solubility and dissolution rate of this pesticide (2); the enhanced transport of low polarity compounds as p,p´-DDT by CDs has been proposed as a method for cleanup of contaminated soils (3).

Chlorpyrifos (CPF), is an insecticide widely used which has a very bad odor, is volatile and presents a very low solubility in water (2mg/L at 25°C). Its encapsulation with cyclodextrins could improve these aspects, allowing the obtention of less problematic formulations. The principal objective of the present work is to investigate the possibility of obtaining complexes between β- and HP-β-CD and chlorpyrifos in order to improve principally the aqueous solubility and dissolution properties of this pesticide, and also the reduction of its volatility and bad odor.

2. Experimental

2.1. MATERIALS

CPF (purity 99 %) was supplied by Riedel-de Haën (Seelze, Germany) and β-CD and HP-β–CD by Roquette (Lestrem, France).

2.2. METHODS

2.2.1. *Preparation of samples*

Physical mixtures (PM) were prepared by simple mixing of both components in the 1:1 molar proportion in order to serve as reference.

Kneading method: a PM of pesticide and β- and HP-β-CDs respectively, were ground together for 45 min in a mortar in order to obtain a homogenous paste.

Sealed heating method: the PM were sealed in aluminium containers and then heated at 60 °C for 48 h.

Grinding method: PM were ground in a spectro mill (Retsch Mülhe). This process was performed during 180 min.

2.2.2. *Phase solubility study*

The solubility studies were carried out according to the method reported by Higuchi and Connors. Solutions containing various concentrations of ß-CD (0.001 - 0.014 M) and HP-β-CD (0.001 - 0.1 M) were added to excess amounts of pesticide (10 mg). The flasks were sealed and shaken at 25 °C for one week. After equilibrium, the samples were filtered with syringe through a 0.22 μm Millipore cellulose nitrate membrane filter, and analyzed spectrophotometrically at 230 nm using a Hitachi U-2000 spectrophotometer.

2.2.3. *Differential scanning calorimetry study (DSC)*

DSC analysis was carried out in a Mettler apparatus equipped with FP85 furnace, FP80 HT temperature control unit and FP89 HT software. Samples of 10 mg were put into aluminium pans. The determinations were performed under static air atmosphere, at a heating rate of 10 °C / min, in the temperature range 25 to 80 °C.

2.2.4. *Hot Stage Microscopy (HSM)*

Physical changes in the samples during heating were monitored using a Mettler FP82 HT HSM apparatus. A small amount of the samples was placed on a glass slide with coverglass and heated at 10 °C/min in the temperature range from 25 to 80 °C.

2.2.5. *X-Ray Diffractometry (XRD)*

X-ray powder diffraction diagrams of different samples were obtained using a X-ray diffractometer Siemens, model Kristalloflex D-500. The measuring conditions are as follows: Ni-filtered CuKα radiation, 36 kV, 26 mA, scanning speed 1° (2Θ) / min, chart speed of 1 cm/min and adequate sensibility, usually 4 x 10^4 counts per second.

2.2.6. *Dissolution rate study*

The dissolution rate studies were performed according to the USP 23 paddle method using a Turu Grau equipment, model D-6. The dissolution medium was deionized water (500 mL), the stirring speed 50 rpm and the temperature maintained at 37 ± 0.5 °C. Aliquots (3 mL) were withdrawn at various time intervals using a syringe and analyzed spectrophotometrically.

3. Results and Discussion

3.1. SOLUBILITY STUDY

The phase solubility diagrams obtained can be classified as type A_L according to Higuchi and Connors´ classification. Since the diagrams have a slope <1, it was assumed that the solubility increase was due to the formation of a stoichiometric 1:1 complex in solution. Consequently, the coprecipitation method, which is employed only for preparation of complexes with a B_S type solubility curve, cannot be applied. The presence of cyclodextrins enhances CPF aqueous apparent solubility about 15 times higher in 0.014 M solution of β-CD and 20 times higher for 0.1 M solutions of HP-β-CD. The apparent stability constants (K_c= 275 and 41 mM^{-1} for the β- and HP-β-CD respectively), were calculated from

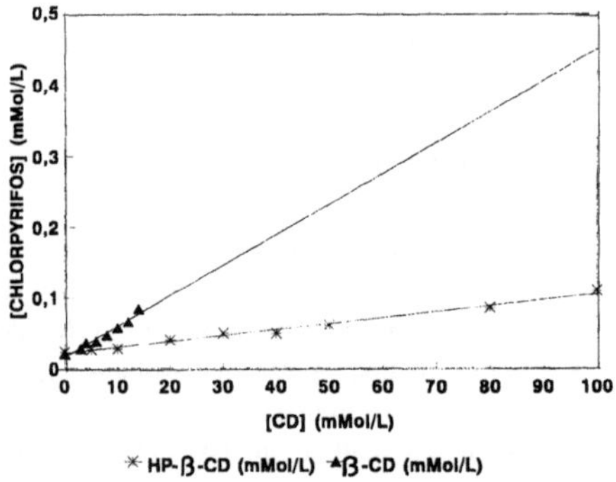

Figure 1. Phase solubility diagrams

the slope of the solubility diagrams.

3.2. DIFFERENTIAL SCANNING CALORIMETRY STUDY (DSC)

The DSC curve of CPF was characterized by the presence of an endothermic peak at 50 °C, corresponding to its melting. In the thermograms with both CDs corresponding to the kneaded and ground samples, the endothermic effect of CPF is present but showing a lower fusion enthalpy that the corresponding PM. These results could be interpreted on the basis of a partial amorphization or complex formation of pesticide into CD cavity. In contrast, this peak disappears in the case of sealed heating samples. These results indicate that these systems are true inclusion complexes.

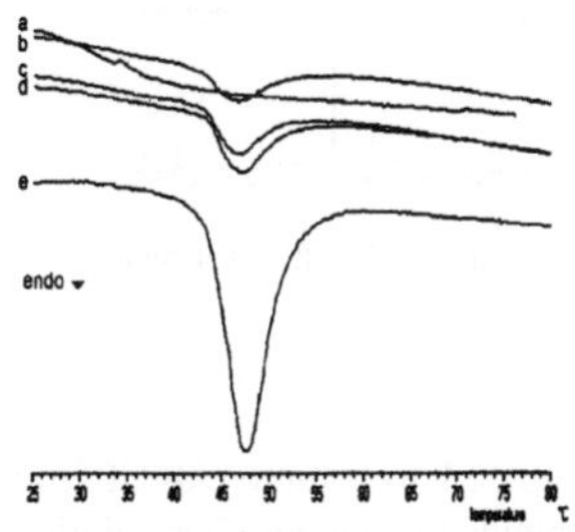

Fig. 2. DSC thermograms of CPF-β-CD samples: a)Sealed heating, b)Kneaded, c)Ground, d)PM, e)CPF.

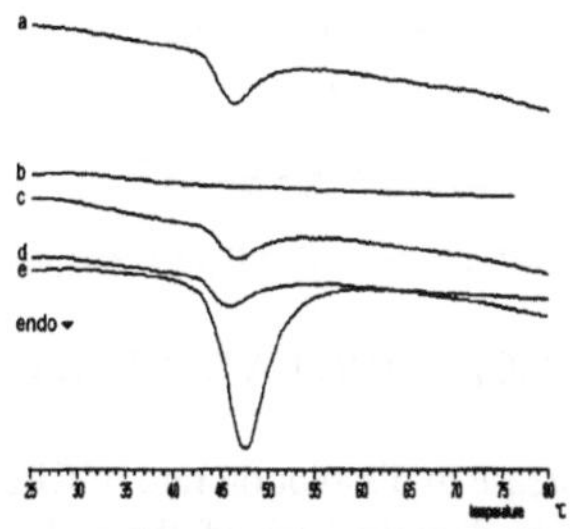

Fig. 3. DSC thermograms of CPF-HPβ-CD samples: a)PM, b)Sealed heating, c)Ground, d)Kneaded, e)CPF

3.3. HOT STAGE MICROSCOPY (HSM)

The feasibility of formation of an inclusion complex was corroborated in the different samples by HSM study. The melting process of CPF was observed by HSM for the PM, kneaded and ground products, but not for the sealed heating samples, confirming by direct observation the results obtained by DSC. It can be thus deduced the reality of the formation of inclusion complexes in the case of using sealed heating method.

3.4. X-RAY DIFFRACTOMETRY (XRD)

In the PM of β-CD and CPF only a small diffraction peak at 3.19 Å (the more intense for pure CPF) can be observed between the others corresponding to β-CD. The kneaded system shows the decreasing of β-CD diffraction peaks due its amorphization, and it makes easier to observe the CPF peaks, which appear much less crystalline than the pure compound. The sealed heating system shows almost exclusively the diffraction peaks corresponding to β-CD, but CPF is present in this system as it was observed by IR spectroscopy (not shown), so CPF has lost its crystallinity after the process. The system ground shows diffraction peaks corresponding to CPF and β-CD, but many β-CD peaks are double, what means that a part of its crystalline structure is being affected probably by the inclusion of CPF, and other part remains without alteration.

In the system HP-β-CD - CPF, the XRD pattern of PM is very similar to those obtained after kneading and grinding, in which the crystallinity of CPF is not very well defined. However, the pattern corresponding to the sealed heating shows very clear and well defined diffraction peaks of CPF, but with different intensity relations in comparison to pure CPF pattern. It indicates the CPF recrystallization with a different orientation in the HP-β-CD-CPF system.

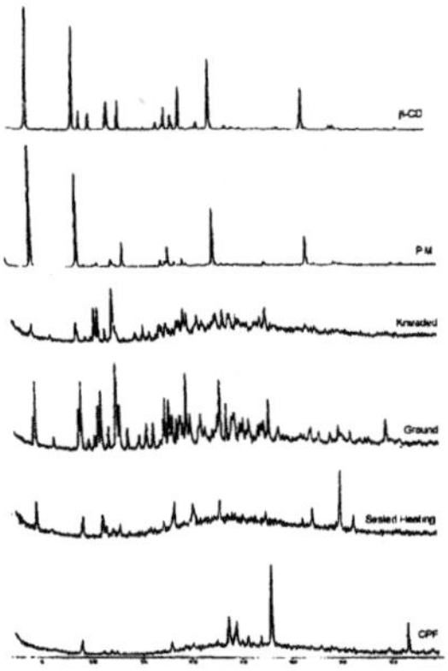

Figure 4. XRD diagrams of CPF-β-CD samples.

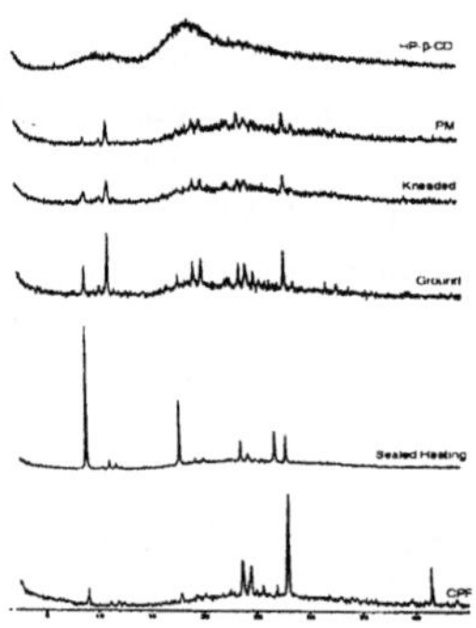

Figure 5. XRD diagrams of CPF-β-CD samples.

3.5. DISSOLUTION RATE STUDY

The increase in dissolution rate of CPF from the PM as compared to commercial CPF may be explained on the basis of the higher solubility of the pesticide in aqueous CDs solutions. It is evident that CPF dissolves faster from the different systems than from physical mixtures. This enhancement in the dissolution rate may be attributed to the inclusion or amorphization of CPF after the treatment, confirmed by DSC and XRD.

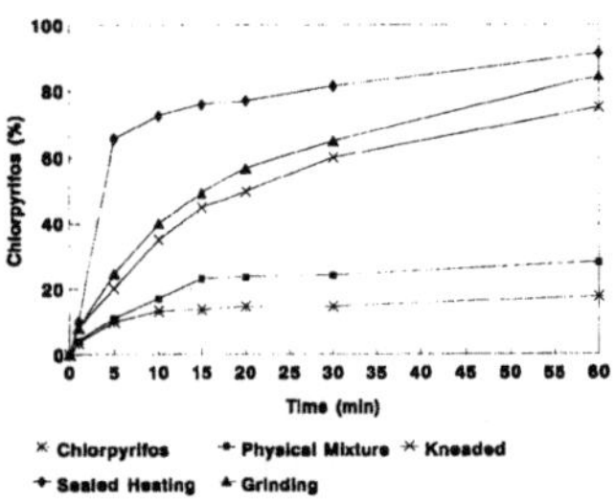

Fig. 6. Dissolution profiles of CPF-β-CD systems

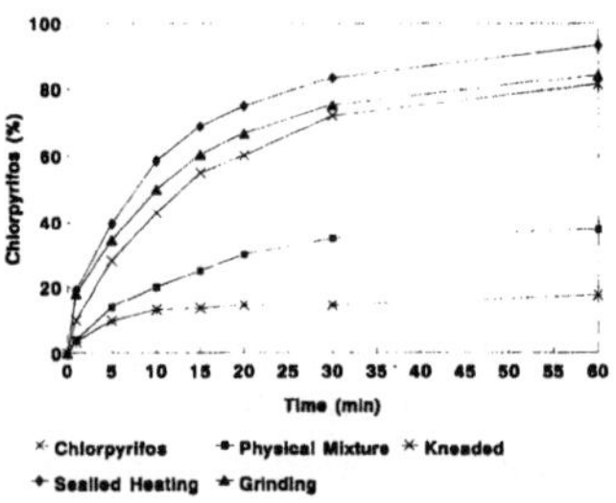

Fig. 7. Dissolution profiles of CPF-HP-β-CD systems

4. Conclusions

From these studies, it is clearly observed the complexation of CPF with β and HP-β-CD in aqueous medium. The preparation of solid complexes of CPF was only completely obtained by the sealed heating method and partially by kneaded and grinding method, as deduced by different characterization techniques. Finally the dissolution study showed that the complexes presented a notable improvement of pesticide dissolution, that can be attributed to the amorphous state or inclusion complexation of CPF into the CD cavity.

REFERENCES

1. Kamiya, M., Nakamura, K. and Sasaki, C. (1994) Inclusion effect of CDs on photodegradation rate of parathion and paraoxon in aquatic medium. *Chemosphere*, 28, 1961-1966.
2. Ginés, J.M. Pérez-Martínez, J.I., Arias, M.J. Moyano, J.R., Morillo, E., Ruiz-Conde, A. and Sánchez-Soto, P.J. (1996) Inclusion of the herbicide 2,4-D with ß-CD by different procesing methods. *Chemosphere*, 33, 321-334.
3. Wang, X. and Brusseau, M.L. (1993) Solubilization of some low-polarity organic compound by HP-β-CD. *Environ. Sci. Technol.*, 27, 2821-2825,
4. Higuchi, T. and Connors, K.A. (1965) Phase-solubility techniques. *Adv. Anal. Chem. Instr.* 4, 117-212.

APPLICATION OF RANDOMLY METHYLATED β-CYCLODEXTRIN FOR BIOREMEDIATION OF CONTAMINATED SOIL

MOLNÁR, M. [2], FENYVESI, É. [1], GRUIZ, K. [2] AND SZEJTLI, J. [1]
[1] *Cyclolab Ltd., 1117 Budapest, Dombóvári út 5-7, HUNGARY*
[2] *Technical University of Budapest, Dept. of Agric. Chem. Technology, 1111 Budapest, Gellért tér. 4. HUNGARY*

1. Introduction

The biodegradation of hydrocarbons in the soil is a complex process, whose quantitative and qualitative aspects depend also on the bioavailability and the toxic effect of the contaminants (1).
Cyclodextrins (CD) are typical „host" molecules and due to their favourable structural character form inclusion complexes with a wide variety of organic compounds (2). This kind of „molecular encapsulation" may affect on bioavailability of the contaminants during biodegradation and also modify toxicity on soil microbes, plants and animals.
The influence of randomly methylated β-cylodextrin (RAMEB) on the biodegradation process and on the toxic effect of contaminated soil was investigated by chemical and biological analyses in details.

2. Materials and methods

Large scale (40 kg) laboratory experiments were carried out for bioremediation of transformer oil contaminated soils. A clean forest soil was used in the experiment, deriving from the Buda Hills. The forest soil was polluted artificially with 30 000 ppm (3%) transformer oil. The effect of RAMEB both on the uncontaminated and on the oil contaminated forest soil was investigated. RAMEB was applied in the soil reactor in the concentration of 5 g/kg soil and 10 g/kg soil. The soil was treated in a natural aerobic system, amended with inorganic nutrients and optimal humidity was ensured.
Biodegradation process was followed by the chemical and biological investigations of the solid and gaseous fractions of the soil. The degradation of the oily contaminant and the microbial activity of the soil was followed by measuring the living cell number, the oil content and the oil-degrading cell concentration in the soil. Oil content was determined after Soxhlet extraction of air dried soil samples by gravimetry and Fourier transfrom infrared (FT-IR) spectrometry. The CO_2 and O_2 content of the soil gas was measured continuously during the bioremediation.
FT-IR spectra were measured on liquid samples with resolution of 4 cm^{-1} using FT-IR Perkin-Elmer Instrument System 2000, from 128 scans. The effect of RAMEB on soil toxicity was investigated by bioassays.

Direct contact biotests were used to simulate the interaction with the complex system, between contaminants, soil matrix and testorganisms. Organisms of three different trophic level were applied in the ecotoxicological tests :

- Bacterial : ***Photobacterium phosphoreum*** bioluminescence inhibition test,
- Plant : *Sinapis alba* and *Lepidum sativum* germination and root elongation test,
- Animal : *Folsomia candida* (Collembola) mortality test.

These tests were developed in our laboratory by the modification of DIN, HS and other standard methods (4,5,6,7).

3. Results and discussion

A laboratory experiment was performed to study the influence of RAMEB on the biodegradation process in a transformer oil contaminated soil.
Randomly methylated ß-cyclodextrin as an additive in the bioremediation technology resulted in three different effects.
•1. Enhanced the biodegradation of oily contaminant in the beginning of the process. By forming a complex with RAMEB hydrocarbons presumably desorbed from fine soil particles and are made available to microorganisms.
•2. Decreased the toxicity of the hydrocarbon contaminants. This retard effect of the cyclodextrin is due to the characteristics of the formed complex. After encapsulation the partition of the contaminant between solid- and water phase has been changed; a more advantageous equilibrium concentration is connected with an increased desorption.
•3. Increased the availability of many of the nutrients of the soil resulting in higher activity and production. This effect may be measured also in case of uncontaminated soil.

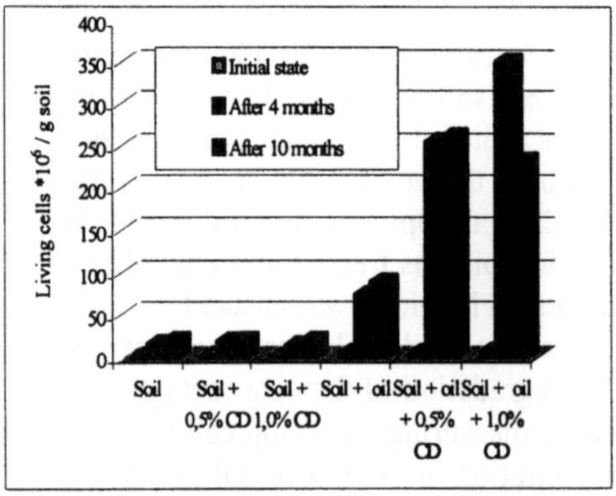

Figure 1. Changes in living cell number during bioremediation

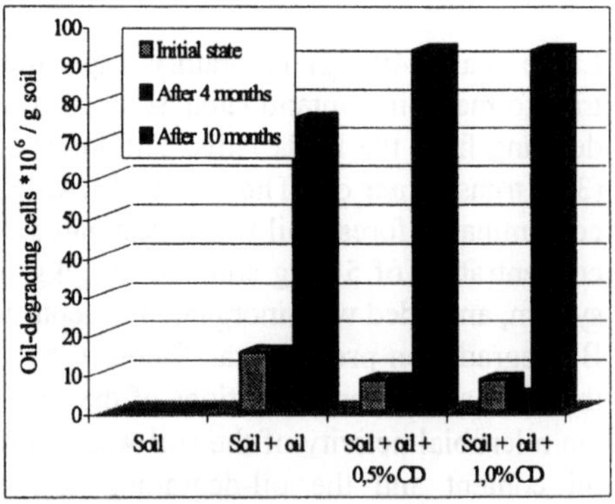

Figure 2. Changes in oil-degrading cell number during bioremediation

Addition of RAMEB positively influenced both living cell number and oil-degrading cell number of soil (Fig.1. and Fig.2.). The RAMEB-treatment led to significantly increased microbial activity. The adaptation of microbes occured on the expiration of 4 months.

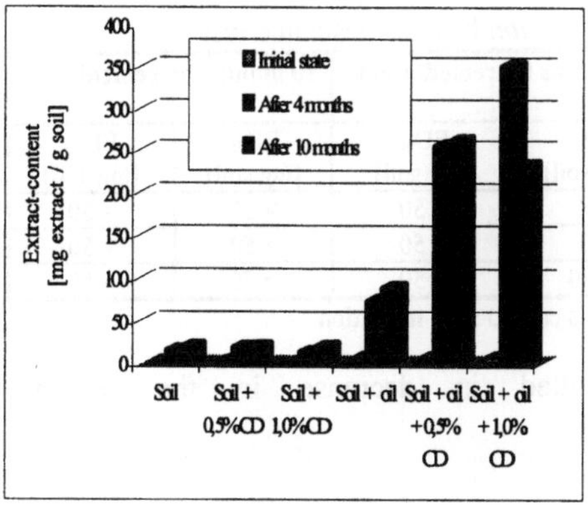

Figure 3. Changes in extract content during bioremediation

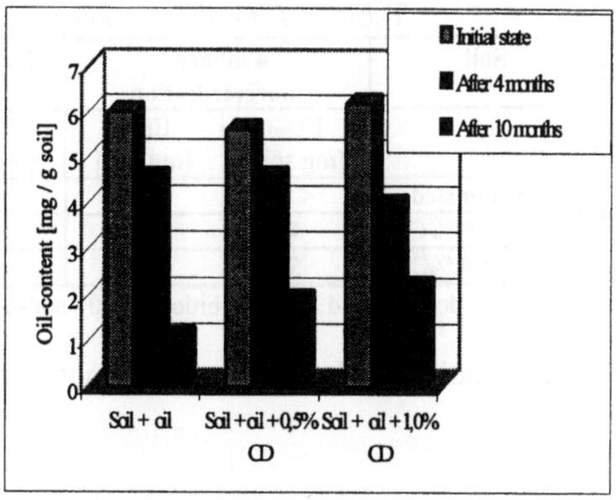

Figure 4. Changes in oil content during bioremediation

The figure illustrates an increase in the amount of extractable organic matter after 4 and 10 months of bioremediation (Fig.3.). Considerable decomposition of humus content of the soil was characteristic. The enhanced microbial activity resulted in a higher extract content, made it more available for chemical and biological processes. This phenomenon was observed in uncontaminated soil as well, but it is very characteristic in the polluted soils and even more significant in RAMEB-treated soil during bioremediation.

On the basis of FT-IR spectra we determined the oil-content of soil (Fig.4.). These results give more detailed information about the oil content of the extract. In spite of increased amount of extractables the amount of biodegradable hydrocarbons decreased during bioremediation. The explanation may be the considerably increased activity of soil microbes, which affects in case of RAMEB-treated soil not only the contamination but the humus compounds of the soil.

Table 1. *Sinapis alba* root - and shoot - elongation test

Soil	4 months bioremediation				8 months bioremediation			
	ED_{20root} [g soil]	ED_{50root} [g soil]	$ED_{20shoot}$ [g soil]	$ED_{50shoot}$ [g soil]	ED_{20root} [g soil]	ED_{50root} [g soil]	$ED_{20shoot}$ [g soil]	$ED_{50shoot}$ [g soil]
Contaminated	1.2	< 0.625	4.0	0.8	1.7	< 0.625	4.8	1.2
Cont. + 0.5 % CD	1.6	< 0.625	4.2	1.1	1.8	< 0.625	6.4	1.8
Cont. + 1.0 %CD	3.4	0.8	10.0	2.8	3.4	< 0.625	> 10.0	4.4

ED20 - soil-dosis caused 20% inhibition ED50 - soil-dosis caused 50% inhibition

Plant inhibition decreased (growth increased) in contaminated soil treated with RAMEB as a consequence of better nutrient availability and of decreased toxicity.

Table 2. *Photobacterium phosphoreum* bioluminescence test

Soil	4 months bioremediation		8 months bioremediation		10 months bioremediation	
	ED_{20} [mg soil]	ED_{50} [mg soil]	ED_{20} [mg soil]	ED_{50} [mg soil]	ED_{20} [mg soil]	ED_{50} [mg soil]
Contaminated	< 3	5	18	> 50	> 50	> 50
Cont. + 0.5 % CD	4	11	30	> 50	> 50	> 50
Cont. + 1.0 %CD	5	10	> 50	> 50	> 50	> 50

ED20 - soil-dosis caused 20% inhibition ED50 - soil-dosis caused 50% inhibition

Encapsulation of the contaminant resulted in decrease in the toxicity

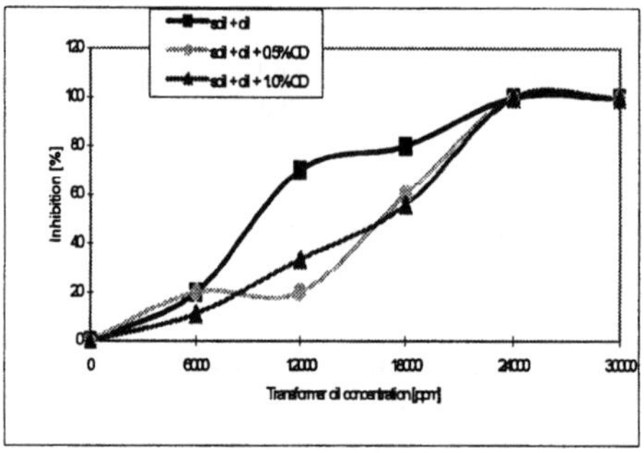

Figure 5. Toxicity for Collembola in the presence of cyclodextrin (after 10 months)

According to our former investigations *Folsomia candida* mortality biotest is sensitive to oily contaminants in general. The figure shows that transformer oil is toxic to the *Folsomia candida* (Fig.5.). Significant difference in toxicity was observed between RAMEB-treated and untreated soil samples. Randomly methylated β-cylodextrin was found to reduce the toxic effect of transformer oil.

5. Conclusion

In the complex system (soil – oily contamination – RAMEB) two contrary process take place :

- an increase in the amount of total extractable organic matter ,
- a decrease in the amount of biodegradable compounds included the hydrocarbon contamination.

RAMEB - treatment reduces the toxic effect of transformer oil and enhances the biodegradation of transformer oil contamination. The results suggest that randomly methylated β-cylodextrin can be used successfully in soil bioremediation, but further studies are necessary to optimise the applied cyclodextrin concentration, the frequency of RAMEB – treatment and the amount of added inorganic nutrients.

References

1. Gruiz, K., Fenyvesi, É., Kriston, É., Molnár, M. and Horváth, B. (1996) Potential use of cyclodextrins in soil bioremediation, *J. of Inclusion Phenomena and Mol. Recognition*, **25,** pp. 233-236.
2. Szejtli J. (1982) *Cyclodextrins and their Inclusion Complexes*, Akadémiai Kiadó, Budapest. pp. 30-35.
3. Horváth, B., Gruiz, K. and Sára, B. (1996) Ecotoxicological testing of soil by four bacterial biotests, *Toxicological and Environmental Chemistry*, **58**, pp. 223-235.
4. Gruiz, K., Murányi, A., Molnár, M. and Horváth, B. (1998) Risk Assessment of Heavy Metal Contamination in Danube Sediment - *Journal of Water Science and Technology*, in press.
5. Riepert,F. and Kula, C. (1996) Development of laboratory methods for testing effects of chemicals and pesticides on Collembola and earthworms, *Inst. for Ecotox.*, **320**, pp. 8-47.
6. DIN 38412 Inhibitory effect of waste water on *Photobacterium phosphoreum* - German Standard, 1991.
7. MSZ 22902-4 Water toxicological test; Plant seedling test - Hungarian Standard, 1991.

Chapter 6

CYCLODEXTRINS AS CHIRAL SELECTORS

MODIFIED CYCLODEXTRINS AS VERSATILE CHIRAL STATIONARY PHASES IN FLAVOUR CHEMISTRY AND LIFE SCIENCES

A. MOSANDL*, F. PODEBRAD, D. BARTSCHAT, A. KAUNZINGER, S. REICHERT, AND M.WÜST
Institut für Lebensmittelchemie, Biozentrum J.W. Goethe-Universität, Marie-Curie-Str. 9, D-60439 Frankfurt/ Main, Germany, Tel.: +49 69 79829202/03, Fax: +49 69 79829207, e-mail: Mosandl@em.uni-frankfurt.de

Abstract

Modified cyclodextrins as chiral stationary phases in capillary gaschromatography are proved to be powerful tools in the enantioselective analysis of different chiral volatiles, detecting origin specific enantiomeric distributions. In flavour chemistry enantioselective analysis is used for authenticity assessment and biogenesis studies. In medicine, enantio-MDGC/ MS serves to diagnostic and metabolic studies of inherited metabolic diseases.

1. Introduction

Tert-butyl-dimethylsilyl-substituted cyclodextrin derivatives are used as versatile chiral stationary phases in different apparative configurations such as:

- Enantioselective capillary gas chromatography (*enantio-GC)*
- Enantioselective multidimensional gas chromatography (*enantio-MDGC*), employing cutting-techniques from a pre column onto a chiral main column
- Enantioselective multidimensional gas chromatography-mass spectrometry (*enantio-MDGC-MS*), combining high selectivity of MDGC with mass selective detection

In flavour chemistry enantioselective analysis is used for authenticity assessment and biogenesis studies, such as for dill ether [1], 3-butyl-(hexahydro)-phthalide stereoisomers [2] and *cis*/*trans* rose oxides [3].

In medicine, enantio-MDGC(-MS) serves to diagnostic and metabolic studies of inherited metabolic diseases, such as lactic-, glyceric-, 2-hydroxyglutaric aciduria [4] or maple syrup urine disease (*MSUD*) [5].

* corresponding author

2. Experimental

Only types of apparatus are given. For full details, see literature references.

Enantio-GC: Shimadzu GC-14A gas chromatograph with chiral column.
Enantio-MDGC: Siemens SiChromat 2-8 double oven chromatograph with column combination: nonchiral pre column/ chiral main column.
Enantio-MDGC-MS: Siemens SiChromat 2-8 double oven chromatograph with column combination: nonchiral pre column/ chiral main column and mass selective main column detection (ITD 800, Finnigan MAT).
Nonchiral stationary phases: different apolar phases.

Chiral stationary phases:
- heptakis-(2,3-di-O-acetyl-6-O-tert-butyldimethylsilyl)-β-cyclodextrin (DiAc-β-CD)
- heptakis-(2,3-di-O-methyl-6-O-tert-butyldimethylsilyl)-β-cyclodextrin (DiMe-β-CD)
- octakis-(2,3-di-O-butyryl-6-O-tert-butyldimethylsilyl)-γ-cyclodextrin (Dibutyryl-γ-CD)

synthesized and used for laboratory made chiral columns.

3. Results and Discussion

These enantioselective phases have proved to be of unique selectivity and versatility and therefore used in enantioselective gas chromatography analysis, as to be shown.

3.1 ENANTIOSELECTIVE ANALYSIS - AUTHENTICITY ASSESSMENT

From organoleptic standpoint, dill ether (3,6-dimethyl-2,3,3a,4,5,7a-hexahydrobenzo[b]furan (1a) is the most important constituent in dill oil. The enantioselective analysis of dill ether and its *cis*-stereoisomers has been achieved by enantio-GC with Dibutyryl-γ-CD. Additionally the sensory properties of the four stereoisomers are evaluated with GC-olfactometry using the same chiral stationary phase[1]. (see Figure 1)

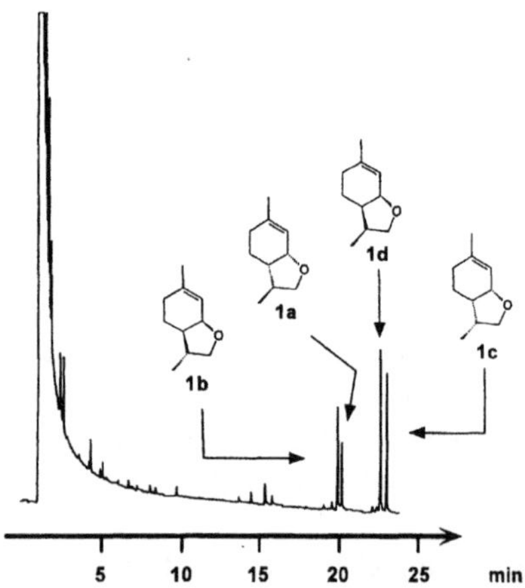

Figure 1. Enantio-GC analysis of dill ether and its *cis*-stereoisomers [1].

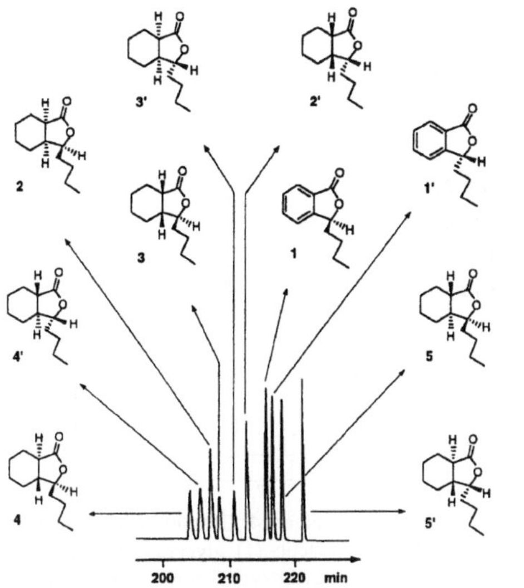

3-Butylphthalide and some hydrogenated derivatives are well-known as character impact flavour compounds of celery, celeriac, and fennel. The simultaneous enantioselective analysis of 3-butyl(hexahydro)phthalides has been realized by enantio-MDGC and DiAc-β-CD as chiral stationary phase in main column separation [2]. (see Figure 2)

Figure 2. Enantio-MDGC analysis of 3-butyl(hexahydro)-phthalides: main column chromatogram [2].

3.2 BIOGENESIS STUDY

The enantiomeric distributions of *cis*- and *trans*-rose oxide ketones in geranium oils were used for proposing a pathway of biogenesis of rose oxide ketones in *Pelargonium* species, including rose oxide as the precursor. Even more the enantioselective analysis offers an interesting approach for the authenticity control of geranium oils [3]. (see Figure 3)

Figure 3. Enantio-MDGC-MS analysis (DiAc-β-CD) of iso-rose oxide (1-4), rose oxide (5-8) and rose oxide ketone (9-12): main column chromatogram [3].

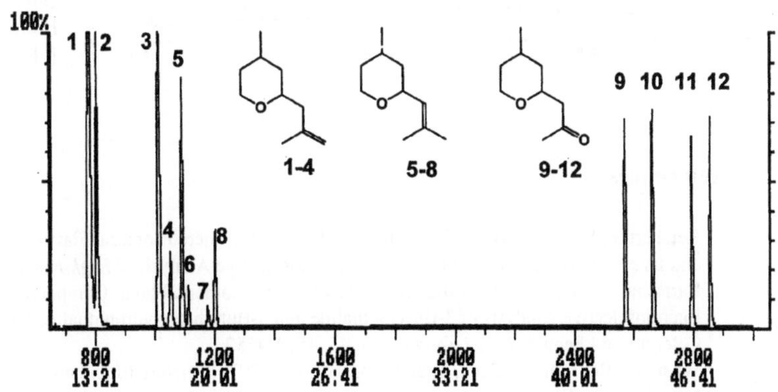

3.3 DIAGNOSTIC AND METABOLIC STUDIES OF INHERITED METABOLIC DISEASE

Urinary organic acid analysis is a vital diagnostic tool in the investigation of patients with suspected inborn errors of metabolism. The presence of chiral metabolites have provided the need for enantioselective analysis [4-6].

The simultaneous stereodifferentiation of lactic-, glyceric- and 2-hydroxyglutaric acid (DiMe-β-CD) is used for diagnosis of patients with D- or L-2-hydroxyglutaric-and D-glyceric aciduria [4,6]. (see Figure 4)

MSUD is an inherited metabolic disorder of the branched-chain amino acids valine, leucine and isoleucine. A specific enzym deficiency leads to accumulation of carboxylic-, α-oxo-, α-hydroxy- and α-amino acids. The simultaneous analysis of metabolites has been achieved by enantio-MDGC-MS with DiMe-β-CD [5]. (see Figure 5)

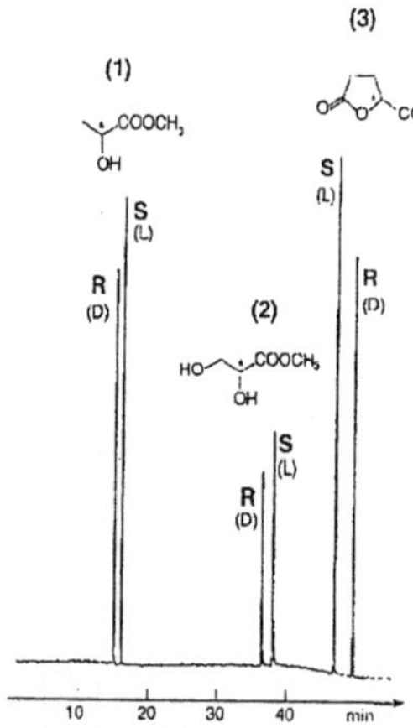

Figure 4: Enantio-MDGC analysis of lactic- (1), glyceric- (2) and 2-hydroxyglutaric acid (3) (methyl-ester derivatives): main column chromatogram [4,6].

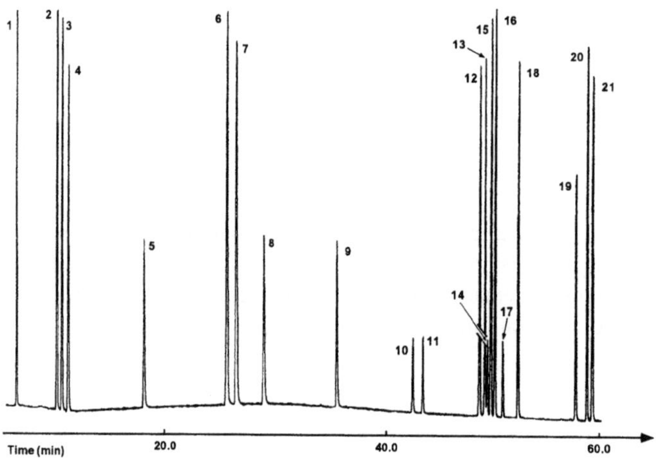

Figure 5. Enantio-MDGC-MS analysis of MSUD metabolites (methylchloroformate derivatives): main column chromatogram (1-4) carboxylic-, (5-8) α-oxo-, (10-17) α-hydroxy- and (18-21) α-amino acids [5].

4. References

[1] Reichert,S., Wüst, M., Beck, T. and Mosandl, A. (1998) Stereoisomeric Flavour Compounds LXXXI: Dill Ether and Its *cis*-Stereoisomers: Synthesis and Enantioselective Analysis, *J High Resolut Chromatogr*, 21, 185-188.

[2] Bartschat,D., Beck, T. and Mosandl, A. (1997) Stereoisomeric Flavor Compounds. 79. Simultaneous Enantioselective Analysis of 3-Butylphthalide and 3-Butylhexahydrophthalide Stereoisomers in Celery, Celeriac, and Fennel, *J Agri Food Chem*, 45, 4554-4557.

[3] Wüst, M., Rexroth, A., Beck, T. and Mosandl, A. (1997) Structure Elucidation of *cis*- and *trans*-Rose Oxide Ketone and its Enantioselective Analysis in Geranium Oils, *Flav Fragr J*, 12, 381-386.

[4] Kaunzinger, A., Rechner, A., Beck, T., Mosandl, A., Sewell, A.C. and Böhles, H. (1996) Chiral Compounds as Indicators of Inherited Metabolic Disease – Simultaneous Stereodifferentiation of Lactic-, 2-Hydroxyglutaric- and Glyceric Acid by Enantioselective cGC, *Enantiomer*, 1, 177-182.

[5] Podebrad, F., Heil, M., Leib, S., Geier, B., Beck, T., Mosandl, A., Sewell, A.C. and Böhles, H. (1997) Analytical Approach in Diagnosis of Inherited Metabolic Disease: Maple Syrup Urine Disease (MSUD) – Simultaneous Analysis of Metabolites in Urine by Enantioselective Multidimensional Capillary Gas Cromatography Mass Spectrometry (Enantio-MDGC-MS), *J High Resolut Chromatogr*, 20, 355-362.

[6] Sewell, A.C., Heil, M., Podebrad, F. and Mosandl, A. (1998) Chiral compounds in metabolism: a look in the molecular mirror, *Eur J Pediatr*, 157, 185-191.

β-CYCLODEXTRINS AS MOLECULAR SCAFFOLDS TO REVERSE THE REGIOSELECTIVITY OF NITRILE OXIDE CYCLOADDITIONS

ADAM G. MEYER,[a] CHRISTOPHER J. EASTON,*,[a] STEPHEN F. LINCOLN[b] AND GREGORY W. SIMPSON[c]
[a]Research School of Chemistry, Australian National University, Canberra, ACT 0200, Australia
[b]Department of Chemistry, University of Adelaide, Adelaide, SA 5005, Australia
[c]CSIRO Molecular Science, Private Bag 10, Clayton South MDC, Clayton, Vic 3169, Australia

Abstract

β-Cyclodextrin has been modified to function as a molecular scaffold such that the regioselectivity of nitrile oxide cycloadditions with unsymmetrical alkynes and alkenes has been reversed.

Introduction

Cyclodextrins (CDs) have been utilized to facilitate organic transformations through the self-assembly of reactants within the annulus. For example, CDs have been used in Diels-Alder cycloadditions to both accelerate the rate of reaction and influence the distribution of products [1]. It has also been reported that CDs control the regioselectivity of nitrile oxide cycloadditions,[2] however these findings are in error [3].

The cycloaddition of nitrile oxides (1,3-dipoles) with alkynes and alkenes (dipolarophiles) generates isoxazoles and isoxazolines, respectively [4]. For reactions involving unsymmetrical dipolarophiles, the product regiochemistry is controlled by steric factors alone and the more hindered end of the dipolarophile becomes attached to the oxygen of the nitrile oxide.

DESIGNING A CYCLODEXTRIN TO ALIGN DIPOLE AND DIPOLAROPHILE

Rather than allowing the dipole and dipolarophile to self-assemble within the annulus of a CD, we imagined that a system could be designed to control the orientation of the reactants and yield a reversal in the regioselectivity of nitrile oxide cycloadditions. To develop this strategy, a dipolarophile would be tethered to β-cyclodextrin (β-CD) *via* substitution of a C-6 hydroxy group. It was anticipated that subsequent formation of an inclusion complex with a nitrile oxide featuring a hydrophobic unit would establish the orientation of the reaction partners for the cycloaddition (*Figure 1*).

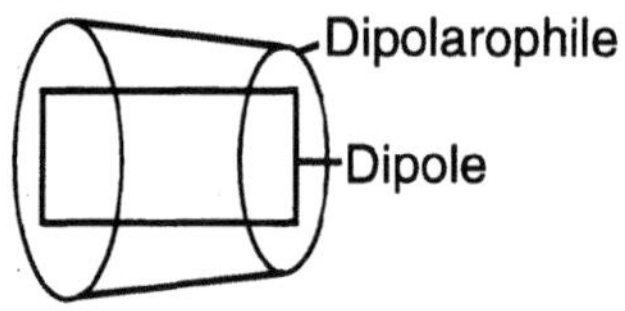

Figure 1

Tethering of an unsymmetrical dipolarophile to β-CD was accomplished by treatment of the amino-substituted CD **1** with propynoyl chloride under basic conditions to give the amide **2** (*Scheme 1*).

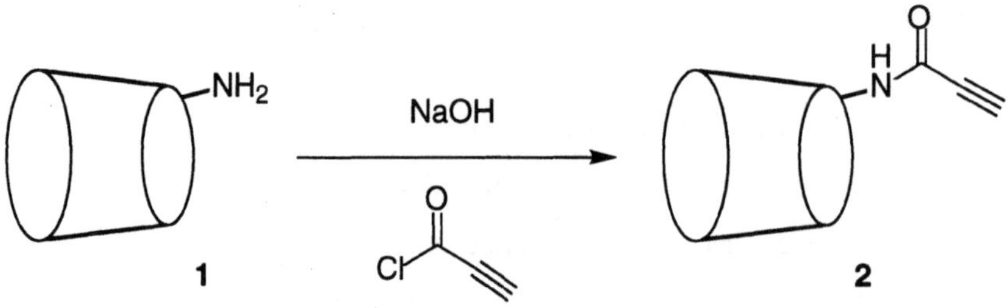

1 **2**

Scheme 1

NITRILE OXIDE CYCLOADDITION OF A TERMINAL ALKYNE TETHERED TO A CYCLODEXTRIN

4-*tert*-Butylbenzonitrile oxide **3** was selected as the dipole since it should form a thermodynamically stable inclusion complex with the amide **2** [5]. The cycloaddition between **2** and **3**, in aqueous solution, reversed the usual regioselectivity of nitrile oxide cycloadditions to terminal alkynes and yielded a 15:1 mixture of the isoxazoles **4** and **5** (*Scheme 2*). To scrutinise the role of the CD annulus, the reaction was repeated in DMF. With the extent of complexation reduced, **4** and **5** were obtained in a ratio of 1:1.5.

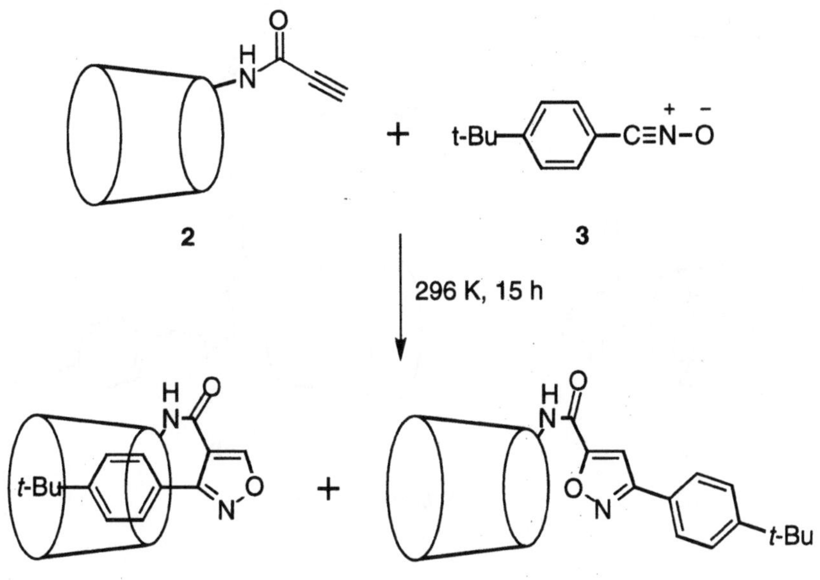

Scheme 2

NITRILE OXIDE CYCLOADDITION OF A TERMINAL ALKENE TETHERED TO A CYCLODEXTRIN

The cycloaddition of 4-*tert*-butylbenzonitrile oxide **3** with the amide **6**, in aqueous solution, has been shown to afford a 2.3:1 mixture of the isoxazolines **7** and **8** (*Scheme 3*) [6]. When this reaction was repeated in DMF, **7** and **8** were produced as a 1:4 mixture.

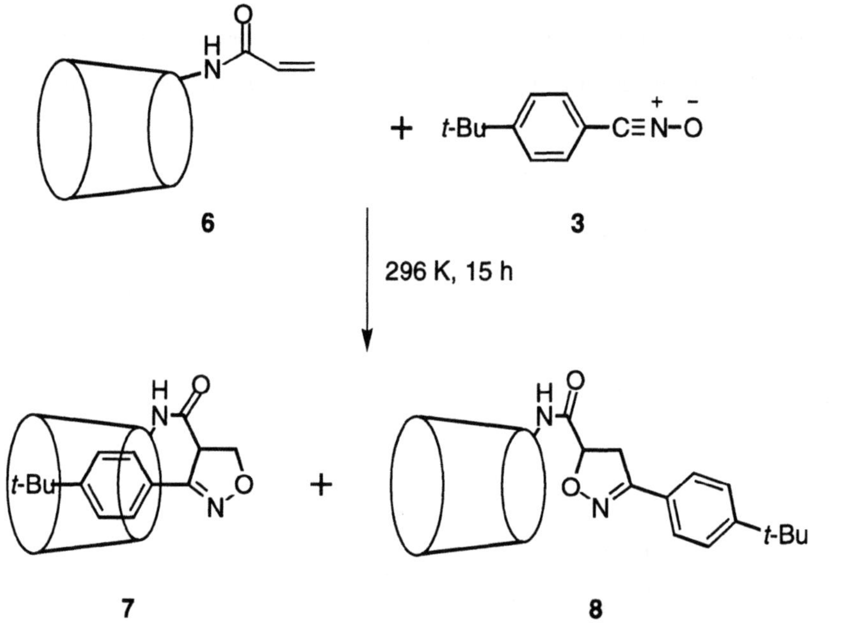

Scheme 3

References

1. (a) Rideout, D. C.; Breslow, R. *J. Am. Chem. Soc.* **1980**, *102*, 7816-7817. (b) Schneider, H.-J.; Sangwan, N. K. *Angew. Chem., Int. Ed. Engl.* **1987**, *26*, 896-897.
2. (a) Rama Rao, K.; Bhanumathi, N.; Srinivasan, T. N.; Sattur, P. B. *Tetrahedron Lett.* **1990**, *31*, 899-902. (b) Rama Rao, K.; Bhanumathi, N.; Sattur, P. B. *Tetrahedron Lett.* **1990**, *31*, 3201-3204. (c) Rama Rao, K. *Pure Appl. Chem.* **1992**, *64*, 1141-1145.
3. Easton, C. J.; Hughes, C. M. M.; Tiekink, E. R. T.; Savage, G. P.; Simpson, G. W. *Tetrahedron Lett.* **1995**, *36*, 629-632.
4. (a) Torssell, K. B. G. *Nitrile oxides, Nitrones and Nitronates in Organic Synthesis: Novel Strategies in Synthesis*; VCH Publishers: New York, 1988. (b) Easton, C. J.; Hughes, C. M. M.; Savage, G. P.; Simpson, G. W. *Adv. Heterocyclic Chem.* **1994**, *60*, 261-327.
5. Matsui, Y.; Nishioka, T.; Fujita, T. *Top. Curr. Chem.* **1985**, *128*, 61-89.
6. Meyer, A. G.; Easton, C. J.; Lincoln, S. F.; Simpson, G. W. *J. Chem. Soc., Chem. Commun.* **1997**, 1517-1518.

ANALYSIS AND CHARACTERISATION OF COMMERCIAL DM-β-CD. INFLUENCE OF MIXTURE COMPOSITION ON CHIRAL SEPARATION IN CAPILLARY ELECTROPHORESIS AND SUPERCRITICAL FLUID CHROMATOGRAPHY

A. SALVADOR[1], E. VARESIO[2], J.L. VEUTHEY[2] and M. DREUX[1]

[1]*Institut de Chimie Organique et Analytique (ICOA), UPRES A-6005, UFR Sciences, Université d'Orléans, BP 6759, F-45067 Orléans, France*

[2]*Laboratoire de Chimie Analytique Pharmaceutique, Université de Genève, 20 bd. D'Yvoy, 1211 Genève 4, Suisse*

1. Introduction

Cyclodextrins (CDs) and their derivatives have found many applications in recent years because of their ability to form inclusion complexes with a large number of molecules. Among different applications, methylated β-CD and more precisely heptakis (2,6-di-O-methyl)-β-CD (DM-β-CD) are often used as chiral discriminating agents in the mobile phase and as background electrolytes in Liquid Chromatography (LC) [1] and Capillary Electrophoresis (CE) [1,2]. However, the purity of commercially available DM-β-CDs is a problem because samples are a mixture of compounds with different numbers of methyl groups and different methylation positions. In this work, composition of three commercial DM-β-CDs was studied by various analytical techniques and the influence of mixture composition on chiral separation was studied by CE and by Supercritical Fluid Chromatography (SFC).

2. Experimental

2.1 APPARATUS

Gas Chromatography (GC) was carried out on a Varian 3400 gas chromatograph (Palo Alto, CA, USA), equipped with a flame ionization detector. The column was a DB5 (15 m x 0.53 mm id, 1.5 µm film). The injector and detector temperature was 210 and 310 °C respectively. The column temperature varied linearly between 130°C to 220 °C at 1°C.min^{-1}. The carrier gas was helium U (Air Liquide, Paris, France). SFC was performed with a Gilson SF3 apparatus (Villiers-le-Bel, France), a Rheodyne 7125 injector with a 20-µl sample loop and a CROCO-CIL™ column oven (CIL-Cluzeau,

Ste-Foy-la-Grande, France). The column temperature was 41°C, and the outlet pressure was 150 bar. SFC detection was performed with an Evaporative Light Scattering Detector (ELSD) or a Varian diode-array detector. CE experiments were carried out on a HP^{3D} CE system (Hewlett-Packard, Waldbronn, Germany) using Hewlett-Packard capillaries of 50 µm I.D. and 64.5 cm total length (56 cm from inlet to detector). Buffer consisted of 60 Mm tris(hydroxymethyl)aminomethane/phosphate/CDs.
Mass Spectrometry (MS) was done with a Perkin-Elmer Sciex API 300 (Foster City, CA, USA) equipped with an ionspray source or heated nebulizer.
The following columns were used in SFC : a porous graphitic carbon column Hypercarb-S column 7 µm (100 x 4.6 mm I.D) from Hypersil (Runcorn, UK) and a Nucleosil NO_2 10 µm (150 x 4.6 mm I.D.) from Shandon (Cheshire, UK).

2.2 REAGENTS

Carbon dioxide (purity 99.7 %, Air Liquide) and methanol (Mallinckrodt Baker, Noisy le sec, France) were of analytical grade. The three DM-β-CDs studied (A, B, C) were commercial products from different sources. Alditol actetate for GC studies was prepared by acid hydrolysis [3].

3. Results and discussion

3.1 ANALYSIS AND CHARACTERISATION OF COMMERCIAL DM-β-CDs

Analysis of the substitution patterns of the three DM-β-CDs was done by the usual GC method after acid hydrolysis, reduction, acetylation and LC method [3] which is less time-consuming (Table 1). Substitution patterns differed greatly for the different samples ; for samples B and C methylation is heterogeneous and the eight possibilities of glucopyranose occurred.

TABLE 1. Determination of substitution patterns by GC : Eight possible arrangements of the glucopyranose unit : [0] no substitution ; [2], [3], [6] monomethylation ; [2,3], [2,6], [3,6] dimethylation ; [2,3,6] trimethylation.

	[2,3,6]	[2,6]	[3,6]	[2,3]	[6]	[2]	[3]	[0]
A	9.9	89	0	0	0.5	0.5	0	0
B	15.3	28.2	9.4	6.6	14.5	14.4	4.4	7.1
C	25.7	28.7	20.7	0.2	23.3	0.6	0.3	0.5

MS of the three samples was also engaged in order to determine the relative percentage of methyl number in each CD ring. As shown in Table 2, methylation is heterogeneous : DM-β-CD A and C are overdimethylated whereas DM-β-CD B is undermethylated compared to pure heptakis-(2,6)-di-O-methyl-β-CD. Analysis of the different DM-β-CDs mixtures with SFC and SFC-MS [4,5]was performed and sample A gave 7 peaks with baseline separation, whereas samples B and C gave only a fingerprint. The 7 components of A were identified with SFC-MS, respectively:18 Me, two of the four isomers of 17 Me, two of the three isomers of 16 Me, 15 Me and 14 Me (data not shown).

TABLE 2 : Determination of number of methyl group (%) by MS

	9 Me	10 Me	11 Me	12 Me	13 Me	14 Me	15 Me	16 Me	17 Me	18 Me
A	0	0	0	0	0	41.6	45.7	10.1	2.5	0
B	2.4	7.8	19.9	30.9	24.8	11.4	2.8	0	0	0
C	0	0	0	0.4	8.6	52.1	25.1	8.0	2.9	1.7

3.2 INFLUENCE OF DM-β-CD COMPOSITION ON CHIRAL DISCRIMINATION

3.2.1 Capillary Electrophoresis

To evaluate chiral recognition of these three DM-β-CDs, the binding constants with amphetamine derivatives have been determined and are reported in Table 3. B and C give the same binding constant values. However, in spite of their similar binding constants, their enantioselectivity varies as shown in Figure 1. Indeed C did not permit separation of MDA whatever the concentration range of DM-β-CD C.

TABLE 3 : *DM-β-CD/amphetamine binding constants (M^{-1}). Amphetamine derivatives : (A), Amphetamine ;(MA), Methamphetamine ; (MDA), 3,4-methylenedioxyamphetamine ; (MDMA), 3,4-methylenedioxymethylamphetamine ; (MDEA) 3,4-methylenedioxyethylamphetamine ; (MDPA),: 3,4-methylenedioxypropylamphetamine.*

	A	A	MA	MA	MDA	MDA	MDMA	MDMA	MDEA	MDEA	MDPA	MDPA
A	93 ± 5	97 ± 5	111 ± 6	119 ± 7	317 ± 18	333 ± 19	376 ± 22	406 ± 32	513 ± 46	554 ± 49	593 ± 58	637 ± 63
B	56 ± 3	57 ± 3	61 ± 3	65 ± 3	136 ± 7	140 ± 7	151 ± 7	158 ± 12	183 ± 16	193 ± 17	196 ± 19	207 ± 20
C	43± 2	44 ± 2	47 ± 2	50 ± 2	139 ± 7	139 ± 7	151 ± 8	159 ± 6	191 ± 10	201 ± 10	210 ± 12	221 ± 13

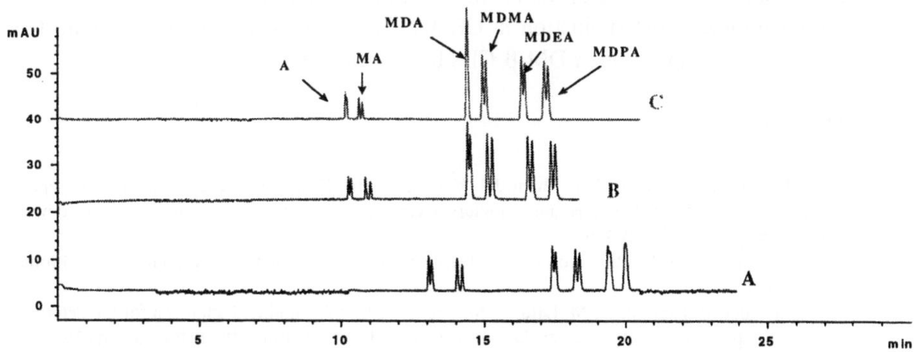

Figure 1 : Influence of DM-β-CD on enantioseparation of amphetamine derivatives. Fused-silica capillary dimensions, 64.5 cm x 50 μm ; electrolyte, 60 mM tris(hydroxymethyl)aminomethane/phosphate + 20 mM DM-β-CD, pH 3 ; applied voltage, 30 kV ; temperature, 25 °C, UV detection, 210 nm.

3.2.2 SFC

Chiral SFC separation is usually performed with CDs derivative based stationary phase. In our laboratory, we have explored the potentiality of DM-β-CD as a chiral discriminating agent in supercritical mobile phase. DM-β-CD was first dissolved in the modifier (MeOH) and then mixed with CO_2. The preliminary results show that the selectivity increases with DM-β-CD concentration and decreases when pressure and temperature of mobile phase increase. As shown in Figure 2, chiral separation occurred in less than 3 minutes. Chiral separation is very sensitive to the composition of DM-β-CD.

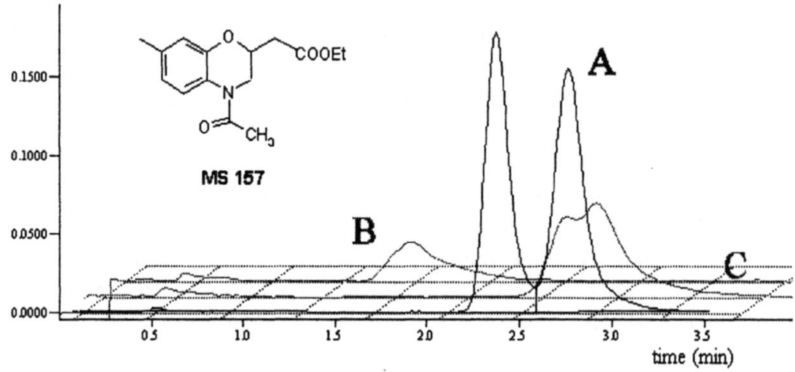

Figure 2 : Influence of DM-β-CD composition on enantioselectivity by SFC . Colonne Hypercarb S (100 x 4.6 mm I.D.), Mobile phase, CO_2/MeOH/DM-β-CD (80 :20 :2, v/v/mM), Flow rate, 3 ml.min-1, outlet pressure 150 bar, Column temperature, 41°C, UV detection at 210 nm.

4. Conclusion

The enantiomeric separation power of methylated β-CD derivatives strongly depends on their composition. Differences in the substitution patterns usually cause differences in the enantiomeric discrimination in CE and SFC. It is therefore necessary to determine the mixture composition of DM-β-CDs before use in CE or SFC.

References

[1] F. Bressolle, M. Audran, T-Y. Pham, J-J. Vallon, (1996), Cyclodextrins and enantiomeric separations of drugs by liquid chromatography and capillary electrophoresis : basic principles and new developments, J. Chromatogr. B, 687, 303-336.

[2] S. Fanali, (1996), Identification of chiral drug isomers by capillary electrophoresis, J. Chromatogr. A, 735, 77-121.

[3] A.Salvador, B. Herbreteau, M. Lafosse, M. Dreux, (1997) A simplified approach for the determination of substitution patterns of methylated cyclodextrins by liquid and subcritical fluid chromatography,Analusis, 25, 263-267.

[4] A. Salvador, B. Herbreteau, M. Dreux, in M. Perrut (Editor), Proceedings of the 5th International Symposium on Supercritical Fluids, Nice, France, March 1998, Vol.2, p 797-801.

[5] A. Salvador, B. Herbreteau, M. Dreux, J. Chromatogr. A., submitted.

PURE MONOPENTENYLATED β-CYCLODEXTRIN AS CHIRAL AGENT : PURITY CHECK BY LC-ELSD AND LC-MS

I. CARON[a], C. ELFAKIR[a], M. DREUX[a], H. LEVEQUE[b], R. DUVAL[b]
[a] *Institut de Chimie Organique et Analytique (ICOA), CNRS UPRES-A 6005, Université d'Orléans, BP 6759, 45067 Orléans Cedex 2, France*
[b]*CHIRALSEP, Parc d'activité de la boissière, 76170 La Frenaye, France*

1. Introduction

Cyclodextrins are extensively used in separative science either as chiral stationary phase bonded to a solid support or as chiral mobile phase for enantiomeric separation in liquid chromatography (LC) and Capillary Electrophoresis (CE).
Mono-2-O-pent-4-enyl derivative of β-cyclodextrin linked via a spacer to silica constitutes an efficient chiral stationary phase used in LC. However, the substitution reaction is not regioselective [1].
Obtaining of high purity monopentenylated β-cyclodextrin is necessary to produce batch to batch reproducible chiral stationary phase and consequently to provide rugged enantiomeric separations. Control of the reaction products is currently made by thin layer chromatography and the substitution patterns are determined by GC-MS analysis of derivative alditol acetates [2, 3].
In this paper, our aim was to develop fast and straightforward liquid chromatographic systems for the analysis of monopentenylated β-cyclodextrins without further derivations.

2. Material

Crude mixture of mono-2-O-pent-4-enyl β-CD, pure mono-2-O-pent-4-enyl β-CD and pure mono-6-O-pent-4-enyl β-CD were synthesised by Chiralsep (La Fresnaye, France).

3. Results and discussion

3.1 LC-ELSD

As previously described [1], the substitution reaction of β-CD to obtain mono-2-O-pent-4-enyl β-CD is not regioselective ; as a result, mono-2-O-pent-4-enyl β-CD is obtained

with small amounts of 3-O and 6-O-pentenyl derivatives as well as di- and tripentenyl derivatives.
The analysis of such reaction mixtures containing compounds with widely differing polarities requires gradient elution. Evaporative light scattering detection, which has proved to be a good choice for cyclodextrin analysis [4-6], has therefore been used.
Figures 1 and 2 show the LC-ELSD elution pattern of the crude mixture of mono-2-O-pent-4-enyl β-CD on a C_{18} silica column or on an NH_2 polymeric column in acetonitrile-water gradient mode. Analyses of β-CD and pure mono-2-O-pent-4-enyl β-CD under these chromatographic conditions show that the elution order is reversed in the two LC systems and that these two products were the main compounds in the crude mixture. Moreover, the 2-OH position was preponderantly substituted.

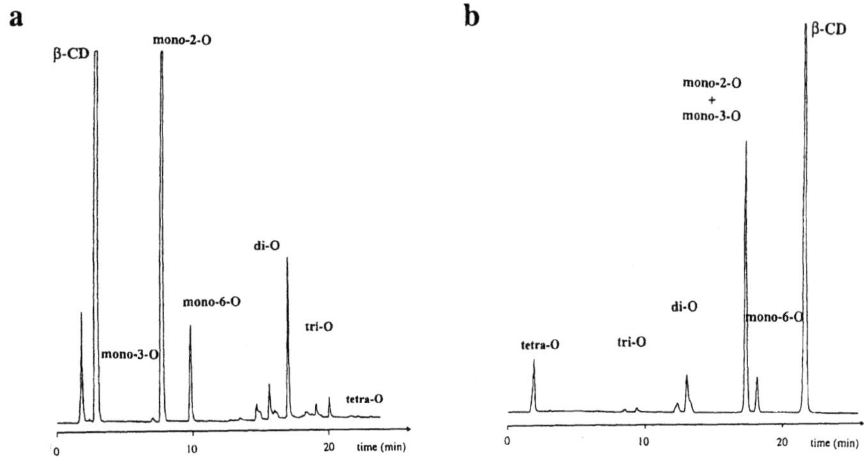

Figure 1. Chromatograms of the crude mixture on : a) Spherisorb ODS 2 (250 x 4.6 mm I.D.) column (gradient elution from 10 % to 20 % ACN in 10 min then to 50 % ACN in 10 min and to 100 % ACN in 5 min) ; b) Polymeric Astec NH_2 (250 x 4.6 mm I.D.) column (gradient elution from 15 % to 30 % water then ACN/water (50 : 50) during 5 min).

3.2 LC-MS

As the LC-ELSD and LC-MS systems have similar chromatographic requirements and acetonitrile-water mobile phases are compatible with mass spectrometry, the direct combination of the LC systems with electrospray mass spectrometry has been carried out to identify each chromatographic peak. Post-column addition of ammonium acetate ions was used for positive ion detection of the sample. $[M+NH_4]^+$ ions were produced at respectively m/z = 1152.4 for β-CD, m/z = 1220.4 for monopentenylated β-CDs, m/z = 1288.5 for dipentenylated β-CDs, m/z = 1356.6 for tripentenylated β-CDs and m/z = 1424.6 for tetrapentenylated β-CDs.
The LC-MS analyses confirmed that di- and tripentenylated derivatives were present in the mixture and revealed that small amounts of tetrapentenylated derivatives were also

present. No overlapping of the chromatographic peaks of the nearest degrees of substitution was observed.
LC-ELSD analysis of pure mono-2-O-pent-4-enyl-β-CD and pure mono-6-O-pent-4-enyl-β-CD and LC-MS analysis of the crude mixture showed that a good resolution between the three positional isomers of mono-O-pent-4-enyl-β-CD (mono-2-O, mono-3-O and mono-6-O) was obtained only on Spherisorb ODS2 column. This analytical chromatographic system can be successfully transposed to the preparative in order to obtain large quantities of high purity monopentenylated β-CD derivatives.

3.3 QUANTITATIVE ANALYSIS

In order to achieve accurate quantitative analysis of monopentenylated β-CDs, calibration curves for β-CD and mono-2-O-pent-4-enyl-β-CD have been carried out on the most efficient chromatographic system (Spherisorb ODS2 column with ACN/water elution gradient).
β-CD and mono-2-O-pent-4-enyl-β-CD showed similar linear ELSD response in logarithmic coordinates as expected [7] with an acceptable correlation coefficient ($R \approx 0.9994$). Consequently, only one calibration curve is necessary to achieve the determination of impurities in the crude mixture. Table I reports the proportion of native β-CD and of the different monofunctionalized β-CDs in the mixture. It confirmed that the 2-OH position was preponderantly substituted but we can see that the reaction yield was low. Quantitative LC-ELSD analysis was also assumed to be suitable for purity determination after purification of the crude mixture.

Table I : Quantitative analysis of β-CD and the three positional isomers of mono-O-pent-4-enyl-β-CD

	Solute				
	β-CD	mono-3-O-pent-4-enyl-β-CD	mono-2-O-pent-4-enyl-β-CD	mono-6-O-pent-4-enyl-β-CD	overpentenylated-β-CD
Percentage (%)	37.5	1.4	28.9	9.3	22.9

Figures 2a and 2b show the separation of racemic dansyl leucine and racemic chlorthalidone on two chiral stationary phases, one obtained with the crude mixture and the second obtained by grafting pure mono-2-O-pent-4-enyl-β-CD. The enantiomeric separation has been found to be highly dependent on the quality of the chiral selector. No separation of the enantiomers of chlorthalidone and a lower selectivity between the enantiomers of dansyl leucine were achieved using the crude mixture as chiral stationary phase.
Thus, it is necessary to purify the crude mixture before grafting it onto functionalyzed silica gel to obtain rugged and successful chiral separations.

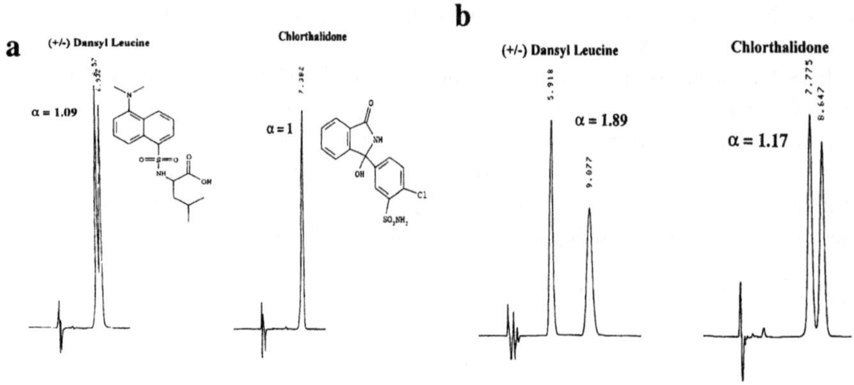

Figure 2. Separation of enantiomers of dansyl leucine and chlorthalidone on a) stationary phase bonded with the crude mixture and b) stationary phase bonded with pure mono-2-O-pent-4-enyl-β-CD ; flow-rate 1 $ml.min^{-1}$; U.V. detection at 254 nm.

4. Conclusion

We can conclude that the LC-ELSD system is suitable for performing a simple and fast control of mono-2-O-pent-4-enyl-β-CD synthesis without further derivations to ensure quality in these products.

For chiral separation, the use of well characterized pentenylated β-CD derivatives, by LC-ELSD and LC-MS, is recommended to achieve better batch to batch reproducibility of chiral stationary phase and in order to evaluate separation mechanisms.

References

1. Ciucanu, I., Kerek, F. (1984) A simple and rapid method for the permethylation of carbohydrates, *Carbohydr. Res.* 131, 209-217.
2. Ciucanu, I., König, W. A. (1994) Immobilization of peralkylated β-cyclodextrin on silica gel for high-performance liquid chromatography, *J. Chromatogr. A* 685, 166-171.
3. Jindrich, J., Pitha, J., Lindberg, B., Seffers, P., Harata, K. (1995) Regioselectivity of alkylation of cyclomaltoheptaose (β-cyclodextrin) and synthesis of its mono-2-O-methyl, -ethyl, -allyl, and -propyl derivatives, *Carbohydr. Res.* 266, 75-80.
4. Caron, I., Salvador, A., Elfakir, C., Herbreteau, B., Dreux, M. (1996) Analysis of partially methylated cyclodextrins by subcritical fluid and liquid chromatography, *J. Chromatogr. A* 746, 103-108.
5. Caron, I., Elfakir, C., Dreux, M. (1997) Advantages of evaporative light scattering detection for the purity control of commercial cyclodextrins, *J. Liq. Chrom. & Rel. Technol.* 20, 1015-1035.
6. Caron, I., Elfakir, C., Dreux, M. (1998) Correlation between the chromatographic properties of silica phenyl packing materials and the retention behaviour of methylated β-cyclodextrins, *Chromatographia* 47, 383-390.
7. Dreux, M., Lafosse, M. (1995), *Carbohydrate Analysis*, in El Rassi, Z. (ed.), J. Chrom. Library, Elsevier, Amsterdam, 58, 515-540.

CYCLODEXTRINS DOMINATE THE CHIRAL CHROMATOGRAPHY

Z. JUVANCZ[1], G. SZEJTLI[2]
1VITUKI Plc., Institute for Water Pollution Control H-1095, Kvassay J. út 1, Budapest, Hungary.
2Cyclolab Ltd., H-1525, P.O.B. 435, Budapest, Hungary

Cyclodextrins (CDs) are dominating as selectors in the majority of chromatographic chiral separations. The share of CDs and CD derivatives in different chiral chromatographic modes (GC, LC, SFC, TLC, CE) are studied both on chiral stationary phases (CSPs) as well as chiral stationary additives (CSAs) in mobile phases. Sources are CHIRBASE (Koppenhoefer et.al.) and CD NEWS LIBRARY DATABASE.
In chiral separation between 1978-1997 CDs are subjects of 1640 publications (scientific papers, reviews, patents, abstracts in symp. proc.). GC and CE are the subject of the most publications among all chromatographic modes (Fig. 1), while CE is the most dynamically developing branch on this field (Fig. 2).

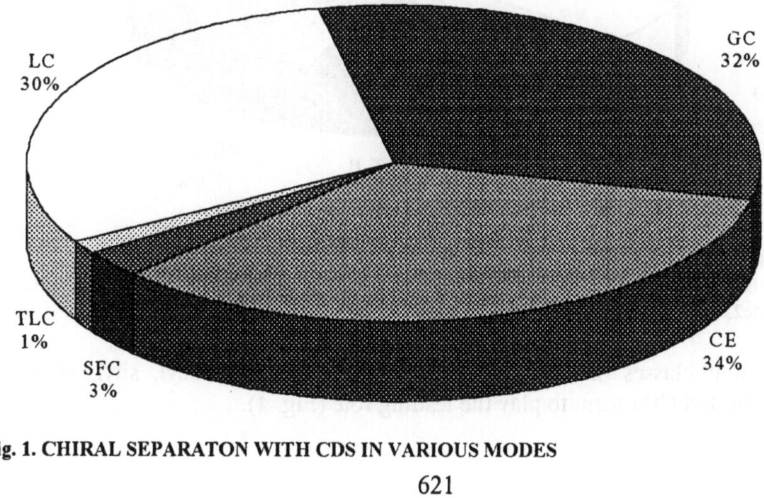

Fig. 1. CHIRAL SEPARATON WITH CDS IN VARIOUS MODES

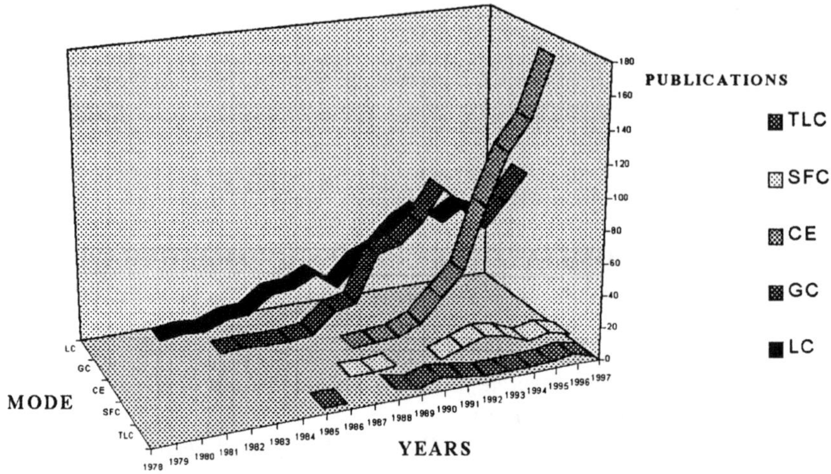

Fig.2. NUMBER OF THE PUBLICATIONS PER YEARS IN DIFFERENT CHROMATOGRAPHIC MODES

GAS CHROMATOGRAPHY
In chiral GC, the leading role of CDs is obvious from the reported chiral separations on different types of CSPs (CD 64%, amide 32%, metal 3%, others 1%). In overwhelming part of the publications derivatized CDs have been used (509 out of 534) in GC. The permethylated βCD is the most popular CSP (266 out of 534 publications), but the share of 6-TMBSi 2,3methyl βCD (49) is increasing rapidly (Fig. 3).

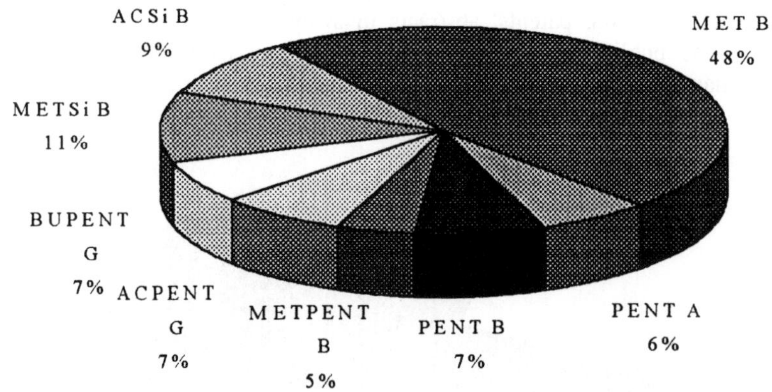

Fig. 3. DISTRIBUTION OF PUBLICATIONS AMONG SELECTED CD BASED CSPS IN GC
Symbols: MET B=perMe βCD; PENT A=perPe αCD; PENT B=perPe βCD; METPENTB=2,6Me3Pe αCD; ACPENTG= 3Ac2,6Pe γCD; METSI B= 6TBDMSi 2,3Me βCD; ACSi B=6TBDMSi 2,3Ac βCD

Among different classes of substituted CDs (native, peralkyl, acyl, silyl, etc.) the peralkyl substituted CDs seem to play the leading role (Fig. 4).

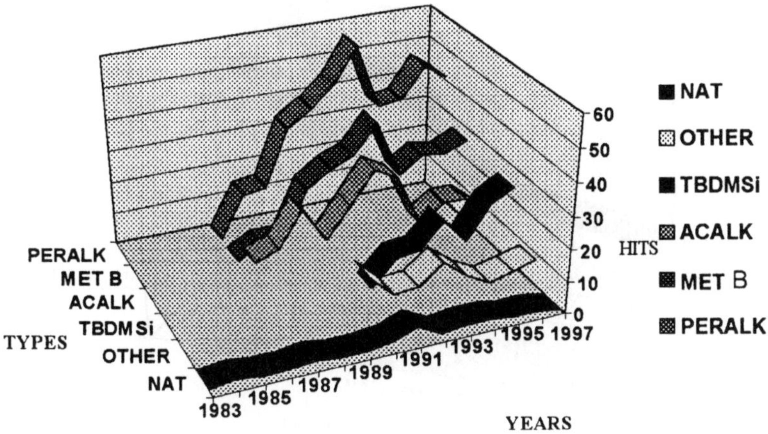

Fig. 4. NUMBERS OF PUBLICATIONS OF VARIOUS CDS PER YEARS IN GC

CAPILLARY ELECTROPHORESIS

The dominant role of CDs is manifested spectacularly in chiral CE (Fig. 5). More than 83% of the publications (581 of 702) refer CDs as CSAs. Among these the mostly used CSAs are the native CDs (34%), followed by the alkyl (25%), ionizable (23%), hydroxyalkyl (15%), others (3%). The applications of ionizable CDs are steeply increasing (Fig. 6). Moreover, the native CDs serve only as basis for comparison in a lot of papers. Fig. 6 shows only a few representatives of the applied CDs, but the tendencies become obvious from these data. A big advantage of ionizable CDs is their separating ability for neutral enantiomers. Almost all publications used water based electrolytes, but 7 papers refer to organic solvent as electrolytes.

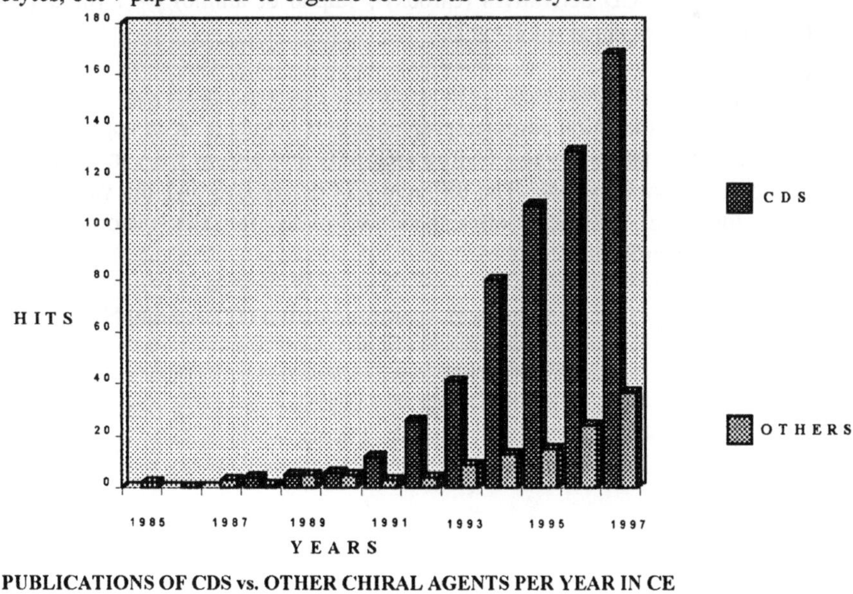

Fig. 5. PUBLICATIONS OF CDS vs. OTHER CHIRAL AGENTS PER YEAR IN CE

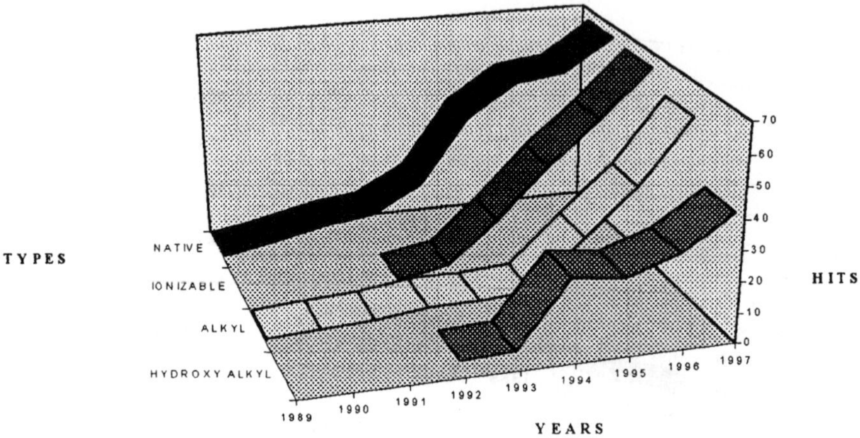

Fig. 6. NUMBER OF PUBLICATIONS OF THE DIFFERENT CDS AS CSAS PER YEARS IN CE

RECOMMENDATION

According to the above mentioned data and the personal experience of the authors, general sequences can be followed for selection of an appropriate chiral selective agent in various chromatographic modes. The probability of success of a chiral separation correlates to the rank of the recommended CDs, which are

in GC: permethylated CD>>6-TMBSi 2,3-methyl βCD>3 butyril-2,3 pentyl γCD> perpentyl αCD. These agents are advisable to solve (15-50%) in an achiral silicone matrix.

in LC: βCD (CSP)> NEC -βCD (CSP)> Hydroxypropyl -βCD (CSA)

in CE: ionizable -βCD derivatives (carboxyl, sulpho or amino) >> Hydroxypropyl -βCD> partially methylated -βCD> βCD

CONCLUSION

The CDs are versatile chiral chromatographic agents having a broad application range. The CDs are the first choice as a CSP in GC, SFC and also as CSA in CE and LC to solve a chiral separation task. The different CD derivatives and the vast number of relevant literature references offer good solutions for the majority of chiral chromatographic problems.

Acknowledgement: This work has been sponsored by the OTKA grant No. T14326.

NEW ASYMMETRIC β-CYCLODEXTRIN DERIVATIVES DESIGNED FOR CHIRAL RECOGNITION

F. DJEDAINI-PILARD[a], M. GOSNAT[a], V. BRUCATO-MAUCLAIRE[c], C.CREMINON[b], J.P. DALBIEZ[a], S. PILARD[d], W. LUIJTEN[d] AND B. PERLY[a]
[a] DRECAM/SCM,[b] DRM/SPI, CEA-Saclay, F-91191 Gif sur Yvette Cedex (France.)[c] CIS bio international, BP 32, F-91192 Gif sur Yvette Cedex (France). [d] Technologie Servier, F-45000 Orléans (France)

1. Introduction

In the continuing challenge of increasing the performances of cyclodextrins (CDs) for various applications, it has been observed that very simple chemical modifications of the CD core lead to very large improvements. A clear illustration is provided by mono-3,6-anhydro-βCD **1**, mono-3,6-anhydro-heptakis-2-*O*-methyl-hexakis-6-*O*-methyl-βCD **2** and mono-3,6-anhydro-heptakis-2,3-*O*-methyl-hexakis-6-*O*-methyl-βCD **3**.

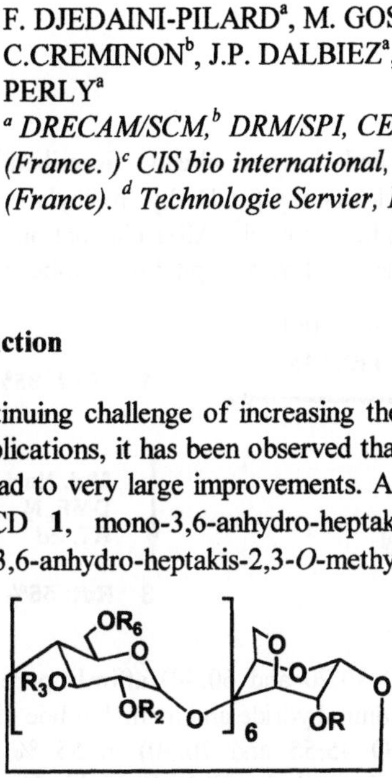

R= R2= R3= R6= H **1**
R= R6= R2= CH3; R3 =H **2**
R= R6= R2= R3= CH3 **3**

These compounds are prepared and purified by HPLC. A structural analysis of **1** alone and with different chiral molecules has been already performed[1-2]. A complete characterization of **2** and **3** has been achieved by high resolution NMR and Mass Spectrometry with electrospray infusion mode and have shown a complete reduction of symmetry.
These three compounds exhibit inclusion properties similar to the parent CD as observed by NMR for a variety of hosts. However, the lack of symmetry induces a very large chiral separation of racemic compounds. Morever they display a strongly increased solubility and solubilization power even at high temperature. The hemolytic character of these three compounds has been also investigated and compared to homogeneous series of pure β-CD derivatives. Finally, it was shown as expected that antibodies raised against β-CD, di-2,6-*O*-methyl-β-CD (DIMEB) and tri-2,3,6-*O*-methyl-β-CD (TRIMEB), respectively, failed to recognize any asymmetric analogue.

2. Materials and Methods

β-CD was obtained from Roquette Frères. Purification was achieved by HPLC (Deltaprep, column μBondapack C_{18}, LSED detection). All NMR experiments were performed using Bruker DRX500 spectrometers operating at 500.13 MHz, after a freeze-drying step and redissolution in deuterium oxide (Euriso-Top, France). ESI-MS analysis were performed on a VG Analytical Tribrid (Micromass, UK) tandem hybrid mass spectrometer with BEqQ geometry. The compounds were individually dissolved in CH_3CN/H_2O 50:50 and infused into the electrospray ion source. Accurate mass measurement has been achieved using polyethylene glycol as the internal reference masses.

3.Results and Discussion

3.1 SYNTHESIS

Synthesis of mono-6-tosyl-6-deoxy-β-CD and **1** were already described[(3,1)]. **2** was obtained in one step from mono-6-tosyl-6-deoxy-β-CD by methylation using a modification of the method reported by Stoddart and al[4]. After elimination of the salts and alkylating agent, the crude product was isolated by precipitation in n-hexane (75%).

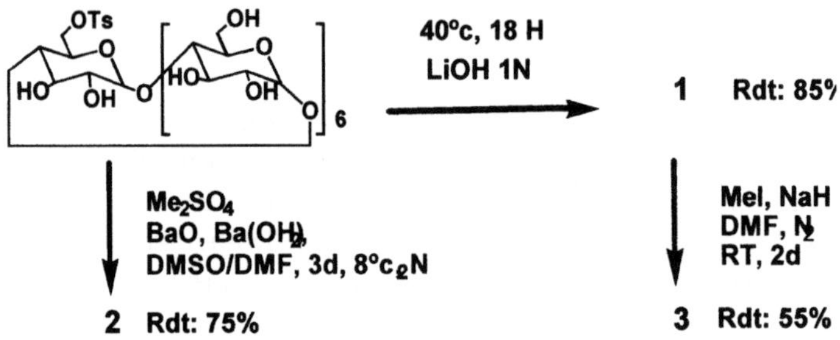

Purification by HPLC (elution CH_3CN-H_2O 30:70 and 60:40) affords pure **2** in 45% yield. Methylation of **1** by treatment[5] with sodium hydride and methyl iodide gives **3** after purification by HPLC (elution CH_3CN-H_2O 45:55 and 70:30) in 55 % yield. The chemical purity was checked by NMR and high resolution ESI-MS (Table 1).

Table 1: Accurate mass data

compound	$[M+Na]^+$	Measured mass	Exact mass	error (ppm)
1	$C_{42}H_{68}O_{34}Na$	1139.341	1139.349	6.9
2	$C_{55}H_{94}O_{34}Na$	1321.542	1321.552	7.7
3	$C_{61}H_{106}O_{34}Na$	1405.613	1405.646	23.8

3.2. SOLUBILITY

The solubility in water *versus* temperature has been determined for **1-3**; the results are displayed in Table 2 and have to be compared with solubility data of β-CD, DIMEB and

TRIMEB under the same conditions[6].

Table 2: solubility (M) in water of β-CD derivatives *vs* temperature

	β-CD	**1**	DIMEB	**2**	TRIMEB	**3**
25°C	0.016	0.350	0.420	0.520	0.220	2.4
90°C	0.200	0.400	0.005	0.530	0.011	3.6

It should be observed that the asymmetric derivatives display a stongly increased solubility in water even at high temperature at the difference to DIMEB and TRIMEB. **2** and **3** indeed do not suffer for the inversed solubility properties of methyled CDs[6]. All these results can be explained by the complete lack of symmetry of compounds **1-3**.

3.3 HEMOLYTIC CHARACTER AND RECOGNITION BY ANTIBODIES

Determination of the hemolytic activity of **1-3** and of the symmetric CDs was achieved using the same experimental method as described elsewhere[7]. Results are expressed as % of the total hemolysis by comparaison with erythrocytes in pure water and displayed in Table 3.

Table 3:concentration (mM) of β-CD derivatives inducing a total hemolysis at 37°C

	β-CD	**1**	DIMEB	**2**	TRIMEB	**3**
C(mM)	10	>40	7.5	>35	5	>30

1-3 were found to be very less hemolytic than the parent β-CD and than the methylated derivatives described as very hemolytic compounds[6]. Anhydro derivatives **1-3** can be considered as non-hemolytic since no hemolysis has been detected at 30 mmol concentration, suggesting that the geometry of the CD does affect the hemolytic properties.

Antibodies were raised against β-CD, DIMEB and TRIMEB, respectively, providing highly sensitive enzyme immunoassays for β-CD[8], DIMEB and TRIMEB respectively. Investigations of cross-reactivity with **1-3** were achieved and the results expressed in term of percentage of cross-reactivity are summarized on Table 4.

Table 4: Relative cross-reactivity (%)

Antibodies	β-CD	**1**	DIMEB	**2**	TRIMEB	**3**
β-CD	100	<0.1	<0.1	<0.1	<0.1	>0.1
DIMEB	<0.1	<0.1	100	<0.1	<0.1	<0.1
TRIMEB	<0.1	<0.1	<0.1	<0.1	100	<0.1

1-3 were poorly recognized by all antibodies. This demonstrates that the modifications introduced by the 3,6-anhydro unit are sufficient to abolish antibody recognition.

3.4 CHIRAL RECOGNITION

An investigation of **2** and **3** by high resolution NMR (COSY, RELAYS, NOESY, HMQC and HMBC) has shown a complete reduction of symmetry and a strong deformation of the cyclodextrin cavity (data not shown). This lack of symmetry induces an improvement of their chiral recognition properties compared to β-CD and even to **1** as observed on the following figure:

Figure 1: ^{1}H NMR spectra (10mM, D2O, 500MHz, 298K) of Dothiepin alone (**a**) and in presence of β-CD (**b**), **1** (**c**) and **2** (**d**).

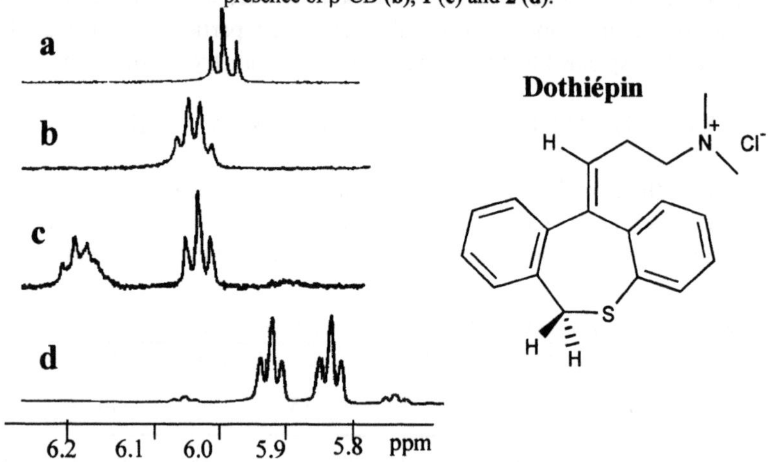

The synthesis of the other asymmetric analogues of β-CD or γ-CD is currently performed in our Laboratory. The characterization and the evaluation of the chiral recognition properties of these new compounds will be assayed by NMR and capillary electrophoresis/ Mass spectrometry.

This work was supported by the European Commission (DGXII) under the FAIR Programme CT95-0300

References

1 F. Djedaïni-Pilard, B. Perly (1993) Patent Fr-93-15470, France
2 M. Gosnat, F. Djedaïni-Pilard, B. Perly ; (1995) *J. Chim. Phys.*, **12**, 1777-1781
3. R.C. Petter, J.S. Salek, C.T. Sikorski, G. Kumaravel, F. Lin (1990) *J. Am. Chem. Soc,* **112**, 3860-3868
4 C.M. Spencer, J.F. Stoddart, R. Zarzycki (1987) *J. Chem. Soc. Perkin trans II*, 1323-1330
5 J. Boger, J. Corcoran, J.M. Lehn (1978) *Helv. Chim. Acta*, 61, 6, 2190-2218
6 K. Uekama, T. Irie (1987) "Cyclodextrins and their industrial uses", D.Duchêne ed, Editions de Santé, Paris, 393-428
7 M. Bost, V. Laine, F. Pilard, A. Gadelle, J. Defaye, B.Perly (1997).*J. Inclusion. Phenom*, **29**, 57-63
8 C. Créminon, F. Pilard, J. Grassi, B. Perly, P. Pradelles (1994) *Carbohydr Res,* **258**, 179-186

Chapter 7

CYCLODEXTRIN COMPLEXES AND ITS CHARACTERIZATION

STUDY OF THE FORMATION OF THE INCLUSION COMPOUND BETWEEN RHODIUM(II) 3-CHLOROCINNAMATE AND β-CYCLODEXTRIN.

MACHADO, F. C.[1,2] ; DE BELLIS, V. M.[2] AND SINISTERRA, R. D.[2]

[1]*Departamento de Química - ICE - Universidade Federal de Juiz de Fora Campus Universitário - Juiz de Fora - MG - Brazil. CEP:36036-330.*
[2]*Departamento de Química- ICEx - Universidade Federal de Minas Gerais Campus Pampulha - Belo Horizonte - MG - Brazil. CEP:31270-901.*

Abstract: The preparation and characterization of the 1:1 inclusion compound of rhodium(II) 3-chlorocinnamate in β-cyclodextrin is reported. Evidence of inclusion was obtained from IR spectroscopic studies, thermal analysis and X-ray powder diffraction results. Given the potential antitumor activity of the rhodium(II) carboxylate and its virtual insolubility in water, its inclusion in β-cyclodextrin opens the possibility for its transference to the aqueous phase.

1. **Introduction**

Since the initial discovery of the antineoplastic activity of *cis*-platin [1], several other complexes containing metals of the platinum group have been prepared and tested for antitumor activity. The activity of rhodium(II) carboxylates as antitumor agents was first demonstrated by Bear and coworkers [2], leading to a new class of anticancer compounds. Rhodium(II) carboxylates have the common tetrabridged structure with a Rh-Rh bond (*Figure 1*).

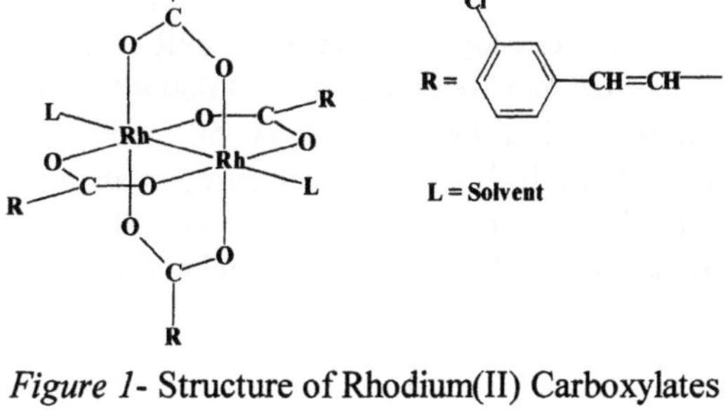

Figure 1- Structure of Rhodium(II) Carboxylates

After a promising beginning, research involving these complexes with antitumor activity diminished, probably due to the toxicity as well as the evident difficulty of solubilization of these compounds in water. It is known that the solubility of sparingly soluble molecules in aqueous medium can be substantially increased by their inclusion in various cyclodextrins [3]. Therefore, we have investigated the possibility of including rhodium(II) carboxylates in cyclodextrins for circumventing this limitation [4]. In the present work we report the preparation of rhodium(II) 3-chlorocinnamate and its inclusion compound with β-cyclodextrin.

2. **Experimental**

2.1.MATERIALS AND PREPARATION OF RHODIUM(II) 3-CHLORO-CINNAMATE

β-cyclodextrin hydrate (β-CD) was purchased from Aldrich and used without further purification. Rhodium(II) 3-chlorocinnamate was prepared by refluxing 1.0 g of $RhCl_3.3H_2O$ and 4.7 g of sodium 3-chlorocinnamate in 100 mL of ethanol for 2.5 hours. The reaction was carried out under dinitrogen atmosphere and the solvent was previously dried according to standard techniques. After cooling down to room temperature, a green solid was collected and dried in a dessicator for 24 hours. The compound was virtually insoluble in water but soluble in most common organic solvents. Chemical analysis revealed the expected carbon and hydrogen content: C, 45.2 (46.3); H, 2.8 (2.6).

2.2.PREPARATION OF THE INCLUSION COMPOUND

This was prepared by adding 0.04 g of solid rhodium(II) 3-chlorocinnamate to an aqueous solution of β-cyclodextrin (0.04 g in 40 mL of water) under constant stirring. The resulting solution was kept at *ca.* 40-50 ^{0}C for 24 hours. The solution was then frozen and lyophilized. A physical mixture of β-CD and rhodium(II) complex, in a 1:1 molar ratio, was also prepared by gently grinding of the components until a homogeneous powder was obtained.

2.3.PHYSICAL MEASUREMENTS

IR spectra were obtained employing a Mattson FTIR 3000 spectrophotometer with the samples recorded as KBr pellets in the 4,000-400 cm^{-1} region. The TG and DTG curves were obtained on a Mettler TG-50 thermobalance of TA-4000 system at a heating rate of 10 $^0C.min^{-1}$, under nitrogen atmosphere. Differential scanning calorimeter (DSC) traces were recorded on a Shimadzu DSC-50 series thermal analysis in the 25-480 ^{0}C range at the scanning rate of 10 $^0C.min^{-1}$, under a 40 sccm flow of dinitrogen. X-ray powder patterns were obtained using a Rigaku Geigerflex diffractometer employing CuK_α radiation.

3. **Results and Discussion**

In general, IR spectra for CD inclusion compounds show only small changes in the absorption bands in comparison to the spectra of the free components, revealing that the host to guest interaction is not particularly strong [5]. IR spectrum of β-CD in the 3,400-3,300 cm^{-1} region shows the characteristic strong and broad band of water, which became sharp in the spectrum of the freeze-dried sample. Also, changes in the relative intensities of modes assigned to the aromatic moiety of the guest, around 1,480 cm^{-1}, are observed in the IR spectrum of the freeze-dried sample in comparison to the spectra of rhodium(II) complex and physical mixture. These results suggest the interaction between the aromatic ring of cinnamate and β-CD cavity.

The TG/DTG curves show a lower thermal stability for the freeze-dried sample when compared to the corresponding physical mixture or to the individual components. The DSC curve of β-CD shows two endotherms at 80 ^{0}C and 320 ^{0}C; the first one corresponds to the lost of H_2O, the latter to decomposition of β-CD. An endothermic peak at 335 ^{0}C is observed in the DSC curve for the rhodium(II) complex. The DSC curve for the physical mixture is just a superposition of the curves observed for the isolated components. A sharp endothermic peak at about 310 ^{0}C is observed in the DSC curve for the freeze - dried sample. In addition, the endothermic peak at 335 ^{0}C due to the decomposition of the rhodium(II) complex does not appear in the DSC curve for the freeze-dried product; it may be attributed to the inclusion complex formation.

The most stringent evidence for the formation of the inclusion compound comes from the X-ray powder diffraction results, as can be seen in *Figure 2*. The diffractogram of the physical mixture is a simple superposition of the diffractograms of β-CD and rhodium(II) carboxylate. The XRD pattern of the freeze-dried sample revealed a new peak at 31^0 (2θ) and other characteristic peaks due to individual components became absent. These results strongly suggest the formation of a new phase that can be associated with the inclusion compound.

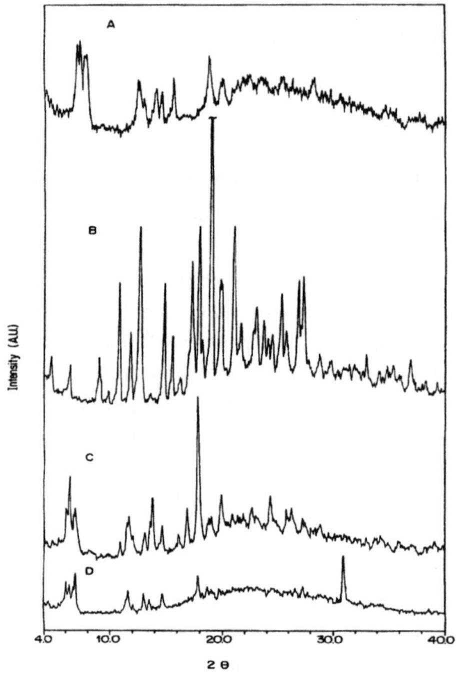

Figure 2 - X-ray diffraction pattern of (A) rhodium(II) 3-chlorocinamate, (B) β-cyclodextrin, (C) physical mixture, (D) inclusion compound.

4. Acknowledgments

The authors are grateful to Conselho Nacional de Desenvolvimento Científico e Tecnológico (CNPq), Coordenação de Aperfeiçoamento de Pessoal de Nível Superior (CAPES) and Fundação de Amparo à Pesquisa do Estado de Minas Gerais (FAPEMIG) for financial support.

5. References

1- Rosenberg, B., Van Camp, L. and Krigas, T. (1965) Inhibition of cell division in *Escherichia coli* by electrolysis products from a platinum electrode, *Nature* 205, 698-699.

2- Hughes, R.G., Bear, J.L. and Kimball, A.P. (1972) Synergistic effect of rhodium acetate and arabinosylcitosine, *Proc. Amer. Assoc. Cancer Res.* 13, 120.

3- Szejtli, J. (1989): *Cyclodextrin Technology*, Kluwer Academic Publishers.

4- Sinisterra, R.D., Najjar, R., Alves, O.L., Santos, P.S., Alves de Carvalho, C.A., Conde da Silva, A.L. (1995) The inclusion of rhodium(II) α-methyl-cinnamate in β-cyclodextrin, *J. Incl. Phen.* 22, 91-98.

5- Saenger, W. (1984): *Inclusion Compounds*, Ed J. L. Atwood, J. E. D. Davies and D. D. MacNicol, *Academic Press*, London, Vol. 2, Ch. 8.

FORMATION OF INCLUSION COMPLEX OF CYCLODEXTRIN IN ORGANIC SOLVENTS AND ALCOHOLS UNDER ANHYDROUS CONDITIONS

H. YOSHII†, T. KOMETANI*, T. FURUTA, Y. WATANABE,
YU-Y. LINKO,** AND P. LINKO**
Department of Biotechnology, Tottori University, Tottori 680, Japan
**Department of Biochemical Engineering, Toyama National College of Technology, Toyama 939, Japan*
***Department of Chemical Technology, Helsinki University of Technology, FIN-02015, P.O.Box 6100 HUT, (Espoo), Finland*

1. Introduction

To prepare a complex between cyclodextrin and a drug, many techniques have been developed such as kneading, grinding, spray drying, roll mixing, and coprecipitation from various solvents. The most popular method is to form the inclusion complex in a solution of cyclodextrin. In this method, the guest compound is dissolved into an aqueous solution of cyclodextrin to form the inclusion complex in a crystalline form. However, since various pharmaceutical drugs have a low solubility in water, it is very important to develop a new method for the effective formation of the inclusion complexes. In this study, we investigated the preparation of inclusion complexes between cyclodextrins and the guest(*d*-limonene, phenyl ethanol, acetophenone or menthol) in two organic solvents, ethanol and methanol. This study presents a simple method of preparing a powder of the inclusion complex with alcohol. A mathematical model for the estimation of the inclusion ratio of the guest compound to CD is also presented with the adsorption of the alcohol on CD by the analogy of substrate inhibition of enzyme kinetics.

2. Materials and Methods

2.1 MATERIALS

Ethanol and methanol were for organic synthesis from Wako Pure Chemical Company (Tokyo). These alcohols contained less than 50 ppm water. Other solvents used were of HPLC grade from Wako Pure Chemical Company. The α-, β-, and γ-cyclodextrins (CD) were obtained from Ensuiko Sugar Refining Co. (Yokohama) Chloroform stabilized by amylene was from Kanto Chemicals Co., Ltd. (Tokyo).

2.2 PREPARATION OF INCLUSION COMPLEX POWDER WITH AN ORGANIC SOLVENT

The inclusion complex powders of the guest compound with CD were prepared as follows in various organic solvents: CD powder (0.5 g), which was dried *in vacuo* at 100 °C for 48 h, was weighed in a 15x130mm glass tube. The mixture of the guest compound (0-20 mmol) and the organic solvent (0.5 ml) were added to the CD powder in the tube, which was vigorously mixed for 1 min in a vortex mixer to make a suspension of the CD and the solvent. The slurry was incubated in a water bath at 30°C for 24 h. The supernatant was removed with a Pasteur pipette after centrifuging at 3500 rev/min (500g) for 15 min, and the wet slurry was dried *in vacuo* at 80°C for 5 h. When the inclusion complex was prepared in ethanol, the molar ratio of ethanol to CD was varied from 0 to 100 in order to investigate the effect of the ethanol amount. The content of the guest compound in the inclusion complex powder was measured by gas chromatography, using the solvent (i.e., chloroform) extraction procedure reported previously[1]. The concentration of the alcohol in the chloroform and water were also analyzed by gas chromatography, by separating on PEG-20M (Shimadzu, Kyoto).

2.3 INCLUSION MODEL IN ETHANOL OR METHANOL

Methanol or ethanol accelerated initially the inclusion of the guest compound to CD. However, the inclusion was inhibited by adding more than the amount of alcohol optimal for the maximum inclusion ratio of the guest, compound, because the alcohol adsorbed CD and interfered with the association of the guest compound into the CD. This scheme resembled that of an enzyme reaction of Michelis-Menten type with substrate inhibition. The encapsulation scheme may be expressed as follows:

$$G + AOH + CD \xleftrightarrow{K_m} G / AOH / CD \xrightarrow{a} Compex \tag{1}$$

$$CD + AOH \xleftrightarrow{K_i} CD \cdot AOH \tag{2}$$

$$CD \cdot AOH + AOH \xleftrightarrow{K_i} CD \cdot (AOH)_2 \tag{3}$$

$$CD \cdot (AOH)_2 + AOH \xleftrightarrow{K_i} CD \cdot (AOH)_3 \tag{4}$$

$$\vdots$$

$$CD \cdot (AOH)_{n-1} + AOH \xleftrightarrow{K_i} CD \cdot (AOH)_n \tag{5}$$

where G is the guest compound and AOH is methanol or ethanol, and *Complex* is the inclusion complex. K_m and K_i are the stability constant and inhibition constant, respectively, and a is the reaction constant of encapsulation. If alcohol is assumed to be adsorbed with the free OH groups in CD, the number of alcohol molecules adsorbed on CD, n, is equal to 18 for α-CD, 21 for β-CD and 24 for γ-CD. The guest compound and alcohol would compete with each other for occupying of the cavity of CD. The adsorption of alcohol on CD reduces the available cavity for the interaction with the guest compound, resulting in a decrease in the inclusion ratio of the guest compound in CD. By using the above encapsulation scheme, the encapsulation ratio of the guest to the CD, R, can be written in analogy to the enzyme kinetics, as

$$R = \frac{a}{1 + \frac{K_m}{S} + \frac{S/K_i(1-(S/K_i)^n)}{(1-S/K_i)}} \tag{6}$$

where S is the molar ratio of alcohol(methanol or ethanol) to CD.

3. Results and Discussion

3.1 SCREENING OF SOLVENT FOR THE FORMATION OF CD COMPLEX.

The inclusion ratios of *d*-limonene into β-CD in various organic solvents, such as ethanol, methanol, acetonitrile, isopropyl alcohol, n-propyl alcohol, toluene, hexane, isopropyl ether, acetone, ethyl acetate, methyl ethyl ketone, tetrahydrofuran, diethyl ether, 1,4-dioxane, 1-butanol, and ethylene glycol dimethyl ether, were investigated. Almost no inclusion complex of *d*-limonene in β-CD was observed in organic solvents except for methanol, ethanol, acetonitrile, isopropyl alcohol and n-propyl alcohol. In the inclusion of *d*-limonene into β-CD, only ethanol and methanol were suitable solvents. Ethanol, methanol and acetonitrile could interact with the hydroxyl groups in the cyclodextrin cavity. Ethanol could be considered to behave as water during the inclusion process. Consequently, it could be assumed that the hydrophobic cavity might allow an incorporation of *d*-limonene to form the inclusion complex without water.

3.2 COMPLEX FORMATION IN ETHANOL

The inclusion ratio of *d*-limonene in β-CD was investigated with anhydrous ethanol The optimum amount of ethanol in the mixture of *d*-limonene and ethanol was investigated by changing molar ratio of ethanol to β-CD in the slurry. Figure 1 shows the inclusion ratios of *d*-limonene or ethanol in β-CD complex powder against the amount of ethanol added to the slurry. With an increase in the amount of ethanol added to β-CD, the inclusion ratio of *d*-limonene increased up to a maximum value at the molar ratio of ethanol to β-CD of about 20, beyond which the inclusion ratio decreased. On the other hand, the inclusion ratio of ethanol to β-CD decreased to a minimum at about molar ratio of 30, and then increased with the increase in the ethanol content in the slurry. When the ethanol content was increased, ethanol displaced *d*-limonene from the cavity, and might form higher order of ethanol-β-CD complexes. The result suggests that ethanol adsorbed on cyclodextrin interacts with the hydroxyl groups of CD, and *d*-limonene could be included easily in β-CD by the adsorption molar ratio of ethanol to β-CD of about 20. The number of ethanol molecules adsorbed on β-CD at the maximum inclusion ratio

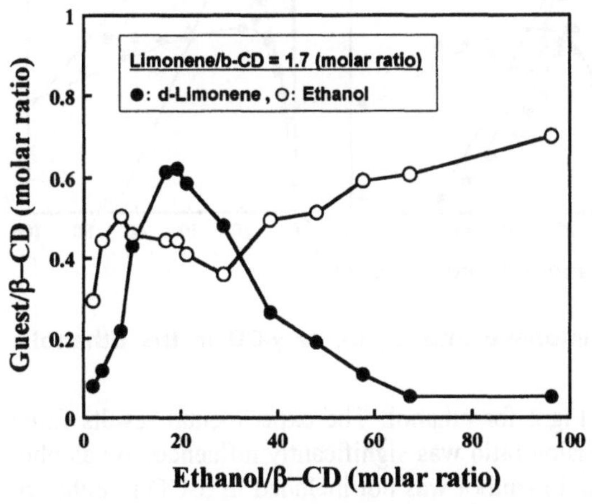

Fig. 1 Encapsulation of Ethanol into β-CD in the Presence or Absence of Water.

was about 21 which is the number of free hydroxyl groups in β-CD. At the higher ethanol content, *d*-limonene might be prevented by ethanol from the formation of an inclusion complex with β-CD. The inclusion ratio decreased as the ethanol content increased over the optimal ethanol addition. This suggests that the adsorption of a large amount of ethanol on CD resulted in the inhibition of the formation of the guest/β-CD complex with an alcohol as a solvent.

3.3 SIMULATION OF THE INCLUSION RATIOS OF VARIOUS GUEST COMPOUNDS WITH THE INCLUSION MODEL

Several guest compounds such as acetophenone, phenyl ethanol, menthol and *d*-limonene were used to form an inclusion complex with α-, β-, or γ-CD in ethanol or methanol. The observed inclusion ratios of the guest compounds were compared with the predicted inclusion ratio by Eq.6, which was based on the alcohol adsorption model on CD. Estimation of the inclusion ratio by Eq.6, in which a, K_m and K_i are used as

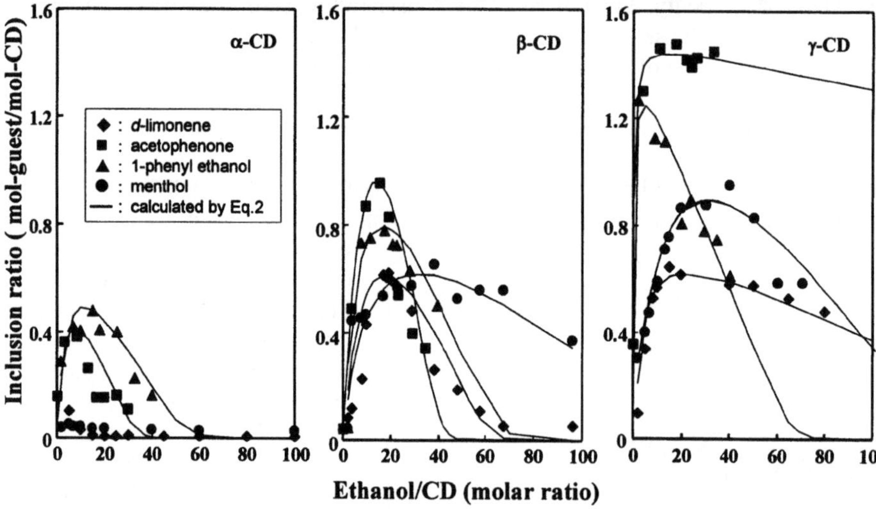

Fig. 2 Inclusion Ratio of *d*-Limonene into α-, β-, or γ-CD in the Ethanol Solvent System.

correlation parameters, is shown in Fig.2 for ethanol. The experimental results fitted well to the calculated lines. The inclusion ratio was significantly influenced by alcohol amount in the solvent. *d*-Limonene and menthol was not included in α-CD in either of ethanol or methanol. The inclusion ratio of the guest compounds increased with an increase in the diameter of the CD cavity.

References

(1) Yoshii, H., Furuta, T., Yasunishi, A.and Hirano, H. (1994) Minimum Number water Molecules required for Inclusion of d-Limonene in the Cyclodextrin Cavity, *J. Biochem.*, **115**, pp.1035-1037

INCLUSION COMPLEXES OF UV FILTERS IN SOLUTION AND IN SOLID STATE

FENYVESI, E. [1], JICSINSZKY, L. [1], SZEJTLI, J. [1], SCHWARZENBACH, R. [2]

[1]*Cyclolab R&D Lab., Dombóvári út 5-7, Budapest, H-1117 Hungary*

[2]*Givaudan - Roure, GH 8600 Dubendorf, Switzerland*

1. Introduction

UV absorbers are often used in pharmaceutical and cosmetic preparations. They are usually sparingly soluble in water which limits their applications. In a previous study we have found that two water-insoluble UV absorbers [3-(4-methylbenzylidene)camphor and 2-hydroxy-4-methoxybenzophenone] can be solubilized by β-cyclodextrin (βCD), random methylated β-CD (RAMEB), γ-cyclodextrin (γCD), and acetyl γ-cyclodextrin (AcγCD) [1]. It was also found that the rate of photofading of aqueous azoreactive dye solutions markedly decreased in the presence of RAMEB-solubilized UV filter. In the present work the solubility of the UV absorber, octyl methoxy cinnamate in aqueous solutions of various cyclodextrins was studied; solid complexes with α-, β- and γ-CD were prepared and the molar ratio of bound guest to CD in the products were calculated on the basis of Evolved Gas Analysis; Molecular Modeling calculations were performed to understand the interaction between the UV filter and the CDs.

2. Experimental

2.1. MATERIALS

UV absorber: octyl methoxy cinnamate (Givaudan-Roure, Schwitzerland).

CDs: α–cyclodextrin (αCD), Lot No. 850902 and β–cyclodextrin (βCD), Lot No. 8709187/14 (Chinoin, Budapest); γ–cyclodextrin (γCD), Lot No. 1006 and random methyl β–cyclodextrin (RAMEB), DS = 1.8 Lot No. 12/2/92 (Wacker Chemie, Munich); acetyl β–cyclodextrin (AcBCD), DS = 6.2, CYL-353/3 and acetyl γ–cyclodextrin (AcGCD), DS = 6.9, CYL-349/1 and hydroxypropyl β–cyclodextrin (HPBCD), DS = 2.7 (Cyclolab, Budapest); maltosyl β–cyclodextrin (MaBCD), DS ≈ 1 Lot No. 93281 (Ensuiko Sugar Refining Co., Yokohama)

2.2. SOLUBILITY ISOTHERMS:

Excess amount of UV absorber was added to 5 mL of the properly diluted cyclodextrin solutions and stirred for 1 hour. The suspensions were left to get the oil separated from the aqueous phase, then aliquots of the aqueous solution were withdrawn carefully, filtered and measured spectrophotometrically.

2.3. PREPARATION OF SOLID COMPLEXES [2]

Kneading method: The proper amount of CD and UV absorber were weighed into a porcelain mortar, and was kneaded for 10 min with a small amount of ethanol-water (1:1) mixture. The product was washed with diethyl ether, and dried on air.

Suspension method: The proper amount of CD was suspended in water at room temperature and the UV absorber was added under rigorous stirring. The stirring was continued for 6 h, then the suspension was freeze dried. The product was washed with diethyl ether and dried on air before analysis.

Coprecipitation method: The proper amount of CD was dissolved in water and the UV absorber was added, then stirred for 2 h and the suspension was centrifuged at 5000 *g* for 15 min The supernatant was decanted, the precipitate was suspended in acetone and filtered through G-3 glass filter. The product was obtained after drying at 60 °C.

2.4. EVOLVED GAS ANALYSIS

Evolved Gas Analysis (EGA) was measured with a Du Pont 916 TEA apparatus, with a heating rate of 8°C/min and nitrogen atmosphere (1.8 L/H).

2.5. COMPUTING METHOD:

***Semiempirical optimizations*:** Performed by HyperChem® 4.5 (Pentium-100). Starting geometries were from semiempirically (AM1) optimized structures (about 400 cycle were required to obtain the optimized geometry, one cycle: 10 min for αCD, 17 min for βCD, and 33 min for γCD).

***Molecular mechanics*:**

Geometry Optimization: HyperChem® 4.5 implemented CHARMM type force field was chosen (Pentium-100). The modified Polak-Ribiere conjugated gradient method was used in the geometry optimization algorithm.

Solution simulation: Performed in a 56*56*56 $Å^3$ periodic box which contains 274-280 molecules of standard TIP3P water molecules (about 10'000 is required to get the optimized geometry, grad <0.05, and ten cycles cost 2.8 min for αCD, 3 min for βCD, and 3.3 min for γCD, respectively).

The presented molecular configurations are chosen from about 200 starting pre-optimized configurations. Positive signs mean less favorable ensembles, and values in shaded boxes printed in italics represent the favorable arrangements of molecules. For molecular dynamics simulations identical conditions for force field and periodic box conditions are used.

3. Results

3.1. SOLUBILITY STUDIES

Solubility isotherms of B_s-type were obtained with β- and γCD. The soluble β- and γCD derivatives and the parent αCD give isotherms of A type, mostly A_P-type except in case of αCD, where A_N-type isotherm (negative deviation from linearity) was obtained.

Methyl β-cyclodextrin is the best solubilizer for the UV-filters: as high as four order of magnitude solubility enhancement was obtained. With hydroxypropyl β-cyclodextrin, maltosyl β-cyclodextrin and acetyl cyclodextrins two - three order of magnitude increase in the solubility was observed. The parent cyclodextrins are less effective: solid complexes were precipitated from their solutions.

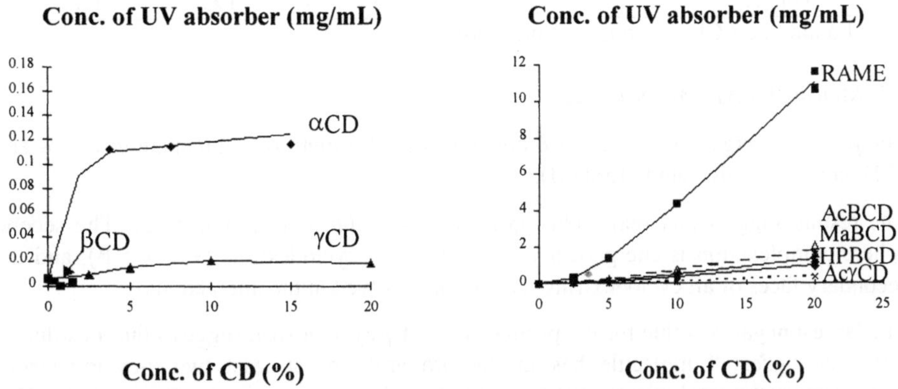

Fig. 1 Solubility of octyl methoxy cinnamate in aqueous CD solutions as a function of CD concentration

3.2. SOLID PRODUCTS

Solid products were prepared with the parent CDs by coprecipitation, suspension and kneading method and characterized with Evolved Gas Analysis (EGA) to obtain the ratio of free to bound guests. Comparing the methods of preparation (Table I) the short kneading showed to be unsuccessful and coprecipitation resulted in higher molar ratio of bound guest, but the yield of preparation was low. With the suspension method the complexes were obtained with 80-90 % yield.

The 1:1 molar ratio of guest/CD was achieved only by γCD, with βCD the 0.4:1 molar ratio could not be surpassed. The very low bound guest/CD molar ratio values in case of α- and βCD suggest that no complex formation takes place.

TABLE I. Conditions of the preparation and characteristics of octyl methoxy cinnamate/CD complexes

Type of CD	Conditions of the preparations			Product			
	Method	stirring time (h)	initial molar ratio (mole/mole)	active ingredient content (%)			calculated molar ratio* (mole/mole)
				total	free	bound	
αCD	susp.	16	1:1	9.9	3.3	6.6	0.2:1
	coprec.	2	1:1	8.5	2.0	6.5	0.2:1
βCD	kneading	0.1	1:1	3.6	n.m.		0.1:1
	susp.	3	1:1	11.3	1.2	10.1	0.4:1
		24	1.2:1	15.2	6.5	8.7	0.4:1
	coprec.	2	1:1	16.7	6.8	9.9	0.4:1
γCD	susp.	24	2.5:1	21.0	3.4	17.6	1:1
	coprec.	2	1:1	20.3	<1	19.3	1:1

* bound guest/CD n.m. not measured

3.3. MOLECULAR MODELING

The relative energy differences of 8 positions of octyl methoxy cinnamate to α-, β- and γCD have been calculated (Table II).

The results suggest that real inclusion *at least in solution* does not occur. The second sphere complexation is energetically favored for all cyclodextrins (γ ≈ α > β), and the secondary faces of all cyclodextrins seem to be involved in the interactions.

The largest negative value for the primary face of γ-cyclodextrin suggests that in solution state one "guest " molecule has in the proximity of at least two γ-cyclodextrins. Summarizing the total energy differences γ-cyclodextrin seems to be the most effective cyclodextrin for the complexation of octyl methoxy cinnamate.

TABLE II Relative Differences in the Total Energies [Kcal/mole] of Solution State Simulation of octyl methoxy cinnamate /Cyclodextrin/Water Systems

	αCD		βCD		γCD	
	primary side	secondary side	primary side	secondary side	primary side	secondary side
separated (>20 Å)	0.0	0.0	0.0	0.0	0.0	0.0
second sphere complex	15.4	*-15.4*	2.9	*-2.9*	15.2	*-15.2*
UV sensitive (middle) part	45.1	*-19.6*	25.7	*-1.9*	*-25.3*	61.6
aromatic ring	62.3	38.6	24.7	6.6	46.7	3.0
alkyl chain	0.5	13.8	63.3	32.4	-0.4	32.1
Sum of Differences	125.5		150.8		117.7	

The significant, energetically favored values are shaded.

Optimizing the structure of the solid complexes we found that in case of γCD the real inclusion complex represents significantly high negative changes in the total energy compared to the separated molecules (-26.8 and -29.2 kcal/mole for the position when alkyl chain and when the aromatic ring, respectively are included in the cavity), some what lower in case of βCD (-13.8 and -15.7 kcal/mole, resp.), while in case of αCD especially the position with the aromatic ring inside is unfavorable (-9.3 and 3.9 kcal/mole, resp.).

4. Conclusions

On the example of octyl methoxy cinnamate it was shown that cyclodextrins (especially the RAMEB) enhance the solubility of UV absorbers. The parent β- and γCD form insoluble complex which precipitates from the solutions. The solid products prepared by suspension or coprecipitation methods contained also not-bound (free) active ingredient. The molar ratio of the bound guest/CD was 1, 0.4 and 0.2 for γ-, β- and αCD, resp. The molecular calculations suggest that in solution not real inclusion takes place, second sphere complexes represent the lowest energy state. Optimizing the structure of the solid complexes only γCD was found to form real inclusion complex.

ACKNOWLEDGEMENT

Thanks are due to Dr. Csaba Novák for the EGA measurements. The technical assistance of Zs. Nagy is greatly acknowledged. The work was partially supported by a grant of Hungarian Research Fund (OTKA T 022003).

REFERENCES

1. Rémi, E., Fenyvesi, É., Rusznák, I., Vig, A.: (1996) The action of lipophilic UV absorbers - solubilized by cyclodextrin - on photofading of aqueous solution of azo reactive dyes *J. Inclusion Phenom. Mol. Recognit. Chem.*, 25(1-3), 203-207

2. Szente, L. (1996) Preparation of cyclodextrin complexes, in Szejtli, J., Osa, T. (ed.) *Comprehensive Supramolecular Chemistry*, Volume 3, 243-252. Elsevier, Oxford, UK.

ENCAPSULATION OF PORPHYRINS BY γ-CYCLODEXTRIN

ASAO NAKAMURA and JÜRGEN-HINRICH FUHRHOP, *Institut für Organische Chemie, Freie Universität Berlin, Takustraße 3, 14195 Berlin, Germany*

1. Introduction

The cavity of cyclodextrins (CDs) is known to provide a media for regio- or stereoselective reactions. Porphyrins and their metal complexes catalyze several important reactions, e.g. photosensitized oxidation of alkenes. Much more interests should be focused on the reaction systems which contain both cyclodextrins and porphyrins. Several papers have been publishsed for the interactiton between porphyrins and cyclodextrins [1–5]. There is, however, few report on the combination of γ-CD and porphyrins [6]. We have tried to characterize the inclusion complexes of γ-CD with porphyrins. It has clearly been shown that porphyrins are a very fit guest for γ-CD and that a large part of the porphyrin ring is covered in the inclusion complex by the γ-CD molecule.

2. Experimental

Tetrasodium salt of *meso*-Tetra(4-sulfonatophenyl)porphyrin (H_2TSPP) was obtained from Strem. Copper complex of TSPP (CuTSPP) was prepared by heating TSPP with copper(II) chrolide in water.

Electronic absorption spectra were recorded with a Perkin-Elmer Lambda 16 spectrophotometer. Fluorescence spectra were recorded with a Perkin-Elmer MPF-44B spectrofluorometer. Circular dichroism spectra were recorded with a Jasco J-600 spectropolarimeter. Electron paramagnetic resonance (EPR) spectra were measured at X-band frequency (9 GHz) using an instrument equipped with a helium cryostat.

3. Results and Discussion

1. Electronic Absorption Spectra

Aggregation at high concentration is often observed for water-soluble porphyrins in aqueous solutions [7–9]. H_2TSPP and CuTSPP showed very broad B-band (Soret band) at the concentration of 1 mM (Table 1). This band broadening can be assigned to the aggregation of the porphyrins. In less polar solvent than water the band shifted to longer wavelength and the band width was smaller. When γ-CD was added to the aqueous solution, the band became narrower. This implies that the aggregates are dissociated through the inclusion by γ-CD. This gives a strong evidence for the assumption that the porphyrins are included into the cavity of γ-CD not as a dimer but as a monomer.

TABLE 1. Band width (FWHM) of the Soret band for CuTSPP.

Solvent	Concentration / M	λ_{max} / nm	$\Delta\lambda_{1/2}$ / nm
DMSO	5×10^{-6}	419	15
DMSO + water (1:1)	5×10^{-6}	416	14
	1×10^{-3}	416	14
water	5×10^{-6}	412	14
	1×10^{-3}	404	20
γ-CD (20 mM)	5×10^{-6}	414	11
	1×10^{-3}	414	11

2. Electron Paramagnetic Resonance Spectra

EPR spectra verified the assumption that γ-CD helps the dissociation of the aggregates of porphyrins into monomer in aqueous solutions. EPR spectrum for CuTSPP (1 mM) in aqueous solution was distorted from the normal spectrum for monomer (Figure 1A). This distortion will be the effect of the interaction between spins which is possible only in the aggregates. In contrast to this the spectrum for the porphyrin in the presence of γ-CD was nearly the same as those for monomer in organic solvents (Figure B). This result is consistent with the conclusion from the measurment of electronic absorption spectra that the porphyrin makes aggregates in aqueous solutions, and that monomer is stabilized by γ-CD through the inclusion of only the monomer.

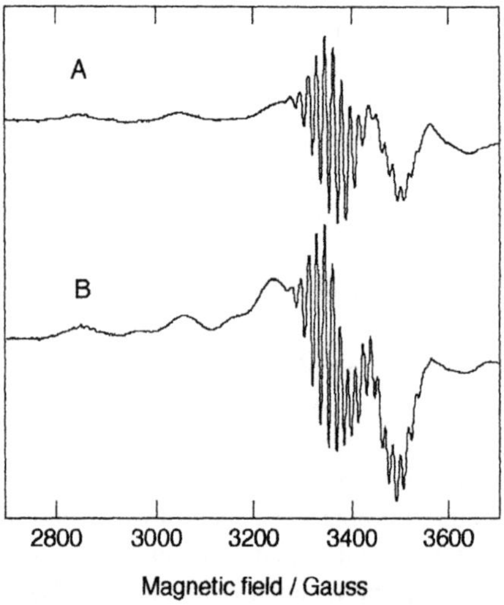

Figure 1. EPR spectra for CuTSPP in water; (A) in the presence of 20 mM of γ-CD; (B) in the absence of γ-CD.

3. Circular Dichroism Spectra

From a primitive molecular modeling using a space-filling model we propose the struc-

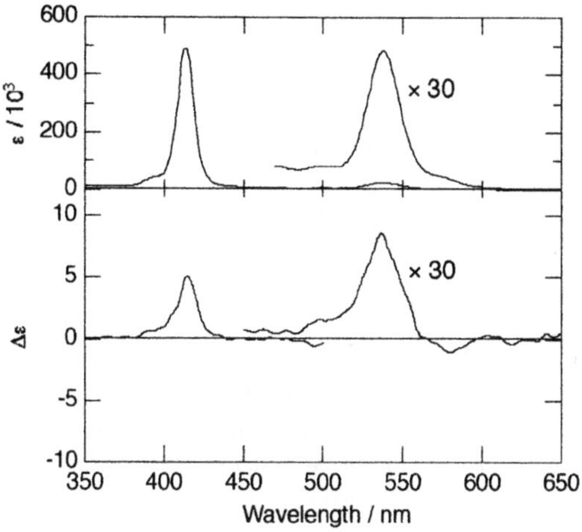

Figure 2. Circular dichrism (lower) and electronic absorption spectra (upper) for the inclusion complex of γ-CD with CuTSPP in a aqueous solution (pH 7).

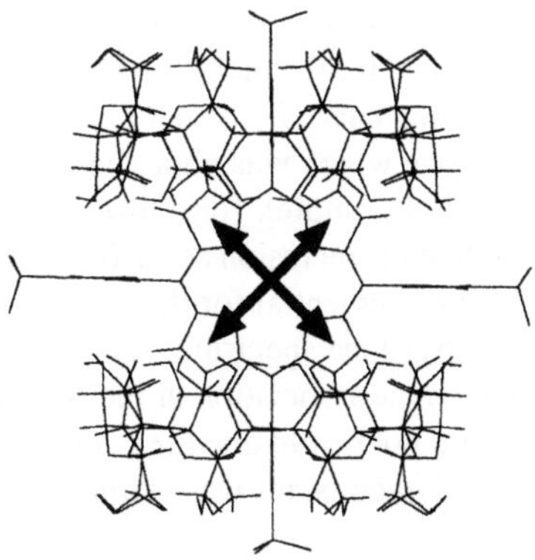

Figure 3. Schematic illustration showing the estimated structure of the inclusion complex of γ-CD with CuTSPP. The position and the direction of the electronic transition moments are shown by arrows.

ture shown in Figure 3 as the structure of the inclusion complex between γ-CD and TPPS. The distance between β-protons on the neighboring pyrrole rings of the porphyrin ring is about 10 Å. The inner radius of the upper rim of the γ-CD cavity is wide enough to accomodate the porphyrin ring of this size.

In this structure model the transition moments for Q- and B-bands are located at 5 Å above the center of the CD cavity. The angle between the transition moments and the rotational symmetry axis of CD is just 45°, By using Kodaka's rule [10] one can predict the sign of the induced circular dichroism spectrum. A weak positive signal is predicted by the theory. In the circular dichroism spectrum for the inclusion complex weak positive bands were observed both at Q- and B-bands (Figure 2). This gives an evidence for the predicted structure.

From this model it is also evident that the binding of more than 2 CD to 1 porphyrin will be impossible. The binding of 2 CD to 1 porphyrin will be the most stable. Another interesting point is that there is a small cleft of the capsule at the center of the inclusion complex. The central metal is exposed to solvents through this cleft.

4. **Fluorescence Spectra**

Fluorescence spectra of H_2TSPP changed largely on the addition of γ-CD. Intensity of the fluorescence was larger and the vibrational subbands were clearer in the presence than in the absence of γ-CD. This means that the environment of the porphyrin is much less polar in the inclusion complex than in the free monomer. Moreover, this implies that not only the peripheral phenyl group but also the substantial part of the porphyrin ring is included into the cavity.

By using the change in the fluorescence intensity we can estimate the association constant for the inclusion complex. The titration curve fited well to the calculated one which was obtained by assuming 1;1 complex formation. This means that the inclusion at the first phenyl group is essential for the change in the fluorescence spectrum. The information on the association of the second CD could not be obtained fro this data. Association constant was estimated at 1600 M^{-1}.

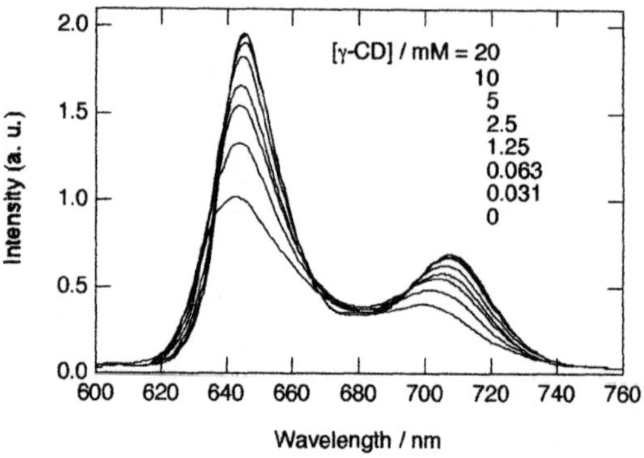

Figure 4. Change in the fluorescence spectrum of H_2TSPP in aqueous solution (pH 7) by increasing the concentration of γ-CD.

Acknowledgement

We would like to thank Dr. Art van der Est and Uwe Krüger for their assistence with the EPR measurements and also for helpful discussions. This work was supported by Alexander von Humboldt Stiftung with a fellowship (to A. N.).

References

1. Mosseri, S., Mialocq, J. C., Perly, B. and Hambright, P. (1991) Porphyrins⇌cyclodextrin. 1. Photooxidation of zinc tetrakis(4-sulfonatophenyl)porphyrin in cyclodextrin cavities: The characterization of ZnTSPP dication. Photolysis, radiolysis, and NMR studies, *J. Phys. Chem.*, *95*, 2196-2203.
2. Mosseri, S., Mialocq, J. C., Perly, B. and Hambright, P. (1991) Porphyrins⇌cyclodextrin. 2. Dissociation, reduction, and proton relaxivity of an iron(III) porphyrin μ-oxo dimer in cyclodextrin solutions, *J. Phys. Chem.*, *95*, 4659-4663.
3. Jiang, T. and Lawrence, D. S. (1995) Suger-coated metalated macrocycles, *J. Am. Chem. Soc.*, *117*, 1857-1858.
4. Manka, J. S. and Lawrence, D. S. (1990) Template-driven self-assembly of a porphyrin-containing supramolecular complex, *J. Am. Chem. Soc.*, *112*, 2440-2442.
5. Willner, I., Adar, E., Goren, Z. and Steinberger, B. (1987) Photosensitized reduction of benzyl and octyl viologens in β-CD aqueous media, *New J. Chem.*, *11*, 769-773.
6. Wu, W. and Stalcup, A. M.(1994) *J. Liq. Chromatogr.*, *17*, 1111.
7. Krishnamurthy, M., Sutter, J. R. and Hambright, P. (1975) Monomer⇌dimer equilibration of water-soluble porphyrins as a function of a co-ordinated metal ion, *J. Chem. Soc., Chem. Commun.*, 13.
8. Pasternack, R. F., Huber, P. R., Boyd, P., Engasser, G., Francesconi, L., Gibbs, E., Fasella, P., Venturo, G. C. and Hinds, L. deC. (1972) On the aggregation of meso-substituted water-soluble porphyrins, *J. Am. Chem. Soc.*, *94*, 4511.
9. Pasternack, R. F., Francesconi, L., Raff, D. and Spiro, E. (1973) Aggregation of nickel(II), copper(II), and zinc(II) derivatives of water-soluble porphyrins, *Inorg. Chem.*, *12*, 2606.
10. Kodaka, M. and Furuta, T. (1989) Induced circular dichroism spectrum of a-cyclodextrin complex with heptylviologen, *Bull. Chem. Soc. Jpn.*, *62*, 1154-1157.

GUEST-DEPENDENT ORDERING OF THE SELF-ASSEMBLED CYCLODEXTRIN INCLUSION COMPLEXES STUDIED BY SCANNING TUNNELING MICROSCOPY

Satoshi Yasuda [1*], Yasuyuki Goto[1], Koji Miyake[1], Kenji Hata[1,2], Jun Sumaoka[3], Akira Harada[4], Makoto Komiyama[3*], and Hidemi Shigekawa[1,2*]

[1]*Institute of Materials Science, and Centerfor Tsukuba Advanced Research Alliance (TAPA), University of Tsukuba, Tsukuba 305-8573, Japan*
[2]*CREST, Japan Science and Technology Corporation (JST)*
[3]*Department of Chemistry and Biotechnology, Graduate School of Engineering, University of Tokyo, Tokyo 113-8656, Japan*
[4]*Department of Macromolecular Science, Faculty of Science, Osaka University, Toyonaka, Osaka 560-0043 Japan*

Scanning tunneling microscopy was performed on the three types of cyclodextrin (CyD) inclusion complexes; (1) α-CyD-penta-ethylene glycol, (2) G1- α -CyD-H_20 (G1: 6-0-a-D-glucopyranosyl), and (3) β-CyD-ACA (ACA: adamantanecarboxylic acid), which were self-assembled on a highly oriented pyrolytic graphite surface. The observed structures were different from those of the three-dimensionally grown crystals determined by X-ray diffraction method, and their structures were different from each other. The obtained results strongly suggests not only that the included guest compounds or the modified CyD structures influence the molecular interactions between CyDs, but also the possibility of the control of the ordering of self-assembly of CyD molecules as a nanoscience technology.

1. INTRODUCTION

Study of the mechanism of the molecular arrangement induced by the self-assemble ordering has been attracting considerable attention because of its importance from both fundamental and practical points of view [1-6]. Since interaction between molecules is reflected on the ordering of self-assembled structures, microscopic analysis of the structures of the assemblies is extremely important to understand the molecular interactions and to develop the nanoscience technology in order to realize the desired materials. In line with the development of the scanning tunneling microscopy (STM), direct observation of self-assembled molecules has been a recent research trend. However, the analysis by STM has mainly been conducted on molecules exhibiting van der Waals interactions; thus study of molecules with anisotropic interaction is lacking [1-6].

Cyclodextrin (CyD) has a torus-shaped cyclic structure and shows selective catalysis in the reactions with various organic materials, for example, by forming inclusion complexes using its cavity. Furthermore, CyD is easily modified chemically, and is expected to play an important role in the field of artificial enzymes and biosensor devices. In addition, CyD shows a strong anisotropic interaction due to hydrogen bonding by the hydroxyl groups existing on both sides of the cavity.

Therefore, study of the self-assembled structures of CyDs is necessary not only for the fundamental understanding of anisotropic molecular interactions, but also for further potential uses of the CyD molecule.

Recently, α- and β-CyD-H_2O complexes ordered on a highly oriented pyrolytic graphite (HOPG) surface were directly observed by STM [7], and as is shown in Fig. 1, they were found to exhibit arrangements different from the cage type crystal structure determined by X-ray diffraction. From analysis using the atom manipulation technique, the new structures were concluded to be caused by the difference in the growth mechanism; CyD crystals are generally grown in the three-dimensional mode from the super-saturated solutions, but on the HOPG surface, CyDs are self-assembled in the two-dimensional growth mode [8]. When molecules are arranged in the two-dimensional growth mode, structural ordering generally becomes more controllable.

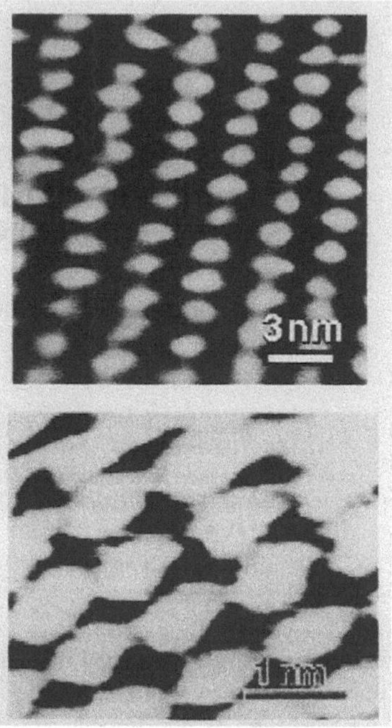

Fig. 1. STM images of (a) α -CyD-H_20 (V_s= 30 mV, I_t = 0.6 nA), and (b) β-CyD-H $_20$ (V_s = 31. l mV, I_t = 0.54nA) on HOPG.

As is describes above, CyD forms inclusion complexes with other molecules, and chemical modification can be easily realized. And the ordering of CyD is governed by the strong anisotropic molecular interactions due to the hydrogen bonding. Therefore, if the molecular interaction can be effectively modified by changing the guest compound or by chemical modification of CyD, control of the growth process and structure of the molecular arrangement is expected to be realized.

In this paper, we present our recent results on the STM study of the guest-dependent ordering of the self-assembled cyclodextrin inclusion complexes.

2. EXPERIMENTAL

In this experiment, we used three types of samples; (1) α-CyD-penta-EG complex (EG: ethylene glycol which has a chain-like structure of 1.5 nm length, (2) G1-α-CyD-H_20 (G1: 6-0-α-D-glucopyranosyl), and (3) β-CyD-ACA (ACA: adamantanecarboxylic acid, which has a spherical structure with the same dimensions as the β-CyD cavity). Their structures are shown schematically in Fig. 2. Specimens were prepared by dropping the above solutions onto a freshly cleaved HOPG surface, and allowing natural drying. STM observations were performed in air at room temperature using a PtIr tip. All images were taken in the constant height mode.

3. RESULTS AND DISCUSSION

3-1. α-CyD-penta-EG

Figure 3 shows an STM image of the α-CyD-penta-EG complexes self-assembled on a HOPG surface. On comparison with the result shown in Fig. 1 (a), the ordering of α-CyD-penta-EG complex

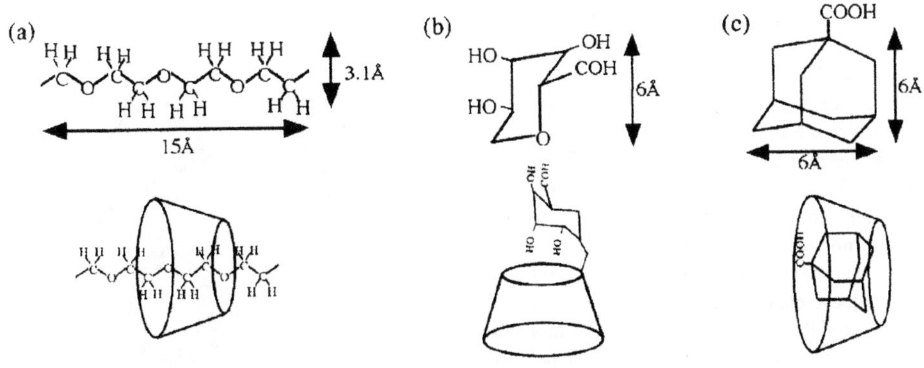

Fig. 2 Schematic structures of guest compounds (upper) and its CyD inclusion complexes (lower).
(a) Penta-EG, (b) Gl-α-CyD, and (c) ACA.

is different from that of the α-CyD-$H_2$0 complex; α -CyD-$H_2$0 complexes formed a channel like structure, while α-CyD-penta-EG complexes formed a structure with a threefold symmetric arrangement.

Fig.3. STM image of α-CyD-penta-EG complexes on HOPG (V_s=-30 mV, I_t=1.0 nA)

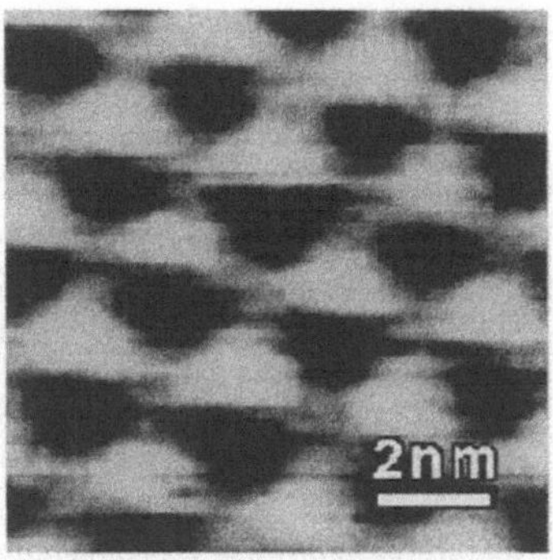

Fig. 4. STM image of α-CyD-H_2O on HOPG (V_s= 30 mV, I_t=1.0 nA)

Since the penta-EG molecule has a structure of 1.5 nm length, it partially projected out of the CyD cavity (Fig. 2(a)) which is expected to affect the structure of the hydroxyl groups of CyD. Therefore, the observed change in arrangement is considered to be caused by the modification of the hydrogen bonding interaction by the existence of the guest compound.

3-2. G1-α-CyD-$H_2$0

Figure 4 shows an STM image of the self-assembled G1-α-CyDs on a HOPG surface, and the arrangement of GI-α-CyDs is similar to that of α-CyD-penta-EG complexes rather than that of α-CyD-$H_2$0. As shown in Fig.2(b), modification of the CyD structure is expected to influence the primary hydroxyl groups, and results in an arrangement different from that of α-CyD-$H_2$0.

The observed results indicates that the chemical modification is also effective for the control of the arrangement by exerting an influence on the hydroxyl groups.

3-3. β-CyD-ACA
ACA self-assembled on a HOPG surface, which is different from that of β-CyD-H_20; β-CyD-H_20 has a more closely packed structure [7]. The β-CyD molecule includes one ACA molecule, which is completely container within the cavity, so that ACA is not expected to influence the hydroxyl groups. Therefore, some other interaction induced by the guest compound is thought to exist. A possible model is that the included ACA molecule changes the strain in the CyD molecule, and induces a change in the dipole moment of the CyD molecule. Dipole-dipole interactions between CyDs results in an important contribution to the mechanism of self-assemble ordering. Further study is in progress.

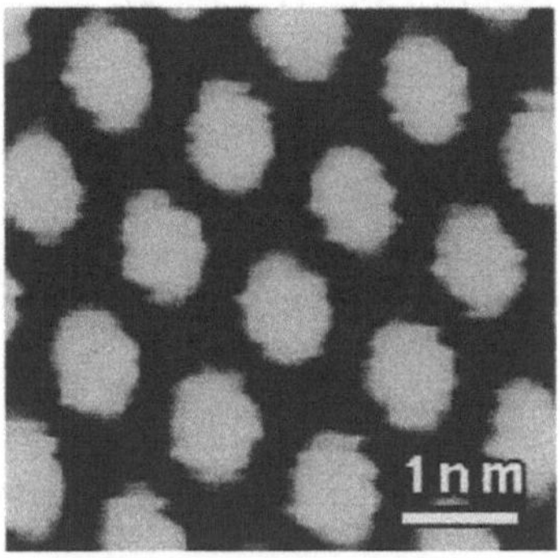

Fig.5 STM image of β-CyD-ACA complexes on HOPG (V_s= - 45 mV, I_t= 2.3 nA).

4. CONCLUSION

We performed STM observations of three types of specimens; (1) α-CyD-penta-EG, (2) G-α-CyD-H_20 and (3) β-CyD-ACA complexes From analysis of the different arrangements observed for these specimens, hydrogen bonding was confirmed to play an important role in the process of self-assembly. In addition, the effect of some other interaction such as the dipole-dipole interaction was considered to exist. These results strongly indicate the possibility of controlling the arrangement of self-assembled CyD inclusion completes by changing the type of guest compound.

REFERENCES

*corresponding authors: yasuda@mat.ims.tsukuba.ac.jp, mkomi@chembio.t.u-tokyo.ac.jp, hidemi@mat.ims.tsukuba.ac.jp, http://www.ims.tsukuba.ac.jp/lab/shigekawa

[1] F. Charra, and J. Cousty: Phys. Rev. Lett. **80** (1998) 1682.
[2] R. Strohmaier, C. Ludwig, J. Petersen, B. Gompf, and W. Eisenmenger: Surf. Sci. **351** (1996) 292. [3] H. Tanaka, T. Nakagawa, and T. Kawai: Surf. Sci. **364** (1996) L575.
[4] Y. Z. Li, M. Chander, J. C. Patrin, J.H. Weaver, L. P. F. Chibante, and R. E. Smalley: Science **253** (1991) 429.
[5] E. 1. Altman, and R. Colton: Surf. Sci. **279** (1992) 49.
[6] Y. Kuk, D. Kim, Y. Suh, K. Park, H. Noh, S. Oh, and S. Kim: Phys. Rev. Lett. **70** (1993) 1948.
[7] H. Shigekawa, T. Morizumi, M. Komiyama, M. Yoshimura, A. Kawazu. and Y. Saito: J. Vac. Sci. Technol. B **9** (1991) 1189.
[8] S. Yasuda, K. Miyake, Y. Goto, M. Ishida, K. Hata, M. Fujita, M. Yoshida, J. Sumaoka, M. Komiyama, and H. Shigekawa: Jpn. J. Appl. Phys. **37** (1998) in print.

THE APPLICATION OF DIFFERENTIAL CONDUCTIVITY METHOD TO THE DETERMINATION OF BINDING CONSTANT OF β-CYCLODEXTRIN WITH SODIUM PERFLUOROOCTANOATE

M.MANABE, H.KAWAMURA, H.KATUURA, and M.SHIOMI
Niihama National College of Technology, Dept. Applied Chemistry and Biotechnology; Niihama, Ehime 792-8580, Japan

1. Introduction

So far, the equivalent conductivity (Λ) has been used in various solution systems for the determination of concentration of ionic species. Some conductivity methods concerning with Λ have also been applied to the cyclodextrin - ionic guest systems for the estimation of binding constant (K).[1,2] In contrast, Manabe et al developed a novel method, "differential conductivity method" and they determined the partition coefficients of 1-alkanols between the bulk water and the micellar phases in the surfactant solution.[3] The differential conductivity defined as the differential of specific conductivity (κ) with concentration (C), i.e., dκ/dC is of much benefit to Λ defined as the quotient, κ/C, when an ionic species is in equilibrium with its modified ones such as complex or aggregate, since the contribution of an aimed species can be extracted by differentiation. In the present study, it was attempted to apply the novel method to β-cyclodextrin - sodium perfluorooctanoate system in order to estimate K.

2. Experimental

Materials. β-Cyclodextrin (CD) supplied by Ensuiko-Sugar Refining Co,.Ltd. was dried in vacuum at around 383K for several hours. Perfluorooctanoic acid from Daikin Industries, Ltd. was neutralized with sodium hydroxide and recrystallized to obtain sodium perfluorooctanoate (SPFO).
Conductivity measurements. A solution with a certain concentration (Cs) of SPFO was prepared as a solvent. A half portion of the solvent was put in a conductivity cell (cell constant; 0.2638 cm^{-1}) and the other portion was used for preparing a CD solution. The CD solution was added little by little in sequence in the conductivity cell. On each addition, the conductivity of the CD+SPFO solution with a given Cs was measured on a conductivity meter (Conductivity Bridge Set, 361B,362B,3962B, Fuso Seisakusyo, Japan) at 25±1/100°C. Solutions were prepared by weight.

3. Results and Discussion

At a given Cs, the specific conductivity (κ) tends to decrease almost linearly with an increase in the CD concentration (Cd) and then, via a clear break, it remains substantially constant, although it decreases very slightly at high Cs, the curve being not shown here. The decrease can be attributed to the complex formation which causes enlargement of the surfactant-ion size (reduction of ionic mobility), and the constancy is due to the completion of complexing. The value of Cd at the break point is in good agreement with Cs, which reflects 1:1 complex formation of CD with SPFO.

Now, a new conductivity, differential conductivity, dκ/dC, is defined and its numerical value is obtained as the slope of a pair of nearest neighbor data, i.e., $(\kappa_2-\kappa_1)/(C_2-C_1)$. The differential conductivity is plotted in Fig.1 against $\overline{Cd}$ (the mean concentration of Cd, $(C_1+C_2)/2$). On the initial addition of CD, dκ/dCd increases slowly at negative values and then goes up suddenly to almost zero in a very narrow Cd region which corresponds to that around the break point in κ vs. Cs curve. The sigmoid curve clearly reflects the characteristic of complex formation. In the low Cd region whiere dκ/dCd is negative (decrease in κ) the complex is formed by the addition of Cd. The constant values nearly zero (no dependence of κ on Cd) in the high Cd region suggests that the nonionic additive (CD) does not interact with any ionic species, after the complexation has been completed. In addition, it is noticed that Cd at the inflection point is close to Cs, manifesting a 1:1 complex formation. In contrast to the sigmoid curve, the concentration dependence of the traditional equivalent conductivity (κ/Cs) changes just asymptotically toward zero, after the break in κ vs. Cs. The distinction can be comprehended from the mathematical analogy of the thermodynamic quantities. That is, the differential conductivity and the equivalent conductivity correspond to the partial molar quantity and the apparent molar one, respectively. Therefore, they are identical to each other only at a limiting dilution. At a certain composition, the former quantity is responsible for only the complex ion formed by the addition of CD even if other ionic species coexists with a constant concentration, although the latter is concerned with all ionic species coexisting. As a result, the outline of the sigmoid curve can be discribed as follows. At first the complex is formed in the negative dκ/dCd region, and then the complex formation has completed at the inflection point which gives the composition of the complex (here, 1:1), and finally, the added CD dissolves irrespective of the complex formation when dκ/dCd is zero. Further, it is supposed that the steeper the slope of the increasing curve, the higher the binding constant, and that the depth of each dκ/dCd from zero line concerns with the complex fraction of added CD.

The sigmoid curve allows to calculate for the binding constant (K) in the following manner. Assuming that a free surfactant (S) and a free CD (D) form 1:1 complex (X), K is expressed as

$$K = Cx / Csf \cdot Cdf \tag{1}$$

where Ci's refer to concentrations of respective species, and they are correlated with each total concentration,

$$Cs = Csf + Cx; \qquad Cd = Cdf + Cx \tag{2}$$

From these equations, Cx can be yielded.

$$Cx = (1/2)\,[(Cs+Cd+1/K) - SQR\{(Cs+Cd+1/K)^2 - 4Cs \cdot Cd\}] \tag{3}$$

If κ is taken to be a function of concentrations of respective ionic species, Csf , Cx, the total differential of κ can be derived as,

$$d\kappa = (d\kappa/dCsf)\cdot dCsf + (d\kappa/dCx)\cdot dCx \quad (4)$$

By dividing Eq.(4) by dCd, Eq.(5) is obtained.

$$d\kappa/dCd = \overline{\kappa}_{sf}\ (dCsf/dCd) \ + \ \overline{\kappa}_{x}\ (dCx/dCd) \quad (5)$$

where $\overline{\kappa}_{sf}$ and $\overline{\kappa}_{x}$ are the differential conductivity of respective ionic species themselves. Inserting Csf of Eq.(2) into Eq.(5),

$$d\kappa/dCd = (\overline{\kappa}_{x}-\overline{\kappa}_{sf})\ (dCx/dCd) \quad (6)$$

where dCx/dCd is derived by differentiating Eq.(4) with Cd at a constant Cs,

$$dCx/dCd = (Cs-Cx)\ /\ (Cs+Cd+1/K - 2Cx) \quad (7)$$

By taking Eqs.(3) and (4) into account, dCx/dCd can be calculated from Cd and Cs if the parameter, K, is suitably assigned. It is noticeable that $d\kappa/dCd$ in Eqs(5) and (6) corresponds to the experimentally determined value in Fig.1, and that it is proportional to dCx/dCd, $\overline{\kappa}_{x}-\overline{\kappa}_{sf}$ being constant. Accordingly, a linear regressive analysis by the least mean squares method between $d\kappa/dCd$ and dCx/dCd provides the most suitable value of parameter, K as illustrated in Fig. 2. The proportional relation of the plots in Fig.2 reveals that the relation in Eq.(6) holds in the present system. A slight deviation from a linear relation seems to be caused by the side reaction mentioned below. In addition, it is emphasized that all plots at different Cs's fall on a single line, which reflects K as well as $\overline{\kappa}_{x}-\overline{\kappa}_{sf}$ depends on neither Cs nor Cd. Obtained values of K are listed in Table 1, together with the slope (A) and the intercept (B) of the linear regressive line of Eq.(6).

TABLE 1. Binding constant for the complex of b-cyclodextrin with sodium perfluoprooctanoate, and the related conductivity parameters

Cs	$K/10^5$	- A	-B
mmol/kg	kg/mol	(mS/cm)/(mol/kg)	
0.999	2.2	12.34	0.271
3.001	1.1	12.12	0.814
4.988	0.94	12.22	1.107
7.001	0.91	11.73	1.471

In a more detailed survey of the values in Table 1, it is found that each parameter depends a little on Cs. First, K tends to decrease with increasing Cs.. The value of K for the same complex reported by Palepu and Richardson [4] is $4.5x10^3$ dm^3/mol which is lower by two order than the present value. They used the equivalent conductivity method proposed by Satake et al [2] and estimated the conductivity difference, 13.3, which corresponds to -A in the same unit. It should be emphasized that even a little difference in the conductivity values affects remarkably to K.

Next, as for B, it should be zero from Eq.(6), but its magnitude increases in proportional to Cs. The line of the calculated value of $d\kappa/dCd$ in neglecting B is drown in Fig.1. It tends to deviate progressively from the experimental values with increasing Cd at high Cs, where the experimental values remains, not zero but negative. These results show that the complexation is accompanied by some side reaction such as the counter-ion binding [5], the effect of the side reaction appears only when Cx becomes high. Accordingly, the most reasonable value of K is the extrapolated value to zero of Cs.

Finally, at a limiting dilution, the differential conductivitis of the complex ion (λ_x) and free surfactant ion (λ_{sf}) which are equal to respective equivalent conductivities can be calculated on the basis of A, $\overline{\kappa}_{sf}$ (=75.0, by measurements for SPFO alone), and λ_{Na} (equivalent conductivity of Na^+ ion, 50.1, literature data), that is, $\lambda_{sf} = \overline{\kappa}_{sf} - \lambda_{Na} = 24.9$, and $\lambda_x = \overline{\kappa}_{sf} + A = 12.6$ in the unit of present differential conductivity. The result that λ_x is close to a half of λ_{sf} is reasonably explained by the enlargement of complex ion.

It is concluded that the differential conductivity gives more detailed information for the complex formation compared to the traditional equivalent conductivity.

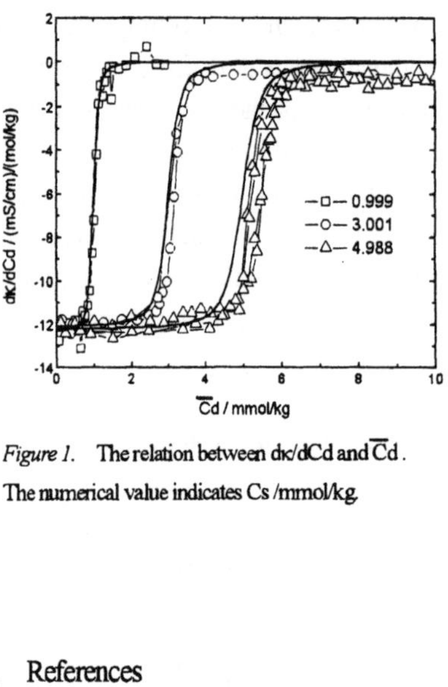

Figure 1. The relation between dκ/dCd and $\overline{Cd}$. The numerical value indicates Cs /mmol/kg.

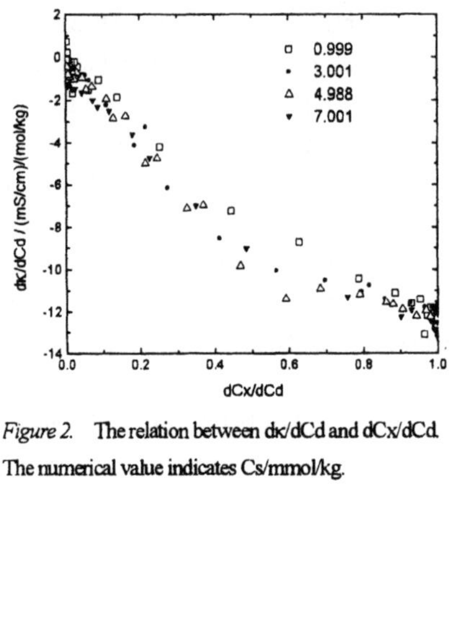

Figure 2. The relation between dκ/dCd and dCx/dCd. The numerical value indicates Cs/mmol/kg.

4. References

1. Okubo,T, Kitano,H., and Ise,N.,(1976) Conductometric study on association of cyclodextrin with colloidal electrolytes, J. Phys. Chem., 80, 2661-2664.
2. Satake,I., Yoshida,S.,Hayakawa,K.,Maeda,T., and Kusumoto,Y., (1986) Conductometric determination of the association constant of b-cyclodextrin with amphiphilic ions, Bull. Chem. Soc. Jpn.,59, 3991-3993.
3. Manabe,M.,Kawamura,H.,Yamashita, A., and Tokunaga,S.,(1987) Effect of alkanols on intermicellar concentration and on ionization of micelles, J. Colloid Interface Sci.,115,147-157.
4. Palepu,R., and Richardson,J.E.,(1989) Binding constant of b-cyclodextrin / surfactant inclusion by conductivity measurements, Langmuir, 5,218-221.
5. Macpherson,Y.E., and Palepu,R. (1990) Counterion binding by surfactant / b-cyclodextrin inclusion complexes, J. Inclusion Phenomena and Molecular Recognition in Chemistry, 9, 134-143.

INFLUENCE OF THE GUEST, THE TYPE AND DEGREE OF SUBSTITUTION ON INCLUSION COMPLEX FORMATION OF SUBSTITUTED β-CYCLODEXTRINS

Ágnes Buvári-Barcza and Lajos Barcza
L. Eötvös University, Institute of Inorganic and Analytical Chemistry, H-1518 Budapest, Hungary

The possibility of inclusion complex formation and the stability of complexes formed is highly influenced by the fit of the guest into the cavity of the host, β-cyclodextrin. In the case of „small" guests (like p-nitro-phenol or p-nitrophenolate), the whole molecule is practically included, while only a part of the guest is inserted, when it is „large" (e.g. phenolphthalein) (*Buvári et al., 1988*).

On the other hand, the size of the cavity of cyclodextrin is altered when the hydroxy groups on its rims are substituted: some kind of extension can be assumed. The „extension" as an effect of substitution can mean (i) the protection of the included guest or (ii) steric hindrance against the guest inclusion. The possibility of the formation of new hydrogen bonds (iii) can be added as a third factor, when both the guest and the new substituent (e.g. hydroxypropyl group) posses proton donor or proton acceptor abilities, because these new hydrogen bondings could highly influence the stability of the inclusion complex.

The monotonous decrease of stability constants of hydroxypropyl-β-cyclodextrin - phenolphthalein inclusion complexes as a function of increasing average degree of substitution has been proved, while the stability constants in systems containing p-nitrophenol or p-nitrophenolate guests attain maxima (*Buv↔ri-Barcza, □ et al., 1998*). It seems that the interaction with the larger and less flexible guest phenolphthalein is more sensitive against the steric hindrance (ii) and therefore the stability constants of its inclusion complexes are monotonously decreased. First of all, the three-site interaction between the guest and hydroxypropyl-β-cyclodextrin is disturbed (*Buvári-Barcza, Á. et al., 1996*).

The interactions with the smaller guests seem to be promoted by increasing degree of substitution at relatively lower values where the protecting (i) and the hydrogen bonding (iii) effects exceed the steric hindrance (ii). At higher average degree of substitution, the steric hindrance becomes little by little predominant: the increase of constants stop, then the values start to decrease.

The picture becomes more complicated when the different substitution patterns are considered. The relative increase of the substitution on *O(6)* position of β-cyclodextrin influences the stabilities more, than the substitutions on *O(2)* or *O(3)* positions. In this respect, a similar trend is valid for both types of guests.

The degree of substitution (and the kind of substituent itself) on *O(6)* position seems to influence basically the molecular recognition ability of β-cyclodextrin derivatives (included the chiral selectivity, too).

To characterize the substitution pattern, the ratio of average degrees (DS) of substitution on primary and secondary hydroxy rims (substitution ratio):

$$R_{DS} = R_{DS(6)/RDS(2,3)} = DS(6) / DS(2,3)$$

is introduced and highly recommended.

Acknowledgement: we thank Hungarian Research Foundation (OTKA 19493) for financial support of this work.

References

Buvári, Á., Barcza, L. and Kajtár, M. (1988) *J. Chem. Soc., Perkin Trans. 2*, , 1687-1690
Buvári-Barcza, Á., Kajtár, J., Szente, L. and Barcza, L. (1996) *J. Chem. Soc., Perkin Trans. 2*, 489-491
Buvári-Barcza, Á., Rák, E., M—sz↔ros, □. and Barcza, L. (1998) *J. Incl. Phenom.*, in press

RELIABLE NMR EXPERIMENTS FOR THE DETERMINATION OF THE STRUCTURE OF CYCLODEXTRIN INCLUSION COMPLEXES IN SOLUTION

C. PÉAN , F. DJEDAÏNI-PILARD and B. PERLY
DRECAM/SCM, CEA-Saclay, F-91191 Gif sur Yvette, (France)

ABSTRACT

Nuclear magnetic resonance of proton (^{1}H-NMR) is a powerfull tool to study "host-guest" interactions. Classical "NOESY-type" experiments and especially the ROESY sequence are useful to determine interactions in cyclodextrins domain. But, due to intrinsic problems, the basic ROESY sequence can lead to misinterpretations. We investigate in this work alternative 2D ^{1}H-NMR "ROESY-type" experiments, and estimate their relative advantages on the well-known case of β-CD/PGE2 complex.

1. INTRODUCTION

^{1}H-NMR is probably the method of choice to study the cyclodextrins complexes [(1)]. Thus, with simple 1D spectrum, cyclodextrin/guest interactions can be clearly displaid with studies of the chemical shifts variations of H3 and H5 CD's protons (internal cavities protons), and proove the interaction fact. But better analysis may be performed by using 2D exchange spectroscopy experiments like NOESY [(2)] (Nuclear Overhauser Effect SpectroscopY) or ROESY [(3)] (Rotating frame Overhauser Effect SpectroscopY), which displays "the spacial proximities" (<5Å) beetwen protons, and confirm unambigously a complex formation. Unfortunately, these sequences have some own drawbacks, which limit their results. Thus, with the NOESY sequence, the studies of CD's complexes are often limited, because of the correlation time ($\omega\tau_c$ ~1.12) of the CD complex, and no NOE cross-peaks are evidenced. With the ROESY sequence, another problem is due to intensities of the ROE cross-peaks which are not correlated to inter-atomic distances; HOHAHA effects contribute to edge off these cross-peak's intensities, and consequently, the inter-atomic distances. Recent published sequences like T-ROESY[(4)] and off-resonance ROESY[(5)] greatly decrease the ROESY inconvenients. We purpose to test all these different NMR experiments on the β-CD/prostaglandin E2 (PGE2) complex [(6)], and to compare their respective benefits.

2. MATERIALS AND METHODS

β-CD were gift from ROQUETTE. PGE2 was purchased from Cayman Chemical (Ann Arbor, USA). Phosphate buffer (0.05 mol.l^{-1}, pH 7.4) was prepared in ultra-pure water using standart procedure. The CD/drug mixture was prepared by mixing accurately volume of mother solution of β-CD (5 mmol.l^{-1}, in phosphate buffer) and PGE2 (5 mmol.l^{-1} in alcoolic solution).

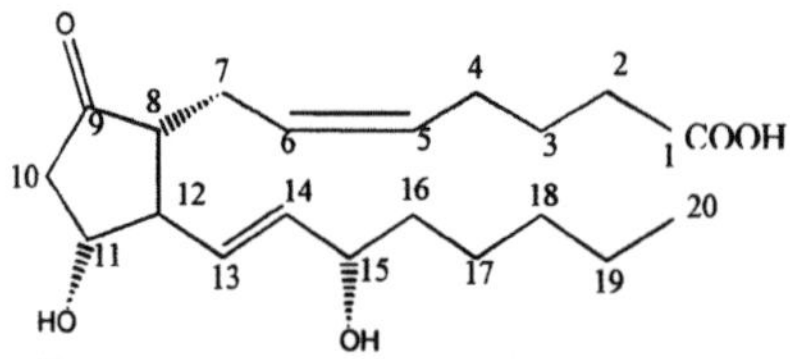

Fig. 1: Molecular structure of protaglandin E2

Samples were homogeneized, freeze-dried, and redissolved in deuterium oxide (Euriso-Top, France). All ^{1}H NMR experiments were performed using a BRUKER DRX 500 spectrometer, at 298K, pH 7.4. These experiments were realized immediately after redissolution of the freeze-dried complex to avoid degradation of PGE2 to PGA2. 2D NMR experiments were performed on the same sample composed of 50/50 proportion β-CD/PGE2. Exactly the same acquisition parameters (receiver gain, mixing time (300ms), scan number, p(90°), ...), and processing parameters were used in the different experiments.

3. RESULTS AND DISCUSSION

3.1 1D-NMR EXPERIMENTS

Previous analysis was made to determine the stoichiometry of the β-CD/PGE2 complex. This study was realized using standart Job plot procedure [(1)]; a serie of 1D ^{1}H-NMR was acquired, where the sum [β-CD]+[PGE2]=cste=5 mmol.l^{-1} was conserved, and by varing the molar ratio (x) of each component. Job plot's (Δ(δ(Hn))*Ctot=f(x)) were traced for diverse protons of PGE2 and for H3 and H5 protons of cyclodextrin, and confirm the 1:1 stoichiometry previously describe in litterature [(6)]. More, treatment of these data with the Benesi-Hildebrand algorithm [(1)] reveal an apparent association constant (Ka) of the complex near 1550 (+/- 50) l.mol^{-1}.

3.2 2D-NMR EXPERIMENTS

NOESY, ROESY, T-ROESY and off-resonance ROESY experiments were performed under identical conditions. As it can be seen in fig.2 a), poor informations were obtained with the NOESY experiment in a reasonnable acquisition time (~12h); it confirms the unfavorable $\omega\tau_c$ value near 1.1. In the pure ROESY spectra (fig.2, b)), many cross-peaks appears beetwen H3 and H5 of β-CD and some protons of the fatty chains of PGE2.

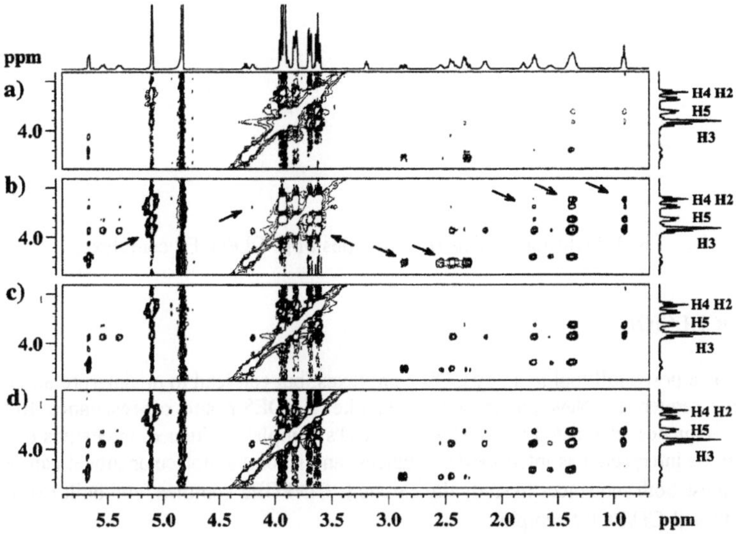

Fig. 2: Extension of the same region (correlation zone of H3 and H5 of cyclodextrin and different protons of PGE2) of the different 2D experiments a) NOESY, b) ROESY, c) T-ROESY, d) off-resonance ROESY. Negative levels are grey -coloured , and all spectra was normalized at the same levels

But concomittely, undesirable cross-peaks appear too (indicated with black arrows on fig.2 b)). These cross-peaks result from spin diffusion effects and TOCSY-ROESY correlations; they lead to misinterpretation of the acquired data and warpe the molecular model issued from cross-peaks interpretation. On the other hand, with T-ROESY (fig.2, c)) and off-resonance ROESY (fig.2, d)) experiments, better results are obtained. HOHAHA effects are greatly diminushed and no TOCSY-ROESY correlations appeared (for example, see the cross-peaks beetwen H2 and H4 of CD and the PGE2 protons, and inter-CD protons signals). More, a better signal-to-noise ratio was obtained with the off-resonance ROESY, compared to T-ROESY. With these both experiments, cross-peaks volume can be interpreted in term of inter-protons distances. So, it can be seen that the more important (number and volume considerations) cross-peaks are obtained with H3 proton of CD and the protons of both lipidic chains of PGE2, compared to H5; this lead to the conclusion that both fatty chain of PGE2 are involved in the complex formation. More, the complex involve the largest rim of the cavity. Futhermore, no cross-peaks were observed with the C5 ring of PGE2 and the CD protons: this demonstrates its non-inclusion in the cavity. A more carreful analysis showed that the largest cross-peaks are obtained with the protons H15, H17, H18, H19,H20 of PGE2 and especially H16 and H16'. It can be interpreted as a preferential complex formation with one of the lipidic chain compared to the other. But in all cases, both chains of PGE2 are not include simultaneously in the CD cavity (fact supported by the absence of inter-chain cross-peaks on the global spectra (data not shown)). So, it can be concluded that we are in presence of two different 1:1 complexes β-CD/PGE2, which involve both lipidic chains of the PGE2. This fact was demonstrated efficiently by using both T-ROESY and off-resonance ROESY 2D-NMR experiments, and two different interaction models for the β-CD/PGE2 complex may be built.

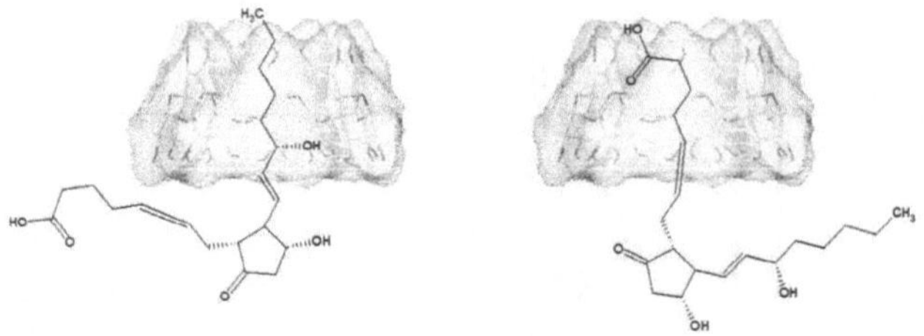

Fig 3: Inclusion models of both types of β–CD/PGE2 complexes

4. CONCLUSION

2D-NMR is a powerfull tool to study CD's complexes, but can lead to misinterpretation by using uneffective sequences. New pulses sequences like T-ROESY and off-resonance ROESY are improved sequences dedicated to the studies of CD's complexes. Indeed, intensities of the cross-peaks may be interpreted as interatomic distances, and rigorous molecular model can be created. We used these both sequences to study the classic β-CD/PGE2 complex, and built two host-guest models of the β-CD/PGE2 complex.

5. REFERENCES

(1) : F. Djedaini, B. Perly *J. Pharm. Sci* **80**, 1157-1161 (1991)
(2) : S. Macura, Y. Huang, D. Suter, R.R. Ernst, *J. Magn. Reson.* **43,** 259-281 (1981)
(3) : A. Bax, D.G. Davis, *J. Magn. Res.* **63**, 207-213 (1985)
(4) : T.L. Hwang, A.J. Shaka, *J. Am. Chem. Soc.* **114**, 3157-3159 (1992)
(5) : H Desvaux, P. Berthault, N. Birlirakis, M. Goldman *J. Magn. Reson. A,* **108,** 219 (1994)
(6) : F. Hirayama, M. Kurihara, K. Uekama, *Chem. Pharm. Bull.* **32**, 4237-4240 (1984)

This work was supported by the European Commission (DGXII) under the FAIR Programme CT95-0300.

^{1}H NMR, CIRCULAR DICHROISM AND UV-VISIBLE SPECTROSCOPIC STUDY OF INCLUSION COMPLEXES FORMATION BETWEEN o-, m-, p-HYDROXYPHENOL AND β-CYCLODEXTRIN

D. LANDY, S. FOURMENTIN, F. ELHOUJHAJI, G. SURPATEANU
Laboratoire de synthèse organique et environnement, Maison de la Recherche sur l'Environnement Industriel de Dunkerque, 59140 Dunkerque (France)

ABSTRACT

Inclusion complex formation between o-, m-, p-hydroxyphenol and β-cyclodextrin (β-CD) in aqueous solution has been investigated by ^{1}H NMR, Circular Dichroism and UV Spectroscopy. The use of an algorithmic treatment to determine formation constants (K_f) of the 1:1 inclusion complexes leads to coherent results between ^{1}H NMR and UV-Visible spectroscopy (classical and competition methods). Circular Dichroism improves the existence of different stoichiometries, since hydrogen bonds lead to pseudo ternary complexes for high hydroxyphenol concentrations.

1. INTRODUCTION

Inclusion complexes between β-CD and substituted aromatics have been studied for a while, but those compounds often exhibit little spectral differences between free and complexed guest, in agreement with relatively low affinity. It has been found [1] that care has to be taken when measuring K_f order of which is close to 10^2. Moreover, if some substituted benzenes like p-nitrophenol [2] have been fully investigated, few recent data are available for this kind of inclusion compound and values of K_f often differ from one technique to another. Little has been done on complex formation between β-CD and hydroxyphenols [3], which are one of P.A.H's metabolites. Thus, the purpose of this study is to obtain coherent and reproductive informations on those compounds from three different techniques.

2. MATERIALS AND METHODS

2.1. Materials

D_2O (>99.9% isotopic purity) was obtained from SDS. β-CD, from Roquette, were used without further purification; Methyl Orange, o-, m-, and p-hydroxyphenol (>99% purity) were purchased from Aldrich. All solutions were prepared in HCl 0.1M in order to keep the ionic strength constant.

2.2. Methods

Spectra were recorded on a Jobin Yvon CD6 spectrometer for circular dichroism, on a Perkin Elmer Lambda 2S spectrometer for UV-Visible absorption and on a Bruker ASPECT 3000 spectrometer operating at 250 MHz for ^{1}H NMR. All spectra were performed at 298 K, just after the solutions were prepared, to avoid excessive oxidation of hydroxyphenols.

3. RESULTS AND DISCUSSION

3.1. Circular dichroism

The concentration generally employed for substituted benzene is nearly equal to 5.10^{-4} M, in order to be compared to UV study and to satisfy the requirement of Benesi-Hilldebrand treatment. For this concentration, spectra exhibit weak signal to noise ratio, which can not allow a precise determination of K_f.
Moreover, as depicted in Fig.1 which shows the intensity of dichroic absorption in function of m-hydroxyphenol concentration, the existence of more than one kind of complex is demonstrated. Since those complexes exhibit very different signals, K_f can not be evaluated by the use of circular dichroism : whatever hydroxyphenol concentration is used, the signal corresponds to spectral response of several species, even if one is predominant.

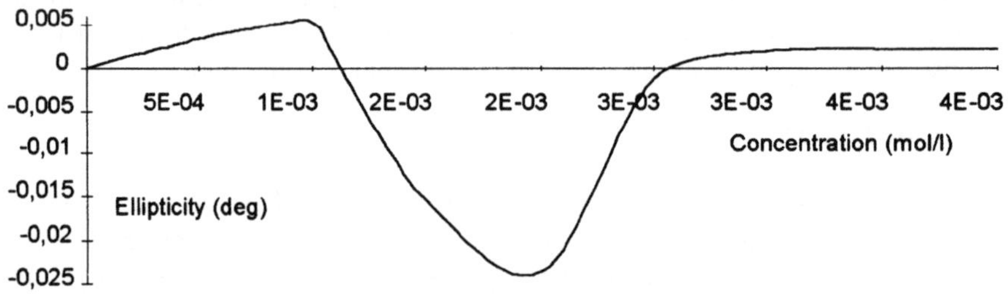

Fig. 1. Dichroic absorption in function of m-hydroxyphenol concentration, in aqueous solution (pH = 1), in the presence of 1.10^{-2} M β-CD.

As hydroxyphenol concentration becomes greater, noise increases in the region of the absorption : this may be due to hydrogen bonds between hydroxyphenol, thus forming flexible molecular systems. Since their respective molecular dimensions preclude the inclusion of more than one hydroxyphenol in the cavity of the β-CD, complexes of higher stoichiometries must be the result of hydrogen bonds between species. These hydrogen bonds seem to take place between 1:1 inclusion compound and another one in the case of m-hydroxyphenol, and between 1:1 inclusion compound and an external hydroxyphenol, whatever diol is concerned.
If hydrogen bonds greatly modify spectral parameters for dichroic absorption, changes due to this phenomena in UV absorption and ^{1}H NMR are intended to be negligible, thus providing two ways for measuring K_f.

3.2. UV-Visible spectroscopic study

3.2.1. *UV absorption*

The main analytical difficulty when studying this kind of complexes lies in the very low spectral differences, which increases the relative importance of experimental errors. The constancy of the guest concentration over the different samples is a critical point : stock solution of β-CD has to be prepared from stock solution of hydroxyphenol and samples with different β-CD concentrations are then obtained by mixing different volumes of the two previous solutions. β-CD, even prepared below solubility limits, undergoes self association, resulting in a very slight and not reproducible absorption. Thus, taking solution with β-CD alone as reference leads to an imprecise obtention of spectra in presence of hydroxyphenol. To avoid this difficulty, we have used spectra's derivatives, since β-CD absorption is nearly constant over short wavelength range. This assertion is confirmed by isobestic points, excellent definition of which could not be reached with absorption curves.

Since Benesi-Hilldebrand treatment lies on approximations and may lead to incorrect values, especially for low formation constants, we have used an algorithmic treatment. If we assume a 1:1 inclusion mechanism, absorption and complex concentration are described as follows :

$$A=\varepsilon_{hyd}*[hyd]+\varepsilon_{comp}*[comp] \quad (1)$$

$$[comp]= 0.5[(1/K_f+[CD]_T+[Hyd]_T)^2-4[CD]_T*[Hyd]_T]^{1/2}+0.5(1/K_f+[CD]_T+[Hyd]_T) \quad (2)$$

where ε stands for molar absorptivity, hyd for hydroxyphenol, comp for complex and $_T$ for total. Then, for a given value of K_f, [comp] is known and ε_{comp} may be calculated from (1) for each β-CD concentration. Standard deviation over ε_{comp} has then to be minimised relative to K_f. $[Hyd]_T$ was kept constant at 5.10^{-4} M while β-CD concentrations were set at 2, 4, 6, 8 and 10.10^{-3} M.

3.2.2. *Competition method*

The same algorithmic treatment applied to methyl-orange (MO) leads to formation constant equal to 243 ± 10 M^{-1}. Since spectral modification between free and complexed MO goes up to 33% (ΔA=0.3), and that absorption maxima of MO (λ_{max}=508nm) and hydroxyphenols (λ_{max}<292nm) are far enough, we applied the competition method.

MO and hydroxyphenol concentrations are set at 2.10^{-5} M and 5.10^{-3}M respectively, while β-CD concentration to be used depends on K_f order and is set to 4.10^{-3} M to obtain a maximal absorption difference.

K_f obtained by mean of the two methods are summarised in Table 1.

Table1. Formation constants for inclusion complexes between hydroxyphenols and β-CD, determined by UV-visible spectroscopy and ^{1}H NMR in aqueous solution. Standard deviations < 5%.

Method	o-hydroxyphenol	m-hydroxyphenol	p-hydroxyphenol
Direct UV absorption	50	85	91
Visible competition method	42	100	110
^{1}H NMR	55	87	95

3.3. ^{1}H NMR study

As it is shown by UV study, complexation results in modifications of the electronic distribution. These modifications occurs not only for hydroxyphenols but also for β-CD, especially for the proton located inside the cavity, where the interactions take place. This suggests that greater displacement of chemical shifts may be expected for molecules interacting strongly with β-CD. It can be seen from Fig.2 that spectral displacement increases from o- to p-hydroxyphenol, as could be predicted from UV data.

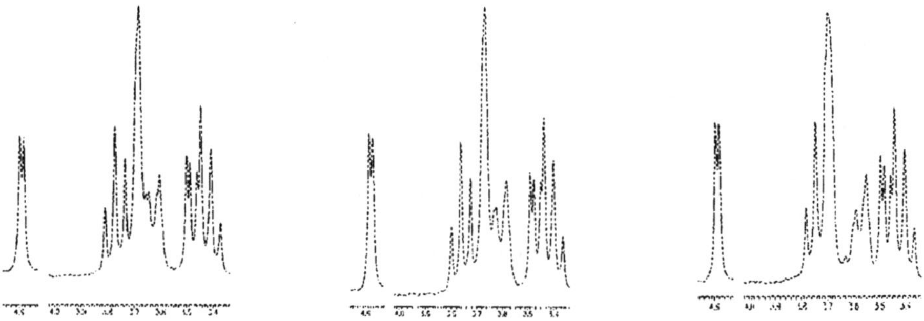

a b c

Fig. 2. ^{1}H NMR spectra in D_2O : 1.10^{-3} M β-CD + 2.10^{-2} M o- (a), m- (b) and p-hydroxyphenol (c).

In order to obtain formation constants we have used the same algorithmic treatment employed for UV study, replacing absorption and molar absorptivities by chemical shifts. The results are presented in Table 1 and are in perfect agreement with those obtained by spectrophotometric studies.

4. CONCLUSION

The combined use of ^{1}H NMR, Circular Dichroism and UV Spectroscopy associated with statistical considerations and algorithmic treatment leads to reliable description of inclusion compounds, even with low affinity. The methodology presented in this study may be useful to prevent erroneous conclusions often obtained when techniques are not confronted. Complexation with β-CD leads to value of formation constants equal to 49, 91 and 99 M^{-1} for o-, m- and p-hydroxyphenol respectively. Existence of pseudo ternary complexes has been demonstrated for high hydroxyphenol concentration.

REFERENCES

[1] Hoenigman, S.M. and Evans, C.E. (1996) Improved accracy and precision in the determination of association constants, Anal. Chem., 68 (18), 3274-3276

[2] Yamamoto, Y., Onda, M., Takahashi, Y.,. Inoue, Y. and Chûjô, R. (1988) Determination of the host-guest geometry in the inclusion complexes of cyclomalto-oligosaccharides with p-nitrophenol in solution, Carbohydr. Res., 182, 41-52

[3] Shimizu, H. (1981) Induced circular dichroism of β-cyclodextrin complexes with o-,m-, and p-disubstitued benzenes, Bull. Chem. Soc. Jpn., 54, 513-519

MOLAR PARTIAL PROPERTIES IN HOST-GUEST SYSTEMS: APPLICATION TO THE INCLUSION COMPLEXES BETWEEN β-CYCLODEXTRIN AND SODIUM ALKANOATES

G. GONZÁLEZ-GAITANO[a], T. SANZ, R. GABARRÓ, J. A. RODRÍGUEZ-CHEDA, M. C. SÁEZ AND G. TARDAJOS[b]

[a] *Departamento de Química y Edafología. Universidad de Navarra. Pamplona 31080 (SPAIN)*

[b] *Departamento de Química-Física I. Universidad Complutense. Madrid 28040 (SPAIN)*

1. Introduction

The formation of an inclusion complex with cyclodextrins involves changes in the degree of hydration of both host and guest molecules. This fact must be reflected in thermodynamic properties as molar partial (or apparent) volumes. To study this phenomenon, we have chosen a homologue series of surfactants, e.g., sodium octanoate (NaO), sodium decanoate (NaD) and sodium dodecanoate (NaL), and the β-CD which is known to form stable complexes with these soaps. The apparent and partial molar volume of the surfactant can be calculated from the density of the solution through

$$v_{\phi,s} = M_s / \rho - (1 + m_{CD} M_{CD})(\rho - \rho_0) / m_s \rho \rho_0 \tag{1}$$

$$v_s = -(\partial V / \partial n_s)_{n_w, n_{CD}} \tag{2}$$

where M_s, m_s and M_{CD}, m_{CD} are the molar masses and molalities for the surfactant and β-CD respectively. Molecular modeling and ^{1}H NMR experiments have been performed to shed light on the microscopic structure of the complexes

2. Experimental

β-CD and sodium alkanoates were from *Sigma* (purity > 99%). All the reactants were used as received without further purification, except the NaL (Ref. 1), and the solutions prepared in bidestilled, deionized and degasified water. The NMR samples were prepared in D_2O as solvent (S.d.S., France, d.d. > 99.9%).

Density measurements have been performed with a technique designed in this laboratory consisting in a new prototype of vibrating tube densimeter[2]. Precision in

density is $1 \cdot 10^{-6}$ $g \cdot cm^{-3}$ and the temperature stability better than 1 mK. The fixed temperature was 25 °C. As a difference with a work by Wilson et al.[3] on the same systems, we did not add buffers to control the pH.
For the recording of the ^{1}H NMR spectra we employed a VARIAN VXR 300S spectrometer operating at 300 MHz. The experiments were done at 20.0 ± 0.1 °C, taking the HDO signal of the solvent as the reference, at 4.63 ppm.
The software employed for the Molecular Mechanics calculations was *Insight II* program[4], implemented in an IRIS 4D/310VGX workstation. The energy minimization of the isolated molecules and complexes was performed with the CVFF forcefield[5], taking the convergence in 0.0001 KcalÅ^{-1} for the RMS of the derivatives. The guests were fitted in the cavity by rigid docking with the refined structures. Cross terms and Morse potentials were considered in the forcefield, together with a distance dependent dielectric constant to account for the solvent effects.

3. Results and Discussion

3.1. MOLAR VOLUMES

The density measurements have been done at fixed molalities of β-CD of 15.32, 12.09 and 9.23 mmol·kg^{-1} for NaO, NaD and NaL respectively. The behavior of the molar volume of the soap when the CD is present is analogous to the observed in cationic surfactants[6]: at infinite dilution the volume is much higher than for the pure surfactant in water and the apparent CMC is reached at higher concentration than the pure surfactant, in an extension apparently equal to the CMC plus the fixed m_{CD}.

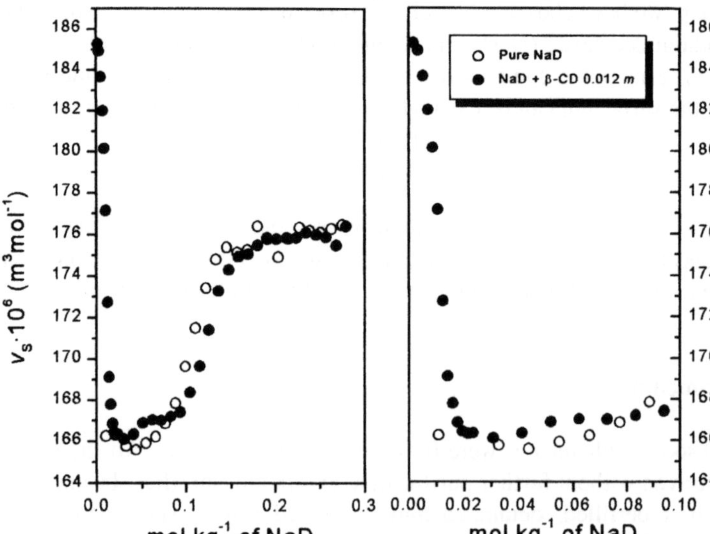

Figure 1. Molar partial volume for sodium decanoate + β-CD

At highest molalities the curves in absence and presence of CD come together. indicating that the complex does not take part into the micelles (Fig. 1). Wilson et al.[3] have measured these systems, but only in the range of low concentrations, bellow the CMC. In Table I are summarized the transfer properties at infinite dilution. The change in the volume of the reaction is the difference between the water expelled from the cavity, which is incorporated to the bulk, and the hole occupied by the part of the surfactant that is included, that is

$$\Delta v_r^0 = v_w^0 n_w - n_{CH_2} v_{CH_2}^0 \quad (3)$$

where v^o_w is the molar volume of water, n_{CH2} the number of CH_2 groups buried into the CD and v^o_{CH2} the volume of a methylene group in water ($15.8 \cdot 10^{-6}$ m^3mol^{-1}). The height of the cavity corresponds to 6.3 CH_2 groups, assuming that the surfactant is in its all-staggered conformation. With this simple scheme, and considering that 6.3 CH_2 groups displace 6.5 water molecules [6], the number of methylene groups of the surfactant that enters can be calculated, resulting in 6.0, 7.0 and 7.9. At zero concentration of surfactant (greatest excess of CD), the most favorable complex will be that of highest stoichiometry possible (CD:S). For the NaO is going to be 1:1, and almost the same for the NaD, but NaL, longer than the other homologues, would permit a small contribution of 2:1 complex. At conditions different than $m_s = 0$, the most favorable stoichiometry will be 1:1. In Table I are the molar ratios, expressed as the quotient $[S]_{comp}/m_{CD}$,. R is nearly 1 in all the cases, although a slight decrease with the chain length is perceived, in the same sense than the transfer volumes.

TABLE I. Transfer parameters for sodium alkanoate + β-CD system at 298.15 K

	$R = [S]_{comp}/m_{CD}$	$v_s^0 \cdot 10^6$ m^3mol^{-1}	$\Delta v_r^0 \cdot 10^6$ m^3mol^{-1}	K $(l \cdot mol^{-1})$
NaO	0.97	150.1	16.8	460
NaD	0.94	185.5	20	1300
NaL	0.91	219	22.4	1700

3.2. ^{1}H NMR AND MOLECULAR MODELING

When the alkanoate is present, upfield shifts of the inner protons, H5 and H3, are observed, whereas the outer H1, H2 and H4, are scarcely shifted. The displacements in the resonances of the guests are not so marked but they move downfield. $\Delta\delta$ differences of the inner H5 of the CD protons (the chemical shift in absence of host or guest molecule minus the observed δ) versus the molar ratio surfactant/CD (Fig. 2) reveal a 1:1 stoichiometry. The binding constants can be calculated from these plots by non linear fit of the data applying the Benesi-Hildebrand method for NMR applications[7] (Table I). Palepu et al.[8], from conductivity measurements, give for this series 370, 740 and 1600 $mol \cdot l^{-1}$, although they report a participation of a 2:1 stoichiometry for NaO that we did not see with our measurements. Anyway, both data are in fair agreement, even though they have been obtained with techniques based in different physicochemical principles.

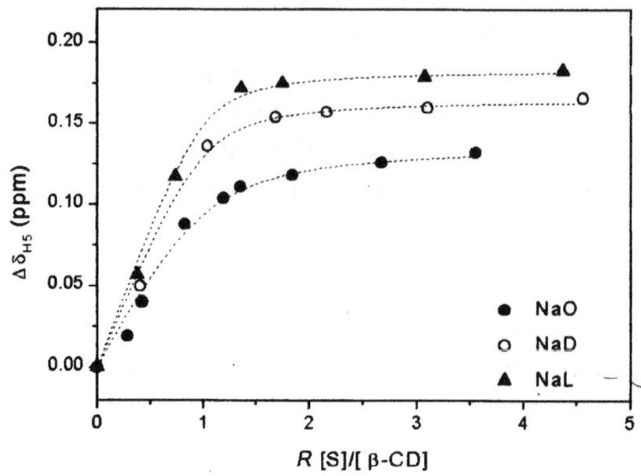

Figure 2. Molar ratio Plot of H5 protons of the β-CD

To find out the molecular structure of the complex. the interaction energies have been calculated. as explained in the Methods section. The host molecule has been approximated from the head towards the wider and narrower rims of the cavity, calculating the interaction energy (van der Waals and electrostatic). The position of minimal interaction energy can be used as the starting point for a minimization, in order to estimate the reaction enthalpy from the difference between the energy of the complex and the energies of the isolated guest and host. Although quite simplistic for giving reliable absolute values, this procedure should permit us to discriminate conformations close in energy.

Thus, the calculated values for the energy of the complexes by this strategy were -85.3, -93.0 and -97.8 $KJmol^{-1}$ for NaO, NaD and NaL, yielding a contribution per methylene group of -3.3 $KJmol^{-1}$. This result is higher than the calculated from the NMR experiments, but the trend is the same. The structure, for NaD with the CD, is shown in Fig. 3. The surfactant is tilted within the cavity, with the possibility of "rattling" inside.

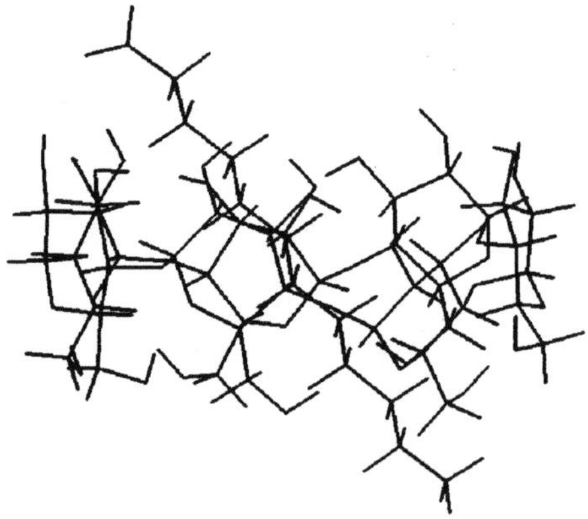

Figure 3. Structure of the complex with NaL

4. References

[1] Rodríguez Cheda, J.A.; Saez díaz, M. C.; Tardajos, G.; González-Gaitano. *Proceedings of the 15th IUPAC Conference on Chemical Thermodynamics.*

[2] Herrero, J.; González-Gaitano, G.; Tardajos, G. (1997) *Rev. Sci. Instrum.* 68, 3835

[3] Wilson, L. D.; Verrall, R. E. (1997) *J. Phys. Chem. 101,* 1970

[4] *Insight II* (3.0.0). San Diego. Biosym Technologies, 1995

[5] Dauber, P.; Roberts, V.A.; Osguthorpe, D.J.; Wolff, J. (1988) *Proteins: Struct., Funct., Genet.* 4, 31

[6] González-Gaitano, G.; Crespo, A.; Compostizo, A.; Tardajos, G. (1997) *J. Phys. Chem. B.* 101, 4413

[7] Bergeron, R.J.; Channing, M.A., Gibeley, G.J.; Pillor, D.M. (1997) *J. Am. Chem. Soc.* 99, 5146

[8] Palepu, R.; Richardson, J. E.; Reinsborough, V. C. (1989) *Langmuir.* 5, 219

THE BINDING OF 2-NAPHTHAMIDES TO β–CYCLODEXTRIN

T.C. WERNER, J. LAROSE AND J.S. ANDERSON
Department of Chemistry
Union College, Schenectady, NY 12308, U.S.A.

1. Introduction

On the basis of fluorescence spectral shifts, Madrid and Mendicuti have recently reported a binding constant (K) of 1965 (+/-159) for the 1:1 complex formed between 2-methylnaphthoate (2-MN) and β–cyclodextrin (β–CD) at 298K [1]. This is considerably larger than the K value reported by Fraiji et al. for the 1:1 complex between 2-acetylnaphthalene (2-AN) and β–CD (581 +/-6) using fluorescence quenching experiments [2]. The former authors suggested that hydrogen bonding by the ester oxygen of 2-MN to the -OH groups on the CD cavity rim might account for this difference in K values. We synthesized 2-naphthamide (2-NA) and its N-methyl (2-MNA) and N,N-dimethyl (2-DMNA) derivatives to see if these guests would also show hydrogen bond stabilization of their 1:1 complexes with β–CD. Using naphthamide spectral shifts in the presence of β–CD, we have measured the K values for the binding of all three naphthamides to β–CD. We have also used molecular modeling of the 1:1 complexes to elucidate our experimental findings. We report herein the results of this work.

2. Materials and Methods

2.1 Material

β–CD was a gift from Cerestar U.S.A., Inc. The naphthamides were synthesized by adding aqueous NH_3 (2NA), CH_3NH_2 (2-MNA) or $(CH_3)_2NH$ (2-DMNA) to 2-naphthoyl chloride (Aldrich Chemical) dissolved in methylene chloride. The products were characterized by IR and NMR.

2.2 Methods

The stock solution of an amide was prepared by stirring some of the solid amide in water for several hours, followed by passage of the solution through 0.2 μ syringe filters (Anotop, Whatman). For fluorescence measurements (2-NA, 2-MNA) in the presence of increasing [β–CD], the amide concentration was adjusted to ensure a maximum absorbance at the exciting wavelength of <0.05. For absorption measurements (2-DMNA) in the presence of increasing [β–CD], the amide concentration was adjusted to ensure a maximum absorbance between 0.2 to 0.8. The concentration of the β–CD stock solution was 0.010 M.

2.3 Calculations

Ab initio calculations (3-21G basis sets) on the naphthalenes were performed using MacSpartan (Wavefunction, Inc.). Once their geometries were optimized, the naphthalenes were docked with β–CD by insertion into the larger (secondary -OH) rim of the cavity using Macromodel (Clark Still) on a SGI Workstation. Non-linear least squares fits of the amide fluorescence data as a function of [β–CD] were performed using Mathematica (Wolfram).

3. Results and Discussion

The results of the *ab initio* MO calculations are given in Table 1. Note that the charges on the amide nitrogens in 2-NA and 2-MNA are more negative than that on the ester oxygen in 2-MN; thus, these amide nitrogens would seem to be more capable than the ester oxygen of hydrogen bonding to the -OH groups on the CD rim. In addition, hydrogens on the amide nitrogens of 2-NA and 2-MNA might also engage in hydrogen bonding with the rim -OH groups.

TABLE 1. Results of *Ab Initio* MO Calculations on Selected Naphthalenes

Naphthalene	*Dipole Moment (D)*	*Dihedral Angle*[1]	*Charge on Ester O*	*Charge on Amide N*
2-NA	3.81	12		-1.06[2],-0.96[3]
2-MNA	4.00	29		-0.66[2],-0.91[3]
2-DMNA	3.68	32		-0.32[2],-0.88[3]
2-AN	3.18	0		
2-MN	1.71	0	-0.52[2],-0.72[3]	

[1]Angle between planes of C=O and ring [2]Mulliken Method [3]Electrostatic Potential Method

Binding constants were determined by using shifts in the fluorescence spectra (2-NA, 2-MNA) or absorption spectra (2-DMNA) as a function of [β–CD]. An example is shown in Figure 1. Absorption data were used for 2-DMNA because its fluorescence is too weak to measure accurately. The K value was extracted from a non-linear least-squares fit of eq.1 [1], where R is the ratio of fluorescence intensities at two wavelengths (360 nm and 369 nm for 2-NA; 358 nm and 344 nm for 2-MNA) or absorbances at two wavelengths (265 nm and 292 nm for 2-DMNA) at a given [β-CD], R_o is the ratio in the absence of β-CD and R_i is the limiting value of R at high [β-CD]. An example fit is shown in Figure 2, and the K values determined in this manner are given in Table 2.

$$R = \frac{R_o + R_i K[CD]}{1 + K[CD]} \quad (1)$$

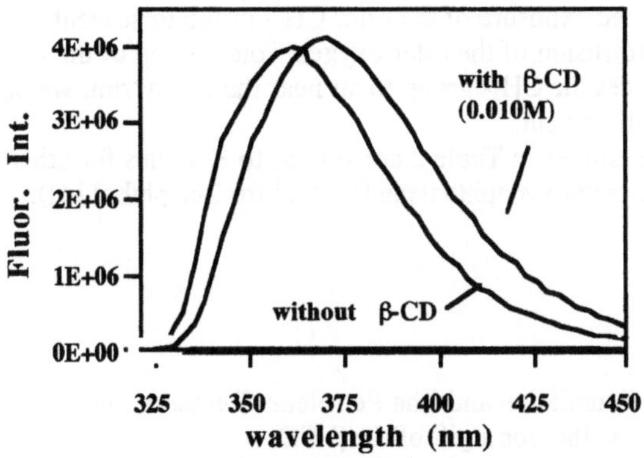

Figure 1. Fluorescence spectra of 2-NA with and without β-CD.

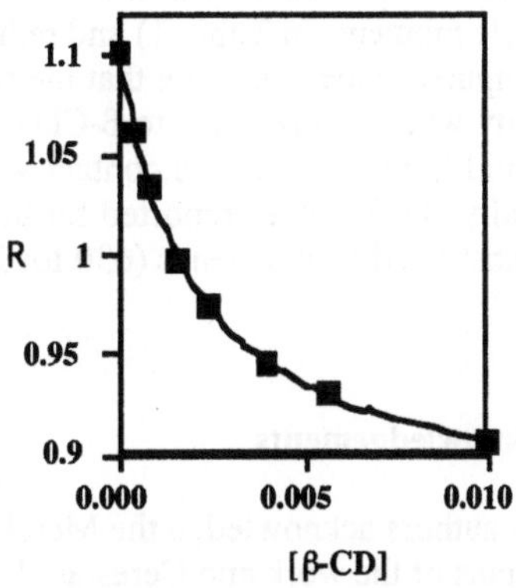

Figure 2. Fit of eq.1 (line) to experimental data (squares) for 2-NA and β-CD

TABLE 2. K Values for 1:1 Complexes Between Selected Naphthalenes and β–CD

Naphthalene	*K Value (+/- std. dev.)*
2-NA	550 (+/- 43)
2-MNA	598 +/- 44)
2-DMNA	653 (+/- 13)
2-AN	581 (+/- 6)[1]
2-MN	1965 (+/- 159)[2]

[1]Reference 2 [2]Reference 1

The data in Table 2 show clearly that K values for all three amides are similar to that of 2-AN, which has no amide nitrogen or ester oxygen, and that methylation of the amide nitrogen does not decrease K. Thus, stabilization of the 1:1 complex by hydrogen bonding from the amide group (either by the nitrogen or its hydrogens) to the CD rim is not significant.

These results are supported by a computer-generated molecular model of the 2-NA:β–CD complex, which shows the amide nitrogen protruding from the top center of the wide end of the CD cavity and not near enough to the rim for the nitrogen, or its hydrogens, to engage in hydrogen bonding interaction with rim -OH groups.

A computer model of the 2-MN:β–CD complex shows that the ester oxygen occupies a similar position to that of the amide nitrogen, indicating that the high K value for this complex is not due to hydrogen bond stabilization involving this oxygen.

We believe that the high K value for the 2-MN:β–CD complex may be due to two factors: the low polarity of 2-MN compared to the other naphthalenes (see calculated dipole moments in Table 1) and reduced exposure of the ester CH_3- group to solvent. Computer models indicate that the protrusion of the ester oxygen from the top of the CD cavity when 2-MN binds to β-CD forces the CH_3- group to sit near the cavity rim, which should limit this group's contact with solvent.

Finally, the K values reported for the amides in Table 2 are similar to K values for other 2-substituted naphthalenes (630 for 2-methoxynaphthalene [3]; 625 for 2-naphthol [4]).

Acknowledgements

The authors acknowledge the Merck Foundation and The Petroleum Research Fund for support of the work and Cerestar U.S.A. Inc. for a gift of the β-CD.

References

1. Madrid, J.M. and Mendicuti, F. (1997) Thermodynamic parameters of the inclusion complexes of 2-methylnaphthoate and α- and β-cyclodextrins, *Appl.Spectrosc.* 51, 1621-1627.

2. Fraiji, Jr., E.K., Cregan, T.R. and Werner, T.C. (1994) Binding of 2-acetylnaphthalene to cyclodextrins studied by fluorescence quenching, *Appl.Spectrosc.* 48, 79-84.

3. Hamai, S. (1982) Association of inclusion compounds of β-cyclodextrin in aqueous solution, *Bull.Chem.Soc.Jpn.* 55, 2721-2729.

4. Yorozu, T., Hoshino, M., Imamura, M. and Shizuka, H. (1982) Photoexcited inclusion complexes of b-naphthol with α-, β-, and γ-cyclodextrins in aqueous solutions, J.Phys.Chem. 86, 4422-4426.

PUBLICATIONS – NEW PRODUCTS; IS THERE ANY CORRELATION?

EDINA BENES, GABRIELLA SZEJTLI
Cyclolab Ltd., H-1525, P.O.B. 435, Budapest, Hungary

Significant expansion of Cyclodextrin literature has been observed in the past years. The average daily number of publications has been increased from the 3.9 in 1995 to 4.7 in 1997.

The propagation of the patents shows the similar trend.

Number of Publications 1981-1997

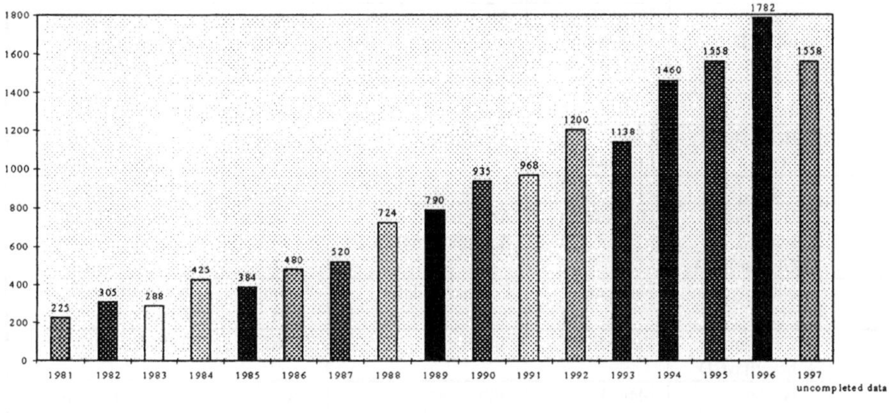

Number of Patent Applications 1981-1997

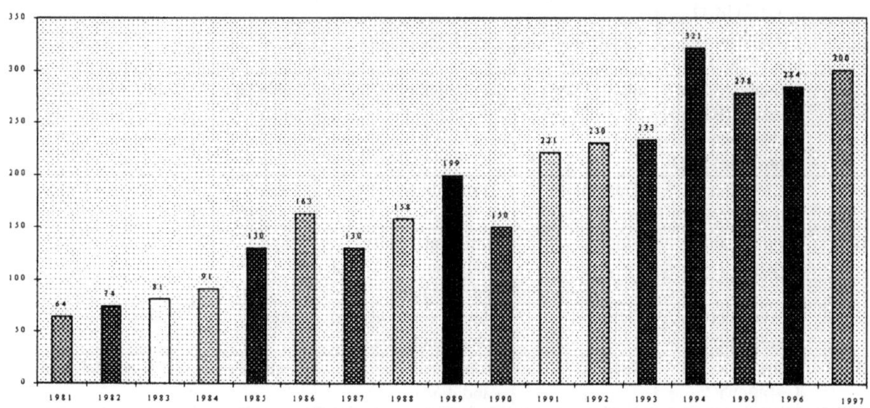

The products approved and introduced on the market until 1998 are presented in the following Tables:

APPROVED AND MARKETED CYCLODEXTRIN/CONTAINING PHARMACEUTICAL PRODUCTS

Publication /patent	Drug/CD	Trade name	Indication	Formulation	Company/country
18/7	PGE_2/βCD	Prostarmon E	Induction of labour	Sublingual tablet	Ono, Japan
64/18	PGE_1/αCD 20 μg/amp.	Prostavasin	Chronic arterial occlusive disease, etc.	Intraarterial	Ono, Japan Schwarz, Germany
64/18	PGE_1/αCD 500 μg/amp.	Prostandin 500	Controlled hypotension during surgery	Infusion	Ono, Japan
15/5	OP-1206/γCD	Opalmon	Buerger's disease	Tablet	Ono, Japan
95/6	Piroxicam/ βCD	Brexin, Cicladol	Antiinflammatory, analgesic	Tablet, sachet, suppository	Chiesi, Masterpharma, Italy
14/4	Garlic oil/βCD	Xund, Tegra, Allidex, Garlessence	Antiartherosclerotic	Dragees	Bipharm, Hermes, Germany Pharmafontana, Hungary CTD, USA
12/1	Benexate/ βCD	Ulgut, Lonmiel	Antiulcerant	Capsules	Teikoku, Japan, Shionogi, Japan.
61/21	Iodine/βCD	Mena-Gargle	Throat disinfectant	Gargling	Kyushin, Japan.
34/3	Dexametha-sone, Glyteer/βCD	Glymesason	Analgesic, Antiinflammatory	Ointment	Fujinaga, Japan.
23/6	Nitroglycerin/ βCD	Nitropen	Coronary dilator	Sublingual tablet	Nippon Kayaku, Japan.
4/1	Cefotiam-hexetil/αCD	Pansporin T	Antibiotics	Tablet	Takeda, Japan.
9/3	New oral cephalosporin (ME 1207)/βCD	Meiact	Antibiotics	Tablet	Meiji Seika, Japan.
4/1	Tiaprofenic acid/βCD	Surgamyl	Analgesic	Tablet	Roussel-Maestrelli, Italy
0	Diphenylhydramine.HCl chlortheo-phylline+βCD	Stada-Travel	travel sickness	Chewing tablet	Stada, Germany
2/0	Chlordiazep-oxide/βCD	Transillium	Tranquilizer	Tablet	Gador, Argentina.
95/6	Piroxicam/-βCD	Flogene	antiinflammatory, analgesic for pediatric use	Liquid	Aché, Brasil
66(28)/6	Hydrocortiso-ne/HPβCD	Dexacort	Mouth wash against aphta, gingivitis, etc.	Liquid	Island
33/5	Itraconazole/ HPβCD	Sporanox	esophageal candidiosis	Liquid	Janssen, Belgium

29/4	Cloramphenicol/ Methyl βCD	Clorocil	eye drop, antibiotic agent	Liquid	Oftalder, Portugal
1/1	Cisapride/-βCD		gastrointestinal mobility stimulant	Rectal suppository	Janssen, Belgium
1/0	Flavopiridol/ HPβCD	Phase II, National Cancer Institute (NCI)	treatment of breast cancer tumors	Intravenous formulation	

APPROVED AND MARKETED CYCLODEXTRIN/CONTAINING COSMETICS AND FOODS

Publications/ patents	Ingredient / CD	Trade name	Product utilization	COMPANY/COUNTRY
7/1	L-carnitin/βCD	Cellutex	cosmetic composition against cellulitis	Regena Ney Cosmetic/ Germany
3/3	tocopherol/βCD	Luminys	color skin cream	Roan S.p.A./Italy
111/92	perfume/βCD	Outdoor fresh Bounce	perfumed fabric softener sheets	Procter & Gamble/USA
28/19	contains βCD	Perfecting Foundation Makeup	CD is a natural carrier for hyaluronic acid, collagen and age protecting system	AVON/USA
5/2	CD absorb unpleasant odor	Activ Complex Plus	Liposomal UV protection Cream	AOK-NERVAL/ Germany
28/26	perfumes/ HPBCD	Vivace/Powder cologne	suspension, long-lasting effect, three different scents *(e.g. water green, shower citrus)*	Shiseido/Japan
3/3	contains CD	Self-Action Tanning Creme	Super Tan for face	Estée Lauder/USA
16/14	contains CD	Klorane	dry shampoo, aerospray and dry powder	Laboratoire Klorane/France
5/2	plant auxin/βCD	Fixpol	tree-wound paste/ wound healing	Reanal Fine Chemicals/ Budapest
85/46 3/3	aroma/βCD	Nestlé Nestea	freeze dried tea: hibiscus, lemon, Japanese green tea	Nihon Shokuhin Kako/Japan
14/9	meat flavour/CD	Steak sauce	mustard taste	Japan
12/11	contains CD	'Nowflash' Vinegar powder		Japan
11/10	flavor/βCD	Flavored salt	natural or synthetic flavor: onion, tarragon, laurel, caraway, smoke, garlic	Compack/Hungary
9/8	aroma/βCD	Tasty like-beef bone	Pet feeding	
24/13	Peppermint oil/βCD	Mr. Etiquet	chewing gum	Kanebo/Japan
		Etiquette Drink	sparkling tablets	Japan
29/16	Reduced cholesterol with CD complexation	Simply Eggs	real whole liquid eggs 80% less cholesterol	Crystal Farms RDC Inc./USA

9/5	Reduced cholesterol content *(Removed by CD)*	Natural Cheese *and* Natural Emmental Cheese	80% less cholesterol of the milk-fat	Entremont M.F./France
11/11	micro-crystalline flavor/βCD	Fruitte filter tea	complex is granulated and mixed to the tea, four different flavors: apple, cinnamon, banana, mango	Cyclolab/Hungary
1/1	Cinnamon flavour/βCD	Dried apple slices	dried apple slices aromatized with cinnamon/CD complex	Bioclean Ltd./ Hungary
11/9	natural lemon-peel flavour/βCD	Lemon-flavoured sugar	flavouring cookies, filling, creams, compotes, fruit soups, etc.	Cyclolab/Hungary
14/9	natural orange-peel flavour/βCD	Orange-flavoured sugar	employed for flavouring cakes, creams, juices etc.	Cyclolab/Hungary
9/5	Contains beer flavour CD complexes	FlavorActiV™	Beer Flavour Standards Kit, for training brewers in beer flavour evaluation	FlavorActiV Ltd./UK
-	alkylisothiocyanates/CD	Wasabi tablets	indoor disinfectant contains: Japanese horse radish extract	Japan
9/6	natural mint oil/CD	*'Hardmint'* Breath manner	long lasting breath freshener tablet	Japan
2/0	contains CD to mask vitamin B odor	High energy fruit juice	vitamin B rich healthy drink	Wakodo/Japan
21/16	natural mint flavour and herb extract stabilized with branched CDs	Flavono	low calorie chewing gum with long lasting flavour	Ensuiko Sugar Co./Japan

This search has revealed that the appearance of many publications (patents) usually forecasts the launching of a new product to the market.

Interesting observation was, however, that the number of publications (patents) significantly increases after the appearance of the product on the market.

These results suggest that the new, useful idea nearly simultaneously emerge at many research centers, companies, but also suggest that the appearance of a new product forcefully stimulates the refinement of product properties, application, etc.

Chapter 8

CYCLODEXTRINS IN FOOD AND COSMETICS

PRELIMINARY STUDY OF DYE SORPTION BY VARIOUS KINDS OF INSOLUBLE SORBENTS CONTAINING beta-CYCLODEXTRIN

M. BACQUET*, T.N.T. PHAN*, G. CRINI**, B.MARTEL* and M. MORCELLET*
*Laboratoire de chimie macromoleculaire URA CNRS 351 Université des Sciences et Technologies de Lille 59655 Villeneuve d'Ascq
**Université de Franche comté - Centre de spectrométrie- La Bouloie 25030 Besançon

Introduction

Interactions between textile dyes and cyclodextrin have been abundantly studied [1,2]. The stability of the inclusion complexes depend : i) on the sterical characteristics of the guest molecule; ii) on the possibility of hydrogen bond type interactions between the host and the guest; iii) on the solvent (polarity, ionic strength...). In our lab, we have prepared two categories of supports carrying β cyclodextrin (β CD) with the objective to use them as sorbents for organic species like textile dyes. In this study, we report the sorption capacity of dyes for these supports which can be described as following : some gels have been synthesised either by crosslinking β CD with epichlorohydrin [3] or by polymerisation of a methacryloyl β CD derivative [4]. On the other hand, β CD has been fixed onto silica beads either by coating or by covalent linkage [4]. We report the results of sorption experiments in which five textile dyes have been tested.

Experimental

The synthesis of the epichlorohydrin crosslinked gels (CCM and CD3) and that of the MAHP gel and functionalized silicas have been described respectively in references 3 and 4. The batch experiments have been carried out in a stopped 100 mL erlenmeyer containing 40 mL of solution and 40 mg of sorbent. The concentration of the dyes was 2.10^{-5} M (excepted for DR 81 = 10^{-5} M). The ionic strength of the solutions was assumed by 1M NaCl. After 24 hours stirring, the absorbance of the supernatant was measured (spectrophotometer UVIKON 900) and the percentage uptake was measured by using the following equation ; $(A_0 - A_{24h}) / Ao$ = % uptake. The methacryloyl (βW7MAHP) and the chlorotriazinyl (βW7MCT) derivatives of βCD were gifts from WACKER Chemie GmbH.

Results and discussion

All the supports reported in table have been tested in batch experiments. Amongst them, some do not contain β CD : carboxymethylcellulose gel (CCM), silica, methacryloyl propylsilane (γMPS) or aminopropylsilane (AMPS) grafted silica that correspond to the intermediate products in the synthesis of the G1 and G4 series, respectively.

Support	Nature	CD or silane* derivatives	CD or silane* content µmole/g
CD3	Gel of epichorhydrin	βCD	132
CCM	Gel of CCM	no	
MAHP gel	Gel	βCD MAHP	714
E3G	Coated silica	βCD MAHP	47.5
E3E		βCD MAHP	86.5
E3I		βCD MAHP	106.4
E3H		βCD MAHP	115.9
G1H1	Grafted silica	γMPS*	42.3*
G1H2		βCD MAHP	98.5
G1D1		γMPS*	135*
G1D2		βCD MAHP	102.9
G1J1		γMPS*	244.5*
G1J2		βCD MAHP	103.2
G1K1		γMPS*	162.6*
G1K2		βCD MAHP	120.5
G4A1		AMPS*	335*
G4A2		βCD MCT	69.3

Table 1 : Nature and chemical composition of the sorbents

The histogram presented in figure 1 shows that all the supports free of β CD, including silica, do have sorption properties, especially towards dyes AB 15 and DR 81. So we can suppose that sorption onto the β CD containing materials are not only due to host guest complexation, but also to physical, electrostatic interactions, and hydrogen bonding. AB 15 that is well sorbed onto almost all the supports is not appropriate to distinguish the presence of β CD in the G1 series (uptake close to 100 % in all cases). On the other hand, in the E3 series, it is possible to observe a correlation between the β CD content of the support and the sorption of DR 81 and AB 15 (see figure 2). Curiously, the reverse tendency has been observed for dyes MB 11 (see figure 2) and MY 30 : their sorption onto the E3 series decreases reversibly with respect to the β CD content. This phenomenon is more obviously observed in the G4 series where the uptake of MB 11 falls from 80 % to ≈ 0 % while β CD is grafted onto the support. This result is rather unexpected for the reason that in a former study, we have shown that MB 11 makes inclusion complex with β CD in solution. From a global point of view, figures 1 shows that MY 30 and RB 7 are adsorbed in a lesser extent than the 3 other dyes. The exception for RB 7 can be observed in the case of the G4 series . Superior results may be due to two possible kinds of interactions between the amino groups of the support and two different parts of the dye molecule : the first possibility can be a complexation with the copper ion present in the chromophore part of the dye; the other possibility is a reaction of grafting with the chlorotriazinyl group that is the reactive functional group of the dye. Amongst the gels, we have observed that non specific interactions occur between CCM and AB 15 or DR 81 that are adsorbed in the same range than onto gel CD 3 where cyclodextrin is present. On the other hand, the efficient sorption of MB 11 and MY 30 onto the former support appears to be due specifically to the presence of β CD. The MAHP gel has rather low results compared to its high β CD content. This can be attributed to the high density of the crosslinking network that probably hinders the accessibility of the majority of the β CD cavities to the substrates.

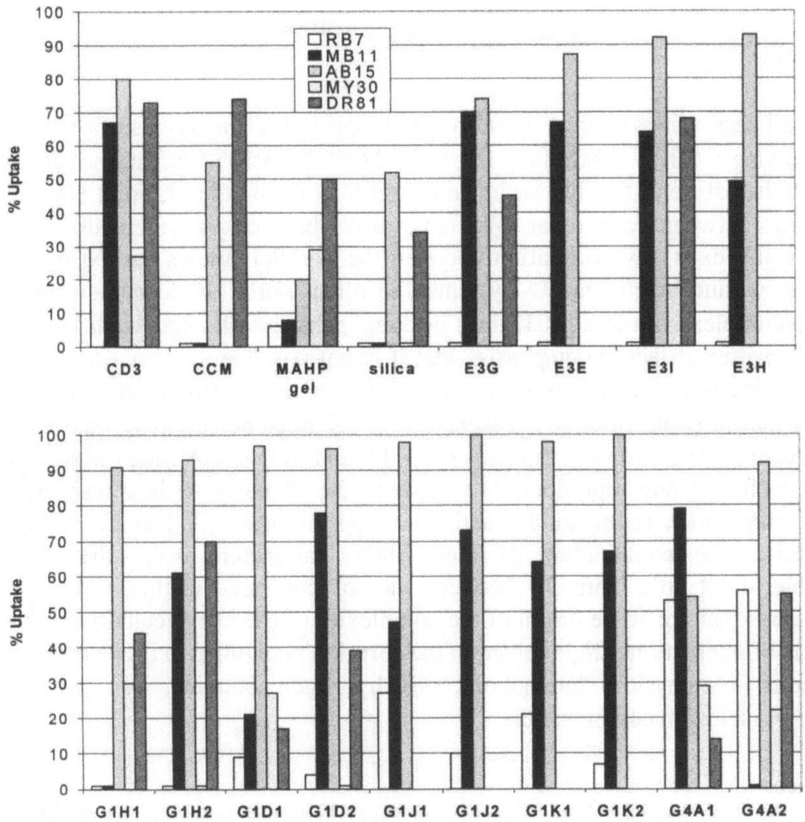

Figure 1 : Sorption uptake of the dyes by the sorbents

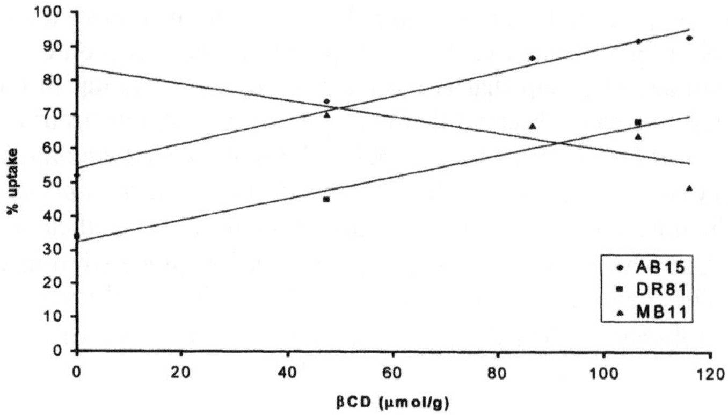

Figure 2 : Adsorption rate against β CD content

Conclusion

The overall decreasing order of sorption with respect to all categories of sorbents is as following : AB 15 > MB 11 ≈ DR 81 > MY 30 ≈ RB 7. In fact, the results vary with the nature of the dye, so it is difficult to compare the different families of sorbents between each other. We have also observed that there does not systematically exist any correlation between the parallel studies carried out in homogeneous and heterogeneous systems ; in other words the formation of an inclusion complex with free β CD does not imply that it will occur with fixed β CD. As a matter of fact we emphasise that it is not possible to carry out such a study by using one dye and to extrapolate to the others. Furthermore, we have observed that in some cases the presence of β CD does not improve, and even makes decrease the sorption capacity (MB 11 and E3 or G4 series systems). Another factor of non reproducibility between the sorbents is their respective ways of preparation (compare E3 and G1 series in figures 1 and 2). As the consequence, the combination of these three parameters make that it is impossible to set up a hierarchy between the sorbents used in this study. The main reasons that are at the origin of the complexity of the interpretation are that five kinds of interactions are involved in the sorption phenomenon i) electrostatic interactions, ii) physical interactions, iii) hydrogen bonding, iv) covalent grafting, v) inclusion complexation.

References

1 Shao Y., Martel B., Morcellet M. and Weltrowski (1996) M. Interactions between β-cyclodextrin and water soluble dyes, *Can. Text. J. 113*, 53-58

2 Shao Y., Martel B., Morcellet M., Weltrowski and Crini G. (1996) Sorption of textile dyes on β cyclodextrin-epichlorohydrin gels, *Proc. 8th Int. Symp. Cyclodextrins*, 571-574

3 Crini G., Bertini S., Torri G., Naggi A., Sforzini D., Vecchi C., Janus L., Leckchiri Y., Morcellet M. Sorption of aromatic compounds in water using insoluble cyclodextrin polymers (1998) in press *J. Appl. Polym. Sci.*

4 Phan T.N.T, Bacquet M., Morcellet M., (1998) Preparation and characterization of sorbents containing βCD derivatives coated or grafted onto silica gel, *Proc. 9th Int. Symp. Cyclodextrins.*

Aknowledgements : special thanks to Mrs A.M. Caze for her skillfull assistance in the sorption experiments. Thanks to the European Community for its financial support through the FAIR program CT 95 0300

RELEASE CHARACTERISTICS OF HYDROPHOBIC FLAVOR ENCAPSULATED IN CYCLODEXTRINS

T. FURUTA, H. YOSHII, I. HASHIMOTO, Y. KATAYAMA, X. LIU, Y-Y. LINKO*, AND P. LINKO*
Department of Biotechnology, Tottori University, Tottori 680-0945 Japan
**Department of Chemical Technology, Helsinki University of Technology, P.O. Box 6100, FIN-02015 HUT, Finland*

1. Introduction

The incorporation of hydrophobic flavors into powders is of great importance in the food and flavouring industries. Several techniques have been applied for encapsulation of hydrophobic flavors. Encapsulants must protect the flavor componds from escaping and chemical change during manufacture and storage. Furthermore, it is often required that flavors should be controllably released during consumption. Encapsulation of hydrophobic liquid flavor in cyclodextrin (CD) is an attractive method for such purposes. This study investigated the microencapsulation of liquid flavors (*d*-limonene, allyl isothiocyanate (AITC), *l*-menthol and β-thujaplicin), and the mechanism of release of these guest compounds included in CD during storage under various temperatures and humidities. An approximation of the first order release rate was applied to obtain the kinetics of AITC and the activation energy of release.

2. Materials and Methods

2.1 MATERIALS

d-Limonene (purity: 95%) was obtained from Nacalai Tesque (Kyoto, Japan) and *l*-menthol and allyl isothiocyanate (AITC) from Wako Chemicals (Tokyo, Japan). α-, β-, and γ-cyclodextrins (CDs) were purchased from Ensuiko Sugar Refining Co. (Tokyo, Japan). β-Thujaplicin (hinokitiol) was a gift from Osaka Organic Chemicals (Osaka, Japan). Other chemicals were of reagent grade.

2.2 PREPARATION OF THE INCLUSION COMPLEX POWDERS

The CD inclusion complex powders of hydrophobic flavors were prepared with a twin-screw kneader. Thirty grams of CD powder were weighed and mixed with each flavor (AITC, *d*-limonene, *l*-menthol or hinokitiol), followed by the addition of distilled water to a moisture content of 50% of dry weight. The mixture was then kneaded in a twin-screw kneader (KRC-S1, Kurimoto Steel Ltd.) at 20 °C for 30 min. The kneaded wet slurry was vacuum dried at 40 °C for 15 h, and 70°C for 5 h. The dried powder was ground and sieved into 80-100 mesh. The amount of the flavor included in CD (inclusion fraction: molar ratio of the guest to CD) was quantified with the same procedure as reported in the previous paper.[1)]

2.3 MEASUREMENT OF RELEASE OF THE INCLUDED GUEST COMPOUND

The release of the guest compound from the inclusion complex powder was measured with the equipment as shown in Fig. 1. The inclusion powder (12) (*ca.* 0.1 g) was weighed in glass bottles (7) (8^{ϕ}x16mm) connected in a series by silicon tubes. The bottles were held in an air bath (6) at 50-70°C. Air was blown through bubbling-bottles (4), which was immersed in a water bath (3), to keep the humidity constant. The air humidity was adjusted by the temperature of the water bath, and monitoring the humidity with a hygrometer (HMP 230, Vaisala). The humid air was blown into the glass bottles (7) through the silicon tube at a rate of 40 ml/min to sweep out the volatiles released from the powder. At prescribed intervals, the glass bottles were taken out, and the remaining amount of the guest compound was measured by gas chromatography asdescribed previously[1].

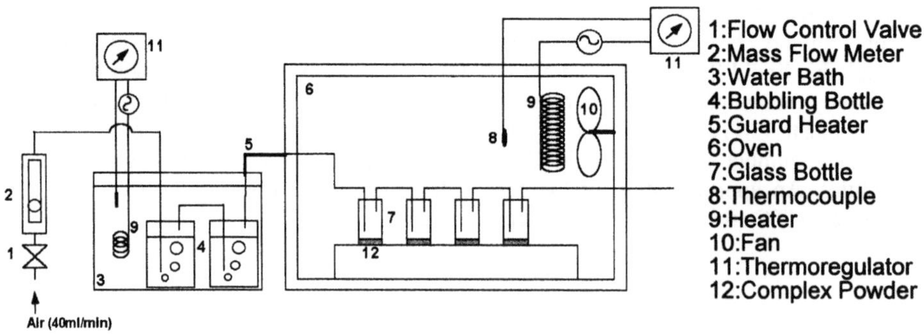

Fig. 1 Experimental equipment for measuring release rate

3. Results and Discussion

Figure 2 shows how the retention of AITC included in β-CD changed as the function of incubation time at 50 °C at different relative humidities. The rate of release of AITC changed markedly with the RH. At zero RH, AITC exhibited an extended release. As the RH was increased, the release of AITC was greately accelerated. At a constant RH, the retention of AITC decrreased rapidly at first, and approached to an equilibrium value (equilibrium retention), which was dependent on the relative humidity. At RH = 90%, the equilibrium retention became zero, and AITC included in β-CD was totally lost. On the contrary, hinokitiol was very difficult to release even at higher RH of above 90% (data were not shown). This indicates that the release characteristics of the flavor compound varied with the combination between the flavor compounds and CDs.

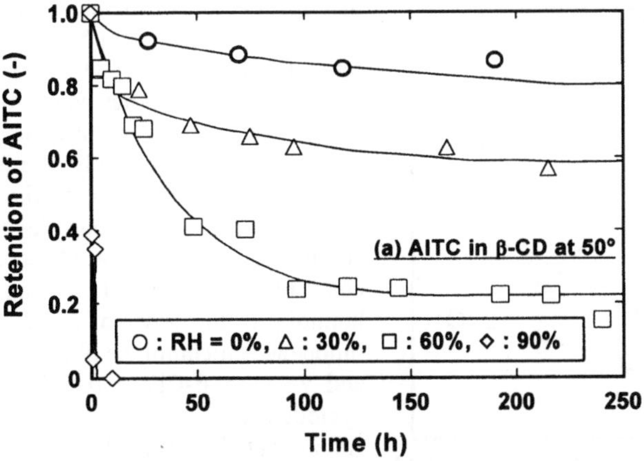

Fig. 2 Releasing behavior of AITC included in CD

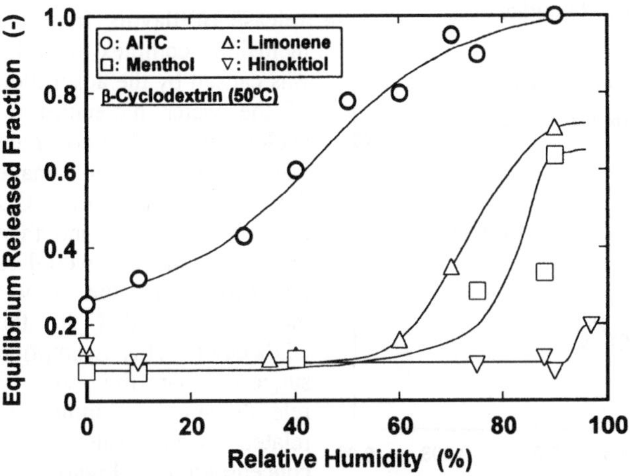

Fig. 3 Equilibrium released fraction of flavors included in β-CD as functions of relative humidity

The fraction of equilibrium release was defined as the difference between the initial retention (= 1.0) and the equilibrium retention at the incubation time of 200 h. The equiribrium release amount of four flavors included in β-CD against the RH is shown in Fig. 3 against RH. AITC exihited a higher value of the equilibrium release than the other three flavor compounds over the total RH range invstigated. Particularly, when the RH was below 60%, less than 15% of flavor was released, except of AITC. The equilibrium release of AITC increased rapidly between the RH = 30 and 60%, followed by a fully release above an RH = 75%. For limonene and menthol, the equilibrium releases were increased markedly at RH above 60%, while being nearly constant at an RH below 60%. Hinokitiol was the most stable flavor included in β-CD. Its equilibrium was about 10% up to the RH = 90%, and 20% of the initial amount could be released at RH = 90%.

The rates of release of AITC included in the CDs were analyzed on the basis of a first order reaction approximation. If the retention of AITC at an incubation time t and the equilibrium retention are denoted by R and R_∞ respectively, the net releasable amount of AITC is R-R_∞. Assuming that the rate of release is linearly proportional to R-R_∞, we obtained the rate of release as,

$$ln\{(R_0 - R_\infty)/(R - R_\infty)\} = kt \tag{1}$$

where R_0 and k are the initial inclusion fraction and the first order release rate constant, respectively. Figure 4 shows the release rate constants of AITC included in the different CDs against the RH. The rates of release of AITC in β-CD and γ-CD showed to be greatly dependent on RH. For γ-CD, the rate of release had a larger value compared to

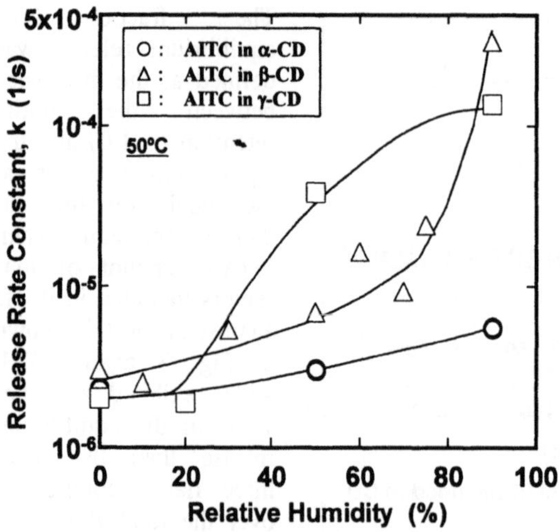

Fig. 4 Release rate constant of AITC included in different CDs against relative humidity

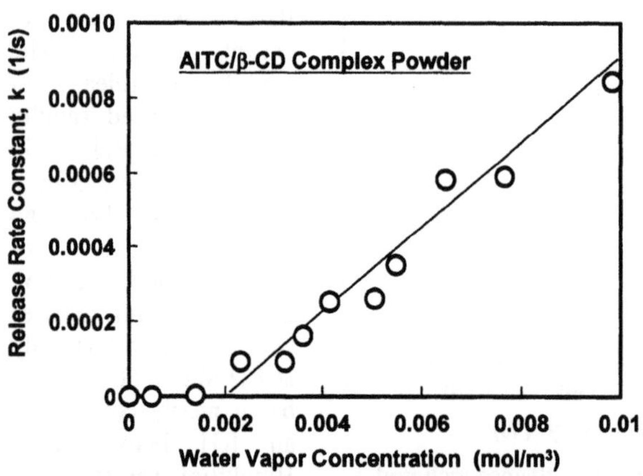

Fig. 5 Release rate constant of AITC included in β-CD as functions of water vapor concentration

that for β-CD. However, as the RH became higher than 50%, the rate of relase for β-CD rapidly increased. On the other hand, for α-CD the rate of release was small amd increased only 3-fold over RH of 0 to 90%. According to Figs 3 and 4 the four flavor compounds were released at a constant temperature and humidity in the order of AITC > limonene = menthol >> hinikitiol, and γ-CD > β-CD > α-CD for cyclodextrin.

As indicated above, the release of flavors from the complexing powder was much influenced by the RH value. This indicates that the release of flavors from the inclusion complex powder was caused by the adsorption of the water molecules on cyclodextrin. The adsorbed water molecules would make the movement of CD molecules flexible, and then the water molecules would be replaced with the flavor molecule included in the CD's cavity. This assumption suggests that the rate of release would be intimately related to the water vapor concentration. Figure 5 shows the correlation of the rate of release of AITC with the concentration of water vapor in the air stream at different temperatures. At low water vapor concentration, the release rate was very low. However, at about 0.002 mol/m^3 of water vapor concentration, the rate of release began to increase linearly with respect to the water vapor concentration. This indicates that the rate release of the flavor compounds was a unique function of the water vapor concentration.

Reference

(1) Yoshii, H., Furuta, T., Yasunishi, A., Linko, Y-Y. and Linko, P. (1996) Oxidation Stability of Eicosapentaenoic and Docosahexaenoic Acid Included in Cyclodextrins, Proceedings of 8th International Cyclodextrin Syposium, Budapest, pp.579-582.

A ^{1}H-NMR STUDY OF INCLUSION COMPOUNDS OF MODEL FOOD FLAVOURS IN β-CYCLODEXTRIN

AIDA MOREIRA DA SILVA,
Department of Food Science and Technology, ESA-Polytechnical Institute of Coimbra, P-3040 Coimbra Portugal
JOSÉ M. A. EMPIS,
Department of Chemical Engineering, Technical University, P-1000 Lisboa Portugal
JOSÉ J.C. TEIXEIRA-DIAS
Department of Chemistry, University of Aveiro, P-3810 Aveiro Portugal

Abstract
A ^{1}H-NMR study of the interactions between β-cyclodextrin (β–CD) and various included flavour molecules (benzaldehyde, vanillin, *trans*-cinnamaldehyde) in aqueous medium is reported. The results confirm that inclusion occurs. Data analysis of data by the continuous variation method shows that all the complexes have 1:1 stoichiometries. Values for the apparent stability constants of the inclusion compounds are estimated and compared with previously reported values.

1. Introduction

β-Cyclodextrin (cyclomalto-heptaose, β-CD) —a short, hollow, truncated cone shaped molecule — is a cyclic oligosaccharide composed by seven α(1-4) linked gluco-pyranose units in normal chair conformation. β-CD interacts with other molecules, which may get into the cavity thus originating inclusion compounds.— is a chiral cyclic oligosaccharide whose natural enantiomer is R-(+). Both in the crystalline hydrate and in aqueous media, the β-CD molecule interacts with water molecules, some of which are removed when a guest of suitable size goes into the cavity. When β-CD interacts with the guest molecule, the difference in chiral properties between host and guest affects the host-guest interaction and the stability of the inclusion complex, In addition, different degrees of fit between the guest enantiomers and the host may influence the degree of hydration of the complex..

In this work, a ^{1}H-NMR study of the interactions between β-CD and benzaldehyde (BNZ), vanillin (VAN) and *trans*- cinnamaldehyde (CIN) in aqueous medium is reported.

2. Methods and materials

β-CD was kindly offered by Wacker-Chemie, Munchen., benzaldehyde (BNZ), vanillin (VAN) and *trans*- cinnamaldehyde (CIN) and D_2O (99.5% isotopic purity) were obtained from Aldrich, Madrid.

Room temperature ^{1}H-NMR spectra were recorded with a 300 MHz General Electric NMR spectrometer,. D_2O was used as solvent, and the residual signal of the non-deuteriated fraction of the solvent (at 24 °C, δ=4.8 ppm) taken as internal reference. Adequate signal-to-noise ratios were obtained by averaging the spectra over 16 or 32 scans.

The H(3) and H(5) protons of β-CD form two inner 'crowns' of hydrogen atoms, in the wider and narrower rims of β-CD, respectively. These 'crowns' of protons have strategic positions for reporting host-guest interactions in the cavity. Both H(3) and H(5) are shifted upfield by the *β-CD.guest* interaction. However, while the H(5) protons give rise to a relatively broad NMR peak and experience a larger shift, the H(3) signal originates a multiplet and its frequency change was difficult to measure due to peak overlapping. [2]. Hence, we used the H(5) NMR signal for probing the *β-CD.guest* interaction. For the solutions containing β-CD and each one of the guests, the H(5) chemical shift differences, Δδ=δ(free)–δ(complexed),

For a 300 MHz spectrometer, and a typical value of the largest observed chemical shift difference ($\Delta\delta_{max}$=0.3), the fast exchange condition (*i.e.*, the exchange rate larger than the reciprocal of the largest observed frequency shift in Hz) implies that inclusion and release of the guest should occur at least 90 times/s. Under these conditions, the frequency of a proton signal is obtained by averaging the frequencies of the free and complexed species, weighted by their mole fractions. From this relationship, one easily arrives at $[C]/[\beta\text{-CD}]_0=\Delta\delta/\Delta\delta_{max}$, that is, Δδ provides a means for measuring the concentration of the inclusion complex, [C] [2].

3. Results and discussion

3.1. STOICHIOMETRY OF THE INCLUSION COMPOUNDS

The complexes stoichiometries were determined using the continuous variation method [1]. 10 mM D_2O solutions of the guest (G) and of β-CD were mixed to constant volume (*i.e.*, the sum of the initial concentrations of β-CD and G remains equal to 10 mM, $[\beta\text{-CD}]_0+[G]_0$=10 mM), and to defined values of $r=[\beta\text{-CD}]_0/([\beta\text{-CD}]_0+[G]_0)$ (r took values from 1/10 to 9/10, in steps of 1/10). By plotting $\Delta\delta.[\beta\text{-CD}]_0$ against r (Δδ is the chemical shift difference for H(5), $\Delta\delta=\delta_{(free\ \beta\text{-CD})}-\delta_{(\beta\text{-CD.G})}$), one obtains maxima at r=0.5 in all cases (Figure 1), pointing to the formation of 1:1 complexes. These 'Job plots' are roughly symmetrical, suggesting that one type of inclusion compounds should be dominant, as competitive formation of complexes would give rise to asymmetric curves [2]. In addition, since both H(3) and H(5) are appreciably shifted and point to the inside of the cavity, it can be inferred that the formed species are inclusion complexes.

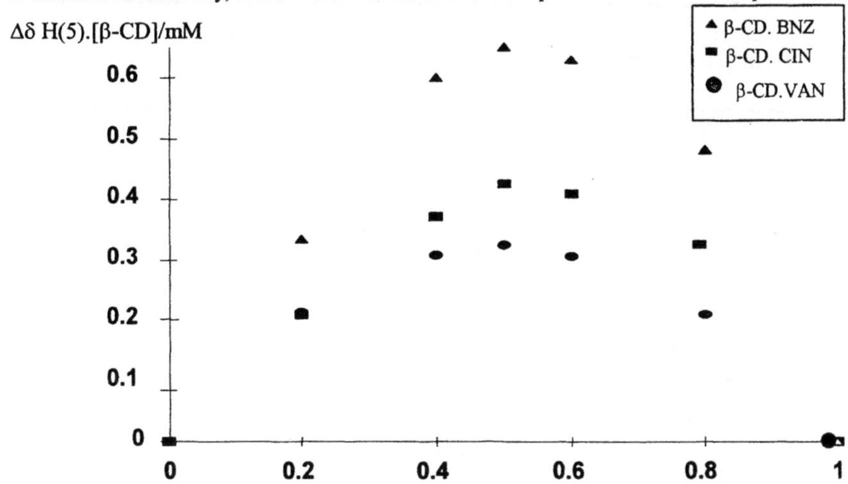

$$r = \frac{[\beta\text{-}CD]_0}{([\beta\text{-}CD]_0+[Guest]_0)}$$

Figure 1- *Continuous variation plots of 300 MHz ^{1}H-NMR spectra, for mixtures of β-CD and Guest, (Benzaldehyde (BNZ). Vanillin (VAN) trans-cinnamaldehyde (CIN)) in D_2O solutions with different values of r in the region of the H(5) signal*

3.2. APPARENT ASSOCIATION CONSTANTS

The equilibrium for the inclusion process in aqueous solution involves hydrated forms of β-CD and G, and represents a substitution of water molecules in the β-CD cavity by the incoming guest molecule. The apparent association constant, K_{app}, measuring the extent of complex formation, was estimated by the Benesi-Hildebrand method [4] (Table 1.).

4. Conclusions

The results herein reported for K_{app} indicate that these systems satisfy the basic requirements (K_{app} in the range 10^2-10^4 M^{-1}) for use in the pharmaceutical and food industries.

In addition, the K_{app} value herein obtained for β-CD.BNZ is of the same order of magnitude of a previously determined value obtained by pulse polarography [2].

Acknowledgements

The authors acknowledge support from Junta de Investigação Científica e Tecnológica (J.N.I.C.T.), Lisboa, to the research unit No.70/94, Molecular Physical Chemistry. A.M.S. thanks MSc.Ana Isabel Rodrigues (INETI, Lisboa) for help provided during acquisition of NMR spectra.

TABLE 1.- *Chemical shift differences for the H(5) protons of β-CD, in mixtures of a D_2O solution of β-CD ($[β\text{-}CD]_0$=0.2 mM) with D_2O solutions of the Guest.*

		Δδ (H–5)		
$[β\text{-}CD]_0$/m	[Guest]/mM	*BNZ*	*VAN*	*CIN*
0.2	9.8	0.263	0.207	0.162
0.2	9	0.259	0.202	0.157
0.2	8	0.252	0.195	0.153
0.2	7	0.250	0.184	0.150
0.2	6	0.246	0.176	0.146
0.2	5	0.242	0.165	0.124
0.2	0	0.000	0.000	0.000

TABLE 2-*Apparent association constants (K_{app}) for the inclusion compounds at room temperature (22°C)*

	K_{app}/M^{-1}	
BNZ	*VAN*	*CIN*
1092	209	256

5. References

1. P. Job, *Ann. Chim.*, **9**, 113, **1928.**

2.H-S, Choi,C-J Chang, A-M Knevel, *Pharmaceutical Research*, , **9**, 4, **1992**

3. G. Jeffrey, W.Saenger, *"Hydrogen bonding in biological* structures", Springer Verlag, Berlin, chapter 18, **1991**

4.. A. Benesi, J. H. Hildebrand, J. Am. Chem. Soc., **71**, 2703, **1949**

β-CYCLODEXTRIN EMULSION: MECANISM OF EMULSIFICATION AND APPLICATION TO THE PREPARATION OF GREEN METALWORKING FLUIDS

L. MENTINK
Research and Development - DIVISION OF APPLICATIONS
ROQUETTE FRÉRES - F 62136 LESTREM

INTRODUCTION:

The use of β-cyclodextrin (BCD) for preparing emulsions or increasing their stability in presence of conventional emulsifying agents is wellknown and mainly described in patent literature (1, 2, 3, 4, 5). In these papers, the purpose was to obtain stable emulsions for cosmetics and food industry. Unfortunately, the effects of BCD, which is not a surfactant, and the properties of the emulsions are not well described in this kind of literature and only a few scientific papers deals with this subject. Shimada et al. (6, 7, 8) consider that the complexes formed between BCD and oils are located at the oil/water interfaces and decrease the interfacial tension, stabilizing by this way the emulsion. Laurent et al. (9, 10) assume that the addition of BCD in an aqueous subphase of trioleine monolayer has a strong effect in its rheological propertie, but a weak effect on the surface pressure. Thus, these authors conclude that BCD could mainly act via a modification of interfacial rheology due to the presence of low soluble complexes at the interface between water and oil.

OBJECTIVE:

Metalworking fluids are used as lubricant and cooling agent during the operations of cutting off, drilling, long turning, tapping, chamfer plugging, grooving, thread cutting of metals.
Metalworking fluids may contain only oily products or also water. When water is present, such fluids are in the form of emulsions, micro-emulsions or solutions.
Conventional metalworking fluids are based on mineral, sulfurized, sulfured or chlorinated oils, which have a very low biodegradability and may be irritant to skin and eyes.
The objective of this work is to formulate new emulsions based on biodegradable components, especially biodegradable oils (natural triglyceride oils or methyl ester oils) and BCD, which are usable and efficient for metal working.

EMULSION PREPARATION:

Emulsions with mineral oil, vegetable triglyceride oils or methyl esters from these oils were prepared as follows:

- dispersion of BCD (KLEPTOSEÒ product from ROQUETTE FRERES) and additives (slushing agent, additional tensive agents, preservatives ...) in water, under mecanical agitation
- addition of the oil in the above dispersion under high shear using a ULTRA-TURRAX or BLENDOR mixer, by increasing progressively the speed to 5000 rpm and maintaining it during 5 minutes.

RESULTS AND DISCUSSION:

1. Emulsifying capacity of BCD:

It appears from a first serie of trials that the emulsifying capacity of BCD depends:

- on the water content in the medium
- on the quantity of BCD used
- on the nature of the oil which must be emulsified
- and on the temperature of emulssification

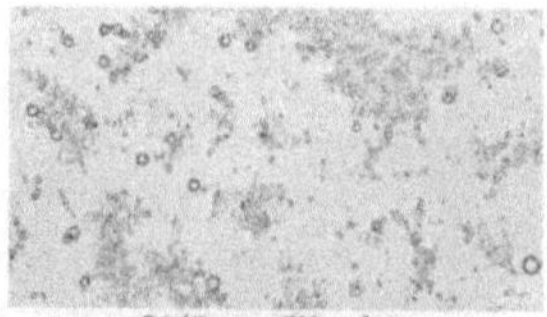

In all cases, if an emulsion is obtained, the continuous phase is water.

2. Emulsion stability:

Stable emulsions can be obtained with all types of oil tested (mineral, triglyceride or methyl ester oils) but the best results are obtained with mineral oils.
In each case, by increasing the BCD content or the oil conten, more stable emulsions are obtained. This phenomenon could be explained by a significant decrease of the average diameter of oil droplets in the emulsions.

Effects of BCD content and oil content on the average diameter of oil droplets

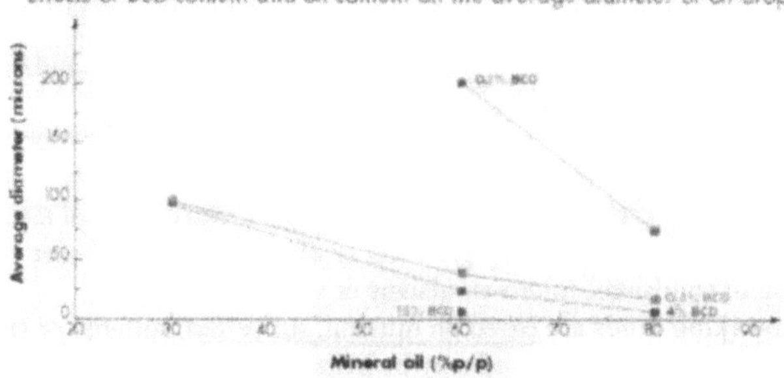

3 Viscosity and surface tension properties:

Metalworking fluids are generally sold in a concentrated form able to be dilued just before using. Concentrates need to have a low viscosity (less than 600 mPas to be pompable at room temperature) and must significatively reduce the surface tension.

Stable emulsions containing only BCD as emulsifier present a too high surface tension and have a too high viscosity.

By adding low quantities of glycols and sorbitian esters, surface tension and viscosity can be easily improved.

Composition (% p/p)							
Recipes	Water (30°F)	KLEPTOSE ®	Vegetable oil	M.E. A.	TWEEN 80	SPAN 85	Hexylen glycol
I	54.67	15	30	0.33	0	0	0
II	53.77	15	30	0.33	0.9	0	0
III	53.17	15	30	0.33	0.9	0.6	0
IV	48.17	15	30	0.33	0.9	0.6	5

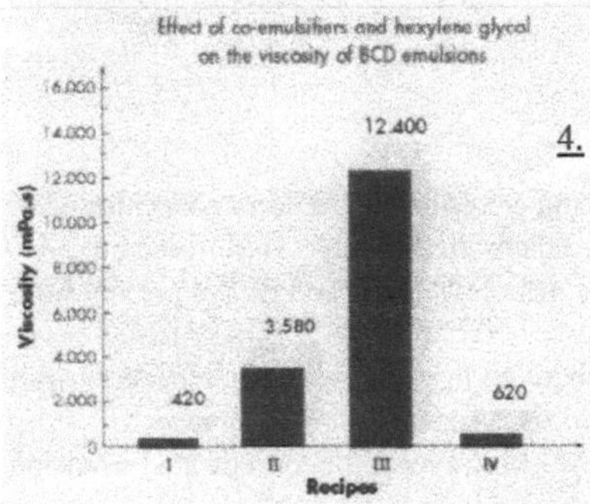

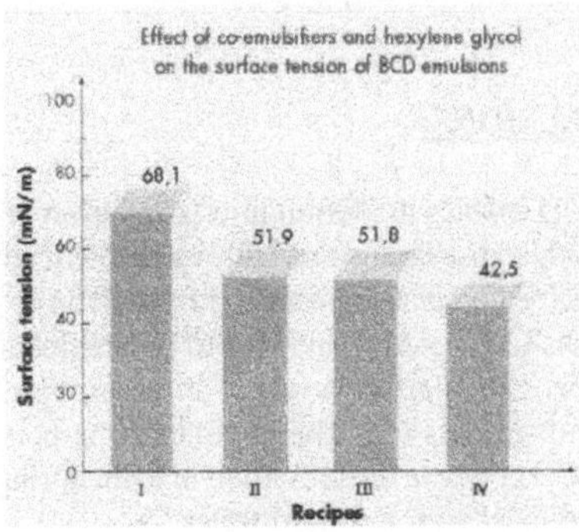

4. Test results in metal working:
The hereunder metalworking fluid with BCD has been tested during slicing in comparion with a conventional cutting emulsion containing mineral oil an slushing, antifrothand extrme-pressure agents. The test are performed on a MANURHIN headstock single-spindle automatic lathe type combinant 42 during a 6 hours productions process.

Tested composition
- Methyl ester oil (METILOIL®ATOCHEM): 4.10%
- β-cyclodextrin (KLEPTOSE®ROQUETTE): 2.00%
- Corrosion inhibitor (LUBRIZOL®5329) 1.00%
- Xanthan gum (RHONE-PULENC): 0.10%
- Tall oil soap (DAUDRUY): 0.06%
- Water: 92.74%

Roughness of the machined parts (PERTHEN M4P analyser)

Roughness in microns (average values)				
	During long turning	During grooving	During cutting off	During plugging (cone 150)
Tested composition	2.5	7.3	2.8	6.1
Control	4.2	6.6	3.8	5.2

Quality of the chips obtained

The quality of the chips with the tested composition is at least equivalent to and often better than the one obtained with the control.

Production index

	Number of machined parts	Production index	Incidents
Tested composition	73	182%	Tool failed during grroving
Control	40	100%	Tool failure during tapping

CONCLUSIONS:

- BCD enables the formulation of stable metal working emulsions in the form concentrated or ready to use emulsions with vegetable triglyceride oils or methyl esters from these oils.
- Such emulsions are easy to prepared an to handle, have low odour and do not present high risks when used, compared to conventional fluids.
- Very good performances of such emulsions are obtained in metalworking, especially with methyl esters oils. The use of extreme-pressure additives seems not necessary.
- BCD prevents the formation of foam during metal working even if the concentrated emulsion is diluted with low hard water.
- Moreover, such emulsions are environment friendly. In fact, they can be formulated with highly biodegradable components and be easily broken down after use by centrifugation at high temperature to recover the oil.

BIBLIOGRAPHY:

1. Sunstar KK. Surfactant free emulsion composition for skin cosmetic comprises cyclodextrins oil, aqueous medium and sodium carboxymethyl cellulose. *Japanese Patent n° 61133138*, 20 June 1996.
2. Kanegafuchi chemical industry Co. Preparation of a new emulsified oil with good flowability - comprises kneaded mixture of oil, water and cyclodextrin followed by mixing in sugar. *Japanese Patent n° 63248433*, 14 October 1988
3. Takeda Chemical Industries, Emulsified food containing oil and fast including cyclodextrin, viscosity increasing agent and protein, *Japanese Patent n° 88041541*, 17 August 1977
4. Teijin LTD. Oil in water emulsufied food containig materials for aqueous and oily phases, viscous material and oil-clarthrated betacyclodextrin as emulsifier, *Japanese Patent n° 52096774*, 13 August 1977
5. Kobayashi Kosei KK. Emulsified cosmetic preparation - using water soluble high molecular compound and cyclodextrin instead of surfactant, *Japanese Patent n° 67031681*, 9 July 1987
6. K. Shimada, Y. Ohe, T. Ohguni, K. Kawano, J. Ishii and T. Nakamura, Emulsifying properties of alpha-beta-gamma-cyclodextrins. *Nippon Shokunin Kogyo Gakkaishi* 38. 16-20 (1991)
7. K. Shimada, K. Fujikawa, K. Yahara and T. Nakamura. Antioxidative properties of xanthan on the autooxidation of soy bean oil in cyclodextrin emulsions. *J. Agric. Food Chem., 40 945-948* (1992)
8. K. Shimada, K. Kawano, J. Ishii and T. Nakamura, Structure of inclusion complexes of cyclodextrins with triglyceride at the vegetable oil/water interface, *J. Food Sc.*, 57, 655-656 (1992)
9. S. Laurent, M.G. Ivanova, D. Pioch, J. Graille and R. Verger, Interactions between betacyclodextrin and insoluble monomolecular films at the argon/water interface/applications to lipase kinetics, *Chem. Phys. Lipids*, 70, 35-42 (1984)
10. S. Laurent, J. Graille, M. Serpelloni and D. Pioch, Study of betacyclodextrin stabilized paraffin oil/water emulsion, *J. Society Cosmetics Chemist.* (1997)

SUBJECT INDEX

MIX
Papier aus verantwortungsvollen Quellen
Paper from responsible sources
FSC® C105338

If you have any concerns about our products,
you can contact us on
ProductSafety@springernature.com

In case Publisher is established outside the EU,
the EU authorized representative is:
Springer Nature Customer Service Center GmbH
Europaplatz 3, 69115 Heidelberg, Germany

Printed by Libri Plureos GmbH
in Hamburg, Germany